# Fluid Mechanics
and
# Hydraulic Machines

# Fluid Mechanics and Hydraulic Machines

**S. C. Gupta**

Professor and Head (Retired)
Department of Mechanical Engineering
Engineering College
Kota, Rajasthan
India

Pearson

ISBN 978-81-7758-364-9

First Impression, 2007
**Current Impression, 2026**



Published by Pearson India Education Services Pvt. Ltd, CIN: U72200TN2005PTC057128.

Head Office: 1st Floor, Berger Tower, Plot No. C-001A/2, Sector 16B, Noida - 201 301, Uttar Pradesh, India.

Registered Office: 6th Floor, Tower A, Unit A and B, International Tech Park, CapitaLand Chennai, 200 Feet Radial Road,Zamin Pallavaram, Old Pallavaram Chennai – 600117, Tamil Nadu, India

Website: in.pearson.com; Email: companysecretary.india@pearson.com

Digitally printed in India by Trinity Academy for Corporate Training Ltd, New Delhi in the year of 2026

*To*
*the next generation of engineers*
*who, through their skill and knowledge, will leave an imprint on the world*
*and change it forever*

*and*

*To*
*my family members*

# Contents

## Symbols Used

| Symbol | Description | SI units |
|---|---|---|
| $A$, $\delta A$ | Surface area | $m^2$ |
| $B$ | Body force | N |
| $a$ | Acoustic velocity | m/s |
| $f$ | Friction coefficient | |
| $F$ | Force | N |
| $Fr$ | Froude number = $U/\sqrt{Lg}$ | |
| $g$ | Earth's gravity | $m/s^2$ |
| $h$, $H$ | Head, distance | m |
| $I$ | Second moment of area | $m^4$ |
| k | Source/sink strength | $m^2/s$ |
| L | Characteristic length | m |
| $m$, $\delta m$ | mass | kg |
| $M$ | Mach number = U/a | |
| $N$ | RPM | rad/s |
| $p$ | Pressure | $N/m^2$ |
| $r, R$ | radial distance | m |
| $Re$ | Reynolds number = UL/n | |
| $t$ | Time | $s$ |
| $T$ | Torque | Nm |
| $u, v, w$ | Velocity components | m/s |
| $U$ | Flow velocity | m/s |
| $V$, $\delta V$ | Volume of control volume | $m^3$ |
| $w, g$ | Specific weight | $N/m^3$ |
| $x$-, $y$-, $z$- | Cartesian coordinates | |

## Greek letters

| Symbol | Description | SI units |
|---|---|---|
| $\omega$ | Angular velocity, Vorticity | rad/s |
| $\Gamma$ | Circulation | $m^2/s$ |
| $\alpha$ | Acceleration | $m/s^2$ |
| $\rho$ | Density | $kg/m^3$ |
| $\tau$ | Shear stress | $N/m^2$ |
| $\mu$ | Dynamic viscosity | kg/ms |
| $\nu$ | Kinematic viscosity | $m^2/s$ |
| $\delta$ | Boundary layer thickness | m |
| $\delta^*$ | Displacement thickness | m |
| $\varepsilon$ | Shear strain rate | $s^{-1}$ |
| $\theta, \phi$ | Angular coordinates | |

# Preface

Mechanics is the science of the action of forces upon material bodies. It studies the behaviour of a substance subjected to external forces and its distortion as a result of the system of forces and couples acting on it. Initially, mechanics was developed for solids. It was later realized that the concepts that explained the effects of forces on a solid body did not hold true for liquids and gases. While the solid bodies stay in a distorted-equilibrium state under the force field and regain their initial shape after the forces are removed, liquids and gases deform, even with a small forces, and do not regain their original position or shape after the removal of the forces. The other difference between solids and fluids is that all the particles of a solid have the same motion, i.e., there is no relative motion amongst the constituents of a solid. On the other hand, the particles constituting a fluid are individual entities and have their own motion. These characteristics complicated the study of motion of the fluid and led to the formulation of the control volume approach, which is different from the Lagrangian approach used in mechanics. For this approach, it is necessary to know either the pressure, velocity and temperature fields at each point, or the overall effect on the control volume, both of which require different types of analysis. These different analyses, of the fluid flowing over the surface or through a passage, are discussed in Chapter 2.

Chapter 1 looks into the properties of different kinds of fluids and the effect of pressure and temperature on them. Chapter 2 discusses how to use fundamental laws, such as those of the conservation of mass, momentum and energy, to develop a mathematical model to provide enough equations to determine the unknowns (velocity, pressure and temperature fields). In the case of compressible fluids, density is also a variable and hence the equation of state is also needed to find the solution. The chapter also explains how to connect the Lagrangian and Eulerian approaches teaching us, in the process, a few important functions like stream function, velocity potential, angular velocity (or vorticity), and circulation, all of which form a sound foundation for mathematical analysis.

Chapter 3 deals with ideal fluid—a fluid having no viscosity, that is, no resistance to motion. Of course, there cannot be such a fluid; this concept is ideal, made only for developing equations that are easily amenable to mathematical tools. Fortunately, in such cases, the equations are linear and allow the concept of superimposition to be used to analyse the flow around any complex geometry. Moreover, both the stream function and the velocity potential satisfy the Caushy–Riemann conditions that we study in complex numbers and functions. This chapter also deals with line integration of the momentum equation and the application of the resulting energy equation (the Bernoulli equation) to the measurement of things like flow through pipes and channels, the time required for emptying of tanks, and flow through orifices.

The application of concepts developed so far is carried further in Chapter 4 through 7. In Chapter 4, flow through pipes and channels is discussed. These two elements are primary; the pipe is used for water distribution and sewage disposal whereas the channel is used for carrying water to remote places. The two flows differ; pressure differences cause the flow through pipes while gravity is responsible for the flow through a channel. In Chapter 5, the formation of the boundary layer around surfaces, and its effect on the flow pattern and analysis techniques, are discussed, This chapter deals with laminar flow whereas Chapter 6 discusses turbulent flow. Difficulties faced in arriving at the closure problem and how to get the one extra relation needed for the solution are discussed. The chapter also debates the mechanism of energy transfer from the main fluid to increase its internal energy, what we call 'the eddy loss'. Chapter 7 deals with the effect of the compressibility of the flow and how it makes the analysis complex. In the process, we learn about the Rayleigh and Fanno lines, formation of shock and how to design a variable area passage like a nozzle or convergent–divergent nozzle for aeroplanes.

Chapter 8 discusses a more logical approach based on the dependence of the objective function (the dependent variables) on the independent variables and by combining these variables to form non-dimensional groups. The reduces the number of groups to be handled, which in turn reduces the required experimentation in the laboratory. The chapter explains how to form these groups and discusses the conditions to be satisfied so the experiment may be carried out in a laboratory based on the model of the given case. From the results thus obtained, the results for the actual situation can be predicted.

Chapters 9 and 10 deals with the basic principles of machines used in day-to-day life, such as hydraulic turbines used for power production, pumps, cranes, presses, lifts, rams, accumulators, and pressure intensifiers. The type of controls needed for these machines and the basic elements of the control used in the industry are also discussed. Finally, the philosophy and the design of experimental setup is discussed in Chapter 11. The appendices provide a quick review of mathematical tools like vectors including gradient, curl and divergence; vector differentiation and integration; tensors; complex numbers; and orthogonal coordinates.

Typical problems from the field have been included to enable students to learn to simplify the complex flows into amenable forms and use the relations learnt in these chapters. Knowledge of mathematical tools is a prerequisite to understanding the subject, and hence the student should be conversant in the use of vectors, tensors, conformal transformation, differentiation and integration, and solution of partial differential equations.

Although many students prefer the complete derivation of results be provided, I have intentionally not carried out their wishes in order to encourage them to work out the relations themselves. After all, the analytical mind is prepared, not imposed upon. However, this book provides guidance so the student, if willing, can reach the next step. In case of difficulty, they should discuss first amongst themselves, with their teacher or may even e-mail me at dr_scg@hotmail.com. Suggestions, too, are welcome.

I owe my sincere thanks and gratitude to all the authors whose names are mentioned in the bibliography and many others whose deliberations on the subject inspired me to learn the subject and put these pages together for young engineers to read. I also thank the editorial board of Pearson Education and all others involved with the project for their tireless efforts and pains. Last, but not least, I owe gratitude to my wife Sudha Gupta who spared me time for this project. I will be remiss if I fail to thank my students, as my interaction with them was invaluable in shaping this book.

S. C. Gupta

CHAPTER 1

# Fluid Statics

OUTLINE

- Introduction
- Continuum Concept
- Pressure Variation and Hydrostablic Forces on Surfaces Submerged in Fluid at Rest
- Metacentric Height
- Pressure Variation in Liquids Confined in Moving Container

## 1.1 Introduction

Every substance in nature is made of matter, which may be organic or inorganic compounds and may exist in different phases, viz. solid, liquid or gaseous or in two or even more phases in equilibrium with each other. These materials, both organic or inorganic, are found in the natural systems like the human or animal body, stars, rivers and so on that we see around us and are also used to develop engineering systems. In this section we shall have a look at it.

### 1.1.1 About the Substance

Substance is the name used for the elements and their compounds. For example, carbon is an element and water is a compound in which the hydrogen and oxygen are combined in the ratio 2:1. There are 112 elements and infinite number of compounds formed from the chemical bondage of different elements. A substance may exist in a single phase, such as solid, liquid or gas, or two phases in equilibrium with each other or all the three phases together that will depend on pressure and temperature of the substance. We may even have mixture of substances that may be homogeneous or heterogeneous. Anyway, the scope of this book is limited to pure substances only and therefore we shall skip discussions about the mixture.

A substance exists in solid, liquid or gaseous phase depending on the gravitational force between the molecules of the substance of which the substance is made of. We learn in kinetic theory that a substance can be thought of made up of molecules that move randomly in all directions with all possible velocities. That is, their motion is chaotic. This behaviour is confirmed by many experimental results; Brownian movement in colloidal solutions is normally cited as an experimentally observed phenomenon in its support. These molecules though extremely small, visible only by a high magnifying power microscope, have mass and hence the molecules attract each other as per the law of gravitation. The force of attraction varies as inverse of square of the distance between the molecules. Thus, in solids the distance between the molecules is so small that the gravitational force does not allow them to escape and the whole substance behaves as solid. The solids therefore are difficult to be compressed by application of force and a large force is needed to do so. When solid is heated, at some temperature, the distance between the molecules increases because of the increase in internal energy and a solid starts behaving differently. First it comes in semi-solid state and finally in the liquid state.

In liquids the distance between the molecules is more than that in solids but not so large that the gravitational force is zero. Because of this, the substance can flow and acquire the shape of the container in which it is stored. However, the molecules in the free liquid surface have enough velocity to leave the free surface and therefore evaporation from the surface goes on though the amount of molecules leaving the surface is small and depends on the temperature of the liquid. At some temperature, called boiling point, more and more molecules acquire velocity high enough to overcome gravitational force between the molecules and thus leave the surface. Thus the liquid changes to gaseous phase. We may also explain the surface tension and capillary effects in similar way. Liquids too, like solids, are difficult to be compressed and their density can be assumed constant. These are often referred as *incompressible* fluids.

On the contrary, the distance between the molecules in gases is large enough for existence of any gravitational force between them and the molecules fly in all directions. The volume occupied by the gas is the volume of the container in which it is stored. Gases are compressible and their density changes with the temperature and pressure. This compressibility introduces many complications in the solution of differential equations to be studied later.

### 1.1.2 Fluid Mechanics

In the initial period of development of science, though much attention was paid on the motion of solid bodies the behaviour of fluids was also under study. This happened because it was a bit easy to treat a whole body as one particle because of the rigid connectivity of all the particles comprising the body, which was not so in the case of fluids where each particle has its own identity and independent motion. Therefore the study of the fluid flow was a much more complex process and its study gained momentum when the control volume approach was thought of. Prior to this, the control mass approach, in which the whole body is taken as a system with no change in its mass with time, could be followed successfully because of the inherent nature of the solid.

We were able to treat it as a single particle while studying the effects of distribution of particles (mass) within it in a simple and easy manner. We shall discuss these aspects in this book.

But we shall first see why do we need such study and what is the scope of fluid mechanics. Let us remember that all the developments are the result of finding a scientific explanation for what we see around us and then to use this knowledge for betterment of our life. We are able to sustain our life because our body has a number of systems, which maintain blood flow throughout the body (cardiovascular system), air circulation to detoxifying the blood resulting from its contact with cells of the body (pulmonary system), excretion of urine to dispose of the waste collected in the blood, and so on. The heart works like a pump, pumping blood, about 5 litres a minute, requiring about 2 W mechanical power supplied by the muscles. The left ventricle of the cardiovascular system sends blood to the lungs for oxygenation and then circulates it to different part of the body supplying the right amount of blood required through a pipe line network made of arteries and millions of capillaries distributed in the body in series and parallel. The blood returns back to the heart, almost at an average rate of 1 litre/min, through veins having reduced oxygen and enhanced carbon dioxide contents. We now know the flow is possible because of pressure difference, which is about $1/6^{th}$ of the atmospheric pressure equivalent to 120 mm of mercury. We note that the cardiovascular system is a closed system as there is no entry or exit point in the system unlike pulmonary system, which has an open system as the lungs breath fresh air and leave carbon dioxide. Humans respire at a normal rate of 12 breaths per minute, inhaling about 0.5 litres of air per breath or 360 litres/h. The lungs squeeze like sponge during exhalation and inflate like a balloon during inhalation. During inhalation, the lungs pressure goes below atmospheric pressure.

Birds have always been a wonder for us and we always dream if we could also fly like them. We know this happens because with every flip of the wing a force is produced whose vertical component keeps the bird afloat in the air and a forwarding force keeps it moving forward. This study has resulted in the development of aeroplanes, jets, scramjets, or rockets.

Rivers, or ocean has attracted the human beings from the start of life on the earth for many reasons. Our rigorous efforts made us a conqueror to ride over the surface waves, to use the flowing energy of water down the slope for transportation and to cross it to reach the desired place. Not only this, we could tame this energy and convert it into useful electrical energy using various man made cyclic systems. Similarly, we are able to convert energy of the wind, tides, etc to electricity.

There is no end to citing examples, especially if we take the man-made systems too, where the study of fluid flow helps in many ways and fluid mechanics is a scientific attempt to answer our curiosities and ambitions.

## 1.2 Continuum Concept

Two methods for analysis of system have been used. In one approach, the system is considered at its molecular level as in kinetic theory of gases. That is, the system is assumed to be made up of molecules in incessant random motion with wide range of velocities. This kind of analytic approach is called *microscopic approach* and has been useful to predict pressure, temperature and internal energy of gases contained in a container in terms of the molecules density (number of molecules per unit volume) and their velocity distribution, expressed by the average, most probable, and root mean square velocity. These relations were obtained using the statistical tools. An important result is obtained for the mean free path, which is the average of the free path defined as the distance travelled by a molecule between its successive collision. This distance will vary and would depend on the proximity of the other molecules around it. The mean free path $\lambda$ is given by the relation, $\lambda = 1/\sigma n$, where $\sigma$ is the collision cross-sectional area = $\pi d^2$, $d$ is the diameter of the molecule and $n$ is the number of molecules in a unit volume.

However, the microscopic analysis is very time consuming and in most of the applications it is not required to go into details of the individual molecule and then taking out the average. When the size of the system is much larger than the mean free path of the molecules, then what happens to the individual molecule is of no consequence and the main thrust lies in the determination of the interaction between the system and its surroundings and also the overall effect left on the system (as well as the surroundings). Such approach is called the *macroscopic approach*. Under such circumstances the matter may be considered as continuous. This is called *continuum concept*, that is, the matter is uniformly and continuously distributed and not made up of molecules with space in between. Thus, we shall follow macroscopic approach in this book.

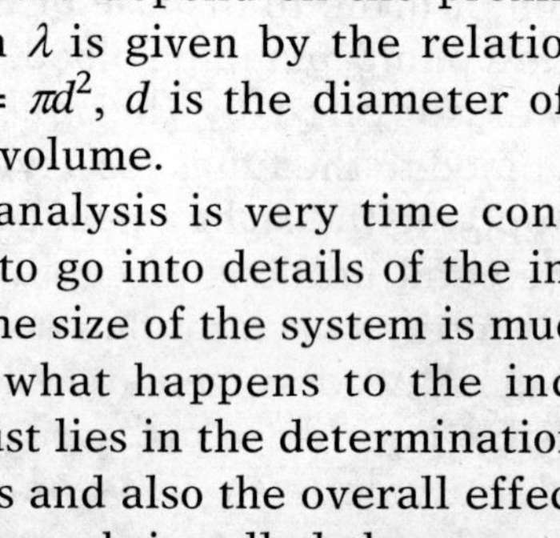


**Fig. 1.2.1** Variation of density with volume

We would like to study behaviour of a system as its size or volume is reduced. Let $\delta V$ be the volume occupied by some mass $\delta m$ of the system at any time. Then their ratio, $\delta m/\delta V$, we know is called the density, $\rho$. Figure. 1.2.1 shows variation of density $\rho$ as volume increases. It is observed that when volume is $\delta V'$ or less, the density changes sharply but the density approaches asymptotically to a value as volume of the system increases; the limiting value is used in day-to-day life as density. This happens because the mater can no longer be considered as continuous when volume of the system has reached $\delta V'$ and the presence of molecules affect the result. This means that we may consider the matter continuous when volume is more than a certain volume, which is related to the mean free path. That is why the concept of *macroscopic approach is applicable only when the smallest size of system is more than the mean free path*. This needs to redefine the definition of pressure, density etc. as the value obtained by taking the limit when area or volume of the element tends to zero. If $\delta A'$ is the smallest area up to which continuum concept holds true then pressure is defined as

$$p = \lim_{\delta A \to \delta A'} \frac{F}{\delta A} \quad \text{and} \quad \rho = \lim_{\delta V \to \delta V'} \frac{\delta m}{\delta V} \tag{1.2.1}$$

Where $F$ is the force acting on surface area $\delta A$ and $\delta m$ is mass of the volume element $\delta V$. Note the difference lies in taking the limit. Here it is not zero as then the continuum concept is not applicable. Density is measured in kg/m$^3$ in SI units whereas pressure in Pascal Pa, which is N/m$^2$. However, MPa, kPa, or bar is commonly used unit for pressure, where 1 Mpa $\equiv 10^6$ Pa, 1 kPa $\equiv 10^3$ Pa, and 1 bar $\equiv 10^5$ Pa. Specific gravity and specific weight are the other ways of expressing density of any substance. Specific gravity is the ratio of the density of the substance with that of water, which is taken as 1000 kg/m$^3$ and *specific weight* is the product of density and gravity, which is usually represented by the symbol $w$ (or $\gamma$). Specific weight of water is 9810 N/m$^3$.[1] For air the density of air at atmospheric pressure at 20°C is 1.24 kg/m$^3$.

Pressure can be given as absolute pressure or how much it is above or below the atmospheric pressure. The positive pressure is called the gauge pressure and the negative pressure the *vacuum*. The atmospheric pressure acts on our whole body but the net force on the body is zero. This atmospheric pressure is about 76 cm of mercury column, which is too large compared to blood pressure of 12 cm of mercury in our body.

The other properties of fluids are dynamic viscosity and thermal conductivity. We learn from Kinetic theory of substances that these properties are related to the movement of molecules from one layer to another in laminar flow case in which there exists some gradient of velocity and temperature across the layers. Molecules from lower velocity layer move with less momentum as compared to those from the higher velocity layer; the effect is dragging of fast moving layer and accelerating of slow speed moving layer. This phenomenon is expressed by the property of the fluid called *dynamic viscosity*, which is given by the relation $\mu = 1/3\, \text{mn}\, \overline{C}\,\lambda$, where $\overline{C}$ is average velocity of the molecules. On macroscopic basis, dynamic viscosity is related to velocity gradient and is given by the Stoke's relation, in a three dimensional flow,

$$\tau \propto (\partial u/\partial y) + (\partial v/\partial x)$$

or,

$$\tau = \mu[(\partial u/\partial y) + (\partial v/\partial x)] \tag{1.2.2}$$

---

[1] Some authors have taken the density of water as 998 and g = 9.806 that gives w = 9787 N/m$^3$; both values have been used in this book. Normally 9810 is commonly used though the other value is more accurate.

where $\mu$ is coefficient of dynamic viscosity, $\tau$ is the shear stress on the surface along which flow is taking place and $u$ and $v$ are the component of velocities along $x$ and $y$ direction of the reference axes assumed to be fixed to the ground. Noting that dimensions for $\tau$ is $ML^{-1}T^{-2}$, for velocity $LT^{-1}$, and for $x$ or $y$ that of L, the dimension of dynamic viscosity is obtained from the principle of dimensional homogeneity to give

$$[\mu] = [\tau]/[\partial u/\partial y]$$

But $$[\partial u/\partial y] = LT^{-1}/L = T^{-1}$$

$\therefore$ $$[\mu] = ML^{-1}T^{-2}/T^{-1} = ML^{-1}T^{-1} \quad (1.2.3)$$

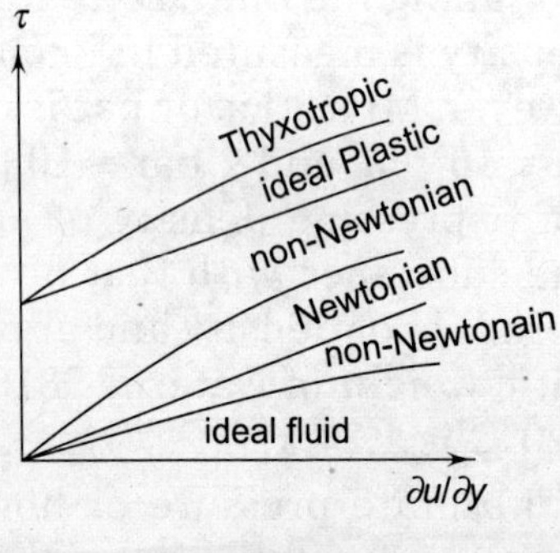


**Fig. 1.2.2** Variation of shear stress with velocity gradient

We have used notation [$\tau$] in Eq. 1.2.3 to represent the dimensions of the quantity $\tau$. Unit of dynamic viscosity in SI system of units is kg/m-s.[2] However, its unit were defined in CGS system in which it is g/cm-s, called *poise*, which is $10^{-1}$ kg/m-s. We may realise that its value will depend on temperature, since the molecular movement is related to temperature and the kind of the fluid. Figure. 1.2.2 shows variation of shear stress at any point in the fluid with velocity gradient for different kinds of the fluids. It is seen that for some fluids, this variation is linear. Such fluids are called *Newtonian* fluids. For other fluids, the variation is non-linear which falls under the category of *non-Newtonian* fluids. Therefore these also include the ones for whom the curves do not start from the origin. This behaviour is observed when the substance is in the plastic stage. Many mathematical models are available to describe the behaviour of the non-linear shear stress with deformation rate (called the constitutive relations) for Newtonian fluids. No such relation, however, can be prescribed for all kinds of non-Newtonian fluids. The power law model, known as the Ostwald – de Waele model, named after the scientist who proposed it, is used for the most common non-Newtonian

[2] $\dfrac{kg}{m-s^2} = \left(kg \times \dfrac{m}{s^2}\right) \times \left(\dfrac{s}{m^2}\right)$ = N–s/m², which is also used as unit for dynamic viscosity.

fluids. According to this model

$$\tau = m\left|\frac{du}{dy}\right|^{n-1}\frac{du}{dy}$$

$$\rightarrow \qquad \mu = m\left|\frac{du}{dy}\right|^{n-1} \tag{1.2.4}$$

where $m$ and $n$ are the flow consistency and flow behaviour index, respectively. We observe that the viscosity in such fluids is function of the deformation rate as well. When $n = 1$, the model is same as for Newtonian fluid. Fluids such as gelatine, blood, milk, etc have $n < 1$ and are called *pseudoplastic fluids*. Whereas the *dilatant fluids* such as sugar in water, aqueous suspension of rice starch, etc have $n > 1$. Table 1.2.1 gives classification according to the relation between shear stress and the rate of angular deformation:

**TABLE 1.2.1** Classification of the fluids

| Shear stress | Kind of fluid |
|---|---|
| $\tau = 0$ | Ideal fluid |
| $\tau = \mu\,(\partial u/\partial y)$ | Newtonial fluid |
| $\tau =$ constat + $\mu\,(\partial u/\partial y)$ | Ideal plastics or Bingham plastics |
| $\tau =$ constat + $\mu\,(\partial u/\partial y)^n$ | Thyxotropic fluids |
| $\tau = \mu\,(\partial u/\partial y)^n$ | Non-Newtonial fluids |

We shall consider only the Newtonian fluids for our discussion in this book. While dealing with the equations, a group $(\mu/\rho)$ often appears and therefore this group of properties of the substance has been given a separate name, the *kinematic viscosity* having unit m$^2$/s in SI units, which is named Stoke, and represented by the symbol $\nu$. We may note that kinematic viscosity has the dimensions $L^2T^{-1}$. Both the dynamic and kinematic viscosities are function of pressure and temperature. But the variation of the viscosity with pressure for liquids is so small that it can be neglected and dynamic viscosity at any temperature $T$ can be obtained from the relation $\mu_T = Ae^{B/T}$, where $A$ and $B$ are the constants. On the other hand, the relation $\mu_T = AT^{1/2}/(1 + B/T)$, in which $A$ and $B$ are constants, gives variation of the viscosity of gases with temperature. We notice that while viscosity of the liquids decreases with temperature, it increases for the gases. Figure. 1.2.3 shows variation of dynamic viscosity for some liquids and gases with temperature.

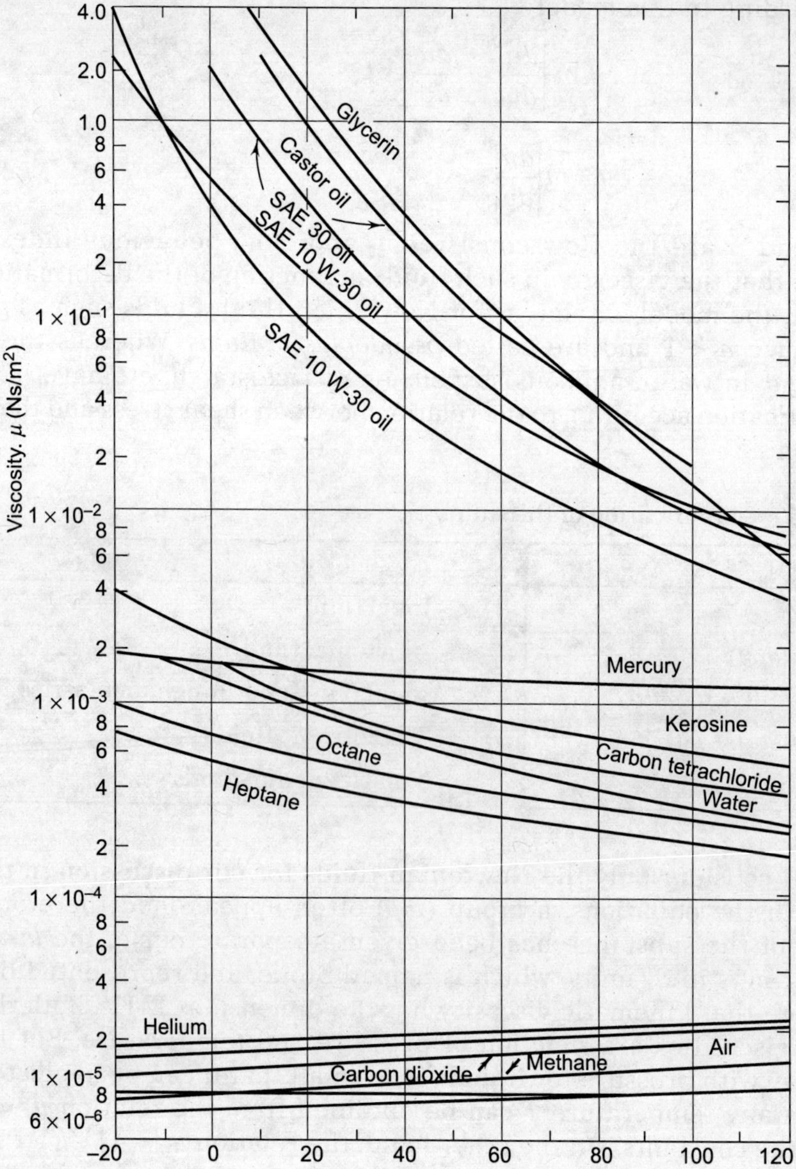


**Fig. 1.2.3** Viscosity of common fluids as a function of temperature

The following empirical relations have also been suggested for variation of viscosity with temperature and pressure:

***For liquids***

$$\mu_t = \frac{\mu_0}{1 + At + Bt^2}$$

and $$\mu_p = \mu_0 \exp[K(p - p_0)] = \mu_0 \exp(Kp_g)$$

where $A$, $B$, and $\mu_0$ are constants in the expression for the dynamic viscosity $\mu_t$ at temperature $t$ °C and depend on the type of the fluid. Here $p_g$ represents the gauge pressure measured above the atmospheric pressure. For water $\mu_0$ = 0.0179 poise, $A$ = 0.03368 and $B$ = 0.000221. This relation shows inverse parabolic variation of the viscosity with temperature that tends to zero at infinite temperature. The other correlation in the Eq. shows variation of viscosity with pressure in which the constant $K$ depends on the nature of the fluid; $p_0$ is the atmospheric pressure. That is we use the gauge pressure in the expression. The increase in viscosity with pressure was expected because the amount of energy required for the relative movement of the molecules is more. The viscosity increases by 10 to 15 percent for a pressure increase of about 75 atmospheres and is doubled for increase from 1 to 1000 atmospheres.

***For gases***

$$\mu_t = \mu_0 [1 + At - Bt^2]$$

For air, $\mu_0 = 1.7 \times 10^{-5}$ Ns/m², $A = 0.56 \times 10^{-7}$, and $B = 0.1189 \times 10^{-9}$.

We note that dynamic viscosity is the result of velocity gradient in the flow region. If the fluid is at rest or moves with uniform velocity then there exists no such gradient and there would not be shear stresses on the face of the element considered in the fluid. However, the pressure that acts normal to the surface would alone be acting. Without going into the proof, we make a statement that the fluids can take only compressive forces and not the tensile forces, unlike a solid. That is pressure is always negative, as per the convention adopted for the solids.

**Example 1.2.1** Determine the kinematic viscosity of a fluid having dynamic viscosity $1.85 \times 10^{-4}$ poise and density 1.208 kg/m³.

**Solution** Given $\mu = 1.85 \times 10^{-4}$ poise

$= 1.85 \times 10^{-4} \times 10^{-1}$

$= 1.85 \times 10^{-5}$ kg/m-s,

and $\rho = 1.208$ kg/m³.

$\therefore \quad \nu = \mu/\rho = 1.531 \times 10^{-5}$ m²/s.

**Example 1.2.2** Determine the density of $H_2$ at a temperature of 40°C and pressure of 2 atm.

**Solution** 1 atm = $1.01325 \times 10^5$ N/m²

$\rightarrow \quad p = 2$ atm $= 2.0265 \times 10^5$ N/m². Also $T = 313$ K

For $H_2$, $R = 8.3143/2$ kJ/kg K = 4157.15 J/kg.

Eq. of state gives, $\rho = p/RT$

$= (2.0265 \times 10^5)/(4157.15 \times 313)$

$= 1.54$ kg/m³

specific weight ($W$) $= \rho g$

$= 1.54 \times 9.81$

$= 15.1074$ N/m³.

**Example 1.2.3** If the velocity distribution in a flow over a surface is given by the relation $u = 2y - 2y^3 + y^4$, determine the shear stress on the surface at $y = 0$ and $y = 0.1$ m if dynamic viscosity of the fluid is $1 \times 10^{-3}$ kg/m-s.

**Solution** Differentiating the given velocity profile with reference to $y$, we get

$$du/dy = 2 - 6y^2 + 4y^3$$

$\therefore$ At $y = 0$, $\tau = \mu(du/dy)_{y=0}$

$$= 1 \times 10^{-3} \times 2$$

$$= 2 \times 10^{-3} \text{ N/m}^2.$$

And at $y = 0.1$ m, $\tau = \mu(du/dy)_{y=0.1}$

$$= 1 \times 10^{-3} \times (2 - 0.06 + 0.0004)$$

$$= 1.994 \times 10^{-3} \text{ N/m}^2.$$

**Example 1.2.4** A plate of area 5.0 m$^2$ is forming interface between two liquids of different viscosities, as shown in Fig. E 1.2.4. The plate is moved with a force 150 N applied at its one end. Determine the velocity with which it moves if the dynamic viscosity of the fluid above it is 0.1 Ns/m$^2$ and the one below it is having three times the dynamic viscosity.

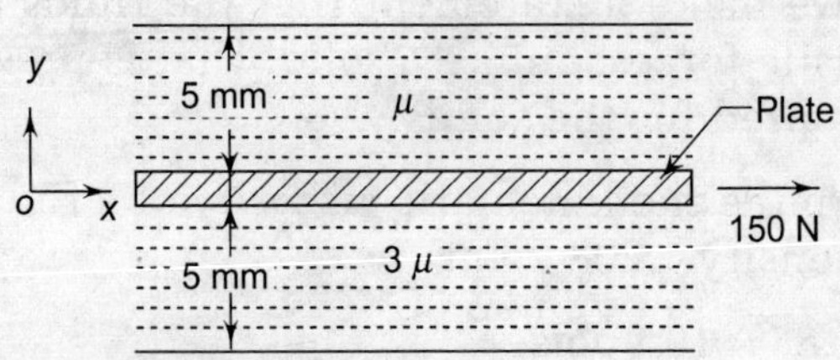


**Fig. E 1.2.4** Sliding of a plate

**Solution** Given. Area of the plate ($A$) = 5.0 m$^2$, Applied force ($F$) = 150 N, Distance of the plate from either surface ($h$) = 5 mm = 0.005 m.

Let $y$ be measured from the plate surface, as shown in the Fig., so that the plate surface is at $y = 0$ and $U$ be the velocity of the plate under equilibrium. Then the total drag force due to the two liquids on the two sides of the plate should be equal to the applied force. This generates the Eq.

$$[\mu(dU/dy)_{y=0} + 3\mu(dU/dy)_{y=0}]A = 150$$

$$\rightarrow \quad [0.1 \times U/0.005 + 3 \times 0.1 \times U/0.005] \times 5 = 150$$

Or, $U = 0.375$ m/s.

**Example 1.2.5** Obtain an expression for the torque on the inner cylinder of the rotating cylinder viscometer. In an experimental set up, $R_1$ = 100 mm, $R_2$ = 108 mm,

$h = 200$ mm, $a = 4.0$ mm, $N = 100$ rpm, and $T = 0.393$ Nm. Determine the dynamic viscosity of the liquid.

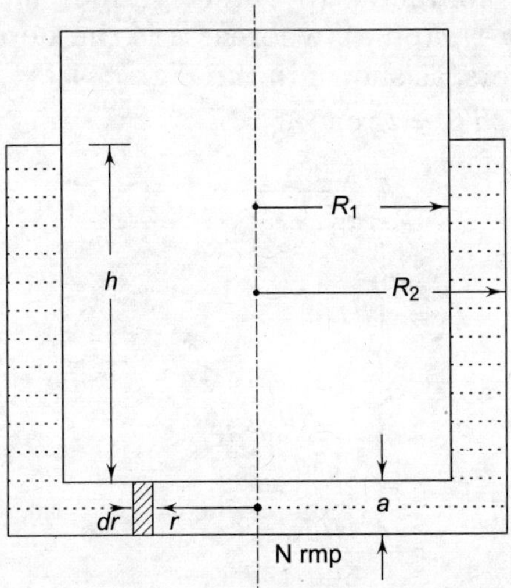


**Fig. E 1.2.5** Rotating cylinder viscometer

**Solution** The viscometer shown in Fig. E1.2.5 consists of two cylinders placed one inside the other with space between the cylinders filled with some fluid whose viscosity is to be determined. The outer cylinder is rotated while the inner cylinder is kept fixed. We can measure the torque applied on the inner cylinder using the transducers or some other method of measurement. (It is also possible to keep the outer cylinder fixed and rotate the inner cylinder.) The fluid layer in contact with the outer cylinder rotates with the cylinder having circumferential velocity $2\pi R_2 N/60$ and the inner layer in contact with the stationary cylinder has zero velocity.

$$\therefore \quad \frac{du}{dy} = \frac{2\pi R_2 N}{60(R_2 - R_1)},$$

$$\text{Shear stress } (\tau_0) = \mu \frac{2\pi R_2 N}{60(R_2 - R_1)}, \text{ and}$$

$$\text{Torque } (T_0) = \text{Area} \times \tau_0 \times R_1$$

$$= (2\pi R_1 h) \times \frac{2\pi R_2 N\mu}{60\,(R_2 - R_1)} \times R_1$$

$$= \frac{\pi^2 R_1^2 R_2 N h\mu}{15(R_2 - R_1)}$$

where $T_0$ is the torque exerted on the inner cylinder due to the liquid contained in the vertical space between the cylinders and rotation of the outer cylinder. The liquid in the bottom of the viscometer will also transfer some torque to the inner cylinder. To determine it, we consider a circular ring of radial width dr at a distance $r$ from the axis of rotation. The layer in contact with outer cylinder has velocity $2\pi r$ N/60 and will induce shear stress = $2\pi r$ N$\mu$/(60 a), whereas a is the depth of liquid between bottom surface of the two cylinders, as shown in the Fig..

$$\therefore \quad dT_1 = \text{Area} \times \tau_0 \times r$$

$$= \frac{\pi^2 N\mu}{15a} r^3 dr,$$

$$T_1 = \int_0^{R_1} dT_1$$

$$= \int_0^{R_1} \frac{\pi^2 N\mu}{15a} r^3 dr$$

$$= \frac{\pi^2 N\mu R_1^4}{60a}$$

and

$$\text{Total torque } (T) = T_0 + T_1$$

$$= \frac{\pi^2 N\mu R_1^2}{15}\left[\frac{R_1^2}{4a} + \frac{R_2 h}{R_2 - R_1}\right]$$

However, $(R_2 - R_1)$ is small and hence the term $\frac{R_1^2}{4a} << \frac{R_2 h}{R_2 - R_1}$. Therefore the first term $T_0$ is often not considered in the calculations. Substituting the values in the above relation, we get $\mu$ = 0.885 kg/ms.

### 1.2.1 Introduction to Stress and Strain

From Eq. 1.2.2, we know that the stress at any point in the flow region is defined by the first derivative of velocity, which is a vector quantity. From vectors we know that a derivative of a vector makes it a tensor of order 2. Thus a stress has nine components being represented in the form

$$\begin{pmatrix} \tau_{xx} & \tau_{xy} & \tau_{xz} \\ \tau_{yx} & \tau_{yy} & \tau_{yz} \\ \tau_{zx} & \tau_{zy} & \tau_{zz} \end{pmatrix} \tag{1.2.4}$$

where $\tau$ is the symbol for stress. Double subscript notation is used with first subscript for the face on which the stress acts and the second subscript for the direction in which it acts. For example, $\tau_{xz}$ is the stress on $x$-face in $z$-direction. Recall the direction normal to the face is used to represent the face. Area is considered positive when the outward drawn normal to the face is towards positive axis. Stress on three faces, normal to which is along the positive direction, is shown in Fig. 1.2.4; the cuboid has been drawn in the fluid region around a point. When the cuboid is shrunk letting its side tend to zero then it gives the stresses at the point.

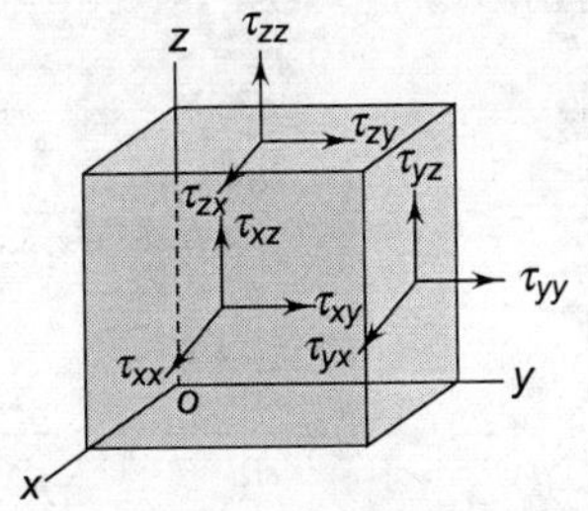


**Fig.1.2.4** Stress representation

The diagonal elements represent the component acting normal to the surface and in many books, $\sigma$ has been used in place of $\tau$. Whereas the off-diagonal components act in the surface and are named shear stress. We may imagine a force acting on the face being resolved along normal to the surface and in the surface; the component in the surface being further resolved along the two mutually perpendicular directions, which normally are parallel to the direction of axes of the reference system used. The diagonal elements, in general, are not equal in magnitude. We define their average as the *thermodynamic pressure* at the point to which the volume element would shrunk in the limit. Thus,

$$p = (\tau_{xx} + \tau_{yy} + \tau_{zz})/3 \tag{1.2.5}$$

In mechanics of solids, we read about the stresses and how are these related to the corresponding strains. We may similarly define strain components for fluids too, representing it by the symbol $\varepsilon$. There is however one significant difference in the behaviour of solid and fluid. When a shear force is applied to a solid, it deforms and stays in deformed position so long the force acts and comes back to original position after removing the force. Fluids, on the contrary, deform continuously even if an infinitesimal shear force acts on it. It is found that the time rate of deformation becomes constant in fluids. Therefore, we use time strain instead of strain. That is, the symbol $\varepsilon$ stands for time rate of strain. It will not be difficult to show that it is nothing but the velocity gradient $\partial u/\partial y$. The corresponding strain rate components are represented as below:

$$\begin{pmatrix} \varepsilon_{xx} & \varepsilon_{xy} & \varepsilon_{xz} \\ \varepsilon_{yx} & \varepsilon_{yy} & \varepsilon_{yz} \\ \varepsilon_{zx} & \varepsilon_{zy} & \varepsilon_{zz} \end{pmatrix} \tag{1.2.6}$$

A fluid is said to be *isotropic* when these relations do not depend on the direction of axes and *homogeneous* when it is independent of the position of the fluid particle in the fluid region. We may also establish that the stress tensor is symmetric, which means the off-diagonal elements are equal. That is, $\tau_{xy} = \tau_{yx}$. In such case, the tensor components are reduced from 9 to 4 — three shear stress viz. $\tau_{xy}$, $\tau_{xz}$, $\tau_{yz}$, and one pressure component.

The Newtonian model is the simplest one, which holds true for a parallel flow. Stokes, an English mathematician, extended it to a general flow, which is described by Eq. 1.2.2 for the shear stresses. For normal stresses, he suggested the following mathematical model

$$\tau_{xx} = -p + 2\mu \frac{\partial u}{\partial x} + \mu' \left[\frac{\partial u}{\partial x} + \frac{\partial v}{\partial y} + \frac{\partial w}{\partial z}\right],$$

$$\tau_{yy} = -p + 2\mu \frac{\partial v}{\partial y} + \mu' \left[\frac{\partial u}{\partial x} + \frac{\partial v}{\partial y} + \frac{\partial w}{\partial z}\right], \text{ and}$$

$$\tau_{zz} = -p + 2\mu \frac{\partial w}{\partial z} + \mu' \left[\frac{\partial u}{\partial x} + \frac{\partial v}{\partial y} + \frac{\partial w}{\partial z}\right], \qquad (1.2.7)$$

where $\mu'$ is another constant, which is determined using Eq. 1.2.5. Substitution of normal stresses from Eq. 1.27 in Eq. 1.2.5 and subsequent simplification gives the average of the normal stresses as

$$-p + 1/3\ (2\mu + 3\mu') \left[\frac{\partial u}{\partial x} + \frac{\partial v}{\partial y} + \frac{\partial w}{\partial z}\right]$$

This will be equal to thermodynamic pressure p (compressive) if

$$(2\mu + 3\mu') \left(\nabla.\vec{U}\right) = 0 \qquad (1.2.8)$$

the negative sign shows that the pressure will always be compressive. We shall later see in Chapter 2 that for incompressible fluids it is zero and hence Eq. 1.2.8 is satisfied automatically. But for any fluid the equation is also satisfied if $2\mu + 3\mu' = 0$, or $\mu' = -2/3\ \mu$. This shows that $\mu'$ is always negative but the cases have been reported for the fluids for which it is positive. However, in some cases this condition though may not be satisfied but the value of the terms in Eq. 1.2.8 is negligibly small and may safely be ignored. Therefore, $\mu'$ can be taken as $-2/3\ \mu$ and then Eq. 1.2.7 takes the form

$$\tau_{xx} = -p + 2\mu \frac{\partial u}{\partial x} - \frac{2}{3}\mu \left[\frac{\partial u}{\partial x} + \frac{\partial v}{\partial y} + \frac{\partial w}{\partial z}\right],$$

$$\tau_{yy} = -p + 2\mu \frac{\partial v}{\partial y} - \frac{2}{3}\mu \left[\frac{\partial u}{\partial x} + \frac{\partial v}{\partial y} + \frac{\partial w}{\partial z}\right], \text{ and}$$

$$\tau_{zz} = -p + 2\mu \frac{\partial w}{\partial z} - \frac{2}{3}\mu \left[\frac{\partial u}{\partial x} + \frac{\partial v}{\partial y} + \frac{\partial w}{\partial z}\right], \qquad (1.2.9)$$

We shall use these constitutive relations later for derivation of Navier-Stokes Eq.; Navier was a French engineer. The Eq. 1.2.9 gives only the normal stresses. The other components similarly were obtained and can be written as (in tensor notation)

$$\tau_{ij} = -p\delta_{ij} + \mu\left(\frac{\partial u_i}{\partial x_j} + \frac{\partial u_j}{\partial x_i}\right) - \frac{2}{3}\mu\delta_{ij}\ \nabla.\vec{U}, \qquad (1.2.10)$$

where $\delta_{ij}$, the Kronecker delta = 1 when $i = j$ and 0 when $i \neq j$. If we take $i = j = 1$ then Eq. 1.2.10 takes the form of the first equation for $\tau_{xx}$ in Eq. 1.2.9. Similarly, we get the other equations of Eq. 1.2.9 by taking $i = j$ equal to 2 and 3, respectively. When $i = 1$ and $j = 2$, Eq. 1.2.10 gives Eq. 1.2.2; substitute $x = 1$ and $y = 2$ in Eq. 1.2.2 to make comparison.

## 1.2.2 Other Properties of Fluids

Another property of importance in fluids is the surface tension. We know a matter is made up from the molecules that move randomly in different directions; the motion is characterised as chaotic motion. Each molecule is surrounded around it by other molecules and is being pulled by a force, called the *cohesive force.* The molecules, which are at a depth more than the molecular diameter from the free surface of the fluid, experience zero force, since the forces from all the sides balance out. However, for the molecules in a thin strip of the size of molecules' diameter near the free surface, the force balance is disturbed as number of molecules above it are less than below it. Thus there is a net force acting on such molecules pulling these molecules to keep them within the fluid surface. It looks like as if there is a membrane on the fluid surface that does not allow the molecules to escape. This phenomenon is called surface tension. If the molecule is to be pushed out of the fluid, as in evaporation process, the molecule is to be given enough kinetic energy to overcome this force of attraction. The amount of energy required to push the molecule just out per unit area of the surface is called *surface tension*, designated by the symbol $\sigma$. However, the surface tension, as used in practice, is defined in terms of the force exerted by the free liquid surface per unit of its length instead of energy term and both are equivalent to each other. Thus surface tension is a property of the fluid that depends on the fluid itself and its temperature. It is easy to recognise that the amount of energy required will depend on the closeness of the molecules, which means more force of attraction and hence more the energy to be supplied or surface tension. This implies that the surface tension would decrease with increase in temperature. When the surface tension is zero, the temperature is called the *critical temperature.*[3]

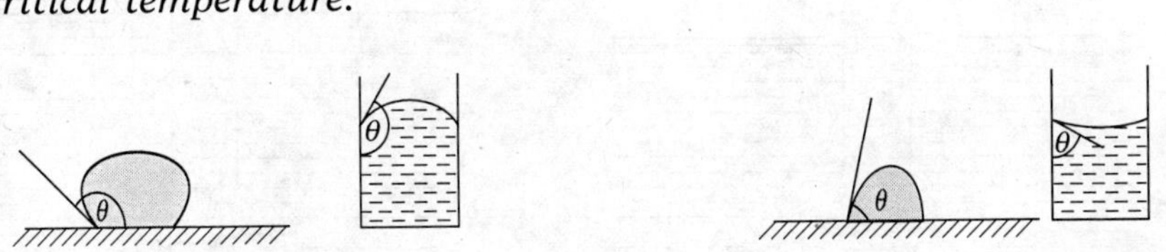

(a) Non-wetting surface, $\theta > \pi/2$ ( Hg-glass) (b) Wetting surface, $\theta < \pi/2$ (water-glass)

**Fig. 1.2.5** Effect of adhesive and cohesive forces on shape of interface

When the two fluids or a fluid and a solid form the interface then molecules of one media pull molecules in the near surface of another media that lie in the region of influence. This force of attraction is called the *adhesive force.* The relative magnitude of cohesive and adhesive forces decides the shape of the interface. Fig. 1.2.5 shows

[3] At this point, liquid and vapour become indistinguishable.

interface between a solid surface and a liquid. In case of mercury, the cohesive force is so strong that the mercury drops are almost spherical, as shown in Fig. 1.2.5 (a). When the size of the drop is large than it is flattened somewhat and does not remain spherical. If we draw a tangent to the interface surface at the point of contact, the line makes an angle $\theta$ with the surface occupied by the drop. The angle $\theta$ is called the *angle of contact* and is around 130° – 150° in the case of glass-mercury interface, instead of 180°, because of impurities. Such fluids are classified as non-wetting liquids. On the other hand adhesive forces are so strong in water that it wets the surface almost. Contaminated water may have a convex surface with $\theta$ upto 25°. Effect of the angle of contact is seen in the meniscus of the free surface of liquid in a tube. We note that while the meniscus is convex in mercury, it is concave for water.

Water in contact with air has surface tension of 0.073 N/m at 20°C; for mercury in contact with air, it is 0.05 N/m at 20°C. Adding some other chemical compounds can modify the surface tension. For example, surface tension of water can be reduced considerably by adding a small quantity of an organic compound such as soap or detergent and can be increased by adding a small quantity of salt such as sodium chloride, etc. Reduction in surface tension allows seaping of water through the pores in th cloth and remove dust particles easily. This explains how do detergent, soap, etc help in washing the clothes.

In normal situations, the surface tension force is significantly small compared to pressure and other forces to be considered in the Eq. of motion of the fluid. But in some applications like rising of liquid in a capillary, bubble mechanics in water boiling, liquid-jet studies and flow through a porous media, the surface tension force is comparable and cannot be neglected. Fig. 1.2.6 shows the forces in some of these cases.

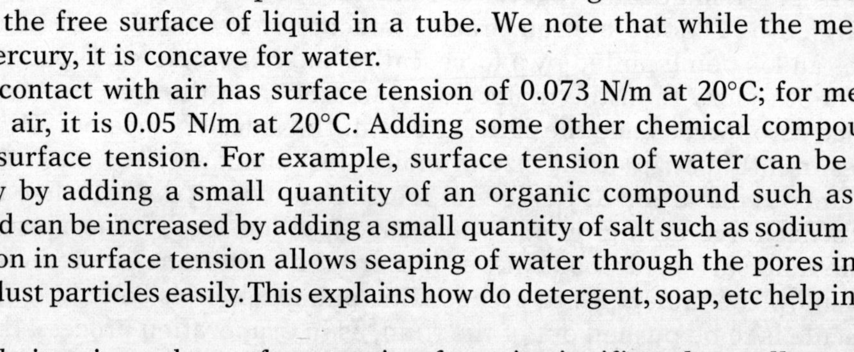

(a) Rise of water in glass tube

(b) Fall of mercury in glass tube

**Fig. 1.2.6** Capillary action in water and mercury

Consider a capillary tube dipped in water (Fig. 1.2.6a) or in mercury (Fig. 1.2.6b). It is observed that the water rises while the mercury level goes down in the capillary tube. This action can be observed when the diameter of the tube is small, less than 1 cm. The water level will be concave, because of acute angle of the contact as discussed above, and the surface tension force $\sigma$ acts in the direction as shown. This force will have a vertical component that forces the water to rise till it is equal to the weight of the water raised in the capillary. For the same reason mercury level goes down the free surface. Balancing the forces, we get

$$\pi d\sigma \cos\theta = (\pi d^2/4)h.\rho g$$

$$\rightarrow \qquad h = 4\sigma\cos\theta/(\rho g d) \tag{1.2.11}$$

where $h$ = height of the liquid column in the capillary tube of diameter $d$ and $\theta$ is the angle of contact.

Now consider a spherical bubble of diameter $d$ enclosing a fluid at pressure $p$. Since the pressure acts in all directions, the net force in any direction would be zero. However, if we take a rim on it then the pressure force above and below it acts in opposite direction and hence a force acts on it, which is the surface tension. The net pressure force responsible for bursting apart of the bubble into two parts is $= p \times$ area perpendicular to the pressure direction $= p \times (\pi d^2/4)$. Equating it to the surface tension force, we get

$$\pi d \times \sigma = p \times (\pi d^2/4) \quad \rightarrow \quad p = 4\sigma/d \tag{1.2.12}$$

Recall if the spherical shell is a metallic one then the stress developed is called hoop stress. If this shell has some thickness such that the two radii are $r_1$ and $r_2$ then

$$p = \sigma\left(\frac{1}{r_1} + \frac{1}{r_2}\right) \tag{1.2.13}$$

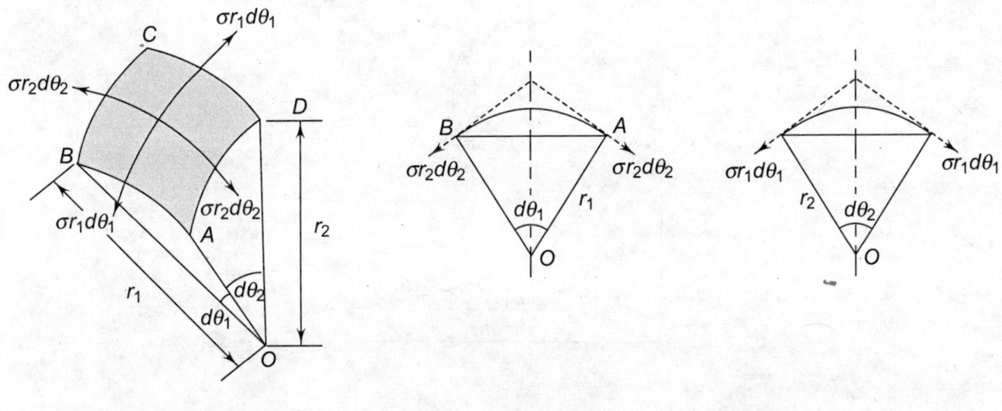


**Fig.1.2.7** Forces acting on a surface separating two immiscible fluids from each other

To get this expression for the fluids, let us consider a curved liquid surface separating the fluid on convex side from the rest of the fluid, as shown in Fig. 1.2.7; $r_1$ and $r_2$ be the radii of curvature of the convex and concave surfaces. Let the angle subtended by these at the centre of curvature $O$ be $d\theta_1$ and $d\theta_2$, respectively. Forces are shown in two-dimensional frame also, which is self-evident. Let $p_1$ and $p_2$ be the pressure on the convex and concave surfaces. For the equilibrium, the vertical forces must balance. This gives the Eq.

$$2\sigma r_2\, d\theta_2 \sin(d\theta_1/2) + 2\sigma r_1\, d\theta_1 \sin(d\theta_2/2)$$

$$= (p_2 - p_1)\, r_1 r_2\, d\theta_1 d\theta_2$$

$$\rightarrow \qquad \Delta p = p_2 - p_1 = \sigma\left(\frac{1}{r_1} + \frac{1}{r_2}\right) \qquad (1.2.14)$$

If we neglect $p_2$ then Eq. 1.2.14 gives Eq. 1.2.13. There is also possibility that a thin liquid surface may be trapped between the same fluids, as in bubble. In this air is present all around the liquid film. In that case the lower surface will also exert the same amount of the force as the upper surface and hence the net force will be doubled. Then Eq. 1.2.14 changes to

$$\Delta p = p_2 - p_1 = \sigma\left(\frac{1}{r_1} + \frac{1}{r_2}\right) \qquad (1.2.15)$$

**Example 1.2.6** Make an analysis of the shape of the water-air interface near a plane wall, as shown in Fig. E 1.2.6. Assume that the slope is small, $1/R = d^2\eta/dx^2$, (where $R$ is the radius of curvature of the interface), and the pressure difference across the interface is balanced by the product of specific weight and interface height as $\Delta p = \rho g \eta$. Boundary conditions: wetting contact angle $\theta = \theta_0$ at $x = 0$ and $\theta = 90°$ as $x \rightarrow \infty$. What is the height $\eta$ at the wall?

**Solution** One of the surface of the interface is plane, i.e. $r_2 = \infty$ whereas for the other surface radius of curvature is $R$. Therefore Eq. 1.2.15 gives

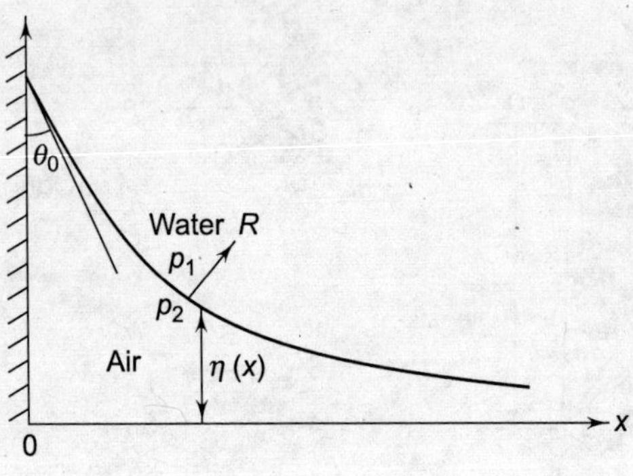


**Fig. E 1.2.6** Water air interface near a wall

$$\Delta p = \frac{\sigma}{R} = \sigma \frac{d^2\eta}{dx^2}$$

Also $\qquad \Delta p = \rho g \eta$ (given)

$$\rightarrow \qquad \frac{d^2\eta}{dx^2} - \frac{\rho g}{\sigma}\eta = 0.$$

Integrating the equation two times with respect to $x$ and using the given boundary conditions, we get

$$\eta = \sqrt{\frac{\sigma}{\rho g}} \cot\theta_0 \exp\left(-\sqrt{\frac{\sigma}{\rho g}}\, x\right)$$

This equation defines the shape of the interface. Substituting $x = 0$, we get

$$\eta_{x=0} = \sqrt{\frac{\sigma}{\rho g}} \cot\theta_0$$

## 1.3 Pressure Variation and Hydrostatic Forces on Surfaces Submerged in Fluid at Rest

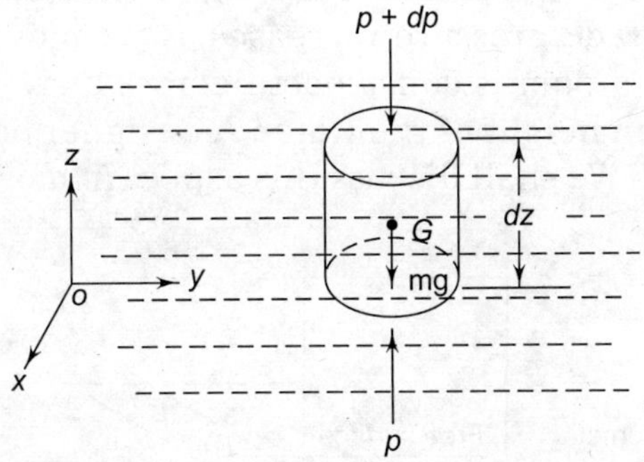


**Fig. 1.3.1** Pressure variation

The pressure force acts at any point of the fluid (which is at rest) that depends on the depth of the point and the gravity field. This can be understood by considering equilibrium of the fluid element, as shown in Fig. 1.3.1. A cylindrical element of cross-sectional area $dA$ and height dz is shown. Then the mass of this fluid element is $m = \rho\, dA\, dz$, where $\rho$ is density of the fluid; the cylinder height is so small that density of the fluid may be considered as constant over its height. Let pressure $p$ acts on its lower face and $p + dp$ on the upper face. Let $z$ axis[4] be taken positive along vertically up direction so that the gravity is along –ve z-direction and its weight mg acts downwards. Balancing the forces in the vertical direction, we get

$$pdA - (p + dp)\, dA = \rho dA\, dzg \quad \rightarrow \quad dp = -\rho g\, dz \tag{1.3.1}$$

This equation means that the pressure at any point depends on the depth of the point from the reference point. It implies that (i) pressure increases if we move down, i.e. pressure increases with the depth, and (ii) the pressure in a horizontal plane would

[4] The coordinate system shown in Fig. 1.3.1 is a right-handed Cartesian coordinate system. It is called right handed because if we rotate a right-handed screw from $x$ towards y the screw will move along $z$ direction. Remember the rotation is to be from $x$ to $y$, $y$ to $z$ or $z$ to $x$ and not the vice versa. In this figure we have taken $y$-$z$ plane in the plane of the paper and $x$ at right angles. We may also take $x$-$y$ in the plane of paper and decide the direction of the third axis by RHS (right-hand screw) rule. We shall follow the Right Hand Coordinate system in this book.

be same and independent of the shape of the container. If we neglect effect of gravity then pressure would be same, a result we read very early in science course and known as *Pascal's* law. This property of the fluid has been effectively utilised in hydraulic press and many other hydraulic machines. Equation 1.3.1 can be integrated if relation between density and height is known. Four cases are possible:

***Case 1***: Density is constant. In this case, integration of Eq. 1.3.1 from ground to any point $z$ above it results in

$$\int_{p_0}^{p} dp = -\rho g \int_{0}^{z} dz$$

$$\rightarrow \quad p = p_0 - \rho g z \tag{1.3.2}$$

Remember that pressure would increase if we go down, since $z$ will be negative in Eq. 1.3.2. For the matter of fact, pressure increases as we move down the water surface in sea, pond or river. Or if a dam is constructed across a river then pressure on its walls increases with increase in height of the water. A designer has to take care of such forces while designing the wall. We shall discuss this aspect in more details in the next section.

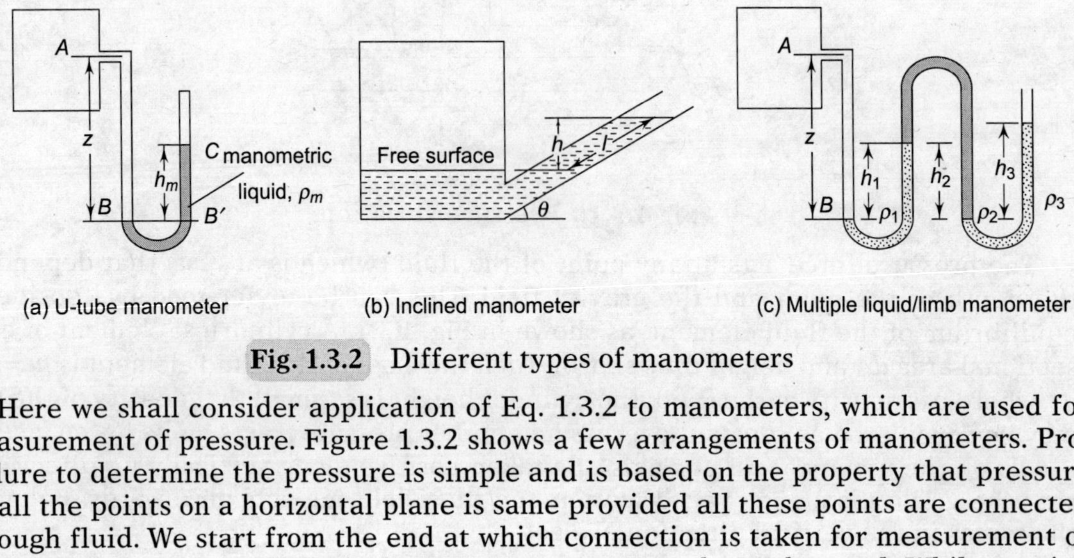


**Fig. 1.3.2** Different types of manometers

Here we shall consider application of Eq. 1.3.2 to manometers, which are used for measurement of pressure. Figure 1.3.2 shows a few arrangements of manometers. Procedure to determine the pressure is simple and is based on the property that pressure on all the points on a horizontal plane is same provided all these points are connected through fluid. We start from the end at which connection is taken for measurement of pressure and traverse through the arrangement to reach another end. While moving across one fluid to another we determine pressure at the interface, which then becomes starting point for another interface. This process continues till we reach the free surface. We also know that the pressure increases while going down and decreases while moving up in the same liquid by a value $\rho gh$, where $h$ is the vertical height between the two points in the same fluid region having density $\rho$. This way an equation is obtained, which is called *manometric equation.*

Let us illustrate the procedure for the arrangement shown in Fig. 1.3.2 (a) for a U-tube manometer, which is often used in laboratories for measurement of pressure. It is connected to the vessel, where pressure is to be measured, at a point A and is filled with some liquid, called the manometric fluid having density $\rho_m$. Due to pressure in the vessel, the manometric liquid takes free surfaces $B$ and $C$, as shown, such that the pressure at horizontal line through $B$ is same in both the limbs. We start from point $A$ and reach the point $C$ following the principle mentioned above to get the manometric equation.

$$p_A + \rho g z - \rho_m g h_m = p_C$$

or, $$p_A - p_C = \rho_m g h_m - \rho g z \approx \rho_m g h_m \tag{1.3.3}$$

where subscripts $A$ and $C$ refer to the points $A$ and $C$, and $\rho$ is the density of the fluid in the vessel that occupies the space above the manometric fluid in the U-tube. Since the $C$ end of the manometer is open to atmosphere, $p_c$ = 1 atm. Therefore, knowing the heights appearing in the equation, $p_A$ can be obtained from this equation. Normally $\rho << \rho_m$, $\rho g z$ term is dropped in general. However, one should justify this.

In case the pressure to be measured is too small then an inclined manometer and a manometric fluid of lower specific gravity may be used so that $l$, the length of the fluid in the inclined limb of the manometer, may be sufficiently large to read accurately. We may also adjust inclination angle of the tube; a decrease will increase the length $l$. You can write easily the manometric equation remembering that while moving along the inclined limb the vertical distance moved is not $l$ but $l \sin \theta$. The inclined manometer is shown in Fig. 1.3.2(b).

On the other hand, we may put a number of U-tubes in series to measure a large pressure. Figure 1.3.2(c) shows two tubes put together. Three fluids of densities $\rho_1$, $\rho_2$, and $\rho_3$ are used; the difference in the meniscus in the two limbs respectively is $h_1$, $h_2$, and $h_3$, respectively. If the pressure at the open end is $p_a$ then the manometric equation is

$$\rho_A + \rho g z - \rho_1 g h_1 + \rho_2 g h_2 - \rho_3 g h_3 = p_a$$

$$\rightarrow \quad p_A - p_a \approx g\,(\rho_1 h_1 - \rho_2 h_2 + \rho_3 h_3)$$

***Case 2****: Isothermal atmosphere.* In atmosphere, density of air decreases as we go up. That is, there exists some relation between density and height of the air column above the sea level. It is known that temperature decreases uniformly in stratosphere upto height of 11 km after which it is constant at that temperature up to 20.1 km; Figure 1.3.3 shows variation of temperature with altitude and Appendix B1 gives details of the standard atmosphere, which of course varies from location to location. For stratosphere, we may assume isothermal condition so that $\rho$ = kp, where $k = 1/RT$, a constant that can be determined knowing temperature of air in that zone. Substituting for $\rho$ in Eq. 1.3.1 and integrating from a point at hight $z_1$ to any point $z$ m further above it, we get

$$\int_{p_1}^{p} dp/p = -kg \int_{z_1}^{z} dz$$

$$\rightarrow \quad \ln\,(p/p_1) = -kg\,(z - z_1)$$

or, $$p/p_1 = e^{-g(z - z_1)/(\mathrm{RT})} \tag{1.3.4}$$

$p_1$ is obtained by consedering pressure, density and temperature variation in the troposhere, which is discussed next.

***Case 3:*** Temperature decreases linearly with altitude $z$, as in troposphere. We know that the temperature decreases uniformly with altitude near the ground surface upto 11 km. Let the temperature at any height $z$ from the ground be $T = T_0 - m\,z$, where $m$ is slope of the temperature curve and represents the change in temperature per metre of the height from the ground, it is called *temperature lapse rate.* Temperature lapse rate upto 11 km is 0.0065°C/m. Assuming the air to obey the perfect gas equation, the density at any height $z$ is given by $\rho = p/\{R\,(T_0 - m\,z)\}$. Substituting it in Eq. 1.3.1 and integrating the Eq. from a point on the ground surface to some point at height $z$, we get

$$\int_{p_0}^{p} \frac{dp}{p} = -\int_{0}^{z} \frac{g\,dz}{R(T_0 - mz)}$$

$$\rightarrow \qquad \ln\frac{p}{p_0} = \frac{g}{mR}\ln\left(\frac{T_0 - mz}{T_0}\right)$$

or,

$$p = p_0\left(1 - \frac{mz}{T_0}\right)^{g/mR}$$

$$= p_0\left(\frac{T}{T_0}\right)^{g/mR} \qquad (1.3.5)$$

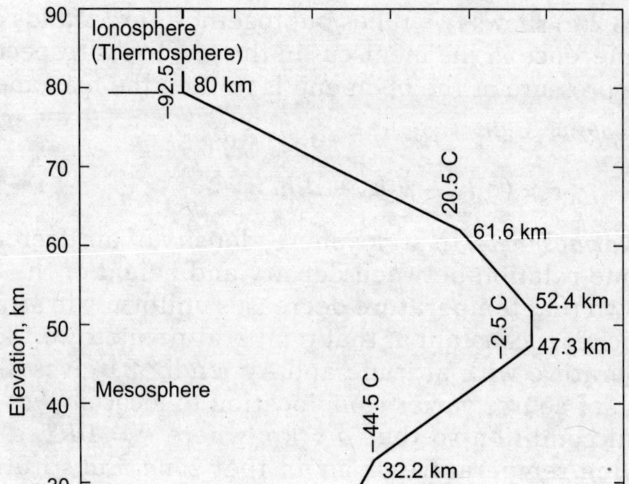


**Fig.1.3.3** Variation of air temperature with altitude

The air in contact with the earth surface becomes hot because of heat convection and radiation from the Sun and rises above where the temperature and pressure decrease. Air being bad conductor of heat gets cooled adiabatically at a cooling rate of about 0.01 °C. This rate of cooling is more than the natural temperature lapse rate in the troposphere. Therefore its density increases and air falls back to the earth thus bringing stability in atmosphere. If however, it is not so then the atmosphere is unstable that results in thunderstorm.

***Case 4**: Isentropic conditions.* In many processes the time required for a process to occur, such as propagation of wave in air, is so small that the process may be assumed to be isentropic for which $p = k\rho^{\gamma}$, where $k$ is a constant $= p_0/\rho_0^{\gamma}$. Substituting it in Eq. 1.3.1, using the equation of state and integrating we get after simplification

$$\int_{p_0}^{p} \frac{dp}{p^{1/\gamma}} = -gk^{1/\gamma}\int_0^z dz$$

$$\rightarrow \quad \frac{p}{p_0} = \left[1 - \frac{\gamma-1}{\gamma}\frac{gz}{RT_0}\right]^{\gamma/(\gamma-1)} \tag{1.3.6}$$

We may also find out temperature at any point using the equation of state.

$$p/\rho^{\gamma} = p_0/\rho_0^{\gamma}$$

$$\rightarrow \quad \frac{\rho}{\rho_0} = \left(\frac{p}{p_0}\right)^{1/\gamma},$$

$$\frac{\rho}{\rho_0} = \left[1 - \frac{\gamma-1}{\gamma}\frac{gz}{RT_0}\right]^{\gamma/(\gamma-1)} \quad \text{(Using Eq.1.3.6)}$$

and $$p = \rho RT \quad \text{(Equation of state)}$$

$$\therefore \quad \frac{T}{T_0} = \frac{p}{p_0} \times \frac{\rho_0}{\rho} \tag{1.3.7}$$

$$= \left[1 - \frac{\gamma-1}{\gamma}\frac{gz}{RT_0}\right]$$

Differentiating Eq. 1.3.7 w.r.t. $z$, we get the temperature lapse rate

$$m = \frac{dT}{dz}$$

$$= -\frac{g}{R}\left(\frac{\gamma-1}{\gamma}\right) \tag{1.3.8}$$

**Example 1.3.1** An inverted vertical manometer is connected across two pipes carrying water as shown in Fig. E 1.3.1. The manometer fluid is gasoline of sp. gr. = 0.796. If the pressure of water in pipe $A$ is 8.35 kPa (gauge) what is the pressure of water in pipe $B$?

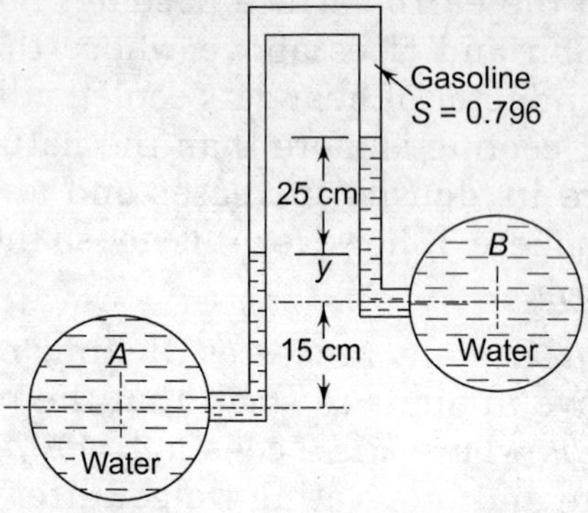


**Fig. E 1.3.1** Inverted manometer

**Solution** Inverted U-tube manometer is used frequently in laboratory when the pressure difference between the two points is small — a manometer fluid of specific gravity less than that of the fluid in pipes is to be used; one such arrangement is shown in Fig. E 1.3.1. We start from one end, say $A$ and reach the other end $B$ to write the manometric Eq.. The equation with $p_A$ and $p_B$ in $kP_a$ is

$$\frac{p_A}{w} - (0.15 + y) - 0.25 \times 0.796 + (0.25 + y) = \frac{p_B}{w}$$

$$\rightarrow \quad \frac{p_B}{w} = \frac{8.35}{9.81} - 0.099 = 0.75 \text{ m of water}$$

and $\quad p_B = 0.75 \times 9.81 = 7.379$ kPa.

In this Eq. $w$ is specific weight of water in kN/m$^3$.

**Example 1.3.2** A glass tube of uniform bore is bent into the form of a square of sides $a$, and filled with equal amounts (volume) of three immiscible liquids of densities $\rho_1$, $\rho_2$, and $\rho_3$. It is known that $\rho_1 < \rho_2 < \rho_3$. If the tube arrangement is filled with the liquids, as shown in the Fig., show that $(2\rho_3 + \rho_1)/3 > \rho_2 > (\rho_3 + 2\rho_1)/3$. If the relative densities of the first and third liquids are 1.0 and 1.2 respectively, find the range of the relative density of the second liquid which makes the above arrangement possible.

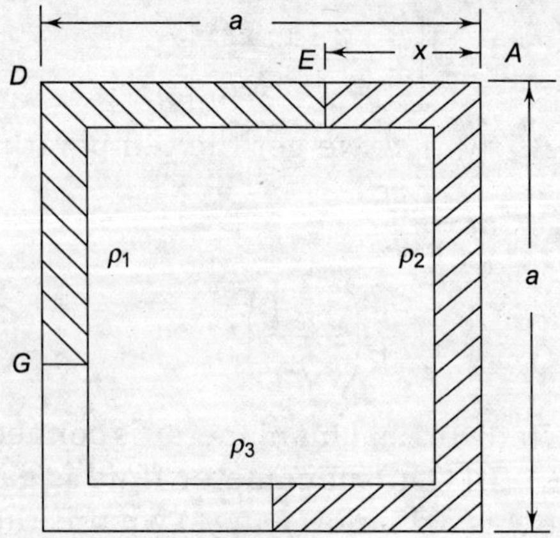


**Fig. E 1.3.2** Three fluids in equilibrium

**Solution** The three liquids in equilibrium with each other are shown in the figure. Let $AE = x$. Since the tube is of same area of cross section and liquids are having same volume, their lengths will also be equal. This gives the following dimensions of the liquids in the tube:

Length of each liquid = 4/3 $a$, since the total length is $4a$.

Liquid 2: $AB = a$, $BF = 4/3\ a - a - x = a/3 - x$.

Liquid 1: $ED = a - x$ and $DG = 4/3\ a - (a - x) = a/3 + x$.

Liquid 3: $CG = a - DG = 2a/3 - x$ and $CF = 4/3\ a - CG = 2a/3 + x$.

For equilibrium, the pressure at the interface $F$ from both the sides must be equal. This condition demands:

Pressure of the column $GD$ + pressure of the column $CG$ = pressure of the column $AB$.

$$\rho_1 g(a/3 + x) + \rho_3 g(2a/3 - x) = \rho_2 ga$$

$$\rightarrow \qquad x = \frac{a}{3}\frac{2\rho_3 + \rho_1 - 3\rho_2}{\rho_3 - \rho_1}$$

But from the given condition of equal lengths of the three liquids, we infer that $0 < x < (4a/3 - a) = a/3$. It implies that

$$0 < \frac{2\rho_3 + \rho_1 - 3\rho_2}{\rho_3 - \rho_1} < 1$$

$$\rightarrow \qquad 2\rho_3 + \rho_1 > 3\rho_2,$$

$$\text{or,} \qquad \rho_2 < (2\rho_3 + \rho_1)/3 \qquad \text{(i)}$$

$$\text{and} \qquad 2\rho_3 + \rho_1 - 3\rho_2 < \rho_3 - \rho_1$$

$$\text{or,} \qquad \rho_2 > (\rho_3 + 2\rho_1)/3 \qquad \text{(ii)}$$

From these two inequalities, we infer $(2\rho_3 + \rho_1)/3 > \rho_2 > (\rho_3 + 2\rho_1)/3$.

Substitution of $\rho_1$ and $\rho_3$ in the inequalities gives after simplification: $1.0667 < \rho_2 < 1.1333$.

### 1.3.1 Pressure Force on a Plane Surface

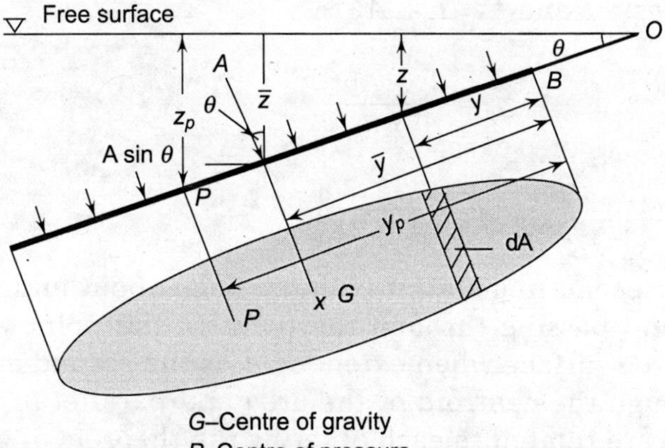


**Fig. 1.3.4** Plane surface immersed in liquid

Let us consider a plane surface *AB* in liquid, as shown in Fig. 1.3.4, immersed such that it makes an angle $\theta$ with the free surface. The surface, seen along normal to *AB*, is also shown just below it. Arrows show pressure acting on the surface, which is perpendicular to the surface because the fluid is at rest. Magnitude of pressure would change along its length, since depth of the surface from the free surface is increasing; a free surface is represented by the inverted triangle symbol ∇. Pressure at any depth $z$ is given by $\rho gz$. Force acting on area $dA$ of the strip considered at depth $z$ is $\delta F = \rho gz\, dA = \rho gy \sin\theta.\, dA$, where $y$ and $\theta$ are shown in the Fig.. Summation of these forces gives the total force acting on the surface. Taking moment about the point $O$, we can locate position of the force acting on the surface. This procedure yields:

$$F = \int_A \delta F = \int_A \rho gz\, dA$$

$$= \int_A \rho gy \sin\theta\, dA$$

$$= \rho g\bar{y} \sin\theta A = \rho g\bar{z} A \tag{1.3.9}$$

where $\bar{z}$ is depth of centre of gravity from the free surface. Taking moments about $O$, we get

$$F \cdot y_p = \int_A \delta F . y = \int_A \rho gy^2 \sin\theta\, dA$$

$$\rightarrow \qquad y_p = \frac{\int_A y^2\, dA}{\int_A y\, dA} = \frac{\text{Second moment of area about } O}{\text{First moment of area about } O} = \frac{I_o}{A\bar{y}}$$

But $\qquad z = y \sin\theta$ and $I_o = I_c + A\bar{y}^2$

$$\therefore \qquad y_p = \frac{I_c + A\bar{y}^2}{A\bar{y}}$$

and

$$z_p = y_p \sin\theta = \bar{z} + \frac{\sin^2\theta}{A\bar{z}} \tag{1.3.10}$$

Here $I_0$ is the second moment of plane surface about an axis perpendicular to the plane of paper and passing through the point $O$, the point where the plane surface would meet the free surface when extended. $I_c$ is the second moment of area about the axis passing through the centroid of the area and parallel to the axis thorough $O$ and the two moments are related to each other by the theorem of parallel axes. Table 1.3.1 gives the second moment of area about the axis passing through the centroid.

**TABLE 1.3.1** Geometric properties of some lamina

| Lamina | Area | Distance of centroid | | Moment of inertia about the centroidal axis XX |
|---|---|---|---|---|
| | | From bottom | From top | |
| Rectangle 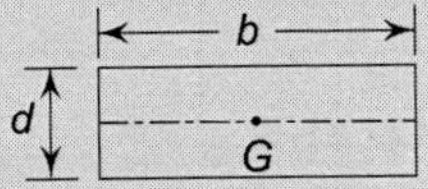 | $bd$ | $d/2$ | $d/2$ | $bd^3/12$ |
| Triangle 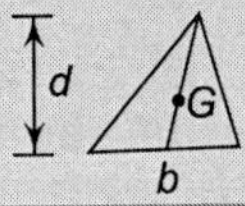 | $bd/2$ | $d/3$ | $2d/3$ | $bd3/36$ |
| Circle 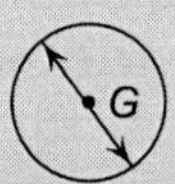 | $\pi d^2/4$ | $d/2$ | $d/2$ | $\pi d^4/64$ |
| Trapezoid 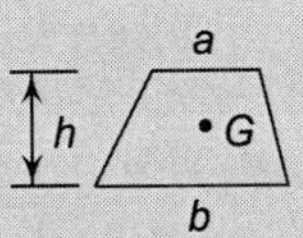 | $(a+b)h/2$ | $\dfrac{b+2a}{b+a}\dfrac{h}{3}$ | $\dfrac{a+2b}{b+a}\dfrac{h}{3}$ | $\dfrac{a^2+4ab+b^2}{a+b}\cdot\dfrac{h^3}{36}$ |

Equation 1.3.10 tells that the *centre of pressure*, the point on the surface where the total force is assumed to act, lies below the centre of gravity, since all the quantities are positive. This was expected since the pressure force on downside is more and hence its moment will be more.

We get some useful information from these two equations that are summarised below.

(i) The component of the force on the plane surface can be resolved along and perpendicular to the free surface. These components are $F \sin \theta$ and $F \cos \theta$, respectively. Therefore, $F_x = \rho g\bar{z}(A\sin\theta)$. But A sin $\theta$, as seen in the figure, is the projection of area A of the surface along the free surface, that is, it gives projected area of A on the vertical plane. We may similarly determine the point of application of the force. Thus to find out the horizontal force along the $x$-direction, we resolve the inclined surface along the vertical direction and then determine the magnitude and the point of action of the horizontal force on this vertical surface.

(ii) Similarly, the force in the vertical direction can be determined by taking the horizontal projection passing through the centroid. This gives the vertical force = $\rho g\bar{z} \times$ (area of the horizontal plane). You may notice that this is the weight of the fluid contained in the cylinder over the horizontal projected area, passing through the centroid of the surface, upto the free surface.

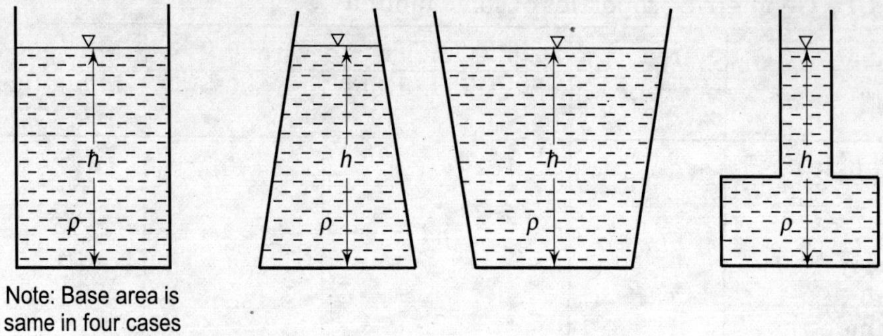


**Fig.1.3.5** Force at the base of different shaped beakers

(iii) An issue related to the weight supported over a horizontal surface is about the force on the base of beakers having different shapes, as shown in Fig. 1.3.5. These beakers have the same base area and are filled by the same liquid upto the same height. We know the hydrostatic force on the base in all these cases will be same though the actual weight of the fluid will be different. The weight of the liquid above the base contained in a cylinder above it is called the virtual weight and this weight is the vertical force on the beaker's base.

We now know that the force on any surface can be determined by determining the force on its horizontal and vertical projection; the horizontal projection is taken to pass through the centroid. The pressure on the vertical projection increases with depth as shown in Fig. 1.3.6 (a). When the surface is fully dipped in a single fluid, the pressure variation is along the slant line. In this figure, *AD* represents the projection with ends *A* and *D* at a depth of $z_1$ and $z_2$, respectively. The pressure at these edges would be $AB = \rho g z_1$ and $CD = \rho g z_2$. The pressure distribution is a trapezoid whose area is $\rho g\ (z_1 + z_2)(z_2 - z_1)/2 = \rho g h + \bar{z}$, where $h = z_2 - z_1$ = height of projected area and $\bar{z} = (z_1 + z_2)/2$ is depth of *CG* from free surface. This is the magnitude of the pressure force per unit width of the vertical surface *AD* acting on it. This is an important observation in the sense that the computation time can be saved when there are layers of the fluid one over the other. We may also show, in the same way that the centre of pressure lies at the centroid of the trapezoid.

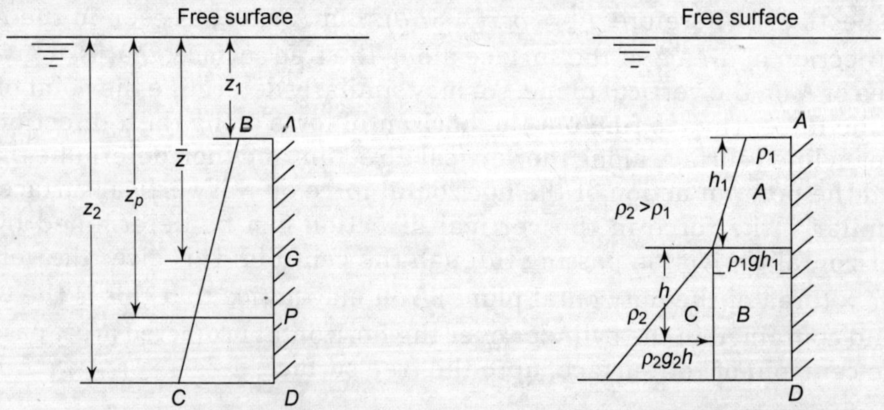


(a) Plate immersed in single fluid (b) Plate immersed in two fluids

**Fig.1.3.6** Vertical surface in fluid

However, when the surface lies in two or more fluids then slope of the pressure curve would change at the interface. Figure 1.3.6 (b) shows pressure variation over the vertical surface dipped in two fluids having densities $\rho_1$ and $\rho_2$, $\rho_2 > \rho_1$. Pressure varies linearly with the depth from the free surface of the top liquid till the liquid interface is reached and there after it remains constant. However, the slope of the pressure curve is changed at the interface and the second fluid adds its contribution to the pressure that varies linearly with the depth from the interface. The pressure variation is as shown in the figure. We may divide this area into three parts *A*, *B* and *C* whose area and centroid can easily be determined. The sum of the three areas gives the total force. To determine the centre of pressure, we take moment about the point *D* and equate it to the moment of the total force. While taking moment we may take the distance of the centre of gravity from the vertical wall or the horizontal line passing through *D* thus giving exact location of the centre of pressure. This problem can also be handled by considering the surface submerged totally in liquid at the bottom and converting the height of the liquids above into the equivalent head of this liquid. This procedure is explained in Example 1.3.6. This is one of the many methods that have been used. One can also find out the pressure force by assuming a fluid of density $\rho_1$ over the entire surface and another fluid of density $(\rho_2 - \rho_1)$ over the portion dipped in fluid of density $\rho_2$. Pressure force due to these fluids is determined as if other fluid is not present and then it is summed up.

**Example 1.3.3** Figure E1.3.3 shows the vertical trapezoidal surface that stores water up to height *h m*. Determine the centre of pressure and the total force on the surface.

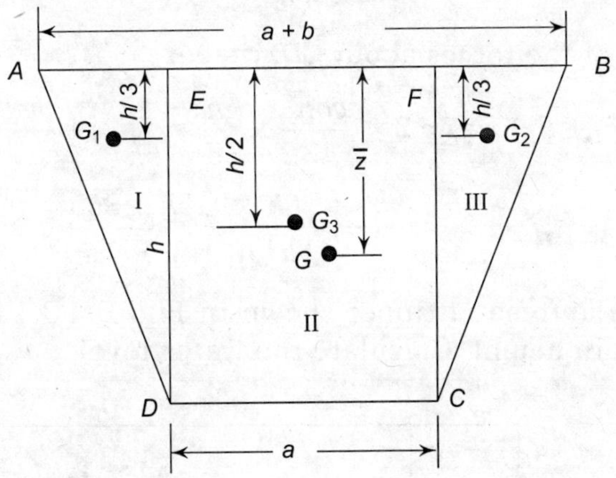


**Fig. E 1.3.3** Vertical trapezoidal surface

**Solution** The surface in Fig. E1.3.3 shows the cross-section of the wall which stores water. The total force and its point of application is given by Eqs. 1.3.9 and 1.3.10, which need the cross-sectional area and the depth of the centroid of the area. This can be calculated by dividing the area into two triangles *ADE* and *BCF* and a rectangle *DEFC*. Their properties are obtained from Table 1.3.1. Area of the triangle is $\frac{1}{2} \times b/2 \times h$ and the centroid is at $h/3$ from *AB*.; $G_1$ and $G_2$ are the centroid. Area of the rectangle is $a \times h$ and the depth of the centroid $G_3$ is $h/2$ from *AB*. Therefore, the total area of the trapezoid is

$$A = 2 \times ½ \times b/2 \times h + a \times h$$
$$= h\,(2a + b)/2 \qquad \text{(A)}$$

To find the depth $\bar{z}$ of $G$ from the free surface, take moments of areas about $AB$. This gives

$$\frac{h}{2}(2a + b)\ \bar{z}$$
$$= ah \times \frac{h}{2} + 2 \times (1/2 \times b/2 \times h \times h/3)$$
$$\rightarrow \qquad \bar{z} = \frac{3a+b}{2a+b}\frac{h}{3} \qquad \text{(B)}$$
$$F_I = w\,\frac{h}{3} \times \left(\frac{1}{2} \times \frac{h}{2} \times h\right) = \frac{wbh^2}{12} = F_{III}$$
$$F_{II} = w \times \frac{h}{2} \times (a \times b) = \frac{wabh}{2}$$
$$z_{pI} = z_{p_{III}} = \frac{h}{3} + \frac{1}{\left(\frac{1}{2} \times \frac{b}{2} \times h\right) \times \frac{h}{3}} = \frac{h}{3} + \frac{12}{bh^2}$$
$$z_{pII} = \frac{h}{2} + \frac{1}{a \times h \times \frac{h}{2}} = \frac{h}{2} + \frac{2}{ah^2}$$

Taking moments of the forces about $CD$, we get

$$(2 \times F_I + F_{II}) \times z_p = 2 \times \frac{wbh^2}{12} \times \left(\frac{h}{3} + \frac{12}{bh^2}\right) + \frac{wabh}{2} \times \left(\frac{h}{2} + \frac{2}{ah^2}\right)$$
$$\Rightarrow \qquad z_p = \frac{bh^4 + 36h + 9abh + 36b}{6bh^2\ (h + 3a)}$$

**Example 1.3.4** The automatic tipper shown in Fig. E 1.3.4 operates when the water level reaches a certain height. Calculate the water level.

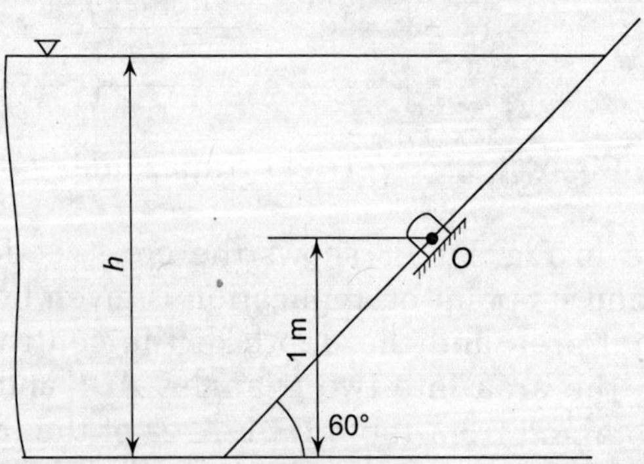


**Fig. E 1.3.4** Tipper gate

**Solution** Tipper gate operates automatically as soon as a particular water level reaches a certain height upstream. Thus it is possible to store water in the canal upto a certain height. If the water level increases, the gate opens and water flows down the canal. Let $h$ is the height of the water level at that moment. We can see that the centre of pressure must lie above the hinge $O$ at such a distance that its moment is more than the moment of the weight of the gate. Since the weight is not given, the centre of pressure must lie at or above $O$. Let us consider a unit span[5] of the tipper. Then

$$\bar{z} = h/2, A = h/\sin 60 \times 1$$

$$= 1.16h$$

$$I_c = \frac{1}{12} \times 1 \times (1.16h)^3$$

$$\rightarrow \quad z_p = h/2 + 0.13h^3 \times \sin^2 60/(1.16h \times h/2)$$

$$= 0.67h.$$

But the hinge at which C.P. would lie is at a depth of $(h - 1)$m. Equating it to $z_p$ obtained above, we get

$$h = 3.03 \text{ m}.$$

**Example 1.3.5** A pipeline 4 m in diameter contains a gate valve. The pressure at the centre of the pipe is 200 kPa. If the pipe is full of oil of specific gravity 0.87, find the force exerted by oil on the gate and the position of the centre of pressure.

**Solution** A gate valve may be considered as plane circular plate at whose centre the oil pressure is 200 kPa. This pressure at the centre is equivalent to a liquid head $z = 200 \times 1000/(870 \times 9.81) = 23.43$ m. That is, the centre of gravity of the surface is at a depth of 23.43 m, which is $\bar{z}$ in Eqs. 1.3.9 and 1.3.10. Further area $A = \pi/4 \times 4^2$ and $I_c = \pi/4 \times 4^4$.

$$\therefore \quad F = 870 \times 9.81 \times 23.43 = 2512.87 \text{ kN}$$

and

$$z_p = 23.43 + \frac{\dfrac{\pi \times 4^4}{64}}{\dfrac{\pi \times 4^2}{4} \times 23.43}$$

$$= 23.43 + 0.043 \text{ m}.$$

That is, the centre of pressure lies below the gate valve centre at a depth of 0.043 m.

[5] The term span refers to the dimension in the direction normal to the plane of paper and shall be used very frequently as we shall deal most of the time a two-dimensional flow.

**Example 1.3.6** Find the total pressure and the depth of the centre of pressure for the vertical lamina shown in Fig. E1.3.6. The upper triangle is dipped in a liquid of specific weight 500 N/m$^3$ and the lower one in a liquid of specific weight 1000 N/m$^3$. Both the triangles are equilateral and equal in size.

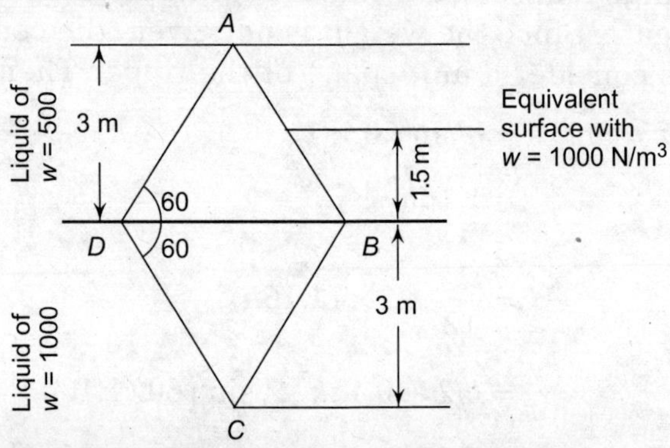


**Fig. E 1.3.6** Centre of pressure

**Solution** We shall attempt the solution using the concept of *equivalent head.* Further, the area and the second moment of area about the axis through the C.G are given by ½ $b \times h$, $bh^3/36$, respectively, where $b$ and $h$ are the side and the height of the triangle. The height of the triangle is given to be 3 m from which we get the side of the triangle $b = 3/\sin 60 = 2\sqrt{3}$ m. For the upper triangle

$$\bar{z}_1 = 2/3 \times 3 = 2 \text{ m}$$

$$A = \frac{1}{2} \times 2\sqrt{3} \times 3 = 3\sqrt{3}$$

and $$I_c = 2\sqrt{3} \times 3^3/36 = 3\sqrt{3}/2$$

∴ $$F_1 = 500 \times 3\sqrt{3} \times 2 = 3000\sqrt{3} \text{ N}$$

and $$z_{p,1} = 2 + 3\sqrt{3}/(2 \times 3\sqrt{3} \times 2) = 2.25 \text{ m}$$

We may replace the upper liquid by the heavy liquid having its free surface at $3 \times 500/1000 = 1.5$ m. That is, we assume the lower triangle is at a depth of 1.5 m from the free surface of the liquid having specific weight 1000 N/m$^3$. We use Eqs. 1.3.9 and 1.3.10 to give for the lower triangle

$$\bar{z}_2 = 1.5 + 1/3 \times 3 = 2.5 \text{ m}, A = \frac{1}{2} \times 2\sqrt{3} \times 3 = 3\sqrt{3}$$

and $$I_c = 2\sqrt{3} \times 3^3/36 = 3\sqrt{3}/2$$

$$\therefore \qquad F_2 = 1000 \times 3\sqrt{3} \times 2.5 = 7500\sqrt{3}\text{ N}$$

and $$z_{p,\,2} = 2.5 + 3\sqrt{3}/(2 \times 3\sqrt{3} \times 2.5) = 2.70\text{ m.}$$

We note that the equivalent free surface is at a depth of 3 – 1.5 = 1.5 m of the original given condition. Therefore the centre of pressure of the lower triangle is at a depth of 1.5 + 2.7 = 4.2 m from the original free surface. We may now proceed to find out the total force and the point of its application by taking moments about the free surface. This gives

$$F = F_1 + F_2 = 18186.533\text{ N}$$

and $$F \times z_p = F_1 \times z_{p,\,1} + F_2 \times z_{p,\,2}$$

or, $$18186.5 \times z_p = 3000\sqrt{3} \times 2.25 + 7500\sqrt{3} \times 4.2$$

$\rightarrow$ $$z_p = 3.642\text{ m.}$$

**Example 1.3.7** Determine the pressure force exerted by a liquid of specific weight $w$ and the point of its application on a vertical parabolic lamina shown in Fig. E 1.3.7.

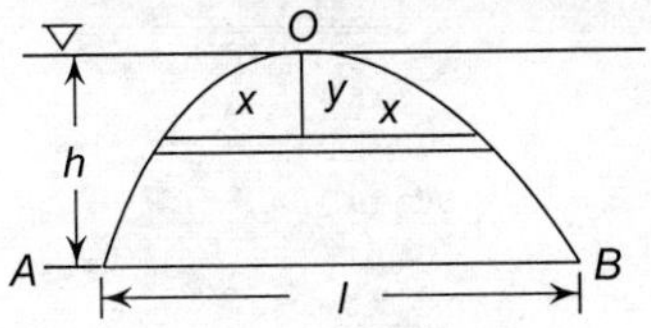


**Fig. E 1.3.7** Force on a parabolic lamina

**Solution** Area of a parabola = 2/3 × area of the corresponding rectangle = $2/3 \times l \times h$ = $2lh/3$. Depth of the centroid from the free surface is = 3/5 $h$. With this information we can find out the hydrostatic force on the lamina $F = w \times 2lh/3 \times 3/5 \times h = 2\,w\,l\,h^2/5$. To find out the point of application of the force, consider the origin of the coordinate axes at $O$ with $y$ along the down side and a strip of width $dy$ at a depth $y$ from the free surface. The parabola is represented by the Eq. $y = 4h\,(x/l)^2$; you may note that when $y = h$, $x = \pm\,l/2$. we shall find out the force on this strip, then take the moment about the free surface and finally sum up these moments to get the total moment of the force. This moment should be equal to the moment of the total hydrostatic force on the lamina. This procedure gives

$$dA = 2x\,dy,\; dF = w.2x\,dy.y$$

$$= \frac{wl}{\sqrt{h}}\,y^{3/2}dy$$

$\rightarrow$ $$F = \int_0^h dF = \frac{2}{5}wlh^2 \text{ and}$$

$$M = \int_0^h \frac{wl}{\sqrt{h}} y^{3/2} xy\, dy = \frac{2}{7} wlh^3$$

But $$M = F \times z_p \to z_p = 5h/7.$$

**Example 1.3.8** A hemispherical depression of diameter 1.2 m exists on one of the vertical sides of a tank. If the tank contains water to an elevation of 3.0 m above the centre of the hemisphere, calculate the forces acting on it.

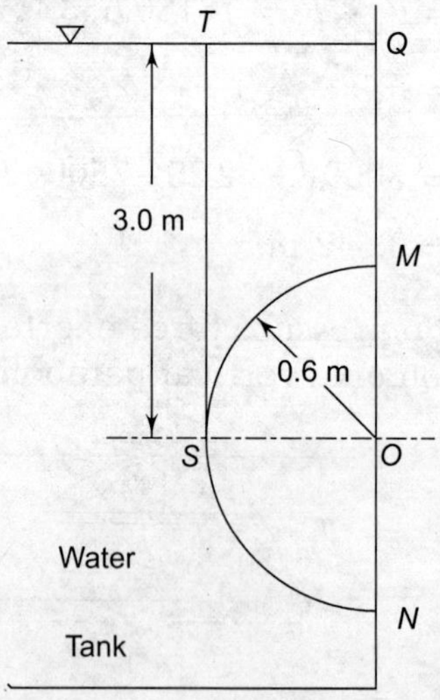


**Fig. E 1.3.8** Forces on a hemispherical projection in a tank

**Solution**

Vertical force on the projection = weight of water *NSTQ* – weight of water *MSTQ*
= weight of water contained in the hemisphere

or. $$F_V = w \times 1/2 \times \left(4/3\ \pi/R^3\right)$$
$$= 9.81 \times 1/2 \times \left(4/3\ \pi(1.2)\right)^3$$
$$= 35.5 \text{ kN}.$$

Horizontal force = force on the vertical circular section *MN*

$$F_H = w \times \pi R^2 \times 3 = 132.9 \text{ kN}.$$

**Example 1.3.9** The gates of a lock gate 5 m wide and 5 m high include an angle of 120° when closed. Each gate is held on two hinges, one placed at the top and the other at the bottom of the gate. If the water levels are 4.5 m and 3 m on the upstream and downstream sides respectively, determine the magnitude of the forces on the hinges due to the water pressure.

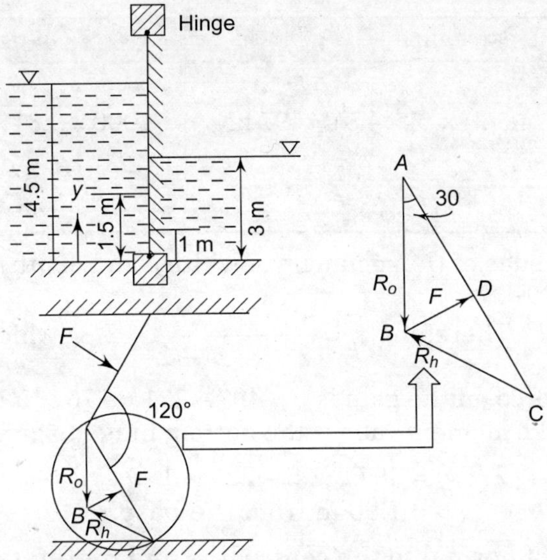


**Fig. E 1.3.9** Lock gate

**Solution** A lock gate is used in sea to maintain different water levels in a dockyard on its two sides. The ship is brought to the dock area from the sea or vice versa by operating one or the other gate to maintain water level as per the situation. The level of the ship can be lowered or raised accordingly. This can also be used to store water during the tides; while tide is over then the gates can be closed not to allow the water to go back to the sea. Lock gate has been used to connect two seas, which are having water at different levels, by a canal carrying the lock gate at both the ends. A canal is constructed across the land separating two seas and lock gate is placed at each end of the canal. When the ship is to be moved from sea at higher level then the gate close to this sea is opened and other gate is closed. This raises water level in the canal to bring it to the level of the sea. Then the ship is moved. Now this gate is closed and when the ship is to move out, the other gate is opened bringing its level to that of this sea so that the ship can easily move.

Figure E1.3.9 shows two views of the lock gate. Water is at a level of 4.5 m on the upstream side and 3 m on the downstream side. Thus there is a net force $F$ on each gate whose close position is shown in the plan. Because of this force, the edges of both the gates are pressed against each other resulting in a reaction force of magnitude $R_o$ acting on each gate in the direction shown. If we consider one gate then there are two forces namely $F$ and $R_0$. For equilibrium of the gate, a third force $R_h$ must act on it which is provided by the hinges. We may determine the force $F$ by determining the force and its position that acts on each side of the gate due to hydrostatic pressure. The gate may be considered as rectangular plate having width 5 m. Calculations can be made in the following tabular way using Eqs. 1.3.9 and 1.3.10 and Table 1.3.1, $y$ is measured from the bottom of the gate.

| | Upstream | Downstream |
|---|---|---|
| $h$ | 4.5 m | 3 m |
| $F$ | $1000 \times 9.81 \times (4.5/2) \times (5 \times 4.5) = 497$ kN | $1000 \times 9.81 \times (3/2) \times (5 \times 3) = 221$ kN |
| $z_p$ | $(2/3) \times 4.5 = 3$ m | $(2/3) \times 3 = 2$ m |
| $y_p$ | $4.5 - 3 = 1.5$ m | $3 - 2 = 1$ m |

Note that when the edge of the rectangle of width $b$ and depth $l$ is in free surface then A = $bl$, $\bar{z} = l/2$ and $I_c = \frac{1}{12}bl^3$. Therefore, $z_p = \frac{1}{2} + \frac{1}{12}\frac{bl^3}{bl.l/2} = 2/3l$, which has been taken in above calculations. The net force on the gate is $F = 497 - 221 = 276$ kN. To locate the height of this force from the base, take moments about the bottom hinge to give

$$276\,y = 497 \times 1.5 - 221 \times 1$$

$$\rightarrow \quad y = 1.90 \text{ m from the base.}$$

The force $R_o$ will act normal to the centre line and therefore makes an angle 30° with the gate. Thus the forces $F$, $R_o$ and $R_h$ should meet at a point, $B$. The triangle of forces can be drawn, which is shown enlarged. We note that the triangles $BCD$ and $ABD$ are congruent because the side $AD = DC$, the side $BD$ is common to the two triangles and $\angle ADB = 90°$. Therefore $\angle BCD = 30°$ and $R_o = R_h$. Resolving the forces in the direction of $F$, we get

$$F = 2R_h \sin 30 \quad \rightarrow R_h = 276 \text{ kN}.$$

Not only the three forces meet at a point, they should also lie in the same plane. Therefore we conclude that the net force exerted by the hinges is 276 kN which acts in a plane at a distance of 1.90 m from the base. Let the force in the top and bottom hinges be $R_T$ and $R_B$, respectively. Taking moment of the hinge forces about the bottom hinge, we get

$$R_T \times 5 = 276 \times 1.9$$

$$\rightarrow \quad R_T = 105 \text{ kN and } R_B = 276 - 105 = 171 \text{ kN}.$$

**Example 1.3.10** The hydrostatic water pressure acts only on one side and to a depth of 10 m from the top of a dock gate that is reinforced with three horizontal beams. Locate the position of these beams so that each of these is equally loaded.

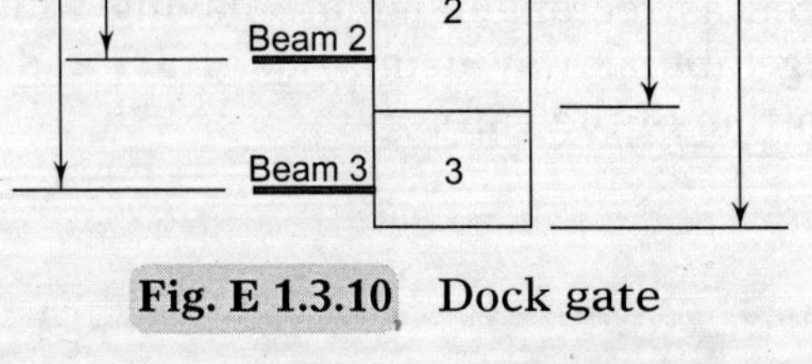


**Fig. E 1.3.10** Dock gate

**Solution** Water is upto a height of $h_3$ on one side of the dock gate, which is shown in Fig. E 1.3.10. We divide the vertical height in three sections such that the hydrostatic force is equal on each of these. We have learnt how to determine the forces and the location of these forces. We may start ab-initio for which we consider an area element of height *dh* at a depth *h*. We can easily get the forces on these sections; these are

$$F_1 = \int_0^{h_1} whdh = \frac{w}{2}h_1^2, \qquad F_2 = \int_{h_1}^{h_2} whdh = \frac{w}{2}(h_2^2 - h_1^2) \qquad \text{and } F_3 = \int_{h_2}^{h_3} whdh = \frac{w}{2}(h_3^2 - h_2^2)$$

The total force on the dock gate can also be determined similarly to give $F = \frac{w}{2}h_3^2$.

For equal force on each section,

$$\frac{w}{2}h_1^2 = \frac{w}{2}(h_2^2 - h_1^2) = \frac{w}{2}(h_3^2 - h_2^2) = \frac{1}{3}\frac{w}{2}h_3^2$$

We know $h_3$ = 10 m (given). Therefore the above equations can be solved to give $h_1$ = 5.77 m and $h_2$ = 8.16 m. Substituting these values, the force on any section is $\frac{w}{2}h_1^2$ = 163.3 kN.

Let us now determine depth of centre of pressure for the second section. The depth of the centre of pressure for the first section is 2/3 $h_1$ = 3.846 m and that of both the first and second section is 2/3 × $h_2$ whereas the total force on the combined section is $\frac{w}{2}h_2^2$. We know the moment of the total force about the free surface is equal to the sum of the force in the individual sections. This gives the depth $h_{2,p}$ of the centre of pressure of the second section from the free surface

$$\frac{w}{2}h_2^2 \times \frac{2}{3}h_2$$

$$= \frac{w}{2}h_1^2 \times \frac{2}{3}h_1 + \frac{w}{2} \times \left(h_2^2 - h_1^2\right)h_{2,p}$$

$$\rightarrow \qquad h_{2,p} = 7.4 \text{ m.}$$

Similarly we may find out the depth of the centre of pressure of the third section, which comes out 9.1 m. At these depths we have to provide horizontal beams that can withstand a load of 163.3 kN.

**Example 1.3.11** A 4 m square gate provided in an oil tank is hinged at its top edge, as shown in Fig. E1.3.11. The tank contains gasoline of specific gravity 0.7 upto a height of 2 m above the top edge of the gate. The space above the oil is subjected to a negative pressure of 6865 N/m$^2$. Determine the necessary vertical pull to be applied at the lower edge to open the gate.

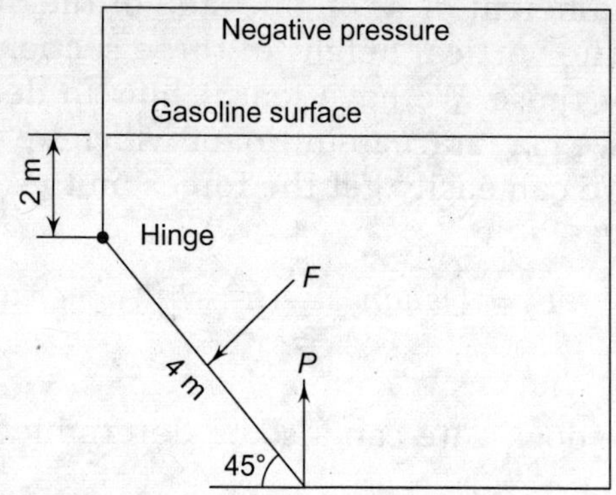


**Fig. E 1.3.11** Negative pressure at the free surface

**Solution** In this problem the pressure at the free surface is negative, which can be handled by decreasing the height of the gasoline in the tank by some height of the gasoline equivalent to the negative pressure. The equivalent height $h$ is given by the relation $h = p/w = 6865/(700 \times 9.81) = 1$ m of gasoline. Now, rest of the process of calculating the hydrostatic force $F$ and its point of application on a plane surface is straightforward.

$$\bar{z} = (2 - 1) + (4/2) \sin 45 = 2.414 \text{ m}, \quad A = 4 \times 4 = 16 \text{ m}^2,$$

$$F = wA\bar{z} = 700 \times 9.81 \times 16 \times 2.414$$

$$= 265150 \text{ N and}$$

$$I_c = \frac{1}{12} \times 4 \times 4^3, \theta = 45°$$

$$\rightarrow \qquad z_p = \bar{z} + \frac{I_c \sin^2 \theta}{A\bar{z}} = 2.804 \text{ m.}$$

Depth of the centre of pressure from the hinge is 2.804 – 1 = 1.804 m. Now taking moments about the hinge (this will eliminate the moment of the force acting on the hinge),

$$P \times 4 \sin 45 = F \times 1.804/\sin 45 \quad \rightarrow \quad P = 239.3 \text{ kN.}$$

**Example 1.3.12** (a) A masonry dam of trapezoidal section is 9 m high with a top width of 1 m and a bottom width of 6 m, the water face of the dam being vertical. If the depth of water impounded is 8 m, find the extreme pressure intensities at the base.

(b) If the height of the dam is increased to 11 m by adding an additional 1 m × 2 m deep rectangular section and the depth of water impounded is raised to 9 m, what would be the extreme pressure intensities at the base? Masonry and water weigh 22000 N/m$^3$ and 9810 N/m$^3$, respectively. What should be minimum coefficient of friction between the masonry and the ground not to allow slip of the dam?

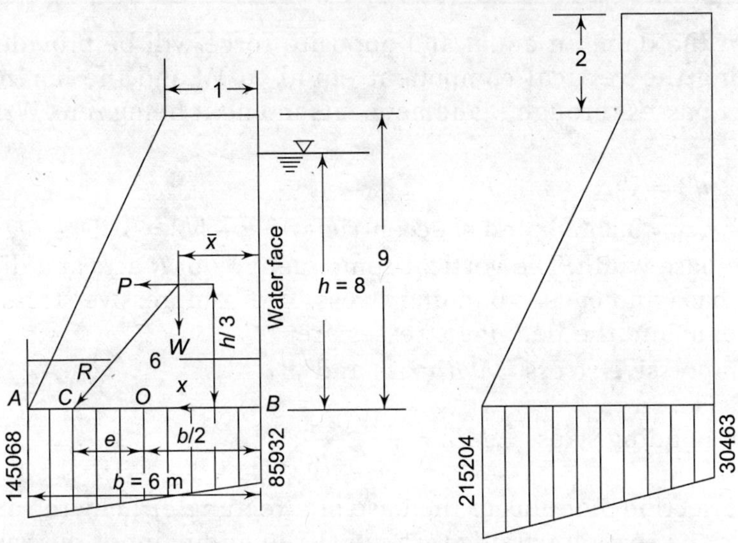


**Fig. E 1.3.12** Pressure intensity at the dam base

**Solution** Section of the dam is shown in Fig. E1.3.12. The various forces that act on the structure are: (i) the water pressure force on the water side of the dam that acts at the centre of pressure and (ii) the weight of the dam. The water force generates a overturning couple resulting in high pressure at the dam base on the far-side corner A and also tries to move the dam in the direction of the force. Therefore the dam is likely to fail (i) by sliding on the soil on which the dam is made, (ii) by overturning, (iii) tensile stress developed at the nearby end B of the dam and (iv) due to excessive compressive stress at the point A. We have to ensure that the dam is safe under all the conditions.

We shall make calculations for 1 m span of the dam. We can easily get the following values:[6]

Weight of dam, $W = \left(\frac{1+6}{2}\right) \times 9 \times 22000 = 693\text{kN}$; $\bar{x} = \frac{1^2 + 1 \times 6 + 6^2}{3(1+6)} = 2.048$ m. $\bar{x}$ is the distance of C.G. from the vertical face.

Hydraulic force, $P = \frac{wh^2}{2} = \frac{9810 \times 8^2}{2} = 313.92\text{kN}$ $\quad z_p = 8/3$ m from the bottom

The centre of gravity and centre of pressure will be at different points but in the same plane. Their resultant *R*, which meets the base of the dam at point *C* at a distance *x* from the base corner *B*, can be determined by taking moments about the point *B*. For

[6] Here the relations from geometry have been used. If you do not remember these relations then the trapezoid can be thought of as if made from a rectangle of size 1 m × 9 m and a triangle of size 5 m base and 9 m height. Area of these simple sections is determined to give volume and weight. Taking moment about the vertical face of the dam determines × position of the centre of gravity.

equilibrium of the dam, an equal and opposite force will be provided by the base of the dam having the vertical component equal to $W$ and the horizontal component equal to $P$ that passes through $B$ and hence its moment being zero. Writing the moment Eq., we get

$$W\bar{x} + P \times h/3 = W.x$$

$\rightarrow$ $x = 3.256$ m and $e$, eccentricity $= x - b/2 = 3.256 - 6/2 = 0.256$ m,

where $b$ is the base width. The vertical component $W$ of $R$ acts at a distance $e$ from the centre of the base and causes bending stress. The compressive stress due to weight of the masonry dam and the bending stresses are

Compressive stress $= W/(b \times 1)$ and

$$\text{maximum bending stress} = M/Z = \frac{W.e}{1 \times b^2/6},$$

where $Z$ is the section modulus of the base of a rectangle, along the axis perpendicular to the plane of the section passing through $O$, the mid-point of the base of the dam. We note that the minimum stress is at $B$ and maximum at $A$. The stress at point $A$ should not become negative and at $B$ not to exceed the safe compressive limit. This gives the condition $\frac{W}{b} - \frac{6We}{b^2} \geq 0$ or, $e \leq b/6$

In the question, however, we have to determine the stress intensity, which comes out to be

$$\frac{W}{b} \pm \frac{6We}{b^2} = 145 \text{ kN/m}^2 \text{ (max) and } 85.93 \text{ kN/m}^2 \text{ (min)}.$$

We note that the minimum stress is more than zero and hence the dam is safe at both the ends $A$ and $B$, since the compressive stress, RCC can take is much higher. We may also check for slipping of the dam because of the hydrostatic force. This requires $P \leq \mu W$, where $\mu$ is the coefficient of friction at the surface of the base. For the given problem, $\mu$ must be $\geq P/W = 313920/693000 = 0.453$ for no-slippage of the dam.

(b) When the height of the dam is increased to 11m then the above procedure can be repeated. The difference comes only in determining the weight of the masonry.

Weight of dam, $W$ = weight of trapeziod + weight of rectangle

$$= 693 + 44 = 737 \text{ kN};$$

$$\bar{x}_1 = 2.048 \text{ m and } \bar{x}_2 = 0.5 \text{ m}$$

$$\rightarrow \quad \bar{x} = \frac{693 \times 2.048 + 44 \times 0.5}{737} = 1.956 \text{ m from vertical face}$$

$$\text{Hydraulic force, } P = \frac{wh^2}{2} = \frac{9810 \times 9^2}{2}$$

$$= 397.305 \text{ kN } z_p = 9/3 = 3 \text{ m from the bottom}$$

$$z = 3.752 \text{ m and } e = z - b/2 = 0.752 \text{ m.}$$

Extremne pressure intensity = 215.204 kN/m$_2$(max) and 30.463kN/m$^2$ (min)

## 1.3.2 Pressure Force on a Curved Surface

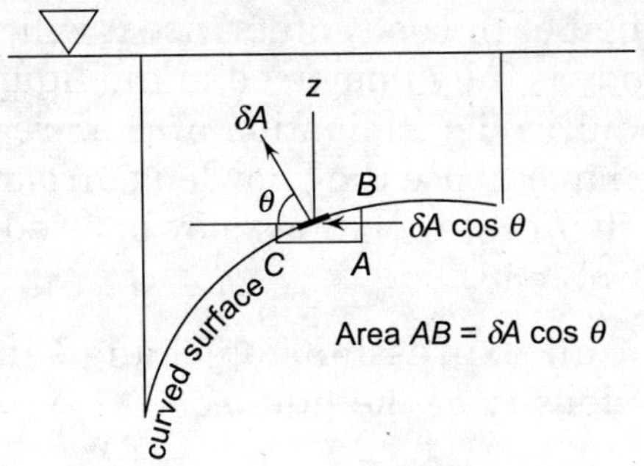


**Fig. 1.3.7** Hydrostatic force on curved surface

Let us now consider a curved surface immersed in a liquid of density $\rho$, as shown in Fig. 1.3.7. In this case the pressure does not only change in the magnitude but also in the direction along the surface. Consider an elemental area $\delta A$ at a depth $z$ from the free surface where $p = \rho gz$ and it acts in a direction making an angle $\theta$ with the $y$-direction. The force in $y$-direction is $F_y = p \times (\delta A \cos \theta)$. But $\delta A \cos \theta$ is the projected area of $\delta A$ on the vertical plane. Similarly, $\delta A \sin \theta$ is its projected area on the horizontal plane. Therefore we may take the projected area of the curved surface on the horizontal and vertical planes and determine the force on these plane surfaces, as we did in the case of plane surfaces.

It is interesting to note that the area of projection of the closed curved surfaces on such planes would be zero, since the projection of surface on one side of the vertical plane passing through the centroid points in one direction and that on the other side in other direction. These projected areas are equal in magnitude but opposite in direction. As far as horizontal force is concerned, the product of pressure and the elemental area at the same depth on both sides of the vertical plane would also be same and the $y$-component of the force would be zero. Hence, the sum would be zero. However, while concluding like this for the vertical force, product of the pressure and area element on both sides of the horizontal plane passing through the centroid will not cancel out as the pressure at these two area elements would be different due to their different depths from the free surface. The force on the lower-half portion will be greater than that on the upper part.

Consider the vertical component of the pressure force on element $\delta A$; it is $w\ z \times \delta A \sin \theta$ equal to weight of the liquid over the projected area of $\delta A$ on the $x$-$y$ face passing through the element. Now summing this force all over the surface, the sum will be equal to the weight of the volume of the liquid displaced by the closed surface. This can be easily understood by dividing the whole surface by a horizontal plane passing through the centre of gravity. The force on the upper portion acts downwards while that on the bottom side it is upwards and these forces will equal to the weight supported by these curved surfaces upto the free surface of the liquid.

This is the *Archimedes principle.* As a result of this vertical force, acting in the upward direction, the weight of the body is reduced by the amount equal to this force, which is called *buoyant force* or the force of buoyancy. The force acts at the centroid of the displaced liquid. The body will float and dip partially in a liquid if the buoyancy force is more than the weight of the body otherwise it will sink. The maximum buoyancy force will act when the body is fully immersed in the liquid. The vertical component of the force on the body requires determination of the area above the surface upto the free surface for which we have to use the knowledge from the geometry and divide the whole area in such a way that these relations may be used effectively. This is explained with the help of solved problems.

**Example 1.3.13** Obtain an expression for change in height of immersion of a hydrometer and relative density of the liquid.

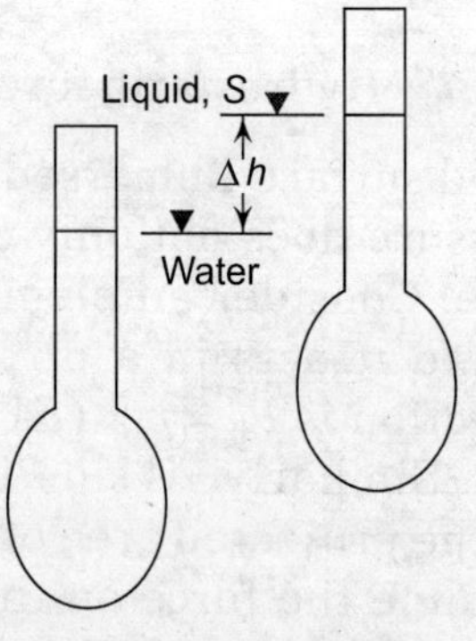


**Fig. E 1.3.13** Hydrometer

**Solution** Hydrometer is used for measurement of specific gravity of a liquid. It consists of a bulb with a long neck fitted with scale. Initially it is immersed in water and it floats; reading at the free surface of the water is taken. Now it is dipped in the liquid whose density is to be obtained experimentally. If the liquid is heavier then the neck of the hydrometer would go up. Let $\Delta h$ be the difference in reading in the two cases, $S$ the specific gravity of the liquid, $V$ the volume of the water displaced by the hydrometer, $W$ the weight of the hydrometer, and $A$ the cross sectional area of the neck. We assume that the bulb of the hydrometer is always under the free surface of the liquid and the neck is peeping out. Since in both the cases, the hydrometer is floating, we may write that the buoyant force is equal to the weight of the hydrometer. This gives

$$W = wV, \text{ and } W = wS\,(V - A\Delta h)$$

$$\rightarrow \qquad \Delta h = \frac{V}{A}\left(1 - \frac{1}{S}\right)$$

or,

$$S = \frac{1}{1 - A\Delta h/V}$$

For a given hydrometer $A$ and $V$ are fixed, knowing $\Delta h$, $S$ can be calculated. This relation is not linear and hence it is difficult to put a scale on the neck of the hydrometer.

**Example 1.3.14** Fig. E 1.3.14 shows the section of a tank which is 2 m long and is full of water under pressure at 19.62 kN/m$^2$. A cylinder of radius 1 m is resting on the tank at one of its corners as shown. Find the horizontal and vertical components of the force acting on the curved surface *ABC* of the cylinder. Neglect the weight of the cylinder and assume it is to be empty.

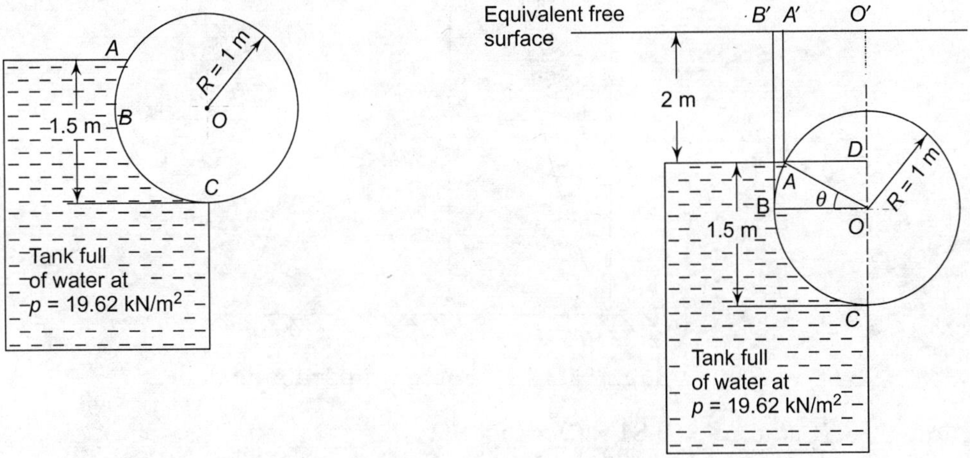


**Fig. E 1.3.14** Pressure on curve *ABC*

**Solution** Given $OA = OC = 1$ m and cylinder rests at one corner $\Rightarrow CD = 1.5$ m. Therefore, $OD = 1.5 - 1 = 0.5$ m.

$\rightarrow \sin\theta = ½$, or $\theta = 30°$ and $AD = OA \cos 30 = 0.886$.

We may add a water column of height h corresponding to pressure 19.62 kN/m$^2$ in the tank, which is wh = 19.62 × 1000 → $h$ = 2 m. this is shown in Fig. E 1.3.14 (b). To find horizontal component of the force, we take projection of the surface *ABC* on the vertical plane, which is a rectangle of height $CD$ = 1.5 m and width 2 m, the length of the cylinder, with its top surface 2 m below the free surface.

$$P_x = wA\bar{z} = 9.81 \times (1.5 \times 2) \times (2 + 1.5/2)$$
$$= 80.9325 \text{ kN}.$$

$P_y$ = Upward force on $BC$ – downward force on $AB$
= Weight of water above $BC$ – Weight of water above $AB$
= Weight of water in the block $A'ABCO'$
= $w \times$ area of block $A'ABCO' \times$ length of cylinder.
= 9.81 × {Rectangle $A'ADO'$ + triangle $OAD$ + sector $OABC$} × length of cylinder
= $9.81 \times \{1 \cos 30 \times 2 + ½ \times 0.5 \times AD + (120/360) \times \pi \times 1^2\} \times 2$
= 58.776 kN.

**Example 1.3.15** Figure E 1.3.15 shows the section of a dam. The water face of the dam is parabolic following the law $y = x^2/4$. Find the horizontal and vertical component of the force exerted by water on the dam per metre length. Find also the resultant force on the dam per metre length.

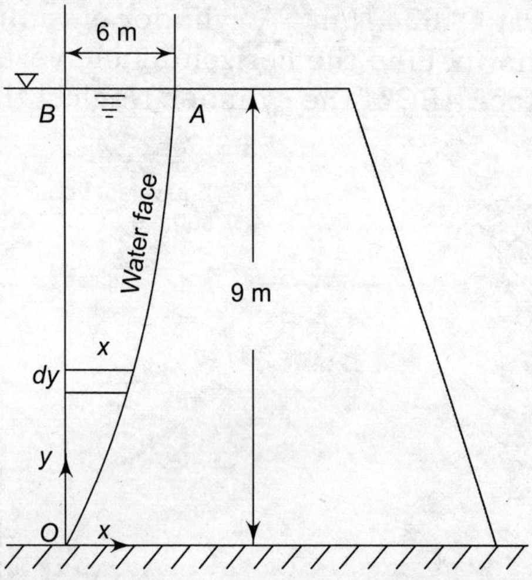


**Fig. E 1.3.15** Force on parabolic dam

**Solution** $P_x = wA\bar{z} = 9.81 \times (9 \times 1) \times 9/2$

$= 397.305$ kN.

$P_y$ = weight of water over the parabolic profile upto the free surface

$$= 9.81\int_0^9 x\,dy = 353.16\,\text{kN s}$$

$$\text{Resultant force} = \left(P_x^2 + P_y^2\right)^{1/2} = 531.576 \text{ kN and}$$

$$\tan\theta = 353.16/397.305$$

$$\rightarrow \qquad \theta = 41°387 \text{ with the horizontal.}$$

**Example 1.3.16** A masonry weir is of trapezoidal section with a top width of 2.0 m and of height 5 m with upstream slope of 1 vertical in 0.1 horizontal and a down stream slope of 1 vertical in 0.75 horizontal. If the weir has stored water up to its crest on the upstream side and has a tail race of 2 m depth on the downstream side, calculate, per unit length of the weir, (i) the resultant force on the base of the weir, and (ii) the minimum and the maximum stresses on the base of the weir.

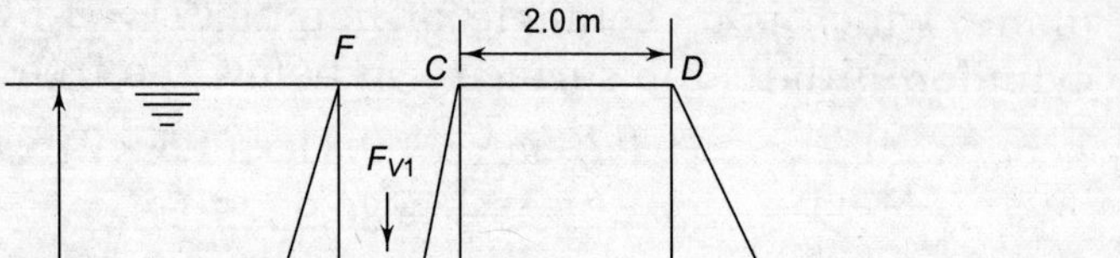


**Fig. E 1.3.16** Maximum and minimum

Assume specific weight of masonry as 22 kN/m$^3$, that of water 9.79 kN/m$^3$ and neglect uplift forces.

**Solution** For convenience, we divide the whole weir region into three geometrical sections, namely triangular section ACF (I), rectangular section (II) and another triangular section (III). The horizontal and vertical forces acting on these sections can be determined; these are tabulated in the following table. Note we are given slope of the faces *AC* and *BD* from which we determine their horizontal projected length, which are 5 × 0.1 = 0.5 and 5 × .75 = 3.75m, respectvely.

**Table Example 1.3.16.** Calculation of forces and their point of application

| Force | Description | Magnitude, kN | | Lever arm | Moment |
|---|---|---|---|---|---|
| | | Horizontal | Vertical | about B, m | KN.m |
| $W_1$ | ½ × (0.5 × 5) × 1 × 22 | | 27.5 | 3.75 + 2 + 0.5/3 = 5.917 | 162.7 |
| $W_2$ | (22 × 5.0) × 1 × 22 | | 220.0 | 3.75 + 212 = 4.75 | 1045.0 |
| $W_3$ | ½ × (3.75 × 5) × 1 × 22 | | 206.25 | 2 × 3.75/3 = 2.5 | 515.6 |
| $F_{V1}$ | ½ × (0.5 × 5) × 1 × 9.79 | | 12.24 | 3.75 + 2 + 2 × 0.5/3 = 6.083 | 74.5 |
| $F_{V2}$ | ½ × (1.5 × 20) × 1 × 9.79 | | 14.69 | 1.5/3 = 0.5 | 7.3 |
| $F_{H1}$ | (1.0 × 5) × 5/2 × 9.79 | 122.38 | | 5/3 = 1.667 | –204.0 |
| $F_{H2}$ | (1.0 × 2) × 2/2 × 9.79 | –19.58 | | 2/3 = 0.667. | 13.1 |
| Sum | 102.8 | | 480.7 | | 1614.2 |

**Note:** The distances in the above table are those which would be needed to determine the moment of the forces about the axis through the point *B*. You should be able to get the data as given in the table. From the table we get

Z, sum of vertical forces = 480.7 kN

and X, sum of horizontal forces = 102.8 kN

$$R\text{, Resultant} = \sqrt{Z^2 + X^2} = \sqrt{480.7^2 + 102.8^2}$$
$$= 491.6 \text{ kN}$$

and $\tan\theta = Z/X$

→ $\theta = 77.93°$

$\theta$ is the angle made by the resultant force on the weir with the horizontal direction and is acing towards the base of the weir, as the vertical force is downwards. Now we may locate the point of the resultant force from point *B* by taking moments of the vertical forces about *B*; (the anticlockwise moment is taken positive); the moments are given in the table. This gives

$$x = \sum M/Z = 1614.2/480 = 3.358 \text{ m.}$$

$b$ = distance of midpoint of weir base from B = 3.75 + 2 + 0.5 = 6.25 m

→ $e$, eccentricity = $x - b/2 = 3.358 - 6.25/2$

$= 0.233$ m.

The maximum and the minimum stresses are:

$$\sigma = \frac{Z}{b}\left(1 \pm \frac{6e}{b}\right) = \frac{480.7}{6.25}\left(1 \pm \frac{6 \times 0.233}{6.25}\right)$$

→ $\sigma_{max} = 94.1$ kPa and $\sigma_{min} = 59.7$ kPa

**Example 1.3.17** A sector gate, as shown in Fig. E 1.3.17 under the equilibrium conditions, is of radius 4 m and length 5 m. It controls the flow of water in a horizontal channel. Determine the total force on the gate.

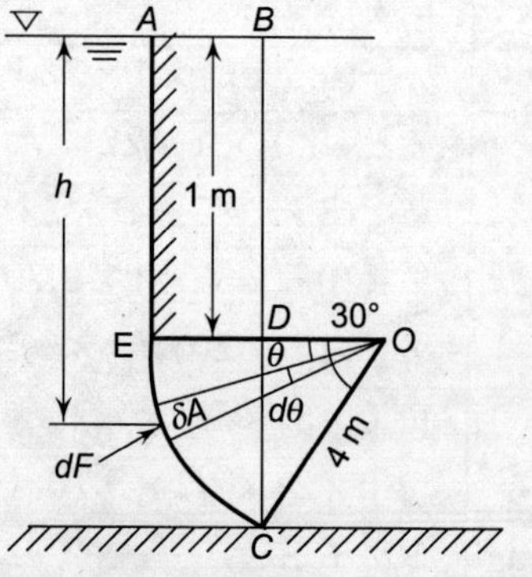


**Fig. E 1.3.17** Forces on radial gate

**Solution** From the figure, $CD = OC \sin 30 = 2$ m and length of the gate is 5 m (given). Also $OD = 4 \cos 30 = 4\sqrt{3}/2$ and $ED = 4 - OD = 4(1 - \sqrt{3}/2)$

Therefore, the horizontal component of the force is the force on vertical rectangular plate CD of width 5 m.

$\therefore \quad P_x = 9.81 \times (2 \times 5) \times (1 + 2/2) = 196.2$ kN

and $z_p = 2 + \dfrac{5 \times 2^3}{12 \times (5 \times 2) \times 2} = 2.167$ m.

The vertical component can be determined by determining the weight of the water in the block *ABDCEA*. But

Volume *ABDCEA* = volume *ABDEA* + volume of sector *OCE* – volume of triangle *OCD*

$= \{(1 \times ED) + (30/360) \times (\pi/4) \times 4^2 - ½ \times OD \times CD\} \times 5$

→ $P_y = w \times$ Volume $ABDCEA = 61.8$ kN.

The resultant force $= \sqrt{(P_x^2 + P_y^2)} = 205.7$ kN.

We obtain its point of action from the vertical wall AE by taking moment of areas about AE.

*Alternative procedure.* We may determine the forces by considering an element $\delta A$ at an angle $\theta$ subtending angle $d\theta$, as shown in Fig.. Depth of this element from the free surface is $h = 1 + 4 \sin\theta$ and $\delta A = 4\, d\theta \times 5 = 20\, d\theta$. The horizontal and the vertical components of the force acting on this element are $\delta F \cos\theta$ and $\theta F \sin\theta$, respectively. integration w.r.t. $\theta$ from $\theta = 0$ to $30° = \pi/6$ radians gives

$$P_x = \int_0^{30} 9.81 \times (20 d\theta) \times h \times \cos\theta = 196.2 \text{ kN}$$

and $$P_y = \int_0^{30} 9.81 \times (20\, d\theta) \times h \times \sin\theta = 61.8 \text{ kN}$$

**Example 1.3.18** A steel rectangular block of specific gravity 7.85 floats at mercury water interface such that its side in water and mercury is $a$ and $b$ respectively. Determine the ratio $a/b$ if specific gravity of mercury is 13.57.

**Solution** For floatation of the block at the interface, the weight of the block should be equal to the sum of the weight of the displaced mercury and water. Therefore, if $A$ is the cross-sectional area of the block,

$$\mathrm{A} \times (a + b) \times 7.85 \times g = \mathrm{A} \times [\, b \times 13.57 + a] \times 1000 \times g$$

$$\rightarrow \qquad a/b = 0.835.$$

## 1.4 Metacentric Height

Stability study of a floating body is of importance because of the vessels floating in the sea, aeroplanes flying in the air and many other such applications. By stability of a body we mean that if any temporary extraneous displacement acts on the body then it should return back to the original position from its disturbed position. If so then the bodyis said to be in stable equilibrium. There are also possibilities that the body maynot return to the same position and takes some other position; it is said to be in *unstable equilibrium*. Or the body may stay in any position and is in neutral or *metastable equilibrium*.

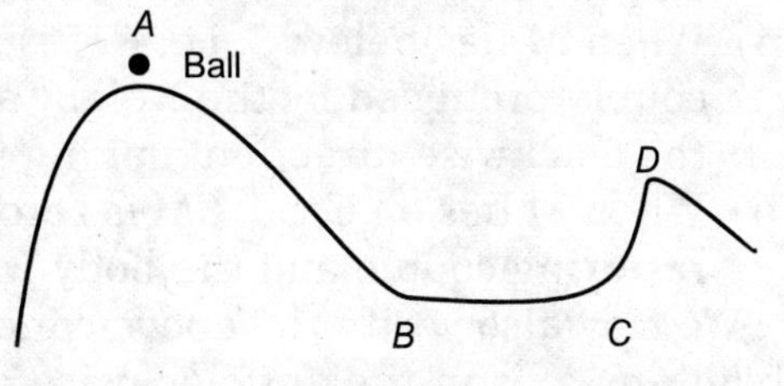


**Fig. 1.4.1** Different positions of a ball

To understand equilibrium, let us imagine a ball placed in position $A$ at the top of a hill, as shown in Fig. 1.4.1. An infinitesimal disturbance would displace it from this position on either side. Let us suppose it moves to its right then it will finally stay at any point on the horizontal portion $BC$ of the hill if it does not cross over the point $D$ to move further down. Before coming to rest it oscillates a few number of times in the valley. We conclude that the position $A$ is unstable, since an infinitesimal disturbance will move the position of the ball to some new position. The position of the ball anywhere on $BC$ represents neutral equilibrium situation. However, if instead of a horizontal portion there is a curved surface at $BC$ then the ball will always stay at the bottom of the curve and it corresponds to the stable equilibrium position. We also observe that position $A$ has higher energy compared to a point on $BC$. In other words the ball will take

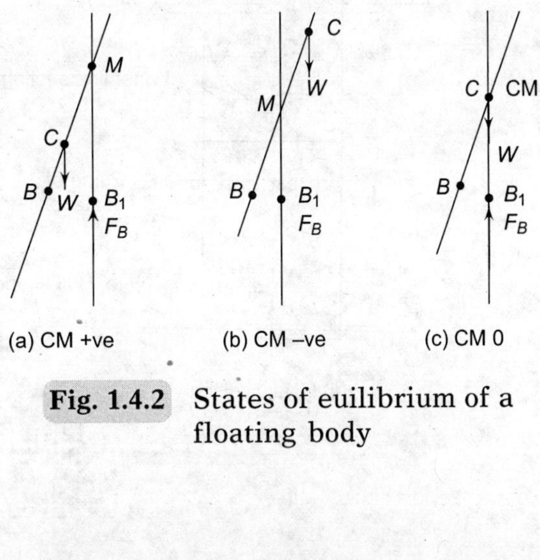


**Fig. 1.4.2** States of euilibrium of a floating body

the position of stable equilibrium where it has minimum energy. This can be generalised for any system and for linear or angular motion.

Now let us consider equilibrium of a floating body having weight $W$ that acts at its centre of gravity $C$. The portion of the body dipped in the liquid will displace it. As a result, buoyancy force $F_B$ acts on the body at the centre of gravity of the displaced liquid. The point $B$ at which the buoyancy force acts is called the *centre of buoyancy*. Initially, in the equilibrium position of the body, the point $B$ will lie vertically below $C$, say at $B$; draw a line $BC$ joining $B$ with $C$. Let the body be displaced clockwise by some angle so that the line $BC$ takes the inclined position, as shown in the Fig.. In this position, the buoyancy force acts at the point $B_1$; the vertical line through $B_1$ meets the line $BC$ at point $M$. The point $M$ is called *metacentre* and the distance $CM$ *metacentric height*. Three possibilities arise, which are:

(i) When $M$ lies above $C$, i.e. $CM$ is positive. In this case, see Fig. 1.4.2 (a), the weight and the buoyancy force will generate anticlockwise couple, called restoring couple, and the body would return back to the original position.

(ii) When $M$ lies below $C$, i.e. $CM$ is negative. From Fig. 1.4.2 (b), we observe that the couple produced by the weight and the force of buoyancy tilts the body further in the clockwise direction making it unstable.

(iii) When $M$ lies on $C$, i.e. $CM$ is zero. In this case, as shown in Fig. 1.4.2(c), there is no restoring couple and the body will stay in the displaced position.

We may also verify the above conclusions if the initial disturbance rotates the body about any axis in the anticlockwise direction. *So we conclude that for the stability of a floating body, the metacentre should lie above the centroid of the body.* The metacentric height depends on the shape and the size of the floating body.

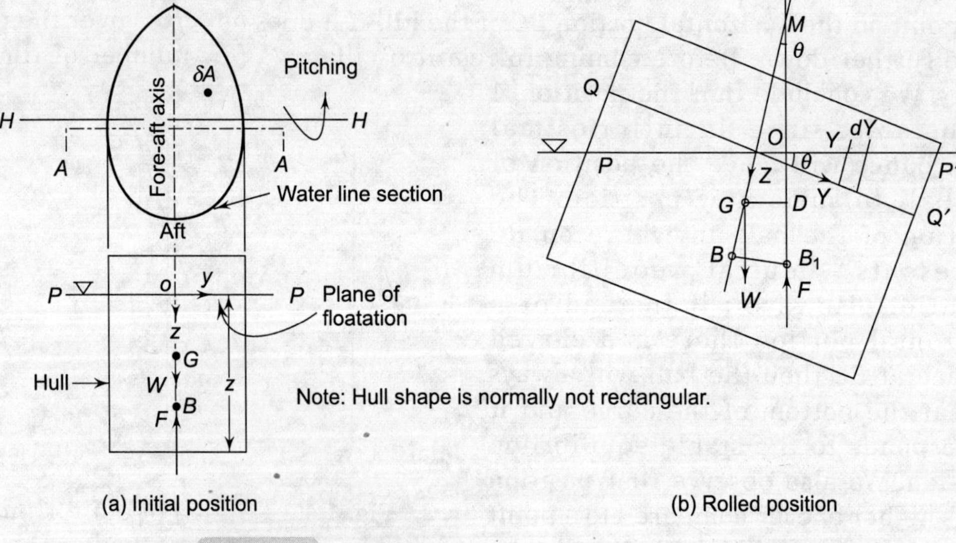


**Fig. 1.4.3** Oscillation of a floating body

We know that when a body is displaced from its stable equilibrium position, it returns back to its initial position after oscillating for a couple of times about its mean position. We would wish to bring the body back to its equilibrium position with minimum oscillations. Normally the floating body of interest is a ship or a boat and therefore we will explain the concept with reference to it though it is applicable to any floating body. Figure 1.4.3 (a) shows the plan and elevation, taken at the section *AA*, of a ship and the name of its different portions. The vessel can oscillate about the fore-aft axis, called *rolling*, or about the horizontal axis *HH*, called *pitching*. The vessel can also oscillate about the third axis, perpendicular to the plane of the board, which is called *yawing*. Moreover, the vessel contains the rotating parts whose axis turns with such motion and in turn Coriolis force acts which changes one kind of oscillation to another kind of oscillation. Pitching occurs when the wave front hits the fore or aft of the ship. The Coriolis component will change it to rolling or yawing. However, the rolling motion is generally considered in the study of oscillations of the vessel because moment of inertia about the fore-aft axis is minimum due to large length and height as compared to the width of the ship.

Figure 1.4.3 (b) shows the rolled position of the vessel at any time $t$. In this position, portion $OP'Q'$ of the ship goes into and $OPQ$ comes out of water. As a result, the new centre of buoyancy is at $B_1$ where force of buoyancy $F = W$ acts, where $W$ is the weight of the ship. The line $BG$ meets the vertical line through $B_1$ at $M$ at an angle $\theta$. The restoring couple due to the weight $W$ of the body is $W \times GD = W.GM \sin\theta \approx W.GM\,\theta$, since $\theta$ is small. Therefore the Eq. of motion by the de Alembert's principle is

$$I_c = \frac{d^2\theta}{dt^2} = -W\,GM\,\theta, \text{ where } I_c = \frac{W}{g}k^2$$

$$\rightarrow \qquad \frac{d^2\theta}{dt^2} + \frac{g \cdot GM}{k^2}\theta = 0.$$

or, $$\theta = A\sin\alpha t + B\cos\alpha t, \text{ where } \alpha = \sqrt{\frac{g \cdot GM}{k^2}}$$

Here $k$ is the radius of gyration of the water line section[7] about the fore-aft axis. If $T$ is the periodic time of the oscillations of the body then we can use the conditions $\theta = 0$ at $t = 0$ and $t = T/2$ to determine the constants $A$ and $B$. This gives

$$B = 0 \text{ and } A\sin(\alpha T/2) = 0$$

But $A \neq 0$, $\therefore\ \alpha T/2 = \pi$

$$\rightarrow \qquad T = 2\pi\sqrt{\frac{k^2}{g \cdot GM}} \tag{1.4.1}$$

The periodic time can be reduced by increasing the metacentric height or reducing the radius of gyration. We would like that vibration of the ship is dampened quickly and periodic time of vibration be small so that passengers are not disturbed.

[7] The ship does not dip into water. The water line section is that section of the ship that lies in the level of water.

### 1.4.1 Determination of Metacentric Height

Metacentric height plays a vital role in stability of the floating bodies and its oscillation frequency; its value depends on the distribution of the material, shape and the size of the floating body. Therefore we would like to determine it in terms of geometrical shape and dimensions of the floating body so that it may be used as useful information while designing such bodies, like ship, boat, etc. Figure 1.4.3 (a) shows two views of the body. The top view is the plan of the deck and the bottom one shows the elevation with the water surface at $PP$, which is the plane of floatation. Let the body rolls clockwise by an angle $\theta$, as shown enlarged in Fig. (b). As a result, the portion $OPQ$ comes out of the liquid whereas the portion $OP'Q'$ goes in and displaces more liquid. As such the point of application of the buoyant force is displaced to the point $B_1$, to the right. The original water surface line $PP$ takes the position $QQ'$ intersecting the new water line $PP'$ at the point $O$. we take $O$ as the origin of the coordinates with $OY$ and $OZ$ as the axes, as shown in the Fig.; $OZ$ is vertically down. Further the total volume of the displaced liquid will not change during the rolling of the body and therefore the two areas $OPQ$ and $OP'Q'$ will be equal.

To determine the position of the centre of buoyancy, consider an area element $\delta A$, as shown in the plan, in the plane of the floatation at a distance $y$ from the origin where the height of the body is $z$. This displaces volume of liquid equal to $(\delta A\, z)$. Taking moments about the X-Z plane passing through $O$ gives the $y$-position of the centre of gravity of the water displaced. In the original undisturbed position of the body, we get

$$V \times y_B = \int (z\delta A) y \tag{1.4.2}$$

In the displaced position, $z$ has increased by $(y \tan\theta)$ as a result of the portion of the body going into the liquid for the element shown in the Fig.. In this case, the new position of centre of buoyancy is

$$V \times y_{B_1} = \int (z + y\tan\theta)\delta A y \tag{1.4.3}$$

Shift $BB_1$ is then obtained by taking difference of the two position of the centre of buoyancy to give

$$BB_1 = V \times (y_B - y_{B_1}) \tag{1.4.4}$$

$$= \int y^2 \tan\theta\, \delta A$$

$$= \tan\theta \int y^2 \delta A$$

But $BB_1 = BM \tan\theta$ and $\int y^2 \delta A = I_{xx}$, the second moment of area about the fore-aft axis in the water line surface.

$$\therefore \quad BM = \frac{I_{xx}}{V} = \frac{\text{Second moment of area of the plane of floating about fore - aft axis}}{\text{Immersed volume}} \tag{1.4.5}$$

Thus, as expected, the metacentric height depends on the second moment of area, i.e. the distribution and the size of the floating body. We have so far not considered the loading of the ship that will affect the second moment of area and therefore plays a vital role in the stability of the vessel. The problem is more aggravated when it carries liquids, as in tankers, in which case the free surface of the liquid in its container changes depending on the acceleration of the ship. Thus, all these factors contribute towards the stability and proper care needs to be taken to ensure stability of the ship in all the conditions not only for the rolling but for the other types of ship motion such as pitching or yawing too.

### Experimental Determination of Metacentric Height

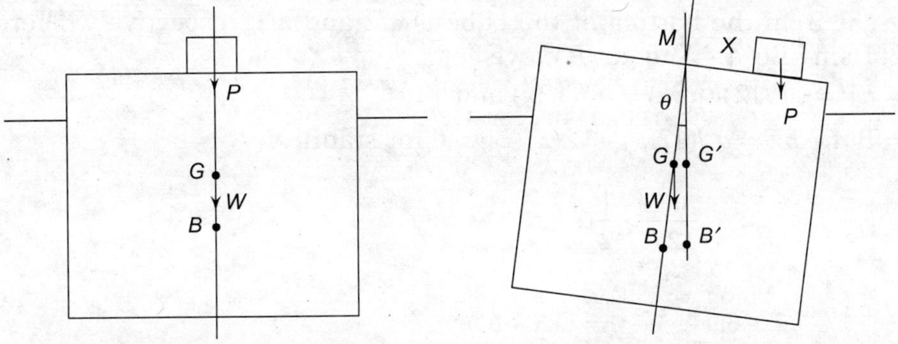


**Fig. 1.4.4** Experimental determination of metacentric height

To determine the metacentric height, a simple experiment is performed in which a weight $P$ is placed on the floating body of weight $W$ such that forces P, W and buoyancy force $F$ are in vertical line. Then the weight $P$ is moved to one side by a distance $X$ that heels the floating body. The body acquires the equilibrium position since the centre of gravity of $P$ and $W$ also shifts to the new point $G'$ such that once again the point of buoyancy and the gravity lie vertically in line. In this new position, the vertical line will intersect the original line $BG$ at the point $M$ in the water line. This point is the metacentre. Let the angle of heel of the body is $\theta$, which can be measured through experiment. The shift in G is due to movement of $P$, we can write

$$P\,X = W \times GG' = W\,GM \tan\theta$$

$$\rightarrow \qquad GM = PX \cot\theta / W \qquad (1.4.6)$$

The metacentric height can thus be determined.

**Example 1.4.1** A cube of side a and relative density $S$ ($S < 1$) floats in water. Determine the conditions for its stability if it is given an angular tilt.

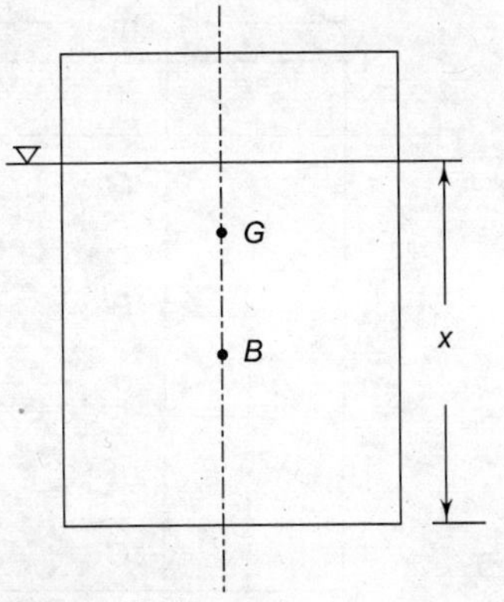


**Fig. E 1.4.1** Stability of a cube

**Solution** Let $x$ be the height of the cube immersed in water. By Archimedes principle,

$$a^3 S w = a^2 \times w \times x$$

$$\rightarrow \qquad x = aS \tag{A}$$

Let $G$ and $B$ represent the centre of gravity of the cube and the displaced water; their height from the bottom of the cube is $a/2$ and $x/2$, respectively. Therefore, $BG = (a - x)/2$. Using Eq. 1.4.5 to get $BM$, we get

$BM = I/V = (a^4/12)/(a^2 x) = a^2/(12x)$ and

$GM = BM - BG = a^2/(12 x) - ½ (a - x) > 0$ for stability

or, $$\frac{a^2}{12x} > \frac{1}{2}(a - x)$$

or, $$1 > 6\left(\frac{x}{a} - \frac{x^2}{a^2}\right) = 6(S - S^2)$$

or, $$S^2 - S + 1/6 > 0$$

or, $$(S - 0.789)\ (S - 0.211) > 0.$$

We get the factors 0.789 and 0.211 taking equality sign. This condition is satisfied if $S > 0.789$ or $< 0.211$; i.e. the specific gravity of the cube should be less than 0.211 or greater than 0.789. If the specific gravity is within these limits then the cube will be unstable. In case the cube has any of these two values then it is in neutral equilibrium.

**Example 1.4.2** A cylinder of 1.5 m diameter and 2 m long floats in seawater with its axis vertical. The base of the cylinder is 1.5 m below the water surface. Find the weight of the cylinder and the position of gravity if it is 0.3 m below the metacentre. Seawater weighs 10055 N/m$^3$.

**Solution** Weight of the cylinder = weight of the seawater displaced

$$= (\pi/4) \times 1.5^2 \times 1.5 \times 10055 = 26653 \text{ N}$$

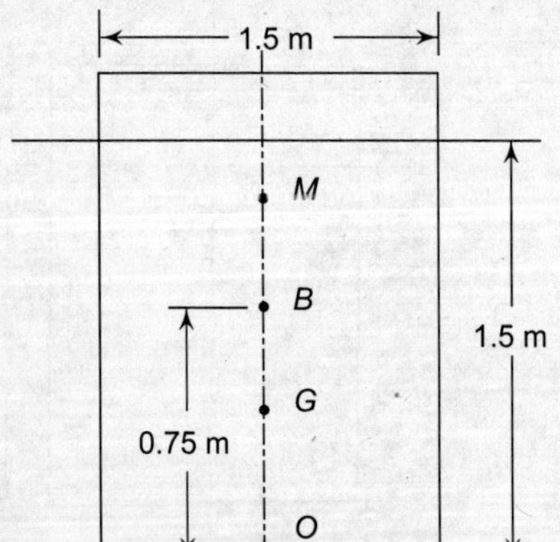


**Fig. E 1.4.2** Weight of cylinder

Figure E 1.4.2 shows the position of the cylinder with $O$, $G$, $B$, $M$ representing the centre of the base, centre of gravity, centre of buoyancy and the metacentre, respectively.

$$OB = 1.5/2 = 0.75\ m \text{ and } BM = \frac{I}{V} = \frac{\pi d^4/64}{(\pi d^2/4)\times 1.5} = 0.09 \text{ m.}$$

But $OM = OB + BM = 0.84$ m and also $= OG + MG$

$\rightarrow$ $OG = 0.54$ m, since $MG = 3$ m (given).

**Example 1.4.3** A barge in the shape of a rectangular block which is 8 m wide, 12.8 m long and 3 m deep floats in water with a draft of 1.8 m. The centre of gravity of the barge is 0.3 m above the water surface. State whether the barge is in stable equilibrium. Calculate the righting moment when the barge heels by 10°.

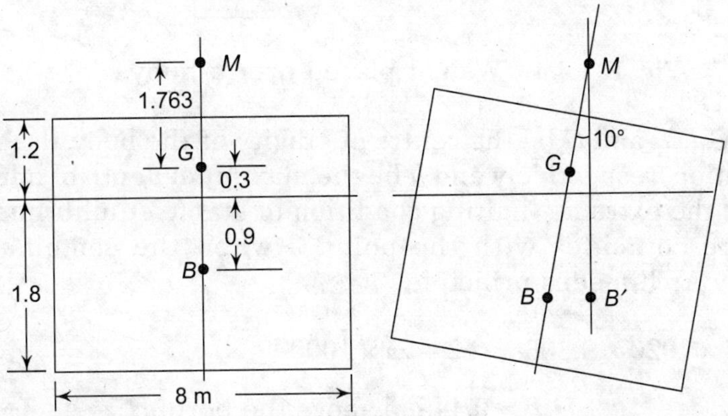


**Fig. E 1.4.3** Righting moment and stability of the range

**Solution** Centre of buoyancy will lie below the free surface at a distance = 1.8/2 = 0.9 m.

Therefore, $BG = 0.9 + 0.3 = 1.2$ m. Further, $BM = \dfrac{12.8\times 8^3/12}{(8\times 12.8\times 1.8)} = 2.963$ m and $GM = 1.763$ m. We observe that $BM > BG$ and hence the barge is in stable equilibrium. The righting moment for 10° heel is given by

$$F_b \times GM \sin 10, \text{ where } F_b = 9.81\times 8\times 12.8\times 1.8 = 1808 \text{ kN}$$

$\rightarrow$ righting moment = 553.56 kNm.

**Example 1.4.4** A cylindrical buoy 2 m in diameter and 1.5 m in height weighs 12 kN and floats in saltwater of density 1020 kg/m$^3$. Centre of gravity of buoy is 0.65 m above O. If a load of 2 kN is placed on its top symmetrically, find the maximum height of the centre of gravity of this load above the bottom if the buoy is to remain in stable equilibrium.

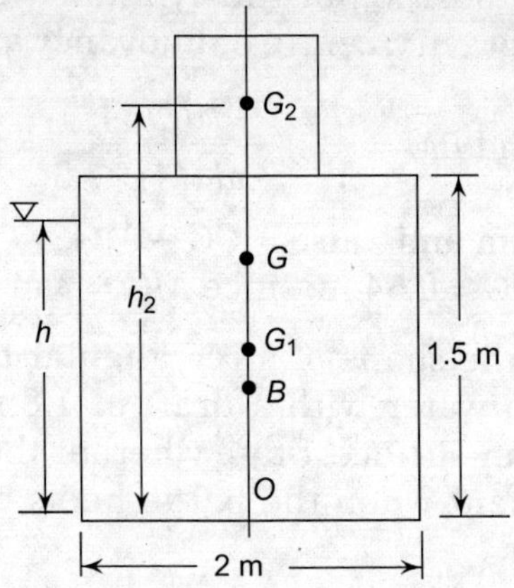


**Fig. E 1.4.4** Weight placed over a buoy

**Solution** Let $G_1$, $G_2$ and $G$ be the centre of gravity of the buoy, the additional load and of the combination, respectively and $h$ be the maximum depth of immersion of the buoy in water under the extreme limiting condition of stable equilibrium. This occurs when the metacentre coincides with the point $G$ when the combination is in neutral equilibrium. By Archimedes principle

$$\frac{\pi \times 2^2}{4} \times h \times 1020 \times 9.806 = (12 + 2) \times 1000$$

$$\rightarrow \qquad h = 0.445 \text{ m above the bottom}$$

Buoyancy force acts at $h/2 = 0.223$ m above the bottom of the buoy. Also

$$BM = BG = \frac{I}{V} = \frac{\pi 2^4/64}{(\pi 2^2/4) \times h} = 0.562 \text{ m}$$

and $\qquad OG = OB + BG = 0.785$ m.

Taking moments about the base, we get

$$2 \times h_2 + 12 \times 0.65 = 14 \times OG$$

$$\rightarrow \qquad h_2 = 1.595 \text{ m.}$$

**Example 1.4.5** A ship displaces 49250 kN of seawater of density 1025 kg/m$^3$. The second moment of area of the water line section about the axis through the centre is 120000 m$^4$ and the centre of buoyancy is 2 m below the centre of gravity. If the radius of gyration is 3.7m, calculate the period of oscillation.

**Solution** The metacentric height is obtained from Eq. 1.4.5 and the time of oscillation from Eq. 1.4.1.

$$GM = \frac{I}{V} - BG = \frac{120000}{(49250 \times 1000)/(1.25 \times 9.806)} - 2 = 0.449 \text{ m}$$

$$T = 2\pi \sqrt{\frac{3.7^2}{9.806 \times 0.449}} = 11.081 \text{ s}$$

**Example 1.4.6** An 80 mm diameter composite cylinder consists of an 80 mm diameter 20 mm thick metallic plate having specific gravity 4.0 attached at the bottom of an 80 mm diameter wooden cylinder of specific gravity 0.8. Find the limits of the length of the wooden portion so that the composite cylinder floats in stable equilibrium in water with its axis vertical.

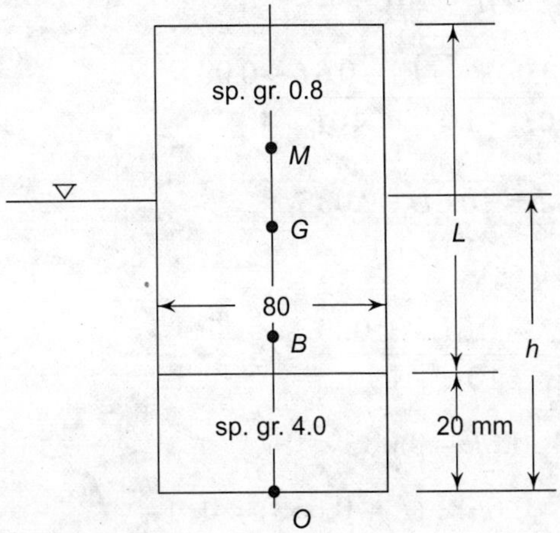


**Fig. E 1.4.6** Stability of cylinder

**Solution** The minimum height of the wooden portion is decided so that the composite cylinder should float for which the weight of the cylinder must be less than or equal to the weight of the liquid displaced. Let $L$ is the height of the wooden portion. Then

$$\frac{\pi(0.08)^2}{4}[0.02\times 4+0.8\times L]\leq\frac{\pi(0.08)^2}{4}[0.02+L]\times 1$$

$\rightarrow$ $$L\geq 0.3 \text{ m.}$$

Thus the wooden portion should be more than 0.3 m so that the cylinder floats in water. This will remain in stable equilibrium condition till the metacentric height is positive, which condition determines the maximum length of the wooden portion. We determine position of $G$ by taking moments of weights about the point $O$ to give

$$OG=\frac{\pi(0.08)^2[0.02\times 4\times 0.01+0.8\times L\times(0.5L+0.02]/4}{\pi(0.08)^2\times(0.02\times 4+0.08\times L)/4}=\frac{5L^2+0.2L+0.01}{10L+1} \quad \text{(A)}$$

Since the cylinder is floating, its weight should be equal to the weight of the displaced liquid. This gives another equation

$$\frac{\pi(0.08)^2}{4}[0.02\times 4+0.8\times L]=\frac{\pi(0.08)^2}{4}h\times 1$$

$\rightarrow$ $$h = 0.08\,(10\,L+1) \quad \text{(B)}$$

But $OB = h/2$. Therefore, $BG = OG - OB = \dfrac{L^2 - 0.6L - 0.03}{10L + 1}$ (C)

$$BM = \frac{I}{V} = \frac{\pi(0.08)^4 \times 4}{64 \times \pi(0.08)^2 \times h} = \frac{0.005}{10L+1}$$ after substituting for $h$ from Equation B.

$$\therefore\ GM = BM - BG = \frac{0.005}{10L+1} - \frac{L^2 - 0.6L - 0.03}{10L+1}$$

$$= -\frac{L^2 - 0.6L - 0.035}{10L+1} > 0$$

or,
$$\frac{L^2 - 0.6L - 0.035}{10L + 1} < 0$$

or,
$$L^2 - 0.6L - 0.035 < 0$$

or,
$$(L - 0.653)\,(L + 0.053) < 0.$$

This condition is satisfied only when $L < 0.653$ m. Thus the limits of the wooden portion are 0.35 m and 0.653 m.

## 1.5 Pressure Variation in Liquids Confined in Moving Container

Tankers are used for shipment of liquid in sea or road transports where the carrier is sometimes accelerated, decelerated or moves with constant acceleration. In such case the free surface of the liquid in the tanker changes and acquires equilibrium state. As discussed in last section, this affects the stability of the vessel. Therefore, we would now discuss how to determine the free surface and its effects on the pressure variation. The other kind of container motion of importance is its rotation about any vertical axis passing through the liquid. The common feature in these kind of motions $P$ is that laws of fluid statics are applicable.

### 1.5.1 Container Moving with Constant Acceleration

Figure 1.5.1 shows a vessel containing liquid moving with constant acceleration $\alpha$ having components $a_y$ and $a_z$ along the $y$ and $z$ axes with the origin in the free surface. Because of the inertia of the liquid and its property to move relative to the surface, the free surface will shift and would take inclined position at an angle $\theta$ with the horizontal direction. This slope will depend on the magnitude and direction of the acceleration. We know the effect of inertia is to oppose the main motion as a result when the vessel accelerates, the liquid will go back resulting in rising of level on the rear end and lowering on the front end of the vessel, as shown in the figure. For constant acceleration the free surface takes the inclined position and stays there. The pressure at any point can be obtained by treating the fluid at rest. That is, the vessel moves but not the fluid, which remains static with respect to the vessel; effect of the vessel movement is restricted to change in slope of the free surface. For this reason this subject is studied with fluid statics. If the vessel moves with no acceleration then the free surface will remain horizontal.

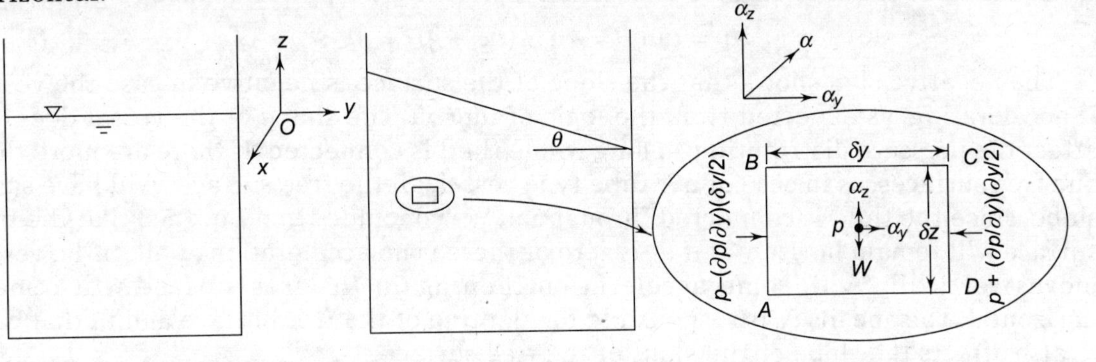


**Fig. 1.5.1** Liquid in a container subjected to a constant acceleration

We may determine the angle $\theta$ by considering a fluid element *ABCD* whose enlarged view is shown in Fig. 1.5.1. Let $p$ is the pressure at the centre of the element considered whose sides are $\delta y$ and $\delta z$. Because the fluid is continuous, pressure at the faces *AB* and *CD* can be written using the Taylor series while dropping the higher order terms; the pressures are shown in the Fig. for $y$-direction and similarly we may write along the $z$-direction. Using Newton's second law of motion, we can write the following equation of motions along $y$ and $z$ directions for the element taking unit width in $x$-direction.

$$\left(p - \frac{\partial p}{\partial y}\frac{\delta y}{2}\right)\delta z - \left(p + \frac{\partial p}{\partial y}\frac{\delta y}{2}\right)\delta z = \rho\,(\delta_y \times \delta_z \times 1)\alpha_y$$

$$\rightarrow \qquad \frac{\partial_p}{\partial_y} = -\,\rho\alpha_y \qquad (1.5.1a)$$

$$\left(p - \frac{\partial p}{\partial z}\frac{\delta p}{2}\right)\delta y - \left(p + \frac{\partial p}{\partial z}\frac{\delta z}{2}\right)\delta y - \rho g\,(\delta y \times \delta z \times 1) = \rho\,(\delta y \times \delta z \times 1)\,a_z$$

$$\rightarrow \qquad \frac{\partial p}{\partial_z} = -\rho(\alpha_z + g) \qquad (1.5.1b)$$

We can integrate Eq. 1.5.1a to give $p$ at any depth.

$$\rightarrow \qquad p = -\rho\alpha_y\, y + f(z)$$

$$\therefore \qquad \frac{\partial p}{\partial z} = f'(z) = -\rho(a_z + g) \text{ (from second Eq.).}$$

Integration with respect to $z$ gives $f$. Thus

$$f = -\rho(\alpha_z + g)z + p_0,$$

where $p_0$ is the constant of integration, the pressure at the free surface. This gives

$$\therefore \quad p = -\rho\,[\alpha_y\, y + (\alpha_z + g) z] + p_0, \tag{1.5.2}$$

Equation of the free surface is obtained by letting $p = p_0$ in Eq. 1.5.2 to give

$$z/y = \tan\theta = -\,a_y\,/\,(a_z + g) \tag{1.5.3}$$

The negative sign shows that the slope of the surface is negative in case the vessel is accelerating, as expected from the logic of inertia. The shape of the vessel does not affect the pressure distribution so long whole fluid is connected. If there are more than one free surfaces, as in case of a *U* tube type vessel, all the free surfaces will have same slope, since the fluid is connected throughout. We conclude from Eq. 1.5.3 that the free surface will remain horizontal if $a_y$ is zero or there is no acceleration at all. If the vessel moves vertically with some acceleration then again the free surface will remain horizontal. This means $a_y$ is responsible for slanting of the free surface and in that case $a_z$ also affects the slope or the slant of the free surface.

**Example 1.5.1** A rectangular tank 3 m long, 2 m wide and 2 m deep contains water to a depth of 1.25 m. If it is accelerated horizontally at 3 m/s$^2$ in the direction of its length, find (a) (i) the inclination of the water surface with the horizontal, (ii) depth of water at the two ends, (iii) total force on the sides at the two ends of the tank and (iv) net horizontal force on the fluid mass.

(b) What maximum acceleration is permissible in this case?

(c) How much water would spill out if the tank is accelerated at 8 m/s$^2$?

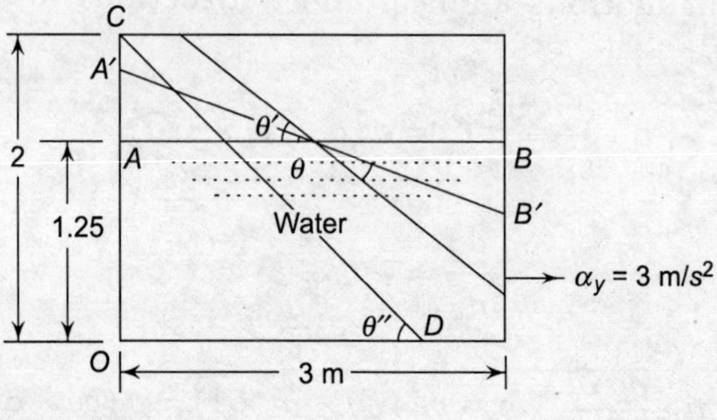


**Fig. E 1.5.1** Horizontal acceleration of a tank

**Solution** (a) Equation 1.5.3 gives the slope of the free surface; the angle $\theta$ is $\tan^{-1}$ (3/9.81) = 17°.

$\therefore AA' = BB' = (3/2)\tan\theta = 0.4587$ m.

Depth of water at the rear end = 1.5 + 0.4587 = 1.7087 m and

Hydrostatic force on rear side of the tank = $w\,A\,\bar{z}$ = 9810 × (2 × 1.7087) × 1.7087/2 = 28.642 kN

Depth of water at the front end = 1.5 – 0.4587 = 0.7913 m

Hydrostatic force on front side of the tank = 9810 × (2 × 0.7913) × 0.7913 /2 = 6.143 kN.

∴ Net horizontal force = 28.642 – 6.143 = 22.499 kN, opposite to the direction of $\alpha_y$.

(b) The tank can be accelerated further till the water level at the rear side just touches the top of the tank. (Further increase in acceleration will spill out the water from the tank.) For this condition, the slope of the free surface is tan $\theta'$ = (2 – 1.25)/1.5 = ½. This gives $a_y = g \times \tan \theta' = 4.905$ m/s$^2$.

(c) In this case tan $\theta''$ = 8/9.81, and the free surface will be *CD*. As seen in the Fig., there is no water on the front end. From the triangle *OCD*, *OD* = *OC*/tan $\theta''$ = 2.4525 m.

∴ Volume of water in the tank = ½ × *OD* × *OC* × width of tank = ½ × 2.4525 × 2 × 2 = 4.905 m$^3$.

Volume of water originally present = 3 × 1.25 × 2 = 7.5 m$^3$.

→ Volume of water spilled out from the tank = 7.5 – 4.905 = 2.595 m$^3$.

**Example 1.5.2** The U-tube shown in the figure is moved horizontally with an acceleration of *g*/3 m/s$^2$. Find the difference of water level in its two limbs.

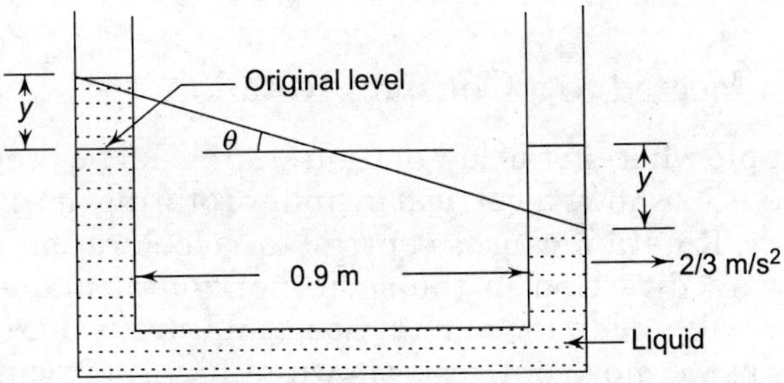


**Fig. E1.5.2** Acceleration of U-tube

**Solution** Slope of the free surface in the U-tube is tan $\theta = (g/3)/g = 1/3$. Also from the figure, the slope is $2y/0.9$, where $y$ is the displacement of the liquid in the limb due to acceleration. Equating the two, we get, $y$ = 0.15 m. Therefore, difference of water level in the two limbs is 2 × $y$ = 0.3 m.

**Example 1.5.3** A tank shown in Fig. E1.5.3 contains oil of specific gravity 0.8. If it is given an acceleration of 5 m/s$^2$ along a 30° inclined plane in the upward direction, determine the slope of the free surface and pressure at *b*.

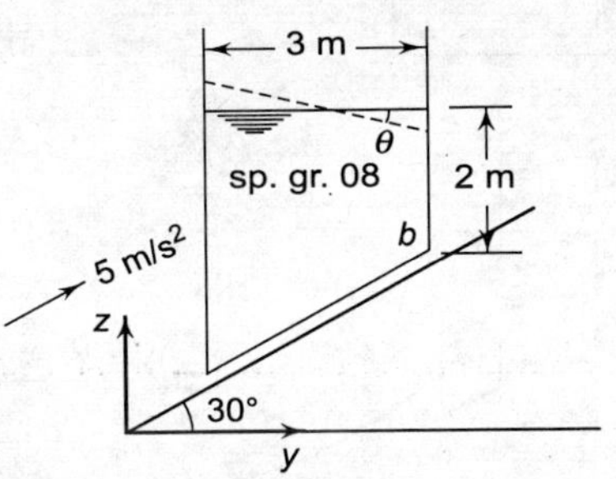


**Fig.E 1.5.3** Acceleration on inclined plane

**Solution** Components of the acceleration along the horizontal and vertical directions are

$$a_y = 5 \cos 30 = 4.33 \text{ and } a_z = 5 \sin 30 = 2.5 \text{ m/s}^2.$$

Slope of the water surface is then $\tan \theta = a_y/(a_z + g) = 0.3519$

$\rightarrow$ $\theta = 19.38°$.

Depth at $b = 2 - (3/2) \tan \theta = 1.472$ m.

The pressure at $b$ now can be determined using the equation for static fluid. However, there is acceleration in the z-direction too that would add to the pressure. We may use Eq. 1.5.2 without considering the first term in $y$ as effect of $a_y$ has already been accounted while using the depth calculated above. Moreover, while going down, $z$ is negative. With these comments, we get

$$p_b/w = (1 + a_z/g) \times 1.472 = 1.847 \text{ m}.$$

or, $$p_b = 1.847 \times 9787 \times 0.8 = 14.463 \text{ kN/m}^2 \text{ (gauge)}.$$

## 1.5.2 Container Subjected to a Constant Rotation

This is another example wherein the law of fluid statics can be used to determine pressure at any depth. Figure 1.5.2 shows a vertical cylinder rotating at some axis, not necessarily the axis of symmetry. Rotation causes centripetal acceleration that increases pressure along radially outward direction in the same horizontal plane. This implies that the liquid's free surface will no longer remain horizontal but would take a shape such that its height increases as we move out, as is shown in Fig. 1.5.2, with minimum depth at $O$, the point where the axis of rotation meets the free surface. The fluid rotates as if it is a rigid body and such motion is called *forced vortex* motion. We take the origin at the base of the cylinder where the axis of rotation meets and the radial and z-directions as shown in the figure.

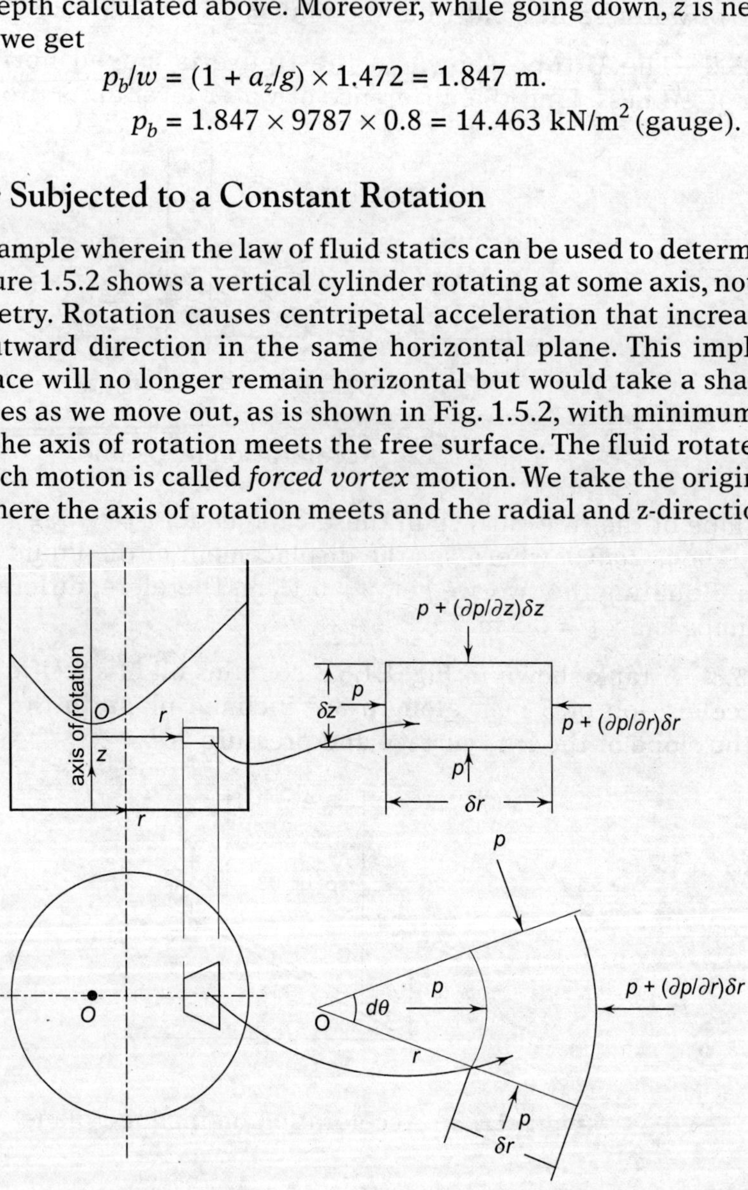


**Fig. 1.5.2** Constant velocity rotation of liquid

Consider an element of radial length $\delta r$ rotating with angular velocity $\omega$ and subtending angle $\delta\theta$ at the axis of rotation; its two enlarged views are also shown with dimensions and the pressure acting at various faces. We use Taylor series to get pressure at other faces. We may balance the forces in the radial and axial directions to give the following equations:

For $r$ – direction: $p.\ r\delta\theta.\delta z - \left(p + \dfrac{\partial p}{\partial r}\delta r\right).\ (r + \delta r).\ \delta\theta.\delta z + 2p\ \delta r\ \dfrac{\delta\theta}{2}\ \delta z = -\rho\omega^2 r.\delta r.\delta\theta.\delta z$

$\rightarrow \qquad \partial p/\partial r = \rho\omega^2 r$, and

For $z$ – direction: $p.r\delta\theta.\delta r - \left(p + \dfrac{\partial p}{\partial z}\delta z\right).\ r\delta\theta.\delta r - \rho g(r\delta\theta.\delta r.\delta z) = 0$

$$\rightarrow \qquad \partial p/\partial z = -\rho g \tag{1.5.4}$$

We may integrate these equations, as explained in section 1.5.1, to give pressure at any point within the fluid. We get

$$p = -\rho g\ (z - z_0) + 1/2\ \rho\omega^2 r^2 + p_0 \tag{1.5.5}$$

We have used the condition $z = z_0$, $p = p_0$ on the free surface at the point $O$, where $z_0$ is the height of the point $O$ from the origin. If we let $p = p_0$ in Eq. 1.5.5 then we get the equation for the free surface as

$$z = z_0 + (\omega^2 r^2)/(2g) \tag{1.5.6}$$

which is the equation of a paraboloid of revolution with vertex at $O$. If the axis of rotation coincides with the axis of the cylinder then the paraboloid will be symmetrical about its axis. If we substitute $z = z_R$ at $r = R$ then Eq. 1.5.6 can be written as

$$z_R - z_0 = (\omega^2 R^2)/(2g) \tag{1.5.7}$$

Finally recalling that the volume of a paraboloid of revolution is equal to half that of the circumscribing cylinder. i.e;

$$\text{volume of paraboloid of revolution} = 1/2\ \pi R^2\ (z_R - z_0).$$

When the cylinder is not rotating then the free surface is horizontal. If $z_1$ is the height from the origin then volume of this additional liquid shall be equal to the volume of paraboloid of revolution. This gives

$$\pi R^2\ (z_1 - z_0) = \pi R^2 (z_R - z_0)/2$$

$$\rightarrow \qquad z_1 = (z_R + z_0)/2$$

$$\text{or,} \qquad z_0 = 2z_1 - z_R.$$

We may express Eq. 1.5.7 in terms of $z_1$

$$z_R = z_0 + \frac{\omega^2 R^2}{2g} = (2z_1 - z_R) + \frac{\omega^2 R^2}{2g}$$

$$\text{or,} \qquad z_R = z_1 + \frac{\omega^2 R^2}{4g} \text{ and } z_0 = z_1 - \frac{\omega^2 R^2}{4g} \tag{1.5.8}$$

It is observed that the rise near the surface is equal to the fall in level at the centre from the original level. If we increase the angular velocity then the water may spill out from the cylinder brim if its height is not sufficient. Slope of the free surface at any radius $r$ can be obtained by differentiating Eq. 1.5.6. with respect to $r$. This gives

$$dz/dr = \omega^2 r/g \tag{1.5.9}$$

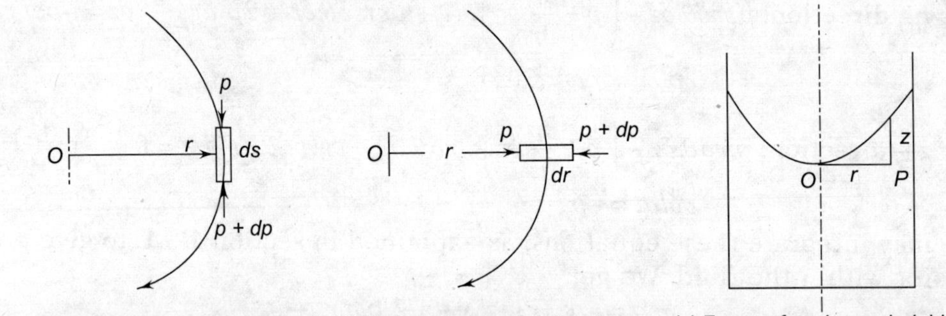


(a) Variation along streamline (b) Variation along radial direction (c) Free surface is paraboloid

**Fig. 1.5.3** Pressure variation along stream line and radial direction

In fact, forced vortex flow is one case of a more general flow along the curved streamline in which the velocity does not change on the same streamline and there exists radial acceleration only. However, in any flow along a curved stream line, the velocity may also change along the streamline. Figure 1.5.3 shows a streamline with $r$ as radius of curvature having the centre of curvature at $O$. To determine variation of pressure along the streamline, consider a cylindrical element of cross sectional area da along it. Let $v$ be the velocity then $v^2/r$ and $v\ dv/ds$ will be the radial and the stream wise acceleration respectively at the point under consideration. Equilibrium of the element demands

Resultant force on the element along the streamline = mass × acceleration

or, $p\,da - (p + dp)\,da = (w/g)\,da \times ds \times v\,dv/ds \rightarrow dp/dv = -(w/g)\,v$,

Where $w$ is the specific weight of the fluid. Integrating this equation in the same horizontal plane, we get

$$\frac{p}{w} + \frac{v^2}{2g} = C \text{ (constant)} \tag{1.5.10}$$

We shall learn in the following chapters that Eq. 1.5.10 is the Bernoulli's equation and shows that the sum of the pressure head and kinetic head remains constant along the streamline. On the other hand, if we take the element along the radial direction, as shown in Fig. 1.5.3 (b), then the radial equilibrium leads to

$$(p + dp)\,da - p\,da = (w/g)\,da \times dr \times (v^2/r) \rightarrow dp/dr = w\,v^2/(gr) \tag{1.5.11}$$

If the relation between $v$ and $r$ is known then Eq. 1.5.11 can be integrated. Substituting $v = wr$ in Eq. 1.5.11 and carrying out integration from $r_1$ to $r_2$, we get

$$\frac{p_2}{w} - \frac{p_1}{w} = \frac{(\omega r_2)^2}{2g} - \frac{(\omega r_1)^2}{2g} = \frac{v_2^2}{2g} - \frac{v_1^2}{2g} \tag{1.5.12}$$

You may get this result from Eq. 1.5.5 as well. This shows that both pressure and velocity increases with increase in radius and the change in pressure head is equal to the change in the velocity head in the same horizontal plane. Let us determine pressure at the point $P$ at a depth $z$ from the free surface, as shown in Fig. 1.5.3 (c), where point $P$ lies in the horizontal plane passing through the point $O$. Pressure at the point $P$ from hydrostatics is $p_a + w\,z$ whereas from Eq. 1.5.12, it is $p_a + w\,\omega^2 r^2/(2g)$, because at $O$ the pressure is atmospheric. Equating pressure at the point $P$ obtained in two different ways, we get

$$p_a + w\,\omega^2 r^2/(2g) = p_a + w\,z$$

$$\rightarrow \qquad z = \omega^2 r^2/(2g)$$

which is Eq. 1.5.7, as expected. The origin has been taken at the free surface for which $z_0 = 0$.

**Example 1.5.4** A closed cylindrical vessel, 0.3 m diameter and 0.6 m high, two-third filled with oil of specific gravity 0.8 is rotated about its vertical axis. Determine the speed of rotation when (a) the oil starts touching the lid and (b) the point at the centre of the base is just clear of oil. How the results will change if the cylinder is open at the top?

**Solution** Given for case (a): $z_1 = 0.6 \times 2/3 = 0.4$ m, $z_R = 0.6$ m (see Fig. E1.5.4 a), and $R = 0.15$ m.

Using Eq. 1.5.8, we get $\omega = 18.68$ rad/s. The liquid free surface is shown in Fig. E1.5.1.

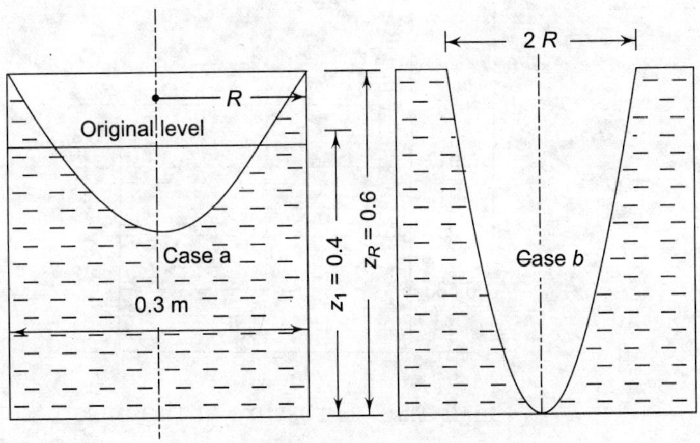


**Fig. E 1.5.4** Rotation of a cylinder

(b) In this case the liquid will collect more and more near the wall as it cannot spill out. Let $R$ be the radius at the cylinder brim when the liquid has touched the bottom of the cylinder. Initially, $1/3^{rd}$ cylinder is not filled with liquid. Therefore, volume of the paraboloid of revolution should be equal to it. This gives

Volume of paraboloid of revolution = volume of unfilled space initially.

or, $1/2\ (\pi R^2 \times 0.6) = 1/3\ \{\pi(0.3/2)^2 \times 0.6$

$\rightarrow$ $R = 0.122$ m.

Now $z_R = 0.6$ m, $z_0 = 0$ m and $R = 0.122$ m,

Eq. 1.5.8 gives $\omega = 28.1$ rad/s.

If the cylinder is open then liquid will spill out as $\omega$ increases beyond 18.68 rad/s. When the liquid touches the bottom of the cylinder then $z_0 = 0$ m, $z_R = 0.6$ m and $R = 0.15$ m. Eq. 1.5.8 gives $\omega = 22.9$ rad/s. Further, the volume of the oil left in the cylinder = ½ $A \times 0.6 = 0.3\ A$, here $A$ is the cross sectional area of the cylinder. In the initial condition, the volume occupied is $A \times 0.6 \times 2/3 = 0.4\ A$.

Therefore, percentage of oil left in the cylinder = $0.3\ A/(0.4A) \times 100 = 75\%$. Thus the effect of removing the lid from the cylinder top is that the oil remaining in the cylinder is decreased by 25% at the time liquid touches cylinder bottom.

**Example 1.5.5** A vertical cylindrical tank 0.5 m diameter and 1.5 m high is closed at the top. It contains water to a depth of 1.2 m, the air above the water surface is at a pressure of 39.24 kPa. Determine the pressure head on the bottom of the tank at the axis and the side, when the tank is rotated at 200 rpm about its axis.

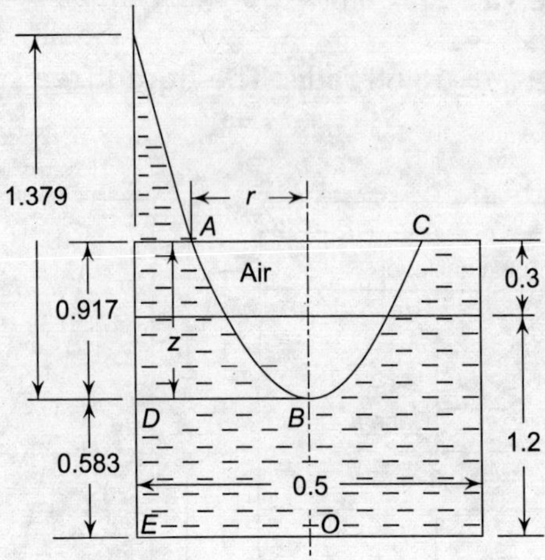


**Fig. E 1.5.5** Pressure at the base

**Solution** When the tank is rotated, the liquid free surface takes parabolic shape $ABC$, as shown in Fig. E1.5.5. Let $r$ is the radius of the point $A$ from the cylinder axis and the origin be at $B$. Volume of the air above the water level before rotation remains same. The volume after and before the rotation are: ½ $\pi r^2 \times z$ and $\pi (0.5)^2 \times 0.3$, where $z$ is the vertical height of paraboloid of rotation and the space above the water surface before rotation is 0.3 m (given). Equating these two, gives

$$r^2 \times z = 0.0375 \tag{A}$$

From Eq. 1.5.7, the equation of the paraboloid of revolution, we get, noting that in this case $z_0 = 0$,

$$z = \left(\frac{2\pi \times 200}{60}\right)^2 \times \frac{r^2}{2 \times 9.81} = 2.357\, r^2 \quad \text{(B)}$$

Solving Eqs. A and B, we get $z = 0.917$ m and $r = 0.202$ m. If $O$ is the mid point of the cylinder base then $OB = 1.5 - 0.917 = 0.583$ m. Further, pressure head at point $B$ due to air pressure (using Eq. $p = \rho gH$) is $= 39.24 \times 1000/9810 = 4$ m of water. Pressure head at point $D$, lying on the cylindrical surface in the horizontal plane passing through the point $B$, due to centrifugal force can now be obtained using Eq. 1.5.12. This gives

$$\frac{p_D}{w} - \frac{p_B}{w} = \frac{\left(\dfrac{2\pi \times 200 \times 0.5}{60}\right)^2}{2 \times 9.81}$$

$$\rightarrow \qquad \frac{p_D}{w} = 5.397\text{m, since } \frac{p_B}{w} = 4\text{m,}$$

Now pressure head at $O$ and $E$ can be determined by using Eq. 1.3.2.
Pressure head at $O$ = pressure head at $B$ + 0.583 = 4 + 0.583 = **4.583** m.
Pressure head at $E$ = pressure head at $D$ + 0.583 = 5.397+ 0.583 = **5.980** m.

## Summary

When the fluid is stationary or moving with constant velocity then the pressure variation with depth of the liquid is given by $-\rho g dz$, where $\rho$ is density of the fluid, $g$ is the acceleration due to gravity and $dz$ is the depth of the point. The negative sign appears because we take $z$-direction positive in the upward direction and $g$ acts in the –ve direction. Integration of the equation from one point to another is possible if the relation between density of the fluid and the height is known. We have considered incompressible fluid and isothermal conditions, linear variation of temperature with altitude or isotropic conditions for compressible fluid to evaluate the pressure variation. Fluid statics leads to the development of manometers used for pressure measurement. The procedure for writing the manometric equation is discussed.

We have then examined the force acting on the plane and the curved surfaces dipped fully or partially in the liquid. The important conclusion is that the surface area can be resolved along the vertical and the horizontal planes to find out the horizontal and the vertical components of the resultant force acting on the surface. The vertical component turns out to be the weight of the liquid displaced by the immersed surface. The centre of pressure is always below the centre of gravity of the area that depends on the second moment of the area, inclination of the surface with the free surface and the depth of the centre of gravity from the free surface. The depth of the centre of pressure is given by Eq. 1.3.10. This study has helped in determining the forces and its position on the hydraulic structures so that these can be designed properly. We have also studied the stability consideration of the dam.

Floating bodies are of prime importance as ships and boats have been used for a long time for transportation. Their stability is of concern to us and it depends on the position of the metacentre vis-a-vis the centre of gravity. It is established that when the metacentre lies above centre of gravity then the body is stable. The other requirement is that once the body is displaced angularly because of some disturbance in the surrounding then it should come back to the original position with minimum number of oscillations. In this respect, we learn that the rolling motion needs attention and the moment of inertia about the fore-aft axis in the water line section should be large.

## PROBLEMS

### Review Questions

1. Differentiate between (i) liquids and gases, (ii) real and ideal fluids, (iii) specific weight and specific gravity, (iv) compressible and incompressible fluids.
2. Define the terms: density, specific gravity, specific volume, and vacuum pressure.
3. What is the difference between dynamic and kinematic viscosity? State their units in different system of units as well as their dimensions.
4. Enunciate Newton's law of viscosity. Explain its importance in fluid motion. What is the effect of temperature on viscosity of water and gas and explain the difference in their behaviour with temperature?
5. Cite examples where surface tension plays a prominent role.
6. 1 litre of crude oil weighs 9.6 N. Calculate its specific weight and specific gravity of the oil.
7. Explain atmospheric pressure. What is the value of atmospheric pressure head in terms of mercury column and in terms of water column? Explain how atmospheric pressure varies with altitude?
8. Differentiate between: (i) absolute and gauge pressure, (ii) simple manometer and differential manometer, and (iii) piezometer and pressure gauge.
9. What do you mean by vacuum pressure?
10. What is a manometer? How are they classified?
11. What are the common liquids used in manometers? What conditions should it satisfy before we choose a manometric liquid?
12. Suggest manometric arrangements to measure a pressure difference of (i) 50 N/m$^2$, 0 kN/m$^2$, and (iii) 100 kN/m$^2$ between the two sections of a pipe carrying a liquid? How will the arrangement differ if the pipe carries air?
13. What considerations govern the diameter of the glass tube to be used in a manometer?
14. What will happen if two limbs of U-tube manometer are of different diameters?

15. What is the difference between U-tube differential manometers and inverted U-tube differential manometers? Where are they used?
16. Describe briefly with sketches the various methods used for measuring pressure exerted by fluids.
17. What do you understand by hydrostatic equation? With the help of this equation, derive the expression for the total thrust on a submerged plane area and the buoyant force acting on it.
18. The diameter of a ram and the plunger of a hydraulic press are 300 mm and 50 mm respectively. find the load that can be lifted with a force of 400 N.
19. What do you understand by the term, ' temperature lapse rate'? Obtain an expression for the temperature Lapse rate. Derive an expression for the pressure at a height Z form the sea level for a static air when the compression of the air is assumed isothermal. The pressure and temperature at sea level are $p_0$ and $T_0$ respectively.
20. What is hydrostatic pressure distribution? Give two examples each where pressure distribution is hydrostatic and the pressure distribution is non-hydrostatic.
21. What do you understand by the total pressure and centre of pressure?
22. Prove that the centre of pressure of a completely submerged plane surface is always below the centre of gravity of the submerged surface or at the most coincides with it. When will that happen?
23. Prove that the total pressure exerted by a static liquid on an inclined submerged plane surface is the same as the force exerted on a vertical plane surface as long as the depth of the centre of gravity of the surface is unaltered.
24. Explain how would you find the resultant pressure on a curved surface immersed in a liquid?
25. Why are different techniques adopted to calculate the total force and its point of application, in case of plane and curves immersed surfaces?
26. What are the conditions of equilibrium of a floating body and a submerged body?
27. Define the terms metacentre, centre of buoyancy, metacentric height, gauge pressure and absolute pressure.
28. A water body is subjected to acceleration in the vertically upward direction. At what acceleration will the pressure difference between two points, separated by a vertical distance h, be zero?
29. What is the Archimedes principle? A body floats in between two fluids of specific weights $\gamma_1$ and $\gamma_2$. What will be the expression for buoyancy force?

## Problems for Practice

1. Determine the torque and power required to turn a 10 cm long, 5 cm diameter shaft at 500 revolutions per minute in a 5.1 cm diameter concentric bearing flooded with a lubricating oil of viscosity 100 centipoise. [0.1029 Nm, 5.387 W]
2. A thrust bearing consists of a 10 cm diameter pad rotating on another pad separated by an oil film, $\mu = 80$ cP, by 1.5 mm. Compute the power dissipated in the bearing if it rotates at 100 rpm. [0.0574 W]

3. A skater weighing 800 N skates at 54 km/hr on ice at 0°C. The average skating area supporting him is 10 $cm^2$ and the effective dynamic coefficient of friction between the skates and the ice is 0.02. If there is actually a thin film of water between the skate and the ice, determine its average thickness. Assume water has dynamic viscosity at 0°C of 1 cP.
4. A U-tube is made up of two capillaries of diameters 1 mm and 1.5 mm respectively. The U tube is kept vertically and partially filled with water of surface tension 0.0075 kg/m and zero contact angle. Calculate the difference in the level of the menisci caused by the capillarity. [0.68 mm]
5. Calculate the gauge pressure at $A$, $B$, $C$, and $D$ in the Fig. in $N/m^2$, $kg/cm^2$, and m of W.G.

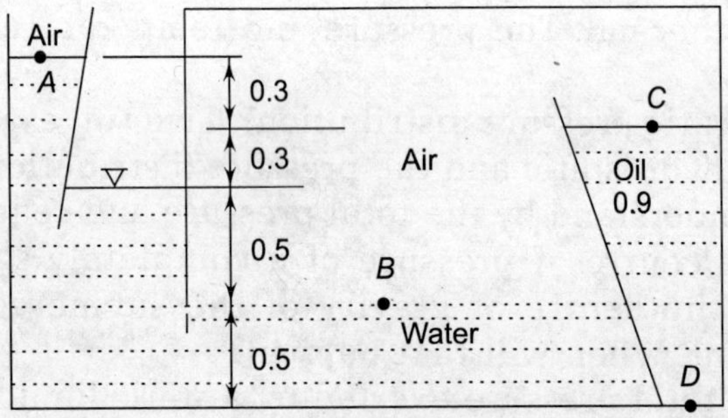


**Fig. P 1.5** Pressure at different points

6. In order to measure the pressure in a pipe containing an oil of specific gravity 0.8, four U-tubes are connected in series as shown in Fig P1.6. The gauging liquid in all the U-tubes is mercury of specific gravity 13.6. Between the columns of mercury the liquid in the U-tubes is water. Find the pressure at the centre of the pipe.

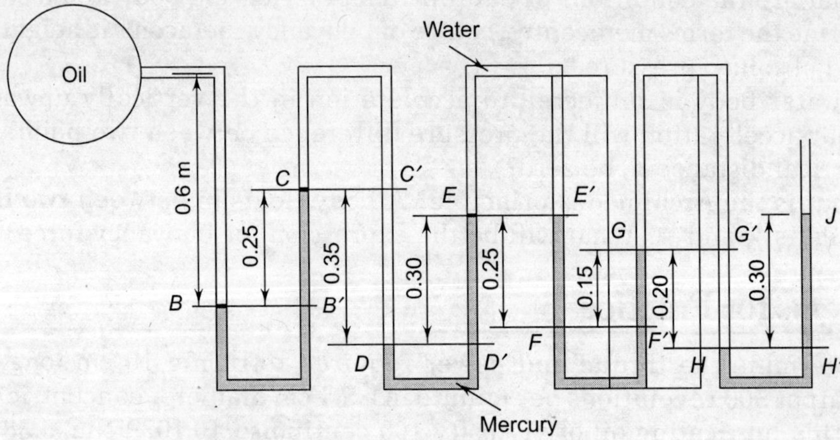


**Fig. P 1.6** Multi-tube manometer

7. In the arrangement shown in Fig. P 1.7, find the difference of pressure of water in the pipes *A* and *B*.

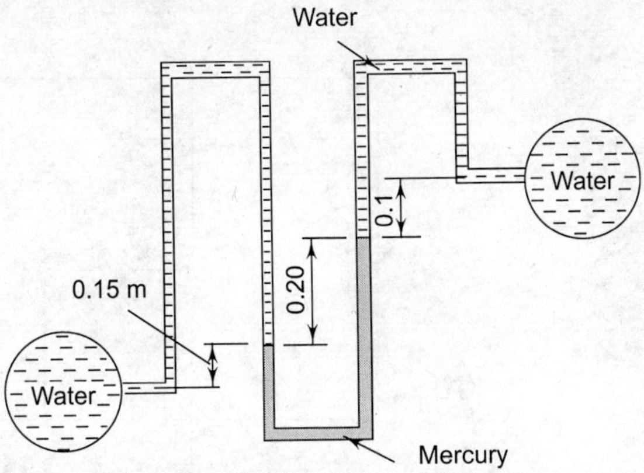


**Fig.P 1.7** Differential manometer

8. Find the pressure difference between the containers *A* and *B* shown in Fig. P 1.8.

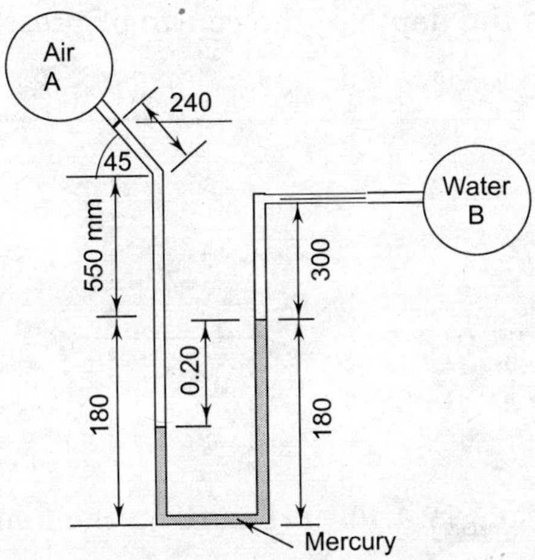


**Fig. P 1.8** Pressure difference

9. Determine the total pressure and the depth of the centre of pressure for the vertical lamina shown in Fig. P 1.9.

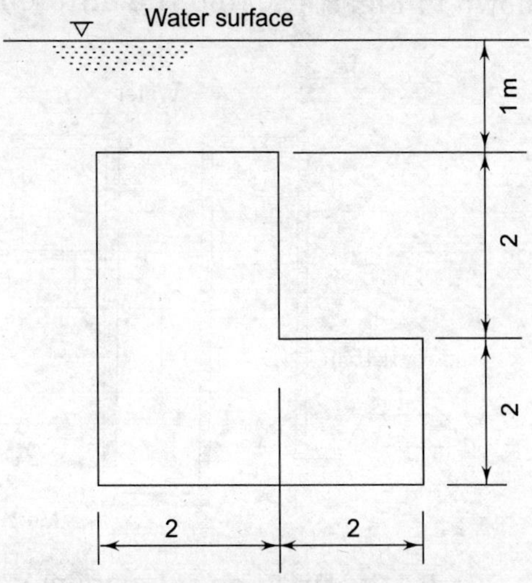


**Fig. P 1.9** Pressure on the lamina

10. Figure P 1.10 shows a vertical semicircular lamina. Determine (i) the total pressure on the lamina and (ii) the depth of the centre of pressure for the lamina.

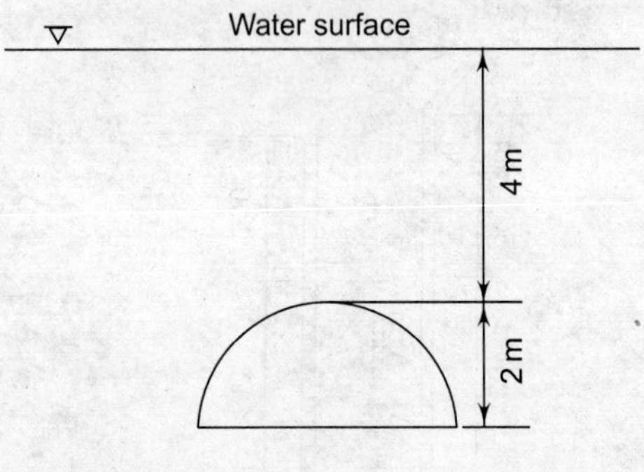


**Fig. P 1.10** Pressure on the lamina

11. A square plate of size $2a$ is placed vertically with its upper edge horizontal and at a depth $a$ below the liquid surface. A square hole of size $a$ is cut out of the plate as shown in Fig. P 1.11. Find the total pressure on one face of the plate and the depth of the centre of pressure.

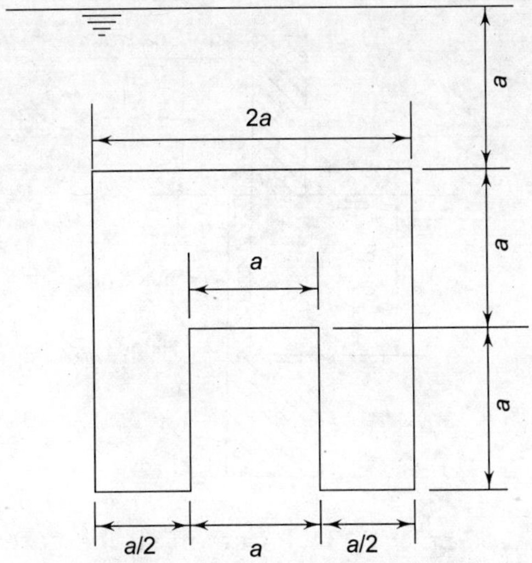


**Fig. P 1.11** Pressure on the lamina

12. A rectangular gate shown in Fig. P 1.12 should tip automatically when the water rises above the certain level. Determine that level in terms of $h$. [1.732 m]

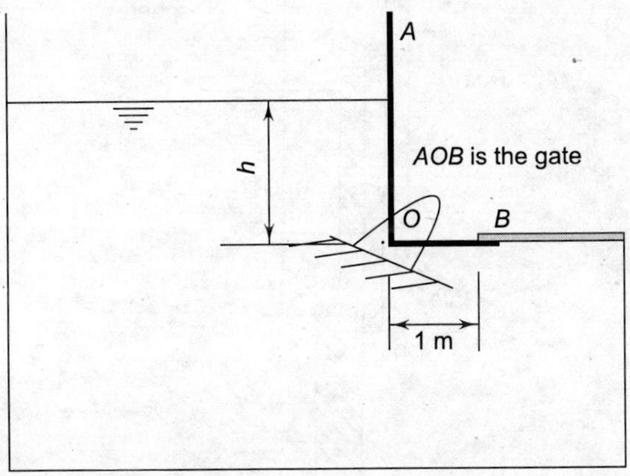


**Fig. P 1.12** Height of water for gate to trip

13. A sluice gate, shown in Fig. P 1.13, is 2 m wide and 1.2 m high. It is hinged at the bottom. On the upstream side there is seawater extending to a height of 1.6 m above the top of the gate, and on the downstream side there is fresh water up to the top of the gate. Find (i) the resultant pressure force acting on the gate, (ii) the position of the centre of pressure, and (c) the least force acting at the top of the gate, which will open the gate. Fresh water and sea water weigh 9.81 kN/m$^3$ and 10.05 kN/m$^3$, respectively.

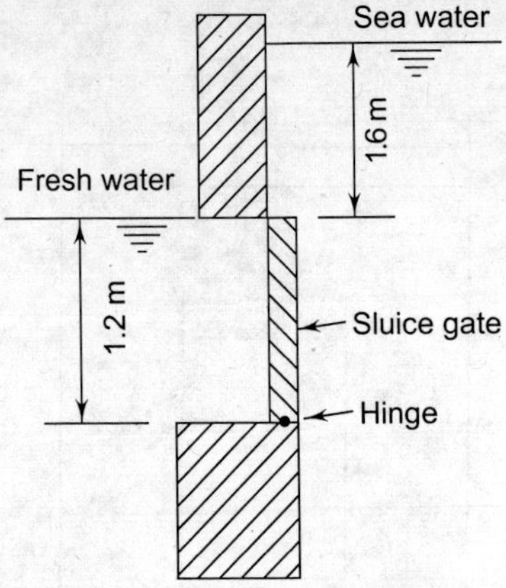


**Fig. P 1.13** Force on sluice gate

14. The gate $AB$ hinged at $B$, as shown in Fig. P 1.14, is 4.5 m wide and is connected by a rod and pulley to a sphere of concrete. Find the diameter of the sphere so that the gate remains just closed. Take specific gravity of the concrete = 2.4. [3.221 m]

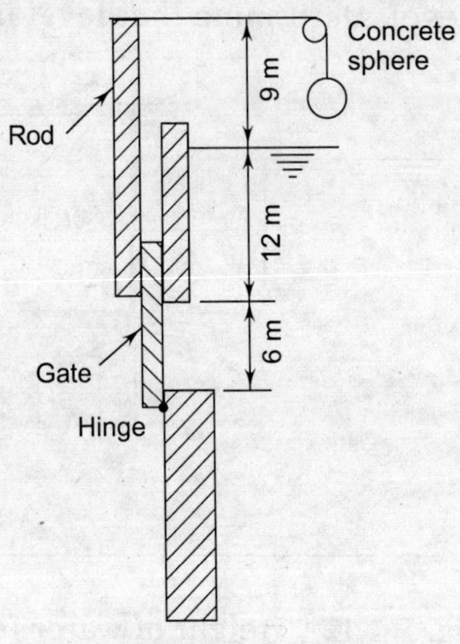


**Fig. P 1.14** Force on gate

15. A slab $AB$ 4 m × 2 m supports water as shown in Fig. P 1.15. The slab is hinged at the bottom $A$. The middle point $D$ of $AB$ is connected to a prop $DE$. If the slab is at 60° with the horizontal and the prop is at 45° with the slab, calculate the thrust in the prop. [64 kN]

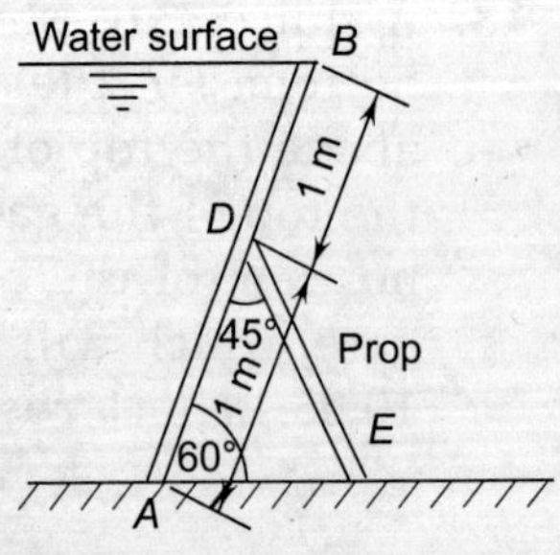


**Fig. P 1.15** Thrust on prop

16. A tank $ABCDE$ contains water up to a depth of 1 m and is 2 m wide. The curve $AB$ is defined by $z = x^2$ and curve $CD$ is a quadrant of a circle of radius

0.3 m. Calculate the magnitude of the resultant forces on surface *AB* and *CD*. Find also the angle that the resultant force makes with the horizontal. [*AB*: 16.29 kN, 52.96°; CD: 7 kN, 42°]

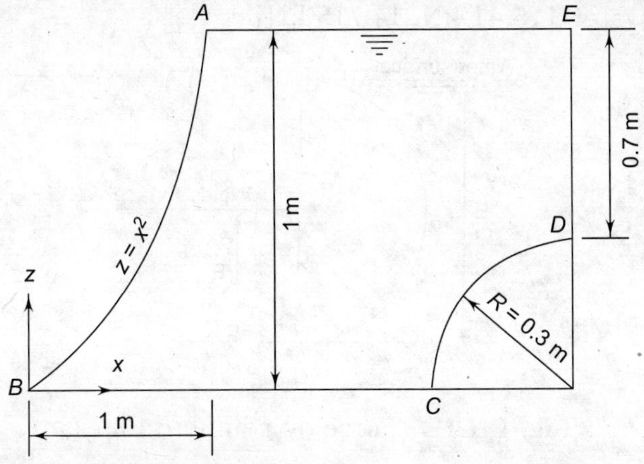


**Fig. P 1.16** Force on curved surfaces

17. Find the horizontal and vertical components of the force on the radial gate shown in Fig. P 1.17. What force *F* is required to open the gate? Ignore the weight of the gate. The gate is 2 m wide. [196. 2 kN, 218.6 kN, 0 kN]

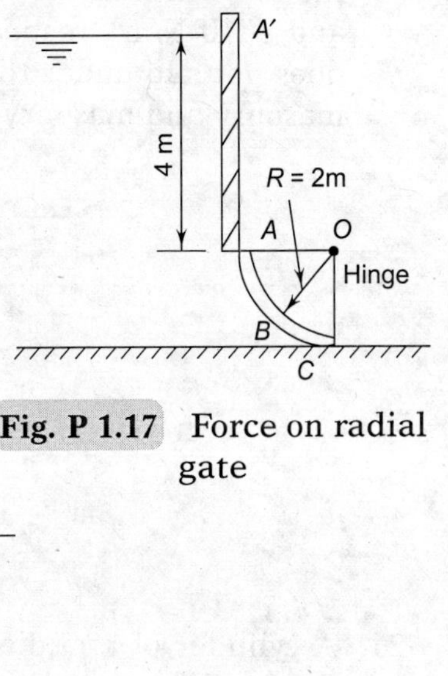


**Fig. P 1.17** Force on radial gate

18. Find the horizontal and the vertical components of the hydrostatic forces on the segmental gate shown in the Fig. P 1.18. The gate is 3m long. [42.376 kN, 12.1 kN]

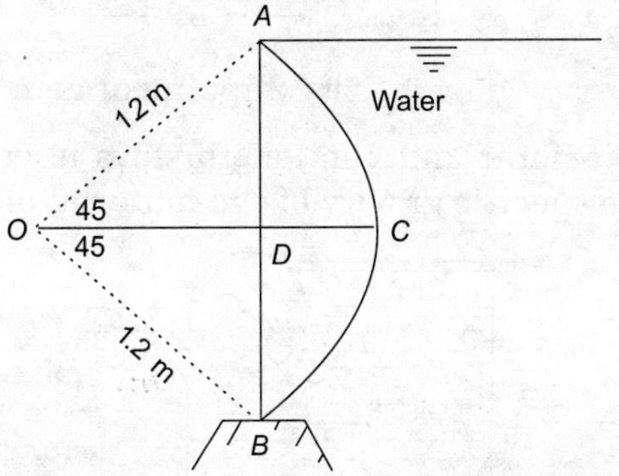


**Fig. P 1.18** Force on segmental gate

19. Figure P 1.19 shows a semi-circular gate *ABC* retaining water on its two sides. The face *ABC* is cylindrical while the face *AC* is a flat vertical surface. Calculate per metre run of the gate, and the vertical and horizontal components of the force exerted by the water. [15.41 kN, 14.715 kN]

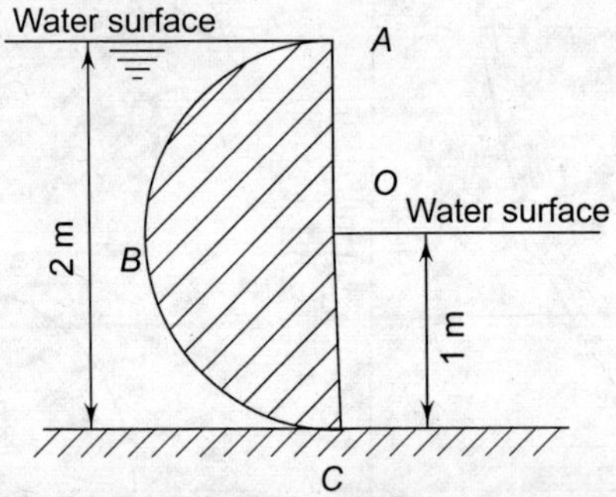


**Fig. P 1.19** Force on semi-circular gate

20. A dam section is 8 m high, the maximum depth of water impounded being 7.5 m, the top width of the section is 1 m, the weight of the masonry and water is 22000 and 9810 N/m$^3$, respectively. Find the minimum width required so that the dam does not slip under the hydrostatic force. Take the coefficient of friction between masonry and masonry as 0.5. The water face of the dam is vertical. [4.48 m]

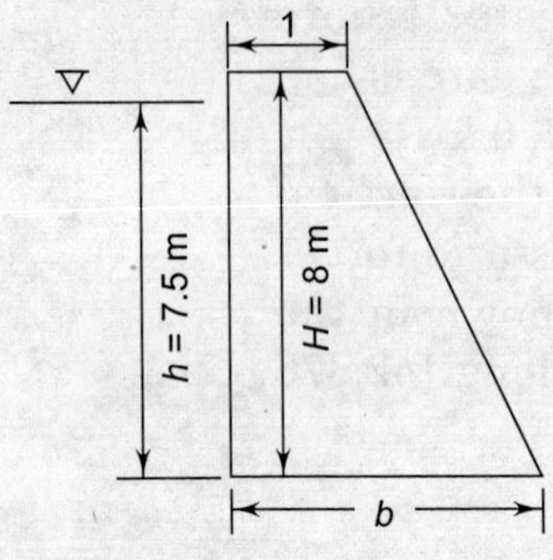


**Fig. P 1.20** Slipping of dam

21. A cylinder of 1 m diameter and 2 m length stays in equilibrium as shown in Fig. P 1.21. Calculate the specific gravity of the material of the cylinder. [797.3 kg/m$^3$]

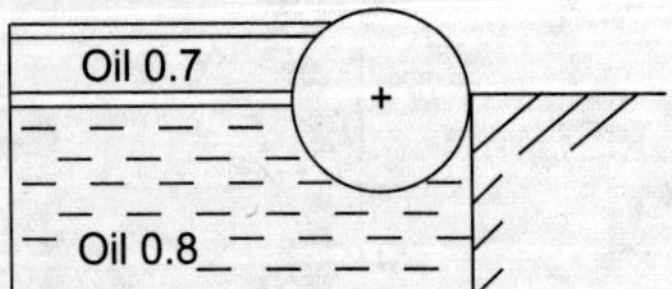


**Fig. P 1.21** Cylinder in equilibrium

22. A hollow cylinder open at both the ends has internal diameter of 30 cm, wall thickness of 15 cm and a length of 90 cm. If it weighs 625 N, find whether the cylinder would be stable while floating in water with its axis vertical. [Unstable].
23. A wooden log of 0.6 m diameter and 5 m length is floating in river water. Find the depth of the wooden log in the water when the specific gravity of the log is 0.7. [0.395 m.]
24. Find the density of a metallic body, which floats at the interface of mercury of specific gravity 13.6 and water such that 40% of its volume is submerged in mercury. [6040 kg/m$^3$.]
25. A block of wood of specific gravity 0.7 floats in water. Determine the metacentric height of the block if its size is 2 m $\times$ 1 m $\times$ 0.8 m. [0.0288 m.]
26. A body has the cylindrical upper portion of 3 m diameter and 1.8 m deep. The lower portion is a curved one, which displaces a volume of 0.5 m$^3$ of water. The centre of buoyancy of the curved portion is at a distance of 4.95 m below the top of the cylinder. The centre of gravity of the whole body is 1.2 m below the top of the cylinder. The total displacement of the water is 3.9 tonnes. Determine the metacentric height of the body. [0.596 m].
27. A rectangular pontoon 10.0 m long, 7 m broad and 2.5 m deep weighs 686.7 kN. It carries on its upper deck an empty boiler of 5 m diameter weighing 588.6 kN. The centre of gravity of the boiler and the pontoon are at their respective centres along a vertical line. Find the metacentric height. Weight density of seawater is 10.104 kN/m$^3$. [0.12 m].
28. A solid cone floats in water with its apex downwards. Determine the least apex angle of cone for stable equilibrium. The specific gravity of the material of the cone is given 0.8. [31°.]
29. A rectangular tank is moving horizontally in the direction of its length with a constant acceleration of 0.4 m/s$^2$. The length, width and depth of the tank are 6 m, 2.5 m, and 2 m respectively. If the depth of water in the tank is 1 m and tank is open at the top then calculate: (i) the angle of the water surface to the horizontal, (ii) the maximum and minimum pressure intensity at the bottom, and (iii) the total force due to water acting on each end of the tank. [13° 44.6', 17008.5 N/m$^2$, 2611.4 N/m$^2$, 868.95 N, 36861.8 N]
30. The rectangular of the above problem contains water to a depth of 1.5 m. Find the horizontal acceleration which may be imparted to the tank in the direction of its length so that (i) the spilling of water from the tank is just on the verge of taking place, (ii) the front bottom corner of the tank is just exposed, and (iii) the bottom of the tank is exposed up to its mid-point. [1.636 m/s$^2$, 49050 N, 12262.5 N, 3.27 m/s$^2$, 49050 N]
31. A rectangular tank of length 6 m, width 2.5 m, and height 2 m is completely filled with water when at rest. The tank is open at the top and is subjected to a horizontal constant linear acceleration of 2.4 m/s$^2$ in the direction of its length. Find the volume of the water spilled from the tank. [11.007 m$^3$.]
32. A ship displaces 49250 kN of sea water of density 1025 kg/m$^3$. The second moment of inertia of the water line section about axis through the centre is 12000 m4 and the centre of buoyancy is 2 m below the centre of gravity. If the radius of gyration is 3.7 m, calculate the period of oscillation. [11.08 s]

CHAPTER

2

# Mathematical Modelling of Flow

## OUTLINE

- Introduction
- Formulation of the Problem
- Kinematics
- Conservation Laws
- Nature of the Equations and Required Boundary Conditions

In this chapter we shall discuss mathematical modelling of the flow of fluids inside a duct or over the surface using three basic laws of nature, namely the conservation of mass, momentum and energy. To make the understanding of the analysis easy we shall also study various ways a flow can be specified and the different approaches that are in use for the mathematical description of the flow. In the process we find that the equations so developed are non-linear partial differential equations of order 2 and hence require boundary conditions to find the unique solution that depend on the nature of these equations. We introduce different types of boundary problems that occur in physical flows. If the fluid is compressible then these equations are coupled too and are to be solved taken together. In this case we shall also need the equation of state to give as many equations as the number of unknowns are. If by chance, the temperature range is wide for the entire flow region then the effect of the temperature on the properties of the fluid such as thermal conductivity and viscosity is also to be taken care of. For this we need the constitutive relations for these properties expressing these as function of temperature. Often such relations are non-linear and even empirical relations, which may be different in different range of temperature. The flow may be unsteady too but the unsteady flow is not studied in this book and hence it is not discussed.

## 2.1 Introduction

We considered a few specific cases of motion of fluid in the last chapter. In general, a flow can take place in any direction with varying velocity, pressure, temperature, density and acceleration. This requires an in-depth study of the motion to develop as many equations as the number of unknowns are. Velocity is a vector quantity and has three components. Temperature and density are scalars whereas pressure is a pseudo-vector[1] thus totalling to three unknowns. Once the velocity field is known then it is an easy job to determine acceleration field and it is no more unknown. So the number of unknowns in the flow problems is 3 + 3 = 6 and we need six equations to get the unique solution. We get these equations from the laws of nature that give five equations because mass and energy are scalar while momentum is a vector quantity. The sixth equation is obtained from the behaviour of the fluid. For gases this equation is the state equation whereas for liquids it is $\rho = constant$.

We study in mechanics that force is related to the rate of change of momentum, and therefore the shear and normal stresses, discussed in Chapter 1, will have to be considered. This means that the momentum equation will contain terms having dynamic or kinematic viscosity. Similarly, in equation of state gas constant appears. We learn in thermodynamics that these constants of the fluid are functions of temperature and therefore their values would change from place to place as temperature changes in the flow field though this variation may be too small. However, for accuracy, these variations need to be taken into account and we need relations expressing these properties as functions of temperature. Such relations are called *constitutive relations*.

Lastly, it is known to us that when a hot surface is placed in air then the air rises and the velocity field acquired by the air determines the amount of heat transferred from the hot surface; the more the temperature difference between the surface and the surrounding air, the more is the air velocity and the heat transfer rate. Thus, there is both energy and momentum transfer and the equations will have terms representing such transfers. This couples the momentum and energy equations, which means that temperature field should be known to determine velocity field and vice versa.

To summarise we have at least six simultaneous equations (neglecting constitutive relations), which are coupled and involve partial derivatives of second order (we shall see this later). These equations are non-linear and therefore the solution of all the equations had posed a great challenge to mathematicians. Procedure for digital solutions has also been persuaded and we have finite difference, finite element and finite volume methods, covered under the regime of computational fluid dynamics. In all these solutions the grid formed is too close and the number of nodes in a simple problem may run into tens of thousands. Therefore, even a supercomputer takes a lot of computation time. We shall restrict ourselves in this book to the development of equations and their analytical solutions for simple problems only.

---

[1] Pseudo-vector is a variable, which can have one of the two directions as compared to a vector that can have any direction. For example, surface area is represented by a normal drawn normal to the surface; it is taken positive when drawn outwards and negative in the opposite direction. Thus, there are only two directions for it. The same is true for pressure. Therefore, it is considered as a scalar for all practical purposes.

### 2.1.1 Coordinate System

For mathematical formulation, we first need to define a reference from which measurements are to be made for the independent variables that include space and time coordinates. As far as time coordinate is concerned, the flows are assumed steady so that time is not involved in the equation. Even if the flow is not steady, we attempt to make it steady by shifting the coordinate system suitably. For example, imagine you are standing on a river bank and a boat is approaching you. As the boat comes closer, you notice the flow pattern starts changing and after it has gone away downstream the flow pattern resumes the original one. That is, the flow is unsteady for some period. However, if you are sitting in the boat then for you the flow pattern will remain same all the time. Thus, if we fix the reference to the boat, the flow can be turned to steady flow. You may think of many other such examples.

Regarding space coordinates, both orthogonal and non-orthogonal systems are in use; the former is often used. An orthogonal system is the one in which all the coordinates are mutually perpendicular to each other. A few commonly used orthogonal systems in three-dimensional space are $(x, y, z)$, $(r, \theta, z)$, $(r, \theta, \phi)$, etc; these are commonly known as Cartesian, cylindrical polar and spherical coordinates. The Cartesian coordinates have linear axes whereas the other two have the curvilinear axes for $\theta$ or $\theta$ and $\phi$. These systems are shown in Fig. 2.1.1; $O$ is the origin of the coordinates. Distances are taken positive when measured in the direction of the axes. For measurement of angles, a reference line is needed and the angle, by convention, are measured positive in the anticlockwise direction from the reference axis. We note that because of the angular coordinates, the coordinate has become curvilinear. Further, the circumference in cylindrical coordinates subtending angle $d\theta$ at the centre is $r\,d\theta$. For a spherical coordinate system, the area $dA = r^2 \sin\theta\, d\theta\, d\phi$ (see Fig. 2.1.1).

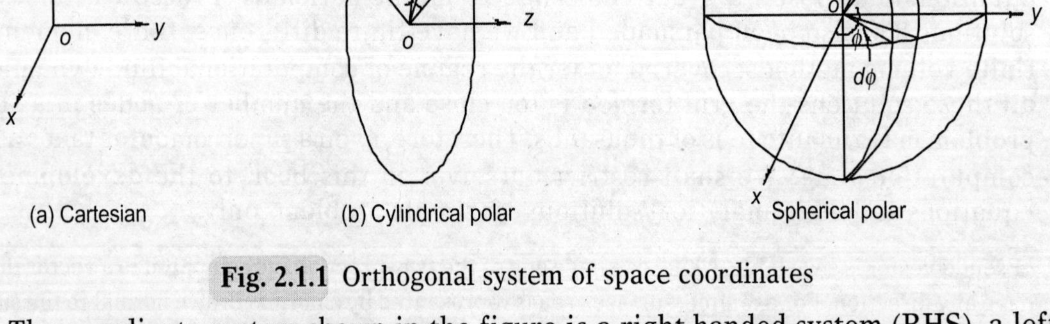


**Fig. 2.1.1** Orthogonal system of space coordinates

The coordinate system shown in the figure is a right-handed system (RHS), a left-handed system can also be used. Moreover, any of the axes can be taken as $x$-axis and once this is fixed the other axis is named depending on the type of system used. For a

RHS, if we call $y$-axis of Fig. 2.1.1 (a) as $x$-axis then $z$-axis will become $y$ and $x$ will become $z$-axis, respectively. Normally we take $z$-axis as vertical axis as it corresponds to height or depth of fluid. While dealing with two-dimensional flow, we take $x$-$y$ plane with $y$-axis as vertical axis.

The origin of the reference system can be fixed anywhere. That is, it can be in the flow region or outside it. It may or may not be fixed within the boundaries of the flow. The origin may be fixed or moving with the flow. Here an inertial frame is of importance and is defined as the one which remains stationary. The frame fixed with the ground is normally taken as inertial frame though we know that this would also move with earth as the latter is not stationary and moves around the Sun.

We have talked of three coordinate systems. In fact, there are many more orthogonal systems having even $n$-dimensions but we shall restrict ourselves to a three-dimensional space only. But why do we need different kind of coordinate systems? The reason is that the mathematical formulation should be simple. To understand, take a circle with its centre as origin of coordinate system. Let a point on this circle has the coordinates $(x, y)$ in Cartesian and $(r_1, \theta)$ in the cylindrical coordinate systems. Then we have learnt in coordinate geometry that $r_1 = \sqrt{x^2 + y^2}$ . Let us specify temperature on its periphery. If $T_1$ is the temperature, then we may write at $r = \sqrt{x^2 + y^2}$, $T = T_1$ or at $r = r_1$, $T = T_1$. Naturally, the latter is much simplified. For this reason, we prefer cylindrical polar co-ordinate in this case.

## 2.2 Formulation of the Problem

We are given a flow for which the velocity field, pressure field, and temperature field distribution is to be determined at all the points lying in the flow region for the specified boundary and initial conditions. We need to determine the velocity field as this determines how much power would be required to maintain the flow or temperature field to know the region where the temperature is maximum so that it can be taken care of while deciding the material of the container through which or over which the flow is taking place. Moreover, we find that fluid enters at some sections and leaves from other sections of the flow region. In fact, the flow may enter at or leave from more than one section. Thus, the flow region can be shown on paper by a control surface surrounding the flow region, which is being crossed at different points by the flowing mass; at some sections the fluid enters and at the other sections of the surface the fluid leaves the flow region. It is a usual convention that the surface is shown by an outward drawn normal to the surface. If $U$ is the velocity of the fluid passing through surface $dS$ then mass flow rate is = $\rho(U.dS)$, the dot product of velocity and surface area. The integral over the surface gives the total mass flow rate to or from the flow region. For steady flow this will be zero, since mass of the region does not change with time.

### 2.2.1 Types of Flow

From a close look at the flow pattern, we can group these in different categories and even show the flow in the flow region with the help of curves. Therefore, we devote some time in the following sections on these concepts as these are very helpful in the development of the subject.

#### 2.2.1.1 Path, Stream and Streak Lines

***Path line.*** Fluid is made of particles that move from entrance to the exit of the flow region. Each particle has its own journey path. If we locate position of the (selected) particle at different times and join such points then a curve is formed. This curve is called *path line.* A path line can be easily obtained in laboratory by choosing some reflective material having density as that of the fluid and inserting a pellet (small sized leaf) made from it at the entrance of the flow to represent the fluid particle. At successive intervals, the flow is photographed which gives the position of the pellet. These points, once obtained, are joined by a curve to give the path line. It is noted that the fluid particles that pass through the same point in the flow region at different times form a loop in the path line. Moreover, the path lines for different particles may pass through the same point in the flow region and hence the different path lines may intersect at a point. Figure 2.2.1 (a) shows different path lines.

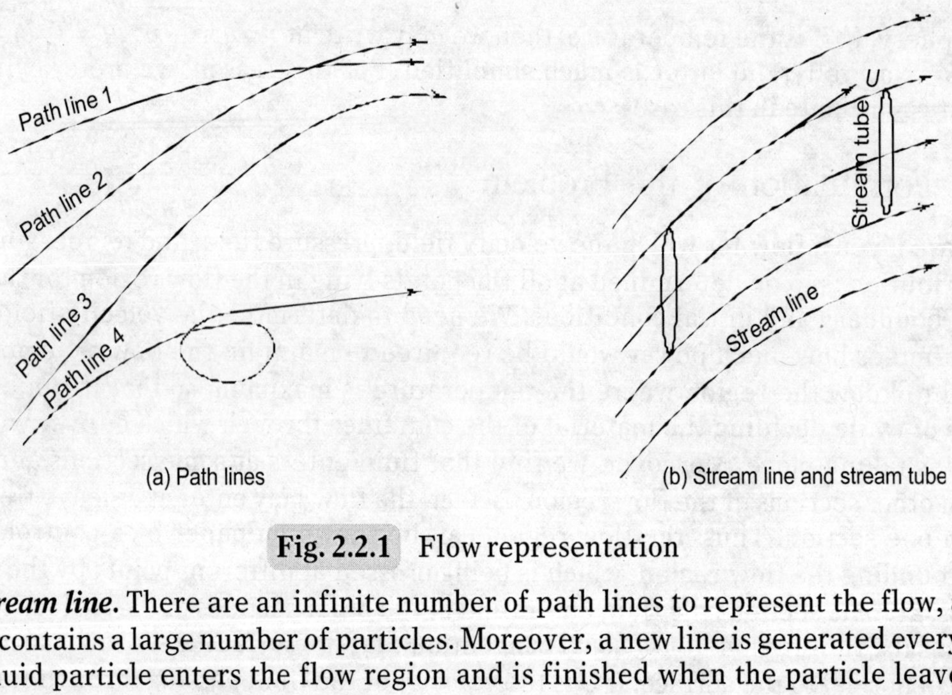


(a) Path lines (b) Stream line and stream tube

**Fig. 2.2.1** Flow representation

***Stream line.*** There are an infinite number of path lines to represent the flow, as the fluid contains a large number of particles. Moreover, a new line is generated every time the fluid particle enters the flow region and is finished when the particle leaves the region. Therefore, we think of an alternate way to represent the flow as in the case of magnetic field; a tangent drawn at any point on the magnetic lines of force gives the direction of force at that point. Similarly, velocity (both magnitude and direction) at any point in the flow of a fluid is measured with an instrument called yaw meter and is represented by a vector drawn at that point. Now we draw a line such that the velocity vector, whose origin lies on it, is tangential to it. Like this a number of such lines can be drawn in the flow region. These lines are called *stream lines.* Hence, a stream line is an imaginary line drawn in the flow region such that a tangent drawn to it at any of its point gives the direction of the flow at that point. A few stream lines are shown in Fig. 2.2.1 (b) in a flow region. Also shown is a stream tube, which is a collection or bundle of all the stream lines passing through the closed curve drawn in the flow region. Stream filament is also used as alternate name for stream tube when variation of the velocity across its cross section is negligible.

Since velocity vector is always tangential to a stream line, there will be no flow across the stream tube and the flow will take place only along it. This helps in determining the equation of the stream line. At any point on the stream line, the velocity vector $\vec{U}$ and the length vector $\vec{dl}$ are parallel and therefore their cross product is zero. As a result, the equation of stream line is

$$\vec{U} \times \vec{dl} = 0$$

or, $i(v\,dz - w\,dy) + j(w\,dx - u\,dz) + k(u\,dy - v\,dx) = 0$ (2.2.1)

♦ $dx/u = dy/v = dz/w$ (in Cartesian coordinates)

where $u$, $v$, and $w$ are the three components of velocity vector $U$ along $x$, $y$ and $z$ axes respectively.[2] Note that the equation in vector form is true for any coordinate system – orthogonal or non-orthogonal system. However, the second equation is true only for the Cartesian system of coordinates. This shows the power of the vector equation.

**Example 2.2.1** The velocity distribution in a three-dimensional flow is given by: $u = -x$, $v = 2y$ and $w = (3 - z)$. Find the equation of the streamline that passes through the point (1, 1, 1).

Streamline is defined by Eq. 2.2.1. Substituting the values of velocity components, we get

$$-\frac{dx}{x} = \frac{dy}{2y} = \frac{dz}{3 - z}$$

Integration of these equations give $-\ln x = 1/2 \ln y + A$, where constant of integration $A$ is obtained by substituting $x = 1$ and $y = 1$ so that the streamline passes through the point (1, 1, 1). This gives $A = 0$ and $x = 1/\sqrt{y}$. Similarly, we may take the other two equations in $dx$ and $dz$ and integrate to give $-\ln x = -\ln (3 - z) + B$ and $B = \ln 2$ for the streamline through the given point. Or, $x = (3 - z)/2$.

Combining these two equations, the equation of the streamline passing through the point (1, 1, 1) is

$$x = 1/\sqrt{y} = (3 - z)/2.$$

***Streak line.*** While visualising the flow pattern in laboratory, the common practice is to inject a dye at some point and then observe in what way the colour pattern appears in the flow. This technique is shown in Fig. 2.2.2 (a), which was employed by Reynolds to find difference in laminar and turbulent flow patterns. Dye is injected at different

[2] A vector is zero only if its components are zero. The components in expanded form in Cartesian coordinates are given. Equating these to zero gives the equation of the stream line.

points in the pipe cross section and movement of the dye is then observed as the particles passing through these points move in the flow. The different-coloured line in the flow then is the locus of all the dye particles emanating from the same point in the flow region. Thus, the common characteristic in this type of curve or line is the point in the flow region at which the dye particle is added. Such line is called *streak line.* Thus a streak line is locus of all the fluid particles passing through a point in the flow region. Naturally, there will be a streak line for each point in the flow region.

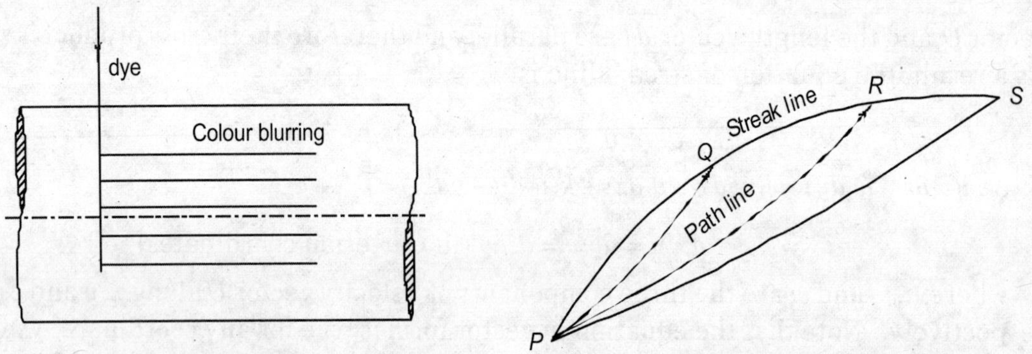


(a) Reynolds experiment (b) Path and streak lines

**Fig. 2.2.2** Streak lines

Now consider one such point $P$ in the flow region of an unsteady flow, as shown in Fig. 2.2.2(b). The particle at some time $t = 0$ passing through this point will have its path line say $PQ$. The next particle will have different velocity due to the unsteady nature of the flow. This means it moves along some other path, say $PR$, and so on. Thus we locate the points such as $Q, R, S$, etc. at any time, which represents the position of all the particles that are coming from the point $P$. By definition, the locus of these points is the streak line at that point of time.

The three lines defined just above are representing different conditions and need not be same. However, for steady flow the three lines merge into one line that is the streak line as well as the stream line. Thus for steady flow the streamline, streak line and path line coincide.

### 2.2.1.2 Steady and Non-steady Flow

When the flow is such that any of the fields like velocity, pressure or temperature do not change with time at a given point, anywhere in the flow region, then the flow is called steady flow otherwise it is an unsteady flow. *Here the key point to be noted is the point.* That means if we go from one point to another then the variable may change but it would not change with time at the same point. In other words, if we add pellet to the flow and take its photograph at different times then the flow pattern would be exactly identical in steady flow. This concept has mathematical importance because then we may write $\partial/\partial t \ldots 0$, where $t$ is the time coordinate used in mathematical formulation of the flow.

### 2.2.1.3 Uniform and Non-uniform Flow

A flow is uniform if the pressure, velocity or temperature fields are not a function of any of the space coordinates. That is, $\partial /\partial x = f/fy = f/fz = 0$ simultaneously. Note that $\partial /\partial x, \partial /\partial y,$ *and* $\partial /\partial z$ are $x, y$ and $z$ components of  (pronounced as del), that is,  $= \partial / \partial \vec{r} = i\partial/\partial x + j\partial/\partial y + \partial/\partial z$, where $\vec{r}$ is the position vector having components $(x, y, z)$ in Cartesian coordinates. For a non-uniform flow, at least one of these variables, if not all of them, will vary with space coordinate.

### 2.2.1.4 Laminar and Turbulent Flow

We mentioned Reynolds' experiment in Section 2.2.1.1. This is shown in Fig. 2.2.2 (a) where the flow is taking place through a transparent pipe in which dye is injected at a certain cross section at different points. He started the experiment with a very low flow speed, increased it slowly and slowly and observed the changes in the streak line pattern. With increasing speed, he observed that (i) initially the flow is along straight line, which transforms later to a wavy form at some distance, (ii) the length of straight line decreases with rising speed and (iii) at certain flow speed, the colour of the dye is mixed with the fluid and the streak lines are not visible. He concluded that at low speed the flow is along the layers or lamina and these layers do not mix with each other. Such a flow has been named as *laminar flow* in which there is transportation of molecules only across the layers. However, the whole liquid becomes coloured, he argued, because the chunk of mass also moves in addition to motion of molecules causing mixing of the fluid. Motion of the mass chunk is irregular and he described such flow as *turbulent flow*. We shall see in the chapter on turbulent flow that analytical solution is possible for the turbulent flow only after using some experimental results because the number of unknowns is always one more than the number of equations formed for analysis. On the contrary, there is no such problem for laminar flows. In mathematics, such problems are described as open- and close-ended problems.

### 2.2.1.5 Dimensionality of the Flow

Dimensionality of the flow plays a significant role in analytical solutions. Let us consider a flow of incompressible fluid through a circular pipe. Figure 2.2.3 shows development of the flow regime in the pipe. Because of the viscosity of the fluid, the layer in contact with the pipe surface takes its speed (zero in this case). As we move towards the centre of the pipe, velocity increases. Velocity profile is shown at three sections *AA*, *BB* and *CC* in the figure. We observe that velocity in the centre goes on increasing because of formation of boundary layer, a layer in which velocity increases from the pipe surface in the normal direction till it is equal to that in the main flow. Boundary layer develops all around the pipe surface and continues till the entire boundary layer meets at the centre of the pipe, which is shown at section *CC*.

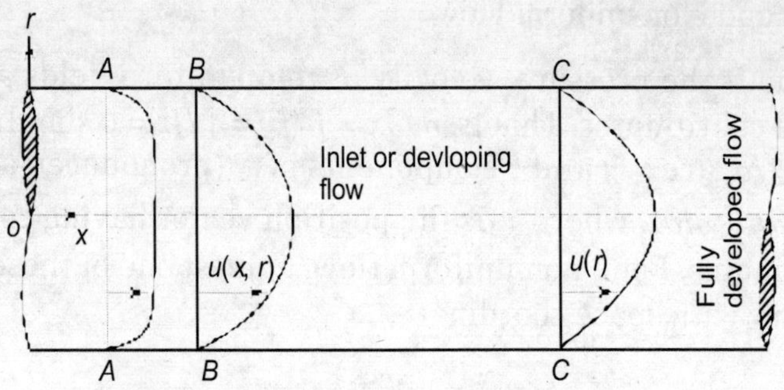


**Fig. 2.2.3** Development of flow in pipe

The figure also shows the coordinate system with *o* as origin and *ox* and or as the axes along and perpendicular to the pipe axis, respectively; the flow is along the positive direction of *x*-axis. We note that in the region up to the section *CC*, velocity component *u* is a function of both the *x* and *r* coordinates and is shown as $u\,(x, r)$ in the figure. This region is called the inlet or *developing flow* region. Moreover, in this region the velocity profile also changes along *x*-direction, as can be seen in the figure. However, from section *CC* onwards there is no change in the velocity profile and the flow is called *fully developed* flow. In this region, *u* does not change with *x* while it does change with *r* with maximum value at the centre and zero at the pipe surface.

Let us consider another variable pressure. Experimental measurements show that variation of pressure on a cross section is negligible and we may assume pressure at any cross section as not to be a function of *r*. Nevertheless, the flow is possible only if the pressure at the inlet is more than that at the outlet of the pipe. This means that pressure changes along *x*-direction and may be expressed as $p(x)$.

At the moment consider the developing flow region. In this region, *u* depends on *x* and *r* while *p* depends on *x* alone. Thus the independent variables *x* and *r* are involved, numbering two. On the other hand, in the fully developed flow region, *u* depends on *r* while *p* depends on *x* alone but in the totality, the two independent variables *x* and *r* are involved. Thus, in both the cases the flow field depends on *x* and *r*. Hence the flow is two-dimensional. Now we have a clear perception regarding how to determine the dimensionality of the flow and what does it mean. We may express the dimensionality of the flow by determining the number of independent variables needed to determine the entire flow field. A flow in pipe is called *symmetric flow* when the flow variables do not change with the circumferential coordinate $\theta$.

### 2.2.1.6 Internal and External Flows

The fluid may flow inside the boundaries of a conduit or over the surface and accordingly are classified as internal and external flows. Pressure is required to maintain the internal flows, whereas gravity is mainly responsible for external flows. The flow through a pipe in the water distribution system in a city is the internal flow and in rivers or man-made canals is external flow.

#### 2.2.1.7 Compressible and Incompressible Flows

This classification refers to the nature of the fluid used. The liquids are often considered as incompressible fluids since the density does not change significantly whereas gases are called compressible fluids as density changes with pressure and temperature. Pressure and density changes in fluids are related by the *bulk compressibility modulus*, defined as $dp/(d\rho/\rho)$. Thus, liquids have infinite bulk compressibility modulus. If the bulk compressibility modulus is independent of temperature, then the liquid is called *barotropic*. Though the liquid is considered as incompressible yet a little compressibility of the liquid is responsible for the water hammer observed in pipes when the tap or the valve is closed suddenly. The sound is as if someone is hammering the pipe. This is caused because of propagation and reflection of the acoustic waves in the confined region. To understand this, consider a very long pipe carrying tonnes of water with some velocity and imagine the valve is suddenly closed (that is, some disturbance or an acoustic wave has been created). Water has enormous momentum and is brought to rest suddenly. Thus, pressure increases. Once the fluid comes to rest, water will start flowing back (upstream) because of the higher pressure in the valve. This continues and because of inertia the pressure now increases upstream till the whole of water comes to rest due to the building up of pressure. Now the whole process is again reversed setting oscillations shifting a large bulk of water to and fro a number of times till these are dampened by internal losses. This phenomenon is called *water hammer*.

The other phenomenon in incompressible fluids is cavitation. When the pressure of liquid decreases to less than the saturation pressure at a temperature then its vapours are formed which are trapped in the liquid film and can be seen as bubbles. These bubbles collapse when they travel to a region of high pressure forming a local cavity to which the fluid rush in from the surroundings to fill the space so created. This results in enormous rise in pressure locally to damage the surface in contact. This process is called *cavitation*. Thus, we notice that incompressibility of fluid while is a bonus to make analytical solution relatively comfortable warns us to take care for its detrimental effects too.

As mentioned above, the gases are normally considered as compressible. But it can also be treated as incompressible fluid if the Mach number is less than 0.3 and there is no heat transfer to the flow region (as then the maximum density variation may be around 5%).

### 2.2.2 Formulation Approach

There are a number of ways in which a problem can be modelled in mathematical form. For example, we may be interested in the distribution of the various fields in the flow region or only in the overall results applicable to the region. We may recognise the fluid particles and follow them for the time they move into the region and leave it, or fix a control volume in the flow region and do not bother for the identity of the particles entering in or leaving the control volume. Before modelling the flow problem, let us explore various approaches that are available with us on need basis.

### 2.2.2.1 Lagrangian and Eulerian Approach

Lagrangian approach is based on following the particle as it moves in space. A point is specified (with reference to the reference frame, normally assumed to be fixed with the ground) at time $t = 0$, from which the position of the particle is expressed as function of time. The first and second differentiation of displacement gives velocity and acceleration of the particle respectively at any position. Thus, all the variables are function of the fixed reference $\vec{r}_0$ and $t$, where $\vec{r}_0$ is the position vector of the fixed point with reference to the reference frame and $t$ is time measured from the instant the particle is at $\vec{r}_0$. Thus, the position of the particle $\vec{r}$ at any time is expressed in Lagrangian approach as, $\vec{r} = f(\vec{r}_0, t)$. This gives

$$\vec{U} = d\vec{r}/dt$$

and $$\vec{\alpha} = \text{acceleration} = d^2\vec{r}/dt^2.$$

Lagrangian approach has been used successfully in solids, since the solid can be idealised as a single particle and then we have developed the methods to find the effect of distribution of the particles within the solid. This model works very well without much complexity. However, in case of the fluids, each particle has its own identity and does not behave like other fluid particles. Therefore, each particle has to be considered separately and then the statistical method is used to find out the average values as is done in kinetic theory of gases. Nevertheless, there is an alternate much simpler solution to this kind of situation, since the behaviour of each particle is of no significance.

Euler suggested that a control volume be considered in the flow region across whose surface fluid particles enter or leave. Once the reference frame and the control volume are decided then each point in the control volume has unique position vector $\vec{r}$ and as the fluid moves, it passes through these different points. This means that in this approach[3]

$$\vec{U} = f(\vec{r}, t)$$

$$\vec{\alpha} = d\vec{U}/dt$$

$$= f\vec{U}/ft + f\vec{U}/f\vec{r}.f\vec{r}/ft$$

$$= f\vec{U}/ft + (\vec{U}.\ \ )\vec{U} \qquad (2.2.2)$$

because $\partial/\partial r$ is operator and position of vectors in a dot product is interchangeable. In Eq. 2.2.2, $d/dt$ is an operator, operating on $\vec{U}$ called *spatial, substantive, material, or total derivative* and is usually represented as $D/Dt$ to avoid confusion with Lagrangian operator $d/dt$; $\partial/\partial t$ is the *local derivative* whereas $\vec{U}.$ is the *convective derivative*, since it takes into account convection of the fluid particles in the control volume. Recall from vectors, for Cartesian coordinates

$$\vec{U} = i\,u + j\,v + k\,w$$

and $$= i\partial/\partial x + j\partial/\partial y + k\partial/\partial z.$$

[3] In vector algebra, we read $\vec{A}\,.\,\vec{B} = \vec{B}\,.\,\vec{A}$ and $\vec{A} \leftrightarrow \vec{B}$ $\quad \vec{B} \leftrightarrow \vec{A}$ that is the vectors can interchange their position in dot product (scalar product) and not in cross product (vector product).

$$\therefore \qquad \vec{U}. \;= u\partial/\partial x + v\partial/\partial y + w\partial/\partial z. \tag{2.2.3}$$

It is seen that the convective derivative is non-linear, since it is a product of the function with its derivative and involves partial derivative of first order. Equation 2.2.2 takes the following forms in the Cartesian and cylindrical coordinate systems.

Certesian Coordinates

$$\alpha_x = \frac{Du}{Dt} = \frac{fu}{ft} + u\frac{fu}{fx} + v\frac{fu}{fy} + w\frac{fu}{fz}$$

$$\alpha_y = \frac{Dv}{Dt} = \frac{fv}{ft} + u\frac{fv}{fx} + v\frac{fv}{fy} + w\frac{fv}{fz}$$

$$\alpha_z = \frac{Dw}{Dt} = \frac{fw}{ft} + u\frac{fw}{fx} + v\frac{fw}{fy} + w\frac{fw}{fz} \tag{2.2.4}$$

Cylinderical Polar Coordinates:

$$\alpha_r = \frac{Du_r}{Dt} = \frac{fu_r}{ft} + u_r\frac{fu_r}{fr} + \frac{u_\theta}{r}\frac{fu_r}{f\theta} + u_z\frac{fu_r}{fz} - \frac{u_\theta^2}{r},$$

$$\alpha_\theta = \frac{Du_\theta}{Dt} = \frac{fu_\theta}{ft} + u_r\frac{fu_\theta}{fr} + \frac{u_\theta}{r}\frac{fu_\theta}{f\theta} + u_z\frac{fu_\theta}{fz} + \frac{u_r\,u_\theta}{r}$$

$$\alpha_z = \frac{Du_z}{Dt} = \frac{fu_z}{ft} + u_r\frac{fu_z}{fr} + \frac{u_\theta}{r}\frac{fu_z}{f\theta} + u_z\frac{fu_z}{fz}. \tag{2.2.5}$$

In the above equation, $(u, v, w)$ are the velocity components along $(x, y, z)$ in Cartesian coordinates whereas $(u_r, u_\theta, u_z)$ are the velocity components along $r, \theta, z$ directions in cylindrical polar coordinate systems; $(u_r, u_\theta, u_z)$ are used as velocity components because the equation gives the idea of the direction in which the velocity is considered. However, $u, v$ and $w$ can also be used to represent three velocity components.

In polar coordinates, terms $u_\theta^2/r$ and $u_\theta u_r/r$ appear. The first term is the centripetal acceleration that acts along $r$-direction and the second one is the Coriolis acceleration that acts along the tangential direction. These terms are introduced because of the curvilinear nature of the $\theta$ coordinate that makes the unit vectors along $r$ and $\theta$ directions to change, as mentioned in Appendix *A*.

### 2.2.2.2 Differential and Integral Approach

We may develop mathematical model considering an infinitesimal control sub-volume within the control volume and shrink it to a point by taking in limit its linear dimension to zero to make the results applicable at the point. For example, to find out the mass flow rate through a point, we may consider a rectangular surface (at the point) of dimensions say $\delta x$ and $\delta y$, normal to the flow direction, and determine the mass flow rate across this surface. Now we determine the limit of this mass flow rate as $\delta x$♦ $0$ and $\delta y$♦ $0$. Such formulation results in differential equation and hence known as differential approach. Such an approach is followed when a designer is interested to know the velocity, pressure, and temperature fields throughout the flow region or we want to have totality of the picture of the flow in the region. However, for engineers, the overall effect matters more than the total solution of the flow and in that case we may consider the mass, momentum and energy efflux from the surface of the control volume and relate it to the time rate of change of these quantities for the control volume. Such an approach leads to integral equations whence the name.

## 2.3 Kinematics

Kinematics is the study of the fluid flow, which deals with fluid motion in terms of displacement, velocity and acceleration without any consideration of the forces that cause the motion. To broaden up the scope of mathematical analysis, concepts like line integral and complex numbers are also taken into consideration along with the Eulerian approach.

### 2.3.1 Stream Function

Mathematicians have mainly developed the subject of Fluid Mechanics. Mass conservation leads to continuity equation (derived later in this chapter). Similarly other conservation principles for energy and momentum give other set of Eq., as mentioned before. One of the methods of solving these equations, the mathematicians thought of, was to reduce the number of equations by introducing a function of space coordinates that satisfy the continuity equation. This somehow could only be done for a two-dimensional flow. The function $\psi(x, y)$ in $x - y$ plane is related to the velocity components for the incompressible fluid as follows:

$$u = \partial\psi/\partial y \text{ and } v = -\,\partial\psi/\partial x \qquad (2.3.1)$$

Velocity field given by Eq. 2.3.1 satisfies continuity Eq. 2.4.3 (discussed later) because the function $\psi$ is continuous function and therefore $f^2/_{fx\,fy} = f^2/_{fy\,fx}$.[4] This function is called *stream function*. We shall use this definition because positive $u$ implies positive $\partial\psi/\partial y$ that means $\psi$ increases along positive $y$. Thus when we plot for $\psi = C$ lines then the value of $\psi$ will increase along positive $y$ direction. The name stream function was given because along the line of constant stream function, substituting for the partial derivatives of $\psi$ from Eq. 2.3.1, we get

$$d\psi = (\partial\psi/\partial x)\,dx + (\partial\psi/\partial y)\,dy = 0.$$

or, $$dx/u = dy/v,$$

which is the equation of a streamline. Thus $\psi = C$ lines represent the streamlines in the flow region. Having established that the streamlines are constant stream function line, we may give some physical significance to $\psi$, which hitherto was only a mathematical function. Consider the control volume, as shown in Fig. 2.3.1, formed by the

[4] Note that continuity equation will also be satisfied if the negative sign appears in the expression for $u$ and a positive sign for $v$.

two streamlines having stream function values $\psi$ and $\psi + d\psi$, the two arbitrary curves *AB* and *CD* drawn across the streamlines, and two *z*-planes spaced unity in *z*-direction. By definition of the stream lies, no mass can flow across these lines, the volume[5] entering from *AB* must be equal to that leaving at *CD* for a steady flow.

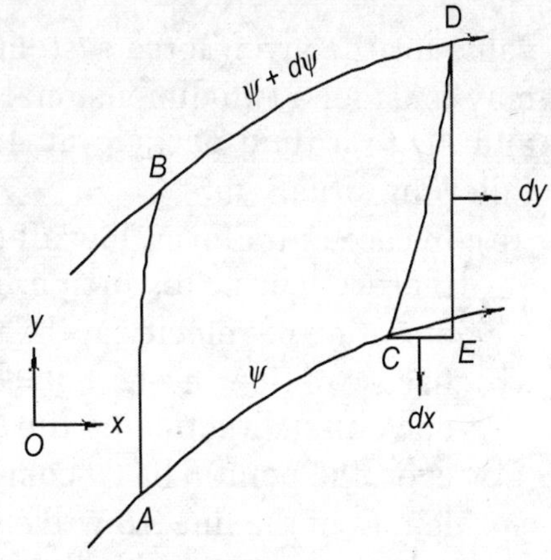


**Fig. 2.3.1** Streamlines are constant lines

But the volume flow rate across *CD* can be expressed in terms of velocity components $(u, v)$, taken positive along positive directions of the $x$ and $y$-axes. Let *CE* and *DE* be the projected areas $dx$ and $dy$ respectively of the surface *CD* of width unity in *z*-direction. According to the convention, these areas are positive along the outward drawn normal, as shown in the figure. Thus volume flow rate across *ED*, *CE* and *CD* is $u\ dy$, $-v\ dx$ and $u\ dy - v\ dx$, respectively; the negative sign with $v$ is because the vectors $v$ and $dx$ are at an angle of 180°. Substituting for $u$ and $v$ from Eq. 2.3.1, the volume flow rate through the control volume is

$$(\partial\psi / \partial y)\ dy + (\partial\psi / \partial x)\ dx = d\psi.$$

That is, the stream function is not just a mathematical function but its difference for the two streamlines represent the volume of the liquid flowing through the area formed by the two stream lines having unit width in a two-dimensional flow. It should be noted that the stream function is defined for a steady, two-dimensional flow. We define separate stream functions for a general flow, one for each plane, to take care for three-dimensional nature of the flow. Thus there are three stream functions, which need not be same.

Though we assumed incompressible flow in our discussions but this is not necessary; the following equation define the stream function for a compressible flow:

$$u = \frac{\rho_0}{\rho}\frac{f\psi}{fy} \text{ and } v = \frac{\rho_0}{\rho}\frac{f\psi}{fx} \tag{2.3.2}$$

where $\rho_0$ is the reference velocity used in the manner as given in Equation (2.3.2). You may verify that the velocity components defined by Eq. 2.3.2 satisfy Eq. 2.4.1 (discussed later). This concept of the steam function can also be extended to unsteady or non-steady flow but then the stream function itself will be function of time too, in addition.

## 2.3.2 Vorticity and Velocity Potential

Like solids, the fluid also deform. In solids deformation takes place and acquires an equilibrium position whereas in the fluids deformation goes on continuously so long the force is acting on it. It is found, however, that the rate of deformation takes a stable

[5] Volume is used because for the incompressible fluid density is constant and therefore constancy of mass also implies constancy of volume.

value for the given force system acting on the fluid element. To study this effect, we may consider a two-dimensional flow and attach two mutually perpendicular lines *AB* and *AD* to a fluid particle at *A*. Depending on the velocity field, we will have the following situations:

(a) In case of uniform flow all the fluid particles would move the same distance and there will be no distortion. Thus the two lines *AB* and *AD* will move parallel in the direction of velocity at the point *A*.

(b) In case of flow in which $u = u(x)$ and $v = v(y)$ then all the points on the line *AB* will move in $y$-direction by the same distance since $v$ is same at all these points. However, the point *B* has $u$-component different from that of the point *A* and therefore length of the line *AB* will change. We may argue similarly for the line *AD*. Thus in this case, there will be linear strain rate but the angles between the lines will not change.

(c) Generally, in a flow the velocity components will change with both $x$ and $y$ coordinates. In such case, the lines will change not only in their length but they will not be parallel anymore.

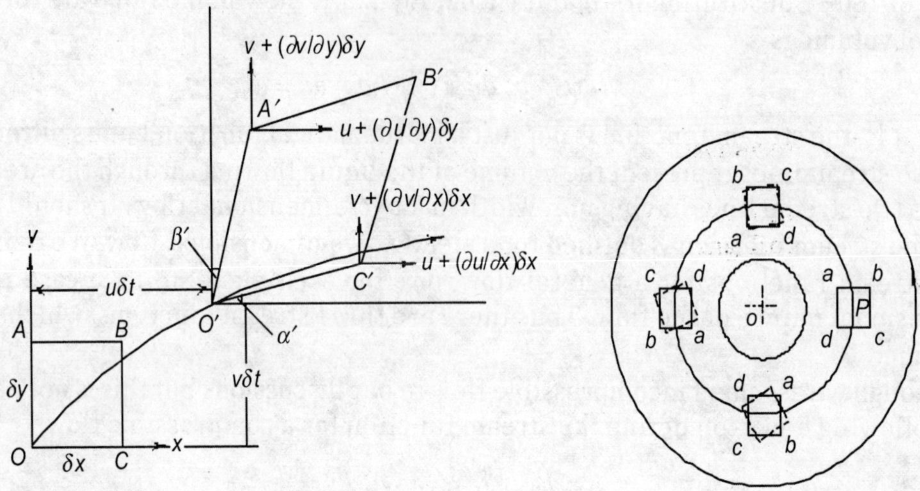


(a) Deformation of fluid element (b) Rigid body rotation

**Fig. 2.3.2** Kinematics of the flow

We shall demonstrate the procedure of calculation of the deformation for two-dimensional, incompressible fluid in Cartesian coordinates. Figure 2.3.2 (a) shows a general motion of the fluid particles in a steady, two-dimensional flow of an incompressible fluid. A fluid element has been shown by a rectangle *OABC* at point *O* at time $t = 0$. This element moves along the streamline and occupies position $O'A'B'C'$ after time $\delta t$ travelling a distance $u\,\delta t$ and $v\,\delta t$ along $x$ and $y$ direction from *O*. ($\delta t$ is assumed so small that $u$ and $v$ at point *O* be taken constant during the travel from *O* to *O′*. (This however, is not necessary condition and the velocity components may even change.) The element will no longer remain a rectangle, as mentioned at (c) above, and would be distorted, as shown in the figure. We may express velocity components at the points *A′* and *C′* by using the Taylor series in terms of their values at point *O′*, which is $(u, v)$.

(Remember for point $C'$, the $x$ coordinate is changing and therefore derivative is with respect to $x$ while it is $y$ coordinate for the point $A'$.) The velocity components at these points are shown in the figure.

In Physics we read that longitudinal strain in a solid is the ratio of change in length of a rod, as a result of the force, with its original length whereas the shear strain is defined in terms of the angle of deformation of a rectangular element. We extend these concepts for the fluid too; keeping in mind that time rate is to be evaluated in this case. To find the linear strain rate along $x$-axis, we may determine $u$. This component at the point $C'$ gives relative displacement of $C'$ with respect to $O'$ in unit time along $x$-direction; dividing it by the initial length $\delta x$ gives the desired result. Similarly, we may consider rotation of the line $OC$ about $O$ in unit time for determining $\alpha$. For this the $v$-component of velocity of $C'$ with respect to $O'$ is responsible that can be determined from the velocities given in the figure. Once the angle of rotation $\alpha$ and $\beta$ of the two lines about the point $O'$ has been obtained in unit time then the shear strain rate $\varepsilon_{xy}$ produced in $x$-$y$ plane is $\alpha - \beta$ and can be determined in terms of the velocity gradients.[6] This procedure is then repeated for other planes. Or, we simply rotate the dependent and independent variables in cyclic order to get these relations in other planes. The procedure is quite simple and the results are given below.

$$\dot{\varepsilon}_{xx} = \frac{u + (fu/fx)\delta x - u}{\delta x} = \frac{\delta u}{\delta x}.$$

Similarly $$\dot{\varepsilon}_{yy} = \frac{\delta v}{\delta y} \text{ and } \dot{\varepsilon}_{zz} = \frac{\delta w}{\delta z} \tag{2.3.3a}$$

$$\dot{\alpha} = \frac{v + (fv/fx)\delta x - v}{\delta x} = \frac{\delta v}{\delta x},$$

$$\dot{\beta} = -\frac{u + (fu/fy)\delta y - u}{\delta y} = -\frac{\delta u}{\delta y}$$

$\therefore$ $$\dot{\varepsilon}_{xy} = \dot{\alpha} - \dot{\beta} = \frac{\delta v}{\delta x} + \frac{\delta u}{\delta y} \tag{2.3.3b}$$

$\alpha$ is taken positive as it is in anticlockwise direction and a negative sign is in the expression for $\beta$, as this angle is along clockwise direction. We may also find out the dilatation of the original volume element of unit size. Due to linear strain, sides will change in unit time to $1 + \varepsilon_{xx}$, $1 + \varepsilon_{yy}$, and $1 + \varepsilon_{zz}$ So the change in volume in unit time to the first order is $\varepsilon_{xx} + \varepsilon_{yy} + \varepsilon_{zz}$. Recall from Chapter 1, these three terms are the diagonal elements of the strain tensor, therefore, their sum represents the dilatation of the volume element. Further, the terms appearing in Eq. 2.3.3 (b) are the off-diagonal elements of this tensor and their sum gives the shear strain rate.

In addition, the point $C$ has transverse and the point $A$ has the lateral displacement with respect to Point $O$; there is angular rotation of the line elements about the axis of

[6] $\varepsilon_{xy}$ has been used for $\dot{\varepsilon}_{xy}$, because of convenience only. But keep in mind that such a quantity is on unit time basis.

the particle. Therefore it is possible to introduce the concept of angular rotation of the fluid particle and the angular velocity, $\omega_z$, of the fluid particle, which Stokes defined as the average of the angular velocities of the two mutually perpendicular lines that are perpendicular to the line about which angular velocity is to be determined. Equation 2.3.3 gives the angular velocity of the two lines.

$$\omega_z = \frac{1}{2}(\dot{\alpha} + \dot{\beta})$$

$$= \frac{1}{2}\left(\frac{\delta v}{\delta x} - \frac{\delta u}{\delta y}\right) \tag{2.3.4}$$

Equation 2.3.4 gives the angular velocity in two dimensional flow cases in $z$ direction and is represented by the symbol $\omega_z$. You will recognise it to be curl $\vec{U}$ for two-dimensional flow. Direction of the angular velocity vector is decided by the right hand screw rule. We may similarly find out the other components of angular velocity about the other axes. In general, we may write for angular velocity $\vec{\omega}$

$$\vec{\omega} = \frac{1}{2}\text{curl } \vec{U} \text{ and } \vec{\Omega} = \text{curl } \vec{U}\text{, where}$$

$$\text{Curl } \vec{U} = \quad \leftrightarrow \vec{U} = \begin{vmatrix} i & j & k \\ f/fx & f/fy & f/fz \\ u & v & w \end{vmatrix} \text{ in (Cartesian)}$$

or

$$= \frac{1}{r}\begin{vmatrix} e_r & re_\theta & e_z \\ f/fr & f/f\theta & f/fz \\ u_r & ru_\theta & u_z \end{vmatrix} \text{ in (Cylindrical)}$$

$$\blacklozenge \ \omega_x = \frac{1}{2}\left(\frac{\delta w}{\delta y} - \frac{\delta v}{\delta z}\right), \ \omega_y = \frac{1}{2}\left(\frac{\delta u}{\delta z} - \frac{\delta w}{\delta x}\right),$$

and

$$\omega_z = \frac{1}{2}\left(\frac{\delta v}{\delta x} - \frac{\delta u}{\delta y}\right) \text{ (Cartesian)} \tag{2.3.5}$$

$$\omega_r = \frac{1}{2}\left(\frac{\delta u_r}{r\delta\theta} - \frac{1}{r}\frac{\delta(u_\theta\, r)}{\delta z}\right),$$

$$\omega_\theta = \frac{1}{2}\left(\frac{\delta u_r}{\delta z} - \frac{\delta u_z}{\delta r}\right),$$

and

$$\omega_z = \frac{1}{2r}\left(\frac{\delta(ru_\theta)}{\delta r} - \frac{\delta u_r}{\delta\theta}\right) \text{ (Cylindrical)}$$

We note that divergence of velocity vector gives the total mass efflux rate through the point and the cross product (also called the curl) gives the angular velocity of the fluid particle at that point. In fact, this result is not specific to the fluid motions only. These relations appear in magnetism, electricity and other fields also where conservative field is dealt with. *Vorticity* $\vec{\Omega}$ is used to represent twice the angular velocity just to avoid half factor.

In general, a flow will take place along a stream line, since it has translational motion. The fluid particles will rotate about the axis through its own mass centre if Curl $\vec{U}$   0. Here emphasis is laid on the own axis of the particle, which is different from the axis passing through the mass centre of the whole fluid in the control volume. Figure 2.3.2 (b) shows rotation of fluid particles around the axis passing through mass centre *O* of the water contained in a cylinder that is rotating about its axis through *O*. In this case, the whole fluid is rotating about this axis but the individual fluid particles do not rotate about their own axes. This can be observed by placing a wooden block, marked abcd at its four corners, anywhere in the liquid, say at *P*. Being lighter, it will float over the water surface. From the figure we observe that the block is rotating about the cylinder axis but not about its own axis. Had there been any angular motion of the fluid particle about its own axis, it might have taken the position shown by dotted lines in the figure, i.e. the fluid particles do not have any angular velocity. Such a motion is called rigid body rotation and is not being considered in our discussions. For a 2D flow, $\partial u/\partial y = -\ \partial v/\partial x$ then there will be no shear of the fluid element and the particle will have pure rotation, whereas there will be no rotation if $\partial u/\partial y = \partial v/\partial x$. When vorticity of the flow is zero then the flow is called *irrotational* flow. In such a case it is possible to define, like stream function, another scalar function $\phi(\vec{r})$ of the space coordinates for steady flow, that is,

$$\vec{U} = \ \ \phi.$$

That is, in Cartesian coordinates, $u = \partial\phi / \partial x$, $v = \partial\phi / \partial y$, and $w = \partial\phi / \partial z$ (2.3.6)

$\phi$ is called the *velocity potential* and is defined only for the irrotational flow such that its gradient gives the velocity field in the flow region. Like stream function we may associate physical meaning to velocity potential too. As in the case of electrical or magnetic potential which is defined at a point as the amount of work done in bringing a positive charge or a north pole from infinity to the point under consideration in presence of respective external field. In this case, the external field is Earth's gravity and the change in height is related to the potential energy. For any change $dh$ in height of a mass $m$ in the Earth's field, the work done is given as

$$\delta W = mg\, dh$$

$$= -\, d(PE) = \vec{F}\, .\, dh\, \vec{k}$$

$$♦\ \vec{F} = -\frac{d}{dh}(PE)\vec{k}$$

$$\text{or, in general, } \vec{F} = - \ \ (PE)$$

This proves our statement. We further note from the analogy between the expression for the force and the velocity potential that velocity vector is synonymous with the force vector in this simile.

For constant $\phi$ and $\psi$ lines, we may write

$$d\phi = (\partial\phi / \partial x)dx + (\partial\phi / \partial y)\, dy = u\, dx + v\, dy = 0.$$

$$\blacklozenge \quad (dy / dx)_{\phi = \text{const}} = -\, u / v, \text{ and}$$

$$d\psi = (\partial\psi / \partial x)\, dx + (\partial\psi / \partial y)\, dy = u\, dx - v\, dy = 0.$$

$$\blacklozenge \quad (dy / dx)_{\psi = \text{const}} = v / u.$$

$$\therefore \quad (dy / dx)_{\phi = \text{const}} \times (dy / dx)_{\psi = \text{const}} = -1 \tag{2.3.7}$$

Equation 2.3.7 shows that the constant $\phi$ and $\psi$ lines are perpendicular to each other at a point in the flow, since *dy*/*dx* is slope of a curve at the point.[7] The constant stream function and the velocity potential lines generate a mesh called the flow net. Figure 2.3.3 shows one such flow net. Flow net was used successfully to determine the solution in many cases for inviscid fluid.

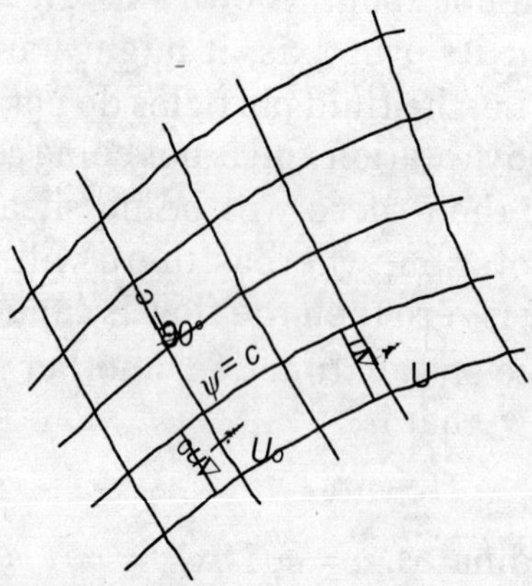


**Fig. 2.3.3** Flow net

The fact that the streamlines and potential lines are perpendicular to each other has been used in fluid mechanics and heat transfer to determine graphically and experimentally the amount of discharge or heat transfer, respectively. We may draw the flow net keeping in mind the following points:

1. The boundaries of the flow region act as limiting streamlines.
2. If $\Delta n_0$ are $\Delta n$ the spacing in a stream tube at two sections where $U_0$ and $U$ are the velocities, then by continuity equation $\Delta n_0\, U_0 = \Delta n\, U$. Or, $\Delta n / \Delta n_0 = U_0 / U$. This means in uniform flow, the streamlines would be parallel and we may draw these lines at equal distance. However, in non-uniform flow the streamlines would not be equally spaced throughout the flow region, since the spacing between the streamlines is inversely proportional to the velocity.
3. We may now draw potential lines keeping distance between the successive potential lines same as that between the streamlines at the point of starting. While drawing the potential lines, we try to maintain these lines perpendicular to the streamlines at the point of intersection.

[7] In Physics, a field is shown as lines such as magnetic lines whereas the equi-potential lines are perpendicular to these lines everywhere in the field. That explains why the name potential was given to the scalar function $\phi$.

4. In uniform flow, this process works well and the intersection of the consecutive streamlines and potential lines produce fairly good square. But it is not so in non-uniform flow. A check for the lines to be perpendicular is that for such square the two diagonals are of equal length.

Let us substitute for $u$ and $v$ from Eq. 2.3.2 in Eq. 2.3.4, we get for irrotational flow,

$$\frac{f^2\psi}{fx^2} + \frac{f^2\psi}{fy^2} = 0.$$

or,
$$^2\psi = 0,$$

where $^2 = \frac{f^2}{fx^2} + \frac{f^2}{fy^2}$ is a Laplacian operator. We may also substitute for $u$, $v$ and $w$ in terms of $\phi$ in the continuity equation and show that $^2\phi = 0$, where $^2$ is defined in Cartesian and cylindrical polar coordinates as follows:

$$^2 = \frac{f^2}{fx^2} + \frac{f^2}{fy^2} + \frac{f^2}{fz^2} \text{ (Cartesian coordinates)} \qquad (2.3.8)$$

$$^2 = \frac{f^2}{fr^2} + \frac{1}{r}\frac{f}{fr} + \frac{1}{r^2}\frac{f^2}{f\theta^2} + \frac{f^2}{fz^2} \text{ (Cylindrical coordinates)}$$

Laplacian operator is linear operator of second order and appears in the equation from other fields of Physics too. Since the stream function and the velocity potential are satisfying the Laplacian equation, an irrotational flow is also called the *potential flow*. We shall analyse in Chapter 3 a few potential flows and use the property of linearity of the Laplacian equation. Recall if there are $n$ solutions to a linear equation then their combinations is also the solution. This is known as the principle of superimposition.

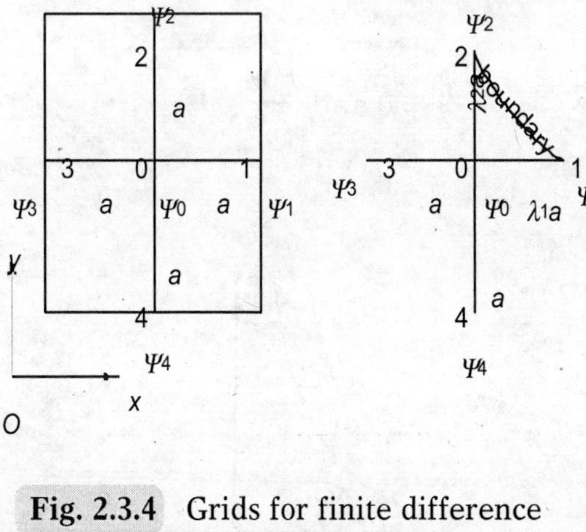


**Fig. 2.3.4** Grids for finite difference

We may also use the central finite difference to plot the flow net. Finite-difference technique is very suitable for elliptic *PDE*, like the Laplace equation. In this technique, we use a grid of points around the point at which the first and second derivatives are to be determined, as shown in Fig. 2.3.4. We have to determine the second derivative of $\psi$ with respect to $x$ and $y$ around the point $O$. We choose the points 1, 2, 3, and 4 around it at a distance $a$; choice of $a$ depends on the variation of these derivatives around the point $O$ and obtained by experimentation. However, $a$ is quite small. Such kind of grid is generated throughout the flow region and the grid is known as symmetric grid. Problem is faced when the grid so formed is close to the boundary as then the grid points lying on the boundary may not be at the same distance $a$, as shown in Fig. 2.3.4(b). Now we shall express Eq. 2.3.8 in the finite difference form using central difference relations.

$$\frac{f^2\psi}{fx^2} = \frac{\psi_1 - 2\psi_0 + \psi_3}{2a} \text{ and } \frac{f^2\psi}{fy^2} = \frac{\psi_2 - 2\psi_0 + \psi_4}{2a}.$$

$$\therefore \quad \frac{f^2\psi}{fx^2} + \frac{f^2\psi}{fy^2} = \frac{\psi_1 + \psi_2 + \psi_3 + \psi_4 - 4\psi_0}{2a} = 0.$$

or, $$\psi_0 = \frac{\psi_1 + \psi_2 + \psi_3 + \psi_4}{4} \quad \text{(Symmetrical grid)}$$

and $$\psi_0 = \frac{\psi_1 / \lambda_1 + \psi_2 + \lambda_2 / \psi_3 + \psi_4}{1/\lambda_1 + 1/\lambda_2 + 2} \quad \text{(Unsymmetrical grid)}$$

where $\lambda_1$ and $\lambda_2$ are shown in Fig. 2.3.4 (b). Procedure for solving these equations is to choose the values of $\psi$ at four points so that the equations are satisfied. Relaxation method is used for solving these Eq. which is not covered here.

**Example 2.3.1** Does the stream function $\psi = xy$ represent a physically possible flow? Determine the velocity components at (1, 2). Is this flow rotational?

**Solution** The flow is physically possible if $\psi$ satisfies the Laplace equation. For this flow

$$\frac{f^2\psi}{fx^2} = 0 \text{ and } \frac{f^2\psi}{fy^2} = 0$$

$$\blacklozenge \quad {}^2\psi = 0.$$

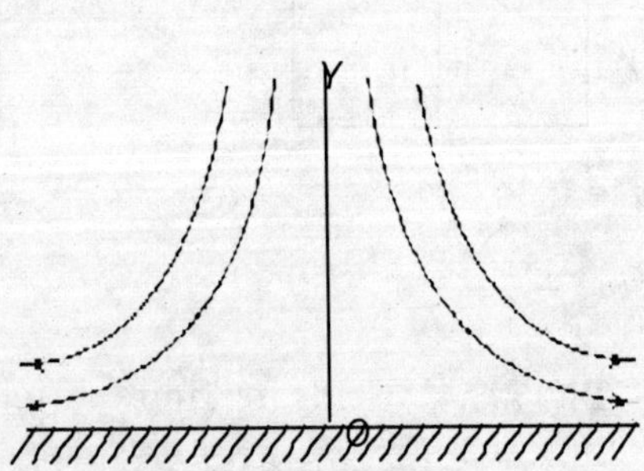


**Fig. E 2.3.1** Possibility of a flow

Hence the flow is physically possible. It is observed from Fig. E 2.3.1. that the constant stream function lines $\psi = xy = C$ are rectangular hyperbola. We will observe such streamlines when a water jet moves in a direction perpendicular to the plate. As water approaches the plate, it is diverted towards the edge, as is seen in the figure. The velocity components at any point, by Eq. 2.3.1, are

$$u = \partial\psi / \partial y = x \qquad \text{and } v = -\,\partial\psi / \partial x = -\,y$$

$\therefore$ Velocity at the point (1, 2) is : $u = 1$ and $v = -\,2$ units.

We further see that $\partial u / \partial y = 0$ and $\partial v / \partial x = 0$. Therefore the angular velocity is also zero. Or in other words the flow is irrotational. Velocity potential therefore exists for this flow.

**Example 2.3.2** What type of motion do the following velocity components $u = a + by - cz$, $v = d - bx - ez$, and $w = f + cx - ey$, where $a$, $b$, etc. are constant, constitute?

**Solution** Since there is no term in time $t$, the flow is steady. The constants do not depend on the density and hence the fluid is incompressible. It also satisfies the continuity equation because

$$\partial u / \partial x = 0,\ \partial v / \partial y = 0,\ \text{and } \partial w / \partial z = 0 \ \blacklozenge \ \ .\vec{U} = 0.$$

Hence the flow is possible. We may determine the angular velocity components to check if it is irrotational flow?

$$\Omega = \text{curl } \vec{U} = \begin{vmatrix} i & j & k \\ f/fx & f/fy & f/fz \\ u & v & w \end{vmatrix}$$

$$= -\,e\hat{i} - c\hat{j} - b\hat{k} \quad 0$$

Hence the flow is rotational for which the angular velocity is same at all the points. We further determine the shear and longitudinal strain rates; these are

$$\varepsilon_{xy} = \tfrac{1}{2}\,(\partial v / \partial x + \partial u / \partial y) = 0 \qquad \text{and similarly } \varepsilon_{yz} = \varepsilon_{zx} = 0,$$

and $\qquad \varepsilon_{xx} = \partial u / \partial x = 0 \qquad$ and similarly $\varepsilon_{yy} = \varepsilon_{zz} = 0.$

The flow is thus rotational with no distortion having constant angular velocity.

**Example 2.3.3** Determine the type of the flow represented by the velocity components $u = x - 4y$ and $v = -\,y - 4x$.

**Solution** Following steps are used for the solution:

**Step 1.** Check whether the flow is possible or not from mass conservation point of view. We note, $\partial u / \partial x = 1$ and $\partial v / \partial y = -1$. Therefore $\partial u / \partial x + \partial v / \partial y = 0$ and hence the flow is possible.

**Step 2.** Determine equation of the stream function. Using Eq. 2.3.1,

$$u = \partial\psi / \partial y = x - 4y\ .$$

On integration with respect to $y$ while keeping $x$ constant, we get

$$\psi = xy - 2\,y^2 + f(x),$$

where $f(x)$ is a constant of integration when $x$ is treated as constant during integration.

∴ $$v = -\partial\psi/\partial x = -y - df/dx = -y - 4x \text{ (given)}.$$

or, $$df/dx = 4x.$$

On integration with respect to $x$, we get

$$f(x) = 2\,x^2 + C, \; C \text{ is the constant of integration.}$$

∴ $$\psi = xy - 2\,y^2 + 2\,x^2 + C.$$

Constant $C$ is determined from the value of $\psi$ of the streamline passing through the origin. If we choose this value to be zero then $C = 0$ in this case.

**Step 3.** Determine Vorticity of the flow, as explained in Example 2.3.2. You may verify that $\tilde{\Omega} = 0$ for the flow and hence it is a potential flow and $\phi$ can be determined in the same manner as we did for finding out $\psi$ in Step 2.

$$u = \partial\phi/\partial x = x - 4y$$

On integration with respect to $x$ keeping $y$ constant,

$$\phi = x^2/2 - 4x\,y + f(y),$$

where $f(y)$ is a constant of integration.

∴ $$v = \partial\phi/\partial y$$

$$= -4x + df/dy = -y - 4x \text{ (given)}.$$

or, $$df/dy = -y.$$

This gives on integration with respect to $x$,

$$f(x) = -y^2/2 + B, \; B \text{ is constant of integration.}$$

∴ $$\phi = x^2/2 - 4x\,y - y^2/2 + B.$$

**Step 4.** Determine the constant of integrations $B$ and $C$ from the given condition. If not given, value of the constants is normally taken as zero

**Note.** We will follow this approach to determine stream function and velocity potential of the flow whenever the velocity field is prescribed for the flow though there are other ways also to integrate the partial differential equation.

**Example 2.3.4** Determine the type of the flow represented by the velocity components $u = y^3/3 + 2x - x^2\,y$ and $v = x\,y^2 - 2y - x^3/3$.

**Solution** Following steps are used for the solutions:

**Step 1.** The velocity field satisfies the Laplace equation for both $\psi$ and $\phi$. Therefore, the flow is possible and is irrotational. Next, we determine the stream function and velocity potential. This time, we shall demonstrate another way of finding out these.

$$u = \partial\psi/\partial y = y^3/3 + 2x - x^2 y \text{ (given)}$$

♦ $$\psi = y^4/12 + 2x\,y - x^2y^2/2 + F(x) \qquad \text{(A)}$$

Similarly we may proceed with $v$-component of the velocity to give on need basis,

$$v = -\partial\psi / \partial y = xy^2 - 2y - x^3/3 \text{ (given)}$$

$$\blacklozenge \qquad \psi = x^4/12 + 2x\,y - x^2y^2/2 + G\,(y) \qquad \text{(B)}$$

Comparing Equations ($A$) and ($B$) we get,

$$F(x) = x^4/12 \text{ and } G\,(y) = y^4/12.$$

$$\therefore \qquad \psi = x^4/12 + y^4/12 + 2x\,y - x^2y^2/2$$

## 2.3.3 Circulation

*Circulation* is defined as the line integral of the velocity field around a closed curve drawn in the flow region and is represented by the symbol $\Gamma$.

$$\Gamma = \int_c \vec{U}\,.\,d\vec{l} = \iint_s \vec{U}\,.\,d\vec{S}, \qquad (2.3.9)$$

$$= \iint_s (\;\leftrightarrow\vec{U}).\,d\vec{s}$$

where $dl$ is the length element along the closed curve $C$ drawn in the flow region. This curve is arbitrary and can be drawn in any manner. It is taken positive when we take the line element along the anticlockwise direction.[8] The line integral can be changed to surface integral using the Stokes theorem. We observe from Eq. 2.3.9 that circulation is related to the vorticity efflux from the control surface in the direction of the normal to the surface. Further it is seen that circulation will be zero for irrotational flow.

To demonstrate the procedure for calculating $\Gamma$ in a simple manner, we take the closed curve to be a rectangle, like the one shown in Fig. 2.3.5 (a). Let $ox$ and $oy$ be the reference axes and the rectangle be of dimensions $\delta x$ and $\delta y$. We shall evaluate the line integral of Eq. 2.3.9 by moving along the path $AB$, $BC$, $CD$ and $DA$, that is in the anticlockwise direction. The length vector points in the direction of the movement. Thus we get

$$\Gamma = u\;\delta x + \{v + (\partial v / \partial x)\;\delta x\}\;\delta y - \{u + (\partial u / \partial y)\;8y\}\;\delta x - v\;\delta y$$

$$= (\partial v / \partial x - \partial u / fy)\delta x\;\delta y$$

$$\text{or,} \qquad \Gamma / (\delta x\;\delta y) = (\partial v / \partial x - \partial u / \partial y) = \omega_z \qquad (2.3.10)$$

[8] In aerodynamics, the convention adopted is opposite to it.

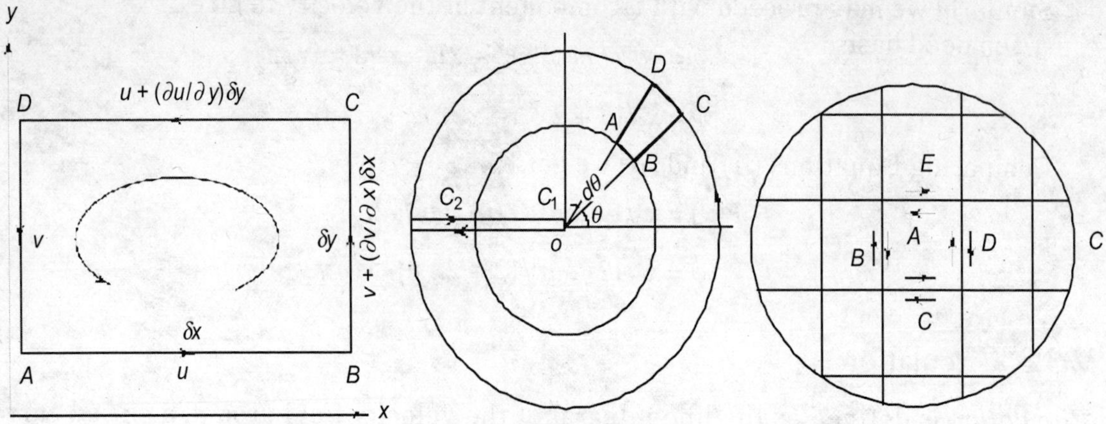


(a) Circulation along a rectangular path (b) Circulation in free vortex (b) Circulation around curve C

**Fig. 2.3.5** Circulation

But $\delta x\ \delta y$ is the area enclosed by the rectangle. Therefore circulation per unit area is the vorticity of the flow. For irrotational flow, we therfore expect the circulation to be zero. Probably, we need some more exploration because at the face of the statement, it looks true. Those who have studied Cauchy's integral theorem, they would be able to see the point because the integral of Eq. 2.3.9 is nothing but Cauchy's integral and the value of the integral depends on the presence of zeros and poles of the integrand in the region bounded by the closed curve. Now consider the curve $C$ drawn in the flow region, as shown in Fig. 2.3.5(c). We can divide this area by drawing equispaced horizontal and vertical lines. Consider 5 adjacent areas $A$, $B$, $C$, $D$, and $E$ and evaluate the circulation in each of these areas. We observe that in the process, contribution from the side common to the two areas cancel out its effect. For example, consider contribution for the line common to $A$ and $E$. For area $A$, we move from right to left and from left to right for area $E$ and therefore these two will cancel out. Similarly for the other sides. Thus ultimately the contribution will come from the area close to the curve. Thus

$$\Gamma = \left(\frac{fv}{fx} - \frac{fu}{fu}\right) dA$$

$$= \int_A \left( \leftrightarrow \vec{U}\right) . \vec{n}\, d A.$$

But by definition: $\Gamma = \oint \vec{U} . d\vec{l}$, therefore

$$\Gamma = \oint \vec{U} . d\vec{l}$$

$$= \int_A \left( \leftrightarrow \vec{U}\right) . \vec{n}\, dA.$$

This equation was extended to a general flow though we started analysis for a two-dimensional flow. This theorem is called *Stokes theorem*.

Let us consider free vortex flow, which is defined by the velocity field

$$u_\theta r = C \qquad \text{and } u_r = 0, \tag{2.3.11}$$

where $u_r$ and $u_\theta$ are the velocity components along the radial and circumferential directions of the flow in the cylindrical polar coordinates. The flow is axisymmetric. It is seen that $r = C$ are the streamline, since $u_r$ is zero and flow is along the circular streamline, in which the circumferential velocity decreases as $r$ increases. Further, using Eq. 2.3.5 for the polar coordinates, we get

$$\vec{\omega} = \frac{1}{r}\begin{vmatrix} e_r & re_\theta \\ f/fr & f/f\theta \\ 0 & C \end{vmatrix} = 0.$$

This shows that the free vortex flow is irrotational. We may also determine the circulation for which the choice of a circular path around the origin of the coordinate system, as shown in Fig. 2.3.5 (b), will be most convenient for carrying out the integration. Using Eq. 2.3.9, the circulation around the circular curve of radius $r$ around the origin is

$$\Gamma = \int_0^{2\pi} u_\theta \leftrightarrow r \, d\theta$$

$$= \int_0^{2\pi} C \, d\theta = 2\pi C,$$

which is not zero, a result contradicting Eq. 2.3.9 according to which circulation ought to be zero for irrotational flow. This is because the point $O$ is a singular point where the velocity is infinite and falls in the region enclosed by the curve; such points are called pole. We may instead choose another path $ABCD$, as shown, for the line integration. Let radii of $AB$ and $CD$ be $r_1$ and $r_2$, respectively and both the arcs subtend an angle $d\theta$ at the centre. In this case, the contribution towards the circulation to the circulation by the portion $BC$ and $DA$ is zero as the radial velocity is zero. Thus,

$$\Gamma = -\, u_{\theta,\, AB} \leftrightarrow r_1 \, d\theta + u_{\theta,\, CD} \leftrightarrow r_2 \, d\theta = c \, d\theta - c d\theta = 0.$$

This was expected, since there is no pole or zero in the region confined by the closed curve. In complex numbers, we learn while studying Cauchy's theorem, that the regions containing the pole and/or zero are called multiply connected regions (else the simply connected regions). To exclude the singular point, it is also possible to split the circular closed curve, by taking two lines $C_1$ close to each other, as shown in Fig. 2.3.5 (b). These lines are running opposite to each other with a small circular curve at the end so as to form a closed path made up of two curves $C$ and $C_1$. The circulation in this case is zero, because the contribution from the two limbs of the curve $C_1$ will cancel out.

In the context of the circulation, we would mention of Kelvin's theorem because of their importance. These are:

(i) Circulation along a closed contour remains constant with the passage of time, and

(ii) A flow, irrotational at some time, must continue to be irrotational thereafter.

**Example 2.3.5** Determine the acceleration and the vorticity of a two-dimensional flow expressed by the velocity field $u = \dfrac{x}{x^2 + y^2}$ and $v = \dfrac{y}{x^2 + y^2}$.

**Solution** The first derivatives of $u$ and $v$ can easily be obtained. These are:

$$\frac{fu}{fx}=\frac{y^2-x^2}{(x^2+y^2)^2},\ \frac{fu}{fy}=\frac{-2xy}{(x^2+y^2)^2},\ \frac{fv}{fx}=\frac{-2xy}{(x^2+y^2)^2},\ \text{and}\ \frac{fv}{fy}=\frac{x^2-y^2}{(x^2+y^2)^2}$$

We note that $\frac{fu}{fx}+\frac{fv}{fy}=0$, which means flow is possible.

$$\therefore \qquad \alpha_x = u\frac{fu}{fx}+v\frac{fu}{fy}=-\frac{x}{(x^2+y^2)^2},$$

$$\alpha_y = u\frac{fv}{fx}+v\frac{fv}{fy}=-\frac{y}{(x^2+y^2)^2},\ \text{and}$$

$$\omega_z=\frac{1}{2}\left(\frac{fv}{fx}-\frac{fu}{fy}\right)=0$$

Flow is irrotational.

## 2.4 Conservation Laws

Conservation laws have contributed a lot in development of science. These include conservation of mass, momentum and energy. In the following sections, we shall learn how to develop a mathematical model for these laws. Analysis is carried out in Cartesian coordinates for convenience of the reader but their derivation, using vector approach, saves time and space.

### 2.4.1 Conservation of Mass — Continuity Equation

The law states that mass cannot be created or destroyed. This law is known to us for millenniums. We shall give mathematical form to the law of mass conservation using different approaches.

#### 2.4.1.1 Differential Approach

In differential approach, the results of the differential equation are applicable to every point in the flow region. The advantage of such approach is the detailed analysis and we know all the parameters of interest at each point. However, the approach is rigorous, time consuming and requires a large computing memory.

(a) ***Control mass approach.***

In thermodynamics, we study closed system and open system, most popularly known as control volume. A closed system consists of the same particles and therefore the law of conservation of mass is implied implicitly in its definition. In addition, the identity of particles also does not change with time in such analysis.

**(b) *Control volume approach.***

In the control volume approach, a fixed region, bounded by the control surface, is marked in which the mass and energy flow in or out of the control surface. Therefore the identity of the particles changes with time. Whether the mass changes or remains constant with time depends on the rate of mass flow across its control surface. If the net mass flow rate was zero then the mass would not change. Under steady conditions both the mass and energy of the control volume remains invariant with time. Subscript $\sigma$ may be used with the quantities to represent it for the control volume. For example, $dE_\sigma$ is the change in energy of the control volume.

Let $\Delta V$ be the volume of the control mass having density $\rho$; $\Delta V$ is so small that density may be considered as same everywhere. Thus the mass conservation can be written as

$$\frac{D}{Dt}(\rho \Delta V) = 0.$$

Here *D/Dt* is the material derivative. Since both *r* and *DV* may be function of time, differentiation gives after simplification

$$\text{or} \quad \frac{D\rho}{Dt} + \frac{\rho}{\Delta V}\frac{D}{Dt}(\Delta V) = \frac{f\rho}{ft} + \vec{U}. \quad \rho + \frac{\rho}{\Delta V}\frac{D}{Dt}(\Delta V) = 0$$

in which *D/Dt* has been replaced from Eq. 2.2.2. The last term $\frac{\rho}{\Delta V}\frac{D}{Dt}(\Delta V)$ is the rate of volumetric dilatation and equals the divergence of the velocity vector at the location of the control mass. Thus the above equation can be written as

$$\frac{f\rho}{ft} + (\vec{U}. \quad )\rho + \rho \quad . \vec{U} = \frac{f\rho}{ft} + \quad .(\rho \vec{U}) = 0 \tag{2.4.1}$$

This equation can also be obtained by considering a particular co-ordinate system. Let us consider a cuboid, as shown in Fig. 2.4.1 (a) having dimensions $\delta x$, $\delta y$ and $\delta z$ along *x*, *y* and *z* directions, respectively, constructed around a point *O* so that when $\delta x$, $\delta y$ and $\delta z$ tend to zero then the *C.V.* shrinks to this point. Let us consider face *OABC*, which we shall refer to as *y* face; *y*-axis is normal to this face. Mass flow across this face is because of the velocity component *v*, which is normal to the *y* face; the other velocity components being parallel to the face, will not make the fluid to flow in or out of the control volume. However, velocity *v* and fluid density $\rho$ will vary, in general, over the surface and we may account for it. We consider these to be constant on this face because such accounting adds to terms of negligible magnitude. Similarly, we may take the density and *v*-component to be same at all the points of the surface *DEFG* ($y + \delta y$ face), at a distance $\delta y$ from the *y* face along *Oy*. However, numerical value of these variables will be different from those at the *y* face.

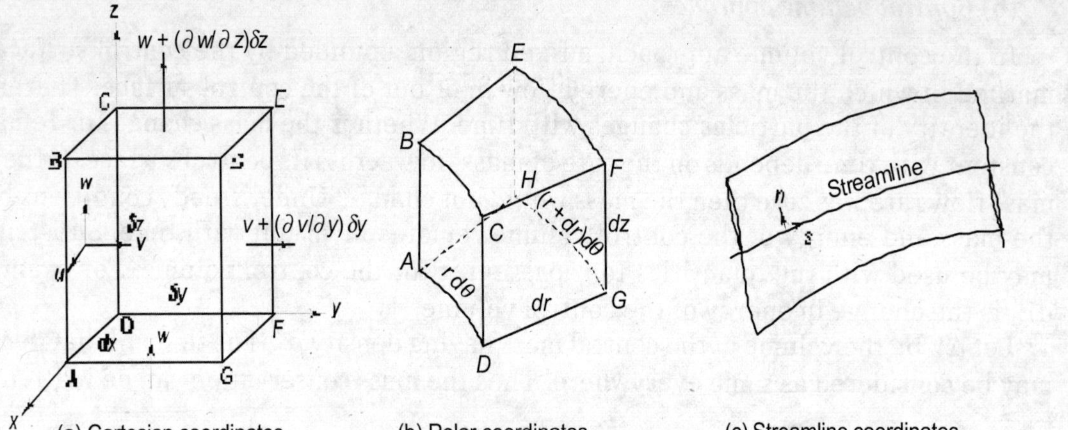


**Fig. 2.4.1** Conservation of mass

Since the fluid is continuous, we may assume that all the flow variables ($\boldsymbol{U}, p, T, \rho$, etc.) will be continuous and have derivatives of all the higher orders so that it may be expressed around the point $O$ at a distance $\delta y$ from it by Taylor series[9]. Let $m_y$ and $m_{y+\delta y}$ be the mass flow rate across the two faces under consideration, which would be in the $y$-direction. Then by the Taylor series,

$$m_{y+\delta y} = m_y + (\partial m_y / \partial y)\, \delta y + \text{terms of higher order} \quad m_y + (\partial m_y / \partial y)\delta y$$

$$\text{or, } m_{y+\delta y} - m_y = \text{net mass flow rate in } y \text{ direction} = (\partial m_y / \partial y)\, \delta y.$$

But $\quad m_y = \rho v.\, \delta x\, \delta z$

♦ $\quad m_{y+\delta y} - m_y = \partial(\rho v) / \partial y . \delta V,$

where $\quad \delta V = \delta x\, .\, \delta y\, .\, \delta z =$ Volume of the cuboid

To find out the net mass flow rate, we may consider faces normal to $x$ and $z$ directions. Keeping in mind that $v$ and $y$ are involved in the above equation, we may write

$$m_{x+\delta x} - m_x = \partial(\rho u) / \partial x.\, \delta V$$

and $\quad m_{z+\delta z} - m_z = \partial(\rho w) / \partial z.\, \delta V.$

Scalar sum of the mass flow rates would give the net mass efflux from the cuboid and should be equal to the time rate of change of its mass, which is $= -\,\partial\,(\rho.\, \delta V) / \partial t$; minus sign comes because a net out flux would decrease the mass and hence the time derivative would be negative. (Remember, 5 cannot be equal to –5, though their absolute value is same). Making mass balance and simplifying, we get

[9] A function $f(x+h)$ at a point $x+h$ in the neighbourhood of $x$ can be expressed in terms of value of the function and its derivatives with respect to $x$ at $x$ by the Taylor series:

$$f(x+h) = f + hf' + (h^2/2!)\, f'' + (h^3/3!)\, f''' + \ldots$$

where a single prime represents first derivative, two primes, the second derivative, and so on, of the function $f(x)$ with respect to $x$. Normally, $h$ is small (since we are taking the limit $h$ ♦ 0), the magnitude of terms $(h^2/2!)\, f''$, $(h^3/3!)\, f'''$, etc is negligibly small as compared to that of the term $hf'$ and are called terms of the order 2, 3, etc, respectively. Therefore in our analysis we retain only the term $h\, f'$ in the infinite series.

$$\frac{f\rho}{ft} + \frac{f(\rho u)}{fx} + \frac{f(\rho v)}{fy} + \frac{f(\rho w)}{fz} = 0, \qquad \frac{f\rho}{ft} + \ .(\rho\vec{U}) = 0, \tag{2.4.1}$$

or
$$\frac{f\rho}{ft} + (\vec{U}.\ )\rho + \rho(\ .\vec{U}) = \frac{D\rho}{Dt} + \rho(\ .\vec{U}) = 0 \tag{2.4.2}$$

For incompressible fluid, density is constant and hence its spatial derivative is zero. Further, $\rho$ cannot be zero. Therefore Eq. 2.4.2 reduces to

$$.\ \vec{U}) = 0,$$

or
$$\left(\frac{fu}{fx} + \frac{fv}{fy} + \frac{fw}{fz}\right) = 0 \text{ (Cartesian coordinates)} \tag{2.4.3}$$

Equation 2.4.1 or 2.4.2 is known as *continuity equation* and we note that it is a non-linear equation as it involves product of $\rho$ and velocity component; the velocity field is called the solenoidal field. However, the equation becomes linear for incompressible fluid, as is seen from Equation 2.4.3. For incompressible fluids, mass conservation also means volume conservation. In the same way, the continuity equation is obtained in cylindrical polar and spherical coordinate system. We reproduce these Eq. below:

$$\frac{f\rho}{ft} + \frac{1}{r}\frac{f(\rho r u_r)}{fr} + \frac{f(\rho u_\theta)}{rf\theta} + \frac{f(\rho u_z)}{fz} = 0, \tag{2.4.4}$$

$$\frac{f\rho}{ft} + \frac{1}{r^2}\frac{f(\rho r^2 u_r)}{fr} + \frac{1}{r\sin\theta}\frac{f(\rho u_\theta \sin\theta)}{f\theta} + \frac{1}{r\sin\theta}\frac{f(\rho u_\phi)}{f\phi} = 0 \tag{2.4.5}$$

We shall derive Eq. 2.4.4 to demonstrate how to form the control volume in polar coordinates, which is shown in Fig. 2.4.1 (b). We take the face *FGDC* of height $dz$ such that *CD* is at *a* distance $r$ and *FG* at $r + dr$ from the origin and rotate it anticlockwise about the vertical line passing through the origin by an angle $d\theta$ so as to occupy the position *ABEH*. This generates curvilinear faces *ABCD* and *EFGH* having areas $rd\theta dz$ and $((r + dr)\, d\theta dz$, respectively. Note the difference in the Cartesian and this system; in the former the area of flow remains constant from one face to another in a direction but changes in this system along the $r$-direction. Therefore, in order to find out the mass flow across $r + dr$ face the area term will also be inside the derivative while taking the variation with $r$ in the Taylor series. This introduces the curvilinear term in Eq. 2.4.4. In spherical coordinates, the curvilinear term will appear in $\phi$ direction as well.

**(c) *Streamline approach.***

In fluid mechanics, yet another commonly used coordinate system is the streamline, in which the streamline and a line perpendicular to it are used as coordinates. These are shown as $s$ and $n$ in Fig. 2.4.1 (c). Such coordinate system is sometimes referred to as the natural coordinate system. In this system, velocity along the streamline direction is shown as $u_s$ and perpendicular to it by $u_n$. Let $A$ be the area of the stream tube, then the general continuity equation and the one for steady flow are

$$\frac{f\rho}{ft} + \frac{1}{A}\frac{f(\rho u_s A)}{fs} = 0$$

and $$\rho u_s A = \text{Constant.} \tag{2.4.6}$$

Importance of the continuity equation lies in the fact that it represents mass conservation and hence for any flow to be possible, this equation has to be satisfied. That is, satisfaction of continuity equation by any velocity field is necessary condition.

The problem that arises while using Eq. 2.4.6, is that the velocity is not uniform over the area over which the mass flow rate is to be determined. We define an *average velocity* such that it gives the actual discharge when used in this equation. Figure 2.4.2 shows how to determine it in rectangular (Cartesian) and cylindrical coordinates. *OA* is the plane surface of area *A* of height *H* and width *B* on which the velocity changes and the average velocity is to be determined, Fig. 2.4.2 (a). We choose an elemental area *dy* at

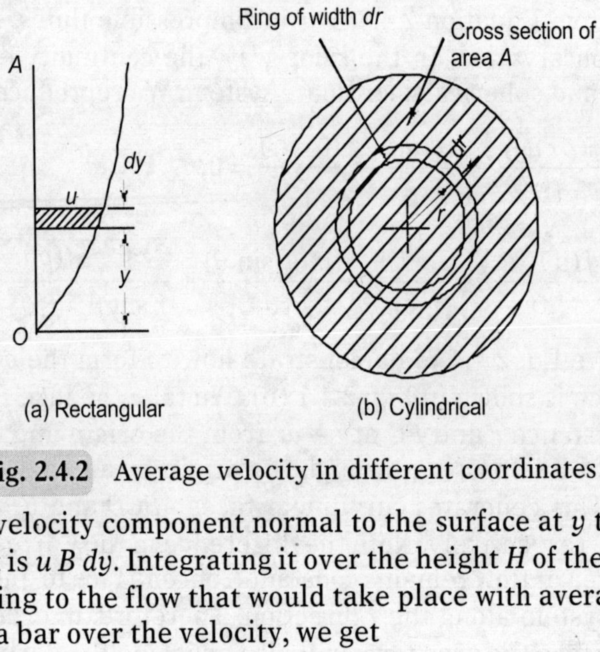


**Fig. 2.4.2** Average velocity in different coordinates

*y* from *O*. If *u* is the velocity component normal to the surface at *y* then the discharge through the element is *u B dy*. Integrating it over the height *H* of the surface (since *B* is constant) and equating to the flow that would take place with average velocity on the plane, indicated by a bar over the velocity, we get

$$\bar{u} = \frac{1}{A}\int_A u\,dA$$

$$= \frac{1}{H}\int_0^H u\,dy \tag{2.4.7a}$$

Similarly, for a circular cross section, we choose a circular ring element of thickness dr at *r* from the centre, as shown in Fig. 2.4.2 (b). This gives the average velocity of the flow through a pipe

$$\bar{u} = \frac{1}{A}\int_0^R 2\pi r u\,dr \tag{2.4.7b}$$

When we determine average velocity, the momentum and the kinetic energy of the mass can also be obtained from this average velocity by using a correction factor called momentum and energy correction factors. These factors are obtained by dividing the actual momentum (or energy) by the momentum (or energy) obtained from average velocity. For example, the momentum of the liquid passing through an element area $dA$ is $\rho U dA \leftrightarrow U = \rho U^2 \, dA$. This, on integration, gives the total momentum. While the momentum on the basis of average velocity is $\rho U_{av}^2 A$. Thus we get equation

$$\alpha \rho A U_{av}^2 = \int_A \rho U^2 \, dA \blacklozenge \alpha \text{ (Momentum correction factor)} = \frac{1}{A} \int_A \left( \frac{U}{U_{av}} \right)^2 dA. \quad (2.4.7c)$$

$$\text{Similarly, } \beta \leftrightarrow \rho A U_{av}^3 = \int_A \rho U^3 \, dA \blacklozenge \beta \text{ (Energy correction factor)} = \frac{1}{A} \int_A \left( \frac{U}{U_{av}} \right)^3 dA. \quad (2.4.7d)$$

You will appreciate the fact that these correction factors had to be introduced because the average velocity is obtained on the basis of mass conservation and not the momentum or energy conservation.

**Example 2.4.1** A pipe diameter changes from 0.5 m to 1.0 m in a length of 1 m. If the discharge through this pipe is $Q$ from the smaller end to the bigger end of the pipe, find out the general expression of the velocity at any section of the pipe. If $Q = 0.785$ m$^3$/s, determine the convective acceleration at 0.5 m from the entrance. Now if the flow increases to 1.570 m$^3$/s in 10 s, find the total acceleration.

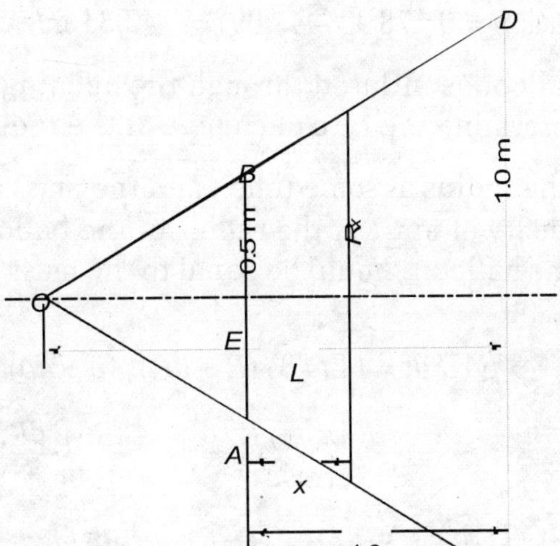


**Fig. E 2.4.1** Flow through a diffuser

**Solution** Figure E 2.4.1 shows the configuration with the various distances. For the triangles $OAB$ and $OCD$, by the law of similar triangles, we may write

$$\frac{L-1}{L} = \frac{0.5}{1.0} \quad \blacklozenge\ L = 2.0 \text{ m and } OE = L - 1 = 1 \text{ m.}$$

The pipe radius $R_x$ at a distance $x$ from the entrance is

$$\frac{R_x}{OE + x} = \frac{0.25}{OE} \quad \blacklozenge\ R_x = 0.25\,(1 + x),$$

$$A \text{ ( flow area)} = \pi\,\{0.25\,(1 + x)\}^2 \text{ and } u = \frac{Q}{A} = \frac{Q}{\pi\,\{0.25\,(1 + x)\}^2}$$

Convective acceleration is given as

$$\alpha_x = u\,\frac{du}{dx}$$

$$= \frac{Q^2}{(0.0625\pi)^2}\,\frac{1}{(1 + x)^2}\left(\frac{-2}{(1 + x)^3}\right).$$

Substituting $x = 0.5$ m, we get acceleration at the point. This gives after simplification, $\alpha_x = -4.209$ m/s$^2$.

When the discharge is increasing the local acceleration would also add to the total acceleration, which is given by $\partial u/\partial t$.

At $\quad x = 0.5,\ R = 0.375\text{m and } A = 0.442 \text{ m}^2.$

$\therefore \quad u_1 = 0.875 / A = 1.776 \text{ and } u_2 = 1.570 / A = 3.552 \text{ m/s.}$

$$\blacklozenge\ \partial u / \partial t = (u_2 - u_1) / 10 = 0.178 \text{ m/s}^2.$$

The total acceleration = 0.178 + (– 4.209) = – 4.031 m/s$^2$.

**Example 2.4.2** A balloon is inflated through the opening of area $A$ with air entering into it at speed $U$. determine the rate of change of the radius of the balloon.

**Solution** Let $R$ be the radius at some time $t$ and the mass added to it in time $dt$ is $\rho AU\,dt$, where $\rho$ is the density of air. Let the radius of the balloon changes to $R + dR$. Then increase in mass of the balloon should be equal to the mass added to it, by conservation of mass. This gives

$$\rho AU\,dt = (4\pi / 3)\,(R + dR)^3\,(\rho + d\rho) - (4\pi / 3)\,R^3\rho$$

$$\cup (4\pi / 3)\,R^3\,d\rho + 4\pi R^2\,\rho\,dR; \text{ neglecting higher order terms.}$$

or, $$4\pi R^2\,\rho dR / dt = \rho AU - (4\pi / 3)\,R^3\,d\rho / dt$$

or, $$\frac{dR}{dt} = \frac{AU}{4\pi R^2} - \frac{R}{3\rho}\,\frac{d\rho}{dt}$$

**Example 2.4.3** Determine the missing velocity component for the possible flow of an incompressible fluid having the other components, $v = ax^3 - by^2 + cz^2$ and $w = bx^3 - cy^2 + az^2x$.

**Solution** A flow is possible only when it satisfies the continuity equation (2.4.3). Substituting the velocity components and simplifying, we get

$$\partial u / \partial x = 2by - 2azx.$$

or,

$$u = \int (fu/fx)\, fx + F(y, z)$$

$$= 2bxy - azx^2 + F(y, z)$$

Note that the integration is with respect to $x$ in which $y$ and $z$ are treated as constant. For this reason we have $F(y, z)$ instead of a constant. This function can be determined if the boundary condition is given and then we shall get a unique velocity field.

### 2.4.1.2 Integral Approach

So far we have discussed the differential approach and expressed the conservation of mass in the differential equation form which are valid at every point in the flow region. For integral approach, we may consider a control volume and determine the net mass efflux from its control surface and equate it to the rate of change of mass of the control volume. This appears to be same as we followed in control volume approach in section 2.4.1.1. However, there is a little difference in the manner of determination of the mass afflux and rate of change of mass of control volume. Consider a control volume, Fig. 2.4.3, with a small area element $d\vec{A}$ on its control surface where velocity is $\vec{U}$. We can determine the mass flow rate across all such area elements, in which the given control surface may be divided, and sum up to give the net mass efflux rate while taking outflow positive and inflow as negative (this comes automatically from our definition of positive directions of area and velocity).

The efflux from the control surface $S$ is $= \iint_S \rho \vec{U}.d\vec{A}$. We may also find out the rate of change of mass of the control volume by dividing the control volume into small volume elements, each of volume $dV$, and sum up their masses to give total mass at a time. Then the rate of accumulation is $= \frac{f}{ft} \iiint_V \rho\, dV$. The mass conservation demands the total rate of change of mass to be zero. This gives the integral form of the continuity equation

$$\frac{f}{ft} \iiint_V \rho\, dV + \iint_S \rho \vec{U}.d\vec{A} = 0 \qquad (2.4.8)$$

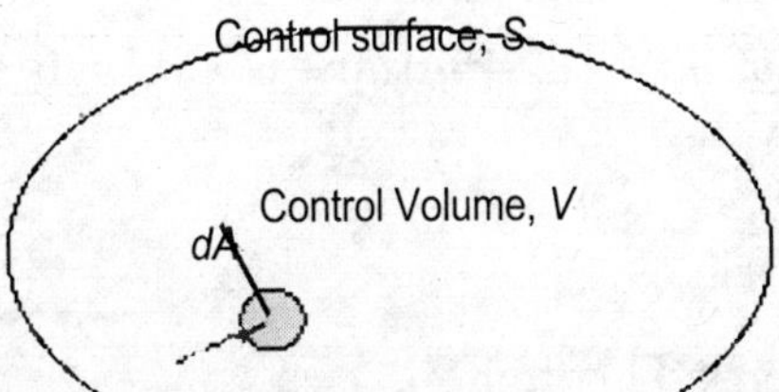


**Fig. 2.4.3** Integral approach

The surface integral in Eq. 2.4.8 can be changed to volume integral using the Gauss divergence theorem. This gives

$$\iiint_V \left[ \frac{f\rho}{ft} + \ .(\rho\vec{U}) \right] dV = 0.$$

or,
$$\frac{f\rho}{ft} + \ .(\rho\vec{U}) = 0,$$

Which is same as in Eq. 2.4.1. This explains that the results is independent of the solution technique followed.

**Example 2.4.4** Water enters a rectangular duct at a rate of 10 m$^3$/s as shown in the figure. Two of the faces of the duct are porous. On the upper face, water is added at a rate shown by parabolic curve, while on the front face water leaves at a rate determined linearly by the distance from the end. The maximum values of both rates are given in the figure. What is the average velocity of water leaving the duct if duct is 1 m long and has a cross sectional area of 0.1 m$^2$?

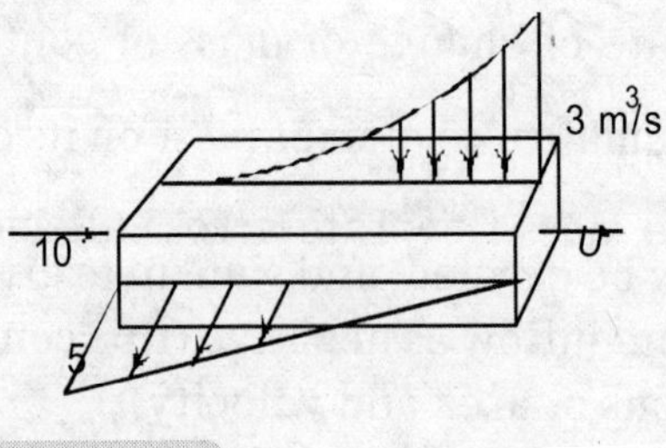


**Fig. E 2.4.4** Mass conservation

**Solution** Let the distance $y$ be measured from the front cross section. For the parabolic curve, we may assume profile to be $q = A + By + Cy^2$, where $A$, $B$ and $C$ are the constant to be determined from the conditions: $y = 0$, $q = dq/dy = 0$ and $y = 1$, $q = 3$ m$^3$/s. This gives $q = 3\,y^2$. Similarly, at the front end the discharge can be represented by the linear equation $q = A + By$ with the conditions that at $y = 0$, $q = 5$ m$^3$/s and at $y = 1$m, $q = 0$. This gives the form $q = 5 - 5y$.

The problem can be solved by using the integral form of the mass balance applied to the control volume made from the interior of the duct of constant cross section and assuming steady conditions so that change in mass of the control volume is zero. Writing the equation, we get

$$-10 - \int_0^1 3y^2\,dy + \int_0^1 (5 - 5y)\,dy + U \leftrightarrow 0.1 = 0$$

$$U = 85 \text{ m / s}.$$ ♦

### 2.4.1.3 Reynolds Transport Theorem

The laws of conservation of mass, momentum and energy are well known in classical physics. These laws were developed for a closed mass and therefore we need to change these for a control volume, which is mostly used in the analysis of flow regions. Consider a control volume at some time $t$, which can also be assumed to be the control mass at that time. This is shown in Fig. 2.4.4 (a) in which the dotted boundary is for the control mass. The control mass moves to a new position at time $t + \delta t$, while the control volume, by definition, is fixed in space, as shown in Fig. 2.4.4 (b). There are three regions II, III and IV; the region III is common to both the control mass and the control volume. That is, the control volume is made of regions II and III and the control mass of the regions III and IV. The control volume is shown displaced from the original position only for clarity in the two situations. In fact, there is no shift of the control volume in the space.

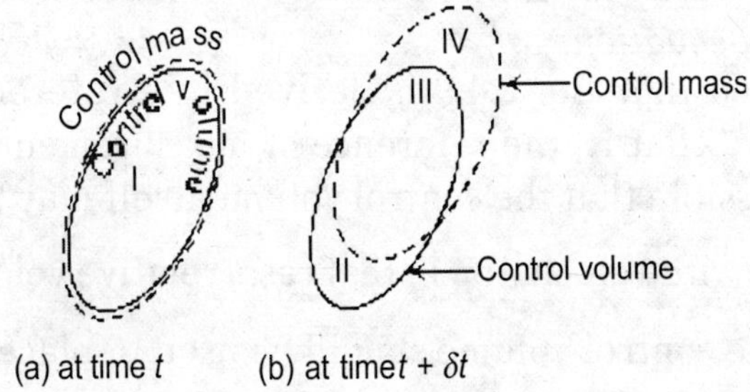


**Fig. 2.4.4** Reynolds transport theorem

Let at any time $t$, $N$ be some property of the control mass for which we want to establish the conservation law in the control volume and $\eta$ be the corresponding specific property throughout the fluid. The increase in property $N$ of the control mass in time $\delta t$ is given by,

$$(N_{t+\delta t} - N_t)_{CM} = \left(\iiint_{III} \eta\rho dV + \iiint_{IV} \eta\rho dV\right)_{t+\delta t} - \left(\iiint_{I} \eta\rho dV\right)_t$$

In this subscript, $CM$ is abbreviation for control mass, $\rho$ is the average density of the fluid in the volume element $dV$, taken in the closed mass, and the integration is over the entire control mass at two different times. We may add and subtract $\left[\iiint_{II} \eta\rho dV\right]_{t+\delta t}$ in the above equation to give

$$\frac{(N_{t+\delta t} - N_t)_{CM}}{\delta t} = \frac{\left[\iiint_{III} \eta\rho dV + \iiint_{II} \eta\rho dV\right]_{t+\delta t} - \left[\iiint_{I} \eta\rho dV\right]_t}{\delta t} + \frac{\left(\iiint_{IV} \eta\rho dV\right)_{t=\delta t}}{\delta t} - \frac{\left(\iiint_{II} \eta\rho dV\right)_{t+\delta t}}{\delta t}$$

Taking its limit as $\delta t$ ♦ 0, the left hand side is the time rate of change of the property $N$ of the control mass. The first term on the right hand side represents the time rate of change of $N$ for the control volume since the two regions II and III at time $t + \delta t$ and region I at time $t$ represent the control volume. The volume integrals over the regions

IV and II are respectively the net efflux and influx of the property from the control surface and can be changed by the Gauss divergence theorem as dot product of the velocity and the surface area at the locations of the mass flow across the surface. Therefore the above equation in the limit $\delta t$ ♦ 0 changes to

$$\left(\frac{dN}{dt}\right)_{CM} = \frac{f}{ft}\iiint_{CV} \eta\rho dV + \iint_{CS} \eta\rho(\vec{U}.d\vec{A}) \qquad (2.4.9)$$

Equation 2.4.9 shows that the time rate of change of property of the control mass is equal to sum of the time rate of change of property of the control volume and the net efflux from its control surface. This property may be mass, momentum or energy, the three quantities for which the law of conservation holds true. This equation is known as *Reynolds transport equation.*

It should be noted that Eq. 2.4.9 is derived in a reference in which the control volume is stationary. That is, the reference of measurement is fixed with the control volume. It is also possible that the control volume itself may be moving with reference to the inertial frame with velocity $U_c$. In that case, relative velocity of flow, $\vec{U}_r = \vec{U} - \vec{U}_c$, with reference to the control volume should be used in place of $\vec{U}$.

### 2.4.1.4 Application of Reynolds Transport Theorem

We shall examine the Reynolds theorem for the three basic laws of conservation of mass, momentum and energy. We may take $\eta$ in Eq. 2.4.9 as 1, $U$ or $U^2$, respectively for specific mass, specific momentum and specific energy.

*Conservation of mass.* Recall that time rate of change of mass of the control mass by its definition is zero. Therefore the law of conservation of mass according to Eq. 2.4.9 is

$$\left(\frac{dN}{dt}\right)_{CM} = \frac{f}{ft}\iiint_{CV} \rho dV + \iint_{CS} \rho(\vec{U}\,.\,d\vec{A}) = 0. \qquad (2.4.10)$$

or, $$\iiint_V \left(\frac{f\rho}{ft} + \quad .(\rho\vec{U})\right) = 0, \text{ (using Gauss divergence theorem)}$$

$$\iiint_V \left(\frac{f\rho}{ft} + \vec{U}\,.\quad \rho + \rho \quad .\vec{U}\right) = 0$$

$$♦ \qquad \frac{D\rho}{Dt} + \rho \quad .\vec{U} = 0 \qquad (2.4.1)$$

Since the integral can be zero only when the integrand is zero everywhere, we note that we get the same continuity equation as obtained by taking the limit shrinking the chosen control volume to a point.

*Conservation of momentum.* We may consider next the law of conservation of momentum in the integral form. The momentum may be linear or angular. The linear momentum is defined as the product of mass and velocity and hence $N$ is $m\vec{U}$ and $\eta$ is

$\vec{U}$, whereas the angular momentum is the moment of the linear momentum about a point or the axis. We shall consider in this book the moment of momentum about an axis, which is called the angular momentum. This is equal to the product of the linear momentum vector acting upon the control surface and the minimum distance between the time of action of the linear momentum and the axis about which the external torque acts on the control volume. The surface force is known as traction force.

Further, we know from the Newton's second law of motion that the rate of change of momentum is equal to the sum of the external forces acting on the control mass and the rate of change of angular momentum is sum of the external torque. This gives the left hand side of Eq. 2.4.9. The external forces are further divided as surface and the body forces because of the nature of the forces. The surface forces are the one, which act upon the surface whereas the body force is due to presence of external field such as gravity, magnetic field, etc. The surface forces are proportional to the surface area whereas the body forces are proportional to the volume of the control volume.

Substituting $N$ and $\eta$ in Eq. 2.4.9, we get the integral form for the linear and angular momentum as

$$\vec{F} = \left(\frac{dm\vec{U}}{dt}\right)_{CM}$$

$$= \frac{f}{ft}\iiint_{CV}\vec{U}\rho\, dV + \iint_{Cs}\vec{U}\rho(\vec{U}\,.\,d\vec{A}) \tag{2.4.11}$$

and

$$\tau = \frac{f}{ft}\iiint_{CV}(\vec{r} \leftarrow \vec{U})\rho dV + \iint_{Cs}(\vec{r} \leftarrow \vec{U})\rho(\vec{U}\,.\,d\vec{A}) \tag{2.4.12}$$

Here $\vec{r}$ is the position vector of the point at which linear momentum is considered from the origin of the axis about which the angular momentum is considered and $\tau$ is the sum of the external torque on the control volume. The traction force is due to the stresses at the control surface of the volume element $dv$ and may be split into the pressure and shear stresses. The pressure $p$ results in a compressive force directing into the fluid and the shear forces result in the force $\vec{F}_{\text{viscous}}$. If $\overset{\blacklozenge}{b}$ represents the net body force per unit mass, then

$$\vec{F} = \iiint_{V}\rho\vec{b}dV - \iint_{S}pd\vec{S} + \vec{F}_{\text{viscous}}$$

Recall that $U$ is the relative velocity of the flow with respect to the control volume and there is no problem if the control volume is inertial or moving with a constant velocity. However, if it is not so then the left hand side of Eq. 2.4.11 for constant mass can be written as

$$\frac{d(m\vec{U}_r)}{dt} = m\frac{d(\vec{U}_r)}{dt}$$

$$= \frac{d(\vec{U} - \vec{U}_c)}{dt}$$

$$= m \frac{d\vec{U}}{dt} - m\vec{\alpha}_c$$

$$= \vec{F} - m\vec{\alpha}_c \quad (2.4.13)$$

where $\vec{\alpha}_c$ is the rectilinear acceleration of the control volume with reference to the fixed coordinate system. We shall consider its application later in the chapter on the theory of jets. When $m$ is also changing then it cannot be taken out of the differential operator, as we did in the above derivation.

**Example 2.4.5** A vane with a turning angle of $\theta$ can move a cart in the same direction as that of the jet of area $A$ moving with the velocity $U$ that strikes the vane, as shown in Fig. E2.4.5. Assume the cart to be stationary and as the jet strikes it, the cart moves at a velocity $u$. Find the expression for $u$ $(t)$ if the cart weighs $W$ $N$. Neglect the friction.

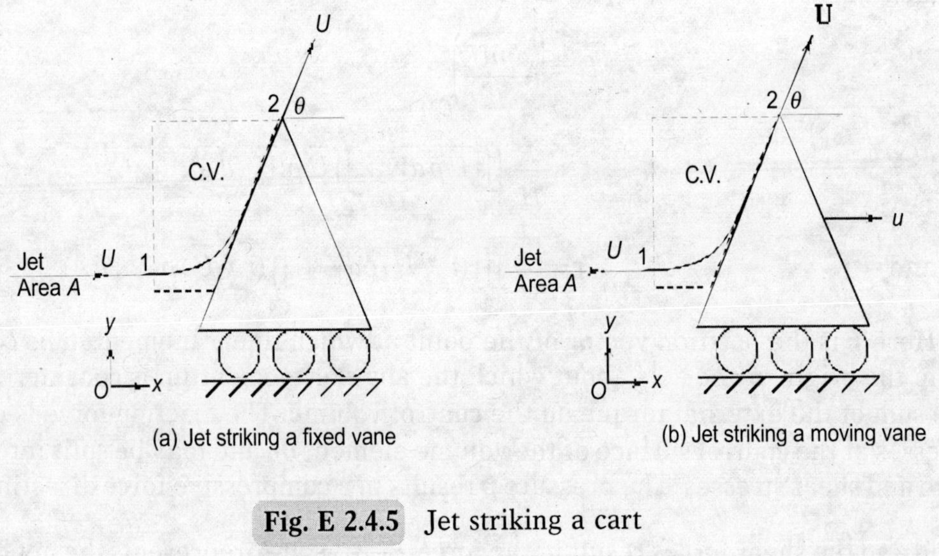


**Fig. E 2.4.5** Jet striking a cart

**Solution** To find out the net force on the cart because of striking of the jet, we may use integral momentum equation. As discussed, it is better to fix the reference axes with the cart so that the fluid appears to come and leave with speed $(U - u)$. Writing it for $x$-direction, we get the force on the cart from Eq. 2.4.11 along $x$-direction. Note $\partial / \partial t$ term is zero, since in the flow with reference axes fixed to vane is steady and $U = U - u$, the control volume shown in Fig. 2.4.4 (b) whose surface is crossed by the fluid at sections 1 and 2, respectively; velocity at section 2 along $x$-axis is $(U - u) \cos \theta$.

$$F_x = A\rho\,(U - u)\,\{(U - u) - (\mathrm{U - u}) \cos \theta\} = A\rho\,(U - u)^2\,(1 - \cos \theta)$$

Invoking the Newton's second law of motion, for cart of weight $W$ Newton, we can write

$$(W/g)\, du\,/\,dt = A\rho\,(U - u)^2\,(1 - \cos \theta)$$

$$\blacklozenge \quad \frac{1}{U - u} = \frac{A\rho g}{W}(1 - \cos\theta) + C$$

where $C$ is the constant of integration. But we know the initial condition that at $t = 0$, $u = 0$, (given). This gives $C = 1/U$. Substituting for $C$, we get

$$\frac{u}{U} = \frac{BUt}{1 + BUt}, \text{ where } B = \frac{A\rho g}{W}(1 - \cos\theta)$$

It can be seen that as $t$ ♦ ×, $u$ ♦ $U$.

**Example 2.4.6** A jet of water strikes a flat plate inclined to it. Work out the distribution of discharge over the plate and the force of impact on it, if (i) the plate is stationary and (ii) the plate is moving with velocity $u$ in the direction of the jet. What happens when the plate is perpendicular to the jet?

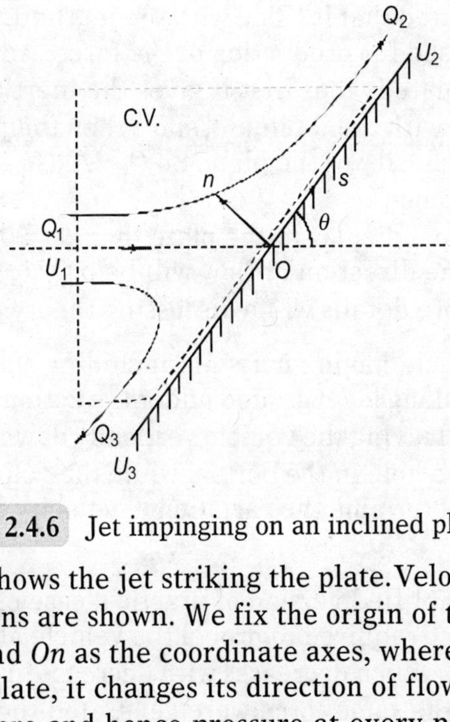


**Fig. E 2.4.6** Jet impinging on an inclined plane

**Solution** Figure E 2.4.6 shows the jet striking the plate. Velocities and the discharge at the inlet and exit sections are shown. We fix the origin of the coordinate system at the point $O$ and take $Os$ and $On$ as the coordinate axes, where $Os$ is along the surface. When the jet strikes the plate, it changes its direction of flow as shown in the figure. Jet is flowing in atmosphere and hence pressure at every point in the jet would be atmospheric. We further assume fluid to be inviscid so that the frictional force between the surface and the fluid is zero. Application of the Bernoulli equation, according to which sum of the potential, pressure and kinetic head is constant, at the two points 1 and 2 or 1 and 3 gives

$$U_2 = U_3 = U_1,$$

neglecting the gravity effect. Since the fluid is inviscid, the force along the s direction will be zero. Taking the control volume as shown by the dotted lines, we can write the mass conservation as well as the integral momentum equation along $s$ and $n$ direction as

$$\rho Q_1 = \rho Q_2 + \rho Q_3 \qquad \blacklozenge \quad Q_1 = Q_2 + Q_3 \tag{A}$$

$$F_s = (\rho\, Q_2\, U_2 - \rho\, Q_3 U_3) - \rho\, Q_1 U_1 \cos\theta = 0$$

♦ $$Q_2 - Q_3 = Q_1 \cos\theta \tag{B}$$

$$F_n = \rho\, Q_1\, U_1 \sin\theta \tag{C}$$

We get Equation (B) because $U_1 = U_2 = U_3$ and $Fs$ is zero, as mentioned earlier; negative sign in the term $\rho\, Q_3 U_3$ is because $U_3$ is in the negative $s$ direction. Solving the two equations, we get

$$Q_2 = Q_1\,(1 + \cos\theta)\,/\,2 \qquad \text{and} \qquad Q_3 = Q_1\,(1 - \cos\theta)\,/\,2$$

Equation (C) gives the force acting on the control volume normal to the surface and hence by Newton's third law a force acts on the surface in the opposite direction. This force can be resolved in the direction of the jet that would cause the plate to move. The component of the force is $\rho Q_1 U_1 \sin^2\theta$. If the plate moves in the direction of the jet then additional space is created that is filled with the jet fluid. This amount does not hit the surface and is not utilised for production of the force. Advantage of fixing the coordinate system with the plate is that first it gives the inertial frame and second it takes care of the space filled with the jet liquid, since then the jet will have relative velocity $U_1 - u$, where $u$ is the velocity of the plate. So $Q_1 = A_1\,(U_1 - u)$ and not $A_1\, U_1$, where $A_1$ is the jet area at the entrance.

When the jet is perpendicular to the plate then $\theta = 90^0$ and $Q_2 = Q_3 = Q_1/2$. Also the force on the plate in the direction of flow will be $\rho\, Q_1 U_1$, which is maximum. We shall study this aspect in more details while discussing theory of fluid machines later.

**Example 2.4.7** A vehicle having mass $M_0$, including that of the fluid it is carrying, moves at a velocity $U_0$ along level ground under the action of a constant force $F$. At time $t = 0$, fluid mass begins leaving the vehicle vertically downwards at a constant rate of $m$ per unit time, through a hole in the bottom while the vehicle continues to move under the action of force $F$. Determine the variation of velocity of the vehicle as a function of time.

**Solution** This is a simplified version of practical case of vehicles and aeroplanes. In both the cases, the force required to propel the vehicle at some velocity $U$ depends on the mass of the vehicle, which decreases with decrease in the mass due to consumption of the fuel. Or else if the same force were generated then the speed would increase. Moreover, the liquid in the given question is falling vertically down and hence does not contribute to any force in the x-direction, the direction of the motion of the vehicle.

Mass at any time $t$ is $M = (M_0 - mt)$.

According to Newton's second law of motion, $F = M\,(dU/dt)$. Substituting for $M$ and integrating the resultant equation, we get

$$U = \frac{F}{m} \ln(M_0 - mt) + C$$

where $C$ is constant of integration to be obtained from the initial condition $t = 0$, $U = U_0$.

$$\therefore \qquad U = \frac{F}{m} \ln \frac{M_0}{(M_0 - mt)} + U_0$$

## 2.4.2 Conservation of Momentum

The conservation of mass gives one equation in ***U*** as mass is scalar quantity and hence cannot determine the velocity field ***U*** that requires three equations. Therefore we need some way to generate the other two equations. Neverthless this principle at least imposes restrains that all the velocity components cannot vary independently. Moreover, when the flow is steady mass leaving the control volume has to enter also into the control volume from some other portion of the control surface.

The other equations are obtained from the conservation of momentum principle. We can change Eq. 2.4.11 into the differential form, as we did for the mass conservation equation by using the divergence theorem to change the surface integral to the volume integral. The procedure would be explained for the *x*-momentum equation that can then be extended similarly for other directions too. Let us rewrite the *x*-component of the equation, which is

$$\vec{F}_x = \iiint_{CV} \rho \vec{b}_x \, dV - \iint_{CS} (p d\vec{A})_x + (\vec{F}_x)_{\text{viscous}}$$

$$= \frac{f}{ft} \iiint_{CV} u \rho \, dV + \iint_{CS} u\rho(\vec{U} \,.\, d\vec{A}) \text{ R.H.S of Equation 2.4.11}$$

But $\qquad \iint_{CS} u\rho(\vec{U} \,.\, d\vec{A}) = \iint_{CS} (\rho u \vec{U}).d\vec{A} = \iiint_{CV} \;.(\rho u \vec{U}) \, dV$

and $\qquad \iint_{CS} (p d\vec{A})_x = \iiint_{CV} (fp/fx) dV$

$$\therefore \qquad \iiint \left[ \frac{f(\rho u)}{ft} + \;.(\rho u \vec{U}) + \frac{fp}{fx} - \rho b_x - (\vec{F}_x)_{\text{viscous}} \right] = 0$$

or, $\qquad \frac{f(\rho u)}{ft} + \;.(\rho u \vec{U}) + \frac{fp}{fx} - \rho b_x - (\vec{F}_x)_{\text{viscous}} = 0.$

We shall now derive the momentum equation using the classical mathematics. We may select volume elements such as shown in Fig. 2.4.1 (a) and consider variation of pressure from one face to another in the same way as we did earlier for velocity components by using the Taylor series. Here we are talking about pressure because shear stress does not exist in ideal fluids. The resultant force on this volume element, in this process, will be

$$-\left( i \frac{fp}{fx} + j \frac{fp}{fy} + k \frac{fp}{fz} \right) dV = - \;\; p \, dV \qquad (2.4.14)$$

The negative sign comes because the pressure acts into the control volume. Let the body force per unit volume be **B** having components $B_x$, $B_y$ and $B_z$, respectively. Normally we consider the gravity as the external field but there are many applications when other fields, such as magnetic or electric field, are also present.

### 2.4.2.1 Euler Equation

Let us now determine the rate of momentum influx to the control volume along $x$, $y$ and $z$ directions. The control volume of size $dx, dy$ and $dz$ along the $x, y$ and $z$-direction of the Cartesian coordinate system is shown in Fig. 2.4.5.[10] The rate of mass flow across $x$ face due to $x$-velocity component $u$ is $\rho udydz$ and the momentum influx rate is $\rho u^2 dydz$. Similarly, the rate of momentum influx due to velocity components $v$ and $z$ with this mass flow rate entering the $x$ face is $\rho uvdydz$ and $\rho uwdydz$, respectively. By Taylor series, the corresponding rate of efflux from the face $x + dx$ can easily be written, which is shown in Fig. 2.4.5. We have not shown the momentum influx and efflux in the figure.

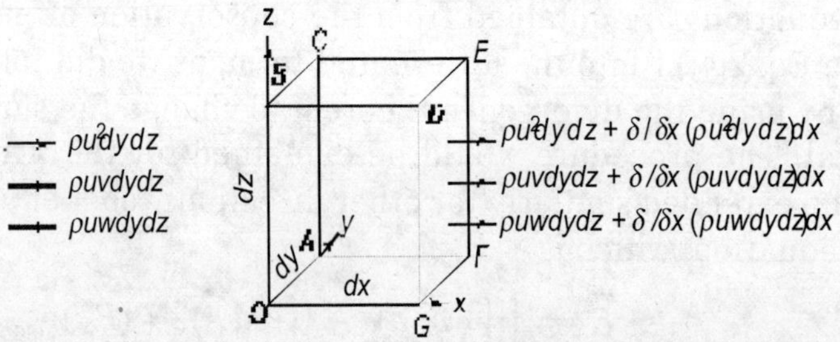


**Fig. 2.4.5** Control volume for derivation of Euler's equation

along the other directions for maintaining the clarity in the figure. Similarly, we may consider the $x$-momentum efflux from the other faces of the control volume. Remember, the velocity component perpendicular to the face at which the calculation is being carried out will appear in the mass flow rate. That is, the mass and the $x$-momentum flow rate across the $y$ face is $\rho vdxdz$ and $(\rho vdxdz)u$, respectively. Hence, the net rate of $x$-momentum efflux from the control volume becomes

$$\left[\frac{f(\rho u^2)}{fx}+\frac{f(\rho uv)}{fy}+\frac{f(\rho uw)}{fx}\right]dV=\left[u\left\{\frac{f(\rho u)}{fx}+\frac{f(\rho v)}{fy}+\frac{f(\rho w)}{fz}\right\}+\rho\left\{u\frac{fu}{fx}+v\frac{fu}{fy}+w\frac{fu}{fz}\right\}\right]dV$$

Similarly the net momentum efflux rate across the control surface in other directions may be obtained from the above equation by changing the variables in cyclic order. Lastly, the time rate of increase in $x$-momentum of the control volume can be written as $\frac{f}{ft}(\rho udV)=\frac{f}{ft}(\rho u)dV$, since $dV$ does not change with time and can be written outside the differential operator. Substituting these in Eq. 2.4.11, we get

$$\frac{f(\rho u)}{ft}+u\left\{\frac{f(\rho u)}{fx}+\frac{f(\rho v)}{fy}+\frac{f(\rho w)}{fz}\right\}+\rho\left\{u\frac{fu}{fx}+v\frac{fu}{fy}+w\frac{fu}{fz}\right\}=-\frac{fp}{fx}+\rho B_x$$

$$\text{or, } u\left\{\frac{f\rho}{ft}+\frac{f(\rho u)}{fx}+\frac{f(\rho v)}{fy}+\frac{f(\rho w)}{fz}\right\}+\rho\left\{\frac{fu}{ft}+u\frac{fu}{fx}+v\frac{fu}{fy}+w\frac{fu}{fz}\right\}=-\frac{fp}{fx}+\rho B_x$$

[10] Note that the coordinate axes chosen are not the same as in Fig. 2.4.1. This has been done intentionally to focus on the fact that these can be chosen in any way we like keeping in mind the left handed or right handed system of coordinates. In this figure we have adopted the RHS.

$$\text{or, } \frac{fu}{ft} + u\frac{fu}{fx} + v\frac{fu}{fy} + w\frac{fu}{fz} = -\frac{1}{\rho}\frac{fp}{fx} + B_x, \tag{2.4.14}$$

Since $\frac{f\rho}{ft} + \frac{f(\rho u)}{fx} + \frac{f(\rho v)}{fy} + \frac{f(\rho w)}{fz}$ is zero by the continuity equations. We may similarly get the momentum equation for $y$ and $z$ directions. The resulting equations in different coordinate systems are:

Cartesian cooriditnates:

$$\begin{aligned} \partial u / \partial t + u\partial u / \partial x + v\partial u / \partial y + w\partial u / \partial z &= -(1/\rho)\, \partial p / \partial x + B_x \\ \partial v / \partial t + u\partial v / \partial x + v\partial v / \partial y + w\partial v / \partial z &= -(1/\rho)\, \partial p / \partial y + B_y \\ \partial w / \partial t + u\partial w / \partial x + v\partial w / \partial y + w\partial w / \partial z &= -(1/\rho)\, \partial p / \partial z + B_z \end{aligned} \tag{2.4.15}$$

Cylinderical Coordinates:

$$\partial u_r / \partial t + u_r \partial u_r / \partial r + (u_\theta / r)\, \partial u / \partial \theta + u_z\, \partial u_r / \partial z - u_\theta^2 / r = -(1/\rho)\, \partial p / \partial r + B_r$$

$$\partial u_\theta / \partial t + u_r \partial u_\theta / \partial r + (u_\theta / r)\, \partial u_\theta / \partial \theta + u_z\, \partial u_\theta / \partial z + u_r u_\theta / r = -(1/\rho)\, \partial p / \partial(r\theta) + B_\theta \tag{2.4.16}$$

$$\partial u_z / \partial t + u_r \partial u_z / \partial r + (u_\theta / r)\, \partial u_z / \partial \theta + u_z\, \partial u_z / \partial z = -(1/\rho)\, \partial p / \partial z + B_y$$

These equations were obtained by Euler and are named after his name. There are now four equations. However, in the process one more unknown, the pressure is also added and the number of equations are equal to the number of unknowns. These equations can be solved to give velocity and the pressure field.

### 2.4.2.2 Navier–Stokes Equation

Euler's equation were obtained for inviscid fluid. But no fluid is inviscid and has some viscosity. Therefore the shear stresses will also act on the control surface and all the three stresses, namely $\tau_{xx}$, $\tau_{yx}$, and $\tau_{zx}$ acting on the $x, y$ and $z$ faces along the $x$ direction will add to the surface force in $x$ direction. Equation 1.2.10 defines these stresses in terms of the velocity gradients. These stresses are reproduced below in the expanded form:

$$\tau_{xx} = -p + 2\mu\frac{fu}{fx} - \frac{2}{3}\mu\left[\frac{fu}{fx} + \frac{fv}{fy} + \frac{fw}{fz}\right],$$

$$\tau_{yy} = -p + 2\mu\frac{fv}{fy} - \frac{2}{3}\mu\left[\frac{fu}{fx} + \frac{fv}{fy} + \frac{fw}{fz}\right],$$

$$\tau_{zz} = -p + 2\mu\frac{fw}{fz} - \frac{2}{3}\mu\left[\frac{fu}{fx} + \frac{fv}{fy} + \frac{fw}{fz}\right] \tag{2.4.17}$$

$$\tau_{xy} = \tau_{yx} = \mu\left(\frac{fu}{fy} + \frac{fv}{fx}\right),$$

$$\tau_{xz} = \tau_{zx} = \mu\left(\frac{fu}{fz} + \frac{fw}{fx}\right),$$

and

$$\tau_{yz} = \tau_{zy} = \mu\left(\frac{fv}{fz} + \frac{fw}{fy}\right)$$

Using Taylor series expansion, we can find out the stresses at the remaining three faces. Thus the total surface force in $x$-direction is

$$dF_{s,x} = \left(\tau_{xx} + \frac{f\tau_{xx}}{fx}dx - \tau_{xx}\right) dy\,dz + \left(\tau_{yx} + \frac{f\tau_{yx}}{fy}dy - \tau_{yx}\right) dx\,dz + \left(\tau_{zx} + \frac{f\tau_{zx}}{fz}dz - \tau_{zx}\right) dx\,dy$$

$$= \left(\frac{f\tau_{xx}}{fx} + \frac{f\tau_{yx}}{fy} + \frac{f\tau_{zx}}{fz}\right) dV \qquad (2.4.18)$$

We may now substitute for stresses from Eq. 2.4.17 in Eq. 2.4.18 to give the surface force in $x$-direction. Similarly, we may get the surface force in $y$ and $z$ directions as well. Rest of the terms of Eq. 2.4.11 were obtained while deriving Euler's equation. Substituting all these values in Eq. 2.4.11 and simplifying the terms, we get

$$\frac{Du}{Dt} = B_x - \frac{1}{\rho}\frac{fp}{fx} + \frac{f}{fx}\left[v\left(2\frac{fu}{fx} - \frac{2}{3}\;.\vec{U}\right)\right] + \frac{f}{fy}\left[v\left(\frac{fu}{fy} + \frac{fv}{fx}\right)\right] + \frac{f}{fz}\left[v\left(\frac{fw}{fx} + \frac{fu}{fz}\right)\right]$$

$$\frac{Dv}{Dt} = B_y - \frac{1}{\rho}\frac{fp}{fy} + \frac{f}{fy}\left[v\left(2\frac{fv}{fy} - \frac{2}{3}\;.\vec{U}\right)\right] + \frac{f}{fx}\left[v\left(\frac{fu}{fy} + \frac{fv}{fx}\right)\right] + \frac{f}{fz}\left[v\left(\frac{fw}{fy} + \frac{fv}{fz}\right)\right] \qquad (2.4.19)$$

$$\frac{Dw}{Dt} = B_z - \frac{1}{\rho}\frac{fp}{fz} + \frac{f}{fz}\left[v\left(2\frac{fw}{fz} - \frac{2}{3}\;.\vec{U}\right)\right] + \frac{f}{fx}\left[v\left(\frac{fu}{fz} + \frac{fw}{fx}\right)\right] + \frac{f}{fy}\left[v\left(\frac{fw}{fy} + \frac{fv}{fz}\right)\right]$$

These equations are called Navier–Stokes Equation. For an incompressible fluid, use of continuity equation $\;.\vec{U} = 0$ gives, on simplification

$$\partial u/\partial t + u\partial u/\partial x + v\partial u/\partial y + w\partial u/\partial z = -(1/\rho)\,\partial p/\partial x + B_x + \nu\;^2u$$

$$\partial v/\partial t + u\partial v/\partial x + v\partial v/\partial y + w\partial v/\partial z = -(1/\rho)\,\partial p/\partial y + B_y + \nu\;^2v \qquad (2.4.20)$$

$$\partial w/\partial t + u\partial w/\partial x + v\partial w/\partial y + w\partial w/\partial z = -(1/\rho)\,\partial p/\partial z + B_z + \nu\;^2w$$

We observe that the N–S equations are non-linear, partial differential equation of order 2. In vector form this equation for incompressible fluid is written as

$$\frac{f\vec{U}}{ft} + \left(\vec{U}\;.\;\right)\vec{U} = -\;(p/\rho) + \vec{B} + \nu\;^2\vec{U} \qquad (2.4.21)$$

Here **B** is the body force per unit mass. If gravity is the only body force acting on the fluid region then the body force per unit mass will act in the vertical down direction and can be expressed at any point $\vec{r}$ as a gradient of the scalar quantity $g$, the acceleration due to gravity, as $\vec{B} = -\;(\vec{g}.\vec{r}) = -gz$, if $z$-axis is chosen along the vertical up-direction. This can be verified by expanding the vector equation in Cartesian coordinates.

### 2.4.2.3 Vorticity Equation

We may get rid of pressure and body force terms in Eq. 2.4.21 by taking its curl, since *curl of gradient of a scalar quantity is zero*. Moreover, we have learnt that curl of velocity vector is the vorticity and therefore such operation will give an equation in vorticity, whose importance we shall discuss later. Choosing *z*-direction along the vertical up-direction and using the vector identity A.6, as given in Appendix A, the convective acceleration terms can be changed to give

$$(\vec{U}.\ )\vec{U} = \ (U^2/2) - \vec{U} \leftrightarrow(\ \leftrightarrow\vec{U})$$

Substituting it in Eq. 2.4.21, we get

$$\frac{f\vec{U}}{ft} - \vec{U}\leftrightarrow\left(\ \leftrightarrow\vec{U}\right) = -\ (U^2/2) - \ (p/\rho) + \vec{B} + \nu\ {}^2\vec{U}$$

$$= -\ (U^2/2 + p/\rho + gz) + \nu\ {}^2\vec{U} \qquad (2.4.22)$$

When the fluid is incompressible and inviscid and the flow is irrotational then we note from this equation that sum of the kinetic, pressure and potential head is constant, which is the Bernoulli equation. We know that curl of vector product of two vectors **A** and **B** is given by

$$\leftrightarrow\left(\vec{A}\leftrightarrow\vec{B}\right) = \vec{A}\left(\ .\vec{B}\right) + \left(\vec{B}.\ \right)\vec{A} - \vec{B}\left(\ .\vec{A}\right) - \left(\vec{A}.\ \right)\vec{B},$$

Taking $\mathbf{A} = \vec{U}, \mathbf{B} = \ \leftrightarrow\vec{U}$, and the continuity equation $\ .\vec{U} = 0$

$$\leftrightarrow\left[\vec{U}\leftrightarrow\left(\ \leftrightarrow\vec{U}\right)\right] = \vec{U}\left[\ .\left(\ \leftrightarrow\vec{U}\right)\right] + \left[\left(\ \leftrightarrow\vec{U}\right).\ \right]\vec{U} - \left(\ \leftrightarrow\vec{U}\right)\left(\ .\vec{U}\right) - \left(\vec{U}.\ \right)\left(\ \leftrightarrow\vec{U}\right)$$

$$= \left[\left(\ \leftrightarrow\vec{U}\right).\ \right]\vec{U} - \left(\vec{U}.\ \right)\left(\ \leftrightarrow\vec{U}\right) = \left(\vec{\omega}.\ \right)\vec{U} - \left(\vec{U}.\ \right)\vec{\omega} \qquad (2.4.23)$$

Taking curl of Eq. 2.4.22 and using Eq. 2.4.23, we get

$$\frac{f\dot{\omega}}{ft} - \left(\dot{\omega}.\ \right)\vec{U} + \left(\vec{U}.\ \right)\dot{\omega} = \nu\ {}^2\dot{\omega} \qquad (2.4.24)$$

We may combine the first and the third term to give total derivative of $\omega$ and take the second term to RHS to give

$$\frac{f\vec{\omega}}{ft} + \left(\vec{U}.\ \right)\dot{\omega} = \left(\vec{\omega}.\ \right)\vec{U} + \nu\ {}^2\dot{\omega} \qquad (2.4.25)$$

Equation 2.2.25 is called the *vorticity transport equation*. If the fluid is initially irrotational and is inviscid then we note from Eq. 2.4.24 that $D\omega/Dt = 0$. That is, the rotation would remain zero thereafter. This result has been obtained for incompressible fluid. For perfect gas the results also hold true when density and pressure gradients are in the same direction. This means that if the flow was initially irrotational then it would remain so at any other time. For viscous fluids, the flow initially irrotational would change to rotational flow because of the velocity gradients near the boundaries. This implies that the vorticity is diffused from the boundaries to the main flow.

**Example 2.4.8** For the study of a free vortex flow that occurs in washbasin, bathtubs, etc. determine the equation for water surface profile. Assume no acceleration in $\theta$ and z direction and in $r$ direction the acceleration is $u_\theta^2 r$, where $u_\theta$ is the velocity in $\theta$ direction at a point $r$ distance from the centre.

**Solution** We shall use the cylindrical polar coordinates for this problem. The momentum equation reduces to the following forms for this flow.

$$-\frac{fp}{fr} = -\rho\frac{u_\theta^2}{r}, \quad -\frac{1}{r}\frac{fp}{f\theta} = 0 \text{ and } \quad -\frac{fp}{fz} = w.$$

The second equation shows that $p$ is not function of $\theta$, i.e. $p = p(r, z)$. Integration of the third equation gives

$$p = -wz + f(r) \blacklozenge \ \partial p/\partial r = df/dr$$

But $\partial p/\partial r = u_\theta^2/r$ (given)

$$\therefore \qquad \frac{df}{dr} = \rho\frac{u_\theta^2}{r} \qquad \text{(A)}$$

Since the irrotational motion of the vortex is maintained without any external torque, the rate of change of momentum is zero (momentum conservation principle). i.e.,

$$\frac{f}{ft}(m\, u_\theta\, r) = 0. \qquad \text{or, } u_\theta\, r = C, \text{ a constant.}$$

Substituting for $u_\theta$ in Equation (A) and integrating it, we get:

$$f = -\frac{\rho C^2}{2r^2} + B,$$

where $B$ is a constant of integration. It is obtained from the condition that at $z = 0$ and as $r \blacklozenge \times, p = p_0$. This gives, $B = p_0$.

$$\therefore \qquad p = -wz - \frac{\rho C^2}{2r^2} + B.$$

For free surface, $p = 0$. we thus get

$$0 = -wz - \frac{\rho C^2}{2r^2} + p_0.$$

or,

$$z = z_0 - \frac{\rho C^2}{2gr^2}, \text{ where } z_0 = p_0/\rho.$$

This is the equation for the free surface profile. We notice that as $r \blacklozenge 0$: $u_\theta \blacklozenge \times$ and $z \blacklozenge -\times$. Obviously, this analysis holds true only in the region away from the origin.

**Example 2.4.9** A plate of mass $M$ is subjected to two equal jets of area of cross section $A$ and moving in opposite directions with the velocity $U_0$, as shown in Fig. E 2.4.9. If at time $t = 0$, the plate is given initial velocity $u_0$, determine its subsequent displacement as a function of time. Assume there is no gravity force acting on the plate.

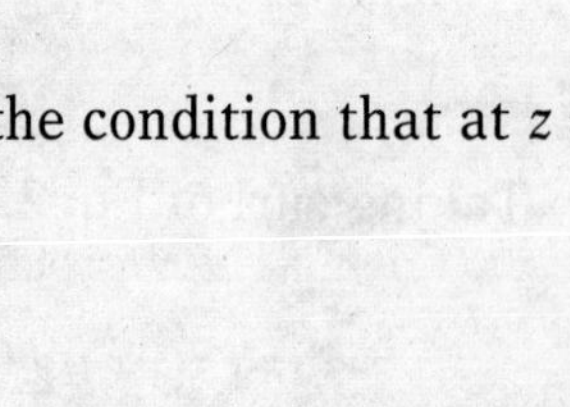


**Fig. E 2.4.9** Two jets striking plate

**Solution** Since the plate is given initial velocity in the + $x$-direction, it will start moving. Let it be at some distance $x$

from the initial position when its velocity is $u$. We fix the reference frame with the plate for simplicity in calculations. Then the force in the $x$ direction due to the jet 1 is $\rho A(U_0 - u)^2$ and that due to the jet 2 is $-\rho A(U_0 + u)^2$; the negative sign with second term is because of the opposite direction of the jet 2. The resultant force is $\rho A\,(U_0 - u^2) - \rho A\,(U_0 + u)^2 = -\,4\rho AU_0 u$. Using Newton's second law of motion

$$du / dt = -\,4\rho AU_0 u / M$$

and
$$u = \frac{dx}{dt} = U_0 \exp\left(-\frac{4\rho AU_0}{M}t\right).$$

Integrating the equation and using the initial condition at $t = 0$, $x = 0$, we get

$$x = \frac{M}{4A\rho}\left[1 - \exp\left(-\frac{4\rho AU_0}{M}t\right)\right].$$

**Example 2.4.10** A small rocket, which has initial weight of 2000 *N*, is to be launched vertically. The rocket consumes 4.0 kg/s of fuel and exhausts it at atmospheric pressure with a velocity of 1000 m/s relative to the rocket. Determine the initial acceleration of the rocket and its velocity after 10 seconds. Assume air resistance to be negligible.

**Solution** Let the mass of the rocket at any time be $M = M_0 - mt$, where $M_0$ is the initial mass of the rocket and $m$ is the rate of the fuel consumption. Since gases are exhausted at atmospheric pressure, there is no pressure thrust. Moreover, the rocket is going vertically up, the gravity force would be acting down. If $\alpha$ is the acceleration of the rocket at time $t$ and $U_0$ is the exhaust velocity with respect to the rocket then

Mass x acceleration = reaction thrust of gases leaving rocket – weight of the body.

or,
$$(M_0 - mt)\,\alpha = m\,(U_0 - 0) - (M_0 - mt)g$$

♦
$$\alpha = \frac{du}{dt} = \frac{mU_0}{M_0 - mt} - g$$

and
$$u = -\,gt - U_0 \ln\left(\frac{M_0}{M_0 - mt}\right),$$

with the initial condition at $t = 0$, $u = 0$. Substituting the data given in the problem, we get: initial acceleration at $t = 0$ is 9.806 m/s$^2$ and at $t = 10$ s, $u = 120.245$ m/s.

### 2.4.3 Conservation of Energy — Energy Equation

It is known for millenniums that the energy cannot be created or destroyed. Or in other words the energy is conserved, which is the first law of thermodynamics, expressed on time rate basis as

$$\frac{DE}{Dt} = \frac{DQ}{Dt} + \frac{DW}{Dt},$$

where
$$\frac{DE}{Dt} = \frac{D}{Dt}\left\{\rho\left(e + U^2/2 - \vec{g}.\vec{r}\right)\right\}$$

$$= \rho\left(\frac{De}{Dt} + U\frac{DU}{Dt} - \vec{g}.\vec{U}\right) \tag{2.4.26}$$

We assume gravity is the only body force acting along –ve $z$-direction and $\vec{r}$ is the displacement of the particle; the term $\vec{g}.\vec{r}$ gives the change in potential energy, and $e$ is the internal energy. All these energies are expressed for a unit volume of the system. Work and heat transfer are the scalar quantiites and represent boundary phenomenon and can be evaluated in the manner we did for finding out the net mass transfer across a finite volume in the flow region. Adopting the procedure and recalling that heat transfer across any surface is given by Fourier law, we have

$$\vec{q} = -k \quad T \text{ (Fourier law), where } \vec{q} = iq_x + jq_y + kq_z$$

$$\text{Net heat transfer} = -\left(fq_x/fx + fq_y/fy + fq_z/fz\right)\delta x\ \delta y\ \delta z$$

or, $$DQ/Dt = -\text{div}\ \vec{q} = \text{div}\ (k \quad T) \tag{2.4.27}$$

where $T$ is the temperature gradient across the surface, which is responsible for the heat transfer and $k$ is the property of the fluid, called thermal conductivity, that depends on the temperature and hence on the space coordinates. Work is done on the surface because of the presence of shear stresses i.e.,

$$w_x = -\left(u\tau_{xx} + v\tau_{xy} + w\tau_{xz}\right)$$

per unit surface area on the $x$-face. We may use Taylor series to determine this on the $x + \delta x$ face. Similar procedure is then carried out at other faces too. This procedure leads ultimately to

$$\frac{DW}{Dt} = \underbrace{\frac{f}{fx}\left(u\tau_{xx} + v\tau_{xy} + w\tau_{xz}\right)}_{x} + \underbrace{\frac{f}{fy}\left(u\tau_{yx} + v\tau_{yy} + w\tau_{yz}\right)}_{y} + \underbrace{\frac{f}{fz}\left(u\tau_{zx} + v\tau_{zy} + w\tau_{zz}\right)}_{z \text{ face contribution}}$$

or, $$\frac{DW}{Dt} = \quad .(\vec{U}.\tau_{ij}) \tag{2.4.28}$$

where $\tau_{ij}$ is the shear stress tensor, defined in Chapter 1. But

$$.(\vec{U}.\tau_{ij}) = \vec{U}.(\quad .\tau_{ij}) + \tau_{ij}\, fU_i/fx_j, (\quad .\tau_{ij}) = p\left(\frac{D\vec{U}}{Dt} - \vec{g}\right) \tag{2.4.29}$$

$$\therefore \quad \vec{U}.(\quad .\tau_{ij}) = \rho\left(U\frac{DU}{Dt} - \vec{g}.\vec{U}\right) \tag{2.4.30}$$

We observe that the two terms in Eq. 2.4.30 are the kinetic and the potential energy terms that cancel out when we make substitution in Eq. 2.4.26 from Eq. 2.4.27, 2.4.28 and 2.4.29. On simplification this gives

$$\rho\frac{De}{Dt} = \text{div}\ (k \quad T) + \tau_{ij}\,\partial u_i/\partial x_j \tag{2.4.31}$$

We may separate out the pressure term from stress tensor for which $i = j$ and then $\partial u_i/\partial x_i = \partial u/\partial x + \partial v/\partial y + \partial w/\partial z = \text{div}\ \vec{U}$. Equation 2.4.31 can be transformed in the following form using the continuity equation and the notation $\tau_{ij}$ for the shear stresses

$$\tau_{ij} = \tau_{ij} - p \text{ (}p\text{ is negative)},$$

$$p \operatorname{div} \overset{\blacklozenge}{U} = -\frac{p}{\rho}\frac{D\rho}{Dt} = \rho\frac{D}{Dt}\left(\frac{p}{\rho}\right) - \frac{D\rho}{Dt} \tag{2.4.32}$$

Substituting from Eq. 2.4.32 in Eq. 2.4.31, we get, after rearrangement of the terms

$$\rho\frac{D}{Dt}(e + \frac{p}{\rho}) = \rho\frac{Dh}{Dt} = \frac{Dp}{Dt} + \operatorname{div}(k \quad T) + \tau_{ij}\,\partial u_i / \partial x_i,\ h = e + \frac{p}{\rho} \tag{2.4.33}$$

where $h$ is the specific enthalpy and represents sum of the internal energy and the flow energy $p/\rho$. The last term is the work done by the shear stresses and represents conversion of the kinetic energy into the internal energy; this term is customarily referred to as the dissipation function, $\Phi$. Using Eq. 1.2.10, we can write the dissipation function in the following form

$$\Phi = \mu\left[\begin{array}{l} 2\left(\frac{fu}{fx}\right)^2 + 2\left(\frac{fu}{fy}\right)^2 + 2\left(\frac{fu}{fz}\right)^2 + \left(\frac{fv}{fx} + \frac{fu}{fy}\right)^2 + \left(\frac{fw}{fy} + \frac{fv}{fz}\right)^2 + \left(\frac{fu}{fz} + \frac{fw}{fx}\right)^2 \\ -\frac{2}{3}\left(\frac{fu}{fx} + \frac{fy}{fy} + \frac{fw}{fz}\right)^2 \end{array}\right] \tag{2.4.34}$$

We expect that the dissipation function must always be positive, as the effect of the viscosity is to change higher grade of energy into the lower grade. The first look of Eq. 2.4.35 agrees with our perceptions because the terms are squared and hence will always be positive. However, this also requires μ  0 and the contribution of the last term in the equation should be less than that of other terms.

Equation 2.4.33 is the general energy equation that takes simple form depending upon the type of fluid. For isotropic and homogeneous fluids, $k$ can be taken as constant as it does not depend on the direction at a point or vary from point to point. Further, the gas may be considered as perfect gas for which the equation of state is applicable. Thus for such fluids, we may use the simplifications

$$k = \text{constant} \qquad dh = c_p\, dT \qquad de = c_v\, dT \tag{2.4.35}$$

If the fluid is taken incompressible, as in liquids, then $\rho$ = constant and Equation 2.4.34 takes the form

$$\rho c_v \frac{DT}{Dt} = k \operatorname{div}(\quad T) + \Phi = k\ {}^2T + \Phi \tag{2.4.36}$$

**Example 2.4.11** Obtain the relationship for thrust and efficiency for a ship propelled by a reaction jet, as shown in Fig. E 2.4.11.

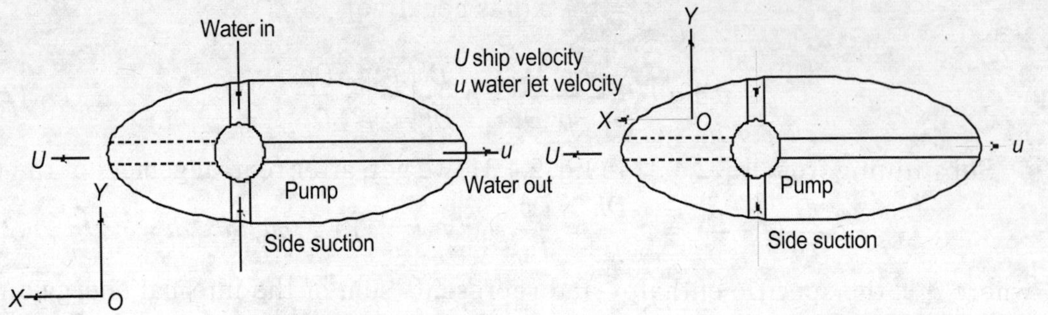


(a) Reference frame attached to ground (b) Reference frame attached to ship

**Fig. E 2.4.11** Ship propelled by a reaction jet

**Solution** Pump takes water from the sides and leaves at high velocity $u$ at the rear end of the ship that generates reaction thrust and propels the ship in forward direction with a velocity $U$. It will be convenient to work with the coordinates fixed to the ship, so that the relative velocity of the jet is $U_r = U + u$ (note $U$ and $u$ are in opposite directions). If $A$ is the area of jet leaving the ship then mass flow rate is $\rho AU_r$, the thrust produced is $\rho AU_r u$ and the power produced is $\rho AU_r uU$. Kinetic energy of the water supplied per second is $1/2\rho AU_r U_r^2$. Therefore efficiency is

$$\eta_1 = \frac{\rho AU_r uU}{1/2\rho AU_r^3} = \frac{2uU}{U_r^2} = \frac{2uU}{(U+u)^2}.$$

$$d\eta/du = 0 \text{ gives, } u = U \text{ and } \eta_{max} = 0.5 = 50\%.$$

It is also possible to take water from the front end of the ship instead through the sides, as shown in the figure by dotted lines. Then the kinetic energy of the water per unit mass is: at inlet $U^2/2$ and at exit $U_r^2/2$.

$$\therefore \qquad \eta_2 = \frac{\rho AU_r uU}{1/2\rho AU_r(U^2 - U_r^2)} = \frac{2uU}{U + U_r}$$

Dividing $\eta_2$ with $\eta_1$, we get

$$\frac{\eta_2}{\eta_1} = \frac{2uU}{2U+u} \leftrightarrow \frac{(U+u)^2}{2Uu} = 4/3 \text{ for the condition } u = U.$$

This means the second arrangement is better. Moreover, it is seen that the second arrangement will have maximum efficiency when $u = 0$ and would decrease with increase in $u$. So $u$ will be fixed from the power requirement considerations.

**Example 2.4.12** Determine the condition for maximum efficiency of a windmill.

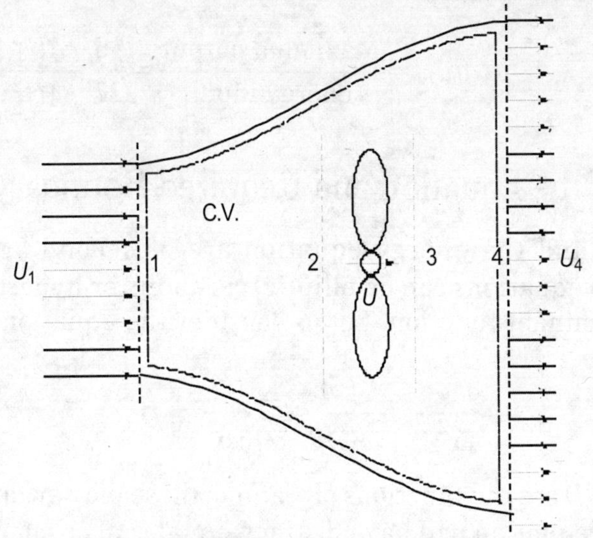


**Fig. E 2.4.12** Wind mill

**Solution** Windmill is used to convert kinetic energy of wind into electric power. It is a propeller turbine. Figure E 2.4.12 shows one such mill placed in a casing in which air enters at section 1 with velocity $U_1$ and leaves at section 4 with velocity $U_4$. The windmill is placed between the sections 2 and 3; the casing section is diffusing. Thus the static pressure continues to rise except in the windmill section where the static pressure would fall. The air exerts a force on the windmill blades and in the process looses momentum thus the integral form of the momentum equation is the right way to handle this problem.

Let $m$ be the mass flow rate of air through the duct. The rate of change of momentum is $m\ (U_1 - U_4)$ and work produced by windmill $= m\ U\ (U_1 - U_4)$, where $U$ is the velocity of flow across the cross-sectional area swept by the windmill. But this work is also equal to change in the kinetic energy of air from entrance to the exit, by the steady flow energy equation, which is $1/2m\ (U_1^2 - U_4^2)$. Equating it to the power developed by the turbine, we get

$$m\ U\ (U_1 - U_4) = 1\ /2m\ (U_1^2 - U_4^2).$$

or, $$U = 1\ /\ 2\ (U_1 + U_4), \qquad m = \rho AU = 1\ /\ 2\ \rho A(U_1 + U_4)$$

and power produced by the windmill $= m \leftrightarrow 1/2\ m\ (U_1^2 - U_4^2)$

$$= 1\ /\ 4\ \rho A\ (U_1 + U_4)\ (U_1^2 - U_4^2).$$

$U_4$ would depend on the power produced by the turbine. For maximum power, first differential with respect to $U_4$ should be zero. This gives: $U_4 = -u_1$ or $U_4 = U_1/3$. But the first solution is not possible. Therefore for maximum power output, $U_4 = U_1/3$ and maximum power $(8\ /\ 27)\ \rho AU_1^3$. Total kinetic energy of air entering the duct is $\frac{1}{2}\ mU_1^2$.

$$\therefore \qquad \eta_{max} = \frac{\text{Maximum output}}{\text{energy input}} = \frac{16}{27}\frac{\rho A U_1^3}{\rho A U_1^3} = 0.593.$$

## 2.5 Nature of the Equation and Required Boundary Conditions

The momentum and the energy equation are of second order involving partial derivatives. Such equations can be elliptic, parabolic or hyperbolic depending on the sign of the discriminant function. Let us consider the equation in $\phi$

$$A\frac{f^2\phi}{fx^2} + B\frac{f^2\phi}{fx\,fy} + C\frac{f^2\phi}{fy^2} = D$$

where $A, B, C$ and $D$ are the functions of $x$ and $y$; these may even be function of $\phi$ and its derivatives. This equation will have distinct, equal real or imaginary roots depending on the value of the discriminant $B^2 - 4AC$. The differential equation is called elliptic, parabolic or hyperbolic, if $B^2 >, =,$ or $< 4AC$, respectively. We have to specify the boundary condition as follows on the surface of the flow region:

Elliptic equation: Complete surface; it is a boundary value problem

Parabolic equation: Boundary conditions must be closed in one direction but remain open at one end of the other direction; it is a mixed initial and boundary value problem

Hyperbolic equation: Boundary conditions must be specified at one portion of the boundary but other boundaries remain open; it is an initial value problem.

In the above classification two terms, the boundary value and the initial value, were used. A boundary value problem is the one in which the value of an unknown function or its derivative is specified on the boundary of the region of interest and the problem is to study the behaviour of the function within the region. In such problems, we start with an assumed solution from one point on one surface to generate solution and in this process reach a point on the other surface. By adjusting the assumed solution, we try to get the solution at other boundary point to match the given function value. On the other hand, in the initial value problem the function is fully defined at one point of the boundary and solution is generated at other points. In unstedy flows there is initial condition problem in which the function is specified all over the region at some known time and then the behaviour of the function with time is studied.

The *N–S* equations for incompressible fluids are elliptic in nature while the boundary layer equations are parabolic. When we study a two-dimensional flow of a compressible flow, the resulting equations are hyperbolic. It is noted that in contrast to the ordinary differential equation, we need a lot more boundary conditions, at each and every point of the surface. While in ordinary differential equation, the number of the boundary conditions required is the order of the equation.

As far as boundary conditions are concerned, we may specify the function or its gradient on the surface. Accordingly, we have three kinds of the boundary conditions in actual applications. The boundary condition will depend on the type of interface of the two fluids or one fluid and another a solid surface. These are:

*First kind*: The function is specified over the surface; for example the no-slip condition $\overline{U} = 0$ on the stationary surface, as in fluid-solid interface.

*Second kind*: The gradient of the function is zero over the surface.

*Third kind*: The gradient of the function is specified over the surface.

***Fluid-solid interface***. At such interface, the fluid takes the velocity and the temperature as that of the surface and is called *no-slip condition*. Thus these boundary conditions are of first kind. We write this condition as

$$\vec{U}_{\text{fluid}} = \vec{U}_{\text{surface}} \tag{2.5.1}$$

However, with regards to temperature, the following type of situations are observed

$$T_{\text{fluid}} = T_{\text{surface}} \text{ (I kind)}, \ \partial T / \partial n = 0 \text{ (II kind) and } \ k \partial T / \partial n = q \text{ (III kind)} \tag{2.5.2}$$

In Eq. 2.5.2, $n$ represents the direction normal to the solid surface. The second kind of boundary condition exists when the interface surface is adiabatic surface, a surface that does not permit heat to flow across the surface. The third type of condition is very common.

***Liquid-gas interface***. At such interface, there must be kinematic equivalence and equality of normal and tangential momentum flux. These can be written in mathematical form taking $s$ as the surface coordinate and $n$ as the normal to the surface, respectively.

$$U_{n.\text{liq}} \cup \partial s / \partial t, \quad p_{\text{liq}} = p_{\text{atm}} \quad \text{and} \quad \partial U / fn|_{\text{liq}} = fT / fn|_{\text{liq}} \cup 0 \tag{2.5.3}$$

This condition assumes negligible evaporation rate of the liquid at the interface.

## Additional Problems

**Example AP 2.1** The fluid in direct contact with a solid boundary has zero velocity; there is no slip at the boundary. Thus the flow over a surface adheres to the plate surface and forms a boundary layer, as shown in Fig. AP 2.1.1. The flow ahead of the plate is uniform with velocity $U = 30\, i$ m/s, the velocity distribution with in the boundary layer ($0 \le y \le \delta$) along *cd* is approximated as $u/U = 2\,(y/\delta) - (y/\delta)^2$. The boundary layer thickness at a location $d$ is $\delta = 5$ mm. The fluid is air with density 1.24 kg/m$^3$. Assuming the plate width perpendicular to the paper to be $w = 0.6$ m, calculate the mass flow rate across surface *bc* of the control volume *abcd*.

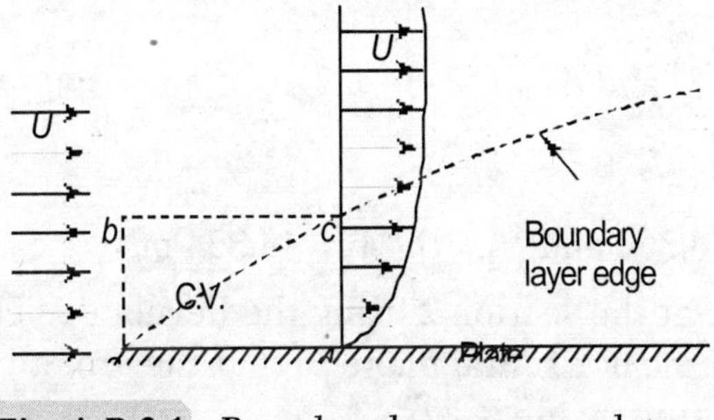


**Fig. A P 2.1** Boundary layer over a plate

**Solution** We can use Eq. 2.4.8 to determine the mass flow across the face *bc* of the control volume *abcd* while noting that the flow is two-dimensional and steady and there is no mass flow across the plate surface *da*. This gives

$$\iint_S \rho \vec{U}.d\vec{A} = \int_{ab} \rho \vec{U}.d\vec{A} + \int_{bc} \rho \vec{U}.d\vec{A} + \int_{cd} \rho \vec{U}.d\vec{A} + \int_{da} \rho \vec{U}.d\vec{A} = 0$$

or,

$$\dot{m}_{bc} = \int_{bc} \rho \vec{U}.d\vec{A} = -\int_{ab} \rho \vec{U}.d\vec{A} - \int_{cd} \rho \vec{U}.d\vec{A},$$

since the mass flow across *da* is zero. For width *b* in *z*-direction, the two integrals on the right hand evaluates to

$\int_{ab} \rho \vec{U}.d\vec{A} = \rho U b \delta$, since $u = U$ across this face and remains constant over the height $\delta$, and

$$\int_{cd} \rho \vec{U}.d\vec{A} = \rho b \int_0^{\delta} U \left\{ \frac{2y}{\delta} - \left(\frac{y}{\delta}\right)^2 \right\} dy = \frac{2}{3} \rho U b \delta.$$

$$\therefore \quad \dot{m}_{bc} = \int_{bc} \rho \vec{U}.d\vec{A} = -\int_{ab} \rho \vec{U}.d\vec{A} - \int_{cd} \rho \vec{U}.d\vec{A} = \tfrac{1}{3} \rho U b \delta$$

$$= 1.3 \leftrightarrow 1.24 \leftrightarrow 30 \leftrightarrow 0.6 \leftrightarrow 0.005 = 0.0372 \text{ kg/s.}$$

**Example AP 2.2** An oil of specific gravity 0.8 flows through a 30 cm diameter by 20 cm reducing bend, which turns the flow through 60°. The rate of flow is 0.45 m³/s and the pressure at inlet, which is 30 cm diameter section, is 122 kPa. Find the force required to keep the bend in position. Neglect loss of energy.

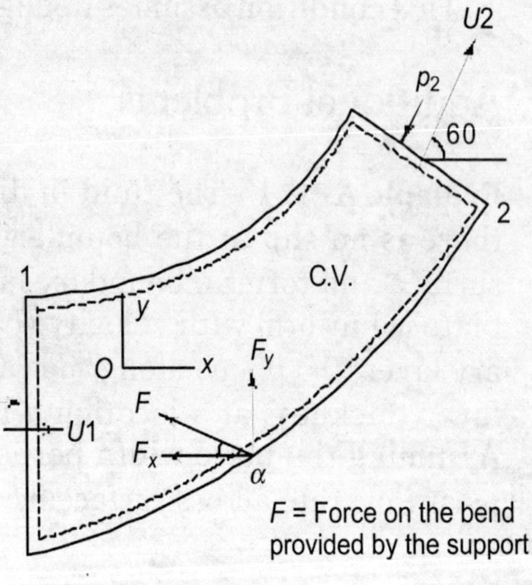


**Fig. AP 2.2** Forces on a bend

**Solution** Let the inlet and the outlet sections be represented by planes 1 and 2, respectively. We can use Eq. 2.2.2, noting that the flow is steady, the velocity at inlet section is known, and the area of cross section is also given. This gives

$$A_1 = (\pi / 4) \leftrightarrow 0.3^2 \text{ and } A_2 = (\pi / 4) \leftrightarrow 0.2^2 \text{ m}^2$$

$$Q = A_1 U_1 = A_2 U_2$$

♦ $U_1 = Q_1 / A_1 = 6.366$ and $U_2 = Q / A_2 = 14.324$ m/s

We get pressure at the section 2 using the Bernoulli's equation and assuming that the bend is lying in the horizontal plane so that the ground takes care of the weight of the bend and the liquid inside it.

$$\frac{p_1}{w} + \frac{U_1^2}{2g} + Z_1 = \frac{p_2}{w} + \frac{U_2^2}{2g} + Z_2$$

♦ $$\frac{p_2}{w} = 7.1536 \text{ m}$$

♦ $$p_2 = 0.8 \times 9810 \times 7.1356 = 5614 \text{ N/m}^2.$$

Let the inner surface of the bend is the control volume and the surface of the bend exerts a force having components $F_x$ and $F_y$ along $x$ and $y$-directions (taken along positive direction of the axes), respectively. Note that $U_2$ will have component $U_2 \cos 60$ along $x$ direction and $U_2 \sin 60$ along the $y$-direction. You can similarly find out the components of pressure along $x$ and $y$-directions keeping in mind that it acts normal to the surface and in the inward side, as shown in the figure. Applying momentum Eq. 2.4.11 in $x$ and $y$ directions, we get

$$p_1 A_1 + F_x - p_2 A_2 \cos 60 = \rho Q (U_2 \cos 60 - U_1)$$

♦ $$F_x = -7455 \text{ N}$$

$$F_y - p_2 A_2 \sin 60 = \rho Q (U_2 \sin 60 - 0)$$

♦ $$F_y = 3039 \text{ N}.$$

This shows that the force components are acting along –ve $x$ and +ve $y$ directions. The bend in turn receives force in opposite direction. If the resultant force makes an angle $\alpha$ with the horizontal direction then

$$F = \sqrt{F_x^2 + F_y^2} = 8050 \text{ N and } \tan \alpha = F_y / F_x$$

♦ $$\alpha = 22.2°$$

Therefore, a force $F$ acting at an angle 22.2° with the –ve $x$-axis direction in the horizontal $x$-$y$ plane, as shown in the figure, will be required to keep the bend in position.

**Example AP 2.3** When it is raining, some people say that it is better to run and some say you should walk to keep drier. Suppose that it is raining straight down at a volume flux rate of $10^{-5}$ m³/s per square metre of ground area, and you have to go 100 m in this rain. Assume an average droplet size of 1 mm³ for which the velocity of rain fall equals to 5 m/s. If a human adult is approximately 2 m high by 1 m wide by 0.5 m deep, analyse the situation to decide whether you should walk at 1 m/s or run at 4 m/s to stay drier.

**Solution** To remain dry, less volume of water should reach the clothes of the man. Therefore, the problem is to determine how much volume of water will cross the control volume around the man who is idealised as a parallelepiped of the dimensions $h = 2$ m, $w = 1$ m and $d = 0.5$ m. When the man runs the rain appears to come at an angle such that the vertical and horizontal components of the velocity are $V_v$ and $V_h$. We are given the volume flow rate per square metre of the ground area. Therefore the volume of the water crossing the control volume per unit time is

$$Q = 10^{-5} \times (\text{top area} \times V_v + \text{front area} \times V_h)$$

The time taken to travel the distance 100 m = $100/V_h$.

∴ Total volume wetting the man = $100 \times 10^{-5} \times (w \times d \times 5 + 1 \times 2 \times V_h)$

Now taking $V_h = 1$ and 4 m/s, the volume of water in two cases comes out to be 0.0009 and 0.000525 m³. Thus, we observe that in the second case when the man moves fast, he is less drenched in rainwater. So running is advisable.

**Example AP 2.4** What is the irrotational velocity field associated with the potential

$$\phi = 3x^2 - 3x + 3y^2 + 16lz^2\, t?$$

Does the flow field satisfy the incompressible continuity equation?

**Solution**

(a) $u = \frac{f\phi}{fx} = 6x - 3, v = \frac{f\phi}{fy} = 6y$, and $w = \frac{f\phi}{fz} = 12t$

♦ $\frac{fu}{fx} = 6, \frac{fv}{fy} = 6$, and $\frac{fw}{fz} = 0.$

(b) The equation of continuity is not satisfied. Therefore the flow is not possible. Therefore function for $\phi$ is fictitious.

**Example AP 2.5** A stream function is given by $\psi = 2x^2y + (3 + t)y^2$. Find the flow rates across the faces of the triangular prism *OAB*, shown in Fig. AP 2.16, having a thickness of 1 unit in the *z* direction at time $t = 1$.

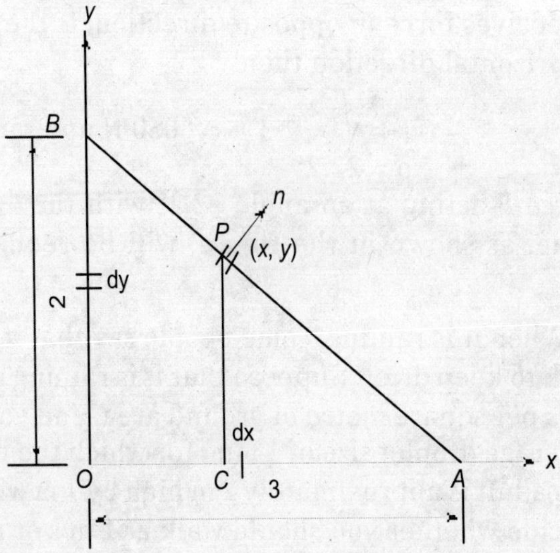


**Fig. AP 2.5** Flow across faces of a triangular prism

(a) $u = \frac{f\psi}{fy} = 2x^2 + 2(3 + t)y, v = -\frac{f\psi}{fx} = -4xy$. We now get velocity at $t = 1$ from these expressions.

(b) Flow across *OB* is due to velocity component *u*, which changes with *y*. We take an element *dy*, as shown in the Fig. Then discharge across *OB* per unit span = $\int_0^2 (u)_{x=0,\, t=1}\, dy = 16$ units. Here we take $x = 0$ for the face *OB*.

(c) We similarly get the discharge through the face *OA* per unit of span. This comes to 0.

(d) Let *n* be the normal to the surface *AB*. Flow across it per unit of its span, using unit vectors *i* and *j* along *x* and *y* directions can be written at $t = 1$s as

$$\int_{AB} (i dy + j dx) . [ui + v j] = \int_{AB} (2x^2 + 8y) dy + \int_{AB} - 4xy\, dx.$$

We need a relation between $y$ and $x$ that comes from the similarity of the triangles *OAB* and *PCA*, which gives:

$$\frac{y}{3-x} = \frac{2}{3},$$

or, $$y = 2/3\,(3-x) \text{ and } dy = -2/3\,dx.$$

Substituting these in the integral and solving it, we get the discharge = 16 units.

*Comment*: The results show as if the flow is steady, since the sum of the flow entering the prism is equal to leaving it. However, we should not be miscarried away by this result as this has been obtained at $t = 1$s. Presence of the term $t$ in the expression for the velocity ensures that the flow is unsteady. It is just a coincidence that at this moment the identity is satisfied.

**Example AP 2.6** The two plates separated by a distance $b$ form a channel, Fig. AP 2.6. One of the plates is porous and the other one is impermeable. A flow takes place within the channel so that the $x$ component of velocity $u$ is a function of $x$ only and its value at the inlet is $u_0$. There is a uniform inflow $v_0$ through the porous wall to the channel so that the velocity component $v$ in the $y$ direction within the channel is a function of $y$ only. Considering the flow to be incompressible, find the expression of $u$ as a function of $x$, and $v$ as a function of $y$.

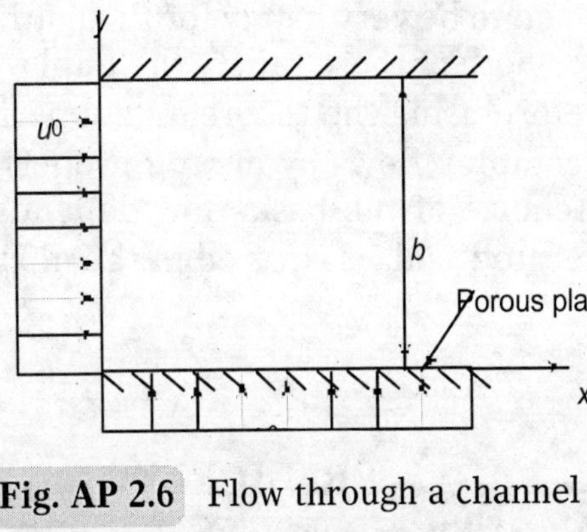


**Fig. AP 2.6** Flow through a channel

**Solution**

(a) The continuity equation at any point is $\frac{fu}{fx} + \frac{fv}{fy} = 0$. Since $u$ is a function of $x$ alone and $v$ of $y$ alone, the partial differentials change into the total differential. This gives: $du/dx = -dv/dy$. This is possible only if both the sides are equal to some constant, say $K$. Now we can integrate these equations separately and use the boundary conditions for unique solution.

(b) Boundary conditions are: $x = 0, u = u_0$; $y = 0, v = v_0$

(c) Result after integration is: $u = u_0 + \frac{v_0}{b}x$ and $v = v_0 - \frac{v_0}{b}y$.

We get $v = 0$ at $y = b$ from this result, which was expected since the plate is impermeable.

## Summary

Any flow can be represented by streamlines and this explains the special role of these lines which are nothing but the constant stream function line. However, in some experimental studies the path line and the streak lines too have the importance. A stream function is a scalar function of the space coordinates for a two-dimensional flow that is related to the velocity components at any point through the mathematical equation (2.3.1). The difference of the stream function of the two nearby streamlines of an incompressible fluid represent the volume flow rate per unit width of the region in the third direction. The stream function satisfies the Laplace equation and hence the law of conservation of mass.

The other important function is the velocity potential, which exists only for the irrotational flows. Such flows are called the potential flows because for such flows the velocity potential exists. On the contrary for a rotational flow, the angular velocity of the rotation is represented by the term vorticity and is given by Eq. 2.3.5. The constant potential lines and the constant stream function lines intersect each other at right angles. For an ideal fluid, these two functions have proved to be very handy for the mathematical analysis of the complex flows, since both of these satisfy the Laplace equation, which is linear in nature. Therefore using the theorem of superimposition, it is possible to compose any complex flows by mixing properly the basic flows of suitable strength; the choice of the basic flow depends on the type of the flow pattern to be generated. This is covered in the next chapter.

# PROBLEMS

## Review Questions

1. Explain the Lagrangian and Eulerian methods of describing the flow.
2. What do you mean by one, two and three-dimensional flow? Explain with some suitable example. Classify the following flows into these categories: (a) flow in a river, (b) flow of water over a spillway, (c) flow through a test section of water tunnel, and (d) flow in a turbo machine.
3. Explain the terms (a) path line, (b) streak line, (c) streamline, and (d) stream tube. Distinguish between these. Are these same for the unsteady flow? Can a path line cross the streamline?

4. Distinguish between (a) steady and unsteady flow, (b) uniform and non-uniform flow, (c) compressible and incompressible fluids, (d) rotational and irrotational flow, and (e) laminar and turbulent flow.
5. All gas flows are compressible and all liquid flows are incompressible. Is this statement true?
6. Define the following and give one practical example for each. (a) Laminar flow, (b) turbulent flow, (c) Steady flow, and (d) uniform flow.
7. Classify the following flow as laminar or turbulent: (a) flow through hypodermic needle, (b) atmospheric wind, (c) flow in a river, and (d) flow of lubricating oil from an oilcan.
8. The end points of all path lines lie on streak line. Is it true?
9. What do you understand by the terms (a) total acceleration, (b) convective acceleration, and (c) local acceleration?
10. Is the continuity equation a conclusion of the principle of conservation of the mass?
11. Is it possible to infer from the continuity equation that the average velocity is inversely proportional to the area of cross-section? Under what conditions is it valid?
12. Justify the statement "But without loosing the generality because such accounting adds to terms of negligible magnitude, we may consider these to be constant on this face" made while deriving continuity equation.
13. For a two-dimensional flow, the discharge per unit time across a line joining two points is equal to the difference between the stream functions at the two points. Comment on the validity of this statement.
14. What is the line of constant stream function? Does this hold true for a three-dimensional flow as well?
15. Write short notes on (a) pure translation, (b) linear deformation, (c) rotation, and (d) angular deformation of the fluid mass.
16. Translation of a fluid element occurs in all types of fluid motion. Is this true?
17. Define the terms velocity potential function and stream function. What are the conditions for flow to be irrotational?
18. Does non-uniform velocity distribution indicate the presence of Vorticity?
19. What do you mean by equipotential line and a line of constant stream function?
20. Describe the use and limitation of the flow nets. Under what conditions when can one draw the flow net?
21. Define the terms (a) vortex flow, (b) forced vortex flow, and (c) free vortex flow. Differentiate between the forced and the free vortex flow.
22. Derive from first principle, the conditions for irrotational flow. Prove that, for potential flow, both the stream function and the velocity potential function satisfy the Laplace equation.
23. How are velocity components prescribed in terms of velocity potential in cylindrical coordinates? What is the corresponding Laplace equation?
24. In terms of potential function and the velocity potential function, express the condition for (a) continuity of flow and (b) irrotational flow.
25. Even though the velocity at a point in a flow is invariant with time, these may be acceleration at the point. How can this be explained?

26. There is no acceleration in steady, uniform flow. Comment.
27. In a two-dimensional flow, show that the line integral from one point to another point is equal to the difference between the velocity potential functions at these two points.
28. Is the motion possible if the fluid is incompressible, inviscid and there is no pressure variation with respect to distances? If it is possible, give equation of motion.
29. Write down the Euler equation for steady, incompressible, irrotational flow along a horizontal line.

## Problems for Practice

1. A two-dimensional flow is described in the Langrangian system as $x = x_0 e^{-kt} + y_0 (1 - e^{-2kt})$ and $y = y_0 e^{kt}$. Find (a) the equation of path line of the particle and (b) the velocity components in Eulerian system. [$(x - y_0)y^2 - x_0 y_0 y + y_0^3 = 0$, $u = - kx + ky (e^{-kt} - e^{-3kt})$, $v = ky$].
2. The velocity in a flow field is given by $u = 3$ m/s and $v = 6$ m/s. Determine the equation of the streamline passing through the origin and the one passing through a point (2, 3). [$y = 2x$; $y = 2x - 1$].
3. The velocity distribution in a three-dimensional flow is given by $u = -x$, $v = 2y$, and $w = -(3 - z)$. Find the equation of the streamline that passes thorough point (1, 1, 1).
4. Determine the equation of a streamline at $t = t_0$ and passing through the point $(x_0, y_0)$ for the velocity field given by: $\vec{U} = (1 + At)i + 2xj$, where $A$ is some constant. How would the equation change for a steady flow? [ $y - y_0 = \dfrac{1}{1 + At_0}(x^2 - x_0^2)$ ].
5. In a steady fluid flow, the velocity components are: $u = 2kx$, $v = 2\,ky$, and $w = -4\,kz$. Find the equation of streamline passing through the point (1, 0, 1). [ $y = 0$ and $z = 1/x^2$].
6. A fluid field is given by $u = 3$ m/s and $v = 6$ m/s. Determine whether this is a possible steady incompressible fluid flow. If so, calculate the acceleration, angular velocity, and velocity at the point (1, 2, 3). [$a = 16i + 12j + 84k$; $w = -1/2\ (7i + k)$; $U = 2i + 12j - 15k$].
7. It is desired to obtain the convective acceleration and the kinematics of a fluid element at the origin (0, 0) in a steady fluid flow from the record of velocity reproduced below.

| $(x, y)$ | (0, 0) | (0.1, 0) | (–0.1, 0) | (0, 0.1) | (0, –0.1) |
|---|---|---|---|---|---|
| Velocity | $10i + 15j$ | $11i + 15.5j$ | $9i + 14.5j$ | $9.4i + 14j$ | $11.6i + 16j$ |

8. The velocity components in a two-dimensional incompressible flow field are expressed as $u = y^3/3 + 2x - x^2 y$ and $v = xy^2 - 2y - x^3/3$. (i) Determine the velocity and acceleration at point $P$ (1, 3). (ii) Is the flow physically possible? If so obtain an expression for the stream function. (iii) What is the discharge between the stream-

lines passing through (1, 3) and (2, 3)? (iv) Is the flow irrotational? If so determine the corresponding potential. and (v) Show that each of the stream and potential functions satisfy the Laplace equation.

9. Check whether the following functions represent a possible flow and determine the nature of the flow: (a) $\phi = \sin(x + y + z)$; (i) $\phi = U \cos\theta / r$.
10. Examine the following velocity components for a possible two-dimensional incompressible flow to be described.

$$u_r = -(U \cos\theta - a^2 U/r^2) \quad \text{and } u_\theta = (U + a^2 U/r^2) \sin\theta$$

11. Which of the following sets of Eq. represent possible two-dimensional incompressible flow?

(a) $u = x + y$; $v = x - y$ (b) $u = x + 2y$; $v = x^2 - y^2$

(c) $u = 4x + y$; $v = x - y^2$ (d) $u = xt + 2y$; $v = x^2 - yt^2$

12. Two components of velocity in an incompressible fluid flow are given by

$$u = x^3 - y^3 \text{ and } v = z^3 - y^3$$

Determine the third component, assuming that the origin is a stagnation point. [$w = 3(y^2 - x^2)$).

13. In a three-dimensional incompressible flow, the $u$ and $v$ velocity components are given by: $u = 1 + x^2 + y^2$ and $v = 1 - 3zy^2$. Determine the $w$ component of the velocity such that $w = 1$ at the origin. [$w = 1 - 2xz + 3yz^2$]
14. What is the irrotational velocity field associated with the potential

$$\phi = 3x^2 - 3x + 3y^2 + 16\,t^2\,t?$$

Does the flow field satisfy the incompressible continuity equation? [no].

15. Find out the average velocity $V$ in a circular pipe of radius $r_0$ and show that the kinetic energy correction factor $= \dfrac{(k+1)^3\,(k+2)^3}{4\,(3k+1)\,(3k+2)}$ for a velocity profile

$$\frac{v}{v_{max}} = \left(1 - \frac{r}{r_0}\right)^k.$$

16. A stream function is given by $\psi = 3x^2\,y + (2 + t)\,y^2$. Find the velocity field and determine its value at a point defined by the position vector $\vec{r} = 1i + 2j - 3k$ when $t = 2$. [22.47 m/s, – 3.27°]
17. Does the velocity potential exist for the 2-dimensional incompressible flow prescribed by $u = x - 4y$ and $v = -(y = 4x)$. If so, determine its form as well as that of the stream function. [$\phi = x^2/2 - y^2/2 - 4xy$ + Constant, $\psi = 2x^2 - 2y^2$ + constant]
18. A cubical air tank, 40 cm side, contains air at 900 kN/m² pressure and 20°C temperature. Air escapes through a 500 mm² area hole at a velocity of 50 m/s. Determine the rate of change of density at an instant when the density of the air in the tank is 7 kg/m³. [–2.73 kg/m³].
19. A 45° reducing pipe-bend in a horizontal plane tapers from 600 mm dia at the inlet to 300 mm dia at the outlet. The pressure at the inlet is 140 kPa gauge and the rate of flow of water through the bend is 0.425 m³/s. Neglecting friction, calcu-

late the net resultant horizontal force exerted by the water on the bend. Assume uniform conditions with straight and parallel streamlines at inlet and outlet and the fluid to be frictionless. [33.23 kN, 13.86° with the negative $x$ direction].

20. Measurement of velocity both in magnitude and direction were made in a vertical plane above the weir, Fig. P 2.20, at distance 0.12 $m$ apart and the following results were recorded: $U_1$ = 4.5 m/s, $\alpha_1$ = 5°; $U_2$ = 4.1 m/s, $\alpha_2$ = 10°; and $U_3$ = 3.8 m/s, $\alpha_3$ = 15°. Calculate the flow rate for unit width of the weir. [1.463 m³/s].
21. Explain whether the velocity components given by $u = 2x^2$ and $v = -6x^2y$ represent a possible steady, two-dimensional, incompressible flow. State whether the flow is irrotational and if so work out the velocity potential function. If rotational, determine the velocity, shear strain and dilation. Proceed to find out the circulation about the circle $x + y^2 - 2ay = 0$.
22. Examine whether the following velocity components represent a rotational or irrotational flow? If the flow is irrotational, calculate the vorticity at point (2, 1). [2.25 m/s/m].
23. Examine whether the following stream functions represent a possible fluid flow. If flow is possible, state whether it is rotational or otherwise: $\psi = x(y^2 - a^2) - x^3/3$.
24. If the velocity field is given by $u = x + y$ and $v = x^3 - y$, find the circulation around a closed contour defined by $x$ =1, $y$ = 0, $y$ = 1, and $x$ = 0. [0].
25. A two-dimensional flow field is given by the velocity components: $u = \dfrac{-y}{x^2+y^2}$ and $v = \dfrac{x}{x^2+y^2}$. Show that it represents a physically possible flow and that the flow field is irrotational. Also calculate the circulation around (a) a square with its corners (1, 0), (2, 0), 2, 1), and (1 ,1); (b) a unit circle with its centre at the origin. Comment on your results.
26. A small jet engine in flight at 30 m/s at a low altitude takes in air through a 0.4 m diameter circular inlet in the front and discharges the exhaust gases through a 0.1 m diameter nozzle at 100 m/s relative to the engine. Neglecting the fuel air ratio, estimate the performance of the engine. [Propulsive efficiency = 46%].

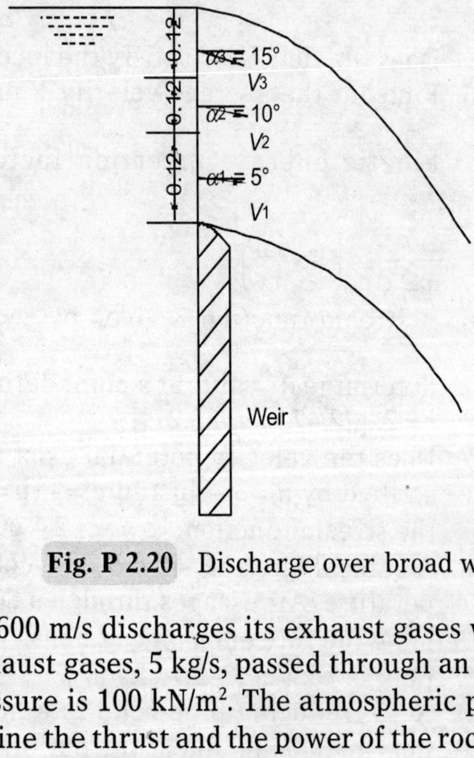


**Fig. P 2.20** Discharge over broad weir

27. A rocket moving forward at a speed of 600 m/s discharges its exhaust gases with an absolute velocity of 200 m/s. The exhaust gases, 5 kg/s, passed through an exit area of 0.06 m² where the absolute pressure is 100 kN/m². The atmospheric pressure at that altitude is 75 kN/m². Determine the thrust and the power of the rocket. [100 N, 160 kW].

28. What force components $F_x$ and $F_y$ are required to hold the black box of Fig. P 2.28 stationary? Assume no mass to be accumulated inside the box and all pressures are at zero gauge. [865.76 N, 258.62 N].

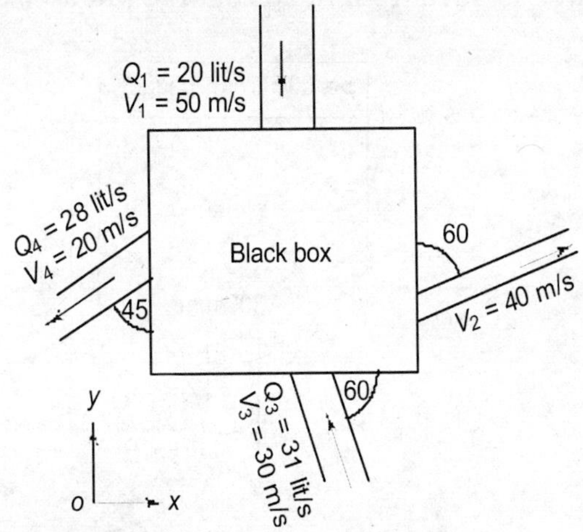


**Fig. P 2.28** Some force on the black box

29. Water flows in a circular pipe. At one section, the diameter is 0.3 m, the static pressure is 260 kPa gauge, the velocity is 3 m/s, and the elevation is 10 m above ground level. The elevation of a section downstream is 0 m, and the pipe diameter is 0.15 m. Find the gauge pressure at the downstream section. Frictional effect may be neglected. Assume density of water to be 999 kg/m³. [290.58 kPa gauge].
30. An ejector pump shown in Fig. P 2.30 uses a high speed water jet issuing from a pipe of area $A_j$ to drag along the surrounding water such that the device works as a pump. If the velocity and pressure profiles are assumed uniform at any section, and the shear stress at the pipe wall is neglected, then show using the momentum theorem

$$p_2 - p_1 = \rho(A_j / A_p)(1 - A_j/A_p)(V_1 - V_j)^2$$

where $p_1$ and $p_2$ are the uniform pressure at the inlet and outlet sections respectively as shown in Fig. P 2.30, $V_j$ is the velocity of the jet, $V_1$ is the velocity of water being dragged by the jet at the inlet and $V_2$ is the velocity of flow at the outlet.

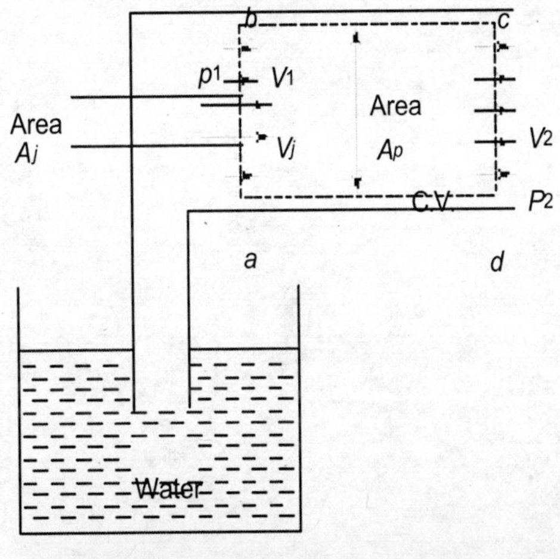


**Fig. P 2.30** An ejector pump

31. Find the net vertical force acting on the circular plate shown in Fig. P 2.31, if the water spreads radially on it. Neglect the weight of water on the plate. Assume the height of free water surface from the discharge plane to remain constant. [79.3 kN].

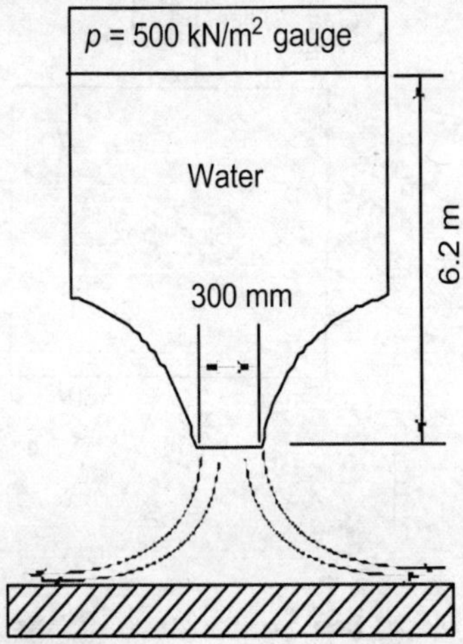


**Fig. P 2.31** Force on plate

32. A jet of water 20 mm diameter leaves the nozzle tip with 15 m/s and is directed vertical upwards. If the jet remains circular, work out the diameter at a height of 5 m above the nozzle tip. Neglect any loss of energy. [23.08 mm].

CHAPTER

3

# Ideal Fluid Flow

## OUTLINE

- Introduction
- Basic Flows
- Superimposition of Basic Flows
- Integration of Euler's Equation
- Modified Bernoulli Equation

## 3.1 Introduction

In the initial period of the development of the subject, assumption of an inviscid and incompressible fluids were made because of less mathematical complications. An inviscid fluid is the one that has no viscosity; this is an ideal assumption because all fluids have some viscosity, however small it may be. On the other hand, incompressibility eliminates the density as one of the variables, thus reducing the number of unknowns. Such fluid is called an ideal fluid. In such case, there will be no shear stress and only the pressure force will act which is normal to the surface, i.e, there will be only diagonal elements in the stress tensor. For such fluids, the order of the momentum equations is reduced to one (Euler's equation) and the momentum equations could be combined into a single equation (Bernoulli's equation). Such an assumption may not give the exact results but at least it can provide the initial approximate value for carrying out further computations, which are normally iterative. Moreover, the flow of such fluids is potential flow for which Laplace equation $\nabla^2\psi = 0$ determines velocity field. We shall study in this chapter how the linearity of this equation helps in analysing complex flows. This concept, in addition to the conformal transformation, helped us in simplifying the study of flow around the complex bodies and the asymmetric aerofoil in terms of flow around a cylinder. But we shall first discuss a few basic flows in the next section to build up the procedure for such studies.

## 3.2 Basic Flows

Uniform flow, source and sink flows, doublet flow and free vortex are simple or basic flows, which can be superimposed over each other to generate a complex flow. A *uniform flow* is a flow having uniform velocity, say $U$, every where in the flow region. The best way to find out stream function and velocity potential for this flow is to align $x$-axis of the coordinate axes with the direction of the flow. In that case, the flow is defined by the velocity field $u = U$ and $v = 0$. Therefore,

$$u = \partial\psi / \partial y = \partial\phi / \partial x = U \quad \text{and} \quad v = -\partial\psi / \partial x = \partial\phi / \partial y = 0$$

$$\rightarrow \quad \psi = Uy \quad \text{and} \quad \phi = Ux \qquad (3.2.1)$$

Thus, streamlines are defined by the straight lines $\psi = Uy$, which are lines parallel to $x$-axis and the constant $\phi$ lines are the lines parallel to $y$-axis. In this description, we have chosen the $x$-axis such that it is along the direction of the flow. If the $x$-axis is making an angle $\alpha$ with the direction of flow then $u = U\cos\alpha$ and $v = U\sin\alpha$. We may similarly determine the equation of the streamline and constant velocity potential lines passing through the point $(r, \theta)$. This gives

$$\psi = U\cos\alpha\, y - U\sin\alpha\, x = Ur\sin(\theta - \alpha), \text{ and}$$

$$\phi = U\cos\alpha\, x + U\sin\alpha\, y = Ur\cos(\theta - \alpha)$$

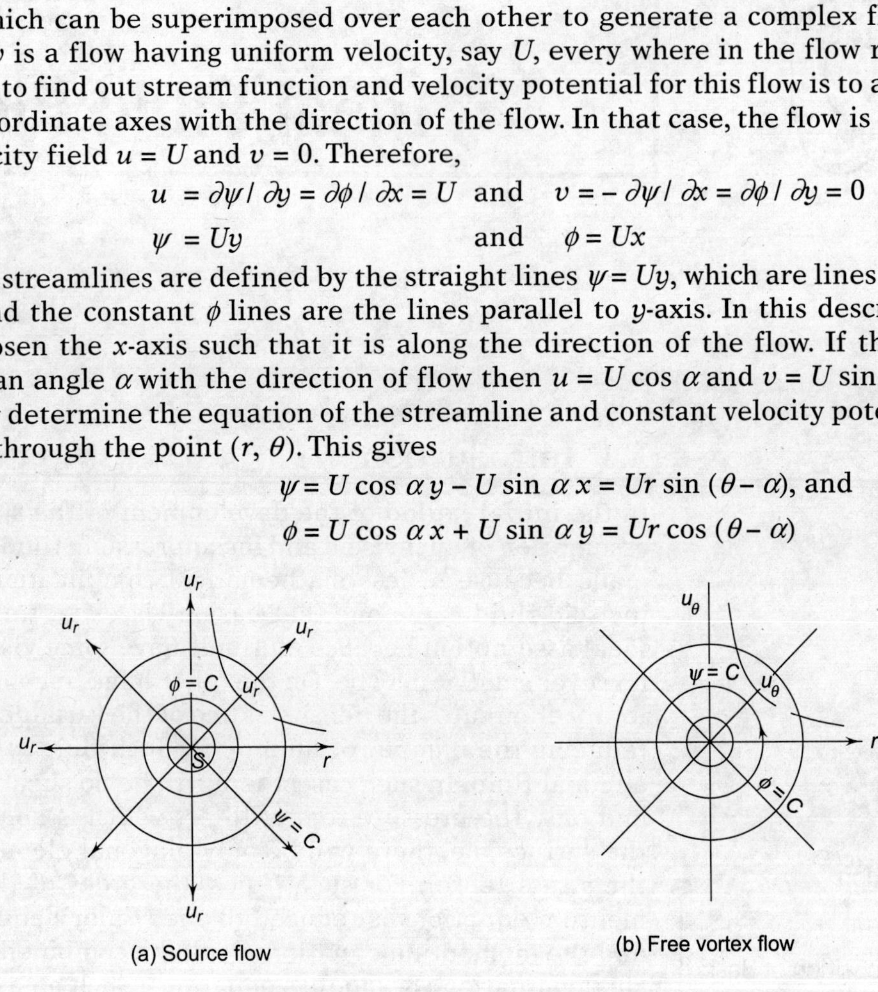


(a) Source flow (b) Free vortex flow

**Fig. 3.2.1** Source and free vortex

A *source or sink flow* is another basic type of flow. Take a porous pipe such that when water flows through it along its length then it can also flow radially at its every section. Figure 3.2.1(a) shows the flow pattern at a cross section of the pipe in which the fluid is shown moving radially with velocity $u_r$ in that direction; there will be no velocity in the $\theta$ direction. Enough amount of fluid is available so that the discharge per unit length of the pipe $q$ does not change along the flow direction and the flow may be considered two-dimensional flow. Such a source is called a *line source.* There is also a point source in which the fluid radiates from a point in all the directions from it. On the contrary, in a sink flow the direction of flow is from the surroundings to the line source *S*. The sink flow is observed in sinks, whence the name. Whirlpool is another example. The velocity $u_r$ at a distance $r$ from the source can be obtained by considering a cylinder of radius $r$ around the pipe axis passing through the source. Let $q$ is the volume flow rate across a unit length of pipe then

$$u_r = \frac{q}{2\pi r} = \frac{k}{r} \text{ and } u_\theta = 0.$$

or,
$$\frac{\partial \psi}{r\partial \theta} = \frac{\partial \phi}{\partial r} = \frac{k}{r} \text{ and } -\frac{\partial \psi}{\partial r} = \frac{\partial \phi}{r\partial \theta} = 0 \tag{3.2.2}$$

$k$ is called the *source strength.* In case of sink flow it is negative since $u_r$ will then be negative by sign convention. On integration and taking constant of integration as zero eq. 3.2.2 gives,

$$\psi = k\theta \text{ and } \phi = k \ln r \tag{3.2.3}$$

Thus, the streamlines are radial lines and the potential lines are concentric circles with centre at the source, the origin of the coordinate system. Moreover, from Eq. 3.2.2, we note that the radial velocity varies hyperbolically with radius, since $u_r\, r$ = constant; this variation is shown in Fig. 3.2.1. It is seen that at the origin, where $r = 0$, the radial velocity becomes infinite. Thus the origin is a singular point, called the pole in mathematical jargon. Singular points are the points where the Laplace equation is not satisfied and are discussed in Appendix A1. For a region enclosing singular points, the circulation around the curve need not be zero and the value depends on the residue at the singular points, as discussed in Chapter 2. In this case, the residue comes to zero, since the velocity component $u_\theta$ is zero along the circular curve around the origin.

Let us consider another basic flow, the *free vortex flow*, which is defined by the velocity field

$$u_\theta = \frac{\Gamma}{2\pi r} = \frac{k}{r} \text{ and } u_r = 0.$$

or,
$$\frac{\partial \psi}{r\partial \theta} = \frac{\partial \phi}{\partial r} = 0 \text{ and } -\frac{\partial \psi}{\partial r} = \frac{\partial \phi}{r\partial \theta} = \frac{k}{r} \tag{3.2.4}$$

where $\Gamma$ is the circulation along a curve drawn in the flow region and is taken positive around the origin when the circulation is in the anticlockwise direction. It is observed that $u_\theta$ decreases hyperbolically as $r$ increases. The stream function and the velocity potential for this flow are

$$\psi = -k \ln r \text{ and } \phi = k\,\theta \tag{3.2.5}$$

Thus the streamlines are concentric circles and the constant potential lines are straight lines, as shown in Fig. 3.2.1(b). We get the impression from the circular streamlines as if this is a kind of flow we observe when a bucket full of water is rotated about the bucket axis (rigid body rotation). A little insight into the two flows shows that in fact it is not so because in the rigid body rotation the tangential velocity $u_\theta$ increases with $r$, whereas it is decreasing in case of free vortex. A free vortex is observed in summer during hurricane, which is nothing but the free vortex flow. This flow has the origin as singular point, which was demonstrated in Chapter 2.

## 3.3 Superimposition of Basic Flows

We learnt in Chapter 2 that both the stream function and the velocity potential satisfy the Laplacian equation for an irrotational flow. The linearity of the operator implies that if different flows are combined together then the stream function of the combined flow will be the sum of the stream function of the individual flow. This is nothing but the principle of superimposition. This has helped us in the study of the complex flows by assuming as if it is made from the superimposition of the basic flows for which we know the stream function and the velocity potential. Once the stream function is known then the velocity field can be obtained from which the pressure field is obtained using the Bernoulli equation. In this section we shall consider a few simple applications to learn the procedure.

The method is applicable to the ideal fluids only, since in real fluids the velocity will be zero over the whole surface and would change along the normal direction. On the contrary, we assume the velocity gradient to be zero on the surface in such analysis. Before we study the procedure a comment will be in order. In the following analysis, we assume presence of some kind of basic flows and then see what kind of streamlines are generated. From their study, we then argue to decide upon the shape of the body around which such flow pattern will be observed when placed in a uniform stream and control the variables to give the desired body profile. However, what we need is to have methods so that we may predict the velocity and pressure distribution for the body placed in a uniform stream for which the velocity of flow is known. Therefore, the following study is indirect way of coming to the point and is called the *inverse problem*.

### 3.3.1 Source–Sink Pair and Doublet

A source and a sink of same strength $k$ may be placed together to generate a flow pattern as shown in Fig. 3.3.1. Let us consider the Cartesian coordinates with $x$-axis along the line joining the source to the sink with the origin at the mid point such that the distance of the source or the sink from the origin is $s$. The streamlines will individually be radial lines emanating from the source and converging towards the sink. Thus, the radial lines from the source will no longer remain straight lines but would bend towards the sink so as to terminate there. To find out the stream function at any point $P$ $(x, y)$ so that the lines $PA$ and $PB$ make an angle $\theta_1$ and $\theta_2$ with the reference $x$-axis, we use the principle of superimposition to give

$$\begin{aligned}\psi &= k\,\theta_1 - k\,\theta_2 \\ &= k\,(\theta_1 - \theta_2) \\ &= -k\,\alpha \text{ and} \\ \phi &= k \ln r_1 - k \ln r_2 \\ &= k \ln (r_1 / r_2), \end{aligned} \tag{3.3.1}$$

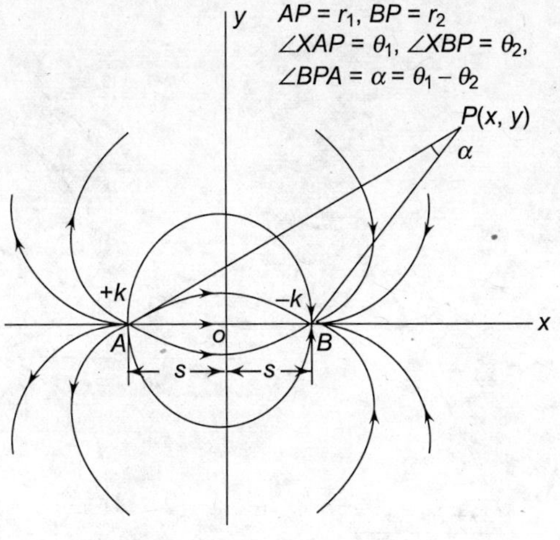


**Fig. 3.3.1** Sink and source pair

where $\alpha = \theta_1 - \theta_2$ and $r_1$ and $r_2$ are the radial distances of the point $P$ from the source and the sink, respectively. We may also express these distance in terms of the Cartesian coordinates $(x, y)$ of the point $P$ by invoking trigonometry.

$$\tan\theta_1 = \frac{y}{x+s}, \quad \tan\theta_2 = \frac{y}{x-s}$$

$$r_1^2 = (x+s)^2 + y^2, \quad r_2^2 = (x-s)^2 + y^2,$$

and

$$\tan\alpha = \tan(\theta_1 - \theta_2)$$

$$= \frac{2ys}{x^2 + y^2 - s^2} \tag{3.3.2}$$

A special flow is formed, called the *doublet flow*, when the source and the sink are brought close to each other such that (i) the distance $\delta s$ between them tends to zero and (ii) the product $\delta s.q$, where $q$ is the volume flow rate per unit length of the line source, remains constant. As the source and sink move towards each other, the equal strength of the two will neutralise each other. To avoid mutual cancellation, the magnitude of the source and sink strength is allowed to grow infinitely large as the distance between them decreases. For this reason, the restriction $\delta s.q$ = constant is laid down.

The product $\delta s$.q is known as the *doublet strength* and is represented by symbol $\mu$. Figure 3.3.2 (a) shows the position of the source and the sink of equal strength $k$, the point $P$ at a distance $r$ from the sink and $r + \delta r$ from the source, and the various angles. As we move the source and sink close to each other, angle $\delta\theta$ becomes small and can be written as $\delta\theta = \tan\delta\theta = \delta s \sin\theta / r$, where $\delta s \sin\theta$ is the length of the perpendicular drawn from the sink, as shown in the figure. For a doublet, the stream function and the potential can be expressed in $r$ and $\theta$, using trigonometry, as follows:

$$\psi = k\theta - k(\theta + \delta\theta) = -k\delta\theta = -\frac{q}{2\pi}\frac{\delta s \sin\theta}{r} = -\frac{\mu \sin\theta}{2\pi r}$$

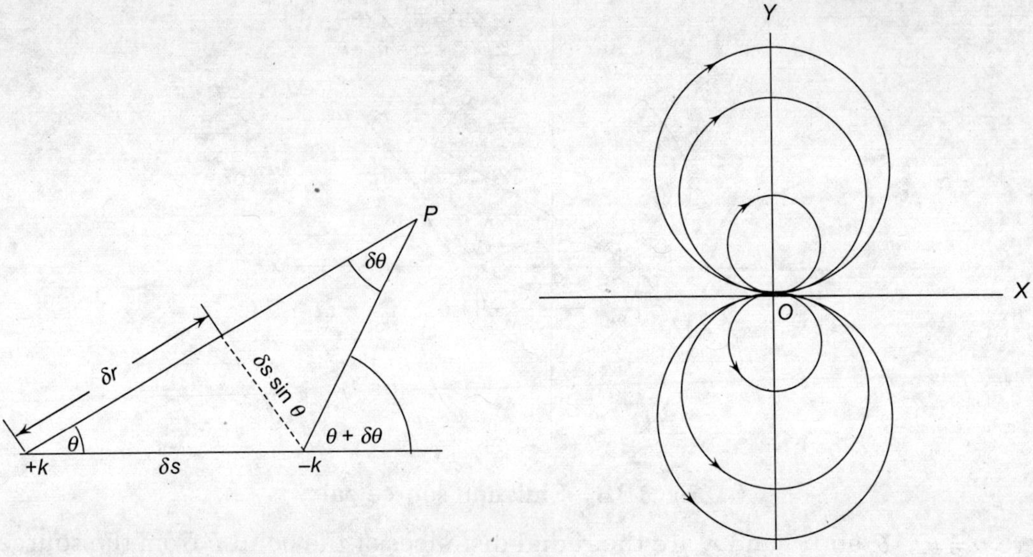


(a) Analysis of doublet (b) Streamlines

**Fig. 3.3.2** Doublet

$$\text{and } \phi = k \ln\left(\frac{r+\delta r}{r}\right) = \frac{k\,\delta r}{r} + \text{smaller terms} \approx \frac{q\,\delta r}{2\pi r} = \frac{q\,\delta s \cos\theta}{2\pi r} = \frac{\mu \cos\theta}{2\pi r} \qquad (3.3.3)$$

We notice that the streamlines are circles with their centre on the $y$-axis and touching the $x$-axis; the direction is from the source to the sink. These are shown in Fig. 3.3.2 (b).

**Example 3.3.1** A point $P$ (0.5, 1) is situated in a flow field of a doublet of strength 5 $m^3/s$. Calculate the velocity at the point and also the stream function. Work out in polar coordinates.

**Solution** Equation 3.3.3 gives the stream function for a doublet. Differentiation with respect to $\theta$ and $r$ gives the radial and tangential velocities.

$$u_r = \frac{1}{r}\frac{\partial \psi}{\partial \theta} = -\frac{\mu \cos\theta}{2\pi r^2}$$

and
$$u_\theta = -\frac{\partial \psi}{\partial_r} = -\frac{\mu \sin\theta}{2\pi r^2}$$

For the given point $P$,

$$r = \sqrt{(0.5^2 + 1^2)} = \sqrt{1.25},$$

$$\sin\theta = y/r = 0.894$$

and
$$\cos\theta = x/r = 0.447.$$

Substituting these values in the expression for the velocity and $\psi$, we get

$u_r = -0.2845$ m/s, $u_\theta = -0.569$ m/s and $\psi = -0.636$ $m^2/s$.

### 3.3.2 Flow Around Half Bodies

A body is called half body if its surface is not closed at one end, for example a bullet, spear, etc. In such a body the area of cross section increases and suddenly at some section it is closed by a vertical surface. When such bodies move in air or any other fluid, we need to study the flow pattern and the velocity distribution so as to calculate the drag and lift force required to move it through the media in the desired direction. We may simulate such flow pattern by placing a source in the uniform flow, as shown in Fig. 3.3.3(a).

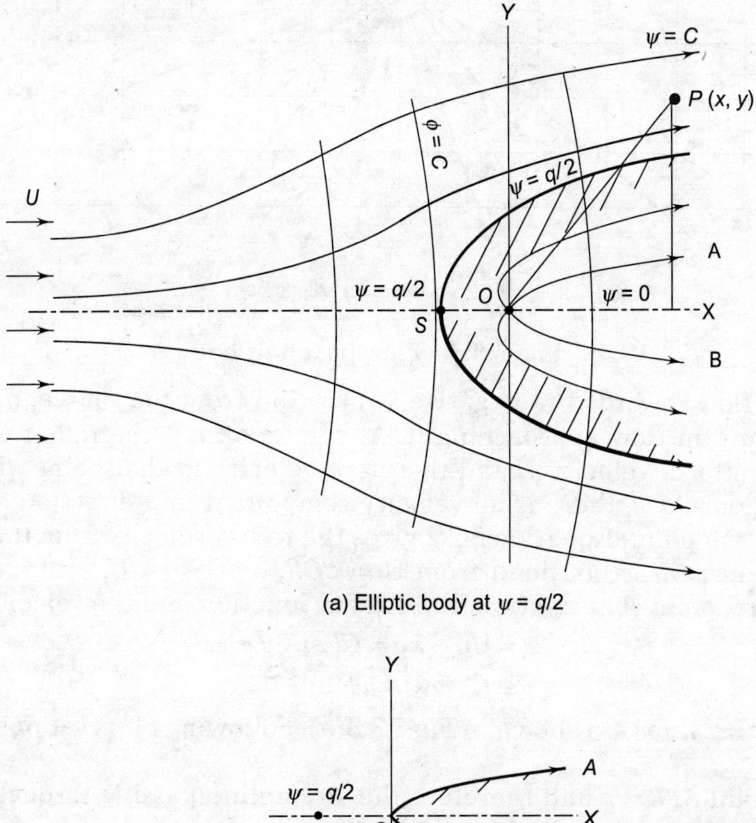


(a) Elliptic body at $\psi = q/2$

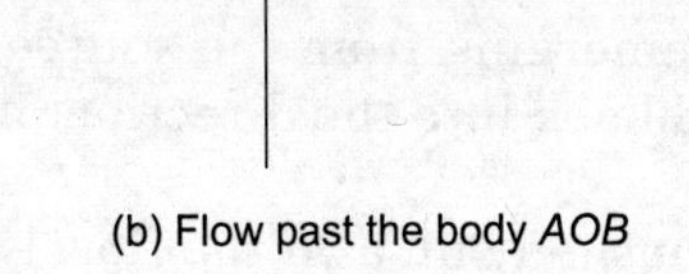


(b) Flow past the body *AOB*

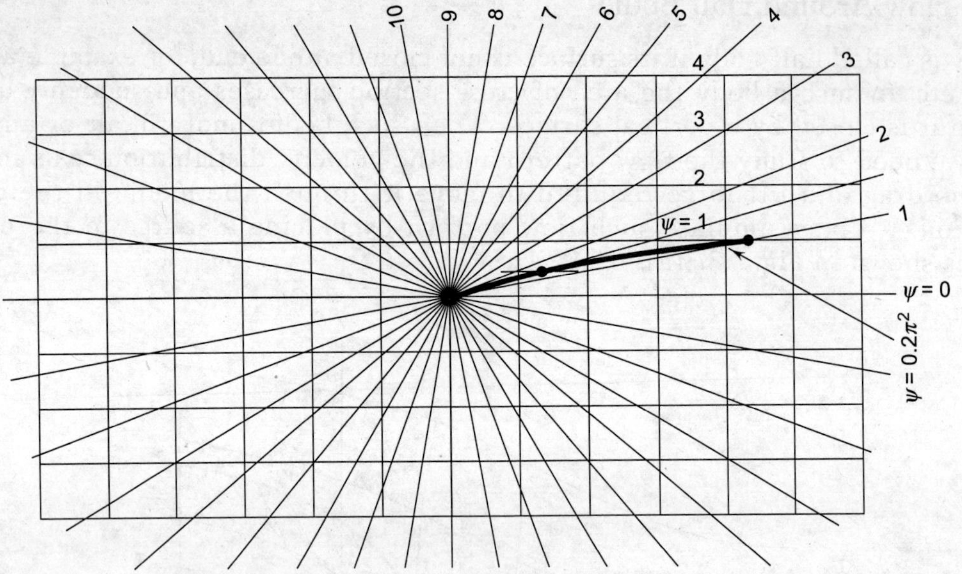


(c) Construction procedure for elliptic body

**Fig. 3.3.3** Flow past half body

The radial flow towards the negative $x$-direction from the source, placed at $O$, will oppose the uniform flow approaching it. At some point $S$, the radial velocity, will be equal to the uniform velocity $U$ and therefore velocity at that point will become zero along $x$-direction. Also, there is no velocity component in $y$-direction, i.e., the point $S$ will be stagnation point. Equation 3.2.2 gives the radial velocity from the source. Therefore, the distance $OS$ is obtained from $U = k/OS$, or $OS = k/U$, where $k$ is the source strength. At any point $P$ in the flow, the stream function and the velocity potential are

$$\psi = Uy + k\theta = Ur \sin \theta + k\theta$$

and $$\phi = Ur \cos \theta + k \ln r \tag{3.3.4}$$

$\psi = C$ and $\phi = C$ lines are shown in Fig.3.3.3(a). Following observations are made from this figure:

(a) At the point $S$, $\theta = \pi$ and therefore the streamline passing through this point has the stream function $= q/2$, since $k = q/(2\pi)$; the stream line $\psi = q/2$ is called the dividing line.

(b) For the radial lines emerging from the source in the direction $\pi/2 < \theta < 3\pi/2$, the radial lines will bend back into the direction of main flow, as shown in the hatched area.

(c) The uniform flow is pushed out, as if the flow is taking place over the surface $\psi = q/2$, since it cannot cross the streamline.

Therefore, the line $\psi = q/2$ has special significance as it divides the flow into two types of flow: (i) the flow in the hatched region and (ii) the flow in the region outside it. If we assume that the body having the shape of $\psi = q/2$ is placed then the flow outside will resemble this flow pattern. We can suitably choose $q$ and $U$ so that the required body profile is produced. We further note that the body is not closed downstream; its $y$-dimension increases asymptotically to $y_{max}$ and hence the name half body. Equation of the half body and its dimensions at different points will be

$$Uy + \frac{q\theta}{2\pi} = \frac{q}{2} \tag{3.3.5}$$

$$\rightarrow \qquad y = r \sin \theta = \frac{q\left(1 - \theta/_{\pi}\right)}{2U} \Rightarrow$$

At $\theta = 0$, $y_{max} = {}^{q}/_{2U}$ (the maximum cordinate) $\quad \theta = \pi/2$, $y = {}^{q}/_{4U}$,

$\theta = \pi$, $y = 0$, and $\quad \theta = 3\pi/2$, $y = -{}^{q}/_{4U}$

The velocity components anywhere in the flow are given by

$$u_r = \frac{\partial \psi}{r\partial \theta} = U \cos \theta + \frac{q}{2\pi r} \text{ and} \tag{3.3.6}$$

$$u_\theta = -\frac{\partial \psi}{\partial r} = -U \sin \theta$$

We note that when $\theta = \pi$ then $u_r$ is zero at a distance $q/(2\pi U) = OS$, as obtained before from the arguments and $u_\theta$ is zero everywhere on this line. Therefore, the point $S$ is the stagnation point. The line $y = 0$ is the streamline for which $\psi = k\theta$ (from Eq. 3.3.4). At point $S$, $\theta = \pi$ and therefore $\psi = q/2$. Further, the streamline $\psi = q/2$ bifurcates into two lines, which meet at infinity downstream. All along this line, the flow is never across it by definition of streamline. Thus we may consider the two streamlines to form a body around which the flow is taking place.

To obtain the shape of the body, we draw the streamlines for the source, which are straight lines through the origin at different angles, and lines parallel to $x$-axis indicating the streamlines for uniform flow. For example, if $q = 0.4\ \pi^2$ m²/s then $k = 0.2\ \pi$. Thus, the streamline at $\theta = 10° = 10/(2\pi)$ has the value $\psi = 1$. Similarly, the lines at angles 20°, 30°, etc will have $\psi$ equal to 2, 3, etc. If the uniform flow is having speed of 1 m/s then $\psi$ = 1 will be line parallel to $x$-axis at 1 m distance from it. The point at which the streamlines for the source and the uniform flow meet will have stream function equal to sum of the corresponding values for the two intersecting lines. We may thus locate points in the flow field where the total stream function is $q/2 = 0.2\ \pi^2 = 1.974$; locus of such points is the profile of the half body around which the flow pattern given by Eq. 3.3.4 will exist. Figure 3.3.3 (c) gives the graphical steps. The stagnation point $(-\ q/(2\pi U), 0)$ is called the leading point of the half body. You may also determine the resultant velocity along the surface of the half body and equate its differential with respect to $\theta$ to zero for finding out the point of maximum velocity. This gives $\theta = \pm 67°$.

There is no reason that we have to consider a half body having $\psi = q/2$ to study the flow around it. We may even consider the flow around the streamlines passing through the origin having stream function zero. In this case the stagnation point will not lie on the surface, rather it would fall in the flow region itself at $S$ at a distance $OS$, determined before, as shown in Fig.. 3.3.3 (b). A look at this body shows that it is a wedge placed in the flow.

In the following sections, we shall avoid repeating the procedure of plotting the streamlines for the individual basic flow and then determine it for the entire flow.

### 3.3.3 Flow Around Elliptic Bodies

There are bodies which have both the leading as well as the trailing edge. That is, the dividing streamline is a closed curve, resembling an ellipse and hence they are called the elliptic bodies. This requires pulling the open end of the streamline on half body back to the centre line, which can be done by adding a sink at some distance from the source. Thus, if a source and sink pair is placed in uniform flow then the flow around the elliptic body is generated.

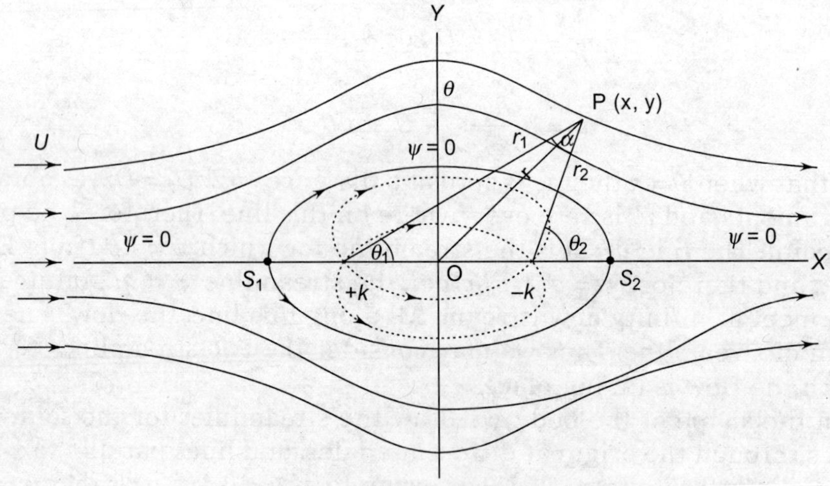

(a) Source and sink placed in uniform flow.

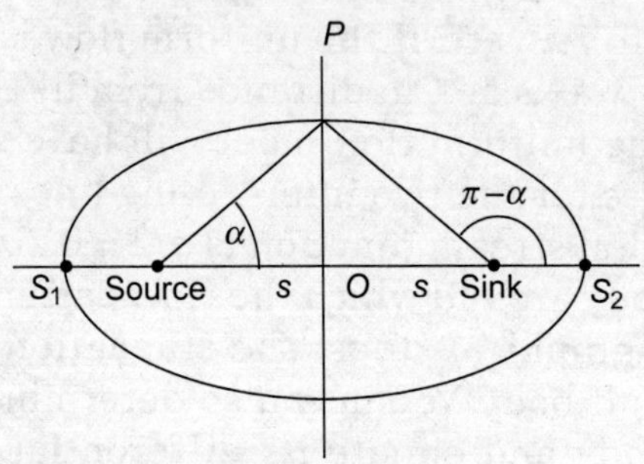

(b) Maximum width of the oval

**Fig. 3.3.4** Flow around elliptic body

Figure 3.3.4 shows uniform flow $U$ superimposed on the source–sink pair of strength $k$ with $O$ as the origin of the coordinate axes. The stream function at any point $P$ $(r, \theta)$ or $(x, y)$, using Eq. 3.3.2, is

$$\psi = Uy - k\alpha = U r \sin\theta + \frac{q}{2\pi}(\theta_1 - \theta_2)$$

$$= Uy - \frac{q}{2\pi}\tan^{-1}\frac{2ys}{x^2 + y^2 - s^2} \tag{3.3.7}$$

This gives $y = \frac{q}{2\pi U}\tan^{-1}\frac{2ys}{x^2 + y^2 - s^2}$ for $\psi = 0$ streamline. This is an equation of Rankine oval, called in common language the elliptic body. We may solve this equation to get the position of the stagnation points $S_1$ and $S_2$. Similarly, we may find out the height of the body at $x = 0$. The length of the major and minor axes will depend on the distance $s$ between the source and the sink, flow velocity $U$ and the source strength $k$. Thus, choosing these parameters suitably, we can generate the desired oval shape.

Alternately, the major and minor axes width can be determined noting that the stagnation points will lie at the $x$-axis and hence at the two stagnation points, $\theta = 0$ or $\pi$. Further, if $r_s$ is the distance of the stagnation point from the origin, then $r_1 = r_s - s$ and $r_2 = r_s + s$. Substituting these values in Eq. 3.3.5, we get half length of the major axis, $r_s = s\sqrt{1 + \frac{2k}{sU}}$.

The half-width of the oval may be determined by considering a point $P$ on the $y$-axis, as shown in Fig. 3.3.4 (b). In this position, the triangles on the two sides of the vertical line $OP$ will be similar. That means if angle $\theta_1 = \alpha$ then $\theta_2 = \pi - \alpha$. The point $P$ lies on the oval body, $\psi = 0$, at a distance $h$ from the origin where $\theta = \pi / 2$. Thus we get from Eq. 3.3.7,

$$0 = Uh + \frac{q}{2\pi}[\alpha - (\pi - \alpha)]$$

or,
$$h = \frac{k}{U}(\pi - 2\alpha)$$

or,
$$\alpha = \frac{\pi}{2} - \frac{Uh}{2k}$$

Also $h = s\tan\alpha = s\cot\left(\frac{Uh}{2k}\right)$.

Solving the transcendental equation we get the value of half-width of the oval.

**Example 3.3.2** Discuss the characteristics of the flow obtained by superimposing of radial inward flow on free vortex flow.

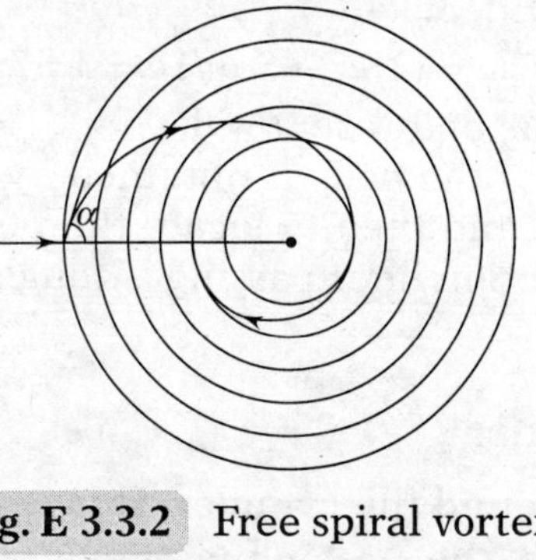


**Fig. E 3.3.2** Free spiral vortex

**Solution** For radial inward flow, $u_r . r = A$, since $Q = 2\pi Br. u_r$, where $B$ is the width of the passage through which the radial flow is taking place and $A$ is a constant. For a free vortex flow, $u_\theta\, r = C$, where $C$ is a constant. The resultant velocity at any point in the resultant flow and the angle between radius vector and the streamline is

$$V = \sqrt{u_r^2 + u_\theta^2}$$

$$= \left(\frac{A^2}{r^2} + \frac{B^2}{r^2}\right)^{1/2}$$

$$= \frac{1}{r}\sqrt{A^2 + B^2}$$

and $$\tan \alpha = \frac{u_\theta}{u_r} = \frac{B}{A}$$

This is the equation for an equiangular spiral, also called as free spiral, as shown in Fig. E3.3.2.

**Example 3.3.3** Is the flow expressed by the velocity potential, $\phi = Ux + K$ in $r$ possible for all $U$ and $K$? Sketch the flow if $U = 10$ m/s and $K = 5$ m²/s.

**Solution** We shall work out in polar coordinates. Replace $x$ with $r \cos \theta$ in the expression for $\phi$ and differentiate with $\theta$ and $r$ to give

$$\frac{\partial \phi}{\partial \theta} = -U_r \sin \theta, \quad \frac{\partial^2 \phi}{\partial \theta^2} = -U_r \cos \theta,$$

$$\frac{\partial \phi}{\partial r} = U \cos \theta + \frac{K}{r} \quad \text{and} \quad \frac{\partial^2 \phi}{\partial r^2} = -\frac{K}{r^2}$$

$$\rightarrow \quad \nabla^2 \phi = \frac{\partial^2 \phi}{\partial r^2} + \frac{1}{r}\frac{\partial \phi}{\partial r} + \frac{1}{r^2}\frac{\partial \phi}{\partial \theta^2} = 0.$$

Therefore the flow is possible and irrotational for all values of $U$ and $K$. A look at the expression for $\phi$ reveals that this is the kind of flow observed when a source of strength $K$ is placed in uniform flow and therefore represents a flow around a half body.

We may obtain stream function also by integrating velocity components or by using the equation derived earlier. This gives

$$\psi = 10\, y + 5\, \theta$$

This equation can be plotted taking the dividing line as $\psi = q/2 = \pi K = \pi \times 5 = 15.7$. For this dividing line, using Eq. 3.3.4,

$$y = (15.7 - 5\, \theta)/10 = 1.57 - 0.5\, \theta.$$

The cardinal points for this body are: $\theta = 0$, $y = y_{max} = 1.57$ m, $\theta = \pi/2$, $y = 0.785$ m and $\theta = \pi$, $y = 0$. The exact shape of the body is obtained by plotting constant $\psi$ lines for the uniform flow and the source and then to locate the points in the field where sum of $\psi$ values of the two stream function lines passing through the point is 15.7. The procedure is explained in section 3.3.2.

## 3.3.4 Flow Around Cylinder

A special case is the flow around the cylinder. This is achieved by placing a doublet in the uniform flow. In this case, the stream function at any point $(r, \theta)$ is

$$\psi = Uy - \frac{\mu \sin\theta}{2\pi r}$$
$$= \left(Ur - \frac{\mu}{2\pi r}\right)\sin\theta \tag{3.3.8}$$

where $\mu$ is the doublet strength and $U$ is the velocity of the uniform flow. For the dividing streamline $\psi = 0$, $r = \sqrt{\mu/(2\pi U)} = R$, we note that the dividing line is the equation of a circle and hence this kind of flow will occur when fluid is flowing over a cylinder of radius $R$. We may now write for $\psi$, $\phi$ and the velocity components in terms of radius $R$ as

$$\psi = U\left(r - \frac{R^2}{r}\right)\sin\theta \text{ and}$$

$$\phi = U\left(r + \frac{R^2}{r}\right)\cos\theta \tag{3.3.9}$$

$$u_r = \frac{\partial\psi}{r\partial\theta} = U\left(1 - \frac{R^2}{r^2}\right)\cos\theta$$

and
$$u_\theta = -\frac{\partial\psi}{\partial r} = -U\left(1 + \frac{R^2}{r^2}\right)\sin\theta \tag{3.3.10}$$

We note from Equation 3.3.10 that at $r = R$, both the radial and the tangential velocities are zero when $\theta$ is 0 or $\pi$. That is, there are two stagnation points, one on the left and the other on the right side at a distance $R$ from the origin. However, at any point $\theta$, $u_r$ is zero everywhere on the cylinder surface ($r = R$) and $u_\theta$ is $-2U \sin\theta$ and hence the velocity is not zero on the cylinder at all other points. The flow around the body is symmetrical about both the $x$ and $y$-axes. The pressure distribution will also be symmetrical; hence there will be no force on the cylinder along the $x$-direction or along the $y$-direction.

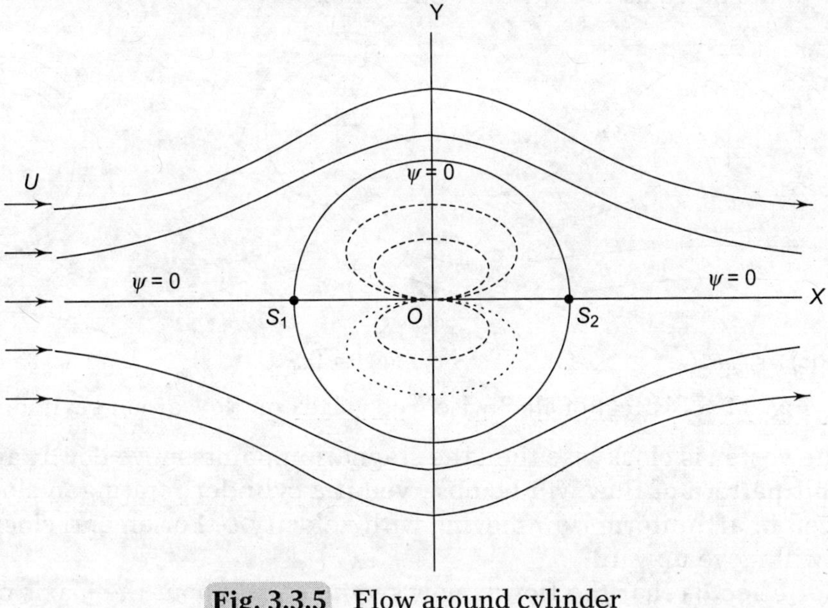


**Fig. 3.3.5** Flow around cylinder

Further, you may verify that the circulation is zero since the vorticity $\vec{\Omega} = \nabla \times \vec{U} = 0$. So we conclude that the inviscid irrotational incompressible flow around a circular cylinder of radius $R$ may be obtained from a combination of a uniform flow $U$ and a doublet flow of strength $\mu = 2\pi UR^2$. This study became important because we could transform other body shapes into a circular cylinder using the conformal mapping techniques.

Let us now study the effect of introducing the clockwise free vortex of strength Γ. The stream function, velocity potential and the velocity components then are:

$$\psi = U\left(r - \frac{R^2}{r}\right)\sin\theta + \frac{\Gamma}{2\pi}\ln\frac{r}{R} \text{ and } \phi = U\left(r + \frac{R^2}{r}\right)\cos\theta + \frac{\Gamma}{2\pi}\theta \tag{3.3.11}$$

$$u_r = \frac{\partial\psi}{r\partial\theta} = U\left(1 - \frac{R^2}{r^2}\right)\cos\theta \text{ and } u_\theta = -\frac{\partial\psi}{\partial r} = -U\left(1 + \frac{R^2}{r^2}\right)\sin\theta - \frac{\Gamma}{2\pi r} \tag{3.3.12}$$

At $r = R$, $\psi = 0$ and therefore the circle represents the stream line. Further, from Eq. 3.3.12, we get at $r = R$, $u_r = 0$ and $u_\theta = -2U\sin\theta - \frac{\Gamma}{2\pi R}$. This gives $\sin\theta = -\Gamma/(4\pi RU)$ for the point where the tangential velocity too is zero. Thus location of the two stagnation points depends on the value of the factor $\Gamma/(4\pi RU)$. If this ratio is less than 1 then there are real values of $\theta$, since $|\sin\theta| \le 1$ and the stagnation points lie below the $x$-axis(because of the negative sign in the expression for $\sin\theta$). But if $\Gamma/(4\pi RU) > 1$ then the stagnation point does not lie on the surface, since $|\sin\theta| > 1$, which has no solution. However, from Eq. 3.3.12, we deduce that $u_r$ is zero at $\theta = -\pi/2$ or $3\pi/2$ when $r > R$. Therefore, the flow pattern in this case looks like as shown in Fig. 3.3.6 (c), which also shows flow patterns for other values of Γ; observe how do the stagnation points move as the ratio $\Gamma(4\pi RU)$ increases.

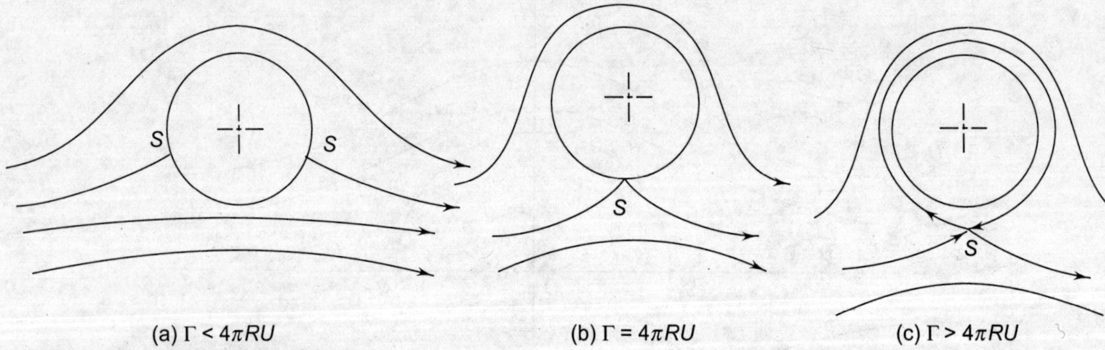


**Fig. 3.3.6** Effect of clockwise free vortex on flow around cylinder

When the vortex is clockwise then the stagnation points move downwards below the cylinder. This pattern of flow will be observed if a cylinder rotating in clockwise direction is placed in an uniform flow moving with velocity $U$. For an anti-clockwise vortex, the points will move upward.

The other effect is that the flow is now symmetrical about the $y$-axis only; the symmetry about the $x$-axis ceases to exist. Since the flow is symmetrical around $y$-axis, the pressure distribution too would be so and hence there is no force on the cylinder in $x$-direction. The force acting on any surface in the direction of flow is called *drag* and the body will be forced to move in the direction of the flow unless some external force holds it in the position. The force acing on any surface in a flow of ideal fluid is due to the pressure distribution, which is found by using the Bernoulli's equation $p/\rho + C^2/2 =$ constant, where $C$ and $p$ are respectively the velocity and pressure at a point on the cylinder. Let $p_s$ be the pressure in the uniform flow field far away from the cylinder where the streamlines are parallel to the $x$-axis. Applying Bernoulli's equation and using Eq. 3.3.10, we get pressure distribution over the cylinder surface as

$$\frac{p}{\rho} + \frac{u_r^2 + u_\theta^2}{2} = \frac{p_s}{\rho} + \frac{U^2}{2}$$

$$\rightarrow \qquad \frac{p - p_s}{\rho} = \left(1 - 4\sin^2\theta\right)\frac{U^2}{2} \qquad (3.3.13)$$

We define dimensionless *pressure coefficient* as

$$C_p = \frac{p - p_s}{\frac{1}{2}\rho U^2} = \left(1 - 4\sin^2\theta\right) \qquad (3.3.14)$$

Distribution of pressure and the dimensionless coefficient of pressure for flow without imposition of free vortex are shown in Fig. 3.3.7. You may see that it is symmetrical about both the $x$ and $y$-axes. $C_p$ is zero at $\theta = \pm 30°$, as shown in the figure. It is positive from –30° to +30° and negative for other angles. The effect of symmetrical pressure distribution is that there is no force on the cylinder either in $x$ or $y$-direction. We should now try to find out the pressure distribution with free vortex and see its effect on the forces acting on the cylindrical surface. Since the symmetry about the $x$-axis is lost, it is expected that there will be a force in the $y$-direction; this force is called *lift*.

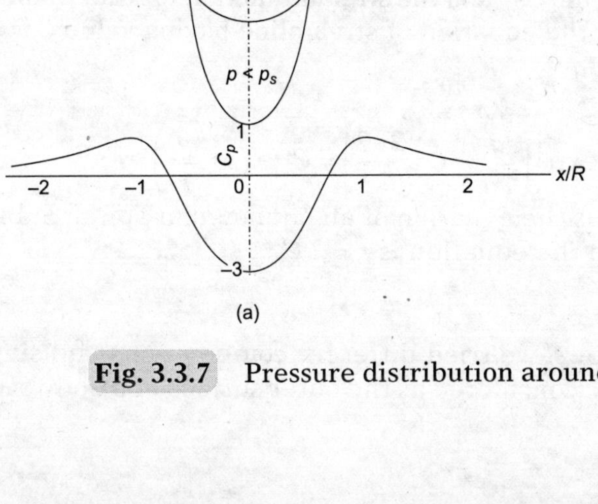

(a)

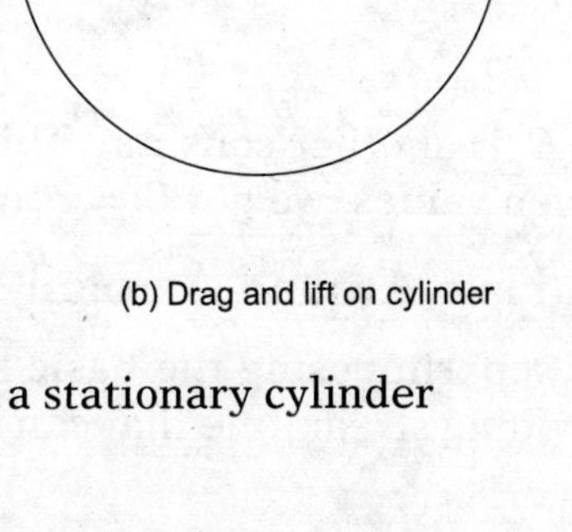

(b) Drag and lift on cylinder

**Fig. 3.3.7** Pressure distribution around a stationary cylinder

If we use values for radial and tangential velocity components from Eq. 3.3.1 in Eq. 3.3.13 then

$$C_p = \frac{p - p_s}{\frac{1}{2}\rho U^2} = 1 - \left( -2\sin\theta - \frac{\Gamma}{2\pi RU} \right)^2 \quad (3.3.15)$$

The pressure force can be resolved along and perpendicular to the direction of uniform flow. Figure 3.3.7(b) shows the pressure force acting on the cylinder arc subtending angle $d\theta$ at the centre. The coefficient of drag is defined as

$$C_D = \frac{\text{Pressure force along the flow direction}}{\frac{1}{2}\rho U^2 \times \text{projected area of cylinder}}$$

$$= \frac{1}{\frac{1}{2}\rho U^2\ (2R \times 1)} \int_0^{2\pi} C_p\ R\cos\theta\, d\theta = 0$$

$$\text{and } C_L = \frac{\text{Pressure force} \perp \text{to the flow direction}}{\frac{1}{2}\rho U^2 \times \text{projected area of cylinder}}$$

$$= \frac{1}{\frac{1}{2}\rho U^2\ (2R \times 1)} \int_0^{2\pi} C_p\ R\sin\theta\, d\theta = \frac{\Gamma}{RU} \quad (3.3.16)$$

where $C_p$ is used from Eq. 3.3.15. *It is noticed that superimposition of vortex (circulation) in the uniform flow over a cylinder produces lift force* and that the lift per unit span of the cylinder is directly proportional to the circulation. This is the *Kutta–Joukowski* theorem. Automobile designers thought of using the concept of rotating cylinders fixed along the side of the vehicle body around its top to produce upward lift so that total downward force on the road be reduced to decrease the frictional drag force.

Another observation is that the drag force is zero, which does not agree with the experimental results. This is known as *De' Alembert's paradox*. The explanation is that pressure force is one source of drag amongst many others. Viscous force is another force that is responsible for this paradox and has been neglected in our above analysis.

**Example 3.3.4** A two-dimensional source of strength $q$ and an anticlockwise vortex of strength $K$ are situated at the origin. Establish the general equation of streamlines. If $q$ = 1.0 m²/s and $k$ = 3.14 m²/s, find the equation of the streamline passing through the point (2, 0) in the polar coordinates.

**Solution** By the principle of superimposition, the stream function for the combined flow is obtained and from this equation the equation of streamline having value $C$ is given as

$$\psi = \frac{q}{2\pi}\theta - \frac{k}{2\pi}\ln r = C$$

$$\rightarrow \qquad r = \exp\left[\left(\frac{q}{2\pi} - C\right)\frac{2\pi\theta}{k}\right] = C_1 \exp\left(\frac{q\theta}{k}\right)$$

where $C_1$ is another constant. This is the equation of an equiangular spiral. Substituting the given values, we get $C_1 = 2$ and the equation is $r = 2\,e^{\theta/\pi}$.

### 3.3.5 Flow Around Aerofoil

While superimposing the basic flows, we tried different combinations and singled out the dividing streamline that can be considered as the outer surface of a body over which the uniform flow takes place. In this process, we could deduce the procedure for half bodies, elliptic and cylindrical bodies. However, we may have some shape, which we may probably not be able to get from superimposition of the basic flows, for example an aerofoil. Therefore we look for some other alternative method to handle such surfaces. The solution comes in the form of conformal mapping, which helps in changing the shape of the given body to a cylinder. For this, we look for theory of complex numbers and form a function $W(z) = \phi(x, y) + i\,\psi(x, y)$, where $\phi$ and $\psi$ are, respectively, the velocity potential and stream function defined in $x$, $y$ plane. Since both $\phi$ and $\psi$ are continuous functions, the function $W(z)$ is analytic function if the differential of $W$ is independent of the way $\Delta z \to 0$. This means we may get $W'(z)$ by varying $x$ by $\Delta x$ and then find out the limit of the change in $W$ when $\Delta x \to 0$. Or, the same procedure be adopted by varying $y$ instead of $x$; the result should be same. Implementation of the procedure gives

$$\mathrm{W}'(z) = \lim_{\Delta x \to 0} \frac{\phi(x + \Delta x, y) - \phi(x, y) + i\psi(x + \Delta x, y) - i\psi(x, y)}{\Delta x} = \frac{\partial \phi}{\partial x} + i\frac{\partial \psi}{\partial x}, \text{ and} \tag{3.3.17}$$

$$\mathrm{W}'(z) = \lim_{\Delta y \to 0} \frac{\phi(x, y + \Delta y) - \phi(x, y) + i\psi(x, y + \Delta y) - i\psi(x, y)}{i\Delta y} = \frac{\partial \psi}{\partial y} - i\frac{\partial \phi}{\partial y} \tag{3.3.18}$$

Comparison of Equation 3.3.17 and 3.3.18 gives

$$\frac{\partial \phi}{\partial x} = \frac{\partial \psi}{\partial y} = u \quad \text{and} \quad -\frac{\partial \psi}{\partial x} = \frac{\partial \phi}{\partial y} = v \tag{3.3.19}$$

Recall that we obtained these relations in the process of introducing the stream function and the velocity potential. These relations, in fact, are the Cauchy–Riemann conditions that are satisfied by any analytical function of a complex number $z$, defined as $z = x + iy$. The flow around the cylinder of radius $R$ in the complex plane $z$ is given by the function

$$W = U\left(z + R^2/z\right) \tag{3.3.20}$$

We may verify Eq. 3.3.20 by writing $W = \phi + i\,\psi$ and $z = x + i\,y$, splitting both sides into real and imaginary parts and then equating the real parts and imaginary parts to give

$$\psi = U\left(1 - \frac{R^2}{r^2}\right)y = U\left(r - \frac{R^2}{r}\right)\sin\theta \text{ and } \phi = U\left(1 + \frac{R^2}{r^2}\right)x = U\left(r + \frac{R^2}{r}\right)\cos\theta \tag{3.3.9}$$

Now let us come back to the original question of how to handle the aerofoil sections. The aerofoils are available in different kinds of sections and therefore each of these has its own function to transform into a circle. We shall determine the transformation function now. This technique is called mapping from $z$ section to $\tilde{z}$ plane, where the aerofoil is drawn in the $z$ plane and the corresponding circle in the $\tilde{z}$ plane.

***Mapping of a circle to a flat plate aerofoil***

Consider the transformation defined by the relation

$$z = \tfrac{1}{2}\left(\tilde{z} + c^2/\tilde{z}\right) \tag{3.3.21}$$

where $c$ is some constant equal to the radius of the circle in $\tilde{z}$ plane and $\tilde{z} = |\tilde{z}|\,e^{i\theta} = |\tilde{z}|\,(\cos\theta + i\sin\theta)$. Substituting for $\tilde{z}$ in Eq. 3.3.21, separating the terms into the real and imaginary parts and then equating them, we get

$$x = \frac{1}{2}\left(|\tilde{z}| + \frac{c^2}{|\tilde{z}|}\right)\cos\theta \quad \text{and} \quad y = \frac{1}{2}\left(|\tilde{z}| - \frac{c^2}{|\tilde{z}|}\right)\sin\theta \tag{3.3.22}$$

If we choose $\tilde{z} = c$ then $x = c \cos \theta$ and $y = 0$. That is, the transformation given by Eq. 3.3.21 results in a flat surface of length $2c$. The centre of the circle is at the origin in this case.

***Mapping of a circle to an ellipse***

If we choose $\tilde{z} = R > c$ in Equation 3.3.22 then

$$x = \frac{1}{2}\left(R + \frac{c^2}{R}\right)\cos\theta \quad \text{and} \quad y = \frac{1}{2}\left(R - \frac{c^2}{R}\right)\sin\theta \tag{3.3.23}$$

We can eliminate $\theta$ from Eq. 3.3.23 by determining $\cos \theta$ and $\sin \theta$, squaring and then adding. This gives

$$\left(\frac{2x}{R + c^2/R}\right)^2 + \left(\frac{2y}{R - c^2/R}\right)^2 = 1,$$

which is the equation of an ellipse having major axis = $R + c^2/R$ and minor axis = $R - c^2/R$. We note that $R/c$ determines the shape of the ellipse.

***Mapping of a circle to an aerofoil with parabolic arc camber***

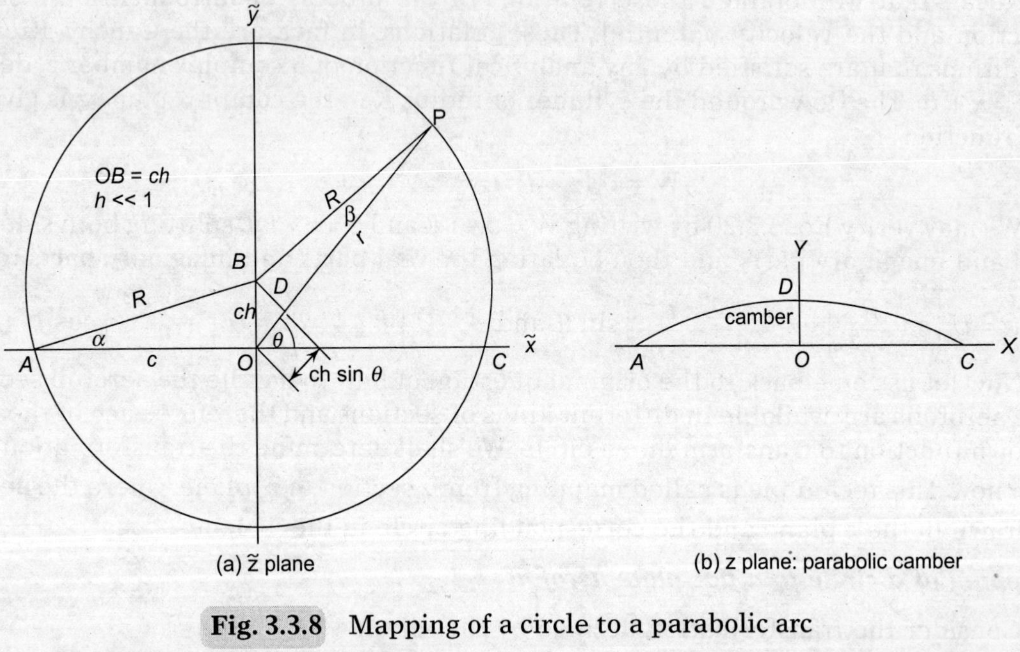


**Fig. 3.3.8** Mapping of a circle to a parabolic arc

We offset the centre of the cylinder of radius $R$ along $\tilde{y}$ by a small vertical distance $OB = ch$ ($h << 1$) from the origin, as shown in Fig. 3.3.8 and choose $c$ in Eq. 3.3.21 as $OA$. Note that now $c$ is not equal to the radius of the circle, which is $R$. Let $\angle OAB = \alpha$ and $\angle OAB = \beta$, where $P(r, \theta)$ is a point on the surface, and origin is at $O$. We draw a perpendicular $BD$ on the radial line $OP$ so that $DP = R \cos \beta$, $BD = R \sin \beta$ and $OD = ch \sin \theta$. We may now write for $r$ considering that the angle $\beta$ is very small so that $\cos \beta = 1$.

$r = OD + DP = ch \sin \theta + R \cos \beta \approx R + ch \sin \theta$ and $\tilde{z} = r e^{i\theta} = r(\cos \theta + i \sin \theta)$.

But $$\frac{R}{c} = \frac{AB}{c} = \frac{\sqrt{OA^2 + OB^2}}{c} = \frac{\sqrt{c^2 + c^2h^2}}{c}$$

$$\rightarrow \quad \frac{r}{c} = \frac{R}{c} + h \sin \theta = \frac{\sqrt{c^2 + c^2h^2}}{c} + h \sin \theta \approx 1 + h \sin \theta$$

and $$\frac{c}{r} = (1 + h \sin \theta)^{-1} \approx 1 - h \sin \theta.$$

neglecting higher order terms of $h \sin \theta$. We may now follow the procedure of separating the terms in Eq. 3.3.21 into real and imaginary parts and use the expressions for $r/c$ and $c/r$, as obtained above. This gives

$$x = C \cos \theta \quad \text{and} \quad y = ch \sin^2\theta = \tfrac{1}{2}\, ch\,(1 - \cos 2\theta) \tag{3.3.24}$$

This is the equation of a parabolic arc with camber $OD$ equal to ch; *camber* is the distance of the point of maxima of the curve from the base line $AC$, called the *chord*. The line $ADC$ is called the camber line. This is shown in Fig. 3.3.8 (b) on $z$ plane.

### *Mapping of a circle to a symmetric aerofoil*

We may move the origin of the cylinder in the horizontal instead of vertical direction by a distance $ch$, in which case $c = OA$ and follow the procedure discussed for parabolic arc to give

$$r = ch \cos \theta + R \cos \beta \approx ch \cos \theta + R = ch \cos \theta + c + ch$$

$$x = C \cos \theta \quad \text{and} \quad y = ch\,(1 + \cos \theta) \sin \theta \tag{3.3.25}$$

We note from Eq. 3.2.25 that the function $y$ is symmetric about $x$-axis as changing $\theta$ to $-\theta$, changes sign of $y$. Therefore this equation represents the aerofoil section symmetric about $x$-axis, as shown in Fig. 3.3.9(b).

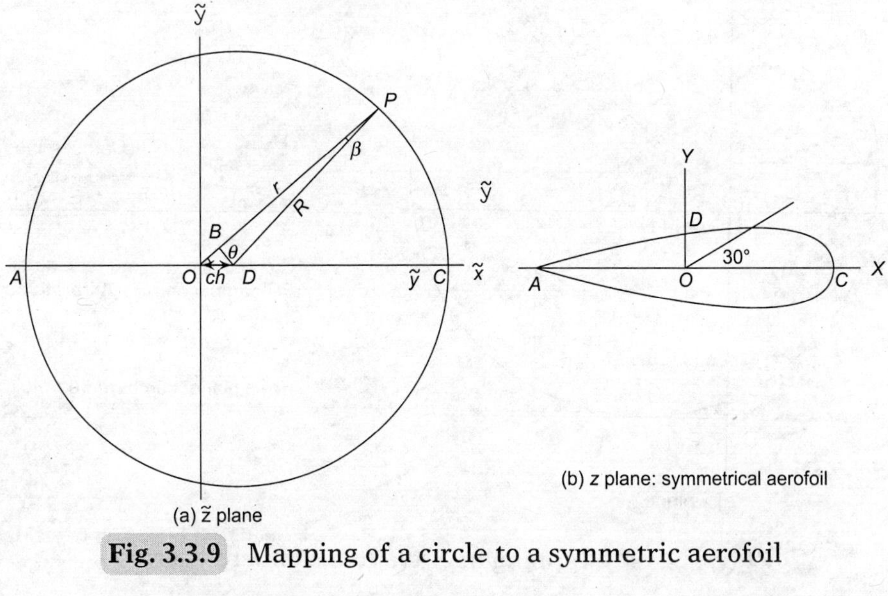


**Fig. 3.3.9** Mapping of a circle to a symmetric aerofoil

Differentiation of $y$ with respect to $\theta$ gives the position of maximum thickness of the aerofoil. This results in $\theta = \pi/3$ and maximum thickness = 2 $y_{max}$ = $(3\sqrt{3}/2)\ ch$. To draw the profile of the foil, let us also get some idea of the slope at $\theta = 0$ and $\pi$. The slope $dy/dx$ is

$$\frac{dy}{dx} = \frac{dy}{d\theta} / \frac{dx}{d\theta} = -h\,\frac{2\cos^2\theta + \cos\theta - 1}{\sin\theta}$$

$$\rightarrow \qquad \left(\frac{dy}{dx}\right)_{\theta=0} = \infty \text{ and } \left(\frac{dy}{dx}\right)_{\theta=\pi} = 0,$$

using L'Hopital's rule in case of $\theta = \pi$. The lower and the upper surfaces of the foil meet tangentially whereas the tangent to the surface at $\theta = 0$ is a vertical line. We may get a few more points at different value of $\theta$ to draw the curve.

***Mapping of a circle to a cambered aerofoil***

Next we move the centre of the circle to any point $B$ such that $OD = ce$ and $BD = ch$, where point $D$ is the projection of the point $B$ on the $\tilde{x}$-axis. Let the line $BE$, parallel to $x$-axis, passes through centre of the circle and $\angle EBP = \gamma$, rest of the construction and symbols are same. In this case, we get the following results using the above procedure

$$r = OF + FG + GP = ce\cos\theta + ch\sin\theta + R\cos\beta$$

$$\approx ce\cos\theta + ch\sin\theta + R \approx ce\cos\theta + ch\sin\theta + c + ce$$

$$\therefore \qquad r/c = 1 + e + e\cos\theta + h\sin\theta,$$

$$c/r = (1 + e + e\cos\theta + h\sin\theta)^{-1}$$

$$= 1 - e - \cos\theta - h\sin\theta \text{ and}$$

$$x = \cos\theta, \qquad y = ce(1 + \cos\theta)\sin\theta + ch\sin^2\theta \qquad (3.3.26)$$

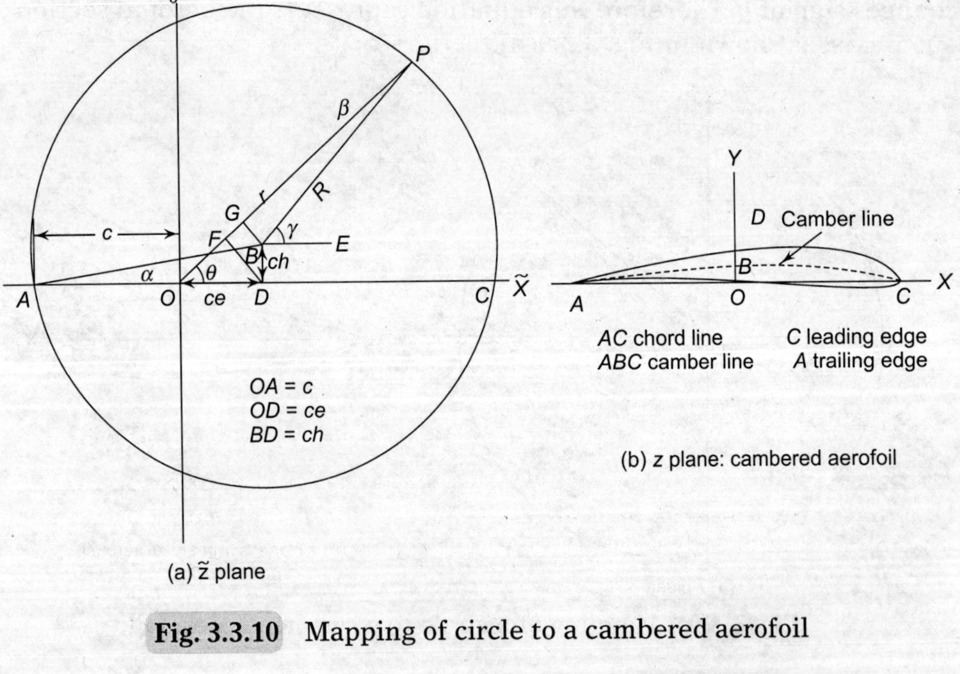


**Fig. 3.3.10** Mapping of circle to a cambered aerofoil

Equation 3.3.26 can be plotted on the $z$-plane using curve tracing techniques and is shown in Fig. 3.3.10(b). We define a curve such that $y$ at any $x$ is ½ $(y_1 + y_2)$, where $y_1$ and $y_2$ are the respective $y$-distances of the upper and the lower surface of the aerofoil at this $x$. This line is called the *camber line* and ½ $(y_1 + y_2)$ is the camber at that $x$. The camber at any $\theta$, from Equation 3.3.26 and definition of camber line, is $ch \sin^2 \theta$, where $\theta$ is the angle made by the radial line passing through the upper surface at a given $x$. We deduce that the camber is maximum when $\theta = 90°$ and is equal to $ch$. We may determine $ch$ knowing the camber of the given aerofoil section at $\theta = 90°$. It is also possible to measure $y$ at two points on the surface and then use Equation 3.3.26 to give two equations in $ch$ and $ce$ whose solution gives these while $c$ itself is half the distance between the leading point $C$ and the trailing edge $A$ of the section.

To generate the cylinder, we need $c$, $ch$ and $ce$ because then points $A$ and $B$ can be located with respect to the origin. Then $BA$ is the radius of the cylinder that we need in the calculation of the lift, velocity and pressure field around the cylinder; which would be around the aerofoil in $x$-$y$ plane. We shall demonstrate the procedure of calculation of the lift in the next section.

Before ending this section, it is worth mentioning that the camber is decided by the vertical shift of the circle from the origin.

## 3.3.6 Lift on Aerofoil

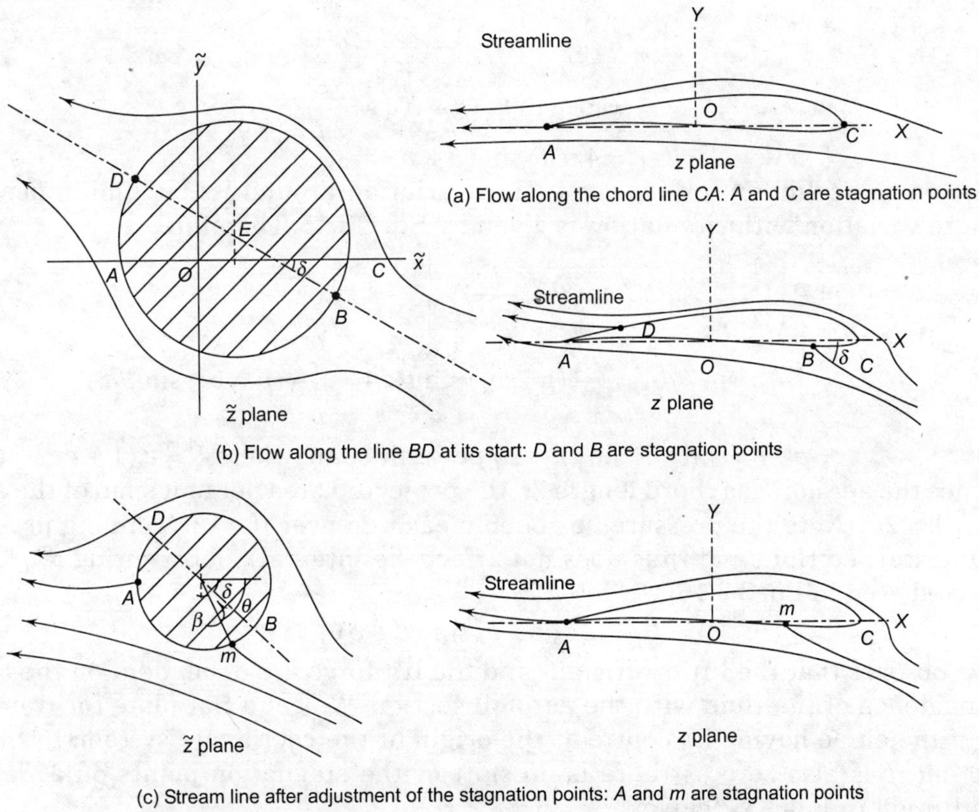


**Fig. 3.3.11** Flow around a general cambered aerofoil

Figure 3.3.11 shows flow around an aerofoil and its conformal transform cylinder for different situations of flow around it. Figure 3.3.11(a) corresponds to the situation when the flow is along its chord line. The stagnation points are at *A* and *C*. Figure 3.3.11(b) shows the flow at the start when the flow direction is along the line as shown. The angle made by this direction with the chord line is called the angle of incidence. Under this condition, the stagnation points are at *B* and *D*.

But the flow will not stabilise and would adjust till the stagnation point *D* coincides with the trailing edge. This will also shift the other stagnation point from *B* to *m*, as shown in Fig. 3.3.11(c) which is caused by the development of clockwise circulation flow around the aerofoil of strength say $\Gamma$. (Recall our discussions from the flow around a rotating cylinder that a clockwise circulation shifts the stagnation points down the cylinder and results in an upward force, the lift.) Let the stagnation points be shifted by angle $-\beta$ (negative sign indicates that the shift is clockwise), then the strength of the circulation can be determined from Eq. 3.3.12 because at $r = R$

$$u_r = 0 \text{ and } u_\theta = -2U\sin\theta - \frac{\Gamma}{2\pi R}$$

The stagnation point '*m*' will lie below the line *BD*, as shown in Fig. 3.3.11(c), that is, at the stagnation point $\theta = -(\beta + \delta)$. This condition gives the strength of the circulation generated by the aerofoil section in order to bring the stagnation point back to the trailing edge.

$$u_m = (u_r^2 + u_\theta^2)^{1/2}$$

$$= 2U\sin\{-(\beta+\delta)\} + \frac{\Gamma}{2\pi R} = 0, \tag{3.3.27}$$

$$\rightarrow \qquad \Gamma = 4\pi UR\sin(\beta+\delta)$$

Lift on the cylinder is due to pressure variation around its circumference where pressure variation with circulation is given by Eq. 3.3.15. Therefore,

$$p - p_s = \rho\left[U^2 - \tfrac{1}{2}\left(2U\sin\theta + \frac{\Gamma}{2\pi R}\right)^2\right]$$

$$L = \oint_A -\rho\sin\theta\, dA = \oint_A -(p-p_s)\sin\theta\, dA = \int_0^{2\pi} -(p-p_s)\sin\theta\, dA$$

$$= \rho U\Gamma = 4\pi U^2 R\sin(\beta+\delta) \text{ per unit span, where } R = c(1+e) \tag{3.3.28}$$

Since the aerofoil has chord length 2*c*, the projected area per unit span of the aerofoil is $2c \times 1 = 2c$. (Note the pressure has been measured over the surrounding pressure of the undisturbed fluid and thus does not affect the integral). Substituting for *L*, *R* and projected area in Eq. 3.3.16, we get

$$C_L = 2\pi(1+e)\sin(\beta+\delta) \tag{3.3.29}$$

We observe that the lift coefficient and the lift force are dependent on the angle of the incidence of the fluid with the aerofoil section. We get a flat plate for transformation of the circle having its centre at the origin of the coordinate system. In this case, $e = 0$ and $\beta$ is also zero as there is no shift of the stagnation points on a flat plate. Substitution in Eq. 3.3.29 gives

$$C_L = 2\pi\delta.$$

The velocity field in the $z$-plane, the plane of the aerofoil, can be determined as follows.

Let us use subscripts $c$ and $a$ to refer to cylinder and the aerofoil planes. From Eq. 3.3.21,

$$\frac{dz}{d\tilde{z}} = \frac{1}{2}\left(1 - \frac{c^2}{\tilde{z}^2}\right) = \frac{1}{2}\left(1 - \frac{c^2}{r^2} e^{-2i\phi}\right)$$

$$= \frac{1}{2}\left(1 - \frac{c^2}{r^2}\cos 2\phi\right) + i\frac{1}{2}\frac{c^2}{r^2}\sin 2\phi$$

$$\therefore \quad \left|\frac{dz}{d\tilde{z}}\right| = \frac{1}{2}\left[\left(1 - \frac{2c^2}{r^2}\cos 2\phi + \frac{c^4}{r^4}\right)\right]^{1/2} \tag{3.3.30}$$

We have learnt that there exists a function $w(\tilde{z}) = \phi(\tilde{z}) + i\,\psi(\tilde{z})$ for which

$$\frac{dw}{d\tilde{z}} = u_c - i\,v_c \qquad \rightarrow C_{\tilde{z}}^2 = \left|\frac{dw}{d\tilde{z}}\right|^2 = u_c^2 + v_c^2$$

Similarly, we get the relations in $z$-plane, the plane of aerofoil

$$\frac{dw}{dz} = u_a - i\,v_a \qquad \rightarrow C_z^2 = \left|\frac{dw}{dz}\right|^2 = u_a^2 + v_a^2$$

This gives

$$C_z = C_{\tilde{z}}\left|\frac{dw/dz}{dw/d\tilde{z}}\right| = \frac{C_{\tilde{z}}}{dz/d\tilde{z}}$$

$$= 2C_{\tilde{z}} \Big/ \left[\left(1 - \frac{2c^2}{r^2}\cos 2\phi + \frac{c^4}{r^4}\right)\right]^{1/2} \tag{3.3.31}$$

**Example 3.3.5** Given the complex potential $w = \ln z^2$, evaluate the stream and velocity potential functions and the components of velocity. Identify the flow pattern.

**Solution** We can separate out the complex function into its real and imaginary parts as follows:

$$w = \phi + i\,\psi = 2\ln z = 2\ln(re^{i\theta}) = 2\ln r + 2i\theta$$

$$\rightarrow \qquad \phi = 2\ln r = \ln(x^2 + y^2) \quad \text{and } \psi = 2\theta = 2\tan^{-1}(y/x).$$

Differentiation with respect to $z$ gives

$$\frac{dw}{dz} = \frac{2}{z} = \frac{2(x - iy)}{(x + iy)(x - iy)}$$

or,

$$u - iv = \frac{2x}{x^2 + y^2} - i\frac{2y}{x^2 + y^2}$$

$$\rightarrow \qquad u = \frac{2x}{x^2 + y^2} \quad \text{and} \quad v = \frac{2y}{x^2 + y^2}$$

We observe from the equation for the streamline that $w = 2 \ln z$ represents a source flow at the origin of the coordinate system with source strength $q = 2 \times 2\pi = 4\pi$ m²/s.

**Example 3.3.6** Figure E3.3.6 shows flow around a cylinder placed in a stream flowing from below; the velocity distribution is also shown both upstream and downstream. the cylinder is of diameter 100 mm and velocity is measured at 100 mm interval. What will be the coefficient of drag $C_D$ for this flow? Take density of the fluid as 1.208 kg/m³.

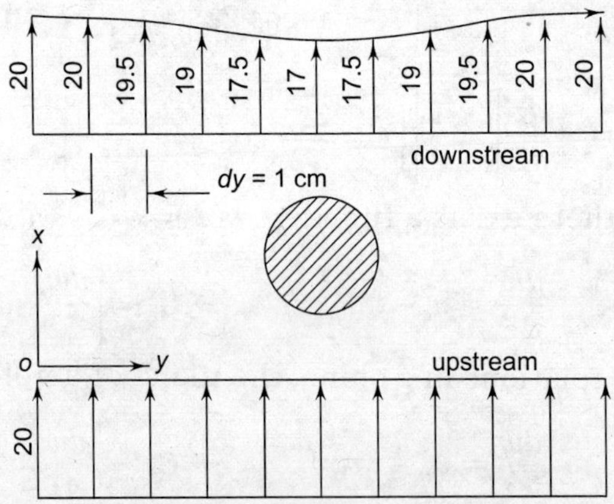


**Fig. E 3.3.6** Flow around cylinder

**Solution** This problem has been chosen to show the laboratory procedure for determining the coefficient of drag from the velocity measurements both upstream and down stream of the body placed in the flow. The procedure is to consider a strip of length $dy$ and then take average velocity in this region for calculation. Let the average velocity for any one such strip at the corresponding positions upstream and downstream be $u_1$ and $u_2$, respectively. The mass flow through the downstream strip will be $\rho u_2 dy$ and the loss of momentum during the flow over the cylinder is $\rho u_2 (u_2 - u_1)\, dy = \rho\,(u_2^2 - u_1 u_2)$. We may divide the entire flow in such strips, determine the momentum loss for each of these and sum all such results to give the entire loss in the momentum. Remember in this question, $dy$ is 0.1 m, $u_1$ is 20 m/s, $\rho = 1.208$ kg/m³ and $u_2$ would change from strip to strip, which is the average of the two values at the extreme ends of the strip. For example, consider the strip nearest to the centre line of the cylinder. The velocities at the extreme ends of the strip are 17 and 17.5 m/s. Therefore, we may take average (17 + 17.5)/2 = 17.25 m/s as the velocity $u_2$ on this strip. We observe the velocity distribution is symmetrical about the vertical axis. Therefore, the above calculations can be made for one side and then multiply it by a factor 2 to give the total loss.

$\therefore$ Loss of momentum $= 2 \times 1.208 \times [(17.25^2 - 17.25 \times 20) + (18.25^2 - 18.25 \times 20) + (19.25^2 - 19.25 \times 20) + (19.75^2 - 19.75 \times 20)] \times 0.1 = -\,23.858$ N

By integral momentum theorem, this loss is caused by the external force acting on the fluid and by Newton's third law the force on the external source is the reaction of this force. Thus the force acting on the cylinder is 23.858 N in the direction of the flow. The coefficient of drag is defined as follows:

$$C_D = \frac{\text{Drag force}}{1/2\rho U^2 \text{ (Projected area perpendicular to flow)}} = \frac{2 \times 23.58}{1.208 \times 20^2 \times (1 \times 0.1)} = 0.988.$$

Recall that the projection of the cylinder of unit width is a rectangle. We observe that the coefficient of drag is quite high.

## 3.4 Integration of Euler's Equation—Bernoulli Equation

For inviscid fluid, the three momentum equations can be combined into one equation. This requires changing the vector momentum equation into a scalar equation and is carried out by line integration of the momentum equations. Recall that the line integral of velocity field resulted in a scalar quantity that we called circulation. Similarly, the momentum equation dealing with forces will result in energy equation, by its definition. This equation was obtained by Euler. However, it is known by the name of Bernoulli who gave this equation centuries ago before the mathematical treatment appeared in literature and was in use for all this period. Here we shall work out in the Cartesian coordinates for an ideal fluid and steady flow and shall assume that the field acting on the fluid is gravity along the negative $z$-direction. We know that $\int \vec{F}.d\vec{l} = F_x\, dx + F_y\, dy + F_z\, dz$. Therefore multiply the $x$, $y$ and $z$-momentum equations, Eq. 2.4.15, by $dx$, $dy$ and $dz$ respectively and sum the resulting equation. That is,

$$u\partial u/\partial x + v\, \partial u/\partial y + w\, \partial u/\partial z = -(1/\rho)\, \partial p/\partial x \qquad \times dx$$

$$u\partial v/\partial x + v\, \partial v/\partial y + w\, \partial v/\partial z = -(1/\rho)\, \partial p/\partial y \qquad \times dy$$

$$u\partial w/\partial x + v\, \partial w/\partial y + w\, \partial w/\partial z = -(1/\rho)\, \partial p/\partial z - g \quad \times dz$$

Multiplying as shown and adding these three equations, we get

$$\text{RHS} = -\frac{1}{\rho}\left(\frac{\partial p}{\partial x}dx + \frac{\partial p}{\partial y}dy + \frac{\partial p}{\partial z}dz\right) - g\,dz = -\frac{dp}{\rho} - g\,dz$$

The left hand side can be changed to a total derivative in two cases, namely, when (i) the line of integration is the stream line and (ii) the flow is irrotational, that is, potential flow. For a streamline

$$\frac{dx}{u} = \frac{dy}{v} = \frac{dz}{w}$$

$$\rightarrow \qquad u\, dy = v\, dx,\ u\, dz = w\, dx, \text{ etc.}$$

We can write the LHS with these relations as

$$\begin{aligned}
\text{LHS} &= [\partial u/\partial x\, (udx) + \partial u/\partial y\, (v\, dx) + \partial u/\partial z\, (w\, dx)] \\
&\quad + [\partial v/\partial x\, (u\, dy) + \partial v/\partial y\, (v\, dy) + \partial v/\partial z\, (w\, dy)] \\
&\quad + [\partial w/\partial x\, (u\, dz) + \partial w/\partial y\, (v\, dz) + \partial w/\partial z\, (w\, dz)] \\
&= [u\partial u/\partial x\, dx + \partial u/\partial y\, u\, dy + \partial u/\partial z\, u\, dz] \\
&\quad + [\partial v/\partial x\, v\, dx + v\, dv/\partial y\, dy + \partial v/\partial z\, v\, dz] \\
&\quad + [\partial w/\partial x\, w\, dx + \partial w/\partial y\, w\, dy + w\partial w/\partial z\, dz] \\
&= u\, du + v\, dv + w\, dw = d(U^2/2)
\end{aligned}$$

Similarly, we may use the relations for irrotational flow. This gives

$$\begin{aligned}\text{LHS} &= u[\partial u/\partial x\, dx + \partial v/\partial x\, dy + \partial w/\partial x\, dz] \\ &\quad + v[\partial u/\partial y\, dx + \partial v/\partial y\, dy + \partial w/\partial y\, dz] \\ &\quad + w[\partial u/\partial z\, dx + \partial v/\partial z\, dy + \partial w/\partial z\, dz] \\ &= u[\partial u/\partial x\, dx + \partial u/\partial y\, dy + \partial u/\partial z\, dz] \\ &\quad + v[\partial v/\partial x\, dx + \partial v/\partial y\, dy + \partial v/\partial z\, dz] \\ &\quad + w[\partial w/\partial x\, dx + \partial w/\partial y\, dy + \partial w/\partial z\, dz] \\ &= u\, du + v\, dv + w\, dz = d(U^2/2)\end{aligned}$$

Therefore the line integral of Euler's equation of momentum gives

$$\int \frac{dp}{\rho} + g\, dz + \frac{dU^2}{2}$$

$$\rightarrow \qquad \int \frac{dp}{\rho} + gz + \frac{U^2}{2} = C \tag{3.4.1}$$

where $C$ is a constant. In case streamline is the line of the integration then the constant would be different for different streamlines; it would be the same, however, for irrotational flow. If the flow is unsteady, the Bernoulli equation is applicable with the difference that the sum of the three energy terms in the equation also becomes function of the time. Recall our comments made about Eq. 2.4.22 that sum of the kinetic, pressure and potential head is constant when the fluid is incompressible and inviscid and the flow is irrotational.

The term $gz$ in Eq. 3.4.1 involves the height of the point from some reference. While applying this equation between the two points 1 and 2, the equation will result in a term $g\,(z_1 - z_2)$. Therefore, it does not matter as to where is the reference line. We may take the datum line as per our wish. It is so taken that one of the $z$ values is zero and other values are positive, but it is not a necessary condition. The integral of pressure term depends on its relation with density of the fluid. Three cases are of consequence and are covered below:

*Incompressible fluid.* In this $\rho$ is constant and can be taken out of the integral sign. Therefore the Bernoulli equation becomes

$$\frac{p}{\rho} + gz + \frac{U^2}{2} = C \tag{3.4.2}$$

*Isothermal flow.* In this case, temperature of the fluid does not change and hence the Bernoulli equation becomes

$$\int \frac{dp}{\rho} = RT \int \frac{d\rho}{\rho} = RT \ln \rho$$

$$\rightarrow \qquad RT \ln \rho + gz + \frac{U^2}{2} = C \tag{3.4.3}$$

*Isentropic flow.* This is a flow during which no heat transfer occurs between the surrounding and the control volume. Such a process is represented by the process equation $p = A\,\rho^{\gamma}$, where $A$ is a constant of proportionality. We may substitute it in Eq. 3.4.1 and carry out the integration.

Bernoulli equation has been obtained on the assumptions that the fluid is ideal and the flow is steady. In fact, we get the same equation from the steady flow energy equation applied to a control volume where there is no heat and work interaction between the system and the surrounding. Therefore, the Bernoulli equation is nothing but the law of conservation of energy. This equation includes the pressure (or flow) and kinetic and potential energy. If we divide the equation by $g$ then each term has the units of head, expressed in meters of the liquid of density $\rho$. Accordingly the terms pressure head, velocity head and potential head (also called the datum head) are used. Sum of the pressure and potential head is called the *piezometric head* and sum of all the heads is named as the *total head*. Thus according to the Bernoulli equation, the total head is constant throughout the flow. We shall delve upon this aspect later in this chapter.

**Example 3.4.1** Figure E3.4.1 shows a pipe discharging water from a tank. The 150 mm diameter pipe is fitted with a nozzle at the outlet. Find the discharge. Also find the pressure heads at the points 1, 2, 3 and 4; ignore the loss of energy head.

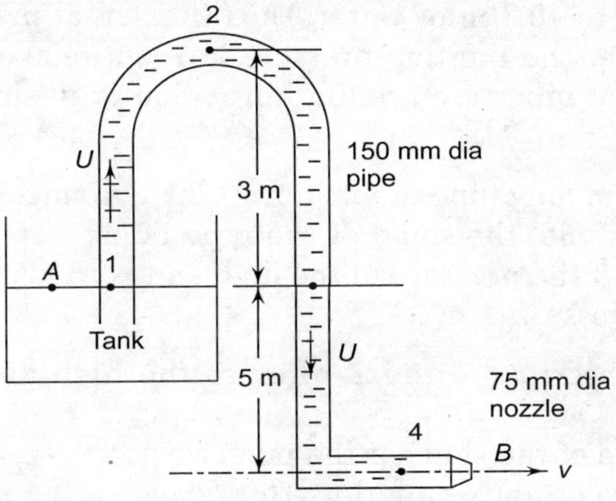


**Fig. E 3.4.1** Pressure at different points in a siphon

**Solution** Since there is no loss of energy head, the total energy at different points will be same. Let $A$ and $B$ be the points in the free surface of water in the tank and at the exit plane of the nozzle respectively. At these two points, pressure will be atmospheric. Further let us measure $z$ from the horizontal line passing through the point 4, $v$ and $U$ be the velocities at the nozzle exit and in the pipe, respectively. We use Bernoulli equation between the points $A$ and $B$ and neglect velocity head at point $A$ because the velocity of water in the free surface is very small due to large area of the tank. This gives

$$\left(z + \frac{p}{w} + \frac{u^2}{2g}\right)_A = \left(z + \frac{p}{w} + \frac{u^2}{2g}\right)_B$$

or,

$$5 + 0 + 0 = 0 + 0 + \frac{v^2}{2g} \rightarrow v = 9.9045 \text{ m/s}$$

The diameter of the pipe is 2 times the diameter of the nozzle exit, the continuity equation gives $U = v/4 = 2.4761$ m/s. Now we may apply Bernoulli equation taking point $A$ and points 1, 2, 3 and 4 in turn to determine pressure at the desired points.

$$5 + 0 + 0 = 5 + \frac{p_1}{w} + \frac{U^2}{2g} \rightarrow \frac{p_1}{w} = -0.3125 \text{ m}$$

Similarly, we get $\qquad \dfrac{p_2}{w} = -3.3125$ m,

$$\frac{p_3}{w} = -0.3125 \text{ m}$$

and $\qquad \dfrac{p_4}{w} = 4.6875$ m.

*Comments* Pressure at any point in the pipe should not decrease to the saturation pressure corresponding to the water temperature else the water in that region will form vapours, which will block the passage and stop the flow. The critical point in the system is the point 2 where it is 3.3125 m below the atmospheric pressure. The atmospheric pressure is equivalent to 10.3 m of water. Thus the actual pressure is 10.3 – 3.3125 = 6.9875 m, which is safe. The limiting pressure at room temperature is around 3 m of water head. This type of pipes are used in hill region or in siphons used for emptying the tank.

**Example 3.4.2** A 400 m long pipe tapers form 1.20 m diameter at the high end to 0.60 m diameter at the low end, the slope of the pipe being 1 in 100. The pipe conveys a discharge of 1.25 m³/s. If the pressure at the high end is 75 kPa, find the pressure at the lower end. Ignore the losses.

**Solution** Let the subscripts 1 and 2 refer to the high and low end of the pipe, respectively.

From the knowledge of the slope of the pipe, we get $z_1 - z_2 = (1/100) \times 400 = 4$ m. We get velocity, from the continuity equation. This gives

$$U_1 = \frac{Q}{(\pi/4)d_1^2} = \frac{4 \times 1.25}{\pi \times 1.2^2} = 1.105 \text{ m/s},$$

$$U_2 = \frac{Q}{(\pi/4)d_2^2} = \frac{4 \times 1.25}{\pi \times 0.6^2} = 4.421 \text{ m/s},$$

$$\frac{U_1^2}{2g} = \frac{1.105^2}{2 \times 9.81} = 0.062 \text{ m}, \ \frac{U_2^2}{2g} = 0.996 \text{ m}, \ \frac{p_1}{w} = \frac{75}{9.81} = 7.645 \text{ m}.$$

Applying Bernoulli equation at the points 1 and 2, we get

$$p_2/w = 10.711 \text{ m} \rightarrow p_2 = 9.81 \times 10.711 = 105.07 \text{ kPa}.$$

**Example 3.4.3** Find out the height $h$ to which the water jet will shoot out in the arrangement shown in the figure. Neglect the loss of energy.

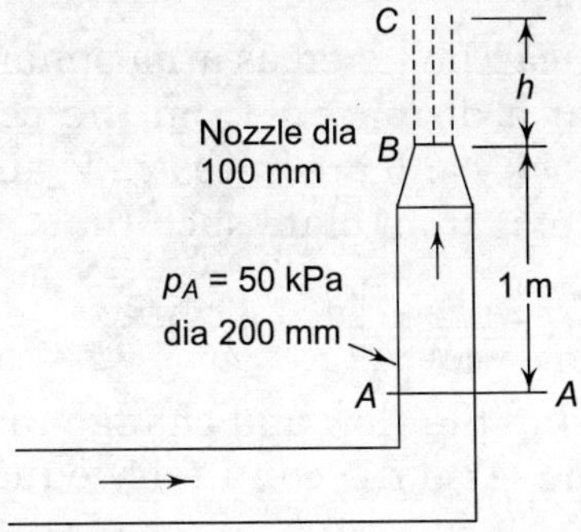


**Fig. E 3.4.3** Shooting height of jet

**Solution** We take line *AA* as the datum for measuring vertical height so that height of the points *B* and *C* are 1 and 1 + *h*. Using continuity and Bernoulli equation, we can write

$$Q = (\pi / 4)\, 0.2^2 \times U_A = (\pi /4)\, 0.1^2 \times U_B$$

$$\rightarrow \qquad U_B = 4U_A \qquad \text{(A)}$$

$$z_B + \frac{U_B^2}{2g} + \frac{p_B}{w} = z_C + \frac{U_C^2}{2g} + \frac{p_C}{w} = z_A + \frac{U_A^2}{2g} + \frac{pA}{w}$$

$$\text{or} \qquad 1 + \frac{U_B^2}{2g} + 0 = (1 + h) + 0 + 0 = 0 + \frac{U_A^2}{2g} + \frac{50}{9.81} \qquad \text{(B)}$$

$$\therefore \qquad \frac{U_B^2}{2g} = h \qquad \text{(C)}$$

Substituting for $U_A$ in terms of $U_B$ from (A) and for $U_B$ from (C) in (B), we get $h = 4.37$m.

**Example 3.4.4** Water flows between two horizontal parallel circular plates each of radius *R* and is discharged into atmosphere. A gap *b* is maintained between the plates. Water enters at the centre of the lower disc through a pipe of radius $r_0$ at velocity $V_0$. The velocity at the exit is $V_R$. Neglecting friction losses and elevation differences, and using Bernoulli equation, derive an expression for the resultant thrust *F* on the upper plate. If $R$ = 30 cm, $r_0$ = 2.5 cm, $b$ = 1.75 cm, $Q$ = 850 l/m, and atmospheric pressure = 101 kPa, find (a) pressure of water in the pipe, (b) pressure of water at a radial distance of 15 cm, and (c) resultant thrust on the upper plate in Newton.

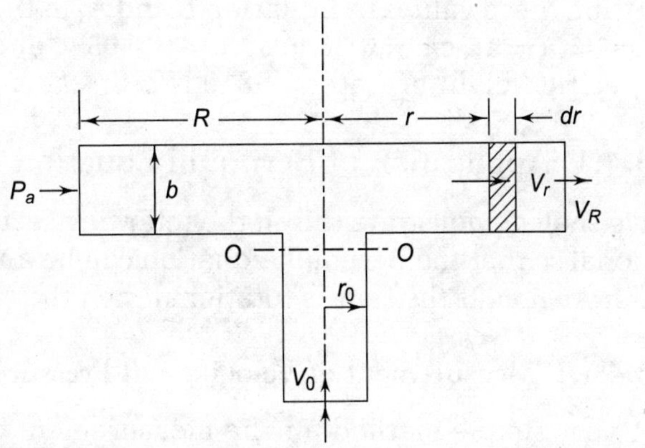


**Fig. E 3.4.4** Radial flow through plates

**Solution** Pressure at the exit can be taken as atmospheric pressure $p_a$. Let us consider a fluid element of width $dr$ at a distance $r$ from the centre line of the arrangement shown in Fig. E3.4.4 where velocity and pressure are $V_r$ and $p_r$, respectively. Let $H$ be the total head at $OO$, at $r$ and the exit from plate such that

$$H = \frac{p_0}{w} + \frac{V_0^2}{2g} = \frac{p_r}{w} + \frac{V_r^2}{2g} = \frac{p_a}{w} + \frac{V_R^2}{2g} \tag{A}$$

as there is no loss of head during the flow and change in potential head between in let and other sections is negligible. Also the continuity equation demands

$$Q = \pi r_0^2 V_0 = 2\pi rb\, V_r = 2\pi Rb V_R$$

$$\rightarrow \qquad V_r = V_R R / r \tag{B}$$

Substituting for the velocity in Equation (A), we get $p_r = w\left(H - \frac{V_R^2}{2g}\frac{R^2}{r^2}\right)$. A force $dF$ acts in vertically upwards direction that is given by $2\pi r \times dr \times p_r$. This on integration from $r = r_0$ to $r_R$ gives the total force on the upper plate in the region $r_0 \le r \le R$. In the central part, pressure on the upper plate will be $p_0$ and therefore the force on this part is $\pi r_0^2 p_0$. Atmospheric pressure acts on the upper side of the plate in the downward direction. Therefore, the total downward force on the plate is

$$F = p_a \pi R^2 - \left[2\pi w \int_{r_0}^{R} \left(Hr - \frac{V_R^2}{2g}\frac{R^2}{r}\right) dr + p_0\, \pi r_0^2\right]$$

$$= p_a \pi R^2 - \left[2\pi w \left\{\frac{H}{2}\left(R^2 - r_0^2\right) - \frac{V_R^2}{2g} R^2 \ln \frac{R}{r_0}\right\} + p_0\, \pi r_0^2\right]$$

We are given: $R = 0.3$ m, $r_0 = 0.025$ m, $b = 0.0175$ m, $Q = 850$ l/m $= 0.85/60$ m³/s. Substituting these values in Equation B and A, in that order, we get, $V_0 = 7.215$ m/s, $V_R = 0.43$ m/s, $V_r$ at $r = 15$ cm $= 0.86$ m/s, $H = 10.305$ w and $p_0 = 75.064$ kPa (absolute); here $w$ is the specific weight of water = 9810 N/m³.

## 3.4.1 Application of Bernoulli Equation

We shall demonstrate this in the following sections where the flow is assumed to be irrotational so that the Bernoulli equation can be applied between any two chosen points in the flow region. If the flow is rotational then the two points are to be taken on a streamline.

### 3.4.1.1 Measurement of Velocity and Pressure in Flow

A very simple method for the measurement of discharge in laboratory would be to collect water in some known time. Dividing the volume so collected in unit time by the area of the flow determines the velocity of flow. This gives the average flow velocity through that area but not the velocity at a point for which we use a probe (sensor) having a very small area. Accurate measurement at a point in the flow region has been a challenging job because any sensor used for the purpose will have some area, however small may it be but cannot be called a point.

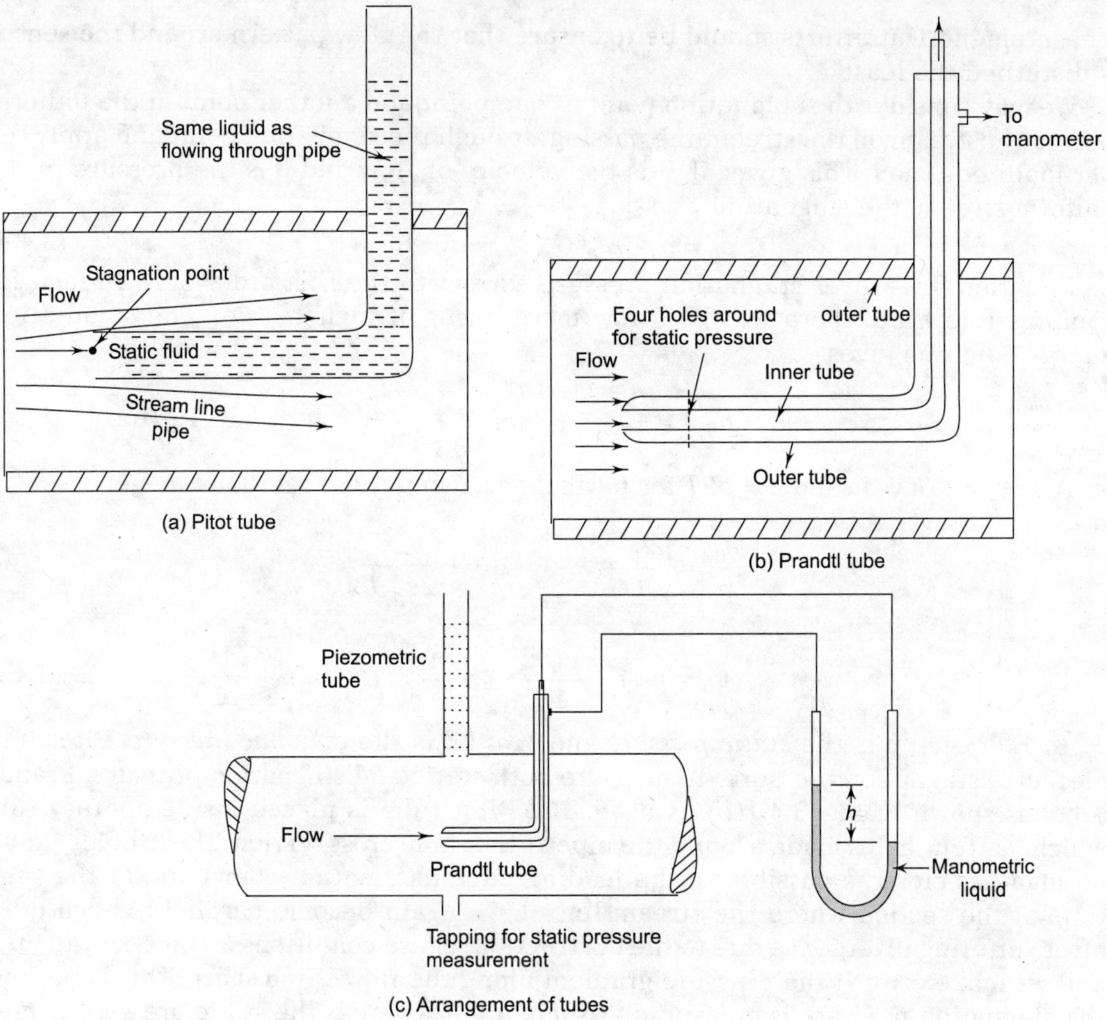


**Fig. 3.4.1** Pressure and velocity measurement

For measuring velocity, in fact we measure the stagnation and the static pressure at a point using the pressure probes, known as stagnation pressure (or Pitot tube) and static pressure tube, respectively. The *stagnation pressure* is defined as the pressure attained by the fluid when it is brought to rest isentropically. It is measured by placing the Pitot tube, as shown in Fig. 3.4.1(a). The Pitot tube is L-shaped with one end placed facing the flow at the point where the stagnation pressure is to be measured and the other end is connected to a manometer. The fluid comes to rest at the points shown in Fig. where velocity is zero. The fluid rises in the tube to a height corresponding to the stagnation pressure. Therefore, the vertical height of the L-shaped tube should be more than this height so that the liquid does not spill out. A few streamlines are also shown. The flow far upstream is uniform and the streamlines are parallel to the pipe axis. The presence of the tube disturbs the streamlines thereby causing some error in the measurement. Our efforts should be to ensure that the flow pattern around the sensor is disturbed the least.

We may consider the stagnation point as one point and another point in the uniform flow far upstream on the streamline passing through the stagnation point and apply the Bernoulli equation. This gives, if $U$ is the velocity of flow and $p$ is the pressure in the uniform stream, the stagnation pressure

$$p_0 = p + \tfrac{1}{2}\rho U^2 \tag{3.4.4}$$

Equation 3.4.4 gives stagnation pressure correct to the first order only. In fact, for compressible fluid, there will be a few more terms of higher order. The steady flow energy equation gives

$$h_0 = h + \frac{U^2}{2} \rightarrow T_0 = T + \frac{U^2}{2c_p}, \quad \text{since } h = c_p T \text{ (for air).}$$

Also $c_p = \gamma R/(\gamma-1)$ and $p = \rho RT$ for perfect gas, using these relations we get

$$\frac{p_0}{p} = \left(\frac{T_0}{T}\right)^{\gamma/(\gamma-1)} = \left(1 + \frac{U^2}{2c_p T}\right)^{\gamma/(\gamma-1)}$$

or,

$$p_0 = p\left(1 + \frac{\rho U^2}{2p} \times \frac{\gamma - 1}{\gamma}\right)^{\gamma/(\gamma-1)} = p + \frac{\rho U^2}{2} + \ldots \tag{3.4.5}$$

In Eq. 3.4.4, $p$ is the *static pressure* and $\tfrac{1}{2}\rho U^2$ is the *dynamic pressure.* Pitot tube measures stagnation pressure. To measure both static and stagnation pressure, Prandtl tube, as shown in Fig. 3.4.1(b), is used. The Pitot tube is placed inside another tube which has four holes made around the pipe at the same cross section. These holes should be made as close as possible to the leading edge (stagnation point) and at the same time at the section where the stream lines have again become parallel to each other after suffering divergence due to the obstruction. These conditions are not easy to meet and we make error if the pressure gradient along the flow is too sharp. This is because the stagnation pressure is measured at some point whereas the static pressure is measured not at the same point but at some distance downstream. Well! We have to accept it and the accuracy depends on the pressure gradient of the flow. In most of the cases the error introduced is tolerable. The Prandtl tube has two tapings, which are connected to the manometer to record difference between the stagnation and static pressure. Figure 3.4.1 (c) shows the arrangement. This difference is nothing but the dynamic head from which the flow velocity can be calculated using Eq. 3.4.4.

Figure 3.4.1(c) also shows another arrangement, called piezometer tube. This is a tube placed at right angles to the flow direction. In a pipe it is placed such that the end fixed in the pipe wall is flushed with the pipe surface so that the tube does not disturb the flow. However, in open channels, it is placed vertically and thus disturbs the flow. The liquid rises in the tube that gives sum of the pressure and datum head. The height of the tube should be enough so that the liquid does not come out from the open end.

#### 3.4.1.2 Flow Through Orifice and Mouthpiece

An orifice is a device, which is fitted at the base or the sides of a vessel to empty it. It may have any shape like triangle, rectangle, square or a circle and are available in variety of sizes. The circular orifice is commonly used. One such sharp-edged orifice of cross sectional area 'a' is shown fitted in the side of a vessel in Fig. 3.4.2; a sharp-edged orifice has the inner edge made sharp and is tapered with increasing diameter at the outer edge. There are other types of orifices too. A square-edged has a straight and parallel passage while a bell-mouthed orifice has a rounded passage and the issuing liquid comes in contact with the entire surface of the orifice. The other types of orifices are also shown in Fig. 3.4.2.

Streamlines of the liquid, while flowing through the orifice, curve so that the minimum area of the jet is at some distance from the orifice plane. This section is called *vena contracta* and the cross sectional area at this section is say $a_c$. Figure 3.4.2(b) shows a square-edged orifice. It is noticed that a vena contracta is formed within the passage in square-edged orifice. The orifice is called thin orifice if the pressure at every point of the jet on the vena contracta is atmospheric, which is the pressure acting on its outside surface. In thick orifice the pressure varies across the cross section. We shall consider a thin orifice for the analysis. We define the following terms:

$$\text{Coefficient of contraction} = C_c = a_c / a \tag{3.4.6}$$

$$\text{Coefficient of velocity} = C_v = u_c / u \tag{3.4.7}$$

Where $u_c$ and $u$ are the velocities at the vena contracta and the orifice, respectively. We may therefore determine actual discharge through the orifice using the continuity equation.

$$Q_a = a_c\, u_c = C_c\, C_v\, (au) = C_d\, Q_{th},$$

where

$$C_d = \frac{a_c \,.\, u_c}{a \,.\, u} = C_c \,.\, C_v \text{ (coefficient of discharge)} \tag{3.4.8}$$

$u$ is obtained by applying Bernoulli equation for which we assume the flow to be irrotational and take two points—one at the free surface and another at the orifice centre. Pressure at these points are equal to atmospheric pressure and choosing the orifice centre line as datum, Bernoulli equation gives $u$, efflux velocity at the orifice = $\sqrt{(2gh)}$, where $h$ is the height of the free surface from the orifice centre line. Thus the theoretical discharge is obtained. Thus is the velocity of free fall of a body under gravity. This discharge through an orifice is given by

$$Q = C_d\, a\sqrt{2gh} \tag{3.4.9}$$

Normally, the coefficient of discharge for a sharp-edged orifice is from 0.59 to 0.68. The coefficient of contraction and velocity range from 0.6 to 0.69 and 0.95 to 0.99, respectively. Coefficient of contraction decreases with its sharpness, orifice size, smoothness of the internal surface of the orifice, etc. Small head and high viscosity improves the coefficient of contraction. For a small circular sharp-edged orifice with water as fluid under moderate head, coefficient of contraction, velocity and discharge are taken as 0.62, 0.98 and 0.61, respectively.

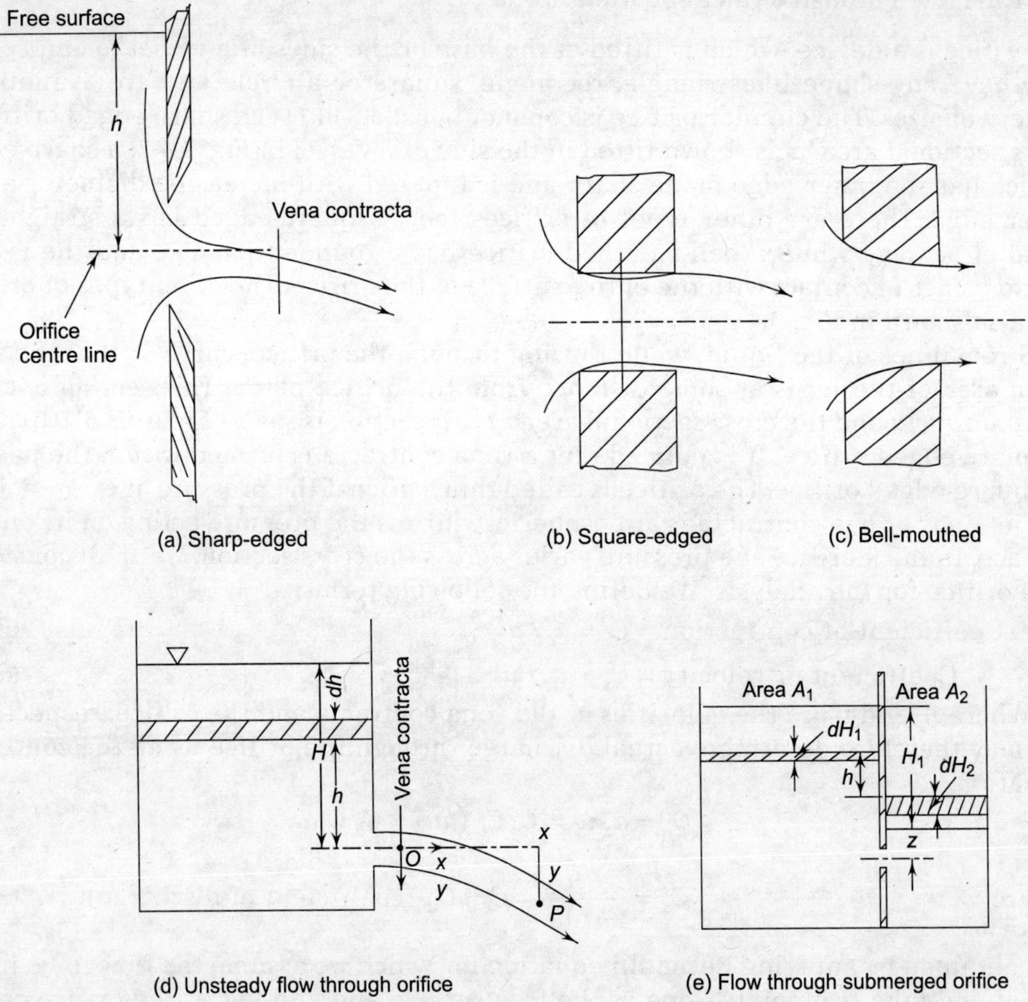


**Fig. 3.4.2** Flow throguh orifice

This value of $C_d$ is increased if a mouthpiece is used. We shall not discuss the mouthpiece in details and would end our discussions by mentioning that the increase is due to the fact that coefficient of contraction is improved significantly in mouthpiece. The following table gives an idea of the coefficient of discharge through some of the orifice and mouthpiece in use.

| Flow through | $C_c$ | $C_v$ |
|---|---|---|
| Orifice | 0.61 | 0.97 to 0.99 |
| External mouthpiece | 0.61 | |
| Converging mouthpiece | 1.0 | 1.0 |
| Borda's (internal) mouthpiece | 0.5 | |

The above relation for discharge was obtained for a thin orifice for which the diameter or thickness of the orifice is negligible as compared to height $h$ of the free surface from its axis. In many cases this may not be true. We will not consider flow through thick orifice.

Coefficient of velocity can be determined experimentally by making an arrangement to measure horizontal and vertical distance of any point on the jet issued from the orifice and making arrangement of water flow to the tank to keep water level at same height from the orifice centre line. Let $Ox$ and $Oy$ be the co-ordinate axes taken at the vena contrata section, the coordinate system is shown in Fig. 3.4.2(d). Let $P$ is the point at $x$, $y$ distance.

When the jet is issued from the orifice, it decreases in its diameter, as shown by measurements, up to a certain section at which the streamlines are parallel and horizontal. This section is called vena contracta section. Let $U$ be the velocity at this section. The jet moves with horizontal velocity $U$ along $x$ and zero velocity in vertical direction at this section. Time taken to travel distance $x = x/U$ and in this time the jet moves down due to gravity by distance $y = ½\, g\, (x/U)^2$. Measuring $x$ and $y$, we can determine $U$ and hence the coefficient of velocity. Thus

$$C_v = \frac{U}{\sqrt{2gh}} = \sqrt{\frac{x^2}{4yH}}. \tag{3.4.10}$$

The coefficient of contraction is obtained by measuring the diameter at the vena contracta.

**Unsteady flow through the orifice**

We mentioned earlier that the orifices are used for emptying of a tank filled with some liquid. When emptying the tank, the level of the liquid above the orifice changes that results in change in the velocity of the jet coming out from the orifice. In such problems, we will calculate the time taken for the water level to fall through some height or completely emptying the tank.

Let $A$ be the area of cross section of the tank at any time $t$ when the height of the free surface above the centre line of the orifice is $h$, which decreases further by $dh$ in time $dt$. Further let $a$ be the cross sectional area of the orifice and $T$ is the time taken for change of head from $H_1$ to $H_2$. We know that the velocity of flow through the orifice is given by $u = \sqrt{(2gh)}$, where $h$ is the head above the orifice centre line. We observe that $h$ is a function of $t$, the time from the start of emptying operation. Then the flow in time $dt$ through the orifice of area $A$ is $A\sqrt{(2gh)}\,dt$ that should be equal to the decrease in volume of the fluid in the tank. Now equating the volume flow rate through the orifice with the decreases in volume in the tank in time $dt$ and integrating the resultant equation,

$$C_d\, a\sqrt{2gh}\, dt = -A\, dh$$

$$\rightarrow \qquad T = -\int_{H_1}^{H_2} \frac{A\,dh}{C_d\, a\sqrt{2gh}} \tag{3.4.11}$$

This equation can be integrated if the relation between $A$ and $h$ is known, that would depend on the contour of the tank. In the above equation, $h$ is the head when the pressure at the free surface is atmospheric. In case, the tank is closed and the pressure above the free surface is different from atmospheric pressure then equivalent head of the liquid corresponding to this pressure can be added to the existing head above the orifice. We shall explain the procedure through example.

Yet another case of unsteady flow is the flow through a submerged orifice, as shown in Fig. 3.4.2(e). The orifice is placed on the wall dividing the two tanks of areas $A_1$ and $A_2$. Let $H_1$ and $H_2$ ($H_1 > H_2$) be the difference in level of the free surface in two tanks at time $t = 0$ and $T$, respectively. Further, $h$ is the difference in head at any time $t$, which changes by $dh$ in time $dt$. Note that we are taking the difference of water level in the two tanks and not the head over the orifice. This can be explained easily by noting that the pressure in the jet coming out of the orifice from left tank will be $p_a + \rho gz$, whereas $z$ is the height of the water surface in the tank $A_2$, and not the atmospheric pressure. So the difference in water levels in the two tanks is responsible for the flow and the flow would cease when it is zero. We may proceed as in the above case of unsteady flow to write the following equation.

$$A_1\, dH_1 = A_2\, dH_2 \rightarrow dH_2 = (A_1 / A_2)\, dH_1$$

and

$$dh = dH_1 + dH_2 = (1 + A_1 / A_2)\, dH_1.$$

or,

$$dH_1 = \frac{dh}{1 + A_1 / A_2}.$$

$U$, velocity through the orifice = $\sqrt{2gh}$

$$dQ = -A_1\, dH_1\, dt = C_d\, a\sqrt{2gh}$$

$$\rightarrow \qquad T = -\frac{C_d a\sqrt{2g}}{A_1}\int \frac{\sqrt{h}\, dh}{1 + A_1 / A_2} \tag{3.4.12}$$

This equation can now be integrated.

**Example 3.4.5** A closed tank has a sharp-edged orifice of 4 cm diameter in one of its walls. The tank is filled with an oil of density 0.82 kg/lit to a depth of 2 m above the centre of the orifice. The air above the oil is under a gauge pressure of 5 kPa. The orifice discharges the liquid into atmosphere. A micrometer contraction gauge gives a jet diameter of 3.15 cm at vena-contracta. The coordinates of a point in the jet with respect to vena-contracta are $x = 1.32$ m and $y = 0.173$ m. Determine the discharge in lit/s. To what value air pressure in the tank should be raised in order to double the discharge through the orifice? .

**Solution** Pressure of air is given as gauge pressure that means we may take atmospheric pressure to be zero. Further density of oil is given on litre basis. We know density of water is 1000 kg/m³ = 1 kg/lit. Therefore, specific gravity of oil = 0.82/1 = 0.82. Further, air pressure is equivalent to $5000/(0.82 \times 1000 \times 9.81) = 0.62$ m of oil. So for the given problem, $H = 2 + 0.62 = 2.62$ m, $x = 1.32$ m and $y = 0.173$ m. Substituting these data in Equation 3.4.10, we get

$$C_v = \sqrt{\frac{x^2}{4yH}} = \sqrt{\frac{1.32^2}{4 \times 0.173 \times 2.6}} = 0.98,$$

$$C_c = \left(\frac{3.15}{4}\right)^2 = 0.62 \text{ and } C_d = C_v \times C_c = 0.6076.$$

$$\therefore Q = C_d \times (\pi/4)\,(0.04)^2 \times \sqrt{2 \times 9.81 \times 2.62} = 5.476 \times 10^{-3}\ \text{m}^3/\text{s} = 5.4761/\text{s}.$$

Since $Q \propto \sqrt{H}$, it will be doubled when the effective head increases four-fold. That is, the new head is $H' = 4 \times 2.62 = 10.48$ m. This means the pressure in the tank should be increased so that its equivalent head is $10.48 - 2 = 8.48$ m; the corresponding pressure is $8.48 \times 0.82 \times 1000 \times 9.81 = 68260$ Pa.

**Example 3.4.6** Two small orifices are located in one of the vertical sidewalls of a tank. The tank bottom is horizontal and rests on a horizontal floor. The centre of the upper orifice is at a depth $h_1$ below the free surface of liquid in the tank. The centre of the lower orifice is at a height of $h_3$ above the floor and $h_2$ below the upper orifice. Show that the liquid jets issuing from the two orifices strike the floor at the same distance if $h_1 = h_3$ irrespective of the type and size of orifices and the magnitude of $h_2$.

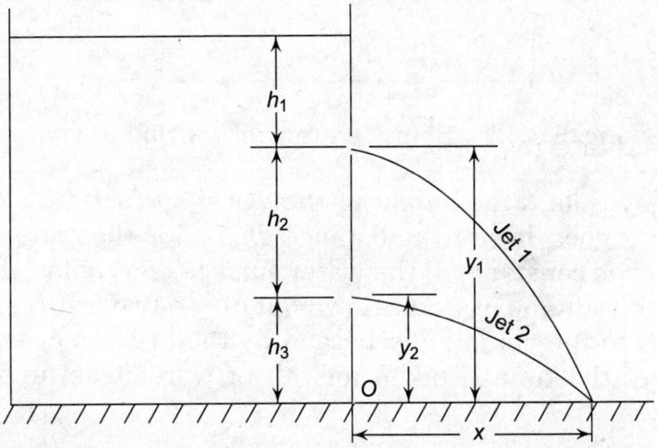


**Fig. E 3.4.6** Jet from two orifices strking at same point on ground

**Solution** Let $t_1$ and $t_2$ be the time taken by jet 1 and 2 respectively to reach the ground. These are obtained by determining the time taken in the vertical fall with zero initial velocity, as explained in the text. Thus

$$h_2 + h_3 = 1/2\, g t_1^2 \qquad \text{and} \qquad h_3 = 1/2\, g t_2^2$$

Further, velocity of the jet from the orifice depends on the head and are given as $V_1 = \sqrt{(2g\,h_1)}$ and $V_2 = \sqrt{\{2g\,(h_1 + h_2)\}}$.

$$\therefore \qquad x = V_1 t_1 = \sqrt{2gh_1}\ \sqrt{2(h_2 + h_3)/g} = [4h_1\,(h_2 + h_3)]^{1/2}$$

and $$x = V_2 t_2 = \sqrt{2g(h_1 + h_2)}\,\sqrt{2h_3/g} = [4h_3(h_1 + h_2)]^{1/2}$$

$$\Rightarrow \quad [4h_1(h_2 + h_3)]^{1/2} = [4\,h_3(h_1 + h_2)]^{1/2} \text{ or, } h_1 = h_3.$$

**Example 3.4.7** A swimming pool is 30 m long and 8 m wide with vertical sides and the bottom at a slope so that the depth of water at the shallow and deep sides are respectively 1 m and 2.5 m, as shown in the figure. If there are two outlets each 0.2 m², one at each of the deep and shallow ends, find the time taken to empty the tank. Take $C_d$ for the orifice = 0.8.

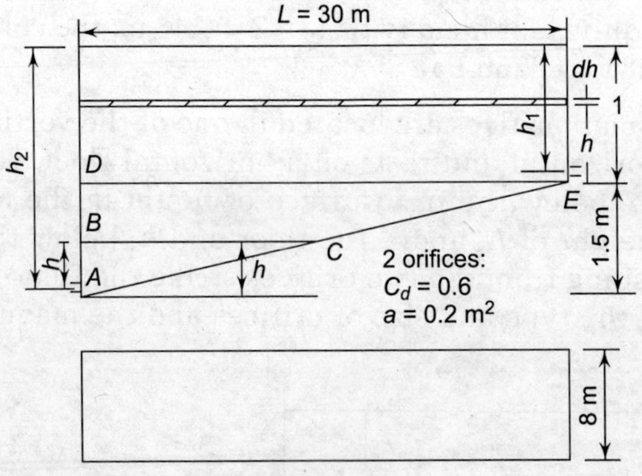


**Fig. E 3.4.7** Emptying time of a swimming pool

**Solution** Let at any time $t$, the height of the water surface from the upper orifice be $h$ m and in time $dt$ it goes down by a distance $dh$. If $A$ is the cross sectional area of the water surface, which is constant till the water surface goes below the first orifice at the shallow end, then the volume of water flown out of the two orifices, each of area $a$ m², is $A.dh$. Equating this to the volume discharged by the orifices gives an equation, which on integration gives the time $t_1$ taken for fall of water level upto the bottom at the shallow end.

$$-A \times dh = C_d\,a\left[\sqrt{2gh} + \sqrt{2g(h+1.5)}\right]dt$$

$$\rightarrow \quad \int_0^{t_1} dt = -\frac{A}{C_d\,a\sqrt{2g}} \times \int_0^1 \frac{dh}{\sqrt{h} + \sqrt{h+1.5}}$$

$$\rightarrow \quad t_1 = \frac{30 \times 8}{0.8 \times 0.2 \times \sqrt{2 \times 9.81}} \times \frac{1}{1.5} \times \frac{2}{3}\left[\left(2.5^{3/2} - 1^{3/2}\right) - 1.5^{3/2}\right] = 167\,\text{s}.$$

Now consider the flow in the lower part of the tank when the length $l$ of the water surface and area of the free surface will decrease because the bottom of the pool is slopping. Let $L$ be the length of the tank and $h_1$ and $h_2$ be the initial depth of two orifices from the free surface then $AD = (h_2 - h_1) = 1.5$ m. If $h$ is the height of the free surface at any time from the bottom orifice then from similar triangles $ABC$ and $ADE$

$$\frac{BC}{DE} = \frac{h}{AD} \Rightarrow BC = hL/AD \text{ and } A = BC.b = \frac{Lbh}{h_2 - h_1}$$

We may proceed once again in the same way as above to get the time $t_2$ taken for emptying the remaining amount of water from the tank.

$$t_2 = \frac{Lb}{C_d a\sqrt{2g}(h_2 - h_1)} \int_0^{h_2 - h_1} \sqrt{h}\, dh$$

$$= \frac{30 \times 8 \times 2/3 \times 1.5^{3/2}}{0.8 \times 0.2 \times \sqrt{2 \times 9.81} \times 1.5} = 276\text{s}.$$

$$\therefore \qquad t = t_1 + t_2 = 443 \text{ s} = 7 \text{ min } 23 \text{ s}.$$

**Example 3.4.8** A cylindrical tank is having a spherical base. The height of cylindrical portion is 5 m and diameter is 4 m. At the bottom of this tank an orifice of diameter 200 mm is fitted. Find the time required for completely emptying the tank. Take $C_d = 0.6$ for the orifice.

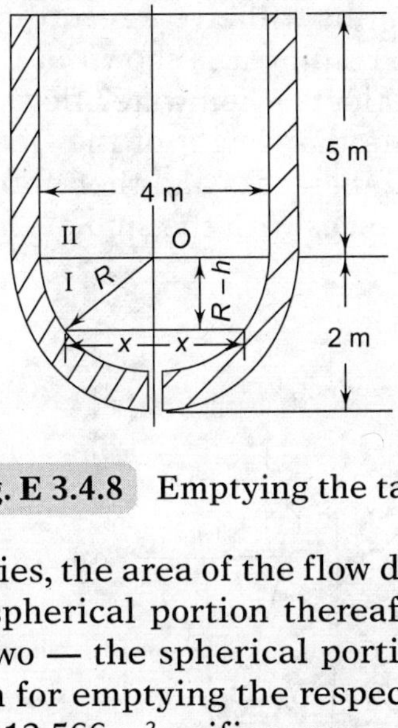


**Fig. E 3.4.8** Emptying the tank

**Solution** While the tank empties, the area of the flow does not change till the level has reached up to the top of the spherical portion thereafter it decreases. Therefore, we divide the whole region into two — the spherical portion I and cylindrical portion II. Let $T_1$ and $T_2$ be the time taken for emptying the respective portions.

For II region: $A = (\pi/4) \times 4^2 = 12.566 \text{ m}^2$, orifice area $a = (\pi/4) \times 40.2^2 \times 10^{-4} = 0.0314 \text{ m}^2$, $H_1 = 2 + 5 = 7$ m and $H_2 = 2$ m. Substituting in Eq. 3.4.11, we get

$$T_2 = -\int_{H_1}^{H_2} \frac{A\,dh}{C_d a\sqrt{2gh}}$$

$$= \frac{2A\left(\sqrt{H_1} - \sqrt{H_2}\right)}{C_d a\sqrt{2g}} = \frac{2 \times 12.566 \times (\sqrt{7} - \sqrt{2})}{0.6 \times 0.0314 \times \sqrt{2 \times 9.81}} = 370.92 \text{ s}.$$

$A$ is constant and could be taken out of the integral sign. For calculating the integral for the spherical portion, take a strip of width $2x$ at a height $h$ from the orifice, as shown in the figure. From geometry $R^2 = x^2 + (R - h)^2 \rightarrow x^2 = 2Rh - h^2$. Therefore, $A = \pi x^2$. Substituting for $A$ in the expression for emptying the time and carrying out the integration, we get

$$T_1 = \int_{H_1}^{H_2} \frac{A\,dh}{C_d\, a\sqrt{2gh}} = \frac{\pi}{C_d\, a\sqrt{2g}}\left[\frac{4}{3}RH_1^{3/2} - \frac{2}{5}H_2^{5/2}\right] = 198.78\text{s}.$$

taking $H_1 = 2$ m and $H_2 = 0$ m. Therefore the total time of emptying $= T_1 + T_2 = 569.7$ s $= 9$ min 29 s.

### 3.4.1.3 Flow Measurement Through Pipes

In laboratories and industries, we need to measure the flow rate, which is named as discharge and is expressed in m³/s or cumec.[1] For quick measurements, water meters used in the houses are satisfactory. It consists of a turbine rotor that rotates by the impulse of water. However, if instead of water air flows through it, as it happens when the water line valve is opened, this will give a recording. A rotameter is also used that gives quick and more accurate results. It is shown in Fig. 3.4.3 and consists of a graduated diverging tube housing a float. When water flows through this tube the drag and the buoyancy force lifts it while the weight of the float makes it fall. With increasing velocity, the upward drag force increases and therefore the float rises up to a distance where the downward force balances with the sum of the upward forces. The scale gives the discharge.

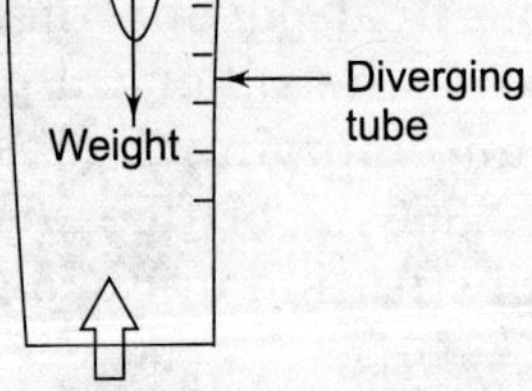


**Fig. 3.4.3** Roatameter

For accurate measurements, velocity is measured at different points and the method discussed for finding out the average velocity is adopted for calculating the discharge. This requires graphical or numerical integration techniques.

[1] Cusec is used as unit for discharge in FPS system of units from where this word cumec is derived for SI system of units.

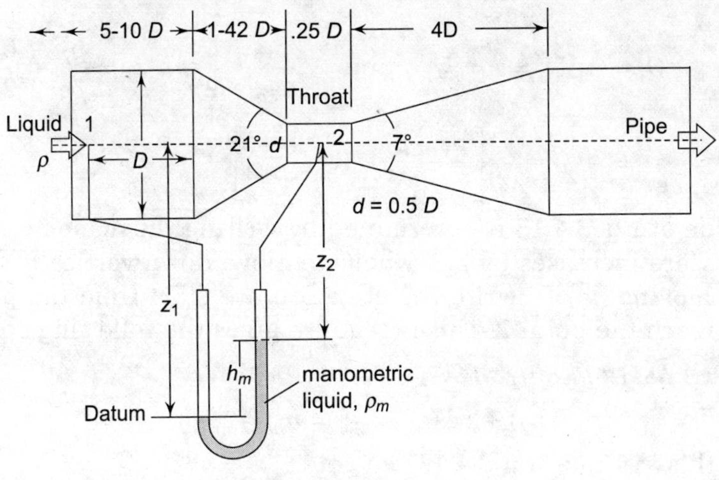


**Fig. 3.4.4** Venturimeter

Now we shall discuss the most common methods used for the measurement. These are the venturimeter, flow nozzle and orifice plate. Figure 3.4.4 shows a venturimeter fitted in a pipe. The pipe may be circular, rectangular or elliptic in shape; the venturimeter will take the shape accordingly. For fitting it with the pipe, flanges are made on both the venturi and the pipe and connected together with bolts and nuts. The venturi consists of a converging part, a throat and a diverging part. The divergence angle is kept small within 5 – 15° so that the flow remains attached to its surface to avoid formation of eddies and loss of total head. There is no such restriction for the converging portion; however, the angle of convergence is kept from 19° to 23°. That is why the diverging portion is lengthier than the converging portion. The head loss in the venturimeter varies from $0.1h$ to $0.2h$, where $h$ is the venturimeter head. For best results, a straight pipe of length 5 to 10 $D$ should precede the venturimeter, where $D$ is the pipe diameter.

We take pressure tapings from two points 1 and 2. Point 1 is $1D$ distance upstream of the throat, where $D$ is the diameter of the main pipe, and the second point 2 is at a section in the throat where the streamlines are almost parallel and a vena contracta is formed. The recommended dimensions are shown in the figure.

We may calculate the discharge through the pipe by applying mass conservation and the Bernoulli equation at the two sections passing through the tapings and assume uniform pressure and velocity on these sections. Thus, the points 1 and 2 in the following Eq. lie at the centre of these sections. We measure the height of any point from some datum that has been taken at the lower meniscus of the manometric fluid, as shown in the figure. (It is not necessary to take this as the datum. In fact, the datum line can be taken at any plane.) Since pressure at point 1 is more than that at the point 2, the level of the manometric liquid in the right limb will be above from that in the left limb. Let the difference in height of the manometric liquid be $h_m$ and the density of the fluid be $\rho_m$. If $A_1$ and $A_2$ represent the areas and $U_1$ and $U_2$ the velocities at the two sections then we can write the following equation using mass conservation and Bernoulli equation:

$$Q = A_1 U_1 = A_2 U_2 \quad \text{and} \quad \frac{p_1}{\rho g} + \frac{U_1^2}{2g} + z_1 = \frac{p_2}{\rho g} + \frac{U_2^2}{2g} + z_2 \qquad (3.4.13)$$

$$\rightarrow \qquad \frac{U_2^2}{2g} = \frac{A_1^2}{A_1^2 - A_2^2}\left[\frac{p_1 - p_2}{\rho g} + (z_1 - z_2)\right]$$

Right hand side of Eq. 3.4.13 is determined by writing the manometric equation. Remember, the pressure increases by $\rho gh$ when we move down vertically a distance $h$ and decreases while moving up. To write the equation, we start from the point 1 along the manometer and reach the point 2. This procedure gives the following equation

$$p_1 + \rho g z_1 - \rho_m g h_m - \rho g(z_1 - h_m) = p_2$$

or,
$$(p_1 - p_2) + \rho g\,(z_1 - z_2) = h_m\,(\rho_m - \rho)$$

Substituting this in Equation 3.4.13, we get

$$U_2 = \left[\frac{A_1^2\, 2g}{A_1^2 - A_2^2}\left\{\frac{p_1 - p_2}{\rho g} + (z_1 - z_2)\right\}\right]^{1/2}$$

$$= \frac{A_1}{\sqrt{A_1^2 - A_2^2}}\sqrt{2gh}, \text{ where } h = h_m\left(\frac{\rho_m}{\rho} - 1\right) \qquad (3.4.14)$$

We note that $\frac{p_1 - p_2}{\rho g} + (z_1 - z_2)$ is the difference of the piezometric heads at the two sections 1 and 2 and therefore should be equal to the level difference of the manometric fluid, equivalent to the head of the flowing liquid. The relation in $h$ and $h_m$ in Eq. 3.4.14 give the equivalence. This means that orientation of the venturimeter will not affect the reading of $h_m$, since it is a measure of the difference of the piezometric head—orientation means that the venturimeter axis may be at any angle with the horizontal line. The venturimeter orientation will effect the pressure at point 2 but not the discharge. Further $\rho_m/\rho = S_m/S$, where $S_m$ and $S$ are respectively the specific gravity of the manometric fluid and the fluid passing through the pipe. Therefore care should be taken while using this term if the fluid flowing through the pipe is different from water.

Bernoulli equation assumes no losses in between the two points under consideration. However, it is not true because of the presence of some viscosity in the liquid. Fortunately, there is not much loss in the converging and throat portions, the losses are mainly in the diverging section. These losses are accounted by introducing coefficient of discharge $C_d$ such that

$$Q_a = C_d\, Q = C_d\, A_2\, U_2$$

$$= \frac{C_d\, A_1\, A_2}{\sqrt{A_1^2 - A_2^2}}\sqrt{2gh} \qquad (3.4.15)$$

Normally diameter $d$ of the throat is much smaller than the pipe diameter $D$ and therefore $A_2 << A_1$. In this case $(A_2/A_1)^2 <<< 1$ and can be neglected as compared to 1. This simplifies Eq. 3.4.15 to

$$\frac{A_1 A_2}{\sqrt{A_1^2 - A_2^2}} = \frac{A_2}{\sqrt{1 - A_2^2 / A_1^2}} \approx A_2 = {}^{\pi}\!/_{4} d^2 \quad \text{and} \quad Q_a = C_d A_2 \sqrt{2gh} \tag{3.4.16}$$

The Venturimeter is suitable for measurement of discharge in industries for a variety of fluid including suspended solids in fluid stream. However, from the experience it is learnt that (i) these are not suitable for pipes of diameters below 7.5 cm, (ii) Venturimeter requires calibration before using it, (iii) less wear and tear and therefore less number of calibrations to check its correctness. The factor $\dfrac{A_2\sqrt{2g}}{\sqrt{1 - A_2^2 / A_1^2}}$ is called the venturi constant and is a constant for the given venturimeter.

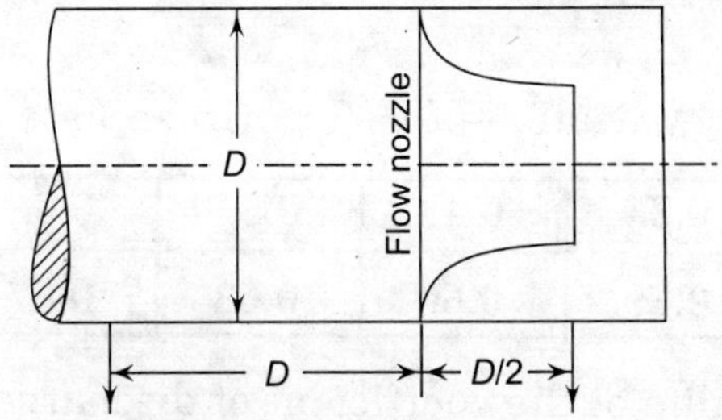


**Fig. 3.4.5** Flow nozzle meter

*Flow nozzle and orifice plates.* These are the variations of the venturimeter. A flow nozzle is a nozzle used for measurement of the discharge and consists of the smooth converging portion only; the flow leaves it into the stream where kinetic energy is converted to static pressure in an uncontrolled manner. Therefore, losses are more and the coefficient of discharge has value ranging from 0.8 to 0.9. Figure 3.4.5 shows the arrangement for the flow nozzle. The flow nozzle can be placed in between the pipes or at the end of the pipe. Recall, the nozzle is also used at the end of the pipe to convert static pressure into kinetic energy to be used elsewhere as in the case of turbines. The exit diameter of the nozzle is less compared to the main pipe diameter and hence Eq. 3.4.16 can be used comfortably. Pressure tapings are taken 1*D* upstream and 0.5*D* downstream, where *D* is the pipe diameter. Nozzle is normally made from gun metal, stainless steel, or monel metal.

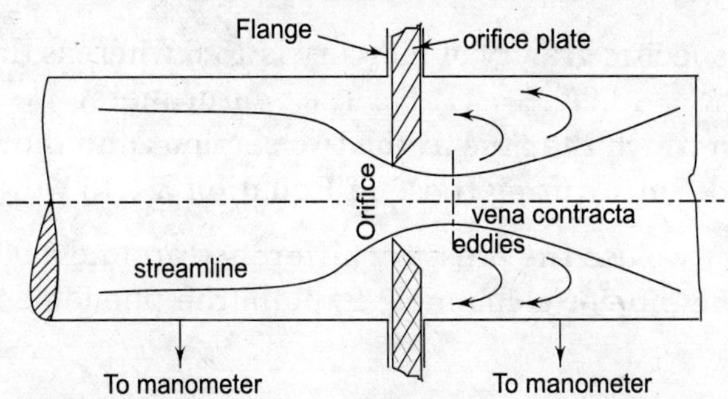


**Fig. 3.4.6** *D* and *D*/2 Orifice plate meter

Arrangement of an orifice plate is shown in Fig. 3.4.6. The orifice plate is a thin plate made from steel, stainless steel, phosphor bronze and other such materials that can withstand the corrosive effect of the flowing medium. It has a concentric circular orifice.Normally diameter of the central hole is about 0.5*D*, where *D* is the pipe diameter. The plate is sandwiched in between the flanges on the connecting pipe, as shown in Fig. 3.4.6. The streamlines bend while passing through the hole resulting in eddy formation downstream and a large pressure drop; pressure loss varies from 40 to 90 percent of the differential pressure across the plate. This results in large loss and therefore the coefficient of discharge varies from 0.6 to 0.63. The following table gives $C_d$ for different $d/D$ ratio, where *d* and *D* are the diameter of the orifice and the pipe respectively.

Variation of $C_d$ for orifice plate

| $d/D$ | 0.2 | 0.3 | 0.4 | 0.5 | 0.6 | 0.7 | 0.8 |
|---|---|---|---|---|---|---|---|
| $C_d$ | 0.58 | 0.59 | 0.60 | 0.63 | 0.67 | 0.72 | 0.78 |

In spite of such poor value of the coefficient of discharge, the orifice meter is popular in industrial use because of its less cost, ease of installation, less space requirements and usable for small diameter pipes, as small as 1.25 cm dia. However, it is susceptible to erosion, corrosion, and scaling. For these reasons, it requires frequent calibration. It tends to clog and hence not suitable for measurements in slurries or fluids carrying solid particles. *D* and *D*/2 pressure taping is very common in practice though other practice like flange tapping is also in use. In former, the tapings are taken at *D* upstream and *D*/2 downstream distance from the plane of the orifice plate. These sections are chosen because of the streamlines becoming almost parallel to each other. Equation 3.4.16 is used to calculate the actual discharge.

**Example 3.4.9** In a vertical pipe conveying oil of specific gravity 0.8, a 300 mm × 150 mm venturimeter is fitted; the direction of flow being upwards. The difference of levels between the throat and entrance sections is 1 m, the oil–mercury differential gauge shows a deflection of 370 mm of mercury. Find the discharge of oil. Neglect losses.

**Solution** We know specific gravity of mercury is 13.6 whereas that of the oil is given to be 0.8. This gives $S_m/S = 13.6/0.8 = 17$. For the venturimeter, we are given $D = 300$ mm and $d = 150$ mm from which the area at the two sections and the venturimeter constant *K* can be determined. Substituting these in Equation 3.4.15 we get $Q = 0.1967$ m$^3$/s.

**Example 3.4.10** May we use the pressure difference around a 90° bend to measure the discharge through the pipe or a channel? Explain the phenomenon and determine the discharge.

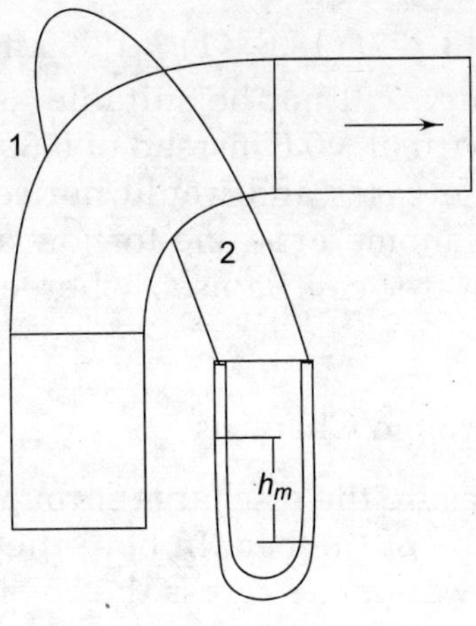


**Fig. E 3.4.10** Elbow meter

**Solution** As a general rule, any device in which the pressure difference exists due to flow of liquid can be used to measure discharge, as the pressure difference would be function of the discharge. In 90° bends, the streamlines are forced to bend and remain no more straight line. This generates centripetal force that is caused due to pressure difference across the bend at the round corners with more pressure on the outer side of the bend. Thus two pressure tapings can be taken, as shown in Fig. E 3.4.10, and connected to a $U$-tube manometer. This is called an *elbow meter*.

Since we are dealing with pipe, cylindrical polar coordinates will be used. Moreover, the flow can be assumed to be steady and axi-symmetrical. That is, $p = p(r)$. Let $U$ is the velocity of the flow in the pipe, then the momentum Eq. result in

$$-\frac{dp}{dr} = -\rho\frac{U^2}{r}, \text{ or } dp = \rho\frac{U^2}{r}dr \rightarrow p_2 - p_1 = \rho U^2 \ln\frac{R_2}{R_1}.$$

or,
$$U = \sqrt{\frac{p_2 - p_1}{\rho\ln(R_2/R_1)}} \quad \text{and} \quad Q = AU = A\sqrt{\frac{p_2 - p_1}{\rho\ln(R_2/R_1)}}$$

where $R_1$ and $R_2$ are the inner and outer radii of curvature of the bend and $A$ is the cross-sectional area of the pipe.

**Example 3.4.11** Water discharge in a 100 mm dia pipe ranges from 1.5 l/s to 20.01 l/s. Recommend the diameter of the orifice of the orifice plate and the manometric arrangements.

**Solution** $C_d$ for an orifice plate depends on the ratio

$$m = d/D.$$

Let $m = 0.5$ at which $C_d = 0.63$ and $d = 50$ mm.

Substituting these values in Eq. 3.4.16 gives the difference $h$ in manometer level of 75.2 mm and 13.377 m of water corresponding to the minimum and maximum discharge through the pipe. If we use mercury in the $U$ tube manometer then these values correspond to

75.2/(13.6 – 1) = 6 mm and 13.377//(13.6 – 1) = 1.069 m of mercury.

Thus both water and mercury will not be suitable as manometric fluid. One alternate is to use an orifice plate with $m = 0.6$ instead of 0.5. You may work out the mercury level difference in the extreme cases and would notice the problem is still there. A better solution is to use two manometers—one for low discharge and another for high discharge; for low discharge, water can be used whereas for high discharge rate mercury will suit.

### 3.4.1.4 Flow Measurement Through Channels

Weir and notch are used to measure the discharge through river and canals in which the flow is taking place due to slope of the bed. In both the methods a restriction is made across the canal. A weir is the wall made across the river. On the other hand, a notch is a cut of certain shape and size made at appropriate place in the wall across the river; the notch may have rectangular, triangular or any other cross section. In both the cases water is stored upstream till it reaches the top of the wall in case of weir and the base of the notch, called the crest. However, the presence of the wall will be felt by the liquid upto certain distance upstream thereafter the flow may be assumed to be uniform having velocity $U$. Water will have some depth over the crest that would depend on the discharge; more the discharge, more is the height of the water. We would prefer this height to be reasonable to allow less error in the measurement for which a Vernier scale can be used. You will realise that a weir will be used only when the discharge rate is very high because the area of flow is large and the head of the water over the top of the wall would be less if discharge were low. For low discharge, a notch will be used so that the height of the liquid over the crest is reasonable to be measured accurately. Thus these flow-measuring devices work on the same principle but their use is decided on the magnitude of the discharge.

*Weir.* It is shown in Fig. 3.4.7. Water flows over the weir. This portion of water is called nappe that falls to the ground and then water flows downstream. Pressure in the region enclosed by the nappe and the weir wall is less than atmospheric pressure. Moreover, the pressure varies a little in the nappe itself but taken as negligible. Thus pressure everywhere in the nappe above the crest is taken as $p_a$, the atmospheric pressure that acts on the free surface also. Let $H$ be the height of free surface over the weir

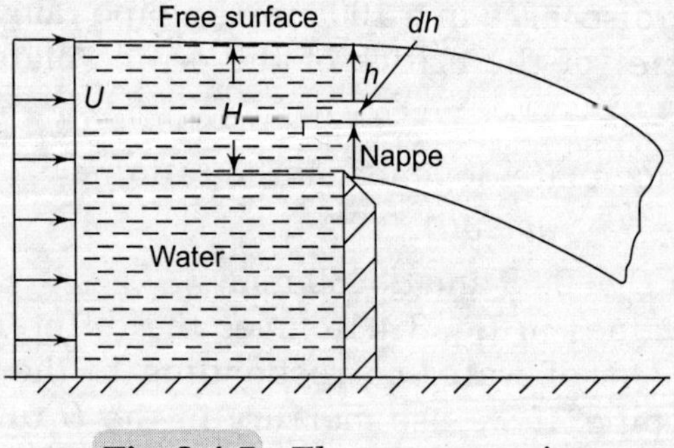


**Fig. 3.4.7** Flow over a weir

crest and consider a fluid element of height $dh$ at a depth $h$ from the free surface where the flow velocity is $u$. Applying the Bernoulli equation at the points in free surface far upstream and in the fluid element, we get (free surface taken as the datum),

$$p_a + U^2/2g = p_a + u^2/2g - gh \quad \rightarrow \quad u = \sqrt{U^2 + 2gh} \tag{3.4.17}$$

$U$ is called the velocity of approach and is normally small because the area of the flow at any section upstream is large and hence often neglected. Moreover, we get theoretical discharge only and a factor, called the discharge coefficient $C_d$ is used to correct it. The actual discharge over the weir is then given by

$$Q_a = C_d \int_0^H u\,b\,dh = \tfrac{2}{3} C_d\, b\sqrt{2g}H^{3/2} \tag{3.4.18}$$

where $b$ is the width of the weir. For a sharp crested weir: $C_d = 0.611 + 0.075\ (H/W)$ for $H/W \le 0.5$, where $W$ is the height of the crest from the ground. In case, the velocity of approach is not negligible then we may take $z = U^2 + 2gh$ and carry out the integration of the expression for the discharge. The integration limit will be changed and we shall get

$$Q_a = C_d \int_0^H u\ b\,dh = C_d \int_0^H \sqrt{U^2 + 2gh}\, b\,dh = C_d \int_0^H \sqrt{2g}\left[\frac{U^2}{2g} + h\right]^{1/2} b\,dh$$

Let $\quad z = \dfrac{U^2}{2g} + h \rightarrow dz = dh$

$$\therefore \quad Q_a = C_d \int_{U^2/2g}^{U^2/2g + H} b\,\sqrt{2g}\,\sqrt{z}\,dz$$

$$= \tfrac{2}{3} C_d\, b\sqrt{2g}\left[\left(H + U^2/2g\right)^{3/2} - \left(U^2/2g\right)^{3/2}\right] \tag{3.4.19}$$

Since the river or canal banks confine the water, the water while flowing near the banks is contracted resulting in decrease in the overall width of the flow. Francis observed that the contraction is of the order of 10 percent of the head $H$ over the crest on each side where the water is in contact with some vertical surface. Thus for a weir, $b = b - 0.2H$ and Equation 3.4.19 is modified accordingly. Presence of each vertical surface in the flow path reduces the width by $0.2H$ and should be taken into account.

*Notch*. Notch, as the name suggests, is a cut made in the wall across an open channel. Fig. 3.4.8 shows a rectangular and $V$ notch. Equation 3.4.19 gives the flow rate for a rectangular notch for which $C_d$ is taken as 0.62– 0.633. For a $V$- notch $b$, the width of the water at depth $h$ from the free surface, increases as we move up from the crest to the free surface where the width is $B$. Using property of triangles, $b$ can be expressed in terms of $(H - h)$ and $B$. Simplifying the integration, we get

$$Q = C_d \int_0^H u\,b\,dh = 8/15\ C_d \tan\theta\sqrt{(2g)}\ H^{5/2}$$

$$= 1.42\ H^{5/2}\ (\text{for } \theta = 45° \text{ and } C_d = 0.6) \tag{3.4.20}$$

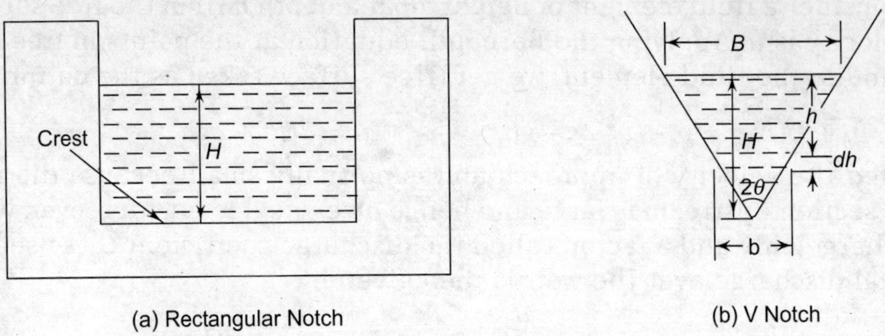


**Fig. 3.4.8** Measurement of discharge in notches

where $\theta$ is half the included angle of the notch. Normally the included angle in $V$ notch used for measurement is 90° and therefore $\theta = 45°$ and $C_d$ is the coefficient of discharge having value 0.6. Substituting these values in the relation gives the constant as 1.42, given in Eq. 3.4.20. Logarithmic differentiation of Eq. 3.4.18 and 3.4.20 with $H$ gives

$$\frac{dQ}{Q} = 1.5\frac{dH}{H} \quad \text{for rectangular notch}$$

and
$$\frac{dQ}{Q} = 2.5\frac{dH}{H} \quad \text{for } V \text{ notch}$$

This shows that for the same percentage error made in measurement of $H$, the error made in discharge is less with rectangular notch. However, $H$ appears in denominator and for lower values of $H$ more is the percentage error made in measurement of $H$. For this reason, a $V$ notch is preferred when the discharge is low as in that case the height over the crest will be sufficient and error made in measurement of $H$ will be less thereby giving more accurate value of discharge. In case, the flow rate is moderate then a combination of $V$ and rectangular notch may be used with the lower portion of $V$ type and upper portion made of rectangular section. For large discharge measurement, rectangular notch will be preferred. If the discharge is very large then a weir itself is used. The added advantage of a $V$ notch is that the coefficient of discharge is almost independent of the height of the free surface over its crest.

**Example 3.4.12** A sharp-edged weir is in the form of a symmetrical trapezium. The horizontal base is 200 mm, the top is 450 mm and the depth is 250 mm as shown in the figure. Find the discharge over the weir when the head of water over the crest is 150 mm. Take $C_d = 0.62$.

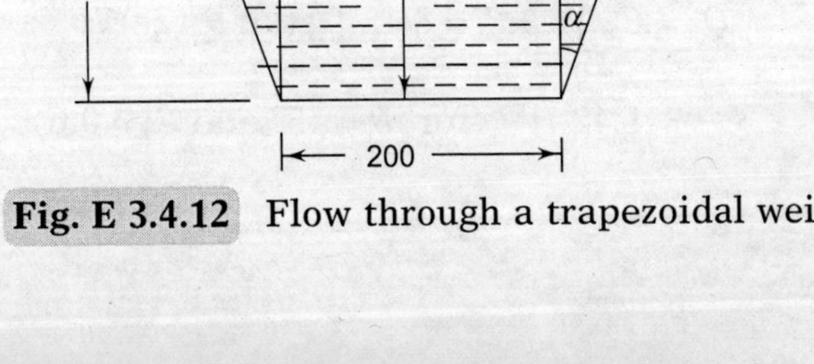


**Fig. E 3.4.12** Flow through a trapezoidal weir

**Solution** The trapezoidal section can be split into a rectangle and a triangle having half included angle $\alpha$ given by tan $\alpha$ = 125/50 = ½ (see Fig. E3.4.12). Equations 3.4.18 and 3.4.20 give discharge over the rectangular and triangular sections, respectively. We note for rectangular weir: $b$ = 200 mm = 0.2 m and $H$ = 0.15 m for both the weirs. Substituting the values in the expression, we get

$$Q_a = 2/3 C_d b\sqrt{2g}H^{3/2} + 8/15 \times C_d \times \sqrt{2g}\,\tan\alpha \times H^{5/2} = 0.0277\ \text{m}^3/\text{s}.$$

**Example 3.4.13** Find the depth and top width of a triangular notch to discharge a maximum quantity of 0.6 m³/s and such that the head shall be 65 mm when the discharge is 0.004 m³/s. Take $C_d$ = 0.62.

**Solution**

Given: $Q$ = 0.004 m³/s when $H$ = 0.065 m.

Equation 3.4.20 for the discharge over a $V$ notch gives

$$0.004 = 8/15 \times 0.62 \times \tan\theta\sqrt{(2g)} \times 0.065^{5/2} \rightarrow \tan\theta = 2.528.$$

Let the head for $Q$ = 0.6 m³/s be $H'$. $C_d$ will be same as it is independent of the head over the crest in $V$ notch. Since only $H$ is changing, we get from Equation 3.4.20

$$0.6/0.004 = (H'/H)^{5/2} \rightarrow H' = 0.482\ \text{m}.$$

Width of the free surface at this height = $2 \times H' \tan\theta$ = 2.436 m.

**Example 3.4.14** Find the rate of flow over a triangular notch of angle 45°, for a head of 0.4 m taking $C_d$ = 0.6. If this flow is wanted within an accuracy of ±2%, what are the limiting values of the head?

**Solution** $\theta$ = 22.5°, $H$ = 0.4 m, $C_d$ = 0.6. Equation 3.4.20 gives

$$Q = 8/15 \times 0.6 \times \sqrt{(2 \times 9.81)} \times \tan(22.5) \times (0.4)^{5/2} = 0.891\ \text{m}^3/\text{s}.$$

Differentiation of Equation 3.4.20 gives $dQ/Q = 2.5\, dH/H$. When the percentage error in $Q$ is ±2, that is $\delta Q/Q$ = 0.02, then

$$dH/H = 0.02/2.5 \rightarrow dH = 0.0032\ \text{m} = 3.2\ \text{mm}.$$

∴ Limiting values of $H$ are 400 ± 3.2 mm.

**Example 3.4.15** A broad-crested weir is built across the entire length of 4 m wide, rectangular channel. The height of weir above the channel floor is 0.3 m and the head of water upstream of the weir is 0.75 m. Assuming $C_d$ = 0.65 and considering the velocity of approach, calculate the discharge of water over the weir. Assume uniform velocity distribution in the nappe.

**Solution** When the width of the weir is large then the lower water surface comes in contact with its surface and the free surface looks like the one shown in Fig. E 3.4.15. Moreover, the depth $h$ of the water level over the crest is significant as compared to $H$, the height of the free surface in the undisturbed flow over the weir crest. Neglecting losses between points 1 and 2 in the free surface, as shown in Fig., and using the Bernoulli equation, we can determine the flow velocty over the weir crest and the discharge $Q$.

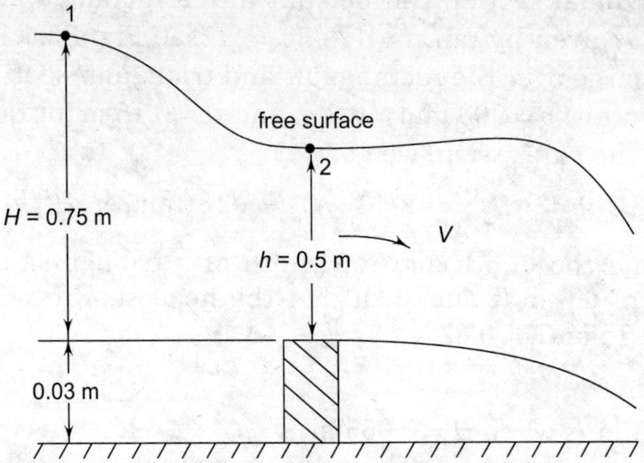


**Fig. E 3.4.15** Flow over broad crested weir

$$\frac{p_1}{w} + \frac{U_1^2}{2g} + z_1 = \frac{p_2}{w} + \frac{U_2^2}{2g} + z_2.$$

or, $$\frac{p_a}{w} + 0 + H = \frac{p_a}{w} + \frac{U_2^2}{2g} + h$$

$\rightarrow$ $$U_2^2 = \sqrt{2g(H-h)} \quad \text{and} \quad Q = C_d\,(L\,h)\sqrt{2g\,(H-h)}.$$

We can find out the condition for maximum discharge by differentiating with respect to $h$ and equating it to zero. This gives $h = 2/3\ H$ and $Q_{max} = 1.7\ C_d\ L\ H^{3/2}$. Here we have neglected the velocity of approach. With this background, we can now proceed to solve the question.

Assume that the condition for the maximum discharge prevails that gives $h = 2/3 \times 0.75 = 0.5$ m. Neglecting the velocity of apprach, the discharge is

$$Q = 1.7 \times 0.65 \times 4 \times 0.75^{3/2} = 2.87 \text{ m}^3/\text{s}.$$

$\therefore$ $U_1$, velocity of approach = $Q/A = 2.87/(4 \times 1.05) = 0.683$ m/s

and $$\frac{U_1^2}{2g} = 0.0238 \text{ m}.$$

While taking into account the velocity of apprach the new discharge is

$$Q' = 1.78 \times C_d \times L \times \left[\left(H + \frac{U_1^2}{2g}\right)^{3/2} - \left(\frac{U_1^2}{2g}\right)^{3/2}\right] = 2.99 \text{ m}^3/\text{s}.$$

[Note: Go back to the Bernoulli equation and do not neglect $U_1$, as done above and then proceed in the same way].

## 3.5 Modified Bernoulli Equation

We mentioned earlier that no fluid is inviscid and hence the shear stresses are always present that result in continuously conversion of kinetic energy into internal energy.

Since the Bernoulli equation does not take into account the internal energy and considers sum of the flow (pressure) energy, kinetic and potential energies only, it becomes essential to account for this conversion. This amounts to say that the sum of the three energies, called total energy $E$, goes on decreasing with the flow direction; the difference of the total energy at the two sections is added to satisfy the Bernoulli equation. This equation is called modified Bernoulli equation and is written as

$$\frac{p_1}{\rho} + gz_1 + \frac{U_1^2}{2} = \frac{p_2}{\rho} + gz_2 + \frac{U_2^2}{2} + \text{energy loss} \tag{3.5.1}$$

This equation can be shown graphically, see Fig. 3.5.1 for a flow from one reservoir to another through pipes of different diameter and length connected in series. We shall learn later that the losses in pipe are classified as major loss due to viscosity and minor losses such as entry loss *ab*, expansion loss *cd*, exit loss ef, etc. The major loss increases linearly with the length of the pipe and is indicated in the Fig. as $h_{f1}$ and $h_{f2}$. To start with, we draw a vertical line *ab* to show the entry loss, an inclined line *bc* whose slope is a measure of the major loss, vertical line *cd* for the expansion loss and continue like this till we reach the last section. The line so obtained is called the *energy gradient line.* Total energy line without considering any losses is also shown and as expected, it is a horizontal line passing through the free surface of the reservoir 1. Thus the vertical height between the total and the energy gradient line gives the loss upto that section from the start. The vertical height from free surface of reservoir 1 to reservoir 2 is the total energy or head loss. In other words, the head available at the end of the pipe is equal to the head at the beginning of the flow minus the head losses. This also means that the flow will take place in the direction in which the energy line is drooping (falling). We shall explore this aspect in chapter 4.

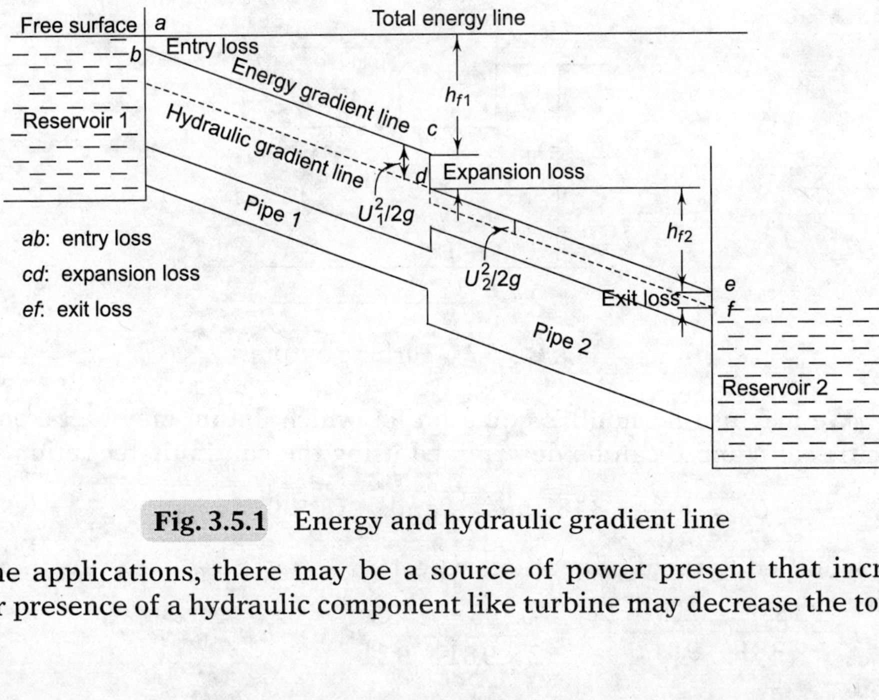


**Fig. 3.5.1** Energy and hydraulic gradient line

In some applications, there may be a source of power present that increases the energy. Or presence of a hydraulic component like turbine may decrease the total energy of the fluid. This can be taken care of by the modified Bernoulli equation by adding or subtracting the corresponding amount of the energy head from the Bernoulli equation. This procedure is illustrated through the following examples.

**Example 3.5.1** The suction and delivery pipe diameter of a pump are 200 mm and 100 mm, respectively. The pump delivers 40 kW power to the fluid; the pressures are 70 and 210 kN/m² at the inlet and outlet of the pump. Find the discharge through the pump if the exit is 0.5 m above the inlet.

**Solution** Let $Q$ be the discharge in m³/s. Then from continuity equation,

$$U_1 = Q/A_1 \text{ and } U_2 = Q/A_2, \text{ where } A_1 = (\pi/4)\,(0.2)^2 \text{ and } A_2 = (\pi/4)\,(0.1)^2.$$

Also 40 kW = $wQH_p/1000$, where $H_p$ is the head developed by the pump in m. This gives $H_p$ in terms of $Q$. The Bernoulli's equation is

$$\frac{U_1^2}{2g} + \frac{p_1}{w} + z_1 = \frac{U_2^2}{2g} + \frac{p_2}{w} + z_2 + H_p.$$

Substituting for the different quantities in the equation and simplifying, we get

$$Q^3 + 0.019\,Q - 0.00527 = 0.$$

This equation is solved by trial and error method, in which we assume some value of $Q$ and see whether LHS is zero or not. If not, we modify the assumed value of $Q$ and repeat the process. This gives, $Q \approx 0.137$ m³/s.

**Example 3.5.2** For a turbine shown in Fig. E 3.5.2, $p_A$ = 2 bar and $p_B$ = – 0.35 bar, $Q$ = 0.4 m³/s. Determine the energy output of the turbine, if diameter of the tube at section $BB$ is 0.8 m and at $A$ 0.4 m.

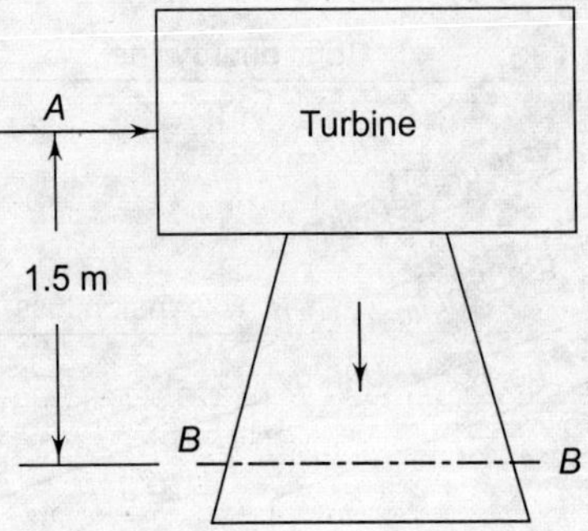


**Fig. E 3.5.2** Turbine output

**Solution** We may use Bernoulli's equation for which datum may be taken at section $BB$. Velocities at $A$ and $B$ can be determined using the continuity equation.

$$U_A = \frac{4Q}{\pi d^2} = 3.185 \text{ m/s} \quad \text{and} \quad U_B = 0.796 \text{ m/s}.$$

Using modified Bernoulli equation, we get

$$\frac{3.185^2}{2 \times 9.81} + \frac{200}{9.81} + 1.5 = \frac{0.796^2}{2 \times 9.81} - \frac{35}{9.81} + 0 + E_m$$

$\rightarrow \qquad E_m = 25.996$ m,

$\therefore$ Energy equivalent of this head = 9.81 × 0.4 × 25.996 = 101.76 kW
and Turbine output = Energy × efficiency = 101.76 × 0.8 = 81.4 kW.

**Example 3.5.3** The diameter of a vertical tapering pipe at sections *A* and *B*, 1.5 m apart are 150 mm and 75 mm respectively, *A* being higher than *B*. For a certain discharge of water down the pipe, the pressure head at *B* is greater than the pressure head at *A* by 0.5 m and the loss of head between the two sections is $h_L$. When the same quantity of water flows up the pipe the pressure head at *B* is greater than the pressure head at *A* by 0.8 m and the loss of head between the sections is 2 $h_L$. Find the discharge and the value of $h_L$.

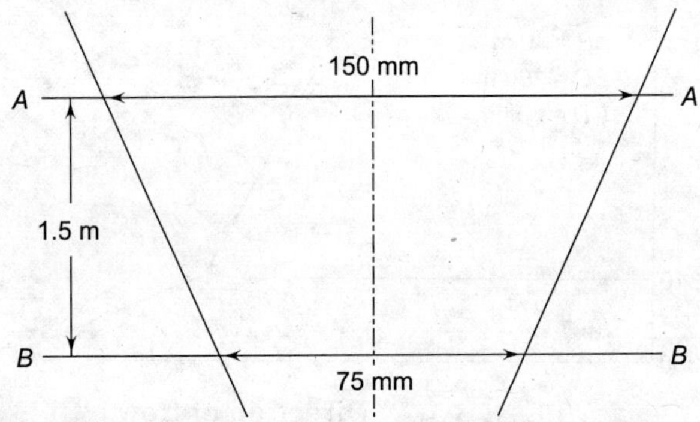


**Fig. E 3.5.3** Discharge through a pipe with head loss

**Solution** Let the subscripts *a* and *b* be used for the sections *AA* and *BB*. Continuity equation at these two sections give

$$Q = (\pi/4)\, d_a^2\, U_a = (\pi/4)\, d_b^2\, U_b \rightarrow U_b = 4U_a.$$

case (i) when the flow is downwards and $p_b/w - p_a/w = 0.5$ m. Bernoulli equation gives

$$\frac{U_a^2}{2g} + \frac{p_a}{w} + z_a = \frac{U_b^2}{2g} + \frac{p_b}{w} + z_b + h_L$$

$$\rightarrow \qquad 15\,\frac{U_a^2}{2g} = 1 - h_L \qquad \text{(A)}$$

case (ii) when the flow is upwards and $p_b/w - p_a/w = 0.8$ m. Bernoulli equation gives

$$\frac{U_a^2}{2g} + \frac{p_a}{w} + z_a = \frac{U_b^2}{2g} + \frac{p_b}{w} + z_b + h_L$$

$$\rightarrow \qquad 15\,\frac{U_a^2}{2g} = 0.7 + 2h_L \qquad \text{(B)}$$

From Equation *A* and *B*, we get $h_L = 0.10$ m and $U_a = 1.085$ m/s.

$$\therefore \qquad Q = (\pi/4)0.15^2 \times 1.085 = 0.0192 \text{ m}^3\text{/s}.$$

**Example 3.5.4** A pipe line *ABCD* of uniform diameter carrying water consists of three straight parts *AB*, *BC* and *CD* of lengths 40 m, 160 m and 60 m respectively. The levels above the datum are: *A* 350m, *B* 345 m, *C* 310 m and *D* 300 m; pressure gauges at *A* and *D* show pressure of 100 and 650 kPa respectively. Determine the direction of flow and the loss of head between the ends of the pipeline. Find also the pressure intensity at a section *E* of the pipe 136 m from *A*.

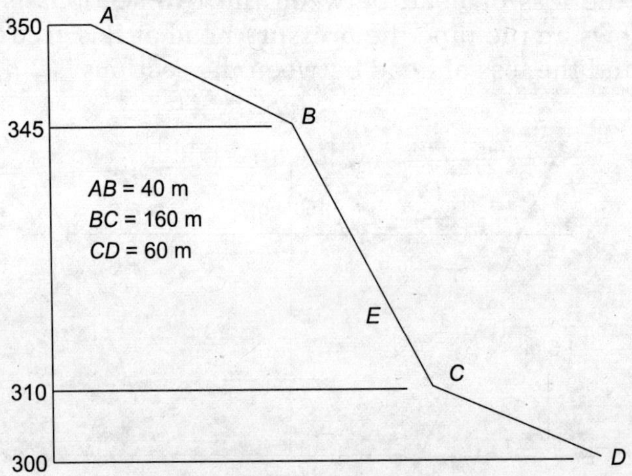


**Fig. E 3.5.4** Direction of flow

**Solution** We are given $p_a/w = 100/9.81 = 10.194$ m and $p_d/w = 650/9.81 = 66.259$ m.

Let $U$ be the flow velocity through the pipe. Bernoulli equation gives the total energy head $E$ at different points.

$$E_a = \frac{U^2}{2g} + \frac{p_a}{w} + z_a = \frac{U^2}{2g} + 10.194 + 350 = 360.194 + \frac{U^2}{2g}.$$

Similarly, $E_d = 366.259 + \dfrac{U^2}{2g}$.

Since energy head at $D$ is greater than that at $A$, the direction of flow is from $D$ to $A$. The total loss of head from $D$ to $A$ is $E_d - E_a = 6.065$ m of water. To find the pressure at $E$, we have to find out the height of the point from datum and the loss from $D$ to $E$, which can be taken proportional to the length of the pipe.

Total length of pipe $DA = 40 + 160 + 60 = 260$ m

and $DE = DA - AE = 124$ m.

The head loss from $D$ to $E$,

$$H_L = (124/260) \times 6.065 = 2.862 \text{ m}.$$

Length $CE = DE - CD = 64$ m.

$\therefore$ height of $E$ above $C = (345 - 310)/160 \times 64 = 14$ m

$\rightarrow$ elevation of $E = z_E = 310 + 14 = 324$ m.

Applying Bernoulli equation between points $D$ and $E$ on simplification gives

$$p_e/w = 37.879 \text{ m and } p_e = 9.81 \times 37.879 = 39.123 \text{ kPa}.$$

**Example 3.5.5** A nozzle whose tip is placed 1 m above the ground level gives a water jet of diameter 7.5 cm and velocity 20 m/s. (a) If the jet leaves the nozzle at an angle of 60° with the horizontal, determine the maximum height that the jet would reach above the ground, the diameter of the jet at this point and the corresponding horizontal distance from the nozzle. (b) If the jet has to reach a window located at a height of 10 m above the ground and at a horizontal distance of 25 m from the nozzle, what should be the initial angle of inclination of the jet? Neglect friction between air and water jet in both cases. [16.3 m, 10.6 cm, 67° or 43°].

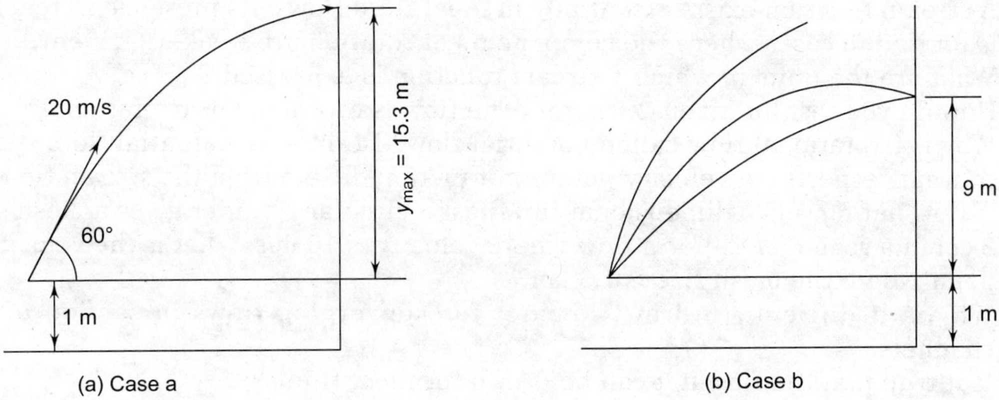


**Fig. E 3.5.5** Trajectory of water

**Solution** Case (a) $y_{max} = \frac{V_y^2}{2g} = \frac{(20 \sin 60)^2}{2 \times 9.81} = 15.3$ m.

The maximum height, above the ground upto which the jet can reach = 15.3 + 1 = **16.3 m**.

At this height the jet will have velocity $V_2$ in horizontal direction only such that

$$V_2 = V_1 \cos 60 = 10 \text{ m/s}.$$

The continuity equation gives:

$$A_1 V_1 = A_2 V_2 \rightarrow d_2^2 = \frac{V_1}{V_2} d_1^2, \text{or, } d_2 = \mathbf{10.6 \text{ cm}}.$$

Case (b) Let the jet direction at the time of issuing it from the nozzle be $\theta$ and time taken by the jet to reach window be $t$ so that $t = x / V_1 \cos \theta$, where $x$ = 25 m, the horizontal distance of the window from the nozzle. Also, the height $y$ of the window from the jet = 9 m that is given by

$$y = V_y t - \frac{1}{2} g t^2 = 20 \sin 60 \times \frac{25}{20 \cos 60} - \frac{9.8}{2} \times \left(\frac{25}{20 \cos 60}\right)^2 \rightarrow \theta = 43° \text{ or } 67° \, 43'.$$

That is, there are two directions when the jet would reach the window placed 25 m away from the jet at a height of 10 m above the ground. Figure 3.5.5(b) shows some of the trajectories of the water jet issued at different angles, including the ones that would reach the window.

# PROBLEMS

## Review Questions

1. Determine the length of the major and minor axes of the Rankine oval.
2. Obtain the relationship between stream function and the velocity components in cylindrical polar coordinates.
3. A stream function exists essentially in ideal fluid flows. Its presence in real fluids is limited to cases where the components of rotation are weak. Comment.
4. What are the units in which a stream function is expressed?
5. How do you test for given $\psi$ whether the flow is rotational or irrotational?
6. Why is irrotational flow called potential flow? Define the potential flow.
7. In what respect(s) a velocity potential function differs from the stream function?
8. A flow net for a two-dimensional flow past a circular cylinder gives a mesh size of 2 cm in a region of uniform flow where velocity is 10 m/s. What is the velocity at a point where the mesh size is 1.5 cm?
9. Why is it difficult to draw flow net for converging flows in section of open channels?
10. Is it true that the flow net can be drawn for ideal fluids?
11. When nothing such as ideal fluid exists, what is the utility of studying flow of ideal fluids?
12. A sphere moves at a constant velocity in still air. Is the flow at a given point steady or unsteady? Draw atypical stream lines, streak line and path line.
13. Explain how the Bernoulli equation which is valid for irrotational flows, can be used to solve real fluid flow problems.
14. The term $p/w$ is given the name flow energy per unit weight of the fluid. Why cannot it be called pressure energy per unit weight like other terms, which are known as kinetic, and potential energies?
15. Steady flow of fluid occurs through a vertical pipe of constant area in a downward direction. Does the velocity increase or decrease in the direction of the flow?
16. Is it possible to have a uniform pressure in a flow along an inclined pipe whose cross section reduces in the downward direction? What is the effect of direction of flow?
17. The velocity head of a free jet of water is 10 m. What is the power of jet if discharge is 0.1 $m^3/s$?
18. Does an ideal fluid require a force to move at a velocity of constant magnitude?
19. If the velocity distribution in a flow section is nearly uniform, will the energy and momentum coefficients be nearly equal? When will they be exactly equal?
20. Distinguish between the surface and the body forces.
21. Discuss the limitations of Bernoulli equation.
22. Derive and plot the pressure distribution over the surface of a circular cylinder kept in a uniform stream of fluid.

23. Derive an expression for the pressure gradient in radial direction in the flow of fluid along a circular path.
24. If the fluid velocity is 5 m/s, what will be the difference of the head between the stagnation and static heads of Pitot tube?
25. The coefficient of a Pitot tube is usually slightly less than 1. Is it due to friction loss? If not, what is the reason.
26. Why should the pressure difference in a venturi meter be always measured between liquid inlet and the throat sections? Why cannot the discharge be computed from the pressure difference between throat and the outlet sections?
27. Is the concept of piezometric head relevant in case of gases? Explain.
28. Why is it necessary to have a straight pipe of 5 to 10 $D$ before installing a venturimeter in it? What happens if a venturi is fitted immediately after a bend in the pipe?
29. The differential head between the inlet and throat of a horizontal venturi meter is $h$ when the discharge is $Q$. What is the discharge when the differential head is $2h$?
30. What is $C_d$ if an orifice meter is considerably less than that of a venturimeter?
31. A Pitot tube inserted at the vena contracta section of a pipe orifice gives a velocity of 15 m/s. the orifice diameter is 4 cm. If $C_d = 0.6$, what is the discharge?
32. If the jet of liquid issuing from an orifice submerges in a body of liquid of the same or different type, does it contract as much as it does in the case of a free orifice jet? What about the trajectory of a submerged jet compared to that of a free jet?
33. What are the two conditions to be satisfied for the Francis formula for contracted weirs to be applicable?
34. Distinguish between static, dynamic and stagnation pressure.

## Problems for Practice

1. A source of strength 120 m$^3$/s/m is located at the origin and is subjected to a uniform flow of 60 m/s in the $x$-direction. Locate the stagnation point and calculate the velocity at (1, 1). What is the value of the stream function on the half body? What is the width of the half body on the $y$-axis and what is it $s$ maximum width?
2. A two-dimensional source of strength $Q$ and an anti-clockwise vortex of strength $K$ are situated at the origin. Establish the general equation of the streamlines. If $Q = 1.0$ m$^2$/s and $K = 3.14$ m$^2$/s, find the equation of the streamline passing through the point ($r = m$, $\theta = 0$). [$r = Ce^{\,Q\theta/K}$, $r = 2e^{\theta/\pi}$]
3. Given the complex potential: $w = z + 1/z$, evaluate the stream and the velocity potential functions in polar coordinates and identify the pattern of the flow.
4. A circular cylinder of radius $a$ is exposed to a uniform stream at velocity $U$, the fluid density is $\rho$. Using Bernoulli equation, compute and plot the pressure distribution over the surface of the cylinder in terms of the non-dimensional pressure $p = (p - p_0)/\left(\frac{1}{2}\rho U^2\right)$. If $R = 1$ m, $U = 2$ m/s, and $p_0 = 1$ bar, determine the maximum and minimum pressure on the cylinder. Assume that the fluid density = 1000 kg/m$^3$. [1.02 bar, 0.94 bar].

5. For the following function in two-dimensional irrotational flow, determine the conjugate functions: (i) $\phi = r^{\pi/\beta} \cos(\pi\theta/\beta)$ (ii) $\psi = A \ln r + m\theta$.
6. The following observations were made for the purpose of calibrating a Pitot tube:

| $U$, m/s | 0.970 | 1.372 | 1.503 | 1.681 | 1.941 |
|---|---|---|---|---|---|
| $H$, cm | 5 | 10 | 12 | 15 | 30 |

Plot $U$ against $\sqrt{h}$ and determine the coefficient for the Pitot tube. [0.988].
7. The head of water over an orifice of diameter 10 cm is 10 m. The water coming out from orifice is collected in a circular tank of diameter 1.5 m. The rise of water level in this tank is 1 m in 25 seconds. Also the coordinate of a point in the jet, measured from vena contracta, is 4.3 m horizontal and 0.5 m vertical. Find, $C_c$, $C_v$, and $C_d$. [0.669, 0.96, 0.643].
8. A 5 cm diameter sharp-edged orifice lies in a horizontal plane and discharges a jet of water, which is directed upward as shown in Fig. P.8. The jet rises to a height of 3.75 m above the plane of the orifice. At what height above the plane of the orifice would jet diameter be equal to the orifice diameter? Assume $C_v$ = 0.972 and $C_c$ = 0.619. Neglect air friction and the height of the vena contracta above the level of the orifice. [2.33 m]

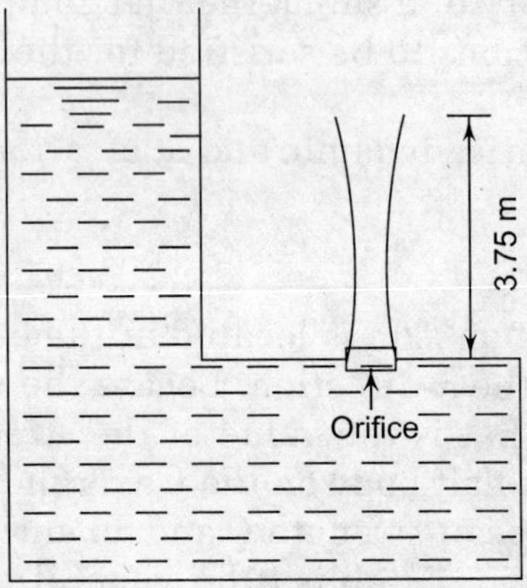


**Fig. P.8** Height in a jet of a given section

9. Two identical orifices are mounted on one side of a vertical tank, Fig. P.9. The height of water above the upper orifice is 3 m. If the jet of water from the two orifices intersect at a horizontal distance of 8 m from the tank, estimate the vertical distance between the two orifices. Calculate the vertical distance of the point of intersection of the jets from the water level in the tank. Assume $C_v$ = 1 for the orifices. [2.33 m, 8.33 m].

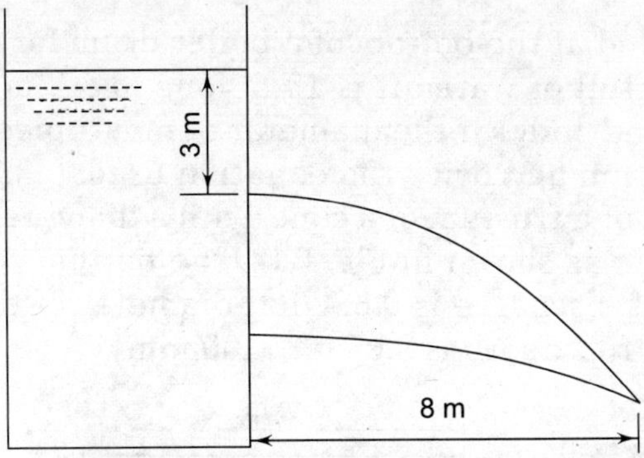


**Fig. P.9** Intersection of two jets

10. An external cylindrical mouth piece 3.5 cm in diameter discharges water under a head of 4 m; the values of $C_v$ and $C_c$ for an equivalent orifice are 0.97 and 0.61. Find the discharge in lps and the head loss. A small hole is drilled into the mouthpiece at vena contracta section, which is connected to a dish of mercury through a vertical piezometer tube as shown in Fig. P.10. To what height will the mercury rise in the piezometer? What is the pressure at vena contracta? [53 cm, 71 kPa (vacuum)].

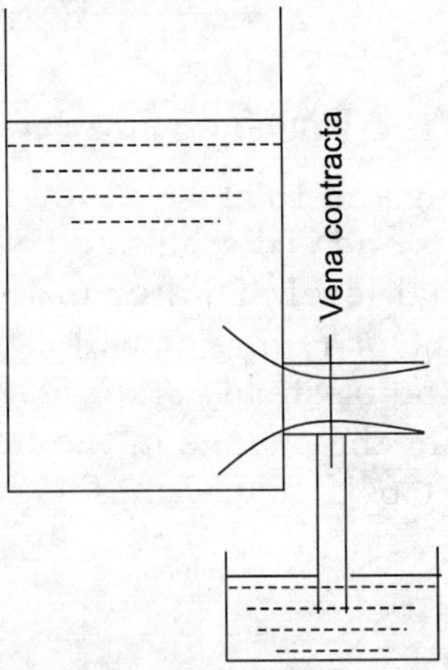


**Fig. P.10** Rise of water in tube at vena contracta

11. An air compressor draws air from the atmosphere through a bell-mouthed entrance calibrated for measuring discharge passing through it in terms of height of water that rises in a single tube manometer installed in the duct, which takes air from the bell-mouth of the compressor. Determine the flow rate of air through the bell-mouth if the rise of water in the manometer tube is 25 cm and the duct has a diameter of 16 cm. For air density = 1.2 kg/m$^3$. [1.28 m$^3$/s]
12. A tank of constant cross-sectional area has orifices each of 9.3 cm$^2$ in area in one of its vertical side, at heights of 6 m and 1.5 m respectively above the bottom of the tank. Calculate the time taken to lower the water level from 9 m to 3.6 m above the bottom. Take $C_d$ = 0.625. [1462 s].

13. A mouthpiece is fitted at the bottom of a boiler drum for emptying it. The drum is horizontal and half full of water. It is 10 m long and 2 m diameter. Find the time required to empty the boiler if the diameter of mouthpiece is 5 cm and its $C_d = 0.8$. Sketch the boiler drum and drive the equation used. [58.3 min].
14. A tank has the form of a frustum of a cone, with a diameter of 2.44 m at the top and 1.22 m at the bottom as shown in Fig. P.14. the bottom contains a circular orifice whose coefficient of discharge is 0.6. What diameter of the orifice will empty the tank in 6 min if the full depth is 3.05 m? [100 mm].

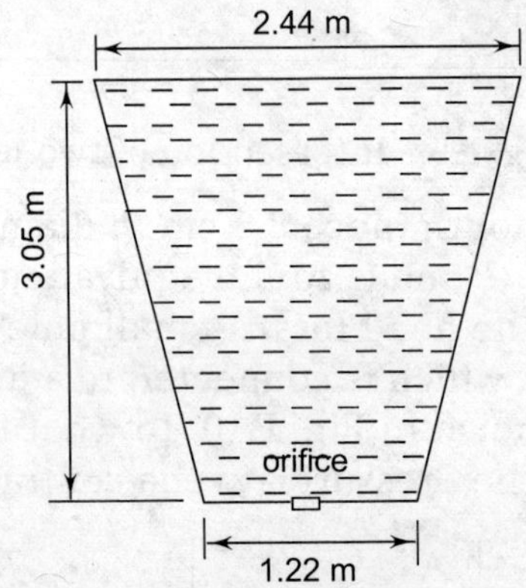


**Fig. P.14** Time for emptying the tank

15. A concrete tank is 10 m long and 16 m wide, and its sides are vertical. Water enters the tank at the rate 0.1 m³/s and is discharged from an orifice of area 0.05 m² at its bottom, see Fig. P.15. Initial level of water in the tank is 5 m. Find whether the liquid level will start rising, or falling or will remain the same. If the liquid level changes then find the value of steady state level to which the liquid will reach. Also find the time taken for the change in the liquid level to be the 60 percent of its total change. Take $C_d = 0.6$. [fall, 0.57 m, 1.41 hrs].

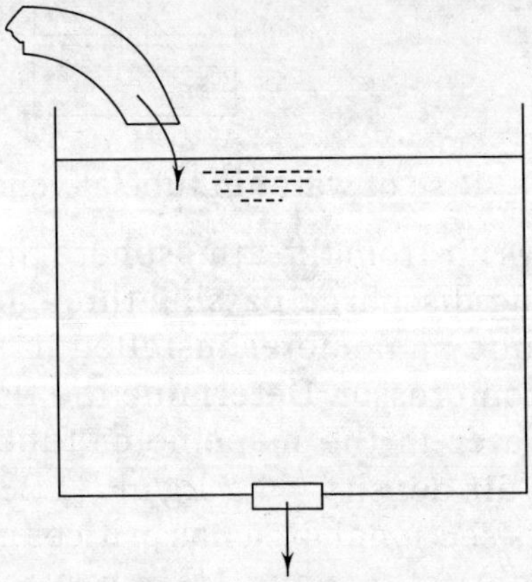

**Fig. P.15** Tank with water in and outflow

16. A pump is to be installed to supply water from a reservoir to a vertical rectangular tank that measures 10 m × 6 m in plan. The desired rise in water level is 4 m and this task is to be accomplished in 30 min. Two arrangements, as shown in Fig. P.16, have been suggested. Based upon the cost of pumping, indicate the arrangement to be preferred.

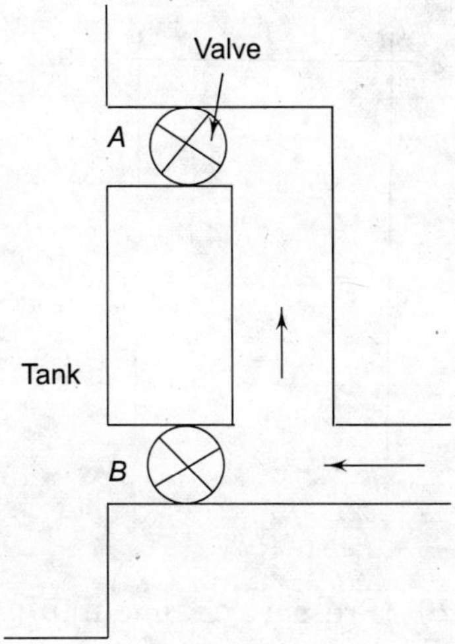


**Fig. P.16** Position of supply in overhead tank

17. During the test on a centrifugal pump, water was drawn steadily from a reservoir through a vertical 30 cm diameter suction pipe, and the pump discharged the water into a horizontal 25 cm diameter delivery pipe. A bourdon pressure gauge just outside the casing on the suction pipe read 0.4 bar gauge while the discharge pressure gauge indicated the pressure of 2.8 bar gauge. A vertical distance of 1.5 m intervened between the pipe centres at the sections where the gauge were attached. If the water discharge through the arrangement was measured as 0.1 $m^3/s$, compute the horse power expanded in running the pump. [33.578 kW].
18. A pipe connects a reservoir to a turbine which discharges water to the tailrace through another pipe. The head loss between the reservoir and the turbine is 10 times the kinetic head in the pipe and that from the turbine to the tailrace is 0.5 times the kinetic head in the pipe. The rate of flow of 1 $m^3/s$ and the pipe diameter in both the case is 1 m. Find the pressure at inlet and exit of the turbine. Take energy level *EL* of turbine = 105 m, *EL* of reservoir = 150, *EL* of tail water = 100 m. [432.54 kPa, – 49.452 kPa, 482 kW].
19. A pipe, 5 cm dia × 250 m long, carries water from station *A* to station *B*, which is located at a level of 10 m height. If the shear stress between the liquid and the pipe wall is 25 $N/m^2$, calculate the pressure change in the pipe and the head loss. [1.98 bar, 10.22 m].

20. The following data pertains to the Y-fitting shown in Fig. P.20. Volumetric inflow to section 1 = 12 l/s, outflow from section 3 = 4 l/s, and pressure gauge reading at section 1 = 25 kPa. Determine the pressure at sections 2 and 3, and the force needed to hold the fitting in position. [ 25.17 kN/m², 24.9 kN/m², 39.71 N, 15.03°].

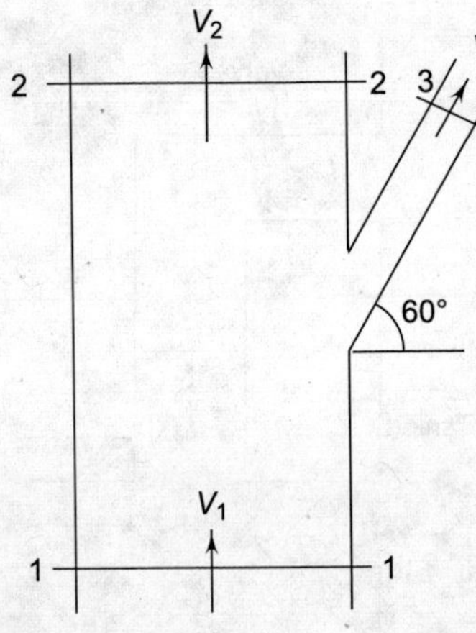


**Fig. P.20** Pressure change in pipe fitting

21. Apply continuity, momentum and energy equation to a propeller in fluid stream and obtain expression for power input and theoretical efficiency. An airoplane with two 2.15 m diameter propellers travels through still air ($\rho$ = 1.22 kg/m³) at 360 km/hr, if discharge through each propeller is 425 m³/s, determine the thrust, power input, theoretical efficiency of propulsion system and the pressure difference across a propeller. [35.51 kN, 4160 kW, 85.38%, 4.892 kPa].
22. A closed tank partially filled with water to a depth of 1 m has an orifice of diameter 1.5 cm at its bottom. The pressure of air above the water surface is raised till the discharge becomes 2 l/s through the orifice of $C_d$ = 0.6. Calculate the pressure of the air. [168.05 kN/m²].
23. A subsonic wind tunnel has a test section 0.5 m × 0.5 m and a settling chamber 1 m × 1 m. A *u*-tube manometer installed between the two locations register 0.1 m difference between the vertical columns. The manometric fluid is is alcohol of specific gravity 0.8. Estimate the test section velocity. [37.32 m/s].
24. Predict the flow rate through a venturimeter, placed at 30° to the horizontal, carrying gasoline of specific gravity 0.8, as shown in Fig. P.24, if the venturimeter is turned into horizontal or vertical position, what will the differential manometer read? [0.00426 m³/s, no change].

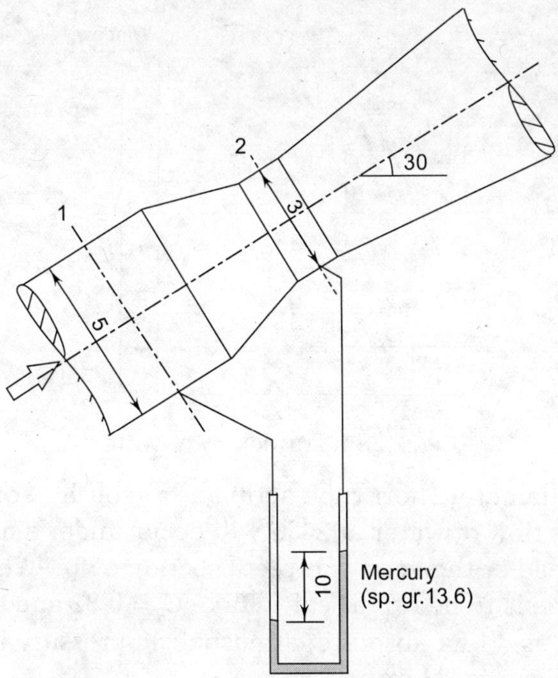


**Fig. P.24** Effect of inclination on flow rate

25. A reservoir having a surface area of 500 m$^2$ is emptied by a 0.5 m wide rectangular weir. How long should it take to empty the reservoir from a height of 3.2 m to 0.1 m above the still? Take $C_d$ = 0.65. [16 m 5 s].
26. The height of water on the upstream and downstream side of a submerged weir of 3 m lengths is 20 cm and 10 cm respectively. If $C_d$ for free and drowned position is 0.6 and 0.8, respectively, find the discharge over the weir. [0.504 m$^3$/s].
27. A broad-crested weir of length 50 m has 50 cm height of water above its crest. Find (a) the maximum discharge. Take $C_d$ = 0.6. Neglect velocity of approach. (b) If the velocity of approach is to be taken into consideration, find the maximum discharge when the channel has a cross-sectional area of 50 m$^2$ on the upstream side. [18.08 m$^3$/s, 18.412 m$^3$/s].
28. A fireman must reach a widow 40 m above the ground, Fig. P.28 with a water jet, from a nozzle of 30 mm diameter discharging 30 kg/s. Assuming the nozzle discharge to be at a height of 2 m from the ground, determine the greatest distance from the building where the fireman can stand so that the jet can reach the window. [142 m].

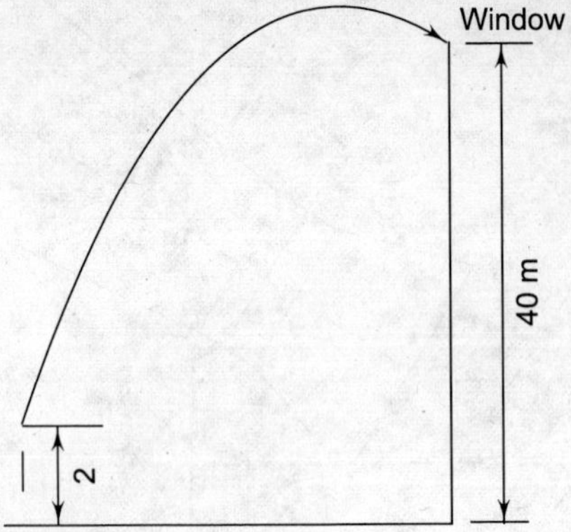


**Fig. P.28** Trajectory of water

29. For the 100 mm diameter short tube acting as a mouthpiece in a tank as shown in Fig. P.29, (a) what flow of water at 24°C will occur under a head of 9.2 m? (b) What is the pressure head at the vena contracta section *C*? (c) What maximum head can be used if the tube is to be full at exit? Take $C_d = 0.82$ and $C_v = 1$. Vapour pressure for water at 24°C is 3 kPa absolute, atmospheric pressure is 101 kPa. [0.084 m$^3$/s, 3.85 m of water vacuum, 23.78 m].

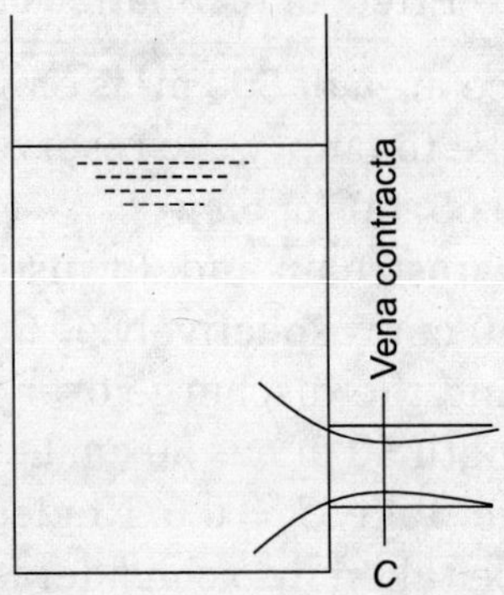


**Fig. P.29** A tank fitted with a mouthpiece

30. A blower fan shown in Fig. P.30 is required to deliver air flow of 90 m$^3$/min at 10 cm W.G. in a 30 cm diameter pipe. Assuming density of the air = 1.2 kg/m$^3$ and blower efficiency = 70 percent, determine the power required to drive the blower. Neglect the compressibility effects and friction loss in the entrance mesh. Assume that the entrance section is large enough to permit taking the inflow velocity of air as zero. [2681 W].

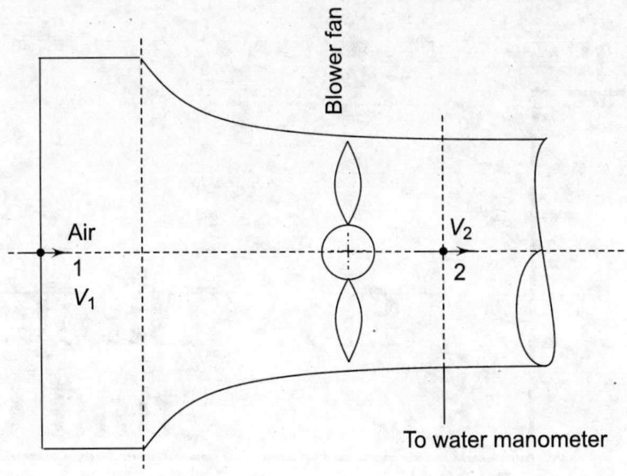


**Fig. P.30** Power required to run a blower fan

31. During testing of a rocket, fuel (liquid hydrocarbon) is supplied at the rate of 1.5 kg/s and liquid oxygen at the rate of 10 kg/s. The gaseous products of combustion flow out of the unit with a high velocity through a 400 cm$^2$ cross-sectional area exit nozzle. Measurements indicate that pressure and temperature at the nozzle exit are 100 kPa and 750 K and the ambient pressure surrounding the stand is 100 kPa. Under these conditions, calculate the nozzle exit velocity and the rocket thrust, i.e. the force necessary to hold the rocket in position. How this force exerted on the test-bed will change if reducing the ambient pressure to 20 kPa simulates the flight conditions? The exhaust gases behave as a perfect gas with $R$ = 600 J/kg K. [1295 m/s$^2$, –1489.5 N, 18092.5 N]
32. A 60 cm diameter pipeline carries oil (specific gravity 0.85) at 82500 m$^3$/day. The friction head loss is 8.5 m per 1000 m of pipe run. It is planned to place pumping station every 20 km along the pipe. Make calculations for the pressure drop in kN/m$^2$ between the pumping stations. [1417.55 kN/m$^2$].
33. When an ideal fluid with ambient velocity $U_0$ flows past a two dimensional circular cylinder, the velocity at the surface of the cylinder is given by $U = 2\ U_0 \sin\theta$. Determine the pressure distribution around the cylinder. Express it in non-dimensional form and plot. Comment on the results obtained.
34. Integrate Euler's equation along a steam line and obtain Bernoulli equation. State all the assumptions.
35. Show that the constant in Bernoulli equation will vary from one stream line to another if the flow is irrotational.

CHAPTER

# 4

# Flow Through Pipe and Channel

## OUTLINE

- Introduction
- Losses in Pipe Flow
- Hydraulic Power Transmitted Through Pipe
- Water Hammer
- Flow in Open Channel

## 4.1 Introduction

In nature water flows in rivers, canals, etc. due to gravity. River or canal beds have some slope that results in a force on the liquid causing it to flow from higher level to lower level. In such systems, the discharge of water depends on the slope and level of water above the bed. However, we can transport water from one place to another through pipes wherein controlling the pressure across the pipe can control the rate of flow. Therefore, flow through pipes has drawn our attention and it is used not only for transmitting liquid but also the gases and even the solid particles. Water or LPG distribution in a city is an example of pipe network. Here our interest is to find out the discharge at the far end of the pipe network and also to determine the head loss so that we can decide about the pressure required at the starting point to send the required amount of water to the farthest corner of the pipe line. We shall study these aspects in details in this chapter.

Normally, a circular pipe is very common in use. But the pipes of other cross sections have also been used. When the fluid flows through the pipe, the pressure decreases along the flow. We express this fact by saying that there is a head loss, which is due to viscosity of the liquid and the surface roughness. The solution of the N–S equations for the flow through a pipe due to pressure difference has been discussed in Chapter 5. But we shall discuss in this chapter a very simple method of determining this head loss—the Darcy's method. We shall also study the other causes of the head loss. This will help in determining how much power is available at the other end of the pipe and how to maximise it.

We need power to maintain flow in a pipe. The amount of power increases with the discharge and the length of the pipe. A huge amount of energy needs to be expanded to transport fluid from one place to another miles away. Therefore, flow by the gravity becomes quite important because we use the natural source of power to maintain the flow. We shall learn how to maximise the discharge with minimum or economic cross section. Moreover a few phenomena are associated with gravity flow, which we shall study. We use canals, which are of great importance for the mankind. There are a number of such man-made canals, like Panama Canal connecting Pacific Ocean with Atlantic Ocean or Egypt canal connecting Mediterranean Sea with the Red Sea, etc. Because of canals it has been possible to store water in hills and transport it down to other places for power generation, supplying drinking water and for agriculture purposes. Because of canals, we are able to think of linking various rivers in the country for better water management. Therefore water flow through a canal is a subject of importance and needs thorough study.

## 4.2 Losses in Pipe Flow

The measurement of pressure at two sections of a pipe carrying fluid shows that it decreases in the flow direction. This loss is associated with the viscous forces. Stanton, Nokuradse and other scientists measured losses in turbulent and laminar flows through pipes with different degree of roughness and Reynolds number. These results were plotted on a log-log graph by Moody and were of great help in the designing of pipe system. See Fig. 4.2 for Moody diagram. The graph gives the friction factor that was introduced by simple analysis by Darcy–Weisbach using a control volume of length $L$ in a circular pipe flow, as shown in Fig. 4.2.1. Flow is supported by pressure $p_1$ while the pressure $p_2$ and the shear force on the wall oppose it. Under steady sate, there is no acceleration and hence inertia force. So pressure and the viscous forces are the only forces acting on the control volume. The force balance required for equilibrium of the element gives the equation.

$$(p_1 - p_2) \times \pi/4\ d^2 = \text{Shear force} = C_D \times (1/2\ \rho U^2) \times \pi\, dl$$

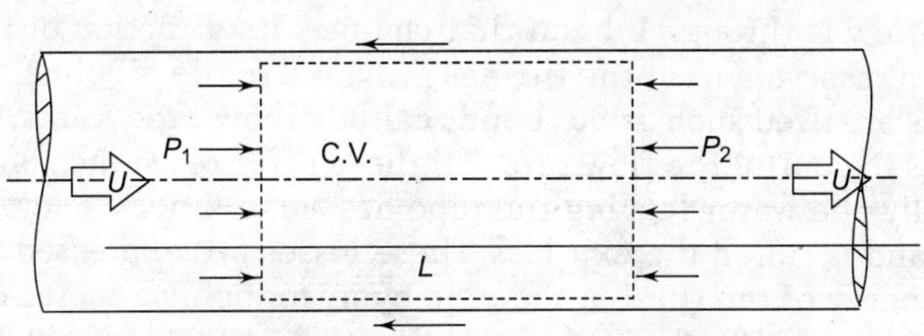


**Fig. 4.2.1** Equilibrium of flow in a pipe

Here $C_D$ is the drag coefficient, $U$ is the flow velocity in pipe of diameter $d$, and $\pi\, dl$ is the surface area on which the viscous force acts. In liquids, it is common practice to express pressure or pressure difference in terms of the height of liquid column, called the pressure head, which is, $h_f = (p_1 - p_2)/w$, where $w$ is the specific weight of the liquid, $= \rho g$. A $U$-tube manometer gives the height of the manometric fluid and is to be changed to the head of the fluid flowing through the pipe using the conversion factor $(\rho_m/\rho - 1)$, where $\rho_m$ and $\rho$ are respectively the densities or specific gravities of the manometric fluid and the fluid flowing through the pipe. Substitute these in the above equation and rewrite it in the following form

$$h_f = \frac{fL}{d} \cdot \frac{U^2}{2g} \approx \frac{fLQ^2}{12d^5} \tag{4.2.1}$$

Here $f$ is called the friction coefficient. In many books factor of 4 is used with $f$; that is, instead of $f$ we use $4f$. This confusion will prevail and only experience will tell to recognise whether $f$ or $4f$ is given. A thumb rule is that if $f < 0.01$ then factor 4 is to be used in Eq. 4.2.1. The trend now is to use Eq. 4.2.1 as such and give value of $f$ accordingly. But you have to be attentive on this issue before using the equation. The last part of the equation is obtained using the equation, $Q = (\pi/4)\, d^2\, U$. However, this equation is in terms of $Q$ and not $U$ and has advantage because in a pipe network, the pipes are in series and hence the velocity will go on changing but the discharge will remain same. Therefore use of this form of the equation makes calculations a bit simple. We shall discuss it later. Figure 4.2.2 shows the Moody diagram. The following empirical relation is also used to calculate $f$.

$$\frac{1}{\sqrt{f}} = -2.0 \log\left(\frac{\varepsilon/D}{3.7} + \frac{2.51}{\mathrm{Re}\sqrt{f}}\right)$$

In distribution system, the pipe length runs into kilometres and therefore this head loss becomes most significant. This is therefore called *major loss*. We shall show through some examples that the other losses are so insignificant in its comparison that those are often not considered in the calculations that too without any appreciable error.

### 4.2.1 Minor Losses

Losses during the flow also occur whenever the pipe diameter increases or decreases suddenly, the flow is through a bend that changes its direction of flow, at the entry to the pipe from a reservoir, or at the fittings provided in the pipeline, etc. Different kinds of pipe fittings are used such as 90° bends called elbows, tee joints, unions, and various types of valves to control the flow rate. All these offer resistance of different degree to the flow. Finally, the water leaving the pipe has some kinetic energy that is lost to the environment and is called the exit loss. These losses are expressed in terms of the specific kinetic energy of the fluid in the pipe by introducing a factor $K$ which is different for different situations. Table 4.2.1 gives the value of factor $K$ for different resistance/obstructions in the pipe system.

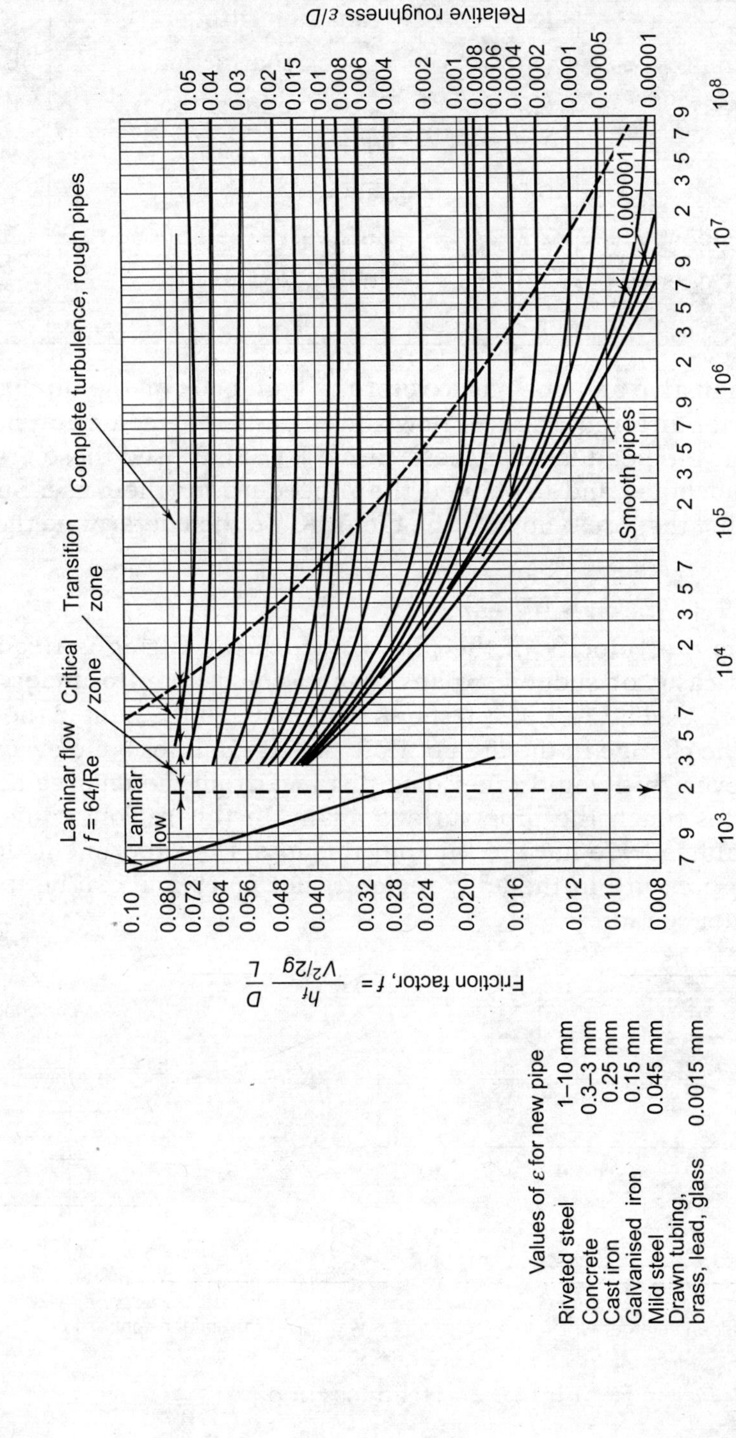


**Fig. 4.2.2** Friction factors for pipes (adapted from Trans. ASME, 66, 672, 1944)

**TABLE 4.2.1** Minor losses

| Transition/fitting | K | Transition/fitting | K |
|---|---|---|---|
| Sudden enlargement | $(1-A_1/A_2)^2$ | Sudden contraction | 0.5 |
| Entry (sharp edged) | 0.5 | Entry (bell-mouthed) | 0 – 0.05 |
| 90° elbow | 1 | 90° bend | 0.6 |
| 180° bend | 0.6 | Tee | 2 |
| Gate valve (fully opened) | 2.2 | Gate valve (half opened) | 5.6 |
| Globe valve (fully opened) | 1.0 | Pump foot valve | 1.5 |
| Exit | 1 | | |

This list is not exhaustive and one has to refer the handbook on the piping system to get the coefficient $K$ for different cases of flow and fittings. Mnemonic graphs have also been developed to get the head loss for these cases. J. Lal has used these graphs in his book on hydraulic machines and explained the procedure for their use. Such graphs save time in calculating the loss. Knowing the factor $K$, we then determine the head loss from the equation.

$$h_{\text{min or loss}} = K\,U^2/(2g) \tag{4.2.2}$$

In this equation, $U$ is the velocity in the pipe in which the fitting is fixed. We shall discuss the first two cases of sudden expansion and contraction to understand what should be $U$ in these cases. Figure 4.2.2 (a) shows a sudden enlargement in the pipe cross-section. Because of the change in the area of flow the streamlines bend away to adhere to the surface. However, this would take some distance from the enlargement section before the streamlines touch the pipe surface again. In this region, eddy forms and results in loss of total head. We have to find out this loss. The experiments have shown that the variation in pressure in the eddy region is not much and can be taken as the pressure $p_1$ at the inlet section.

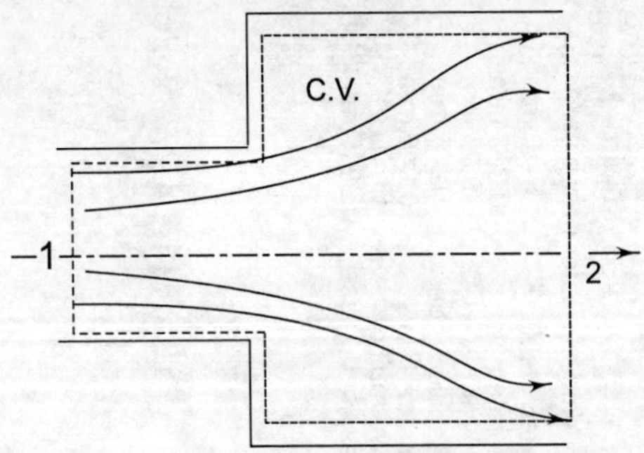


(a) Sudden enlargement

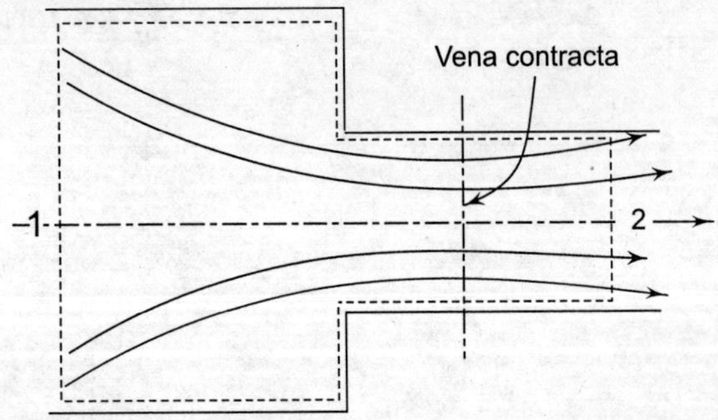


(b) Sudden contraction

**Fig. 4.2.2** Head loss in pipe

We consider the control volume shown by the dotted lines in the figure with sections 1 and 2 at the inlet and outlet respectively. We can use the integral form of mass, momentum and energy Eq.s to give

$$Q = A_1\,U_1 = A_2\,U_2 \tag{A}$$

$$p_1A_1 + p_1\,(A_2 - A_1) - p_2A_2 = \rho\,Q\,(U_2 - U_1)$$

or
$$(p_1 - p_2)\,A_2 = \rho\,Q(U_2 - U_1) \tag{B}$$

and
$$\frac{p_1}{\rho g} + \frac{U_1^2}{2g} = \frac{p_2}{\rho g} + \frac{U_2^2}{2g} + h_L \tag{C}$$

In above equations, $A$ is the cross-sectional area, $h_L$ is the total head loss and rest of the symbols have usual meaning. We have used the experimental observation while writing Equation (B) that pressure in eddy region is equal to $p_1$. Substituting for $U_2$ from (A) in (B) and then $(p_1 - p_2)$ from (B) in (C), we get after simplification

$$h_L = \left(1 - \frac{A_1}{A_2}\right)^2 \frac{U_1^2}{2g} \tag{4.2.3}$$

It is noticed from Eq. 4.2.3 that the head loss is inversely proportional to the ratio of $A_1/A_2$. This expression can also be used to determine the head loss in case of sudden contraction. Figure 4.2.2(b) shows the sudden contraction. The streamlines bend inwards and continue to do so till some section after which these lines move out to attach to the pipe surface. The section where the area of the stream tubes is minimum is called the vena contracta. The experimental results indicate that the loss occurs mostly in the flow after the vena contracta. That is, we determine the loss as if the flow is expanding from area $A_c$ at the vena contracta to area $A_2$ of the pipe at the section 2 of the control volume as shown in Fig..

∴
$$C_c = \frac{A_c}{A_2} = 0.62 \text{ and } U_c = \frac{U_2}{0.62}$$

→
$$h_L = \left(1 - \frac{A_c}{A_2}\right)^2 \frac{U_c^2}{2g} = 038\,\frac{U_2^2}{2g}.$$

This loss is obtained using $C_c = 0.62$ and neglecting the losses prior to the vena contracta section. In practice instead of taking the factor 0.38, we take it as 0.5. Therefore, the head loss in sudden contraction is given by the relation

$$h_L = 0.5\frac{U_2^2}{2g} \tag{4.2.4}$$

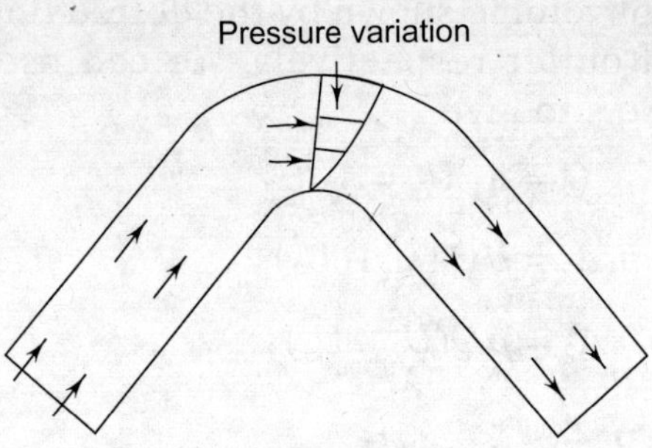


**Fig. 4.2.3** Head loss in bend

The entry loss can also be considered as sudden contraction unless a bell mouthed entry is made for smooth entry of the fluid into the pipe from the reservoir. Naturally, for a smooth entry the loss will be less, as can be seen in the Table 4.2.1. On the other hand the exit loss is due to kinetic energy and therefore the factor K is 1. In situations, as shown in Fig. 4.2.3, the streamlines bend and in the process form a vena contracta after the inner corner thus leaving some space where the eddies circulate and cause head loss. The amount of head loss depends on how smooth or sharp the flow direction is changed. We expect more loss in case of a sharp change. In general, in confined flows the head is lost whenever the streamlines are forced to bend. There is pressure variation also across the curved section, as shown in the Fig. that increases from the inner to the outer edge; this may result in secondary flows adding to the loss further.

All the above losses are known as *minor loss.*

**Example 4.2.1** Water flows from a reservoir through a series of pipes joined as shown in Fig. E 4.2.1. If the exit is 20 m below the free level in the reservoir, calculate the discharge. How much error will be introduced if we neglect the minor losses for the calculation of the discharge?

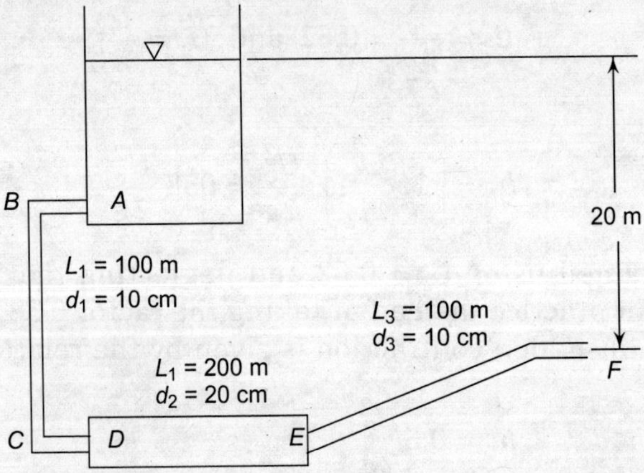


**Fig. E 4.2.1** Error in discharge neglecting minor losses

**Solution** We are given the difference of level at the free surface in the reservoir and the end of the pipeline. Because of this difference, flow will start and would ultimately acquire steady state when the total head loss equals this value. This loss consists of the major loss given by Eq. 4.2.1 and the minor loss obtained from Eq. 4.2.2 in which the factor $K$ is given by Table 4.2.1. The minor loss will occur at (i) entrance to the pipe, (ii) two right angle bends in first pipe, (iii) sudden expansion, (iv) sudden contraction, and (v) exit loss. The continuity equation gives velocity in different pipes in terms of say velocity in pipe 3, which is also the exit velocity. Writing these, we get

$$A_1 U_1 = A_2 U_2 = A_3 U_3 = Q$$

$$\rightarrow \quad U_1 = U_3 = 4\,U_2,$$

since $\quad d_1 = d_3 = 10$ cm and $d_2 = 20$ cm (given).

$$20 = \underbrace{\frac{f}{2g}\left[\frac{L_1U_1^2}{d_1}+\frac{L_2U_2^2}{d_2}+\frac{L_3U_3^2}{d_3}\right]}_{\text{Major loss}}+\underbrace{\left[0.5U_1^2+U_1^2+U_1^2+\left(1-\frac{A_1}{A_2}\right)^2U_1^2+0.5U_2^2+U_3^2\right]}_{\text{Minor loss}}$$

$$= \frac{U_3^2}{2g}\,[20 + 1.25 + 20.0 + 0.5 + 1.0 + 1.0 + 0.56 + 0.03 + 1/0]$$

$$\rightarrow \quad U_3 = 2.94 \text{ m/s and } Q = A_3\,U_3 = 0.0231 \text{ m}^3\text{/s}.$$

Since the pipeline is made up from pipes of different diameters connected in series, use of Eq. 4.2.1 in terms of $Q$ will be handy if we neglect the minor losses. This eliminates determination of velocities in the individual pipe.

$$20 = \frac{fQ_{\text{app.}}^2}{12}\left[\frac{L_1}{d_1^5}+\frac{L_2}{d_2^5}+\frac{L_3}{d_3^5}\right] \rightarrow Q_{\text{app.}} = 0.0242 \text{ m}^3\text{/s}.$$

$$\text{Error} = \frac{0.0242-0.0231}{0.0231} = 4.8\%.$$

Error made by neglecting minor loss is small. In this case length of the pipe is not large, otherwise in actual situation lengths would be of much larger size. In such situation, the error will be still less. On the other hand by ignoring the minor loss there will be significant saving in time. We also get freedom from accounting all such losses.

**Example 4.2.2** Water flows through pipes of diameter 8 mm and 4 mm discharging into a third pipe of diameter 10 mm. The discharge in 8 mm pipe is 6.2 cc/s. Find out as to what should be the maximum discharge in this pipe to make the flow critical. Take critical value of Reynolds number as 2300. State also for this condition the type of flow in the 8 mm and 4 mm pipes. Take kinematic viscosity of water equal to $6.56 \times 10^{-7}$ m$^2$/s.

**Solution**

Given: Let us represent pipes of diameters 8 mm, 4 mm, and 10 mm by subscripts 1, 2, and 3, respectively. $\nu = 6.56 \times 10^{-7}$ m$^2$/s, $d_1 = 8$ mm, $d_2 = 4$ mm $d_3 = 10$ mm, and $\text{Re}_c = 2300$.

Let $U_3$ be the velocity in the third pipe when the flow in it is critical. Then

$$Re_c = \frac{U_3 \times 0.01}{6.56 \times 10^{-7}} = 2300$$

$\rightarrow$ $U_3 = 0.1509$ m/s and $Q_3 = 0.75 \times (0.01)^2 \times 0.1509 = 1.185 \times 10^{-5}$ m$^3$/s.

Using continuity equation,

$Q_3 = Q_1 + Q_2 \rightarrow Q_2 = 1.185 \times 10^{-5} - 6.2 \times 10^{-6} = 0.565 \times 10^{-5}$ m$^3$/s.

$\therefore$ $U_2 = Q_2/A_2 = 0.4496$ m/s and $Re_2 = 0.4496 \times 0.004/6.56 \times 10^{-7} = 2741$.

This means the flow in pipe 2 is turbulent, since Reynolds number is more than the critical Reynolds number for laminar flow. We may similarly find out Reynolds number in pipe 1. This comes out to 1504. So the flow in this pipe is laminar.

**Example 4.2.3** Oil of specific gravity 0.85 is pumped through a horizontal pipe, 150 mm in diameter and 1220 m long. The discharge is 22.5 l/s when the pump, which has an efficiency of 65%, takes 7.35 kW to drive it. Find the viscosity of the oil. Take $f = 16/\text{Re}$.

**Solution**

Given: $d = 0.15$ m, $L = 1220$ m, $Q = 22.5 \times 10^{-3} = 0.0225$ m$^3$/s, $\rho = 850$ kg/m$^3$, $P = 7.35$ kW, $\eta_o = 0.65$.

Note that $f = 16/\text{Re}$ implies a factor of 4 in Equation 4.2.1.

Power required to maintain a flow of $Q$ m$^3$/s in a pipe, in which head loss is $h_f$, is given by

$$P = w\,Q\,h_f/\eta_o$$

$$\rightarrow \qquad h_f = 0.65 \times 7.35/(9.81 \times 0.025) = 25.46 \text{ m.}$$

$$U, \text{Flow velocity} = Q/A = 0.0225/(0.75 \times 0.15^2) = 1.273$$

But, $$h_f = \frac{4 \times f \times 1220 \times 1.273^2}{9.81 \times 0.15} \rightarrow f = 0.0095 = \frac{16}{\text{Re}}. \text{ Or, } R_e = 1684.$$

and $$\text{Re} = \frac{1.273 \times 0.15 \times 850}{\mu} \rightarrow \mu = 0.0964 \text{ Ns/m}^2.$$

**Example 4.2.4** A pipe of 250 mm diameter, 1500 m long is laid at a slope of 1 in 200 for the first half-length and at a slope of 1 in 150 m for the rest of the length. The pressure at the upper and lower ends of the pipe is 100 kPa and 50 kPa, respectively. Find the discharge through the pipe. Take $f = 0.008$.

**Solution** Slope of 1 in 200 means the vertical rise in the level of the pipe is 1 m after moving a horizontal distance of 200 m. From this information, the height of different points on the pipe from the lower end can be determined. We get the height of the upper end of the pipe = 750/200 + 750/150 = 8.75 m. Let subscripts 1 and 2 are used for lower and the upper end of the pipe. Then $z_2 - z_1 = 8.75$ m. Moreover, $U_1 = U_2$ because the diameter of the pipe is same at the two ends. In addition

$$\frac{p_2}{w} = \frac{100}{9.81} = 10.194 \text{ m}$$

and $$\frac{p_1}{w} = \frac{50}{9.81} = 5.097 \text{ m.}$$

Applying Bernoulli equation at the two points, we get

$$h_f = \left(\frac{p_1}{w} + \frac{U_1^2}{2g} + z_1\right) - \left(\frac{p_2}{w} + \frac{U_2^2}{2g} + z_2\right) = (5.097 - 10.194) - 8.75 = -13.847 \text{ m.}$$

The negative sign shows that the flow is from the end 2 to end 1 and not vice versa, as we took while writing the Bernoulli equation. Now using Eq. 4.2.1, we can get the discharge as

$$h_f = \frac{4fLQ^2}{12d^5} = \frac{4 \times 0.008 \times Q^2}{12 \times 0.2^5} \rightarrow Q = 0.0449 \text{ m}^3\text{/s.}$$

**Example 4.2.5** Water flows through a pipe *AB* 15 cm $\phi$ and 400 m long. The point *B* is 20 m above point *A*. The discharge is 0.02 m$^3$/s from *A* to *B*. find the pressure at *A* if the pressure at *B* is 200 kPa. Take $f = 0.024$. Suppose after 12 years of service the friction is doubled, what would be the discharge if the pressure at *A* and *B* remain unchanged?

**Solution** The value of *f* suggests that Eq. 4.2.1 is to be used as such. The purpose of this question is to show the effect of rusting of the pipe on the discharge during the service. Let subscripts 1 and 2 refer to the points *A* and *B* of the pipe. We are given: $z_2 - z_1 = 20$ m, $p_2/w = 200/9.81 = 20.387$ m, $Q = 0.02$ m$^3$/s.

$$\therefore \quad h_f = \frac{0.024 \times 400 \times (0.02)^2}{12 \times (0.15)^5} = 41.8 \text{ m.}$$

Applying modified Bernoulli equation between points 1 and 2, we get

$$\frac{p_1}{w} = \frac{p_2}{w} + (z_2 - z_1) + h_f = 20.387 + 20 + 4.18 = 44.567 \text{ m.}$$

$$\rightarrow \quad p_1 = 9.81 \times 44.567 = 437.2 \text{ kPa.}$$

Let *Q* is the new discharge rate after 12 years. Since there is no change in pressure at the either ends, the head loss too is unchanged, since the total head loss is equivalent to pressure head loss in constant diameter pipe laid horizontally. Equation 4.2.1 can now be used with $f = 2 \times 0.024$ to give *Q*. This gives $Q = 0.014$ m$^3$/s. Hence the discharge has decreased.

**Example 4.2.6** Figure E 4.2.6 shows two reservoirs connected by a 300 mm diameter 1 km long pipe, the difference of water levels in the tanks *A* and *B* is 15 m. The pipe passes over a hill at *C*. Find the length of the pipe from the reservoir *A* to *C* so that the absolute pressure head at *C* is 2.5 m. The height of *C* above the water surface of the reservoir *A* is 5 m. Take $f = 0.0075$ and neglect all the minor losses. Determine the discharge also.

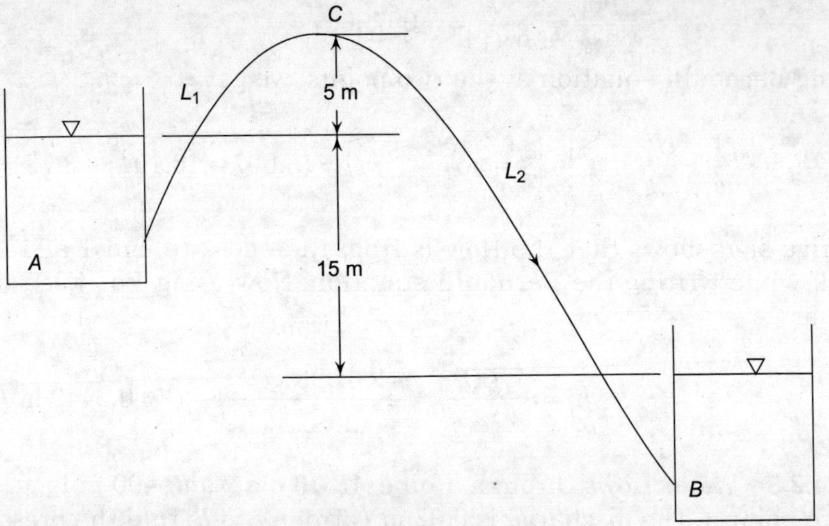


**Fig. E 4.2.6** Limit of the upper point in pipe distribution system

**Solution** Let $Q$ be the discharge through the pipe of $d$ = 0.3 m, $L$ = 1000 m with $h_f$ = head difference in levels of $A$ and $B$ = 15 m. From Eq. 4.2.1, we get

$$h_f = \frac{fLQ^2}{12d^5} \rightarrow Q^2 = \frac{15 \times 12 \times (0.3)^5}{4 \times 0.0075 \times 1000} \text{ or, } Q = 0.1213 \text{ m}^3/\text{s}.$$

$$\rightarrow \quad U = Q/A = 0.1213/(0.75 \times 0.3^2) = 1.716 \text{ m/s and } U^2/(2g) = 0.15 \text{ m}.$$

$$h_{f,AC} = (L_1/1000) \times 15 = 0.015L_1 \text{ m}.$$

Major loss $h_{f,AC}$ is proportional to the length. Therefore this relation is used to get it from loss in 1000 m long pipe Applying modified Bernoulli equation between two points one in free surface of reservoir $A$ and other at $C$ and taking datum from the free surface in reservoir $A$, we get

$$10.3 + 0 + 0 = 2.5 + 0.15 + 5 + 0.015\, L_1 \rightarrow L_1 = 176.67 \text{ m}.$$

**Example 4.2.7** Calculate the power that must be applied to pump 65 l/s of water from the low-level reservoir to the high level reservoir shown in Fig. E 4.2.7. Neglect all minor losses. Find also the maximum dependable flow, which can be pumped through this system. Take $f$ = 0.006.

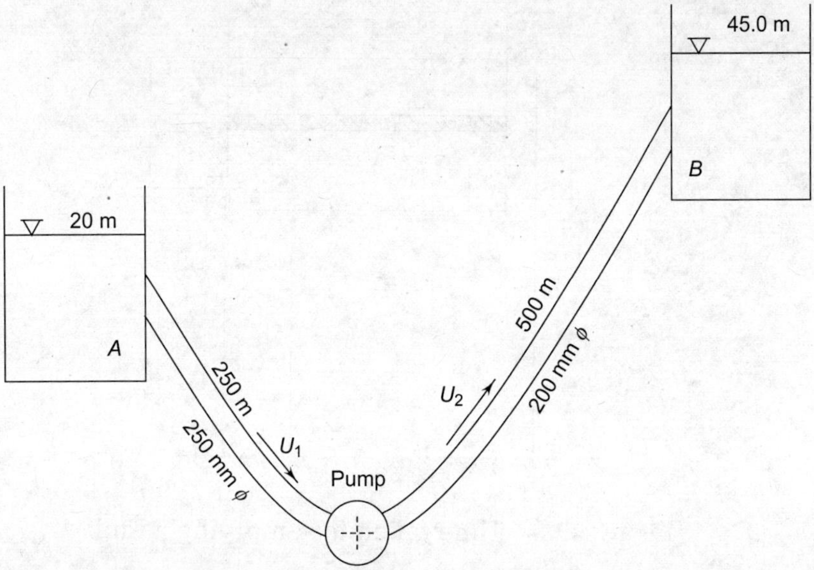


**Fig. E 4.2.7** Power required in pumping water

**Solution** It is a practice to write the level of any point from the reference datum line. For example, height of the free surface in reservoir *A* from reference line is written as 20 m. The pump has to lift water from a level of 20 to 45 m and to supply head to overcome the frictional losses during the flow. Using Eq. 4.2.1, we get

$$h_f = \frac{4fLQ^2}{12d^5} = \frac{4 \times 0.006 \times 0.065^2}{12}\left[\frac{250}{0.25^5} + \frac{500}{0.2^5}\right] = 15.23 \text{ m.}$$

$\therefore$ Head supplied by the pump, $H = (45 - 20) + 15.23 = 40.23$ m.

$P$, Power required = $w\,Q\,H = 9.81 \times 0.065 \times 40.23 = 25.67$ kW.

**Example 4.2.8** A reservoir 10 m × 5 m cross-section contains water up to a depth of 3 m. The water level is 25 m above the ground. It is desired to empty the tank through a 15 cm diameter opening near the bottom of the reservoir. Is it faster to empty it through the opening as an orifice or through a straight vertical pipe reaching the ground? Take $C_D$ of the orifice as 0.6 and $f$ for the pipe as 0.02.

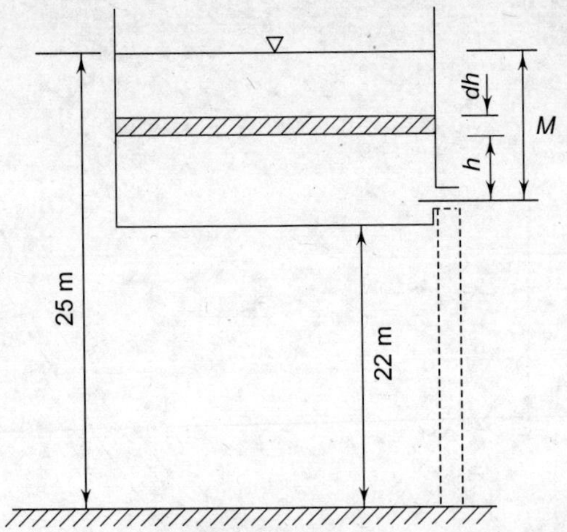


**Fig. E 4.2.8** Time taken for emptying a tank

**Solution** Let at any time $h$ be the height of water level from the opening in the tank , it goes down by $dh$ in time $dt$, $A$ and $a$ are respectively the area of cross-section of the tank and the orifice. When using an orifice the velocity of flow is $\sqrt{(2gh)}$ and we can write the following equation

$$-A dh = C_d \, a\sqrt{2gh} \; dt \rightarrow T = \int_0^3 \frac{A dh}{C_d a\sqrt{2gh}} = 3687.4 \text{ s.}$$

While using pipe, there will be losses in head at (i) entry, (ii) one bend, (iii) exit loss, and (iv) friction loss. The factor $K$ for the minor losses can be taken as 2. Since the pipe is going to the ground, we may measure height of the free surface from the ground. Let $h$ is the height at any time which goes down by $dh$ in time $dt$ and would change from 25 m to 22 m. Then using Eq. 4.2.1 and volume flow from the tank = volume of water flowing out of the pipe in time $dt$, we get

$$h = \left(2 + \frac{fL}{d}\right)\frac{U^2}{2g}, -A \, dt = a \, U \rightarrow \int_0^T dt = \int_{22}^{25} \frac{a}{A}\sqrt{\frac{2gh}{2 + fL/d}} \quad \text{or } T = 879.5 \text{ s.}$$

We note that the time taken for emptying the tank using the pipe is considerably reduced. This is mainly because the head causing flow is increased in this case.

### 4.2.2 Series and Parallel Arrangement

Pipes are connected together in different ways. These pipes may be of same or different diameter. The pipes of same diameter are connected because these are available in certain length. But we are talking of situations where pipes of different diameters are to be used. These pipes may be connected one after another, called the series arrangement. Or two reservoirs may be connected by a number of pipes running between them. This is the parallel arrangement. In any pipe system, both the arrangements can be found. We shall now examine the features of these different arrangements. Figure 4.2.4 shows these arrangements. For simplicity only the lengths of the pipes in parallel arrangement have been shown of equal length but it is not necessary.

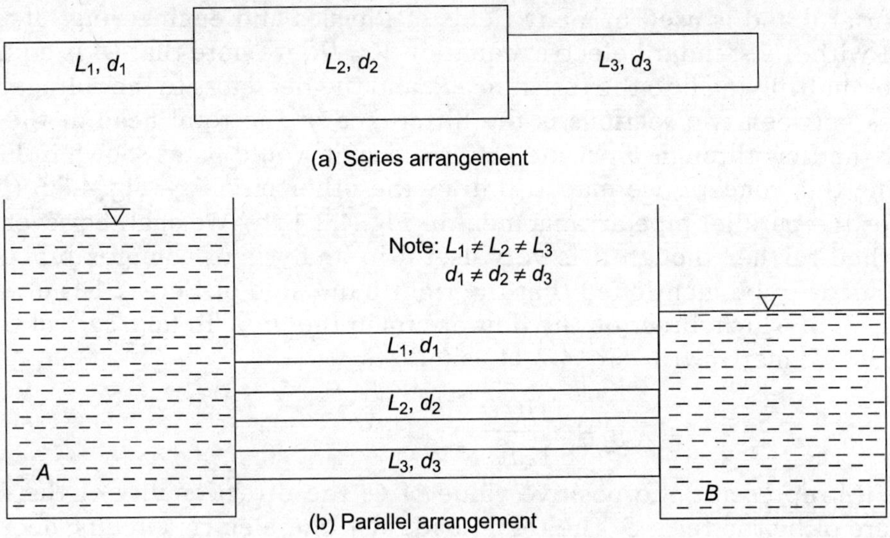


**Fig. 4.2.4** Different arrangements of pipes in pipe network

We observe from Fig. 4.2.4 (a) that the flow in all the pipes remain same and the total loss is sum of loses from entrance to exit. That is, if there are $n$ pipes in series then

$$Q = Q_1 = Q_2 = Q_3 = \ldots \text{ and } h_f = \sum_{i=1}^{n} h_{fi} \tag{4.2.5}$$

On the other hand, in parallel arrangement, the discharge will distribute in the branching lines in such a manner that the head loss in all these pipes is same. Thus for $n$ pipes,

$$h_{f1} = h_{f2} = h_{f3} = \ldots \text{ and } Q = \sum_{i=1}^{\eta} Q_i \tag{4.2.6}$$

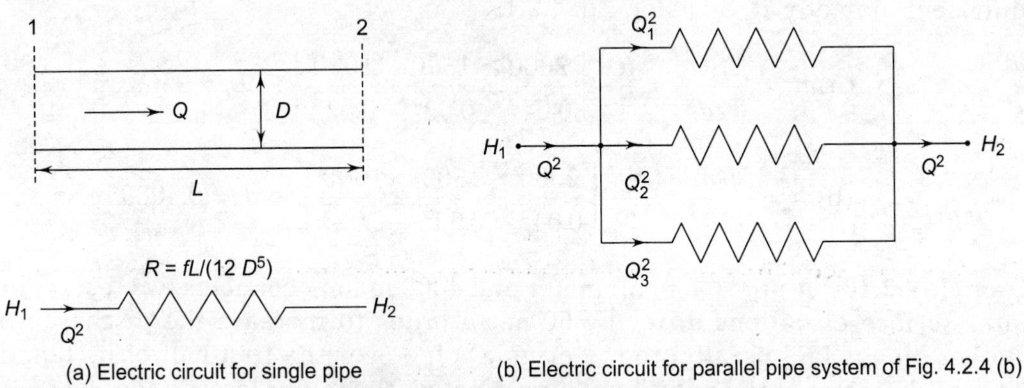


**Fig. 4.2.5** Electrical analogy

In analysis of the pipe line system, the analogy concept is quite useful. This is an attempt in which the pipe lining system is represented by an electric circuit, which can then be easily simplified using circuit analysis techniques. This technique has found to be very useful and is used in many fields of Physics and engineering. If we compare Eq. 4.2.1 with the standard electric equation $V = IR$, we note that $Q^2$ is equivalent to $I$, the factor multiplying it to the resistance $R$, and the head loss to the voltage. Recall that head loss between two sections is the difference of the total head at these sections. Thus for the flow through pipe, the electric circuit would be as shown in Fig. 4.2.5 (a). Extending this concept, we may also draw the other circuits—Fig. 4.2.5 (b) has been drawn for the parallel pipe arrangement of Fig. 4.2.4 (b). We shall however not pursue this method further though it is very useful in analysis of complex piping system. A remark needs to be mentioned that the right hand side of Eq. 4.2.1 is always positive irrespective of which direction the flow occurs in the pipe. To take care of the direction of the flow, we may rewrite Eq. 4.2.11 as follows:

$$h_f = \frac{fL|Q|Q}{12d^5} \tag{4.2.7}$$

Here $|Q|$ represents the positive value of $Q$; the directionality of the flow is then taken care of by the term $Q$. The total head, just like electric circuits, decreases when we move in the direction of flow and increases in the opposite direction.

**Example 4.2.9** The pipe system consists of three pipes $AB$ (2000 m and 0.04m $\phi$), $BC$ (1500 m and 0.03 m $\phi$), and $CD$ (1000 m and 0.02 m $\phi$) connected in series. Determine (a) the equivalent length of a single pipe whose diameter is 30 cm, and (b) the equivalent diameter of a single pipe whose length is 4500 m.

**Solution** When pipes of different diameters are connected in sseries then these can be replaced by (a) a single pipe of dia $d$ of equivalent length $L_e$ or (b) a pipe of length $L$ and equivalent diameter $d_e$. The equivalent length or diameter is decided such that the resistance to flow is equal to the actual resistance. In this problem, we shall ignore the minor losses. However, these can also be considered if required in the manner discussed earlier. We can write the following equations to get equivalent length $L_e$ or equivalent diameter $d_e$ by using Eq. 4.2.1.

$$\text{(a) } h_f = \frac{fL_eQ^2}{12d^5} = \frac{fQ^2}{12}\left[\frac{2000}{0.04^5} + \frac{1500}{0.03^5} + \frac{1000}{0.02^5}\right] \rightarrow L_e = 9568.36 \text{ m.}$$

$$\text{(b) } h_f = \frac{fQ^2}{12d_e^5} = \frac{fQ^2}{12}\left[\frac{2000}{0.04^5} + \frac{1500}{0.03^5} + \frac{1000}{0.02^5}\right] \rightarrow d_e = 0.4089 \text{ m.}$$

**Example 4.2.10** A pipe 0.3 m diameter and 4000 m long connects two reservoirs whose water surface elevations differ by 60 m. In order to increase the discharge by 25%, another pipe is laid parallel to the original pipe over the first half of its length. Compute the diameter of the additional pipe. Neglect the minor losses and take $f = 0.032$ for all the pipes.

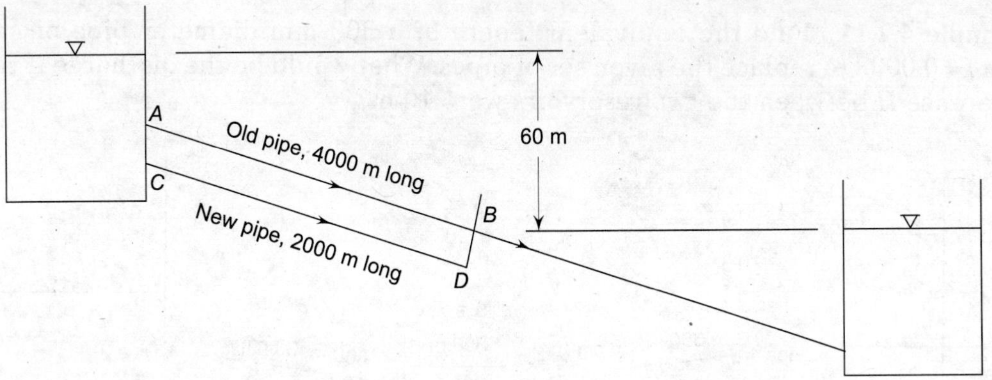


**Fig. E 4.2.10** Augmentation of discharge by placing a pipe in parallel to the old pipe

**Solution** Laying a pipe in parallel increases the total discharge because the discharge through individual pipe and therefore the resistance to the flow is decreased. We shall not go into the discussions of deciding as to in which portion of the original pipe, the new pipe is to be laid or what length it should run in parallel. Here we shall just make calculations to demonstrate the statement. Figure E 4.2.10 shows the arrangement. Let $Q$ be the total discharge in the first case. Then

$$\frac{0.032 \times 4000 \times Q^2}{12 \times 0.3^5} = 60 \rightarrow Q = 0.1174 \text{ m}^3/\text{s}$$

and $$Q_{new} = 1.25 \times 0.1174 = 0.1468 \text{ m}^3/\text{s}.$$

Because the pipes *AB* and *CD* are in parallel, the head loss will be the same. The total head loss in the old pipe will now be sum of the loss in the first half and the second half, which gives one equation. The other equation is obtained by balancing the volume discharge at point *B*.

$$Q_1 + Q_2 = Q_{new} = 0.1468 \text{ and}$$

$$\frac{0.032 \times 2000}{12}\left[\frac{Q_1^2}{0.3^5} + \frac{Q_{new}^2}{0.3^5}\right] = 60$$

$$\rightarrow \qquad Q^2_{\ new} = 0.0273375 - Q_1^2 \qquad \text{(A)}$$

Solving these equations we get, $Q_1 = 0.0777$ and $Q_2 = 0.0691$ m³/s. This shows that the discharge in the old pipe is decreased but the total discharge is increased. Equating the head loss in pipes *AB* and *CD*, we get

$$h_f = \frac{fLQ_1^2}{12 \times 0.03^5} = \frac{fLQ_2^2}{12 \times d^5} \rightarrow d = 28.6 \text{ cm}.$$

Let us decrease $Q_1$ to zero then from Equation *A*, $Q_{new} = 0.1653$ which means increase in discharge by $(0.1653 - 0.1174)/0.1174 = 40.84$ or, 40.84%. For this condition, the diameter of the new pipe can be determined as above.

**Example 4.2.11** Find the equivalent length of a 300 mm diameter pipe of cast iron with $f = 0.0003$ to replace the given set of pipes. What would be the discharge if the head difference $H$ between the two reservoirs were 10 m?

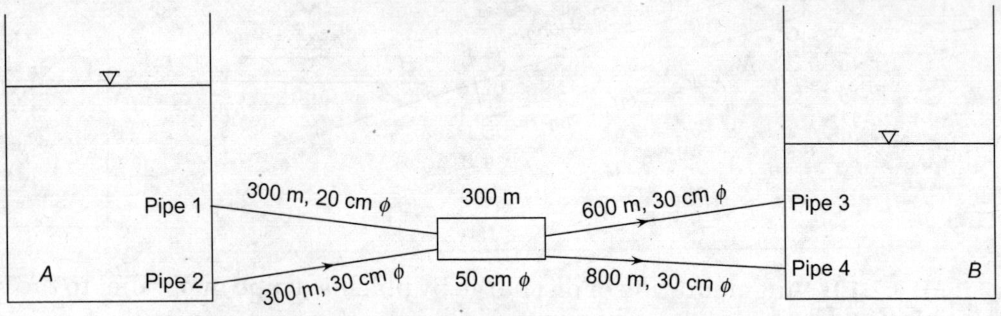


**Fig. E 4.2.11** Equivalent length of pipe

**Solution** Let the subscripts 1, 2, 3, and 4 refer to the pipes 1, 2, 3 and 4, respectively and no subscript for the central pipe. For the given configuration, we can write the following equations using (i) conservation of mass or volume, (ii) head loss in the parallel pipes is same and (iii) the total head loss between the reservoirs is 10 m.

$$Q_1 + Q_2 = Q = Q_3 + Q_4 \tag{A}$$

$$h_{f1} = h_{f2} \qquad \text{and } h_{f3} = h_{f4} \tag{B}$$

$$h_{f1} + h_f + h_{f4} = 10 \tag{C}$$

Using Eq. 4.2.1 for head loss and Equation (B), we get

$$\frac{300 f Q_1^2}{12 \times 0.2^5} = \frac{300 f Q_2^2}{12 \times 0.3^5} \rightarrow Q_2 = 2.756\, Q_1$$

$$\frac{600 f Q_3^2}{12 \times 0.3^5} = \frac{800 f Q_4^2}{12 \times 0.3^5} \rightarrow Q_3 = 1.1547\, Q_4$$

Substituting for $Q_2$ and $Q_3$ in Equation (A), on simplification we get

$$Q_1 = 0.266\, Q, \quad Q_2 = 0.733\, Q, \quad Q_3 = 0.536\, Q, \text{ and } \; Q_4 = 0.464\, Q$$

Now using Equation (C), we get

$$\frac{300 f Q_1^2}{12 \times 0.2^5} + \frac{300 f Q^2}{12 \times 0.5^5} + \frac{800 f Q_4^2}{12 \times 0.3^5} = \frac{f L_e Q^2}{12 \times 0.3^5} \rightarrow L_e = 356.8 \text{ m.}$$

Using Equation 4.2.1, we can determine the discharge, which comes out to be 1.65 m/s.

**Example 4.2.12** A pipe system is shown in Fig. E4.2.12. Determine the pressure in N/m$^2$ on the suction and the discharge side of the pump and the power required to run the pump if the pump and the motor efficiencies are each 70%. Use the relation for head loss in nozzle, $h_L = \frac{1 - C_v^2}{C_v^2}\frac{U_n^2}{2g}$, where $C_v$ is the coefficient of velocity and $U_n$ is the velocity at the nozzle exit. Show the energy and hydraulic gradient line also. Consider all losses.

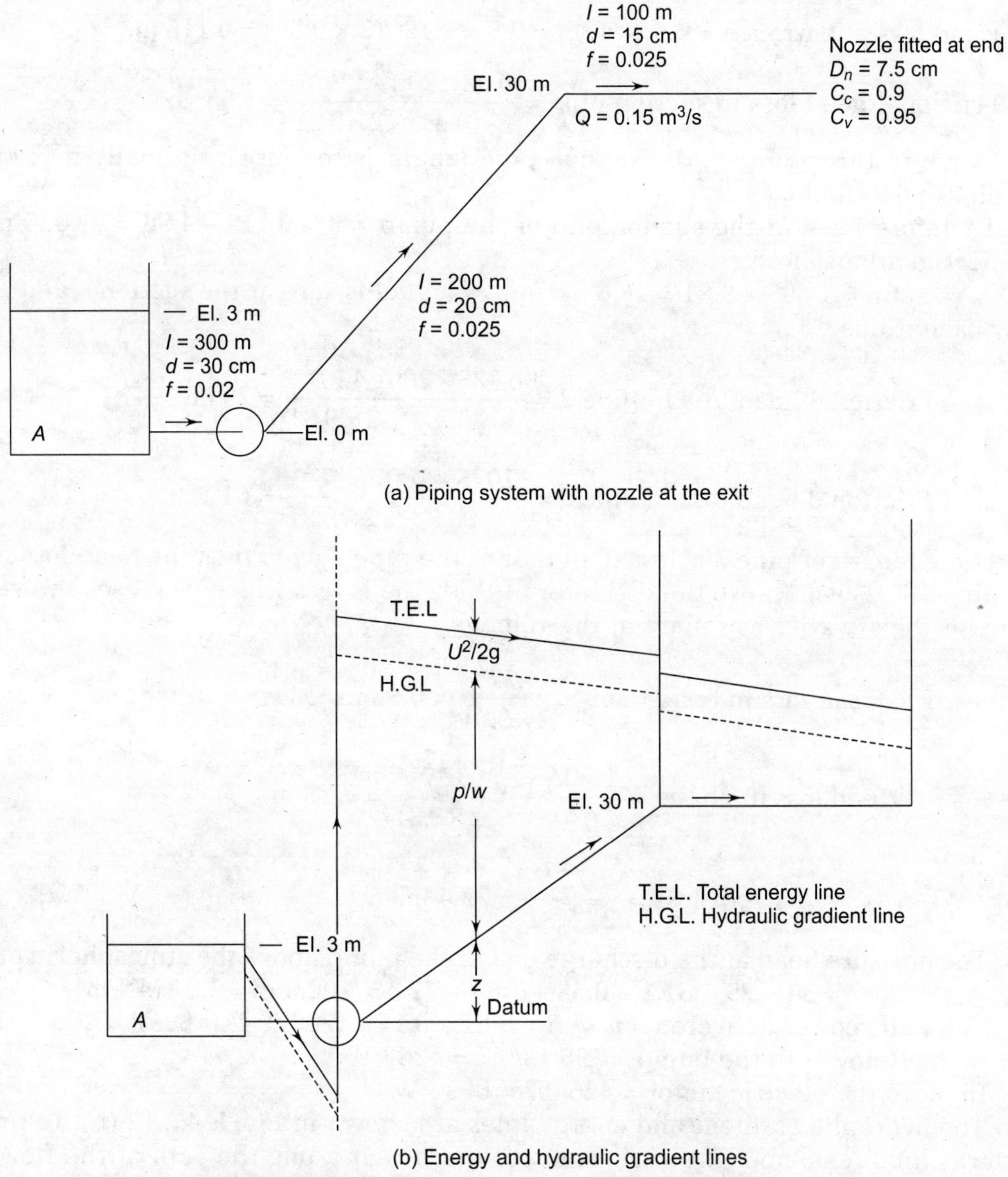


**Fig. E 4.2.12** Suction and discharge pressure across pump

**Solution** This example is chosen because in pipe installation, specially on the suction side, the minor losses can not be neglected. There are three pipes, which we shall represent by subscripts 1, 2 and 3, starting from the reservoir end pipe. From the given data, we can determine velocity in these pipes. These are:

$$U_1 = \frac{Q}{\pi/4d_1^2} = \frac{0.15}{\pi/4 \times 0.3^2} = 2.12. \text{ Similarly, } U_2 = 4.77,\ U_3 = 8.48, \text{ and } U_n = 37.7 \text{ m/s.}$$

Head loss at entrance = 0.5 $U_1^2/(2g)$ = 0.5 × $2.12^2/(2 \times 9.81)$ = 0.115 m.

$$\text{Frictional head loss in suction pipe} = \frac{0.02 \times 200}{0.3}\frac{2.12^2}{2 \times 9.81} = 4.58 \text{ m.}$$

[Note: in this problem, the suction pipe length is too large. Normally it is kept as small as possible.]

Therefore head at the suction end of the pump = 3 – 0.115 – 4.58 = –1.695 m with respect to atmosphere.

∴ Vacuum = 1.695 × 9.81 = 16.63 kN/m$^2$. Vacuum is safe at the suction side to cause any cavitation.

$$\text{Frictional head loss in pipe 2} = \frac{0.025 \times 200}{0.2}\frac{4.77^2}{2 \times 9.81} = 29 \text{ m.}$$

$$\text{Frictional head loss in pipe 3} = \frac{0.025 \times 100}{0.15}\frac{8.48^2}{2 \times 9.81} = 61.1 \text{ m.}$$

[Note length of pipe 3 is less than that of the pipe 2 even then the head loss is more in pipe 3. This is because the diameter of this pipe is less which increases the velocity. The head loss varies as square of the velocity.]

$$\text{Head loss in bend} = 0.5 \times \frac{4.77^2}{2 \times 9.81} = 0.58 \text{ m.}$$

$$\text{Head loss in nozzle} = \frac{1 - 0.95^2}{0.95^2} \times \frac{37.7^2}{2 \times 9.81} = 7.83 \text{ m.}$$

$$\text{Exit loss} = \frac{37.7^2}{2 \times 9.81} = 72.4 \text{ m.}$$

The pressure head at the discharge end of the pump above the atmospheric pressure
= 30 + 29 + 61.1 + 0.58 + 7.83 + 72.4 = 200.9 m = 1.97 MN/m$^2$.

The waterpower delivered = $w\,Q\,H$ = 9.81 × 0.15 × {200.9 – (–1.695)} = 298.13 kW.

∴ Input power to the pump = 198.13/0.7 = 426 kW

Input to the electric motor = 426/0.7 = 609 kW.

The hydraulic gradient and energy lines are shown in Fig. E 4.2.12 (b). To draw the energy line, remember the various losses that occur along the path of the flow. These include the major and minor losses. However, if a pump was placed in the water circuit then the energy line would shoot up by a distance equivalent to the energy input expressed in terms of *m* of head of the liquid.

Let us start from the reservoir. There is entry loss and so we go vertically down from EL 3m by a distance equivalent to this loss on the scale initially showing 3 m level; this has been taken at the free surface. The loss in the pipe is linear with distance and hence the energy line will be drooping straight up to the centre of the pump where the energy line goes up vertically as explained above. We may proceed further till we reach the other end of the water circuit and complete the energy line. To get the hydraulic gradient line, we decrease the velocity head from the energy line at each section.

**Example 4.2.13** Three reservoirs *A*, *B* and *C* are connected by a pipe system, as shown in the figure. If the discharge from the reservoir *A* is 0.05 cumec, determine the rate of flow into or from the reservoirs *B* and *C*. Find also the *RL* of water level in the reservoir *C*. Take $f = 0.02$.

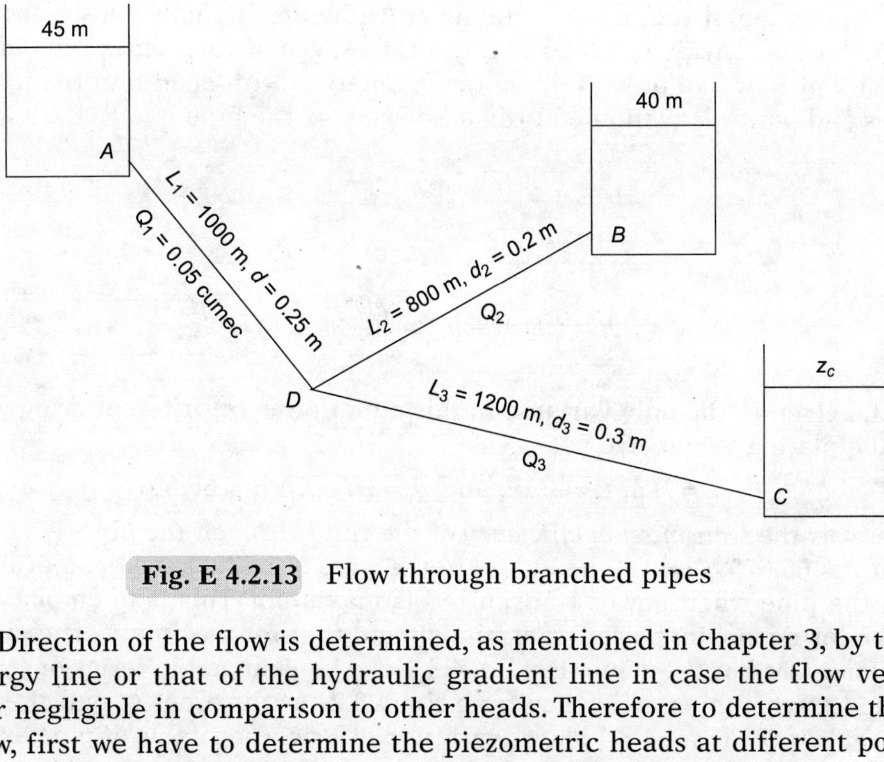


**Fig. E 4.2.13** Flow through branched pipes

**Solution** Direction of the flow is determined, as mentioned in chapter 3, by the slope of the energy line or that of the hydraulic gradient line in case the flow velocity is constant or negligible in comparison to other heads. Therefore to determine the direction of flow, first we have to determine the piezometric heads at different points. We shall ignore the minor losses in our calculations, since the pipes are long.

$$\text{Piezometric head at } D = z_D + \frac{p_D}{w} = z_A - h_{f1} = 45 - \frac{0.032 \times 1000 \times 0.05^2}{12 \times 0.25^5} = 38.23 \text{ m.}$$

Piezometric head at $B = 40$ m > 38.23. This means the flow will occur from *B* to *D*. Using Eq. 4.2.1

$$h_{f2} = 40 - 38.23 = \frac{0.032 \times 800 \times Q_2^2}{12 \times 0.2^5} \rightarrow Q_2 = 0.016 \text{ m}^3\text{/s.}$$

Applying continuity equation at point $D$, we get: $Q_3 = Q_1 - Q_2 = 0.066$ m$^3$/s. From this head loss in pipe 3 is

$$h_{f3} = \frac{0.032 \times 1200 \times 0.066^2}{12 \times 0.3^5} = 5.69 \text{ m.}$$

$$\therefore \quad z_C = E_D - h_{f3} = 38.23 - 5.69 = 32.54 \text{ m.}$$

## 4.3 Hydraulic Power Transmitted Through Pipe

Pipes are used not only for transmitting fluid but in this way power too is transmitted from one place to another. We would therefore like to design the pipe so that the head loss due to friction of the pipe is minimum; we ignore the minor losses while making such calculations. This is justified because the length of such pipes runs in km. If $U$ is the velocity of flow in a pipe of diameter $D$, length $L$ with head $H$ at the inlet then the head loss and power $P$ available at the other end of the pipe is

$$h_f = \frac{fL}{D} \cdot \frac{U^2}{2g}$$

and

$$P = wQ(H - h_f) = \frac{\pi D^2}{4} wU\left(H - \frac{fL}{D} \cdot \frac{U^2}{2g}\right) \tag{4.3.1}$$

For the given $H$, the only variable in this equation is $U$. Differentiating with respect to $U$ and equating to zero, we get

$$h_f = H/3,\ P = 2/3H, \text{ and } \eta_t = (H - h_f)/H = 66.6\% \tag{4.3.2}$$

Here, $\eta_t$ is the *transmission efficiency* of the fluid through the pipe. We note that the efficiency is 66.67% and the head loss in the pipe is 1/3$^{rd}$ of the head available at the inlet to the pipe when power transmitted is maximum. This is to emphasize that the condition for maximum power transmission and that for maximum efficiency of power transmission are not the same. For Eq. 4.3.1 can be changed in terms of $Q$ and we may determine transmission efficiency from Eq. 4.3.2 for different $Q$, and then plot these results against $Q$. Figure 4.3.1 shows variation of Power available at pipe exit, transmission efficiency and head loss. We observe that efficiency is maximum at no flow and zero at the maximum flow condition. On the other hand power transmitted is zero at both these conditions and is maximum somewhere when head loss is 1/3rd of the inlet total head $H_1$. This means when discharge rate is less than this value, the transmission efficiency is more than 66.67%. This is so because the head loss increases parabolically with $Q$.

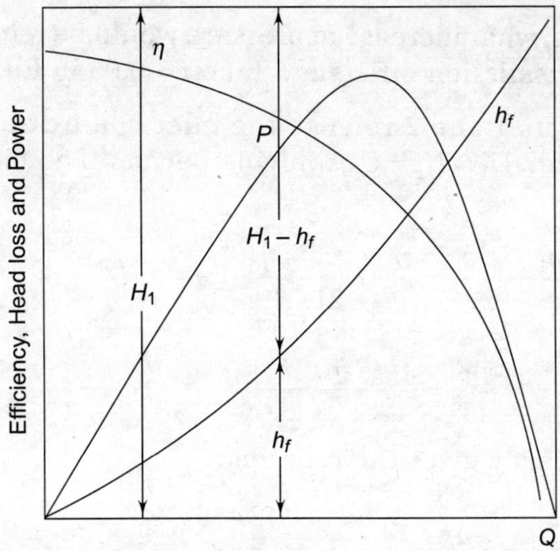


**Fig. 4.3.1** Hydraulic power transmission

A common application in hydraulic power sector is for carrying water from water reservoirs to the turbine house where a nozzle is fitted at the pipe end. Such a pipe is called penstock. Let $d_n$ and $U_n$ be the diameter and velocity at the exit of the nozzle. Neglecting the losses in the nozzle and using the continuity equation, we can write

$$H = \frac{fL}{d}\frac{U^2}{2g} + \frac{U_n^2}{2g} \text{ and } Ud^2 = U_n\, d_n^2$$

$$\rightarrow \qquad U_n^2 = \frac{2gH}{1 + \dfrac{fL}{d}\dfrac{d_n^4}{d^4}} \tag{4.3.3}$$

$$\text{Power available in nozzle jet} = wQ\frac{U_n^2}{2g} = w\pi d_n^2\, U_n \frac{U_n^2}{2g} = K\frac{d_n^2}{(d^5 + fLd_n^4)^{3/2}}.$$

$K$ is a constant for given arrangement that depends on diameter and length of the pipe. Thus jet power is the function of the nozzle diameter. In order to maximise it, its derivative with respect to $d_n$ be equated to zero. This gives

$$d_n = \left(\frac{d^5}{2fL}\right)^{1/4} \tag{4.3.4}$$

**Example 4.3.1** Three horizontal pipes each of 15 cm diameter have been laid in parallel for a length of 5 km from hydraulic pump house to the factory. If a constant pressure of 5 Mpa is maintained at the inlet, (a) make calculations for the power delivered to the factory, assume the transmission efficiency is 95%. (b) If one of the pipes becomes unavailable during service and the same power is required to be transmitted at the same delivery pressure, what increase in pressure would be required at the pump house? Also calculate the transmission efficiency. Take $f = 0.03$ in Eq. 4.2.1.

**Solution** Let subscripts 1 and 2 refer to the inlet and exit end of the pipe. Since the pipe is of same diameter, $U_1 = U_2 = U$. Applying Bernoulli equation at the points 1 and 2, we get

$$\frac{p_1}{w} + \frac{U^2}{2g} = \frac{p_2}{w} + \frac{U^2}{2g} + \frac{fL}{d}\frac{U^2}{2g}$$

$$\rightarrow \qquad \frac{p_1}{w} = \frac{p_2}{w} + \frac{fL}{d}\frac{U^2}{2g} \qquad \text{(A)}$$

Transmission efficiency gives the relation

$$\eta_t = \frac{p_2/w + U_2^2/2g}{p_1/w + U_1^2/2g} = 0.95 \qquad \text{(B)}$$

We are given $p_1/w = 5000/9.81$ and $f = 0.03$. Substituting these values in Equations (A) and (B) we get,

$$U = 0.707 \text{ m/s}.$$

$$\therefore \qquad Q = 3 \times \pi/4\ (0.15)^2 \times 0.707 = 0.0375 \text{ m}^3\text{/s},$$

$$H_1 = 5000/9.81 + (0.707)^2/(2 \times 9.81) = 509.7 \text{ m},$$

and $$P_1 = wQH_1 = 9.81 \times 0.0375 \times 509.7 = 187.506 \text{ kW}.$$

$$H_2 = H_1 - \frac{fLU^2}{2g} = 484.20 \text{ m},$$

and $$P_2 = wQH_2 = 9.81 \times 0.0375 \times 484.2 = 178.13 \text{ kW}.$$

Since the power requirement and head at the factory end is to be kept same and same discharge $Q$ is to be supplied by two pipes, the velocity in each pipe (using $Q = AU$), will be $U = 1.06$ m/s.

$$h_f = \frac{fL}{d}\frac{U^2}{2g} = \frac{0.03 \times 5000}{0.15} \times \frac{(1.06)^2}{2 \times 9.81} = 57.27 \text{ m},$$

$$H_1 = H_2 + h_f + \frac{U^2}{2g} = 484.2 + 57.27 + \frac{(1.06)^2}{2 \times 9.81} = 541.53 \text{ m}.$$

$$\frac{p_1}{w} = H_1 - \frac{U^2}{2g} = 541.47$$

$$\rightarrow \qquad p_1 = 9810 \times 541.47 \times 10^{-6} = 5.31 \text{ MPa}.$$

Necessary increase in pressure = 5.31 – 5 = 0.31 Mpa.

Efficiency of transmission = $H_2/H_1$ = 89.4%.

**Example 4.3.2** A rubber hose, 1.2 cm diameter and 18 m long, is connected to a garden tap, which is fitted to a main where the water pressure is 560 kPa (gauge). Calculate the exit velocity and the ratio of the kinetic energy of the issuing jets for the following conditions: (i) exit diameter is restricted to 0.3 cm by a nozzle arrangement; (ii) water flows unrestricted into the atmosphere. Take the following data for the loss coefficients: for tap 7.5, for nozzle 0.1. Hose is held horizontally for which $f = 0.024$. Neglect the flow velocity in the main. Take specific weight of water = 10 kN/m$^3$.

**Solution** Let subscripts 1 and 2 refer to the inlet and exit end of the pipe and 3 to the section after the nozzle, respectively when the nozzle is fitted to the pipe. Since pressure at section 1 is given as gauge pressure and that at section 3 is atmospheric, which is zero gauge, and that the velocity of flow is negligible (given), application of Bernoulli equation between the points 1 and 3 gives

$$\frac{p_1}{w} = \frac{U_3^2}{2g} + h_f, \text{ where } h_f = 7.5\frac{U_2^2}{2g} + 0.1\frac{U_3^2}{2g} + \frac{0.024 \times 18}{0.012}\frac{U_2^2}{2 \times 9.81} + \frac{U_3^2}{2g}$$

or,
$$\frac{560}{10} = \frac{U_2^2}{2 \times 9.81}\left[1.1 \times \frac{U_3^2}{U_2^2} + 7.5 + 36\right]$$

But $\quad d_2^2\, U_2 = d_3^2\, U_3$ (Continuity equation)

$\rightarrow \quad U_2 = 1.84,\ U_3 = 29.44$ m/s and $Q = \pi/4\,(0.003)^2 \times 29.44 = 2.08 \times 10^{-4}$ m$^3$/s.

When water flows unrestricted then Bernoulli equation when applied between the points 1 and 2 gives

$$\frac{p_1}{w} = \frac{U_2^2}{2g} + h_f, \text{ where } h_f = 7.5\,\frac{U_2^2}{2g} + \frac{0.024 \times 18}{0.012}\frac{U_2^2}{2 \times 9.81} + \frac{U_2^2}{2g}$$

$\rightarrow \quad U_2 = 4.97$ m/s and $Q' = \pi/4(0.012)^2 \times 4.97 = 5.61 \times 10^{-4}$ m$^3$/s

$$\therefore \quad \frac{\text{Energy with nozzle}}{\text{Energy without nozzle}} = \frac{1/2\rho Q U_3^2}{1/2\rho Q' U_2^2} = \frac{2.08 \times 10^{-4} \times (29.44)^2}{5.61 \times 10^{-4} \times (4.97)^2} = 13.$$

## 4.4 Water Hammer

When the flow of the water in a pipe is stopped suddenly for some reasons then the momentum of the water in the pipe is changed to pressure at the end where the pipe is closed. This raises pressure locally till the water comes to rest and then water starts moving back. The water moves upstream, comes to rest and moves again downstream. This process of to and fro motion of water confined in the pipe continues till the ripples (pressure waves) die out. The result is the generation of sound, which we can hear clearly. You too must have heard such sound when the bathroom tap is closed instantaneously; more is the speed of closure more is the sound intensity. The pressure reached in pipe may be of several orders to cause bursting of the pipe. Therefore this phenomenon is to be checked, more so in case of Penstock, etc which carry tons of water. It is called *water hammer* or *hammer blow*, because the sound we hear appears as if a hammer has hit the pipe. We may analyse the problem either from the momentum or the energy consideration. We shall delve upon these methods now for (i) gradual closure and (ii) quick closure of the valve, respectively.

***Momentum consideration***. Let $A$ and $L$ be the area of cross section and length of the pipe, $U$ the velocity of water in a rigid pipe that is brought to rest in time $t$ seconds, and $p$ is the rise in pressure after the water comes to rest.

Retardation = $(U - 0)/t = U/t$.

Retarding force = mass × retardation = $(\rho AL) \times U/t$.

This force must be equal to $pA$. Equating the two, we get

$$p = \frac{\rho LU}{t} = \frac{wLU}{gt} \text{ and } h = \frac{p}{w} = \frac{LU}{gt} \tag{4.3.5}$$

This equation shows that for the given pipe system, the rise in pressure depends on the time taken in closing the valve. Less is this time, more is the rise in pressure. This suggests that the valve should be closed slowly. This rise in pressure is based on the assumption that the water is incompressible. However, all the liquids have some compressibility and, therefore, the rise in pressure would be less.

***Energy consideration***. The flowing mass has kinetic energy which is converted into the strain energy of the fluid and the pipe as the fluid comes to rest. This results in circumferential and longitudinal stress in the pipe material. Let us find out these stresses and the strain energy. If $K$ is the bulk modulus of the liquid then

$$\text{The strain energy of the liquid} = \frac{p^2}{2K} \times \text{volume of the liquid in pipe} = \frac{p^2}{2K} \times AL.$$

If $d$ is the diameter of the pipe, $E$ the Young's modulus of the pipe material, and $t$ the thickness of the pipe then the longitudinal ($f_1$), circumferential ($f_2$) stresses and the strain energy of the pipe per unit volume are

$$f_1 = \frac{pd}{4t},\ f_2 = \frac{pd}{2t} \text{ and Strain energy} = \frac{1}{2E}\left[f_1^2 + f_2^2 - 2\mu f_1 f_2\right]$$

Substituting for the stresses, after simplification we get

$$\text{Strain energy} = \frac{1}{2E}\left[\frac{p^2d^2}{16t^2} + \frac{p^2d^2}{4t^2} - 2\mu\frac{p^2d^2}{8t^2}\right].$$

$\mu$ is the Poisson's ratio, which is normally 0.25. Also the volume of the pipe = $\pi dLt$.

Substituting these, we get strain energy of the pipe material = $\dfrac{p^2dLA}{2Et}$

K.E. of the liquid = ½ mass × $U^2$ = ½ × $(\rho AL) \times U^2$.

This must be equal to the sum of strain energy stored in liquid and pipe material. Equating these, after simplification we get

$$\frac{1}{2}\rho ALU^2 = \frac{p^2 dLA}{2Et} + \frac{p^2 LA}{2K}$$

$$\rightarrow \qquad p = U\left[\frac{\rho}{1/K + d/(Et)}\right]^{1/2} \tag{4.3.6}$$

For a rigid pipe, $E$ is equal to infinity and therefore Eq. 4.3.6 gives the same results as obtained from Equation 4.3.5 by the momentum consideration. This was expected too, since the results do not depend on the method of analysis.

## 4.5 Flow in Open Channel

Open channel flow, as mentioned in the introduction, is another way of transporting fluid from one place to another. Canal, irrigation drain, and sewer lines are some examples of man-made hydraulic systems, whose importance is very well known. In such flow, the surface is open to atmosphere with atmospheric pressure on its surface throughout the flow length. The motive force comes from the gravity due to the slope of the bed. The steady state of flow is reached when the component of the weight of the liquid in channel balances the frictional force from the sides and bottom of the channel. Normally the depth of the flow throughout the length of the flow remains constant unless there is some obstruction downstream. For example, dam is constructed across the river to store it. In this case the water level would rise from some point miles away from the gate and the variation in depth will take place gradually. There are situations when depth of flow may change suddenly; we shall discuss these different aspects of the free flow surface in this section.

### 4.5.1 Uniform Flow

In most of the cases the flow can safely be assumed to have same depth throughout the length of the flow. Moreover, the flow is turbulent as the Reynolds number is more than the critical value. Like pipe, we can use hydraulic diameter, $d_h$ (or radius $R_h$)[1] as the characteristic length in the expression for Reynolds number. It is defined as the ratio of the area of cross section $A$ to the wetted perimeter $P$ of the channel cross-section. The concept of hydraulic diameter (or mean hydraulic depth in channel) can be extended to any kind of non-circular geometry. For a circular pipe running full, for example,

$$A = \pi/4\ d^2 \text{ and } P = \pi d \rightarrow d_h = d/4.$$

Since factor ¼ comes in this description of hydraulic diameter, many scientists have taken 4 $d_h$ rather than $d_h$ as the hydraulic diameter so that it is $d$ for the pipe too, which we usually take. A channel may have any cross section, the most commonly used one is

[1] Symbol $m$ is very common in use to represent hydraulic diameter, also called the hydraulic mean depth, in channels and shall be used in this section.

the trapezoidal section though others like rectangular, circular, etc. are also in use. Whereas the circular channel is normally used in sewage lines and the rectangular section is a special case of the trapezoidal section. These sections are shown in Fig. 4.5.1. We shall now determine hydraulic depth for these sections, which shall be used in later sections. In the flowing relations, $b$ is the base width, $d$ is the depth of water in the channel, and $R$ (= $D/2$) is the radius of the circular pipe. These sections are shown in Fig. 4.5.1.

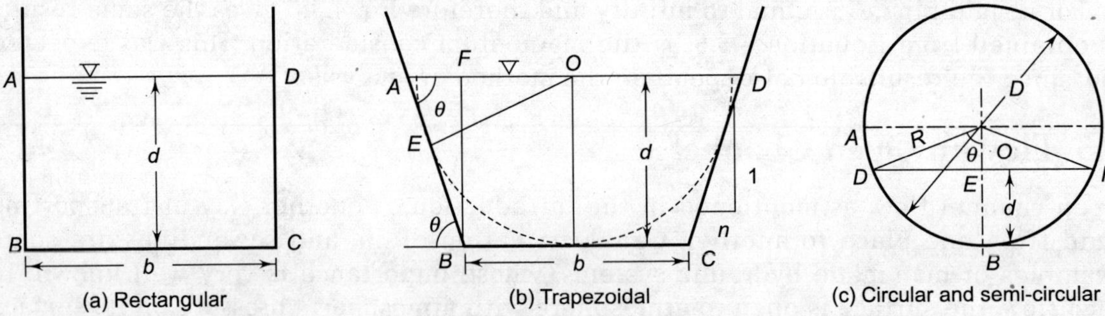


**Fig. 4.5.1** Hydraulic diameter for different cross-sections

*Rectangular section*: $A = b \times d$, and $P = AB + BC + CD = b + 2d$

$$\therefore \quad m = b\, d/(b + 2d) \tag{4.5.1}$$

*Trapezoidal section*: Width of free surface = $b + 2\,nd$, as can be seen from Fig. 4.5.1 (b). Here $1/n$ is the slope of the slant side of the trapezoid. We have taken a symmetric trapezoid, that is, both the sides are slanting by same angle.

$$A = \frac{1}{2}d(b + b + 2nd) = d(b = nd),\ P = b + 2\sqrt{n^2d^2 + d^2}$$

$$\rightarrow \quad m = \frac{d(b+nd)}{b+2\sqrt{n^2d^2+d^2}} \tag{4.5.2}$$

*Semicircular section*:

$$A = \frac{\pi D^2}{8},\ P = \frac{\pi D}{2} \rightarrow m = \frac{D}{4} \tag{4.5.3}$$

*Circular section*:

$$A = \frac{\pi D^2}{4},\ P = \pi d \rightarrow m = \frac{D}{4} \tag{4.5.4}$$

*Circular section partly filled.* Let the circular section is partially filled up to level FD. Join points $D$ and $E$ with $O$, the centre, as shown in Fig. 4.5.1 (c). We may write $OE = R \cos \theta$ and $DE = R \sin \theta$. Then

area $DBF$ = Area of the sector $OBF - 2 \times$ area of the triangle $OED$
and area of the triangle $OED = ½ \times R \cos\theta \times R \sin\theta = 1/4\, R^2 \sin 2\theta$.

Thus area $DBF$,

$$A = R^2\,(\theta - 1/2 \sin 2\theta),\ P = 2RQ \rightarrow m = \frac{R(\theta - 1/2\sin 2\theta)}{2\theta} \qquad (4.5.5)$$

In fact, whatever the shape of the channel may be, we may determine the mean depth in this way. We have to find the area of the cross section of the flow area and the wetted perimeter using the properties of geometric sections and the trigonometric relations.

Now let us examine nature of the flow in a channel. Will it be turbulent or laminar? Consider, for example, a flow through a rectangular channel 1 m wide and 0.5 m deep with water flowing through it at 2 m/s velocity. For water $\mu = 0.001$ kg/ms. Thus for this flow the Reynolds number based on $m$ is

$$A = 1 \times 0.5 = 0.5\ \text{m}^2,\ P = 1 + 2 \times 0.5 = 2\ \text{m} \quad \rightarrow m = 0.5/2 = 0.25\ \text{m}.$$

$$Re = Um\rho/\mu = 2 \times 0.25 \times 1000/0.001 = 500 \times 10^3.$$

This channel is very small as compared to the canals and rivers and the velocity of flow is also small. Even then the flow is turbulent. Therefore we may assume that the flow with free surfaces are turbulent in nature. The effect of turbulence is to make velocity variation at any section, constant except near the bottom and the sides of the channel. Moreover the velocity at the surface is affected by the wind too. Experiments confirm that this variation in velocity is almost negligible and the velocity at a cross-section may be considered constant. Under steady state, the velocity at different sections will also be same. Such flows are called *uniform flow*.

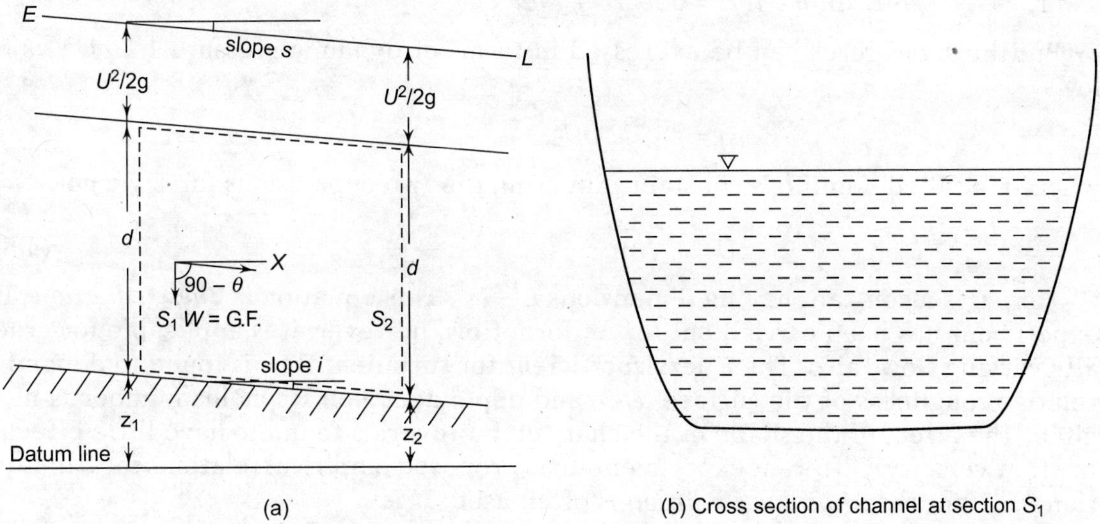


**Fig. 4.5.2** Uniform flow through a channel

Figure 4.5.2 shows uniform flow over a bed having constant slope $i$. Consider a control volume shown by the dotted lines. Applying the modified Bernoulli equation between two points 1 and 2 in the free surface on sections 1 and 2 while measuring height from some datum line, as shown in the Fig., we get

$$\frac{p_1}{w} + \frac{U^2}{2g} + (z_1 + d) = \frac{p_2}{w} + \frac{U^2}{2g} + (z_2 + d) + h_f$$

$$\rightarrow \qquad h_f = \left(\frac{p_1}{w} + z_1\right) - \left(\frac{p_2}{w} + z_2\right) = z_1 - z_2$$

Since velocity is same at all the sections and the pressure at points 1 and 2 is atmospheric pressure, the free surface of the water has the same slope as that of the bed. Moreover, velocity being uniform, the energy line *EL* also has the slope of the bed.

The cross section of the channel at section $S_1$ is shown in the figure. Since the area of cross section may change during the flow, we have considered an average area. Also for a small distance the area may also be considered invariant. Let *A* be the average area of the channel section, *U* the average velocity of flow, *L* be the length of the section $S_1$ $S_2$ *P* be the perimeter of the wetted surface, $i$ = slope of the bed, and $\tau$ = average shear stress at the solid boundaries. The integral momentum equation for *x*-direction, when applied to the control volume, gives

Net external force = rate of change of momentum, which is zero as there is no change in velocity.

$$W \sin \theta - \tau PL = 0$$

or, $$\rho g AL \sin \theta - \tau PL = 0 \rightarrow \tau = \rho g\, mi \qquad (4.5.6)$$

But the shear force can be expressed in terms of dynamic pressure $1/2\ \rho U^2$, as follows:

$$\tau = f \times 1/2\ \rho U^2,$$

where *f* is called Fanning's constant. Equating the two expressions for $\tau$, we get

$$U = C\sqrt{mi} \qquad (4.5.7)$$

Here *C* is a constant having dimensions $L^{1/2}T^{-1}$. This equation is *Chezy equation*. The experiment has been carried out for uniform flow, however, it is applicable for gradually varying flows also. The Chezy coefficient for turbulent flow is found to depend on relative roughness of the surface, $\varepsilon/m$, and dependence on Reynolds number is negligible. The effect of the shape of the channel, however, is found to have little effect on the flow velocity. Different experimentalists proposed empirical relations for *C* in Equation 4.5.7 but the *Manning's relation* is often used. It is

$$C = 1/n\ m^{1/6} \quad \rightarrow U = 1/n\ m^{2/3}\sqrt{i} \qquad (4.5.8)$$

In this equation, *n* is the roughness coefficient whose value depends on the type of the surface and the data is available to get it. Table 4.5.1 gives values of *n* for a few cases.

**TABLE 4.5.1** Values of $n$ in Equation 4.5.8.

| Type of surface | n |
|---|---|
| Smooth cement, planed timber | 0.010 |
| Rough timber, canvas | 0.012 |
| Cast iron, brick work | 0.013 |
| Rubber masonry | 0.018 |
| Firm gravel | 0.02 |
| Canals and rivers in good condition | 0.025 |
| Canals and rivers in bad condition | 0.035 |

The Manning equation takes into account the effect of the shape of the channel because of the presence of the hydraulic mean depth in $C$.

#### 4.5.1.1 Optimum Hydraulic Cross Section

Rectangular, trapezoidal and even semi-circular cross sections have been used for construction of the channel. In the foregoing section, we determined the hydraulic mean depth in a few cases. We also got different relations for the velocity of flow. Therefore the discharge at any section will depend on relation which is used for the determination of the average velocity of the flow in the channel. That is,

$$Q = AC\sqrt{mi} \text{ (from Chezy relation)}$$

or,

$$Q = A/n\, m^{2/3}\sqrt{i} \text{ (from Manning relation)} \tag{4.5.9}$$

In both cases, $Q$ is the function of both $A$ and $m$ for given $i$ that depend on the topography of the area through which the channel would pass. While designing a channel, often we fix the discharge $Q$ for the quantity of water available from the reservoir. Therefore, we would like to construct a channel so that the excavation cost and its lining cost is minimum. The former will depend on the area of cross section of the channel and the later on the wetted perimeter. Therefore, both the area $A$ and the perimeter $P$ should be minimum. From Eq. 4.5.9 we infer that for given $Q$, A is minimum when $m$ is maximum. But $m = A/P$, which implies $P$ is minimum. Consequently *the consideration to get the best section for canals is to minimise the wetted perimeter.* However, it is not so in the case of sewage lines because a pipe of some cross-section is used for conveying the sewage fluid and then we have to find out the maximum discharge that can take place through it and the condition for the same. Therefore we have to consider the whole equation (4.5.9) for optimisation rather than minimising the perimeter.

We have seen that the perimeter and the area of cross section depend on different geometric dimensions, such as width, slanting member length or its slope, etc. Therefore to maximise $Q$, we may differentiate the expression for $P$ or Eq. 4.5.9 partially with respect to one variable at a time, depending upon the criterion we set for optimisation as discussed in last paragraph, and equate it to zero. This gives sufficient number of equations to get the optimal value of these variables. We shall consider two cases, namely flow through the trapezoidal channel and the partly-filled circular pipe.

It should be remembered that our results would also depend on whether the Chezy or the Manning equation is used for the average velocity, since there are empirical relations. However, the difference is not much.

***Best trapezoidal section.*** This is an open channel with free surface exposed to atmosphere. Hence it is designed on the basis of minimum perimeter. Refer Eq. 4.5.2 and eliminate $b$ between $A$ and $P$. This gives

$$A = d(b + nd),\ P = b + 2\sqrt{n^2d^2 + d^2}$$

$$\therefore \qquad P = \frac{A}{d} - nd + 2d\sqrt{n^2 + 1}$$

$$\rightarrow \qquad \frac{dP}{dd} = -\frac{A}{d^2} - n + 2\sqrt{n^2 + 1} = 0$$

$$\text{or,} \qquad -\frac{d(b + nd)}{d^2} - n + 2\sqrt{n^2 + 1} = 0$$

$$\rightarrow \qquad \frac{(b + 2nd)}{2} = d\sqrt{n^2 + 1} \tag{4.5.10}$$

This equation can also be written in the following form

$$m = \frac{A}{P} = \frac{d(b + nd)}{b + 2d\sqrt{n^2 + 1}} = \frac{d(b + nd)}{b + (b + 2nd)} = \frac{d}{2} \tag{4.5.11}$$

We may also express the result of Equation 4.5.10 as the three sides of the trapezoid circumscribing the circle drawn with $O$ as centre and $OE$ as radius, where $O$ is the mid point in the free surface and $OE$ is perpendicular drawn on $AB$ from $O$. We may consider triangles $OEA$ and $ABF$. We observe that these are similar triangles.

$$\therefore \qquad \sin\theta = \frac{BF}{AF} = \frac{OE}{AO} \quad \rightarrow \frac{d}{d\sqrt{n^2 + 1}} = \frac{OE}{(b + 2nd)/2} \quad \text{or, } OE = d,$$

For the best trapezoidal section, the hydraulic mean depth must be half the depth of water in the channel or the three sides of the trapezoid are tangential to the circle drawn with $O$ as centre and $d$ as radius.

We may also determine the slope of its slanting side by differentiating $P$ with respect to $n$. It can be shown that the process gives $n = 1/\sqrt{3}$ or $\theta = 60°$. With this value of $\theta$, Equation 4.5.10 gives

$$b = 2/\sqrt{3}\,d \text{ and } P = b + 2d\sqrt{n^2 + 1} = 6/\sqrt{3}\,d = 3b \tag{4.5.12}$$

This means each sloping side is equal to the bed width.

***Best circular section.*** It was mentioned earlier that a circular section is used in sewage lines and it is required to determine the best depth of flow so that the discharge through it is maximum. Therefore we have to express discharge in terms of the depth and then differentiate it. For this we have many choices to choose from, because the relations for average velocity are empirical.

Figure 4.5.1(c) shows the circular section of radius $R$ (diameter $D$) with water filled upto height $d$ and the centre at $O$. We determined the area of the section filled with water and the wetted perimeter, which is given by Eq. 4.5.5. Using this equation and choosing the Chezy equation, we get

$$Q = AC\sqrt{\frac{A}{P}i} = C\sqrt{\frac{A^3}{P}i}\text{, where } A = R^2\ (\theta - 1/2 \sin 2\theta),\ P = 2R\theta$$

or, $$Q = K(\theta - 1/2 \sin 2\theta)^{3/2}\ \theta^{-1/2}\text{, where } K \text{ is a constant} \tag{4.5.13}$$

Angle $\theta$ is shown in figure; this is the angle made by line $OD$ or $OF$ with the vertical, where $DF$ represents the free water surface. For a given pipe lay out, $C$ and $i$ are constant, which demands $A^3/P$ to be maximum. Here the variable is $\theta$. Differentiation with respect to $\theta$ gives the desired condition.

$$\frac{d}{d\theta}\left(\frac{A^3}{P}\right) = 3\frac{A^2}{P}\frac{dA}{d\theta} - \frac{A^3}{P^2}\frac{dP}{d\theta} = 0,$$

or, $$3P\frac{dA}{d\theta} = A\frac{dP}{d\theta}.$$

$$3 \times 2R\theta \times R^2\ (1 - \cos 2\theta) = R^2\ (\theta - 1/2 \sin 2\theta) \times 2R$$

$$\rightarrow \quad \sin 2\theta - 6\theta \cos 2\theta + 4\theta = 0$$

$$\theta = 154°,\ d = R(1 - \cos\theta) = 0.95\ D \text{ and } m = A/P = 0.2886\ D \tag{4.5.14}$$

Note that the condition for maxima results in a transcendental equation whose solution is obtained by either hit and trial method or plotting $6\theta \cos 2\theta - \sin 2\theta$ and $4\theta$ against $\theta$ on $y$-$x$ plot; the point of intersection of the two curves gives $\theta$. Condition 4.5.13 shows that maximum discharge takes place when the pipe is not full. Substituting $\theta = \pi$ in Eq. 4.5.13 gives the discharge through the pipe when it is running full. From this information, we may calculate ratio of the discharge through full pipe and when discharging maximum. This gives

$$\frac{Q}{Q_{max}} = 0.86 \tag{4.5.15}$$

Circular channel is also designed for maximum flow velocity. Expressing velocity in terms of the pipe parameters and then differentiating it with respect to $\theta$, we obtain desired conditions. For maximum velocity condition this gives

$$\theta = 128°45',\ d = R(1 - \cos\ \theta) = 0.81\ D \text{ and } m = A/P = 0.3048\ D \tag{4.5.16}$$

We notice that water depth or mean hydraulic depth for maximum discharge and maximum velocity are different.

We may similarly proceed to determine conditions for maximum discharge and maximum velocity using the Manning equation for the average velocity. The results obtained are

$$\theta = 151°11', d = R(1 - \cos\theta) = 0.938\,D \text{ and } m = A/P = 0.29\,D \tag{4.5.17}$$

Further, we may similarly obtain the following relations:

$$\frac{Q}{Q_{\text{full}}} = \frac{1}{\pi}(\theta - 1/2\sin 2\theta)^{5/3}\,\theta^{-2/3} \text{ and } \frac{V}{V_{\text{full}}} = (\theta - 1/2\sin 2\theta)^{2/3} \tag{4.5.18}$$

Also from Fig. 4.5.1 (c), we get the following relation between $d$ and $D$

$$\frac{d}{D} = \frac{1}{2} - \frac{1}{2}\cos\theta \tag{4.5.19}$$

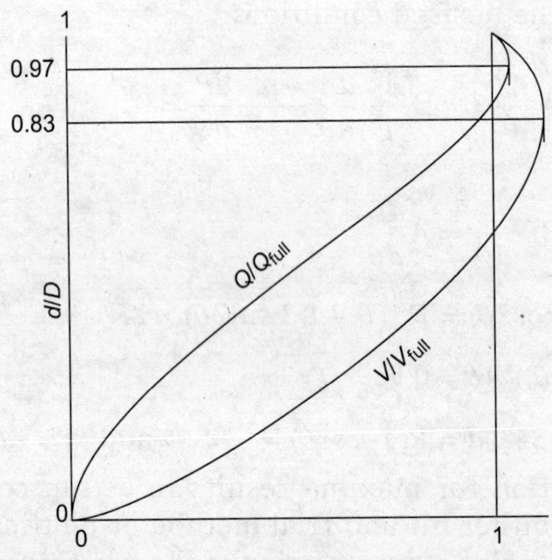


**Fig. 4.5.3** Variation of discharge and velocity with depth

For different $d/D$, $\theta$ is obtained from Eq. 4.5.19 and then Eq. 4.5.18 gives the ratio of discharge and velocity. These are plotted on $x$-$y$ plot; Fig. 4.5.3 shows such plot. These results are obtained on the basis that the Manning constant $n$ remains constant. In real situation, it is not so. The experimental results reveal that the maximum discharge or maximum velocity takes place for the condition $d/D = 0.97$ and $0.83$, respectively.

**Example 4.5.1** Figure 4.5.1 shows a cross section of a channel conveying water at a depth of 9m. The bed slope of the channel is 1 in 1200. Taking Manning constant $n = 0.02$, find discharge.

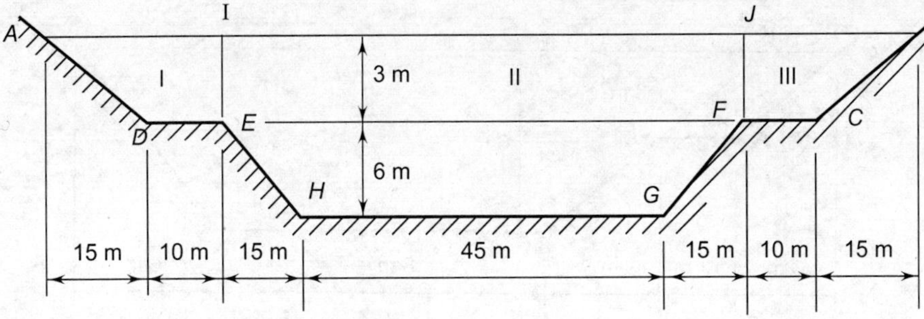


**Fig. E 4.5.1** Discharge through a given cross section

**Solution** To find out the area of discharge, we divide the section into three sections I, II and III, as shown in the figure so that area and perimeter of each section is determinable from its geometric shape. This gives $m = A/P$ for these sections.

Discharge through section I.

$$A_1 = \tfrac{1}{2} \times IE \times (ED + AI) = \tfrac{1}{2} \times 3\,(10 + 25) = 52.50 \text{ m}^2,$$

$$P_1, \text{ wetted perimeter} = ED + DA = 10 + \sqrt{15^2 + 3^2} = 25.297 \text{ m}.$$

$$\rightarrow \qquad m_1 = A_1/P_1 = 2.073 \text{ m}$$

and
$$Q_1 = \frac{A_1}{n}\, m_1^{2/3}\, i^{1/2} = \frac{52.50}{0.022}(2.0753)^{2/3}\left(\frac{1}{1200}\right)^{1/2} = 112.081 \text{ m}^3/\text{s}.$$

Discharge through section II.

$$A_2 = (75 \times 3) + 1/2 \times 6\,(75 + 45) = 585.00 \text{ m}^2,$$

$$P_2, \text{ wetted perimeter} = 45 + 2\sqrt{15^2 + 6^2} = 77.311 \text{ m}.$$

$$\rightarrow \qquad m_2 = A_2/P_2 = 7.5668 \text{ m}$$

and
$$Q_2 = \frac{A_1}{n}\, m_2^{2/3}\, i^{1/2} = \frac{585}{0.022}(7.5668)^{2/3}\left(\frac{1}{1200}\right)^{1/2} = 2958.585 \text{ m}^3/\text{s}.$$

Discharge through section III = Discharge through section I (by symmetry) = 112.081 $m^3$/s.

$\therefore$ Total discharge = $Q_1 + Q_2 + Q_2$ = 3182.747 $m^3$/s.

**Example 4.5.2** The figure shows a cross section of a channel conveying water at a depth of 6 m. The bed slope of the channel is 0.0075. Taking Manning constant $n$ = 0.025, find discharge.

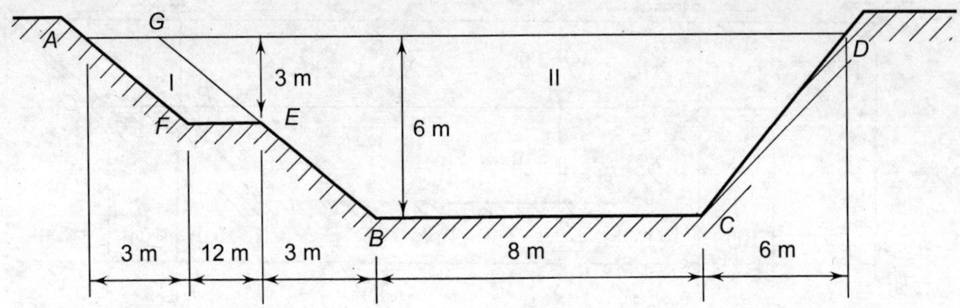


**Fig. E 4.5.2** Discharge through a given cross section

**Solution** To find out the area of discharge, we divide the section into two areas I and II. Note that this problem is different from example 4.5.1 in the sense that the area is divided such that one section is rhombus and the other a trapezoidal. It was also possible to divide it into two trapezoids by drawing a vertical line through point $E$ up to the free surface.

Discharge through section I.

$$A_1 = 12 \times 3 = 36 \text{ m}^2,$$

$$P_1, \text{ wetted perimeter} = 12 + 3\sqrt{2} = 16.243 \text{ m}.$$

$$\rightarrow \quad m_1 = A_1/P_1 = 2.216 \text{ m}$$

and
$$Q_1 = \frac{A_1}{n} m_1^{2/3} i^{1/2} = \frac{36}{0.025}(2.216)^{2/3}(0.0075)^{1/2} = 47.879 \text{ m}^3/\text{s}.$$

Discharge through section II.

$$A_2 = 6(8 + 6) = 84 \text{ m}^2,$$

$$P_2, \text{ wetted perimeter} = 8 + 3\sqrt{2} + 6\sqrt{2} = 20.728 \text{ m}.$$

$$\rightarrow \quad m_2 = A_2/P_2 = 4.052 \text{ m}$$

and
$$Q_2 = \frac{A_2}{n} m_2^{2/3} i^{1/2} = \frac{84}{0.025}(4.052)^{2/3}(0.0075)^{1/2} = 233.874 \text{ m}^3/\text{s}.$$

$$\therefore \quad \text{Total discharge} = Q_1 + Q_2 = 281.753 \text{ m}^3/\text{s}.$$

## 4.5.2 Gradually Varying Flow

In rivers and canals, it is common practice to put an obstruction on the way to store water. As a result the water level starts rising at the obstruction site which affects the water level upstream for some distance. The depth of water decreases upstream from the obstruction site. On the other hand instead of obstruction, the slope of the bed becomes a little more than that required for uniform flow thereby the depth of water would decrease downstream; however, whether the flow will accelerate shall depend on the slope. We shall first consider the later case and assume that the slope is gentle so that the fluid is not accelerating but the depth is gradually decreasing. We shall also assume that the flow is steady and $q$ is constant and that the Chezy and Manning relations for the discharge can be used in such a case. (Recall, though these relations were derived for uniform flow but we mentioned that these are applicable even for non-uniform flows too.) The energy line will now have slope that is different from that of the bed.

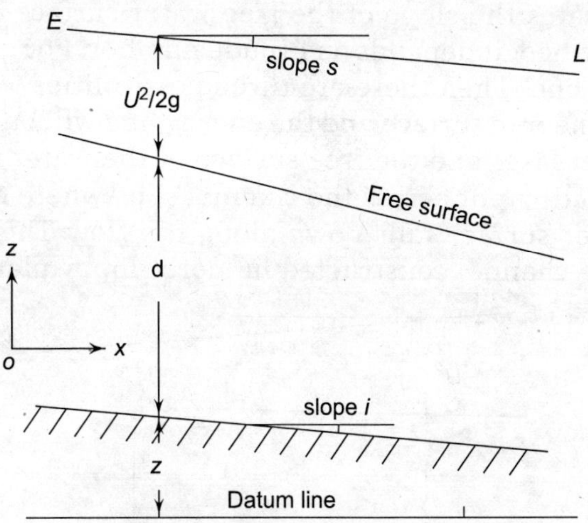


**Fig. 4.5.4** Gradually varying flow

Figure. 4.5.4 shows a gradually varying flow. The total head at any section is

$$E = z + d + \frac{U^2}{2g} = E(x),$$

since $U$ is varying in the direction of flow ($x$–direction). From this equation, slope of the energy line can be determined by differentiating it with respect to $x$ to give

$$\frac{dE}{dx} = \frac{dz}{dx} + \frac{dd}{dx} + \frac{U}{g}\frac{dU}{dx}$$

But $$U = q/d$$

→ $$\frac{dU}{dx} = -\frac{q}{d^2}\frac{dd}{dx} = -\frac{U}{d}\frac{dd}{dx}$$

∴ $$\frac{dE}{dx} = \frac{dz}{dx} + \frac{dd}{dx} - \frac{U^2}{gd}\frac{dd}{dx}$$

here $q$ is discharge per unit width of the channel. Writing $-i$ and $-s$ for the slope of the bed ($dz/dx$) and energy line ($dE/dx$); –ve sign because these are drooping along $x$-direction. We get

$$-s = -i + \frac{dd}{dx}\left(1 - \frac{U^2}{gd}\right)$$

$$\rightarrow \qquad \frac{dd}{dx} = \frac{i-s}{1-Fr^2}, \text{ where } Fr = \frac{U}{\sqrt{gd}} \tag{4.5.20}$$

Equation 4.5.20 relates the slope of the free water surface $dd/dx$ with the slope of the energy line and the bed; it depends on Froude number. For gradual flow the Froude; it number is less than one. Then there are three possibilities:

(i) For $i = s$ slope of the free surface and the energy line will be same as that of the bed.
(ii) For $i > s$ slope increases and the free surface of the water rises downstream; as in case of an obstruction put across the channel somewhere downstream.
(iii) For $i < s$ the water surface falls down along the flow. This situation may exist in flows through the channel constructed in more sloppy planes.

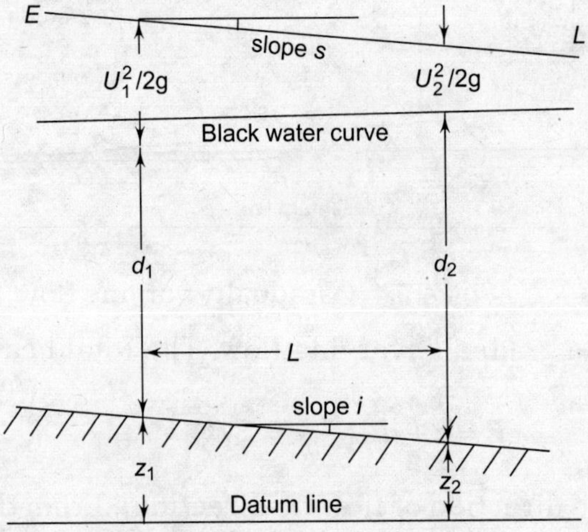


**Fig. 4.5.5** Length of back curve

We will now examine the second case where the water surface is rising. Figure 4.5.5 shows the energy line and the water surface. Effect of the obstruction propagates upstream and after some distance $L$ the water surface depth becomes constant. This portion of the water surface is called the *back water curve* and we would be interested to know the length of this curve. Let us represent these two sections by subscripts 2 and 1, respectively. The figure shows the datum head, depth and the kinetic head at these sections. Applying Bernoulli equation at sections 1 and 2 and assuming head loss $h_L$ between these sections, we get

$$z_1 + d_1 + \frac{U_1^2}{2g} = z_2 + d_2 + \frac{U_2^2}{2g} + h_L.$$

But $z_1 - z_2 = i \times L$, $h_L = s \times L$. Also writing $E = d + U^2/2g$, we get

$$L(i-s) = E_2 - E_1 \text{ or, } L = (E_2 - E_1)/(i-s) \quad (4.5.21)$$

## 4.5.3 Alternate Depth for Mean Depth

We used the mean depth to represent the uniform flow through a channel. We also learnt that at any section the height of the free surface from the bed of the channel is the piezometric head. Therefore the total energy per unit Weight at any section is sum of the depth of water and the kinetic head. This sum is called the *specific energy*, which is defined as the total energy per unit weight of the fluid.

$$E_s = d + \frac{U^2}{2g} = d + \frac{q^2}{2gd^2} \quad (4.5.22)$$

since $q$ = discharge/width of the channel = $U\,d$. We observe from Eq. 4.5.22:

(i) For large depth of flow, contribution from the kinetic term is negligible and the specific energy asymptotically tends to ∞ with depth of the flow.

(ii) On the other hand, the kinetic energy dominates at lower depth of flow when the specific energy tends to ∞ as $d \to 0$. That is, the specific energy goes to infinity asymptotically as we approach $d = 0$.

These two observations help in plotting the specific energy against depth of flow for given $q$ or $E_s$ at a section. Figure 4.5.6(a) shows the variation. Thus the specific energy at a section changes from ∞ to ∞ when flow depth transits from high to low value. It is possible if for some flow depth, its value is minimum. This depth is called the *critical depth*, $d_c$.

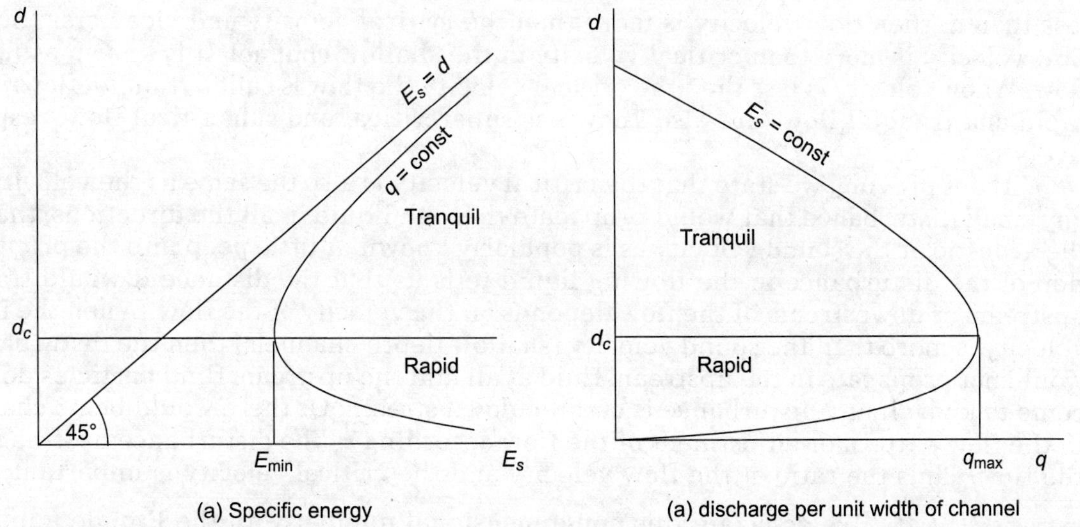


(a) Specific energy (a) discharge per unit width of channel

**Fig. 4.5.6** Variation of specific energy and discharge at a section with depth of low

The critical depth can be obtained by minimising the specific energy while taking $q$ as constant. Applying it to Eq. 4.5.22, we get

$$\frac{dE_s}{dd} = 1 - \frac{q^2}{gd^3} = 0$$

$$\rightarrow \qquad d_c = \left(\frac{q^2}{g}\right)^{1/3} \quad \text{and} \quad E_{\min} = \frac{3}{2}\, d_c \qquad (4.5.23)$$

The critical depth given by Eq. 4.5.23 therefore corresponds to stable state of the flowing liquid. Equation 4.5.22 can be rearranged as $q^2 = 2gd^2(E_s - d)$ and differentiated for given $E_s$ to give condition for maximum discharge. This gives the same result as given by Eq. 4.5.23. That is, when the flow depth is $d_c$ then the specific energy at a section will be minimum and the discharge will be maximum. Figure 4.5.6(b) shows variation of $q$ with $d$ at a section for given $E_s$. The critical velocity may be determined using Eq. 4.5.23.

$$U_c = q/d_c = \frac{\left(gd_c^3\right)^{1/2}}{d_c} = (gd_c)^{1/2} \qquad (4.5.24)$$

We know that a system left to itself acquires a stable state in which the energy is minimum. That is, factor $(g\, d_c)^{1/2}$ can be used to nondimensionalise the velocity. At the critical section, this ratio will be 1. We identify the ratio as Froude number. Accordingly we can also use critical depth as an alternate to the mean depth for non-dimensionalisation. Equation 4.5.22 is cubic in $d$ and hence will have three roots. However, one of these roots is negative and has no physical sanctity. Thus there are two meaningful roots of the equation; one is above and the other below the critical depth. These two values of depth are known as *alternate depth*. Thus when the flow depth is less than $d_c$ then flow velocity is more than the critical velocity and vice versa. Since flow velocity is more than critical velocity in the shallow channel, it is known as *rapid* flow. When velocity is less than the critical velocity the flow is called *tranquil* flow. The rapid and tranquil flows are also known as supercritical and sub-critical flow, respectively.

Without proving, we state that the critical velocity is also the same as the velocity of any small disturbance that would propagate through liquid in all the directions, that is the acoustic or the sound velocity as is popularly known. A little peep into the propagation of the disturbance in the flowing liquid tells us that the distance it would travel upstream or downstream of the flow depends on the velocity of the flow. When the flow velocity is more than the sound velocity (shallow depth channels) then the disturbance would not propagate in the upstream fluid at all and the upstream fluid particles do not come to know if any disturbance is created down stream. Or there would be no change in the flow pattern or adjustment of the flow according to the disturbance upstream of the flow. Thus the ratio of the flow velocity and the critical velocity is important; the ratio is $U/\sqrt{gd_c}$. We associate this non-dimensional number with the Froude number. *This reflects that the free surface flows are influenced by the Froude number.*

### 4.5.3 Non-uniform Flow

When flow starts, it is unsteady and slowly and slowly it acquirs steady state. During this period, the flow parameters adjust and when the gravity force due to slope of the bed is equal to the shear force then it acquires steady state. During the initial adjustment of the flow both the depth and the discharge vary. Whether the flow will be rapid or tranquil under the steady state depends on the initial conditions at the time of discharge from the reservoir. In most of the cases it is rapid flow and we look for means to change it to tranquil flow so that the flow velocity is decreased to save the erosion of the canal sides.

Nevertheless, it is observed that when a rapid flow occurs then it suddenly changes to tranquil flow for some reason and the depth of the flow changes suddenly.

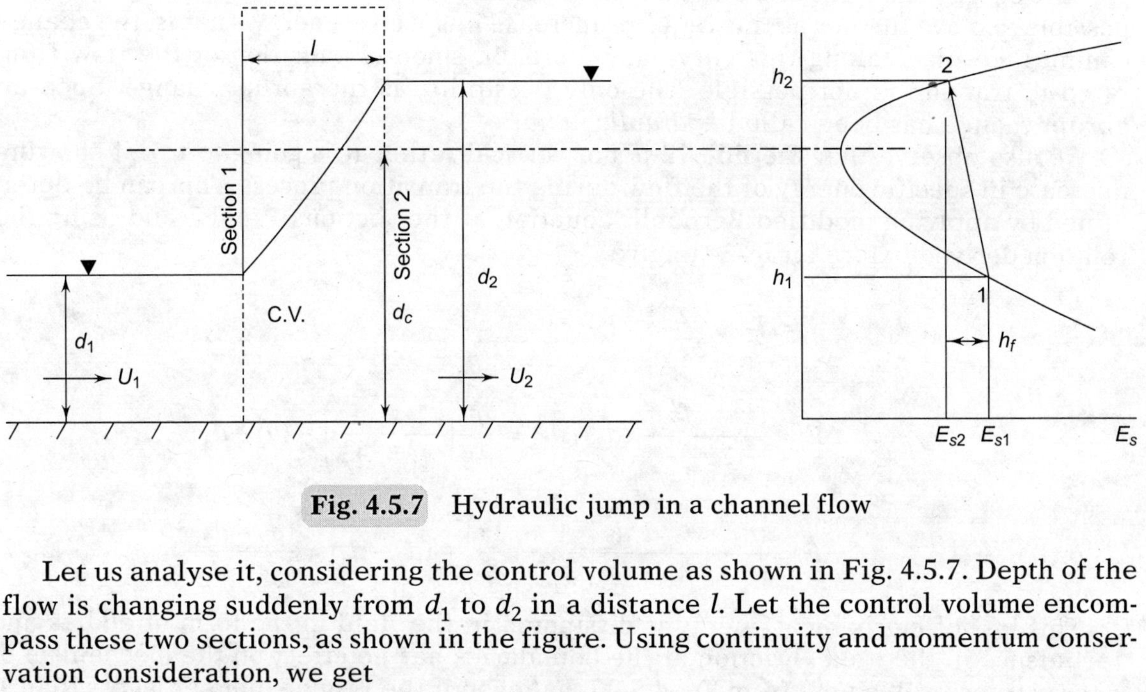


**Fig. 4.5.7** Hydraulic jump in a channel flow

Let us analyse it, considering the control volume as shown in Fig. 4.5.7. Depth of the flow is changing suddenly from $d_1$ to $d_2$ in a distance $l$. Let the control volume encompass these two sections, as shown in the figure. Using continuity and momentum conservation consideration, we get

From continuity equation: $U_1 d_1 = U_2 d_2 = q$

From momentum equation: $\rho g(d_1^2 - d_2^2)/2 = (U_2 - U_1)\rho q$

$$\rightarrow \qquad d_1^2 - d_2^2 = \frac{2q^2}{g}\left(\frac{1}{d_2} - \frac{1}{d_1}\right)$$

or,

$$\frac{q^2}{2g} = \frac{(d_1 + d_2)d_1 d_2}{4}$$

and,

$$d_2 = -\frac{d_1}{2} + \sqrt{\frac{d_1^2}{4} + \frac{2d_1 U_1^2}{g}} \qquad (4.5.25)$$

or, $$\frac{d_2}{d_1} = \frac{1}{2}\left(\sqrt{1+8F_{r1}^2} - 1\right) \tag{4.5.26}$$

There will be one more root of the quadratic equitation in $d_2/d_1$ with negative sign for the discriminate. However, this has no physical relevance, as then $d_2$ is negative; therefore it has not been considered. Equation 4.5.25 has been changed in non-dimensional form where $F_{r1}$ is the Froude number at section 1 before the jump occurs. The process is shown on depth v/s specific energy diagram for a given discharge $q$, in which the points 1 and 2 represent the state before and after the jump; the line 12 shows the direction of the process. Let us consider this change from 1 to 2 to occur along the curve that is drawn for given $q$. In that case the point would move to left of 1, as the flow initially is rapid. This continues till we reach the critical point. During the process specific energy is decreasing and hence the process is possible. However, it will not be possible to move further as this requires increase in specific energy. That is, the change cannot take place along this curve any more, or, smooth transition of the flow from rapid to tranqûil is not possible. The only possibility is the abrupt change. Such an abrupt change has been called *hydraulic jump.*

We also observe that the line 12 is not vertical rather it is going left of it showing decrease in specific energy of the flow during the transition process. This can be determined by applying modified Bernoulli equation at the sections 1 and 2 and using the relation derived before for $q^2/2g$ to give

$$d_1 + \frac{U_1^2}{2g} = d_2 + \frac{U_2^2}{2g} + \Delta E$$

$$\rightarrow \qquad \Delta E = \frac{U_1^2 - U_2^2}{2g} - (d_2 - d_1) = \frac{q^2}{2g}\left(\frac{1}{d_1^2} - \frac{1}{d_2^2}\right) - (d_2 - d_1)$$

or, $$\Delta E = \frac{(d_1 + d_2)d_1 d_2}{4}\left(\frac{1}{d_1^2} - \frac{1}{d_2^2}\right) - (d_2 - d_1) = \frac{(d_2 - d_1)^3}{4d_1 d_2} \tag{4.5.27}$$

This loss of mechanical energy is dissipated in the fluid in the form of eddies and turbulence in the wake. Friction at the boundaries has no effect on the mechanism of energy transfer. We note from Eq. 4.5.26 that even if the Froude number is less than 1, there is increase in depth of the flow as a result of such discontinuity in the flow, a result that does not agree with the experimental observations that jump ocurs when $Fr > 1$. To be specific, such an answer can be made only by the second law analysis of the flow by determining the change in the entropy during the process so that we may infer the possibilities of any process. Hydraulic jump is a useful way to change rapid flow to a tranquil flow. Not only this, the energy of the flow is also reduced in the process.

We may also express $d_2/d_1$ as ratio of the Froude number at the respective sections, using definition of Froude number and Eq. 4.5.27.

$$F_{r1}^2 = \frac{U_1^2}{gd_1} = \frac{q^2}{gd_1^3} \text{ and } F_{r2}^2 = \frac{U_2^2}{gd_2} = \frac{q}{gd_2^3}$$

$$\rightarrow \qquad \frac{F_{r2}^2}{F_{r1}^2} = \left(\frac{d_1}{d_2}\right)^3 = \frac{8}{\left(\sqrt{1+8F_{r1}^2} - 1\right)^3} \qquad (4.5.28)$$

We may now summarise the above discussions.

1. Jump occurs only when the Froude number is more than 1 and in the process the Froude number becomes less than 1.
2. The ratio $d_2/d_1$ is more than 1 and increases with the Froude number at the section 1.
3. $(d_2 - d_1)/d_1$ can be used to represent strength of the jump and depends on $F_{r1}$.
4. Loss of energy is associated with the formation of the jump that depends on the strength of the jump; more the strength more is the loss.
5. Hydraulic jump may be used for (i) dissipation of energy as utilised at the foot of the spillway in a dam, (ii) mixing of the fluids as in chemical plants, or (iii) reduction of the velocity as at the base of the dam where large velocities would otherwise result in scouring.

## PROBLEMS

## Review Questions

### Pipe Flow

1. Differentiate between laminar and turbulent flow. What is the Reynolds number up to which the flow in a pipe is considered as laminar flow? Why there is not a single value for this number?
2. Is it possible to have laminar flow through a pipe at $R_e = 25{,}000$? Under what condition is this possible, if at all?
3. What is the significance of: upper and lower critical Reynolds numbers?
4. Air having a viscosity of 0.00018 Poise and density of 1.2 kg/m$^3$ flows through a pipe of 5 cm diameter. What is the critical velocity?
5. Discuss importance of hydraulic gradient and total energy lines.
6. What is a siphon and where is it used. What precautions have to be taken while using it?
7. Two reservoirs are connected by a siphon. Vaporisation of water is found to take place at the submit of the siphon. What measures have to be taken to remedy this undesirable situation?
8. Is it always necessary that Hydraulic Gradient Line (H.G.L.) be above the pipe axis? Can it not intersect the pipe axis or lie below the pipe axis?
9. The loss of head due to friction in a pipe is 12 m of oil. If the rate of flow of the oil though the pipe is 600 N/s, what is the power loss?
10. What is the minor loss in pipe flows? Under what conditions does a minor loss becomes a major loss?

11. If $C_c = 0.6$, what is the head loss coefficient of a sudden contraction?
12. What precisely is the advantage gained by making enlargement and contraction in pipes gradual instead of their being abrupt?
13. By having rounded pipe entrance and exit instead of sharp-edged entrance and exit, is it possible to reduce the respective losses of head?
14. The friction factor in a laminar flow through a pipe is 0.016 $m^3$/s, what is the Reynolds number?
15. Under what circumstances we shall be justified in neglecting the minor loss? If we do not then what type of problems do we face?
16. What conclusion can be drawn from the curves in Moody's chart with regard to the friction factor of the pipes at high Reynolds number?
17. In a system of parallel pipes, the same hydraulic gradient is valid for all the pipes. Is this true?
18. Out of the series and parallel pipe arrangement, which one would you prefer and why? Support your answer with examples.
19. What is the underlying concept of equivalent length and how does it help the field engineers?
20. When transmitting power through fluid, there is some condition when the efficiency is maximum. Does it imply that efficiency of this pipe line can never be more than this value?
21. Two pipes having the same length and friction factor are connected in parallel. The diameters are 10 cm and 20 cm. If the head loss in the 10 cm pipe is 16 m, what is the head loss in the 20 cm pipe? If the discharge through 10 cm pipe is 0.01 $m^3$/s, what is the discharge through the 20 cm pipe?
22. A pipe has $D = 40$ cm, $L = 100$ m, $f = 0.005$. What is the length of an equivalent pipe which has $D = 20$ cm and $f = 0.008$?
23. The loss of head due to friction in a pipeline is 30 m. For maximum power transmission through the pipe what should be the head at the exit of the pipe?
24. The efficiency of power transmission in a hydraulic system is 80% and the friction head loss in the pipe is 50 m. what should be the supply head?
25. In a hydraulic power transmission system, the supply head is 250 m. A nozzle fitted at the delivery end of the pipe gives a jet whose velocity head is 220 m. What is the efficiency of power transmission? Neglect the nozzle loss.
26. In water hammer analysis, what is the criterion for classifying whether a valve closure is sudden or gradual?
27. Describe the sequence of events that occurs when a valve at the end of a long pipe is abruptly closed. Consider the fluid and pipe to be elastic.
28. Water flowing with a velocity $V$ in a pipe line is brought to rest by closing a valve in time interval $T$. Show that the pressure rises due to valve closure is given by $p = \dfrac{wLV}{gT}$, where $L$ is the length of the pipe upstream of the valve.

**Channel Flow**

29. What are the differences between pipe flow and open channel flow?

30. What are the regimes of flow? Which is the most common regime?
31. What are the conditions under which the maximum velocity in an open channel section occurs a little below the free surface?
32. Justify the criterion used for determining the most economical size for (a) canal and (b) sewage line.
33. If the channel flow is convex in curvature, the pressures across a vertical section are more than those in hydraulic pressure distribution. Is this true?
34. Differentiate between (a) the steady and non-steady flow, (b) uniform and non–uniform flow, (c) gradual and rapid varying flow, and (d) critical, rapid and tranquil flow in an open channel.
35. What is specific energy?
36. What do you understand by the critical depth. Discuss its importance.
37. What is the state of the open channel flow when the Froude number is unity?
38. What is the state of the flow likely to be when the depth is relatively large and the velocity is small in an open channel?
39. Mention three characteristics of a supercritical state of open channel flow.
40. When the specific energy is minimum for a given discharge, what is the depth in the channel?
41. If the depth of flow is less than the critical depth, what is the type of the channel slope? What is the slope if the depth is more than the critical depth?
42. If $E_{min}$ in a rectangular open channel section is 0.75 m, what is the discharge in $m^3/s$?
43. If the critical velocity head in a rectangular channel is 0.75 m, what is the critical depth?
44. What are alternate depths? What are conjugate depths? Can a sequent depth be called a conjugate depth?
45. In an open channel flow, the total energy line, water surface and channel bed are parallel to each other. What type of flow is this?
46. A river with grassy banks has a Manning's $n = 0.05$ and Chezy's $C = 20$. What is its hydraulic radius?
47. What is the condition to be satisfied for the most economical section of an open channel?
48. The cross section of an open channel is a trapezium of the semi hexagonal shape. If the base width = 3 m, what is the depth of flow? What is the hydraulic radius?

## Problems for Practice

1. A 150 mm diameter horizontal pipe conveys water. If the slope of the hydraulic gradient line 1 in 130, find the rate of flow. Take $f = 0.005$. [$1.88 \times 10^{-2}$]

2. Karman–Prandtl equation, $\frac{1}{\sqrt{f}} = 2\ln\frac{R_0}{k} + 1.74$, can determine the friction factor for turbulent flow through rough pipes, where $R_0$ is the pipe radius and $k$ is the average roughness of the pipe. Two reservoirs with a surface level difference of 20 m are to be connected by a 1 m diameter pipe which is 6 km long. What will be the

discharge when a cast iron pipe of roughness $k = 0.3$ mm is used? What will be the percentage increase in the discharge if steel pipe of roughness 0.1 mm replaces the cast iron pipe? Neglect all local losses. [0.823 $m^3$/s, 11.42%]

3. A pipe 100 mm in diameter and 40 m long conveys water at a velocity of 2.5 m/s. If a 200 mm diameter pipe replaces the central 20 m length of pipe, find the saving in head lost. Assume that the changes in section are sudden. Take $f = 0.01$ and coefficient of contraction $C_c = 0.62$. [2.171 m]
4. There is a sudden enlargement in a pipe line from diameter $d_1$ to $d_2$. If the pressure at section 1–1, before the enlargement, and section 2–2, at some distance $L$ from the enlargement section, are equal, find the distance $L$ between the sections, if the diameter ratio is $\sqrt{2}$ and the pipe coefficient $f = 0.005$.
5. Two tanks whose water surface levels differ by 10 m are connected by three pipes of diameters 250 mm, 150 mm, and 200 mm in series. Their lengths are 300 m, 250 m, and 150 m, respectively. Their coefficients of friction are: 0.005, 0.006, and 0.0075, respectively. Find the rate of flow thorough the pipe line considering all losses. Find also the values of the individual losses of head.
6. A pipe line *BCD* 840 m long connects two reservoirs *A* and *E* having a difference of water levels of 20 m. The length *BC* of the pipe line is 300 m and is horizontal at a depth of 5 m below the water level of the reservoir *A* and is of mild steel whose $f = f_1$. The length *CD* is 540 m long and it is made of cast iron whose coefficient of friction is $f_2$. Both the pipes are 250 mm in diameter. A partially closed valve is provided in the part *CD* at a distance of 180 m from the reservoir *E*. If the discharge through the pipe is 92.5 $m^3$/s, the pressure at *C* is 3.77 kPa and the loss of head at the valve is 6 m, find the friction coefficients. [0.0055, 0.006]
7. Two pipes each 100 m long, one of diameter 100 mm and other of diameter 75 mm are provided in parallel connecting two tanks, the difference of water levels of which is 50 m. Find the discharge to the lower tank, if however, a single pipe 100 m ling is provided connecting the two tanks and the discharge to the lower tank is to be increased by 25%, find the diameter of this pipe, for all pipes, take $f = 0.008$. Ignore minor losses. [0.0354 $m^3$/s, 128 mm]
8. Two pipes connected to reservoir discharge water freely to the atmosphere at their outlet ends. The first pipe, which is 150 mm in diameter and 300 m in length, has its outlet end at a depth of 3.8 m below the water level in the reservoir. The second pipe is 200 mm in diameter and 600 m long. If the combined discharge is 50 $m^3$/s, find how much below the reservoir water level, the outlet end of he second pipe is situated? Take $f = 0.0045$. Neglect the minor losses. [1.7765 m]
9. Pipes of 75 mm diameter are to be used to siphon water from a main canal to a branch canal, the difference of water level between the two canals being 15 m, the length from the main canal to the submit of the pipe line is 20 m, the total length of the pipe being 50 m. Determine the number of pipes required to discharge at least 50 l/s to the branch canal. Find also the maximum height of the submit above the water level of the main canal in order that the pressure at the submit may not fall below 20 kPa (absolute). Take barometric pressure as 760 mm of $H_g$. take $f = 0.0075$. Ignore the secondary losses. [3, 1.038 m]

10. A 200 mm diameter pipe line connecting two reservoirs whose difference of level is 11.1 m comprises of 300 m length *AB* and a straight 450 m length *BC* having a valve 150 m from the lower end *C* as shown in Fig. P.10. The pipes *AB* and *BC* have friction factors $f_1$ and $f_2$. When the flow is 2.6 $m^3$ per minute, a gauge at *B* shows a pressure head of 0.51 m of water and the loss of head across the valve is 4.5 m. Find the friction factors. Determine the maximum rate of flow if the loss of head across the valve when fully open is 1.2 m. Also draw the hydraulic gradient for this quantity. Allow for the effect of velocity head in the pipe at the exit but neglect loss of head at a pipe entry. Assume that the horizontal projection of *BC* equals its length. [0.0041, 0.0047, 0.053 $m^3$/s]

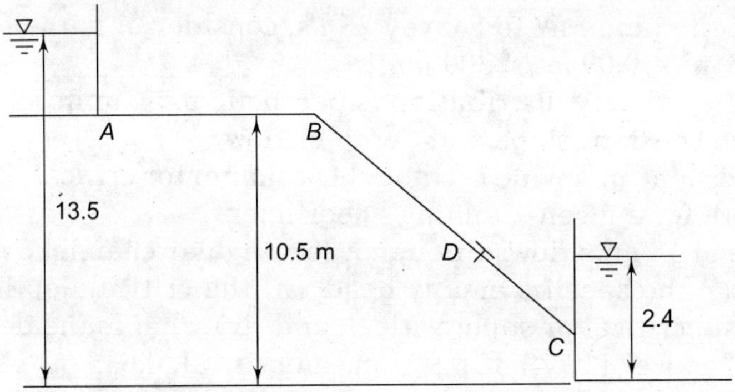


**Fig. P.10** Flow through pipe

11. An existing pipe line 1000 m long consists of three sizes namely 400 mm diameter for the first 400 m, 300 mm diameter for the next 400 m, and 250 mm diameter for the remaining 200 m, neglecting the secondary losses, find the diameter of a uniform pipe 1000 m long to replace the compound pipe. [0.3 m]
12. Two pipes of diameters 400 mm and 200 mm are each 300 m long. When the pipes are connected in series and the discharge through the pipe line is 0.1 $m^3$/s, find the loss of head incurred. What would be the loss of head in the system to pass the same total discharge when the pipes are connected in parallel? Take $f$ = 0.0075 for both the pipes. Comment. [23.96 m, 0.52 m]
13. Two tanks are connected by a pipe system as shown in Fig P.13 find the discharge through each pipe. Take $f$ = 0.008 for all the pipes. Neglect all losses other than the friction loss.
14. A pipe 199 m pong tapers from 100 mm diameter at one end to 50 mm diameter at the other end, and conveys a discharge of 0.008 $m^3$/s of water. Taking $f$ = 0,0075, find the loss of head in the pipe due to friction. [1.19 m]

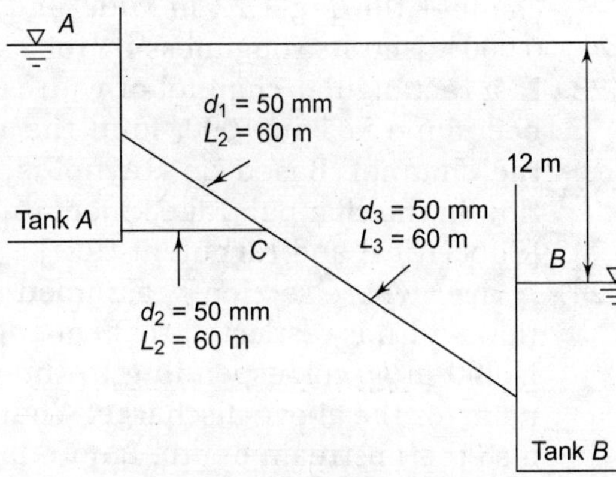


**Fig. P.13** Discharge through pipe network

15. A 250 mm diameter pipe 900 m long connects two reservoirs *A* and *B* having a difference of level of 6, *A* being the higher. Two other reservoirs *C* and *D*, having a difference of level of 12 m, *C* being higher and 3 m below the reservoir *A*, are connected by a 300 mm diameter pipe 1800 m long. In order to increase the quantity of water entering the reservoir *D*, the two pipes are connected by a pipe *MN* 1650 m long which branches from the 250 mm diameter pipe at a point *M* 300 m from *A* and connects with the 300 mm dia pipe at a point *N* 600 m from *D*. The friction coefficient is 0.005 for the 250 mm dia and 300 mm dia pipes and 0.0056 for the pipe *MN*. Determine (a) the quantities of water entering the reservoirs *B* and *D* when 25 litres per second flow through the pipe *MN* and (b) the minimum diameter of the pipe *MN* to convey 25 l/s, consider only the friction losses in the pipes.[0.053 $m^3$/s, 0.09 $m^3$/s, 200 mm]
16. Show that the velocity distribution is parabolic for a uniform flow in a wide open channel with constant slope and depth of flow.
17. Find the bed slope of a wide rectangular channel for critical flow *t* to reach a given critical depth for a given Manning's coefficient.
18. A discharge of 18 $m^3$/s flows through a rectangular channel 6 m wide at a depth of 1.6 m, find (i) the specific energy head, (ii) the critical depth, (iii) state whether the flow is subcritical or supercritical, and (iv) what is the depth alternate to the depth given above? [1.779 m, 0.972 m, subcritical, 0.633 m]
19. A channel 5 m wide conveys a discharge of 10 $m^3$/s of water. Plot graph for the following: (i) static energy head, (ii) kinetic energy head, and (iii) specific energy head.
20. A channel of parabolic section whose section profile has the equation $y = K x^2$ conveys a discharge *Q*. Find the critical depth.
21. A channel of circular cross section 1 m in diameter is half full. If the flow through it is critical flow find the discharge. Find also the critical slope. Take Manning's coefficient = 0.018. [0.771 $m^3$/s, 0.00793]
22. In a hydraulics laboratory, a flow of 0.412 $m^3$/s was measured from a rectangular channel flowing 1.22 m wide and 0.61 m deep. If the slope of the channel was 0.0004, find its roughness factor using Manning's formula. [$n = 0.0163$]
23. In a rectangular channel of width 3 m, the uniform depth of flow is 1.2 m and the bed slope is 1 in 1000. Find the mean shear stress on the sides and bottom of the channel. Based on Reynolds number and Froude number find the type of flow in the channel. Take kinematic viscosity = $10^{-4}$ $m^2$/s and Chezy's coefficient 70. [Subcritical and turbulent flow]
24. A river whose section $p_i$ assumed to be rectangular is 40 m wide. At a bridge the flow width is restricted to 35 m due to bridge piers. The maximum flood discharge is 600 $m^3$/s. corresponding to the condition the depth upstream should be minimum for the above discharge. What type of flow will prevail under the bridge? Find also the upstream depth. Ignore loss of energy head. [Critical, 4/65 m]
25. Uniform flow exists in a 3 m wide channel at a depth of flow of 1. 5 m, the bed slope of the channel is 0.0009. (a) Find the height of the hump in the bed of the channel

to produce critical depth without affecting the upstream depth. (b) Alternatively find the extent to which the width of the channel can be reduced to produce critical depth without affecting the upstream depth. Take Manning's $N = 0.015$. [0.3565 m, 2.079 m]

26. A circular culvert has a capacity of 0.5 $m^3/s$ when running full. Velocity should not be less than 0.7 m/s if the depth of flow is $1/4^{th}$ of the diameter. Assuming uniform flow, find the diameter and the slope, taking Manning's coefficient = 0.012. [0.798 m, 0.0012]

CHAPTER

5

# Laminar Flow

**OUTLINE**

- Introduction–Types of solutions
- Exact Solution of N–S equation
- Low Reynolds Flow
- Boundary Layer
- Integral Solution for Boudary Layer
- Boundary Layer Separation and Control
- Flow over Curved Surfaces

## 5.1 Introduction—Types of Solutions

We discussed laminar flow in Chapter 2 and noticed that flow takes place as if the fluid is made of layers or lamina and the interaction between these adjacent layers is through the transfer of molecules, which results in the various transport properties, namely, the dynamic viscosity, thermal conductivity, diffusion, etc. We also developed the mathematical model for this flow and noted that the resulting equations are (i) non-linear, (ii) partial differential equation of second order, (iii) elliptic, and (iv) set of simultaneous equations. For the solution of these equations, we have four alternatives, namely

1. Exact solution of exact equation
2. Approximate solution of exact equation
3. Exact solution of approximate equation
4. approximate solution of approximate equation

The first kind of solution was possible in some simple cases, such as Couette, Hagan-Poiseuille flows, flow through annulus, etc. Stokes flow and hydrodynamic lubrication are a few examples from the second category of solution. Boundary layer flow fills the third category of the solution. A large category of flows fall into the last category of solutions.

## 5.2 Exact Solution of N–S Equation

There are some simple two-dimensional flows in which the governing equations reduce to total differential equations and their exact solution can be obtained. Flows between two parallel plates in a pipe or between the rotating cylinders are a few to name as example. These flows can be idealised as two-dimensional flows, since the flow in the pipe and annulus is axisymmetric while in the channel two plates of infinite width in $z$ direction form the passage and hence there is no variation in $z$ direction. While the flow in pipe and annulus is because of pressure difference from inlet to outlet, flow through a channel may be due to pressure difference as well as moving the plates relative to each other. We shall now consider the analysis procedure for these three kinds of flows.

### 5.2.1 Couette Flow

Couette flow is a flow of the incompressible fluid passing through a channel, as shown in Fig. 5.2.1, because of both—the pressure difference along its length and the movement of the upper plate with some velocity $U$ while keeping the lower plate fixed. (If the lower plate also moves with $-U$ velocity then the flow developed is shear flow. If the flow is due to pressure gradient only then it is called plane Poiseuille flow.) When the plate moves, the fluid layer in its contact also moves with the same velocity that induces the next layers to move subsequently with some velocity. This process goes on and ultimately all the layers in the flow region move along the flow direction from zero velocity at lower plate (it is stationary) to the velocity of the upper plate $U$. Had there been no pressure difference across the channel, we might expect the variation of the velocity to be linear. Pressure gradient affects the velocity profile in the region. Let the coordinate system be fixed at point $O$ to the lower plate and the height of the channel passage be $h$. We also assume that the plates are long enough along the flow direction so that the flow is fully developed, which means the velocity profile has stabilised and does not change with $x$ direction. We will now investigate the fully developed flow. Equation 2.4.15 governs this flow with the boundary conditions: $y = 0$, $u = v = 0$ and $y = h$, $u = U$, $v = 0$.

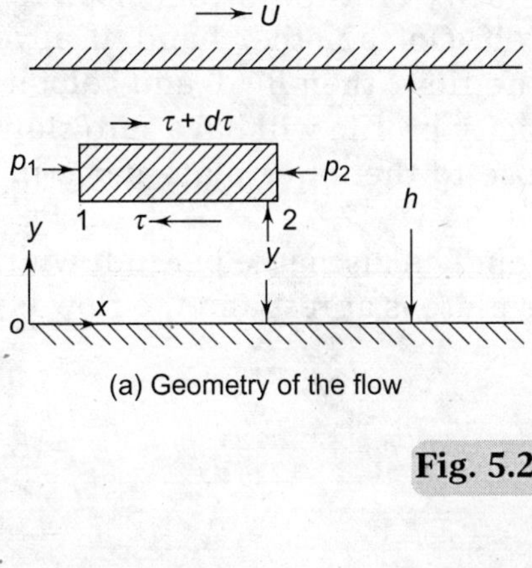


(a) Geometry of the flow

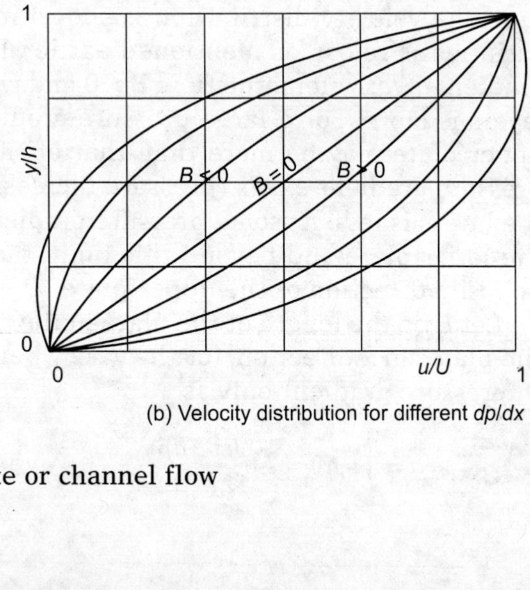


(b) Velocity distribution for different $dp/dx$

**Fig. 5.2.1** Couette or channel flow

For a fully developed flow, $x$ variation of $u$-component of velocity is zero, that is, $\partial u/\partial x = 0$. The continuity equation then gives $\partial v/\partial y = 0$. Its integration with respect to $y$ gives $v = c$. From the boundary condition at $y = o$, $v = 0$, we get constant $c = 0$. Therefore, $v = 0$ everywhere in the fully developed flow, which means its first and higher order derivatives with respect to $x$ or $y$ will be zero. Now consider the momentum equations, which give after dropping the terms having $v$ or its derivatives and $x$-derivatives of $u$

$$\text{X–momentum Eq.}: u\frac{\partial u}{\partial x} + v\frac{\partial u}{\partial y} = -\frac{1}{\rho}\frac{\partial p}{\partial x} + \nu\left(\frac{\partial^2 u}{\partial x^2} + \frac{\partial^2 u}{\partial y^2}\right) \rightarrow -\frac{1}{\rho}\frac{\partial p}{\partial x} + \nu\frac{\partial^2 u}{\partial y^2} = 0$$

$$\text{Y–momentum Eq.}: u\frac{\partial v}{\partial x} + v\frac{\partial v}{\partial y} = -\frac{1}{\rho}\frac{\partial p}{\partial x} + \nu\left(\frac{\partial^2 v}{\partial x^2} + \frac{\partial^2 v}{\partial y^2}\right) \rightarrow -\frac{1}{\rho}\frac{\partial p}{\partial y} = 0$$

The second equation implies that $p$ is not a function of $y$ and hence $\partial p/\partial x$ can be written as $dp/dx$. Therefore $N$–$S$ equation is reduced to

$$d^2u / dy^2 = (1/\mu)\, dp/dx = B'(\text{say}) \tag{5.2.1}$$

Boundary conditios: $u = 0$ at $y = 0$ and $u = U$ at $y = h$

In the last equation, the partial derivatives have been replaced with the total derivative because $u$ is function of $y$ alone. We get an ordinary differential equation of second order in $u$ for which enough number of boundary conditions are given to give an unique solution. So there is unique solution to these equations. Integration of Eq. 5.2.1 two times with respect to $y$ and using the boundary condition gives

$$u = B'y^2/2 + ay + b \qquad \rightarrow \frac{u}{U} = B\frac{y}{h}\left(1 - \frac{y}{h}\right) + \frac{y}{h}, \text{ where } B = -\frac{h^2}{2\mu U}\frac{dp}{dx} \tag{5.2.2}$$

The constant of integrations $a$ and $b$ were obtained from the boundary conditions. Rest is just rearranging the terms and puting them in non-dimensional form. Equation 5.2.2 is parametric with $B$ and hence gives a family of curves for the velocity distribution. The velocity distribution is shown in Fig. 5.2.1(b). We note that when $B = 0$, the variation is linear, as mentioned earlier. For a decreasing pressure (negative pressure gradient $dp/dx$) along the flow, $B > 0$ and we detect that the velocity of the intermediate layers is more. For a large pressure gradient, the velocity of a few layers close to the upper plate may be more than that of the plate itself. On the other hand, if adverse pressure gradient exists (pressure increasing along the flow) then $B < 0$ and velocity of the layers is less. At some pressure gradient, the velocity profile will have zero slope at the lower plate and beyond this limit, the layers close to the lower plate move in the opposite direction of the upper plate.

Let $L$ be the length of the plate in the flow direction. The discharge per unit width of the plates in $x$ direction, for the case when the upper plate is at rest and the flow is due to pressure gradient only, is

$$q = \int_0^h u\,dy = -\frac{h^3}{12\mu}\frac{dp}{dx} \rightarrow \overline{U} = \text{average velocity} = q/h = -\frac{h^2}{12\mu}\frac{dp}{dx} \tag{5.2.3}$$

or, $$\frac{dp}{dx} = -\frac{12\mu\overline{U}}{h^2}. \text{ But } (p_1 - p_2) / L = -dp / dx,$$

$$\therefore \quad p_1 - p_2 = -L\, dp/dx = 12\mu\overline{U}L / h^2 \tag{5.2.4}$$

Equation 5.2.1 can also be obtained by considering a fluid element of length $dx$ and height $dy$ in the flow region at a distance $y$ from the lower plate, as shown in Fig. 5.2.1 (a). The layers above it try to pull it in forward direction while the lower one pulls it back with the result the shear forces act on it in the direction shown in figure, where $\tau = \mu\, du/dy$. Also let $p_1$ and $p_2$ are the pressure at the two ends of the element. Making balance of forces acting on this element, we get

$$(p_1 - p_2)\, dy = d\tau\, dx \qquad \text{or, } dp / dx = d\tau / dy = \mu\, d^2u / dy^2$$

This is the same equation. The two terms refer to the pressure and viscous forces in the *NS* equation. We assume absence of the body force, but then what happens to the inertia force is not known in this kind of analysis.

**Example 5.2.1** For steady flow between parallel plates, $h = 2.0$ mm, $\mu = 0.05$ kg/ms, and $dp/dx = -10$ kN/m$^3$. Determine $q$, maximum shear and maximum velocity.

**Solution** Equation 5.2.2 gives $du/dy$ on differentiation with respect to $y$ from which we get

$$\tau = \mu\, du / dy = -\frac{1}{2}\frac{dp}{dx}(h - y)y$$

Maximum shear occurs at the stationary wall, which is at $y = 0$.

Therefore, $\tau_{max} = -(h/2)\, dp/dx = 10$ N/m$^2$.

Average velocity, $U = -\dfrac{h^2}{12}\dfrac{dp}{dx} = 0.0667$ m/s and $q = Uh = 0.1334 \times 10^{-3}$ per m$^3$

Maximum velocity occurs at the centre of the channel, which is at $y = h/2$. Substituting in Eq. 5.2.2, we get

$$U_{max} = -\frac{1}{8\mu}\frac{dp}{dx}h^2 = \frac{1}{8 \times 0.5} \times 10^4 \times 0.02^2 = 0.10 \text{ m/s}$$

**Example 5.2.2** Water at 60°C flows between two large plates. The lower plate moves to the left at a speed of 0.3 m/s. The plate spacing is 3 mm and the flow is laminar. Determine the pressure gradient required to produce zero net flow at a cross section. Take $\mu = 4.7 \times 10^{-4}$ Ns/m$^2$ at 60°C.

**Solution** This is the Couette flow with the difference that the lower plate is moving instead of the upper plate. Without going into the arguments (you should be able to justify it), we write the governing momentum equation as $\mu\, d^2u/dy^2 = dp/dx$ with the boundary conditions $y = 0$, $u = -U$ and $y = h$, $u = 0$. This gives the solution

$$u = \frac{1}{2\mu}\frac{dp}{dx}(y^2 - yb) + U\left(\frac{y}{b} - 1\right) \quad \text{and}$$

$$Q = \int_0^h u\,dy = -\frac{1}{12\mu}\frac{dp}{dx}h^3 - \frac{Uh}{2} \tag{A}$$

We are given $\mu = 4.7 \times 10^{-4}$ Ns/m², $h = 0.003$ m, and $U = 0.3$ m/s. For $Q = 0$, we get from Equation (A)

$$\frac{dp}{dx} = -\frac{6\mu U}{h^2} = -94.0 \text{ N/m}^2 \text{ per m}$$

A negative pressure gradient indicates that the pressure must decrease in the direction of the plate from the left.

**Example 5.2.3** A continuous belt passing upward through a chemical bath at velocity $U_0$, picks up a liquid film of thickness $h$, density $\rho$ and viscosity $\mu$. Gravity tends to make the liquid drain down, but the movement of the belt keep the fluid from running off completely. Assume that the flow is fully developed and that the atmosphere exerts no shear force at the interface of the film. State clearly the boundary condition to be satisfied by velocity at $y = 0$ and $y = h$. Obtain an expression for the velocity profile.

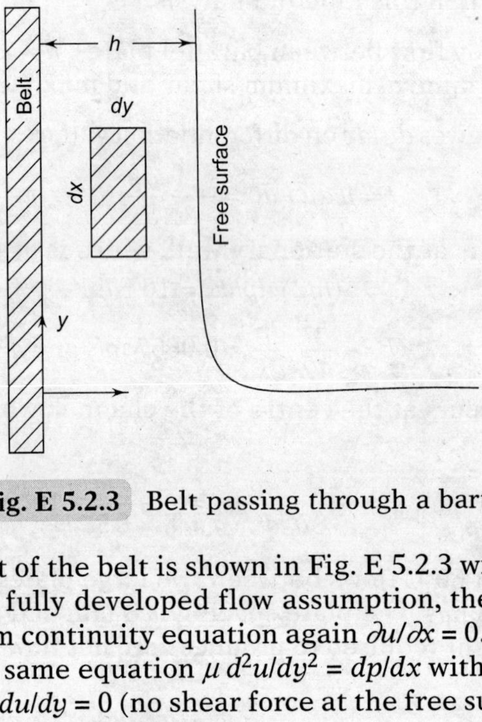


**Fig. E 5.2.3** Belt passing through a barth

**Solution** The movement of the belt is shown in Fig. E 5.2.3 with the coordinate system used for analysis. Under fully developed flow assumption, the velocity component $v$ is zero everywhere and from continuity equation again $\partial u/\partial x = 0$. So we get inertia term to be zero and this leads to same equation $\mu\, d^2u/dy^2 = dp/dx$ with the boundary conditions at $y = 0$, $u = U_0$ and $y = h$, $du/dy = 0$ (no shear force at the free surface). Integration of the equation twice with respect to $y$ and using the boundary conditions gives

$$u = \frac{\rho g y^2}{2\mu} - \frac{\rho g h y}{\mu} + U_0$$

**Example 5.2.4** Kerosene oil flows upwards through inclined plates at the rate of 2.0 l/s per metre width. The distance between the plates is 10 mm and the inclination of plates is 20° with the horizontal. Determine the difference of pressure between the two sections 10 m apart. Take $\rho = 800$ kg/m³ and $\mu = 2x = 10^{-3}$ kg/m s.

**Solution** Given: $h = 10 \times 10^{-3}$ m, $\rho = 800$ kg/m$^3$, $\mu = 2\times 10^{-3}$ kg/ms, $q = 2\times10^{-3}$ m$^3$/s. From this data, we obtain

$$\overline{U} = \frac{q}{h \times 1} = \frac{2\times10^{-3}}{10\times10^{-3}} = 0.2 \text{ m/s and } z_2 - z_1 = 10 \sin 20 = 3.42 \text{ m.}$$

Using Eq. 5.2.4[1], we get the difference of piezometric head at the two sections.

$$\left(\frac{p_1}{\gamma}+z_1\right)-\left(\frac{p_2}{\gamma}+z_2\right) = \frac{12\times2\times10^{-3}\times0.2\times10}{9810\times0.01^2} \rightarrow p_1 - p_2 = 27.31 \text{ kN/m}^2$$

**Example 5.2.5** Laminar flow takes place between the parallel plates 10 mm apart. The plates are inclined at 45° with the horizontal. For oil of viscosity 0.9 kg/ms and mass density 1260 kg/m$^3$, the pressure at two points 1.0 m apart are 80 and 250 kPa when the upper plate moves at 2.0 m/s velocity relative to the lower plate but in opposite direction to flow, see Fig. 5.2.5. Determine (a) velocity distribution, (b) the maximum velocity and (c) shear stress on the opposite plate.

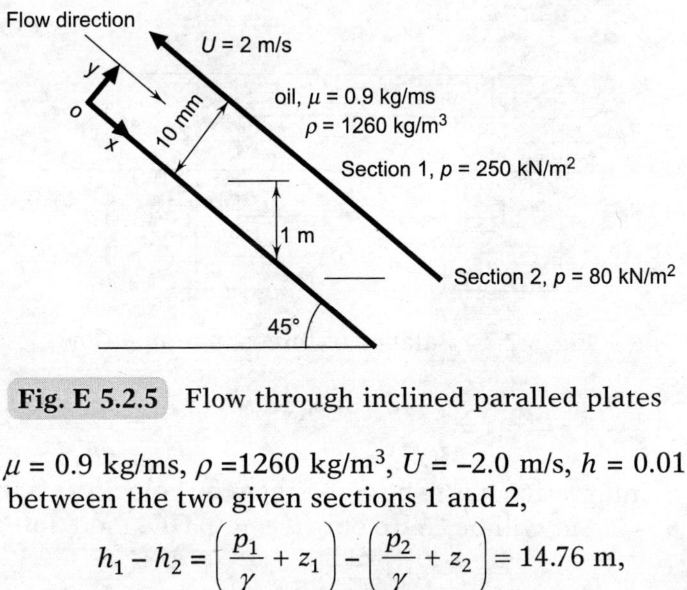


**Fig. E 5.2.5** Flow through inclined paralled plates

**Solution** Given: $\mu = 0.9$ kg/ms, $\rho = 1260$ kg/m$^3$, $U = -2.0$ m/s, $h = 0.01$ m, difference of piezometric head between the two given sections 1 and 2,

$$h_1 - h_2 = \left(\frac{p_1}{\gamma}+z_1\right)-\left(\frac{p_2}{\gamma}+z_2\right) = 14.76 \text{ m,}$$

where $\gamma = 1.26\times9.81 = 12.36$ kN/m$^3$

$L$ = distance between two points = 1/sin 45 = 1.414 m.

Since $h_1 > h_2$, the flow is in downward direction for which $\partial h/\partial x = -14.76/1.414 = -10.438$ and $\partial p/\partial x = \gamma\partial h/\partial x = -129$ kN/m$^2$. Substituting the values in Eq. 5.2.2 and simplifying, we get

$$u = 516.49\, y - 71.65\, y^2.$$

[1] While deriving this equation, the effet of gravity was not considered. Since both pressure and gravity force terms in N-S equtions are linear, these two may be combined to give piezometric head and Eq. 5.2.4 therefore has been used with change in piezometric head in place of pressure head.

Differentiating it with respect to $y$ and equating to zero gives location of the layer from the surface where the velocity is maximum. It gives, $y$ = 3.6 mm = $3.6 \times 10^{-3}$ m. Substituting it in the above equation, we get the maximum velocity = 0.93 m/s.

$$\text{Shear stress on the plate} = \mu \left.\frac{\partial u}{\partial y}\right|_{y=0.01} = -\,824.837\ \text{N/m}^2.$$

## 5.2.2 Hagen–Poiseuille Flow

Hogen and Poiseuille analysed the flow through a circular pipe due to the pressure gradient in the direction of the flow. Naturally, the cylindrical polar coordinate system would be the most suitable for analysis of this flow. The flow through pipe would be axi-symmetric, since there is no variation in velocity or pressure along the circumferential direction. This means $u_\theta = 0$. We further take the pipe axis as $x$-axis. Assuming the fluid to be incompressible and the flow to be steady, you must be able to show from *N–S* equation or balancing the forces in a pipe flow[2], as shown in Fig. 5.2.2, that for such a flow

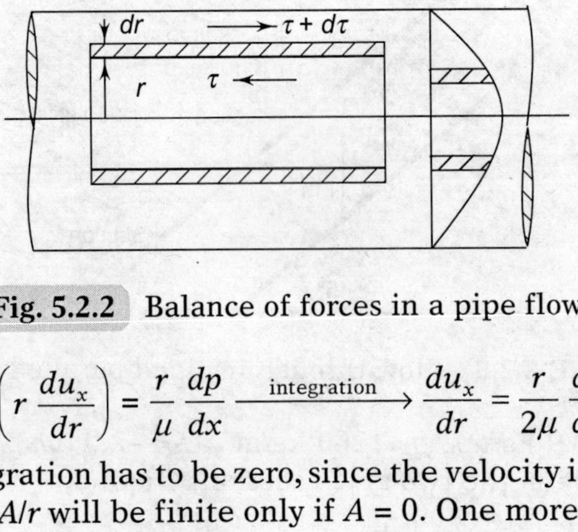


**Fig. 5.2.2** Balance of forces in a pipe flow

$$\frac{d}{dr}\left(r\frac{du_x}{dr}\right) = \frac{r}{\mu}\frac{dp}{dx} \xrightarrow{\text{integration}} \frac{du_x}{dr} = \frac{r}{2\mu}\frac{dp}{dx} + \frac{A}{r} = \frac{r}{2\mu}\frac{dp}{dx}$$

The constant of integration has to be zero, since the velocity is finite at $r = 0$. Note the term $A/r \to \infty$ as $r \to 0$. $A/r$ will be finite only if $A = 0$. One more integration gives

$$u_x = \frac{r^2}{4\mu}\frac{dp}{dx} + c, \tag{5.2.5}$$

where $c$ is constant of integration obtained from the boundary condition at $r = R$, $u = 0$. This gives

$$u_x = \frac{1}{4\mu}\left(-\frac{dp}{dx}\right)\left(R^2 - r^2\right) \tag{5.2.6}$$

Following the procedures of Section 5.2.1, we may determine the average velocity, maximum velocity, or discharge, etc. The results are

---

[2]We have studied the procedure in Section 5.2.1.

$$u_{max} = -\frac{R^2}{4\mu}\frac{dp}{dx} \quad \text{at } r = 0,$$

$$Q = \int_0^R 2\pi r u_x\, dr = -\frac{\pi R^4}{8\mu}\frac{dp}{dx} = \pi R^2 U$$

$$\rightarrow \qquad U = -\frac{R^2}{8\mu}\frac{dp}{dx} = u_{max}/2 \qquad (5.2.7)$$

and
$$-dp/dx = \frac{8\mu Q}{\pi R^4} = \frac{128\mu Q}{\pi D^4} \qquad (5.2.8)$$

If we plot the velocity at any cross section with $r$, we would observe that the velocity is symmetrically distributed around the central axis; the velocity is maximum at the axis and zero on the walls. The velocity distribution is paraboloid of revolution[3].

Some power is required to maintain the flow and can be determined from the product of the force $F$ opposing the flow and the average flow velocity $U$. If pressure at two sections is $p_1$ and $p_2$ then the force $F = (p_1 - p_2)\ \pi R^2$.

$$\therefore \qquad \text{Power} = F \times U = (p_1 - p_2) \times \pi R^2 \times U = (p_1 - p_2)\ Q$$

The total drag force can also be determined and is given by

$$D = \tau_w \times (2\pi RL) = \pi\mu R^2 L\,(-\,dp/dx),$$

since
$$\tau_w = \mu(du/dy)_{y=0} = -\,\mu(du/dr)_{y=0} = (-\,dp/dx) \times R/2.$$

where $y = R - r$ is the distance of the point from the pipe surface.

The drag or skin-friction coefficient therefore is $C_D = \dfrac{\tau_w}{\frac{1}{2}\rho U^2} = \dfrac{16}{\text{Re}}$, where $\text{Re} = \dfrac{\rho UD}{\mu}$

Darcy–Weisbach defined a friction factor by the relation $f = 4\tau_0 / \dfrac{\rho U^2}{2}$, where $\tau_0$ is the shear stress on the pipe wall and $U$ is the average velocity of the flow through the pipe. Though they defined this empirical relation for a turbulent flow yet it is equally applicable to the laminar flow. Thus, $f = 4\ C_f$. Or, $f = 64 / \text{Re}$, where Re is the Reynolds number based on diameter of the pipe and average flow velocity.

For fully developed turbulent flow it is difficult to determine such a relation. Friction factor, as found from carefully carried out experiments, depends on the Reynolds number and some measure of the surface roughness, usually expressed as non-dimensional number made with the hydraulic diameter. Since the surface roughness would vary, we use average surface roughness.

From experimental observations, we find that the flow in the pipe remains laminar up to Re = 2000 beyond which waviness starts developing in the flow and becomes complete turbulent at some higher Reynolds number. However, if we start decreasing the Reynolds number and see the limit at which the flow pattern resembles a laminar flow, it is found that this number is 4000. Thus there exists two critical numbers, called the lower critical and the upper critical numbers. We assume the flow to be laminar till

[3] A paraboloid of revolution is a three-dimensional geometry obtained by rotating a parabola about its axis by 360°. It has volume equal to ½ that of the corresponding circumscribing parallelopiped.

Re < 2000. However, the flow has been assumed to remain laminar even up to Re = 2300. That is, the upper limit of Reynolds number for a laminar flow in a circular pipe is taken from 2000 to 2300.

We defined the momentum and energy correction factors in Chapter 2. We can determine these using the velocity distribution profile given by Eq. 5.2.6. This is simple integration. The results of integration, after using the expression for average velocity as well, are

$$\alpha = 4/3 \text{ and } \beta = 2.$$

Flow through porous media has drawn our attention for a long time. Percolation of water through the pores in the ground, flow of water within the rock from one to another location within the ground, filters used in water filtration plant, pharmaceutical and chemical industries are a few examples of flow through the porous medium which are of immense importance for us. These pores are small enough but matter a lot. The pores can be modelled as circular pipes through which the flow takes place because of gravity or may even be due to pressure difference inside the earth surface or in filters. The large number of such pipes vary in diameter and length and need introduction of some constant to determine the volume flow rate per unit area of solid, expressed in terms of average velocity $\vec{U} = -(k/\mu)\nabla p^\star$.[4] $k$ is called the *permeability* and has the dimensions of length squared. It is the average of the square of the pore diameters. The unit of permeability is Darcy.

**Example 5.2.6** Determine the optimum diameter of the pipe required to carry 100 l/s of crude oil ($\rho = 950$ kg/m$^3$, $\mu = 8 \times 10^{-2}$ kg/ms) and still maintain the laminar flow. Also determine the power required for its transport.

**Solution** Given $Q = 100 \times 10^{-3} = 0.1$ m$^3$/s, $\rho = 950$ kg/m$^3$, $\mu = 8 \times 10^{-2}$ kg/ms

For laminar flow, the Reynolds number for flow through a pipe is around 2000 to 2300; say we take it 2100. This determines the optimal pipe diameter.

$$U = \frac{4Q}{\pi D^2}, \text{Re} = \frac{UD\rho}{\mu} = \frac{4Q\rho}{\pi D\mu} \rightarrow D = \frac{4 \times 0.1 \times 950}{\pi \times 8 \times 10^{-2} \times 2100} = 0.72 \text{ m.}$$

Using Eq. 5.2.8, we get the pressure drop in length $L = 1$ km = 1000 m.

$$\Delta p = \frac{128 \times Q\mu L}{\pi D^4} = \frac{128 \times 0.1 \times 8 \times 10^{-2}\ 1000}{\pi \times 0.72^4} = 1212.73 \text{ N/m}^2$$

and $\qquad P = \Delta p \times Q = 0.1213$ kW

**Example 5.2.7** For steady laminar flow through a conical pipe, as shown in figure, determine an expression for the pressure loss across it.

[4] $P^\star$ is defined as $p - \rho g.R$ and includes the gravity term. This is convenient procedure to get rid of the body term in the *N–S* equation when the fluid is incompressible and the only body force is the gravity force.

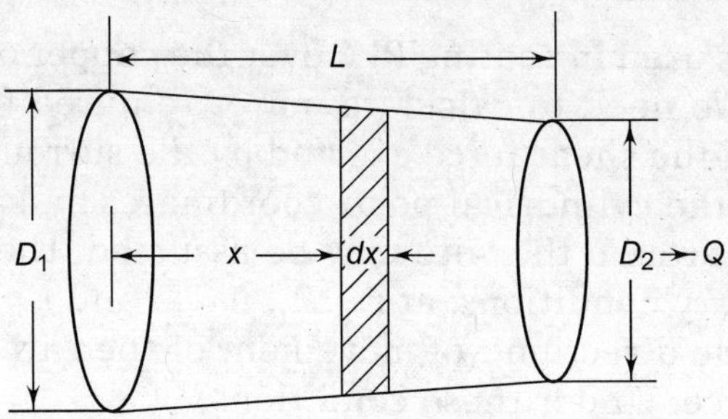


Fig. E 5.2.7 Flow through conical pipe

**Solution** Let $D$ be the diameter of the pipe at a distance $x$ from the entrance then

$$D = D_2 + (D_1 - D_2)(L - x)/L$$

This gives, $dD/dx = -(D_1 - D_2)/L$

For a laminar flow in the pipe, pressure loss is given by Eq. 5.2.8. This gives on integration

$$p_1 - p_2 = \int_{D_1}^{D_2} \frac{128\,\mu Q}{\pi D^4} \frac{dx}{dD}\, dD = \int_{D_1}^{D_2} \frac{128\,\mu Q}{\pi D^4} \frac{L}{D_1 - D_2}\, dD$$

$$= \frac{128\,\mu QL}{3\pi[D_1 - D_2]} \left[\frac{1}{D_2^3} - \frac{1}{D_1^3}\right]$$

**Example 5.2.8** A round wire of radius $R_1$ and length l is coated continuously by drawing it at a uniform velocity $U$ through a pipe of radius $R_2$ containing the coating liquid. If liquid is maintained at constant pressure everywhere, obtain an expression for power required to draw it.

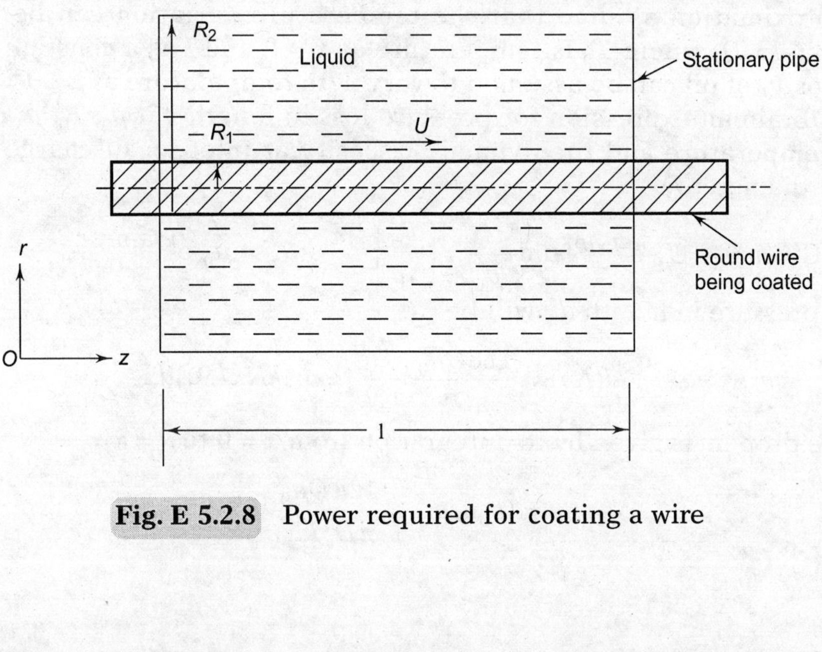


Fig. E 5.2.8 Power required for coating a wire

**Solution** This process is used in coating *PVC* over the copper or aluminium wire, enamelling of the wire, etc. We need to calculate power required for pulling (drawing) the wire that will depend on the shear force exerted by the surrounding liquid on the moving wire. We should use the cylindrical polar coordinates and an axi-symmetric, steady laminar flow conditions around the wire may be assumed. Therefore NS equations can be used with the boundary conditions: at $r = R_1$, $u = U$ and $r = R_2$, $u = 0$, where $u$ is the velocity component in the direction $z$ of movement of the wire. We may further assume that there is no body force. Under these conditions,

$\partial/\partial\theta = \partial/\partial t = \partial/\partial z = 0$, the *NS* equation would reduces to

$$\frac{\partial^2 u}{\partial r^2} + \frac{1}{r}\frac{\partial u}{\partial r} = 0, \text{ since } \mu \neq 0$$

On integration twice with respect to $r$ and using the boundary conditions at $r = R_1$, $u = U$ and at $r = R_2$, $u = 0$, we get

$$u = A \ln r + B = \frac{\mu U}{\ln (R_1/R_2)} \ln (r/R_2)$$

$$\frac{du}{dr} = \frac{U}{\ln (R_1/R_2)}\frac{1}{r}, \quad \tau_0 = \frac{\mu U}{\ln (R_1/R_2)}\left(\frac{1}{r}\right)_{r = R_1} = \frac{\mu U}{\ln(R_1/R_2)}\frac{1}{R_1}$$

$$\text{and } P = \text{work done/s} = (2\pi R_1 l) \times \frac{\mu U}{\ln(R_1/R_2)}\frac{1}{R_1} \times U = \frac{2\pi l \mu U^2}{\ln (R_1/R_2)}$$

**Example 5.2.9** When heated oil is pumped from the oil well to its destination through an non-insulated pipe line, the temperature of oil $T_x$ at any distance $x$ from the inlet decreases along the pipe length due to heat transfer to surrounding environment. As a first approximation assume that the temperature variation can be expresses as $T_x/T_0 = \exp(-kx/D)$ where $T_0$ is temperature at $x = 0$ and $k$ is a constant. Further, dynamic viscosity of oil can be assumed to vary with temperature as $\mu \sim T^{-N}$ where $N$ is a constant. Obtain an expression for pressure loss in length $x$ if $p_0$, $T_0$, $\mu_0$ represent the pressure, temperature and the dynamic viscosity at inlet for discharge $Q$ flowing in the pipe of diameter $D$.

**Solution** Given: $T_x = T_0\, e^{-kx/D}$ and $\frac{\mu_x}{\mu_0} = \left(\frac{T_x}{T_0}\right)^{-N} \rightarrow \mu_x = \mu_0\, e^{kNx/D}$

Loss of pressure in length $dx$ will be

$$d(\Delta p) = \frac{128 Q\mu_0}{\pi D^4}\frac{D}{kN}[\exp(kNx/D)]dx$$

pressure drop $\Delta p$ is given by its integration from $x = 0$ to $x = x$

$$\Delta p = (p_0 - p_x) = \frac{128\, Q\mu_0}{\pi D^3\, kN}[e^{kNx/D} - 1]$$

### 5.2.3 Flow Between Rotating Cylinders

Let us consider some fluid present between two infinite long concentric cylinders and that any or both of them be rotating in any direction with say angular velocities $\omega_1$ and $\omega_2$. In this case, the radial and axial component of velocity will be zero; only the circumferential component of the velocity will exist. We may argue as before and get the governing following equations from *N–S* equations in cylindrical polar coordinate, with the cylinder axis as *x*-axis.

$$\rho\frac{u_\theta^2}{r} = \frac{dp}{dr}$$

and $$\frac{d^2u_\theta}{dr^2} + \frac{1}{r}\frac{du_\theta}{dr} - \frac{u_\theta}{r} = 0$$

The first equation shows that the centrifugal force is caused because of the centripetal force produced by increase in pressure with *r*. The second equation can be integrated with respect to *r* two times to give the solution with two constants of integration, obtained from the boundary conditions $r = R_1$, $u_\theta = \omega_1 R_1$ and $r = R_2$, $u_\theta = \omega_2 R_2$. This gives the circumferential velocity distribution

$$u_\theta = \frac{1}{R_2^2 - R_1^2}\left[\left(\omega_2 R_2^2 - \omega_1 R_1^2\right) r + \frac{R_1^2 R_2^2 (\omega_1 - \omega_2)}{4}\right] \tag{5.2.9}$$

Our interest is to find out the torque acting on the two cylinders as well. This torque is because of the shear force acting on the annulus surfaces by the rotating fluid. Shear stress is related to shear strain rate in cylindrical polar coordinates by the relation

$$\tau_{r\theta} = \mu\left(\frac{\partial u_\theta}{\partial r} + \frac{\partial u_r}{r\partial\theta} - \frac{u_\theta}{r}\right) = \mu\left(\frac{\partial u_\theta}{\partial r} - \frac{u_\theta}{r}\right)$$

$$= \mu r\frac{d}{dr}\left(\frac{u_\theta}{r}\right) = \frac{2\mu}{R_2^2 - R_1^2}(\omega_1 - \omega_2)\frac{R_1^2 - R_2^2}{r^2}$$

$$\rightarrow \qquad \tau_{r\theta}\big|_{r=R_1} = \frac{2\mu R_2^2}{R_2^2 - R_1^2}(\omega_1 - \omega_2) \text{ and } \tau_{r\theta}\big|_{r=R_2} = \frac{2\mu R_1^2}{R_2^2 - R_1^2}(\omega_1 - \omega_2) \tag{5.2.10}$$

If cylinder 1 is at rest ($\omega_1 = 0$) then the torque exerted on the second cylinder is equal to moment of the shear force on it. Using relations from Eq. 5.2.10,

$$T_2 = \frac{2\mu R_1^2\,\omega_2}{R_2^2 - R_1^2} \times 2\pi R_2 L \times R_2 = 4\pi\mu L\frac{R_1^2 R_2^2 \omega_2}{R_2^2 - R_1^2} \tag{5.2.11}$$

The same torque would act on cylinder 1 but in opposite direction. This also means that external torque of this magnitude is to be applied to keep it fixed. This arrangement has been used to determine viscosity of the fluid and was discussed in Chapter 1. F. M. White has discussed how to handle the flow in the space between the two eccentric circular cylinders, called eccentric annulus. The procedure is to use conformal transformation to change it into concentric annulus. The transformation used for the purpose is

$$w = y + iz = M\tan\varsigma/2, \text{ where } \varsigma = \xi + i\eta,\ M = \sqrt{F^2 - a^2} \text{ and } F = (a^2 - b^2 + c^2)/(2c)$$

Here $a$, $b$, and $c$ respectively are the radii of outer and inner cylinders and the eccentricity in their axes. Muskhelishvili proposed this transformation.

## 5.3 Low Reynolds Flow

It is possible to neglect inertia term in case of flows at low Reynolds number, since its magnitude may be very small compared to the viscous force. This case belongs to the second kind of solution—approximate solution of the exact equation. The resulting equations will then be linear in nature and can be handled more comfortably. A few such cases are discussed below.

### 5.3.1 Stokes Flow

Stokes studied flow around a sphere of radius '$a$' placed in a uniform flow having low velocity $U_\infty$, for which the Reynolds number is around 1, such that the inertia terms can be neglected. Only gravity is the body force and can be included with pressure term, represented by $p^\star = p + \rho g z$. *N–S* equation takes the form

$$\mu\nabla^2\vec{U} = \nabla p^\star \text{ with boundary conditions } \vec{U} = 0 \text{ at } \vec{r} = a \text{ and } \vec{U} \to U_\infty \text{ as } \vec{r} \to \infty \quad (5.3.1)$$

Such a flow is axisymmetric and use of spherical coordinates because of the geometry of the object is recommended. This equation requires two times integration and use of the given boundary conditions to get the unique solution. Without going into the details of the solution technique (that you should be able to do) the results are

$$u_r = U_\infty \cos\phi\left(1 - \frac{3a}{2r} + \frac{a^3}{2r^3}\right) \quad \text{and} \quad u_\phi = U_\infty \sin\phi\left(-1 + \frac{3a}{4r} + \frac{5a^3}{4r^3}\right) \quad (5.3.2)$$

$$p^* = p^*_\infty - \frac{3\mu a U_\infty \cos\phi}{2r^2} \quad \text{and} \quad \tau_{\phi r} = \frac{\mu U_\infty \sin\phi}{r}\left(1 - \frac{3a}{4r} + \frac{5a^3}{4r^3}\right) \quad (5.3.3)$$

$p^\star_\infty$ is measured at far off distance in the undisturbed flow. Equation 5.3.3 reveals that the pressure on the front side of the sphere is more as compared to its rear side pressure. Thus, the pressure force acts along the direction of the flow. Viscous force also acts on the surface but in the opposite direction. Both the pressure and the viscous forces can easily be determined. You should try and verify that the viscous force is only 1/3$^{rd}$ of the total drag force on sphere. We have learned how to calculate the pressure force on the surface. Similarly, the viscous force is determined by determining the shear stress from Newton's law of viscosity. The total force on the sphere comes out to

$$F = 6\pi\mu a U_\infty \quad (5.3.4)$$

This relationship is the *Stokes law* that holds true when the Reynolds number is less than 0.1. It helped in experimental determination of the electric charge on an electron in Millikan's oil drop experiment and Boltzman constant using Einstein theory on Brownian movement. Stokes law is also helpful in determining viscosity of fluid in laboratory. A small spherical steel ball is made to fall down under gravity in a tank full of the liquid whose viscosity is to be determined. The gravity moves the ball down which is opposed by the buoyant and the viscous forces. As the fall velocity of the ball increases the later force increases; the net force on the ball is zero when the ball attains constant velocity. The ball then attains the *terminal velocity*, which remains constant during subsequent motion and is measured by recording the time for a known vertical fall. Under these conditions, the equilibrium equation is

$$\frac{4\pi}{3}a^3\rho_s g = \frac{4\pi}{3}a^3\rho_f g + 6\pi\mu a U_\infty \tag{5.3.4.a}$$

We will just make a comment that the drag force does not depend significantly on the shape and orientation of the body in the flow. The analysis carried out on a circular disc of diameter $d$ shows that the drag force is 16 $\mu a\ U_\infty$ and (32/3)$\mu a\ U_\infty$ when placed perpendicular and parallel to the flow direction, respectively. Comparison of their magnitudes confirms our comments.

**Example 5.3.1** A viscometer consisting of a long vertical tank containing the fluid at the desired ambient conditions and a small sphere is employed to determine the viscosity of oil. If a steel sphere of specific gravity 8 and 1 cm diameter falls with a constant velocity of 0.05 m/s through a lubricating oil of specific gravity 0.9, calculate its viscosity.

**Solution** We are given $a$ = 0.5/100 = 0.005 m, $\rho_s$ = 8000, $\rho_f$ = 900 and $U_\infty$ = 0.05 m/s. Substitution in Eq. 5.3.4. (a) gives $\mu$ = 7.74 Ns/m². Let us also check the Reynolds number in this case, which is

$$\text{Re} = \frac{\rho_f U_\infty d}{\mu} = \frac{900 \times 0.05 \times .01}{7.74} = 0.058$$

Value of Re < 1 justifies use of Eq. 5.3.4(a).

## 5.3.2 Hydrodynamic Lubrication

This is another flow that has considerable importance in industry and considers determination of pressure and velocity field in the fluid confined between two non-parallel plates with one plate fixed and the second moving. Moreover, the gap between the plates is much less as compared to other plate dimensions and the inclination between the plates is also very small. Situation in bearings, where the shaft (slipper) rotates in its bearing, can be approximated as fluid confined between the two plates inclined at an angle $\alpha$ to each other having gap $h_1$ and $h_2$ at the ends, as shown in Fig. 5.3.1. This is so because the radius of the shaft is much larger than the gap between the shaft and the bearing.

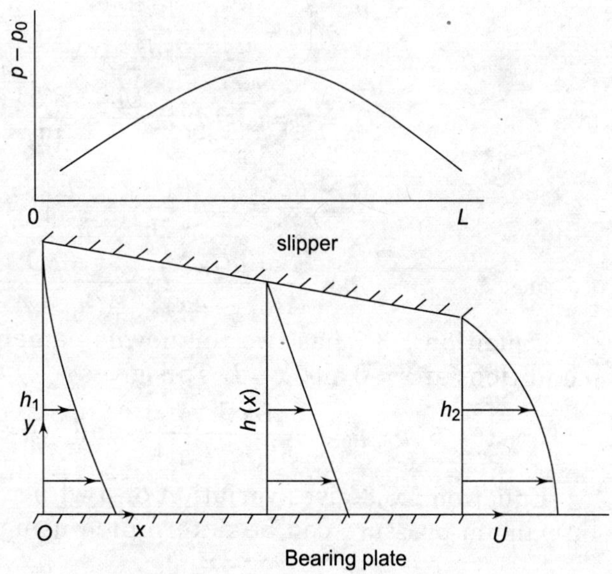


**Fig. 5.3.1** Modelling of a bearing

For analysis, let us take the coordinate system as shown in the figure. Then the gap $h$ at any distance $x$ is $h_1 - (h_1 - h_2)\, x/L$, where $L$ is the length and $h/L << 1$. We assume the lower plate moving with velocity $U$ and the upper plate fixed. Let us have an estimate of the inertia and viscous force to check if the inertia force may be neglected as compared to the viscous force. This is done in the following manner

$$\frac{\text{Inertia force}}{\text{viscous force}} = \frac{\rho u(\partial u / \partial x)}{\mu \partial^2 u / \partial y^2} = \frac{\rho U^2 / L}{\mu U / h^2} = \frac{\rho U L}{\mu}\left(\frac{h}{L}\right)^2 <<< 1$$

Thus the inertia force, though not completely absent as in Couette flow, is negligible in this case. This difference separates this flow from the Couette flow and there is difference too in the boundary conditions, which are:

$$y = 0,\ u = U, \text{ at } y = h\,(x),\ u = 0, \text{ at } x = 0,\ p = p_0 \text{ and at } x = L,\ p = p_0.$$

To be exact, the pressure variation at the inlet and exit sections will not be constant and there would be some variation; this variation however is ignored because of small magnitude of variation owing to small film thickness.

We may also neglect the terms $\partial^2 u/\partial x^2$ as compared to the term $\partial^2 u/\partial y^2$ because of its low magnitude. Further, the $v$ component of velocity and its derivatives will almost be non-existent, the $y$-momentum equation is satisfied automatically. With these simplifications, the momentum equation is reduced to

$$\mu \frac{d^2 u}{dx^2} = \frac{dp}{dx}$$

We infere that in this case too, pressure is not a function of $y$. Like before, we can integrate it with respect to $y$ and get the following results using the given boundary conditions

$$\frac{u}{U} = \left(1 - \frac{y}{h}\right) - \frac{h^2}{2\mu U}\frac{dp}{dx}\left(1 - \frac{y}{h}\right)\frac{y}{h} \tag{5.3.5}$$

$$Q_x = Q = \int_0^{h(x)} u\, dy = \frac{Uh}{2} - \frac{p' h^3}{12\mu}, \text{ where } p' = \frac{dp}{dx}$$

$$\rightarrow \qquad p' = \frac{dp}{dx} = 12\mu\left(\frac{U}{2h^2} - \frac{Q}{h^3}\right) \tag{5.3.6}$$

or,
$$p = \frac{6\mu U}{A(h_1 - Ax)} - \frac{6\mu Q}{A(h_1 - Ax)^2} + B, \text{ where } A = (h_1 - h_2)/L \tag{5.3.7}$$

Equaition 5.3.7 has two unknowns namely $B$ and $Q$, which are obtained using the conditions at $x = 0$ and $x = L$. This gives

$$p - p_0 = \frac{6\mu U x (h - h_2)}{h^2 (h_1 + h_2)} \quad \text{and} \quad Q = \frac{U h_1 h_2}{h_1 + h_2} \tag{5.3.8}$$

Equation 5.3.8 gives variation of $p$ with $x$, which is shown in Fig. 5.3.1. The point of maximum pressure can be determined using calculus. At the point of maximum pressure $dp/dx = 0$. Therefore, Eq. 5.3.5 gives $\frac{u}{U} = 1 - \frac{y}{h}$. Thus, at this section velocity varies linearly.

Because of this pressure, a force $F$ acts in the upward direction on the upper plate that supports it. The force $F$ as well as the shear force $D$ per unit width of the plate can be evaluated as follows

$$F = \int_0^L (p - p_0)dx = \frac{6\mu UL^2}{(h_1 - h_2)^2}\left[\ln\frac{h_1}{h_2} - 2\left\{\frac{h_1 - h_2}{h_1 + h_2}\right\}\right] \tag{5.3.9}$$

$$\tau_0 = -\mu\frac{\partial u}{\partial y}\bigg|_{y=0} = 4\mu\frac{U}{h} - \frac{6\mu Uh_1h_2}{h^2(h_1 + h_2)} \tag{5.3.10}$$

$$D = \int_0^L \tau_0\, dx = \frac{\mu UL}{(h_1 - h_2)}\left[4\ln\frac{h_1}{h_2} - 6\frac{h_1 - h_2}{h_1 + h_2}\right] \tag{5.3.11}$$

We note that $h_1 - h_2$ should be small in order that the supporting force may be large. The Michell thrust bearing works on this principle of hydrodynamic lubrication. As said, this is an idealisation, since the actual curves of the shaft will be curved.

**Example 5.3.2** A journal bearing supports a circular shaft of diameter $D = 10$ cm turning at 3600 rpm. The bearing length $L = 10$ cm. The gap between the shaft and the bearing, $h = 0.1$ mm, is filled with a lubricant whose viscosity is $\mu = 6.7 \times 10^{-5}$ Pas. Calculate the torque applied to the shaft to overcome the friction in the bearing and the power $P$ consumed in the bearing by the friction.

**Solution** Since the gap at the bearing end is not given, we shall assume it to be constant. In that case the problem is changed to that of Couette flow with the shaft as moving plate of length 10 cm. The shear stress exerted on the shaft having radius $R = D/2$ is

$\tau_s = \mu U/h = \mu\omega R/h$ and $T$, torque on the shaft $= \tau_s\,(2\pi RL) \times R$

$$= \pi\mu\omega LD^3/(4h) = \frac{\pi \times 6.7 \times 10^{-5} \times (2\pi \times 3600/60) \times 0.1 \times 0.1^3}{4 \times 10^{-4}} = 1.994 \times 10^{-2}\text{ Nm}$$

$P = T\omega = 7.479$ W

## 5.4 Boundary Layer

When the fluid flows over the surface then the layer of the fluid in contact with the surface acquires the velocity of the surface, a condition called *no-slip* condition. The velocity of the successive layers goes on increasing along normal to the surface, though the rate of increase decreases as we move away from the surface. That is, the velocity plot has maximum slope at the surface and it ultimately reduces to zero at some point away from the surface; theoretically this point will be at infinity. But in reality variation is fast initially and after some distance, a few mm, the velocity of the layers is almost same. In other words, the presence of the surface is felt by the fluid layers up to certain distance normal to the surface and all the layers above it move as if no surface is present.

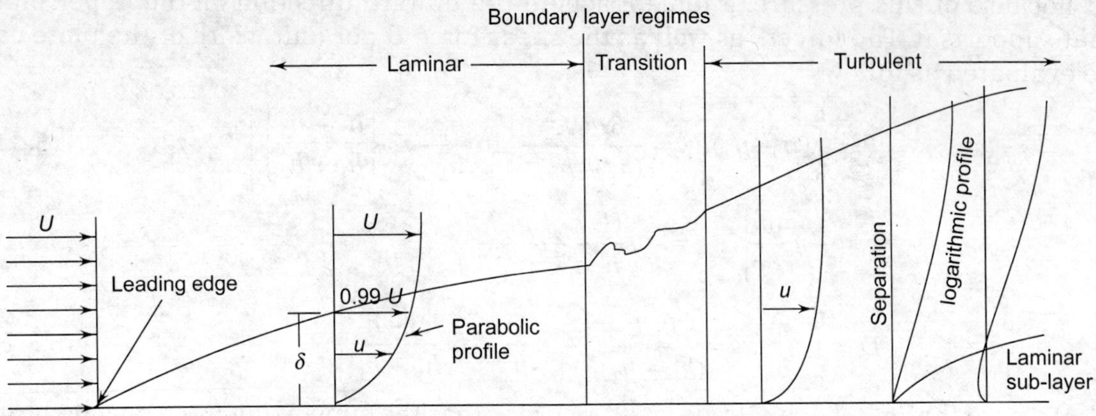


**Fig. 5.4.1** Boundary layer development over a flat plate

Figure 5.4.1 shows the experimental results of Prandtl obtained for a flow over a flat plate. Velocity component $u$ (along the flow direction) was measured at a section in the direction perpendicular to the surface and was plotted at that section. Such measurements were made and plotted at different sections of the plate; the figure shows these at a few sections. These plots support the observations mentioned earlier. Such behaviour is explained recognising the fact that the effect of viscosity is to oppose the motion. Thus the fluid layer velocity is decreased as we approach the surface. We can also explain the growth of the boundary layer from the fact that the slowness of layer decreases the mass flow rate and in order to satisfy the mass conservation, the area of flow should increase and that's why thickening of the layer along the flow direction.

Prandtl argued that the effect of viscosity is limited to some distance normal to the surface and he divided the flow into two zones—the *boundary layer* close to the surface and *potential flow zone* away from the surface. To demarcate the two zones, he took arbitrarily a point at a distance such that the $u$ component at this point is equal to 99 percent of the free stream velocity $U$ and called distance of this point from the surface as the *boundary layer thickness* at that section. Thus he obtained a number of such points at different sections and joined these to give a demarking line, as shown in the figure. We make the following observations:

1. Boundary layer grows along the flow direction. This is because the layers in contact with the surface becomes slow and therefore more area is required to satisfy the mass conservation.
2. The flow within the boundary layer is laminar up to certain distance at which section the Reynolds number is about $3.2 \times 10^5$. At this point, the boundary layer becomes wavy and remains so till the Reynolds number is about $5.5 \times 10^5$. Finally flow in the boundary layer turns to turbulent flow. These two Reynolds number based on the distance $x$ of the section from the leading edge are known in literature as lower and upper critical numbers and the flow between the two sections as *transition region*.
3. As we move down the plate, Reynolds number increases and the slope of the velocity profile near the surface decreases; at some point it is zero. At this point the boundary layer is ready to leave the surface and this point is called the point of

separation. Separation occurs because the fluid particle does not have enough kinetic energy to overcome the frictional force. At the point of separation, $\partial u/\partial n = 0$, where $n$ represents the normal to the surface; for a flat plate $n$ is $y$.

4. Further downstream, there is another sub-layer close to the surface in which velocity is negative. The fluid particles leave the main flow surface and circulate in this sub-layer trying to go with the main flow getting energy from it.
5. Velocity profile in laminar boundary layer is parabolic whereas in turbulent boundary layer it is logarithmic.
6. The boundary layer thickness varies with Reynolds number in the following manner

   For laminar flow $\dfrac{\delta}{x} = \dfrac{5}{\sqrt{\mathrm{Re}_x}}$ and for turbulent flow $\dfrac{\delta}{x} = \dfrac{0.37}{\mathrm{Re}_x^{0.2}}$

### 5.4.1 Displacement, Momentum and Energy Thickness

We discussed formation of the boundary layer zone near the surface placed in the flowing fluid, which has thickness that increases in the direction of the flow. At the end of the boundary layer velocity of the layer is 0.99 $U$, where $U$ is the mainstream velocity. Since the layers in the boundary layer region move slow, there is loss of mass, momentum and energy flux rate. For that reason, we think of an imaginary distance by which the plate may be moved normal to it in an uniform flow field having velocity $U$ such that the decrease in the these quantities (mass, momentum, or energy) due to displacement of the plate is equivalent to that caused by the decrease in velocity in the boundary layer flow.

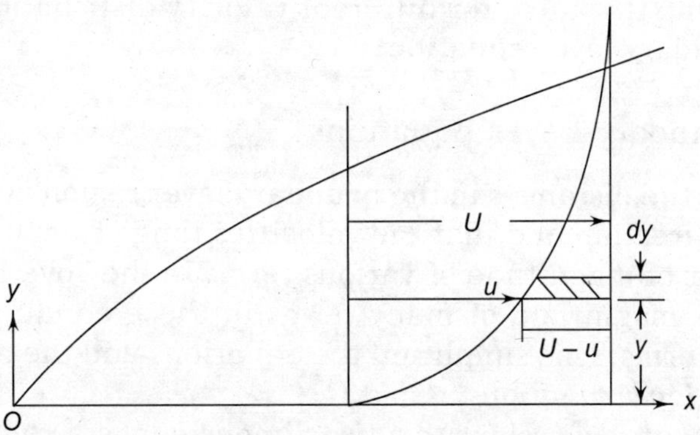


**Fig. 5.4.2** Displacement thickness

Let us consider a boundary layer flow over the flat plate of unit width along $z$-direction, which is shown in Fig. 5.4.2, and a layer of thickness $dy$ at a distance $y$ from the surface. Flow velocity at this layer is $u$ whereas the fluid velocity in the potential flow region is $U$. Thus reduction in mass flow rate through this strip due to loss in velocity is $\rho(U-u)\,dy$, equal to the area of the hatched strip in the figure. The total loss of mass flow rate is then obtained by integrating this term from $y = 0$ to $\infty$. This gives:

$$\text{Loss in mass flow rate} = \rho U\delta^* = \int_0^\infty \rho(U-u)\,dy \rightarrow \delta^* = \int_0^\infty (1-u/U)\,dy \qquad (5.4.1)$$

where $\delta^*$ is the distance by which the plate has been shifted into the flow so that the resulting loss in the mass flow rate is equivalent to that caused by the formation of the boundary layer. This thickness is called *displacement thickness* for the obvious reason. We may now define similarly the momentum and energy thickness, represented by $\delta_m$ and $\delta_e$.[4] Note that the loss in momentum and energy corresponding to the loss in mass $\rho\,(U-u)\,dy$ in the boundary layer is $\rho\,(U-u)\,u\,dy$ and ½ $\rho\,(U-u)\,u^2\,dy$, respectively whereas that by shifting the plate in uniform flow is $\rho U^2\,dy$ and ½ $\rho U^3\,dy$, respectively. Equating these and simplifying we get

$$\delta_m = \int_0^\infty (u/U)\,(1-u/U)\,dy \quad \text{and} \quad \delta_e = \int_0^\infty (u/U)^2\,(1-u/U)\,dy \tag{5.4.2}$$

**Note:** Though the integration is from 0 to $\infty$, the integral will be always zero when $y > \delta$, since the factor $(1 - u/U)$ is negligible. For all practical purpose, the integration limit may be taken up to $\delta$.

These thickness are not arbitrary like the boundary layer thickness and are related to the velocity distribution in boundary layer flow. Dependence of $u/U$ on $\delta$ brings in dependence of various displacement thicknesses on $\delta$; a rough or approximate idea can be obtained from the following table:

| Type of boundary layer | Displacement thickness | Momentum thickness |
|---|---|---|
| Laminar | $\delta/3$ | $\delta/8$ |
| Turbulent | $\delta/8$ | $7\delta/72$ |

We shall determine thickness for different velocity distributions while studying the solution of the boundary layer equation.

## 5.4.2 Prandtl Boundary Layer Equations

Prandtl argued that the distances in the boundary layer region are very small compared to that in the flow direction and that the velocities range from 0 to $U$. So he thought of comparing the order of magnitude of various terms in the governing equations. He observed that relative magnitude of many terms in these equations is negligibly small compared to other terms. This simplified the equations and the resulting equations are known as *boundary layer equations*.

We shall restrict our discussions to a two-dimensional-steady flow of an incompressible fluid over a plate just for simplicity to explain the procedure; in fact it is equally applicable for a three-dimensional flow, compressible fluids, or the flow on a curved surface. We non-dimensionalise the lengths $x$ and $y$ along $x$ and $y$-direction with the length of the plate $L$, the velocity components $u$ and $v$ by the potential flow velocity $U$ and the pressure by the dynamic pressure $\rho U^2$, in order to reduce the governing equations to a non-dimensional form. Many authors prefer to use different symbols for the non-dimensional variables to avoid confusion but still others have retained the same symbols; the form of equation itself speaks off. Let us rewrite the governing equations in the non-dimensional form retaining the old symbols for non-dimensional variables:

[5] We have used symbol $\theta$ also for the momentum thickness in the text.

$$\frac{\partial u}{\partial x} + \frac{\partial v}{\partial y} = 0,$$

$$u\frac{\partial u}{\partial x} + v\frac{\partial v}{\partial y} = \frac{\partial p}{\partial x} + \frac{1}{\text{Re}}\left(\frac{\partial^2 u}{\partial x^2} + \frac{\partial^2 u}{\partial y^2}\right), \text{ and}$$

$$u\frac{\partial v}{\partial x} + v\frac{\partial v}{\partial y} = \frac{\partial p}{\partial y} + \frac{1}{\text{Re}}\left(\frac{\partial^2 v}{\partial x^2} + \frac{\partial^2 v}{\partial y^2}\right), \tag{5.4.3}$$

Re is the Reynolds number = $UL/\nu$ whose appearance reflects that the equation is written in non-dimensional form. For the boundary layer, $x$ varies from 0 to 1 and we say it is of the order 1, written as $x \sim O(1)$. $y << x$ while $y$ varies from 0 to $\delta/L$ and we say $y \sim O(\delta)$. Similarly $u \sim O(1)$. By the order of the magnitude, we do not mean that all the terms in Eq. 5.4.3 necessarily will have the maximum value at the same point. There is no reason for all of them to be maximum at one point only. This raised a grave question about the validity of comparing the order of the terms in the equations, since Eq. 5.4.3 is applicable at a point whereas its terms have maximum magnitude, which are compared in order to delete small terms, at different points at a section. This issue has been dealt with Meksyn in his book. Nonetheless, the closeness of the results obtained from Prandtl boundary layer theory for different flows gives us confidence in his arguments and we shall proceed to discuss his solution.

We may now get the order of various terms in the equations. Order of the derivatives of $u$ are $O(\partial u/\partial x) \sim 1$, $O(\partial u/\partial y) \sim 1/\delta$, $O(\partial^2 u/\partial x^2) \sim 1$, and $O(\partial^2 u/\partial y^2) \sim 1/\delta^2$. For example, take $\partial^2 u/\partial y^2$. Here $u \sim O(1)$ and therefore the order of change in $u$ is also 1 as it changes from 0 to 1. Similarly the order of $y$ is $\delta$ and so is the order of change in $y$ as it changes from 0 to $\delta$. Therefore, $O(\partial^2 u/\partial y^2.) = 1/\delta^2$. This way order of all the terms in the equation can be obtained.

Substituting the order in continuity equation we get, $O(\partial v/\partial y)$ should also be 1 to satisfy this equation since $O(\partial u/\partial x)$ is 1. This means $O(v) \sim \delta$, since $y$ is of the order $\delta$. Now we can determine the order of the derivatives of $v$, these are: $O(\partial v/\partial x) \sim \delta$, $O(\partial v/\partial y) \sim 1$, $O(\partial^2 v/\partial x^2) \sim \delta$, and $O(\partial^2 v/\partial y^2) \sim \delta \times (1/\delta^2) = 1/\delta$. Order of each term has been written below the term in Eq. 5.4.4.

$$\text{X-momentum}: \quad \underset{1\times 1}{u\frac{\partial u}{\partial x}} + \underset{\delta\times 1/\delta}{v\frac{\partial u}{\partial y}} = \frac{\partial p}{\partial x} + \frac{1}{\text{Re}}\left(\underset{1}{\frac{\partial^2 u}{\partial x^2}} + \underset{1/\delta^2}{\frac{\partial^2 u}{\partial y^2}}\right)$$

$$\text{Y-momentum}: \quad \underset{1\times \delta}{u\frac{\partial v}{\partial x}} + \underset{\delta\times 1}{v\frac{\partial v}{\partial y}} = \frac{\partial p}{\partial y} + \frac{1}{\text{Re}}\left(\underset{\delta}{\frac{\partial^2 v}{\partial x^2}} + \underset{1/\delta}{\frac{\partial^2 v}{\partial y^2}}\right) \tag{5.4.4}$$

We observe from the $x$-momentum equation that

(i) $O(\partial^2 u/\partial y^2) >> O(\partial^2 u/\partial x^2)$, since $\partial^2 u/\partial y^2$ is of $O(1/\delta^2)$ whereas $\partial^2 u/\partial x^2$ is of $O(1)$. Therefore $\partial^2 u/\partial x^2$ can be neglected from the viscous force terms.

(ii) The other terms in this equation are of the order 1. This implies that 1/Re shoud be of $O(\delta^2)$ so that viscous force term may be of $O(1)$ so as to retain it in the equation. This means the flow should have high Reynolds number.

Similarly, we may go ahead with the $y$-momentum equation and note that $\partial^2 v/\partial x^2$ may be neglected compared to $\partial^2 v/\partial y^2$ and that $\partial p/\partial y$ should be of the order $\delta$, since all the other terms in the equation are of this order. This is an important conclusion, as it means the variation for the pressure in the $y$-direction can be neglected and pressure at any section will be the pressure in the potential flow zone, which can be determined using any of the methods discussed in Chapter 3. In other words, the *external pressure is impressed on the boundary layer*. This also reduces one unknown as $p$ is now known and is a function of $x$ alone. Lastly, the terms in $y$-momentum equation is of the order $\delta$ whereas in $x$-momentum equation, these are of the order 1. Therefore we may drop $y$-momentum equation even. This results, reverting back to the dimensional coordinates, in

$$\frac{\partial u}{\partial x}+\frac{\partial v}{\partial y}=0, \qquad \text{and} \qquad u\frac{\partial u}{\partial x}+v\frac{\partial v}{\partial y}=-\frac{1}{\rho}\frac{dp}{dx}+\nu\frac{\partial^2 u}{\partial y^2} \tag{5.4.5}$$

These equations are known as Prandtl boundary layer equations, which are parabolic type. These equations however are not applicable in the region close to the leading edge where $y$ is comparable with $x$. Schmidt and Schroeder made the following objections to this simplified derivation of the boundary layer equations. These are

(i) $\partial u/\partial x$ is not of $O$ (1) everywhere when $u$ is increasing rapidly.

(ii) Even if the order of the terms $u\ \partial u/\partial x$ and $v\ \partial u/\partial x$ be one, does this mean that their sum will also be of the same order 1.

(iii) In case of curved surfaces $\partial u_s/\partial n$ need not preserve the sign across the boundary layer.

They too gave a rigorous treatment to the boundary layer flow.

### 5.4.3 Solution of Boundary Layer Equations

We know the ordinary differential equations have many advantages over the partial differential equations of the same order because the number of boundary conditions required to get unique solution is equal to the order of the equation. For a partial differential equation, these are needed at infinite points of the boundary of the surface enclosing the region, depending on the natúre of the equation. Moreover, the subject of solving the *ODE* has been exhausted to a great extent and we have the solution for any equation. Therefore many researchers have made efforts to find out some independent variable to combine two of them, namely $x$ and $y$, into one. This possibility saw the light because the velocity profiles in the boundary layer at different sections are similar. This mean that similar points at different sections have the same non-dimensional $u$-component. The velocity is made non-dimensional using the free stream velocity at that section whereas the similar points at different sections are obtained using a scale factor $g(x)$ for $y$-coordinate. This way, it could be possible to describe a single velocity profile that describes the variation of velocity across the boundary layer at all the sections in the laminar boundary layer. Schlichting has discussed this problem in length and has obtained conditions for the type of flow that can have similar solutions.

Blasius was studying flow over a flat plate. He used the principle of similarity, which demands that the velocity at two points on any two arbitrary sections at $x_1$ and $x_2$ should satisfy the equation

$$\frac{u[x_1, y/g(x_1)]}{U(x_1)} = \frac{u[x_2, y/g(x_2)]}{U(x_2)} \tag{5.4.6}$$

$g(x_1)$ and $g(x_2)$ are the scaling factors for $y$ direction, which, in general, will be different. It is possible to write

$$\frac{u}{U} = f(\eta), \text{ where } \eta = y\sqrt{\frac{U}{\nu x}} = \frac{y}{\delta}, \text{ and } \delta = \sqrt{\frac{\nu x}{U}} \tag{5.4.7}$$

$$\therefore \qquad \psi = \int u\,dy = \int Uf(\eta)\delta\,d\eta = \sqrt{U\nu x}\int f(\eta)\,d\eta = \sqrt{U\nu x}\, f + \text{constant} \tag{5.4.8}$$

constant is taken as zero. We note from Eq. 5.4.7 that value of scaling factor $g$ depends on $x$ value of the section, the fluid property $\nu$ and flow velocity $u$. We now differentiate Eq. 5.4.8 to express different derivatives in the Prandtl b.l. equation as derivative of $\eta$. While differentiating with respect to $x$, we note that $\psi$ is the product of two functions and care must be taken. Equation 5.4.7 gives

$\frac{\partial \eta}{\partial x} = -\frac{1}{2}\frac{y}{\delta}\frac{1}{x}$ and $\frac{\partial \eta}{\partial y} = \frac{1}{\delta}$, respectively. The process yields

$$u = \frac{\partial \psi}{\partial y} = \frac{\partial \psi}{\partial \eta}\frac{\partial \eta}{\partial y} = Uf' \qquad v = -\frac{\partial \psi}{\partial x} = -\frac{\partial \psi}{\partial \eta}\frac{\partial \eta}{\partial x} = \frac{1}{2}\sqrt{\frac{\nu U}{x}}[\eta f' - f] \tag{5.4.9}$$

$$\frac{\partial u}{\partial x} = -\frac{U}{2}\frac{\eta}{x}f'' \qquad \frac{\partial u}{\partial y} = U\sqrt{\frac{U}{\nu x}}\, f'' \tag{5.4.10}$$

$$\text{and} \quad \frac{\partial^2 u}{\partial x^2} = \frac{U^2}{\nu x}f''' \tag{5.4.11}$$

The number of primes over $f$ represents the order of derivative with respect to $\eta$. Substituting these in Eq. 5.4.5 and noting that for a flat plate, $dp/dx = 0$, after simplification we get

$$2f''' + ff'' = 0 \tag{5.4.12}$$

This is a third order non-linear ordinary differential equation. The boundary conditions are:

$$y = 0,\ u = v = 0 \rightarrow \text{at } \eta = 0,\ f = f' = 0.$$

$$\text{At} \qquad \eta \rightarrow \infty,\ u \rightarrow U \rightarrow \text{at } \eta \rightarrow \infty,\ f' \rightarrow 1. \tag{5.4.13}$$

The boundary conditions on $f$ and its derivatives are obtained from Eq. 5.4.9 with the help of Eq. 5.4.13. For example, $u = 0$ means $f' = 0$ and then $v = 0$ in association with $u = 0$ means $f = 0$ and so on. This is a two-point boundary value problem, which is solved by *matching technique.* The solution was given by Blasius and named after him. He obtained a series solution near $\eta = 0$ and an asymptotic solution in the far field ($\eta \rightarrow \infty$). The *power series* near $\eta = 0$ is assumed to be of the form

$$f = A_0 + A_1\eta + \frac{A_2}{2!}\eta^2 + \frac{A_3}{3!}\eta^3 + \frac{A_4}{4!}\eta^4 + \cdots = \sum_{i=0}^{\infty}\frac{A_i}{i!}\eta^i \tag{5.4.14}$$

This give the first, second and third derivatives on differentiation with respect to $\eta$. Substituting these in Eq. 5.4.12, we get

$$A_3 + 2A_4\,\eta + \left(A_2^2 + 2A_5\right)\frac{\eta^2}{2!} + (4A_2 A_3 + 2A_6)\frac{\eta^3}{3!}\cdots = 0 \tag{5.4.15}$$

This equation is satisfied if the coefficient of powers of $\eta$ are zero. This gives:

$$A_3 = A_4 = A_6 = A_7 = 0,\ A_5 = -1/2\,A_2^2,\ A_8 = -11/4\,A_2^3, \text{ etc.}$$

Therefore the series solution near $\eta = 0$ is:

$$f = \sum_{n=0}^{\infty}\left(-\frac{1}{2}\right)\frac{C_n A_2^{n+1}}{(3n+2)!}\eta^{3n+2} \tag{5.4.16}$$

In which $C_0 = 1$, $C_1 = 1$, $C_2, = 11$, $C_3 = 375$, $C_4 = 27897$, $C_5 = 3{,}817{,}137$, etc. We do not need more terms in the series.

Now let us try to find out the asymptotic solution at $\eta \to \infty$. The procedure is to assume an asymptotic solution of the type

$$f = f_1 + f_2 + f_3 + \cdots, \tag{5.4.17}$$

where $f_2 << f_1, f_3 << f_2$, and so on i.e. the successive approximations decrease very fast. We determine functions $f_1$, $f_2$, etc in succession by first approximating the solution to $f = f_1$ and then to $f = f_1 + f_2$, and so on. Using the b.c. at $\eta = \infty$, we may write at a far off point from the surface in the main flow.

$$u/U = f' = f_1' = 1 \to f_1 = \eta - \beta$$

$\beta$ is a constant to be determined; we note from this equation that $f_1'(\infty) = f_1''(\infty) = 0$. Now include second term $f_2$ of the asymptotic solution so that $f = (\eta - \beta) + f_2$ and substitute in Eq. 5.4.12. We get

$$2f_2''' + (\eta - \beta)f_2'' = 0 \xrightarrow{\text{integration}} \ln f_2'' = -\frac{\eta^2}{4} + \frac{\beta\eta}{2} + C = \frac{\eta^2}{4} + \frac{\beta\eta}{2} - \frac{\beta^2}{4} + \ln \text{K}.$$

$C$ is a constant of integration that has been chosen in the manner shown in the equation so that we may rewrite this equation in the following form and then integrate the equation.

$$f_2'' = K\exp\left[-\left(\frac{\eta - \beta}{2}\right)^2\right] \xrightarrow{\text{integration}} f_2' = K\int_{\eta=\infty}^{\eta}\exp\left[-\left(\frac{\eta - \beta}{2}\right)^2\right]d\eta \tag{5.4.18}$$

We note that the integral in Eq. 5.4.18 is zero when $\eta = \infty$. One more integration of Eq. 5.4.18 gives

$$f = f_1 + f_2 = (\eta - \beta) + K\int_{\eta=\infty}^{\eta}d\eta\int_{\eta=\infty}^{\eta}\exp\left[-\left(\frac{\eta - \beta}{2}\right)^2\right]d\eta \tag{5.4.19}$$

We may improve the solution further taking the third term in the asymptotic expansion and proceeding in the similar way. We leave it at this stage. Eq. 5.4.19 has two unknowns $K$ and $\beta$ and the power series 5.4.16 has one unknown $A_2$ that are determined by matching the power series and the asymptotic solutions at some point $\eta = \eta_1$ where both the solutions are applicable. Attempts of Blasius to match the solution and to locate the point led to $A_2 = 0.332$, $\beta = 1.73$ and $K = 0.231$. (In literature $\alpha$ is used for $A_2$.) Blasius solution in the boundary layer over a flat plate is reproduced in the following table and is shown in Fig. 5.4.3.

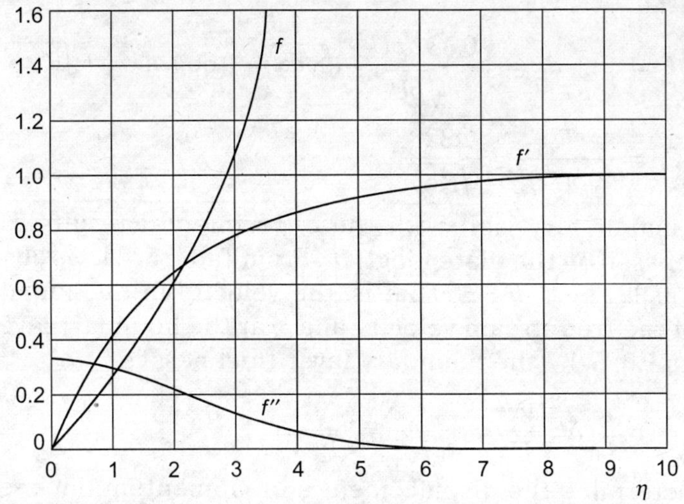


**Fig. 5.4.3** Distribution of the velocity functions across b.l.

**TABLE 5.4.1** Blasius velocity profile

| $\eta$ | $f$ | $f'$ | $f''$ | $\eta$ | $f$ | $f'$ | $f''$ |
|---|---|---|---|---|---|---|---|
| 0 | 0 | 0 | 0.33206 | 2.8 | 1.23099 | 0.81152 | 0.18401 |
| 0.2 | 0.00664 | 0.06641 | 0.33199 | 3.2 | 1.56911 | 0.87609 | 0.13913 |
| 0.4 | 0.02656 | 0.13277 | 0.33147 | 3.6 | 1.92954 | 0.92333 | 0.09809 |
| 0.8 | 0.10611 | 0.26471 | 0.32739 | 4.0 | 2.30576 | 0.95552 | 0.06424 |
| 1.2 | 0.23795 | 0.39378 | 0.31659 | 4.4 | 2.69238 | 0.97587 | 0.03897 |
| 1.6 | 0.42032 | 0.51676 | 0.29667 | 4.8 | 3.08534 | 0.98779 | 0.02187 |
| 2.0 | 0.65003 | 0.62977 | 0.26675 | 5.0 | 3.28329 | 0.99155 | 0.01591 |
| 2.4 | 0.92230 | 0.72899 | 0.22809 | 8.8 | 7.07923 | 1.00000 | 0.00000 |

For engineers, total shear force is of importance. The local shear stress on the wall is

$$\tau_w = \mu\left(\frac{\partial u}{\partial y}\right)_{y=0} = \mu U \frac{\partial}{\partial \eta}\left(f' \frac{\partial \eta}{\partial y}\right)_{\eta=0}$$

$$= \frac{f''(0)\,\rho U^2}{\sqrt{\text{Re}_x}} = \frac{0.332\,\rho U^2}{\sqrt{\text{Re}_x}}, \text{ where } \text{Re}_x = \frac{\rho U x}{\mu} \qquad (5.4.20)$$

The non-dimensional shear stress at any section $x$ is obtained by using a factor ½ $\rho U^2$. Also the shear force at any section and the average shear force on the plate surface of unit width and length $L$ in the direction of flow can be determined. The procedure yields

$$C_{fx} = \frac{\tau_w}{\frac{1}{2}\rho U^2} = \frac{0.664}{\sqrt{\text{Re}_x}} \qquad (5.4.21)$$

$$F = \int_0^L \tau_w \, dx = \int_0^L \frac{0.332\rho U^2}{\sqrt{\rho U}} \sqrt{x dx} = 0.664\rho U^2 L / \sqrt{\text{Re}_L} \tag{5.4.22}$$

$$C_f = \frac{F}{\frac{1}{2}\rho U^2 L} = \frac{1.328}{\sqrt{\text{Re}_L}} \tag{5.4.23}$$

These results match very well with the experimental results. Here $\text{Re}_L$ refer to Reynolds number based on the plate length $L$. From Table 5.4.1, we observe that $f'$, which represents $u/U$, is 0.99155 at $\eta = 5$. That is, the velocity in the boundary layer is more than 99 percent of the free stream velocity and $\eta$ at the boundary layer edge is 5. Using expression for $\eta$ in Eq. 5.4.7 the boundary layer thickness is

$$\delta / \sqrt{\frac{\nu x}{U}} = 5 \rightarrow \frac{\delta}{x} = \frac{5.0}{\sqrt{\text{Re}_x}} \tag{5.4.24}$$

We may also determine the displacement and momentum thickness using Eq. 5.4.1 and 5.4.2.

$$\delta^* = \int_0^\infty (1 - u/U) dy = \sqrt{\frac{\nu x}{U}} \int_0^\infty (1 - f') d\eta = \sqrt{\frac{\nu x}{U}}\, x \lim_{n \to \infty} (\eta - f) = \frac{1.7208\, x}{\sqrt{\text{Re}_x}} \tag{5.4.25}$$

$$\theta = \int_0^\infty \frac{u}{U}\left(1 - \frac{u}{U}\right) dy = \sqrt{\frac{\nu x}{U}} \int_0^\infty f'(1 - f') d\eta = \frac{0.664\, x}{\sqrt{\text{Re}_x}} \tag{5.4.26}$$

We note that the $\delta^* \sim \delta/3$ and $\theta \sim \delta/8$ for a laminar boundary layer, which we mentioned earlier.

## 5.5 Integral Solution for Boundary Layer

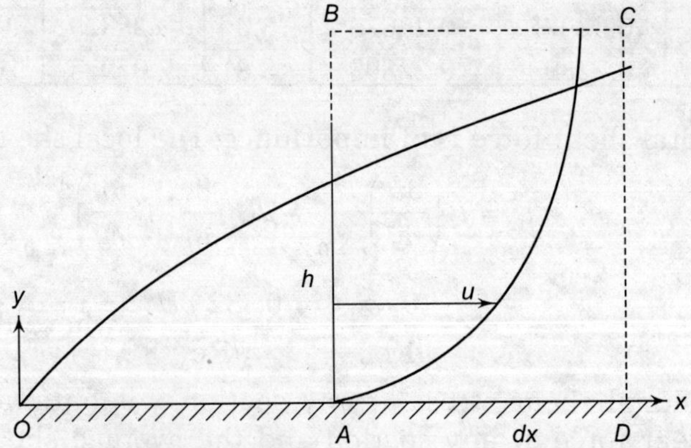


**Fig. 5.5.1** Integral solution

Other group of scientists were trying to find the integral solution for the boundary layer. The procedure of integral solution technique was discussed in Chapter 2. Consider a C.V. as shown in Fig. 5.5.1 in which the face $BC$ is at a height $h$ from the plate so that it lies in the potential flow region of incompressible fluid. Further assume that the flow is steady and 2-dimensional and there is no mass flow across the plate. Since we have learnt the method in details earlier, we proceed directly to write the Steps for solution taking unit width in the normal direction.

Momentum flow rate respectively across section *AB* and *CD* in *x*-direction is

$$= \rho \int_0^h u^2\, dy \;\text{ and }\; \rho \int_0^h u^2\, dy + \rho \frac{d}{dx}\left( \int_0^h u^2\, dy \right) dx$$

Mass flow rate across section *AB* and *CD* = $\rho \int_0^h u dy$ and $\rho \int_0^h u dy + \rho \dfrac{d}{dx}\left( \int_0^h u dy \right) dx$

If $m_{BC}$ is the mass flow rate (cflux) across the section *BC* then mass conservation demands

$$\left\{ \rho \int_0^h u dy + \rho \frac{d}{dx}\left( \int_0^h u dy \right) dx \right\} - \rho \int_0^h u dy + m_{BC} = 0$$

$$\rightarrow \qquad m_{BC} = -\rho \frac{d}{dx}\left( \int_0^h u dy \right) dx$$

Negative sign shows that the mass enters the control volume with velocity *U* and has the momentum. The net *x*-momentum flux per unit time to the control volume is

$$\left\{ \rho \int_0^h u^2 dy + \rho \frac{d}{dx}\left( \int_0^h u^2 dy \right) dx \right\} - \rho \int_0^h u^2 dy - \rho U \frac{d}{dx}\left( \int_0^h u dy \right) dx$$

$$= -\left[ \rho \frac{d}{dx}\left( \int_0^h u(U-u) dy \right) dx - \rho \frac{dU}{dx} \int_0^h u dy \right] dx$$

The external force acting on the control volume in the direction of flow are the shear force at the wall and the pressure force on the surfaces *AB* and *CD* that would be equal to *h*(*dp*/*dx*) *dx*, calculated in the way as discussed in Chapter 2. Applying the integral momentum theorem, we get

$$-\left[ \rho \frac{d}{dx}\left( \int_0^h u(U-u) dy \right) dx - \rho \frac{dU}{dx} \int_0^h u dy \right] dx = -\left[ \tau_w + h \frac{dp}{dx} \right] dx.$$

But $\qquad dp/dx = -\rho U \dfrac{dU}{dx}$ (Bernoulli equation)

$$\rightarrow \qquad h \frac{dp}{dx} = \int_0^h \frac{dp}{dx} dy = -\rho \int_0^h U \frac{dp}{dx} dy = -\rho \frac{dU}{dx} \int_0^h U dy$$

and

$$\tau_w = \rho \frac{d}{dx}(\theta U^2) + \rho U \frac{dU}{dx} \delta^\star \qquad (5.5.1)$$

The integrals have been replaced by using definition of displacement and momentum thickness.[6] Equation 5.5.1 is known as *Von Karman integral equation*[7]. For a plane plate, $dU/dx$ is zero and $U$ is not a function of $x$. Therefore, Eq. 5.5.1 changes to

$$\tau_w = \rho U^2 \frac{d\theta}{dx} \tag{5.5.2}$$

The integrals in Eq. 5.5.1 can be evaluated if velocity variation with $y$ is known. The integral solution is based on choosing this profile and comparing results with experimental results. Thus the accuracy of the solution depends upon the closeness of the chosen velocity profile with the actual velocity distribution. The procedure in explained through solved examples. However, in the following paras Von Karman and Pohlhausen solution is given to stress the complexity involved in case of flow with pressure gradient.

To evaluate the integrals, Pohlhausen suggested choosing velocity profile that would satisfy the boundary conditions at the surface and the boundary layer edge. These boundary conditions are: no slip condition at the surface, continuity of the velocity profile at the b.l. edge, and then the conditions derived with the help of b.l. equations; these are summarised below.

At $\quad y = 0,\ u = v = 0,\ \partial u/\partial y = \tau_w/\mu,\ \partial u/\partial y = -\nu UU',\ \partial^3 u/\partial y^3 = 0$ (5.5.3)

At $\quad y = \delta,\ u = U$ and $\partial u/\partial y = \partial^2 u/\partial y^2 = \partial^3 u/\partial y^3 = \cdots = 0$ (5.5.4)

A word of caution regarding using the boundary conditions. Boundary conditions derived with the help of b.l. equation must be used only when the main boundary conditions are not enough to determine the velocity profile and that too alternately at $y = 0$ and $y = \delta$. These conditions suggest that a polynomial of any degree for $u/U$ in $y/\delta$ may be chosen and the constants in the polynomial can be determined using the boundary conditions. Karman took the velocity profile $u = a + by + cy^2 + dy^3$ having four unknowns for the flow over a flat plate and chose the boundary conditions $u = v = 0$ at $y = 0$ and $u = U$, $\partial u/\partial y = 0$ at $y = \delta$ to determine the constants. This gives, after rearranging the terms,

$$\frac{u}{U} = \frac{3}{2}\frac{y}{\delta} - \frac{1}{2}\left(\frac{y}{\delta}\right)^3$$

$$\rightarrow \qquad \theta = \int_0^\delta \left(1 - \frac{u}{U}\right)\frac{u}{U}\,dy = \frac{39}{280}\,\delta,$$

$$\left(\frac{\partial u}{\partial y}\right)_w = \frac{3U}{2\delta} \text{ and } \tau_w = \frac{3\mu U}{2\delta}$$

Substituting these results in Eq. 5.5.1, we get for flat plate ($dU/dx = 0$)

$$\frac{39}{280}\rho U^2 \frac{d\delta}{dx} = \frac{3\mu U}{2\delta}$$

[6] The Karman equation can also be obtained by integrating the b.l. equations from $y = 0$ to $y = \delta$.

[7] This equation can also be obtained by integration of the *N–S* equations along a curve. This has been done in Appendix A 1.6.

$$\rightarrow \qquad \delta d\delta = \frac{140}{13}\frac{\nu}{U}dx \xrightarrow{\text{integration}} \delta = 4.64\sqrt{\frac{\nu x}{U}} + C.$$

But at $x = 0$ (leading edge), $\delta = 0 \rightarrow C = 0$.

$$\therefore \qquad \frac{\delta}{x} = \frac{4.64}{\sqrt{\text{Re}_x}} \tag{5.5.5}$$

This result differs from Blasius solution for which the numerator was 5. The difference points out that the velocity chosen is deviating from the actual velocity to some extent. One has to try different velocity profiles so that the result may match the experimental results. There is no set procedure to choose the exact velocity distribution at any section. The Von Karman solution of Eq. 5.5.5 is widely accepted.

***Non-zero pressure gradient flows.*** A very large class of flow exists in which case the pressure varies in the direction of the flow; the pressure may increase or decrease. Velocity profile is chosen that might be close to the actual profile in that case. We shall briefly mention the solution proposed by Holstein and Bohlen who took a 4th order polynomial in the variable $\eta$, where $\eta = y/\delta$ and $\delta$ is the boundary layer thickness at a section. They introduced a non-dimensional factor $\lambda$ to represent pressure gradient. $\lambda$ is defined as $\lambda = \frac{\delta^2}{\nu}\frac{dU}{dx}$. The chosen polynomial and the boundary conditions were

$$\frac{u}{U} = a\eta + b\eta^2 + c\eta^3 + d\eta^4$$

$$\text{At} \qquad \eta = 0 : u = 0, v = 0 \rightarrow \frac{\nu}{\delta^2}\frac{\partial^2 u}{\partial \eta^2} = \frac{1}{\rho}\frac{dp}{dx} = -U\frac{dU}{dx} \tag{5.5.6}$$

$$\eta = 1 : u = U \text{ and } \frac{\partial u}{\partial \eta} = \frac{\partial^2 u}{\partial \eta^2} = 0$$

Using the boundary conditions and simplifying the velocity profile we get

$$u/U = F(\eta) + \lambda G(\eta),$$

$$F(\eta) = 2\eta - 2\eta^3 + \eta^4 \text{ and } G(\eta) = \eta(1-\eta)^3/6 \tag{5.5.7}$$

$$\rightarrow \qquad \tau_w = \mu\left(\frac{\partial u}{\partial y}\right)_{\eta=0} = \frac{\mu U}{\delta}\left(2 + \frac{\lambda}{6}\right),$$

$$\frac{\delta^*}{\delta} = \int_0^1 \left(1 - \frac{u}{U}\right) d\eta = \frac{3}{10} - \frac{\lambda}{120} \tag{5.5.8}$$

$$\text{and} \qquad \frac{\theta}{\delta} = \int_0^1 \frac{u}{U}\left(1 - \frac{u}{U}\right) d\eta = \frac{37}{315} - \frac{\lambda}{945} - \frac{\lambda^2}{9072} \tag{5.5.9}$$

$\lambda$ is known as the shape factor as it determines the shape of the velocity profile in the boundary layer for non-zero pressure gradient flows. We now substitute these in Eq. 5.5.1, which generates a complex equation hard enough to be integrated with respect to $x$. Som and Biswas have reproduced the solution that originally is in German. Let us define the non-dimensional parameters

$$L = \frac{\tau_w \theta}{\mu U} = \left(2 + \frac{\lambda}{6}\right)\left(\frac{37}{315} - \frac{\lambda}{945} - \frac{\lambda^2}{9072}\right) = f_1(\mathrm{K}) \tag{5.5.10}$$

$$H = \frac{\delta^*}{\theta} = \frac{3/10 - \lambda/120}{37/315 - \lambda/945 - \lambda^2/9072} = f_2(K)$$

and
$$K = \left(\frac{\theta}{\delta}\right)^2 \lambda$$

Equation 5.5.1 thus gives an equation in $K$ alone. Using these non-dimensionless parameters, the integral momentum equation reduces to

$$U\frac{d\theta}{dx} + \theta(2 + H)\frac{dU}{dx} = \frac{\nu L}{\theta}$$

or,
$$U\frac{d}{dx}\left(\frac{\theta^2}{\nu}\right) = 2\,[L - K(H + 2)] \tag{5.5.11}$$

Walz suggested that right and side of Eq. 5.5.11 can be represented to a fair degree of accuracy by a linear relation in $K$ so that

$$2[L - K(H + 2)] = a - b\,K, \tag{5.5.12}$$

here $a = 0.47$ and $b = 6$. With this simplification the Eq. 5.5.11 turns into

$$\frac{d}{dx}\left(\frac{U\theta^2}{\nu}\right) = a - (b - 1)\frac{U\theta^2}{\nu}\frac{1}{U}\frac{dU}{dx}$$

This gives on integration with the b.c $U = 0$ at $x = 0$

$$\frac{U\theta^2}{\nu} = \frac{a}{U^{b-1}}\int_0^x U^{b-1}\,dx \tag{5.5.13}$$

Finally the momentum thickness is

$$\theta^2 = \frac{0.47\,\nu}{U^6}\int_0^x U^5\,dx \tag{5.5.14}$$

Equation 5.5.14 gives the limit of $\theta$ at $x = 0$ using the L' Hospital rule. The value is

$$[\theta^2]_{x=0} = \frac{0.47\,\nu}{U^6(0)} \tag{5.5.15}$$

**Example 5.5.1** Assuming the velocity distribution in the laminar boundary layer with zero pressure gradient may be expressed as $\frac{u}{U} = \sin\left(\frac{\pi}{2}\frac{y}{\delta}\right)$, where $u$ is the velocity at a distance in the boundary layer, $U$ is the free stream velocity and $\delta$ is the b.l. thickness. Determine the drag coefficient in terms of the Reynolds number.

**Solution** We need momentum thickness in Eq. 5.5.2, which is obtained form Eq. 5.4.2.

$$\theta = \int_0^\delta \frac{u}{U}\left(1 - \frac{u}{U}\right)dy = \left(\frac{4 - \pi}{2\pi}\right)\delta \quad \text{and} \quad \tau_w = \mu\left.\frac{du}{dy}\right|_{y=0} = \frac{\mu U \pi}{2\delta}$$

Substituting in Eq. 5.5.2, we get the equation.

$$U^2\left(\frac{4-\pi}{2\pi}\right)\frac{d\delta}{dx}=\frac{\mu U\pi}{2\delta\rho}\xrightarrow{\text{integration w.r.t. } x}\delta=\sqrt{\frac{2\pi^2}{4-\pi}}\sqrt{\frac{\mu x}{\rho U}}$$

$$D=\int_0^L\tau_w\,dx=\int_0^L\frac{\mu U\pi}{2\delta}\,dx \text{ and } C_D=\frac{D}{\rho U^2L}=\frac{1.31}{\sqrt{\mathrm{Re}_L}}$$

**Example 5.5.2** Water flows with a free stream velocity of 2 m/s past a flat plate of length $L = 10$ cm. Using (i) Blasius solution and (ii) the integral solution with a third order velocity profile, find (a) the thickness of the velocity boundary layer at a location $x = 5$ cm, (b) fluid velocity at the same location at a distance $y = 0.0225$ cm away from the plate, and (c) the drag force per meter of depth into the paper. Take for water, $\rho = 983.1$ kg/m³ and $\nu = 0.4748 \times 10^{-6}$ m²/s.

$$\mathrm{Re}_x=\frac{Ux}{\nu}=\frac{2\times 0.05}{0.4748\times 10^{-6}}=210615$$

(i) Blasius profile.

(a) $\dfrac{\delta}{x}=5\sqrt{\mathrm{Re}_x}\rightarrow \delta=5\times 0.05\times(210615)^{-1/2}=0.05447$ cm

(b) For $x = 5$ cm and $y = 0.225$ cm, $\eta=y\sqrt{\dfrac{U}{\nu x}}=\dfrac{y}{x}\sqrt{\mathrm{Re}_x}=\dfrac{0.0225}{5}\sqrt{210615}=2.065.$

From the Table 5.4.1, $f' = u/U = 0.629$ at this value of $\eta$

$\rightarrow \quad u = 1.258$ m/s

(c) $\mathrm{Re}_L=\dfrac{UL}{\nu}=\dfrac{2\times 01}{0.4748\times 10^{-6}}=421230$. $C_f$ from Eq. 5.4.23 is

$$C_f=\frac{F}{\frac{1}{2}\rho U^2L}=\frac{1.328}{\sqrt{\mathrm{Re}_L}}=\frac{1.328}{\sqrt{421230}}=0.002046\rightarrow$$

$F=0.002046\times \frac{1}{2}\quad \rho U^2L=0.4$ N

(ii) *Integral solution.* Analysis of third order velocity profile is given in the text. Using Eq. 5.5.5, we get $\delta = 0.05055$ cm. Substituting values of $\delta$ and $y$ in the velocity profile, we get

$$\frac{u}{U}=\frac{3}{2}\frac{y}{\delta}-\frac{1}{2}\left(\frac{y}{\delta}\right)^3\rightarrow u=1.25 \text{ m/s}$$

$$\tau_w=\mu\left.\frac{\partial u}{\partial y}\right|_{y=0}=\frac{3}{2}\frac{\mu U}{\delta}=0.323275\frac{\rho U^2}{\mathrm{Re}_x}, \text{ and}$$

Wall shear stress at any $x$ is

$$F=\int_0^L\tau_w\,dy=0.646\rho U^2\sqrt{\frac{\nu L}{U}}=0.39 \text{ N}$$

Results obtained by two approaches are close to each other.

**Example 5.5.3** Water flows between two parallel walls. The velocity is uniform at the entrance and in the core region. Beyond a distance $L_e$ downstream from the entrance, the flow becomes fully developed so that the velocity varies over the entire width $2h$ of the channel. In the boundary layer region, velocity varies as $u = U(x)\,(y/\delta)^3$ where $\delta = A\sqrt{x}$; $A$ is a constant. Determine the acceleration on the axis of symmetry for $0 \le x \le L_e$.

**Solution** We fix the coordinates with the bottom plate at the entrance of the channel measuring $y$ upward. As the flow proceeds, the boundary layer development takes place. let at any $x$ the boundary layer thickness is $\delta$. Let $U_0$ and $U(x)$ be the flow velocity at the entrance and the core region of the channel then the mass flow across the section, by mass conservation, is the mass entering the channel for steady flow gives the equation

$$U_0 h = U(x)\,(h - \delta) + \int_0^{\delta} U\left(\frac{y}{\delta}\right)^3 dy$$

or,

$$U_0 h = U(x)\left(h - \frac{3\delta}{4}\right)$$

We are given $\delta = A\sqrt{x}$ and at $x = L_e$, $\delta = h$ (flow is fully developed). This gives $A$ and hence $\delta$ at any section. Substituting these results gives

$$U(x) = \frac{U_0}{1 - 3/4\sqrt{x/L_e}} \quad \text{and}$$

$$\alpha = \frac{dU}{dx} = \frac{3U_0^2}{8L_e\left[h - \frac{3}{4}\sqrt{\frac{x}{L_e}}\right]^3}\left[\frac{x}{L_e}\right]^{-1/2}$$

The acceleration will be convective only as the flow is steady.

## 5.6 Boundary Layer Separation and Control

In section 5.4, we discussed the reason for formation of boundary layer and learnt that the slope of the velocity profile at the surface is maximum. However, as flow proceeds, the value of the slope will, in general, change depending upon the external conditions. For example, if pressure is increasing, referred as adverse pressure gradient, then the slope of the velocity curve will decrease and it is possible that at some location it becomes zero for large adverse pressure gradient. This means that at that section the layer next to the layer in contact with the surface has zero velocity. Therefore the fluid particle is not able to move forward with the fluid. At the subsequent sections the slope of the velocity profile is negative and the particles at such section would move in the opposite direction. That explains the phenomenon of boundary layer separation and its onset is determined by the condition $\partial u / \partial n = 0$, where $n$ is the direction normal to the surface.

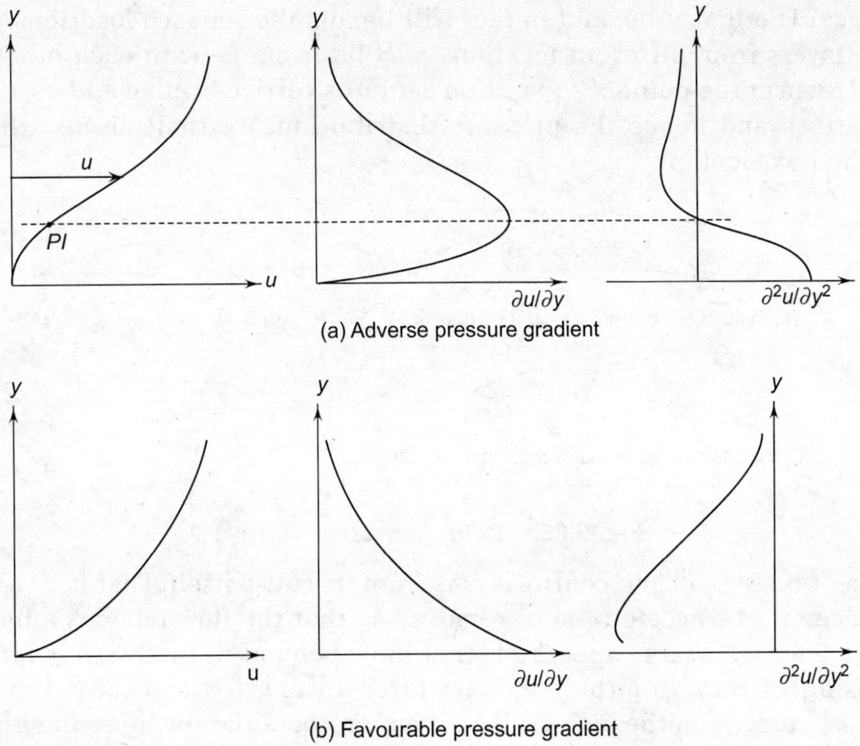


**Fig. 5.6.1** Variation of velocity, its first and second derivative in b.l. region

From b.l. equation also, we may explain the b.l. Separation. Since at the surface, $u = v = 0$, Eq. 5.4.5 gives $\mu(\partial^2 u / \partial y^2) = dp / dx$. Consider the location where the first derivative $\partial u / \partial y$ is zero, that is, the point of separation. In adverse pressure situation ($dp/dx > 0$), the velocity profile in the b.l. is such that (i) the velocity increases from 0 to $U$ joining the main flow smoothly, (ii) the first derivative is zero at the surface and the b.l. edge, which means it must have at least one point of maxima in the b.l. region, and (iii) the second derivative, a measure of the curvature of the velocity curve, is initially positive at the surface and tends to zero at the free surface. At the point of maxima of $\partial u/\partial y$ curve, $\partial^2 u/\partial y^2$ must be zero. Therefore, the curve for $\partial^2 u/\partial y^2$ starts with some positive value, passes through the origin, and then tends to zero again at the free surface. The section, where $\partial^2 u/\partial y^2$ is zero, there exists a point of inflexion on the velocity profile. These observations suggest the kind of the profiles for $u$, $\partial u/\partial y$ and $\partial^2 u/\partial y^2$, as shown in Fig. 5.6.1 (a). These profiles are also shown for favourable pressure gradient in Fig. 5.6.1 (b). You should now be able to explain this type of profile in this case.

We may conclude that the deficiency in kinetic energy of the particles close to the surface, because of viscosity of the fluid, is responsible for the layers to leave the surface. The fluid particles leaving the surface thereafter get energy from the main flow that helps in pushing it forward to join back the main flow. It is not necessary that all the particles or layers will leave the surface from the same section in a three dimensional surface. There may be, and in fact will be, number of such locations on the surface. These layers from different locations may later merge with each other. Thus the flow downstream of the point of separation contains vortices, called eddies, that disturb the flow pattern and hence the pressure distribution. We shall discuss this in more details in the next section.

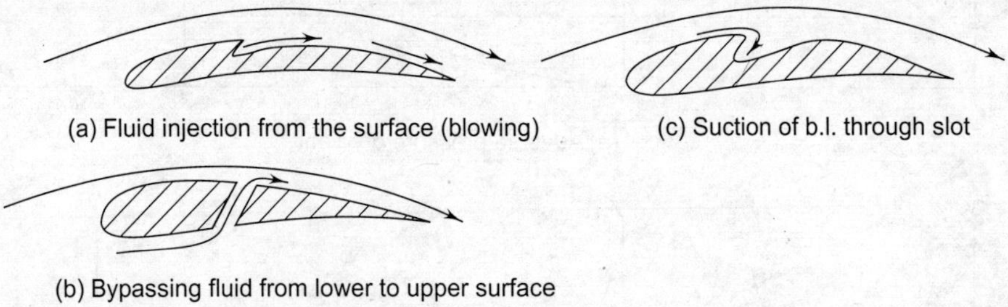


**Fig. 5.6.2** Boundary layer control

The idea of b.l. separation control stems from its cause itself. That is, the layer close to the surface may be accelerated or removed so that the flow remains adhered to the surface. Fig. 5.6.2 shows a few methods that have been used successfully for accelerating, energising, or removing the boundary layer. In the first and second methods, the fluid near the surface on the upper side is accelerated either by injecting high velocity fluid (blowing) through the holes made in the surface or bypassing some of the fluid from other side of the body through the slots made in it. Whatever the method, the slots should be tangential to the surface where the fluid is entering so that it least disturbs the flow pattern. Blowing method in which a high velocity fluid from outside source is injected is a common method used in practice. Yet in another method, the slow moving fluid is sucked through the slots so that the fluid layers remain in contact with the surface.

*Streamlining of the surface.* This method employs a streamlined body so that the adverse pressure gradient may not be large when the fluid flows over its curved surface. Recall that it is one of the important causes of the separation of the boundary layer. We cannot eliminate the possibilities of adverse pressure gradient when the flow is over curved surface, but we can reduce this effect.

## 5.7 Flow over Curved Surfaces

So far mainly we concentrated on the flows over a flat plate. But when the flow takes place over a curved surface then the area of flow changes causing either acceleration or deceleration of the flowing fluid. This affects the boundary layer formation along the surface. While passing over a convex surface, the flow is initially accelerated till it reaches the point after which the area starts increasing resulting in increase in pressure downstream. It is during this journey of the fluid that the boundary layer separates as was discussed above. The point at which the flow will separate and the way it affects the flow downstream depends on the Reynolds number based on the mainstream velocity and some linear characteristic length. This effect is shown in Fig. 5.7.1.

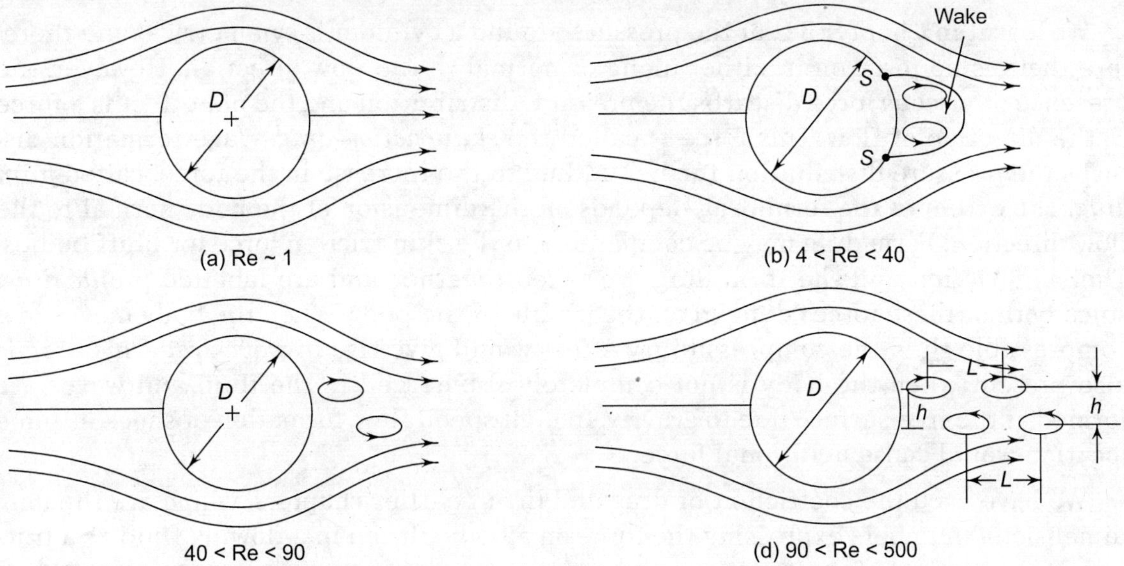


**Fig. 5.7.1** Influence of Reynolds number on wake zone

When Reynolds number is extremely small, around 1, the flow is symmetrical and there is no point of separation, as shown in Fig. 5.7.1 (a). For Reynolds number in between 4 and 40, the flow separates and the point of separations moves upstream along the surface with increasing Re. Thus the flow downstream is divided into two regions—one the *wake zone* in which eddies are present and another the main flow, as seen in Fig. 5.7.1 (b). Flow in the wake zone is symmetrical and steady, the eddies on the upper and lower side rotate in opposite directions, and the size of the eddy grows with Re. After crossing Reynolds number = 40, the wake region flow also becomes unsteady and the wake starts undulating (exhibiting a wave-like phenomenon), resulting in shedding vortices into the stream. At Re 90 or more, the eddies are shed alternately from upper and bottom side with a regular pattern of alternately shed clockwise and anti-clockwise vortices, called the *Von Karman street*, as shown in Fig. 5.7.1 (e). Strouhal number, defined as $fD/U$, expresses frequency $f$ of vortex shedding. Von Karman showed that for stability of the vortex arrangement, $h/L = 0.281$, where $L$ is the distance between two successive vortices and $h$ is the distance between the two rows of vortices; these are shown in Fig. 7.5.1 (d). The alternating breaking away of the vortices produces a periodic force on the cylinder in the transverse direction. If the natural frequency of the cylinder in the transverse direction equals the shedding frequency then resonance occurs. This observation is very useful in the aerodynamic study of stability of suspension bridges, telephone and power transmission lines, etc. as the wind blowing past these, generates vortices with certain frequency. For Reynolds numbers beyond 500, multiple frequencies start showing up and the wake becomes turbulent.

We learnt in Chapter 3 that the pressure around a cylinder is symmetrical and therefore there is no force on it, either along or normal to the flow direction. However, the presence of viscous force disturbs the pressure distribution and the net effect is a force in the direction of flow. This force is called the *skin-friction drag*. Wake formation disturbs the pressure distribution further adding to the increase in the force, called *form drag*. The extent of the form drag depends on the dimension of the body normal to the flow direction. Form drag may be comparable to the skin-friction force for bluff bodies. The skin-friction and the form drag are taken together and are labelled *profile drag*, since both of these forces depend on the profile of the body. When the body moves in a compressible fluid the compressibility effect would give rise to *wave-drag*. This drag is also present when the body is not completely submerged in the fluid and waves are formed at the free surface due to gravity. In high-speed flow, formation of shock at some location would cause additional force.

We have used the coefficient of drag and lift in earlier chapters, which are the non-dimensional numbers expressing the force on a body placed in a flowing fluid as a fraction of the product of dynamic pressure and some area. This area is the area of the body projected on a plane normal to the direction of the force under consideration. The projected area, for example, is $L \times D$ for a cylinder of diameter $D$ and length $L$, $L \times B$ for flat plate of length $L$ and width $B$ placed normal to the flow, $(\pi/4)D^2$ for sphere of diameter $D$, etc. The expression for these two coefficients is reproduced below wherein $D$ is the drag force and $L$ is the lift force on the body.

$$C_D = \frac{D}{\frac{1}{2}\rho U^2 A} \text{ and } C_L = \frac{L}{\frac{1}{2}\rho U^2 A} \tag{5.7.1}$$

These coefficients were obtained experimentally through a very carefully carried out experiments for a very large range of Reynolds number and for flow around a flat plate normal to flow direction, sphere, cylinder and aerofoil; the results are reproduced in Fig. 5.7.2 on a log-log graph for the coefficient of drag. We observe from this result that

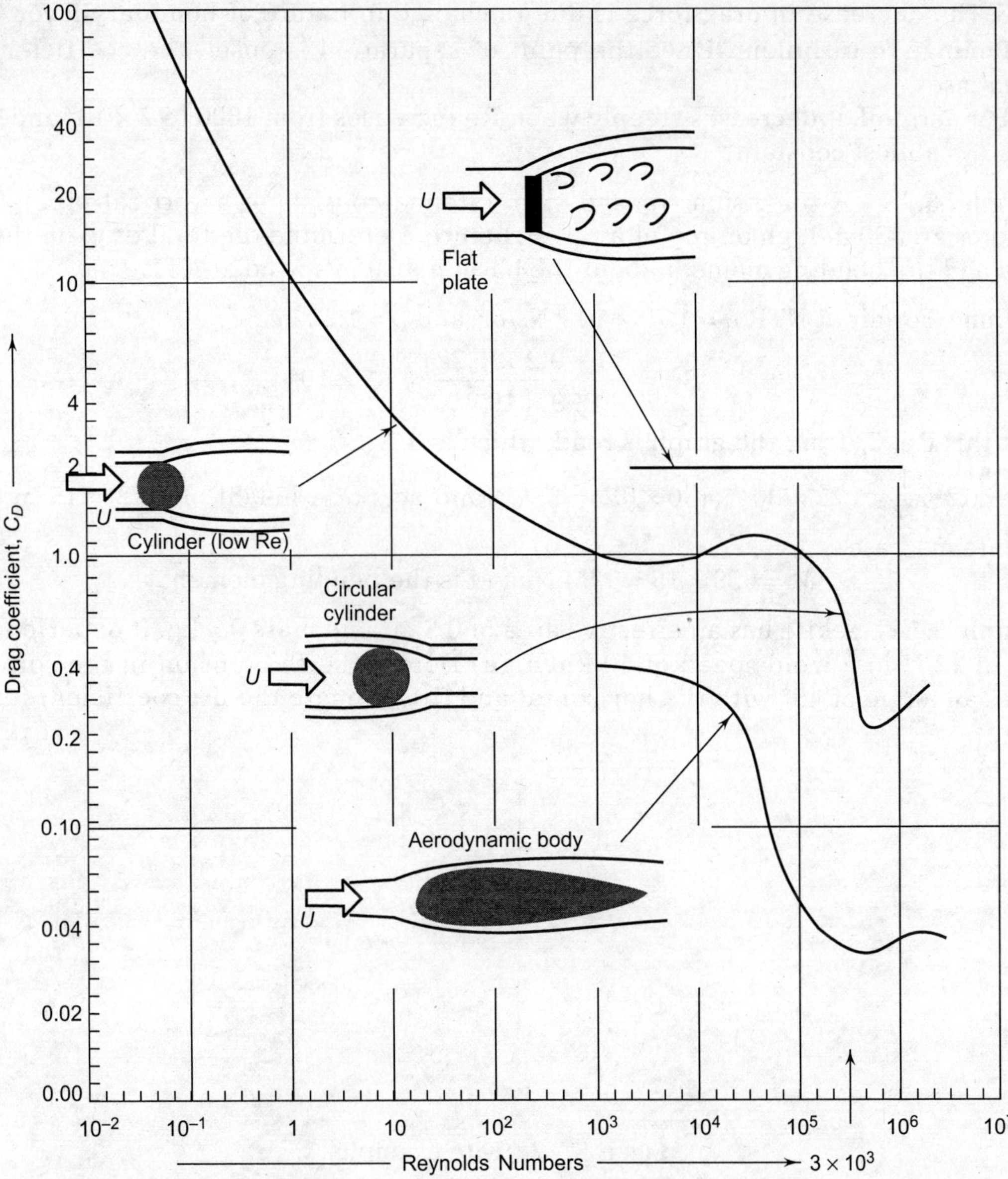


**Fig. 5.7.2** Drag coefficient v/s Reynolds number for two-dimensional bodies

(i) It is maximum for the flat plate and least for the aerofoil. This agrees with our perceptions, since the bluff body disturbs the flow most and a streamlined the least. That is why, we always try to make a body aerodynamically streamlined.

(ii) For a flat plate the coefficient is constant over the entire range of Re.

(iii) For a sphere or cylinder, it decreases monotonically up to Re equal to 2000, remains almost constant up to Re = $2 \times 10^5$, and finally decreases steeply by a factor

5. The decrease in drag force is due to change in nature of boundary layer from laminar to turbulent. Once the point of separation is onset, the coefficient increases.

(iv) For aerofoil, it decreases steeply when Re increases from 1000 to $2 \times 10^5$ and then stays almost constant.

**Example. 5.7.1** A television transmission antenna consists of a vertical pipe 20 cm diameter and 30 m high on top of a tall structure. Determine the total drag on the antenna and the bending moment about the base in a 30 m/s wind at NTP.

**Solution** For air at NTP, $\mu = 1.79 \times 10^{-5}$ Ns/m² and $\rho = 1.22$ kg/m³.

$$\therefore \quad \text{Re} = \frac{30 \times 0.2 \times 1.22}{1.79 \times 10^{-5}} = 4.1 \times 10^5$$

At this Re, $C_D$ from the graph is read, which is 0.2.

$D = 0.2 \times \frac{1}{2} \times 1.22 \times 30^2 \times (30 \times 0.2) = 659$ N and acts at a height of 30/2 = 15 m from the antenna base.

$$\therefore \quad M = 659 \times 15 = 9.9 \text{ kNm; } M \text{ is the bending moment.}$$

**Example 5.7.2** A kite has an effective area of 0.5 m² and mass 0.3 kg. It experiences a drag of 12 N in a wind speed of 40 km/h. (a) Determine the tension in the cord if it makes an angle of 45° with the horizontal and (b) estimate the lift coefficient for the kite.

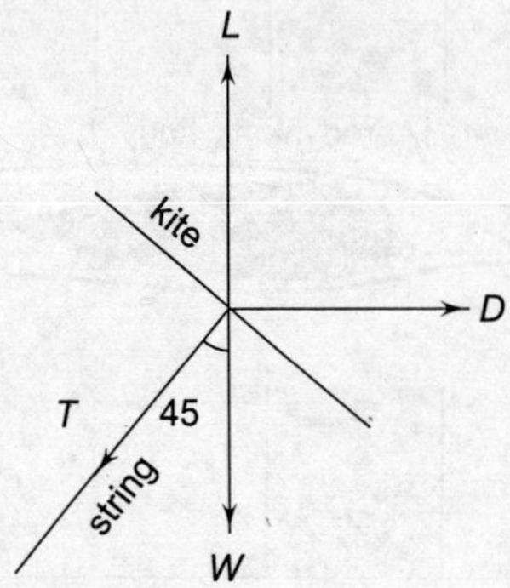


**Fig. E 5.7.2** Kite example

**Solution** $U = 49 \times 1000/3600 = 11.11$ m/s, $W$ = weight of the kite $= 0.3 \times 9.81 = 2.943$ N and $D$ = drag = 12 N. The kite is in equilibrium because of (i) $W$, (ii) $T$, the tension in the string, (iii) the drag force $D$ acting in the direction of wind and (iv) the lift force $L$ produced by the wind flow across the kite. These forces are shown in Fig. E 5.7.2 For equilibrium

$D - T\cos 45 = 0$ and $L - W - T\sin 45 = 0 \rightarrow T = 16.97$ N and $L = 14.94$ N.

$$C_L = \frac{2 \times 14.94}{1.2 \times 11.11^2 \times 0.5} = 0.404$$

# PROBLEMS

## Review Questions

1. What is the difficulty in mathematical solution of viscous flow problems? Is it possible to get exact solutions?
2. What is the contribution of Prof. G.G. Stokes in obtaining the *N–S* equations?
3. In what type of flow is div *U* dropped from the *N–S* equation?
4. What terms are dropped from *N–S* equations if the flow is steady?
5. What are the four types of forces that appear in the final form of the *N–S* equation for steady, incompressible, constant viscosity flow? Write down the mathematical expression of these forces.
6. Mention three of the well-known viscous flow problems for which exact solution of the *N–S* equation are available.
7. List out the approximations of *N–S* equation employed in viscous flow analysis. Justify the approximations in any one case.
8. In laminar flow between parallel plates, why is the velocity maximum at the point of zero shear stress?
9. What type of velocity distribution in laminar flow between parallel plates would give a uniform shear stress across the section?
10. The diameter of a plunger and the radial clearance in a dashpot are 89 mm and 0.5 mm, respectively. If the velocity of the plunger is 2 cm/s, what is the average velocity of the fluid through the clearance?
11. What is the justification in dropping inertial forces and body forces while analysing problems in creeping flows?
12. What is the merit of the hydrostatic lubrication? What is its limitation?
13. The viscous torque in a journal bearing of 10 cm diameter is 10 Nm. All other parameters being same, what is the viscous torque of a 20 cm diameter bearing?
14. Is Prandtl's boundary layer theory developed for analysing highly viscous fluid flows?
15. Prandtl's boundary layer equations are approximated versions of *N–S* equations. Is this true?
16. Perfect fluid flow theory is capable of providing solution to the flow characteristic inside the boundary layer. Comment.
17. What is the most remarkable achievement of boundary layer theory?
18. Viscous friction exercises considerable influence near the boundary. Does it mean that the viscosity increases as the boundary is approached?
19. The shear stress at the boundary of a flat plate decreases with distance from the leading edge. Explain the reason, both physically and mathematically considering the case of a laminar boundary layer as an example.
20. Explain why the boundary shear stress in a turbulent boundary layer is more than in a laminar boundary layer for the same Reynolds number.

21. Justify the formation of a laminar sub layer inside a turbulent boundary layer.
22. A smooth flat plate is covered by a laminar boundary layer over a part of its length followed by a turbulent boundary layer. Will the computed drag be more or less than the two drag if the entire boundary layer is assumed to be turbulent?
23. Distinguish between favourable and adverse pressure gradients.
24. Does the boundary layer get separated only in zones of adverse pressure gradient? Can it not occur in regions where the pressure gradient is favourable?
25. Does the boundary layer separation occur only on curved boundaries and not on flat surfaces?
26. Which type of drag is predominant in the flow past a bluff body?
27. What is the objective of stream lining an immersed body?
28. A circular cylinder is of 10 cm diameter and 50 cm length. It is exposed to an air stream of uniform velocity 10 m/s with its curved surface facing the flow. If $C_d = 1.2$ and density of air = 1.2 kg/m$^3$, what is the drag?
29. If the drag coefficient in Stoke's flow past a sphere is 48, what is the Reynolds number?
30. A ping-pong ball of weight 23.64 milli-Newton and diameter 3.56 cm travels with a velocity of 30 m/s. If the air density is 1.2 kg/m$^3$, what is the lift coefficient?
31. A circular cylinder of 1 m diameter is rotated to have a peripheral velocity of 19 m/s. what is the circulation around the cylinder?
32. A double stagnation point occurs on a circular cylinder of 0.5 m diameter when kept in a uniform flow field of velocity 10 m/s. what is the circulation around the cylinder?
33. The drag of an aeroplane should generally be minimised to achieve economy in fuel consumption. On what occasion is it necessary to have as much drag as possible?
34. Why is it necessary to run aeroplane on the ground for some distance before take-off?

## Problems for Practice

1. Assuming the velocity distribution in the boundary layer as $\frac{u}{U} = \frac{2y}{\delta} - \left(\frac{y}{\delta}\right)^2$, determine the thickness of the boundary layer and the shear stress intensity 0.8 m from the leading edge and the drag force on one side of a plate, 1.2 m long and 1.1 m wide placed in water of velocity 0.2 m/s. Take $\mu = 0.001$ Ns/m$^2$ for water. [10.95 mm, 0.0365 N/m$^2$, 0.0787 N]
2. In a stream of oil of specific gravity 0.95 and kinematic viscosity 0.92 stoke moving at 5.75 m/s, a plate of 500 mm length and 250 mm width is placed parallel to the direction of flow. Calculate he friction drag on the one side of the plate. Find also thickness of the boundary layer and the shear stress at the trailing edge of the plate. [14.747 N, 13.9 mm, 58.99 N/m$^2$]
3. A plate 3 m × 1.5 m is held in water moving at 1.25 m/s parallel to its length. If the flow in the boundary layer is laminar at the leading edge of the plate, (i) find the distance from the leading edge where the boundary layer flow changes from

laminar to turbulent flow, (ii) find the thickness of the boundary layer at this section, and (iii) find the frictional drag on the plate considering both its sides. Take $\mu$ = 0.001 Ns/m$^2$. [400 mm, 2.78 mm, 21.8 N]

4. A 1/60 model of a ship was towed in water at a velocity of 0.4 m/s, the model having a length of 2.5 m and a draft of 0.2 m. Each side of the immersed part of the hull may be taken to be a plate 2.5 m deep. Calculate the frictional drag on the hull and the thickness of the boundary layer at the stern of the model. If the total drag on the model is found to be 1.175 N, find the drag on the prototype. Take $\mu$ = 0.001 Ns/m$^2$. [0.3712 N, 59.5 mm, 1971116 N.]
5. The coefficient of drag due to boundary layer over a thin flat plate 4 m long and 0.3 m wide is given by $C_D = \dfrac{0.455}{(\log R_l)^{2.58}}$, where $R_l$ is the Reynolds number at the length $l$. If the plate is kept parallel to the flow of water having a free stream velocity of 4 m/s, find the total drag force considering both sides of the plate. Take kinematic viscosity of water = $1 \times 10^{-6}$ m$^2$/s.
6. A stream lined automobile with a projected area of 2.75 m$^2$ on a vertical plane at right angles to the direction of motion, runs at 108 km/hr while wind blows at a velocity of 36 km/hr in the opposite direction. Find the drag resistance due to wind action and the power required to overcome the resistance. Take $C_D$ = 0.42. If however, the wind blows in the lateral direction at 36 km/hr at right angles to the direction of motion of the automobile, what would be the drag force tending to topple the vehicle. Area of the vehicle exposed to wind for this case may be taken as 6 m$^2$ and the corresponding values of $C_D$ may be taken as 0.75. Density of air = 1.27 kg/m$^3$. [1173.48 N, 35.2 kW, 285.75 N].
7. At what distance from centre of the tube of radius $r_0$ does the average velocity occur in laminar flow? [0.707 $r_0$]
8. The cup anemometer shown in P 5.7 is placed in wind blowing at a velocity of 54 km/hr. Ignoring bearing friction find the speed of rotation. Take $C_D$ = 0.34 for the convex surface and 1.33 for the concave surface. [35.2 rpm]

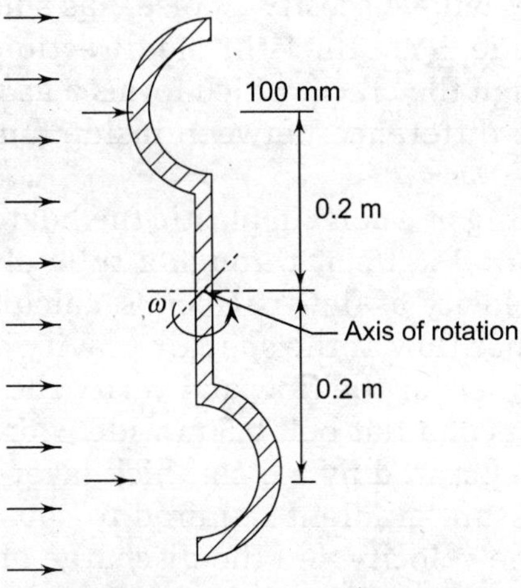


**Fig. P 5.7** Cup anemometer

9. A paratrooper descends in air of density 1.25 kg/m$^3$ with a parachute, which is hemispherical having a diameter of 3.25 m. If the weight of the paratrooper and the parachute is 785 N, find the terminal velocity reached by him. Take $C_D = 0.5$. [17.4 m/s].
10. A vehicle having a projected area of 6 m$^2$ is travelling at 80 km/hr. The resistance to the motion of the vehicle is 2160 N. If 20% and 10% of the total resistance are due to rolling and surface friction respectively and the balance resistance is the form drag, determine the form drag coefficient. Take $\rho = 1.25$ kg/m$^3$ for air. [0.817]
11. A Clark-Y type aerofoil moves at a velocity of 15 m/s through standard atmosphere at an elevation of 2.5 km. The aerofoil has a chord length of 2.25 m and a span of 13.5 m, the angle of attack being 5° 25′. Calculate the weight, which the wing carries, and the power required to drive the aerofoil. Take corresponding to $\alpha = 5° 25'$, $C_L = 0.465$ and $C_D = 0.02$. Density of air = 1.25 kg/m$^3$. [137.9 kN. 815.75 kW]
12. Oil flows through a pipe 15 cm diameter and 650 m in length with a velocity of 0.5 m/s. if the kinematic viscosity of oil is 18.7 stokes; find the power lost in overcoming the friction. Take specific gravity of oil = 0.9. [24.73 W]
13. Oil of specific weight 910 kg/m$^3$ and kinematic viscosity 0.002 m$^2$/s is pumped through a 15 cm dia 300 m long pipe at the rate of 20 tonnes per hour. Show that the flow is viscous and find the head lost. [30 m]
14. Oil is pumped along a horizontal 15 cm dia pipe 200 m long. The specific gravity of the oil is 0.8 and its kinematic viscosity is 1.3 stokes. Flow is laminar so that the fictional coefficient for the pipe will be 16/Re in which Re is the Reynolds number. It has 12 kW motor to drive the pump, which has an efficiency of 65%. Find the quantity of the oil flowing through the pipe in a minute. If the pipe is not horizontal with the other end 10 m above the level of the oil in the pump, find the power required to deliver 0.03 m$^3$/s.
15. Castor oil of viscosity 9.871 poise flows between two plates 4 mm apart. If the difference of pressure between the two sections 100 m apart were 0.025 bar, find the rate of flow between the plates.
16. A container full of oil having viscosity 2 poise, has a horizontal parallel crack in its wall which is 60 cm wide, 5 cm thick, in the direction of flow. Find the flow of oil leaking per hour through the crack, which forms a gap of 0.05 cm between parallel surfaces. The pressure difference between inside and outside faces of the gap is 0.15 bar.
17. In the injection moulding of a certain plastic the heated compound is forced under a pressure difference of 7 kgf/cm$^2$ through a tube of 0.5 cm diameter and 25 cm length. If the mean velocity of flow is 40 cm/s, calculate the viscosity of the compound, assuming laminar flow. If the specific gravity of the compound is 1.5, calculate the Reynolds number for the flow and verify the assumption. [5.365 Ns/m$^2$]
18. A belt conveyor consists of a flat belt 0.5 m wide which slides at a velocity of 4 m/s parallel to a surface separated by a 6 cm thick layer of oil of viscosity 0.25 Ns/m. determine (i) the pressure gradient required to cause no shear stress at the belt surface, (ii) the average velocity and the discharge of oil to be maintained for the above. [–555.55 N/m$^2$, 2.67 m/s, 0.08 m$^3$/s]

19. A coaxial cylinder viscometer consists of the inner stationary cylinder of diameter 10 cm and the outer drum of internal diameter 10.2 cm. When the outer drum rotates at 500 rpm, the torque registered on the torsion meter attached to the inner cylinder is 3 Nm. Determine the viscosity of the fluid filled in the viscometer to a depth of 20 cm. [0.365 Ns/m$^2$]
20. A viscometer consisting of a long vertical tank containing a fluid at the desired ambient conditions and a small sphere is employed for determining the viscosity of oil. If a steel sphere (specific gravity 8.0) of 1 cm diameter falls with a constant velocity of 5 cm/s through the lubricating oil (specific gravity 0.9), calculate the viscosity of the oil.
21. A 2 m diameter cylinder rotates at 1800 rpm in an air stream ($\rho$ = 1.225 kg/m$^3$) of 25 m/s. Estimate the theoretical lift per unit length of the cylinder. [11.55 kN]
22. Experiments were conducted in wind tunnel with a wind speed of 50 km/hr on a float plate 1 m long and 1 m wide. The specific weight of air is 1.15 kg/m$^3$. The plate is kept at such an angle that the coefficients of lift and drag are 0.75 and 0.15 respectively. Determine (i) the lift force, (ii) the drag force, (iii) the resultant force, and (iv) power exerted by the air stream on the plate. [83 N, 16.7 N, 84.7 N, 78.7$^0$, 0.231 kW]
23. Determine the final steady rate at which an air bubble, 0.5 mm dia, rises in a jar of honey with specific gravity 1.3 and dynamic viscosity 0.5 Ns/m$^2$. [0.354 mm/s]
24. What will be the water discharge through 0.60 m thick filter of cross-sectional area 15 m$^2$ if it is made of 0.2 mm diameter uniform sand? Assume the head loss through the filter as 0.2 m, water viscosity 0.001 kg/ms, and porosity of filter 0.5. [0.05445 m$^3$/s]
25. Below are given the wind tunnel measurements of velocity at different elevation from the boundary

| $y$ mm | 1 | 2 | 5 | 7 | 10 | 15 | 0 | 25 | 50 | 75 | 100 | 150 | 200 |
|---|---|---|---|---|---|---|---|---|---|---|---|---|---|
| $u$ m/s | 6.1 | 6.8 | 7.98 | 8.4 | 8.95 | 9.58 | 10.05 | 10.25 | 10.4 | 10.3 | 10.6 | 10.2 | 10.42 |

What is the free stream velocity? Determine $\delta$ the thickness of nominal boundary layer; also determine the exponent $n$ in the equation $u/U = (y/d)^{1/n}$. Is the boundary layer laminar or turbulent? Comment on the observation and justify the increase in subsequent stations for the measurement as we move away from the boundary. [10.38 m/s, 5 mm, 6, turbulent]

26. If velocity distribution in laminar boundary layer over a flat plate is assumed to be given by $u = a + by + cy^2$, determine its form using the necessary boundary conditions. [$u/U = 2(y/\delta) - (y/\delta)^2$]
27. A wind tunnel has a cross-section at its inlet of 1.0 m × 1.0 m and a length of 10 m. Wind at uniform velocity of 15 m/s enters the tunnel at 20°C. Determine the cross-sectional dimension at the end of the test section which will yield zero pressure gradient along its length. Assume velocity distribution in turbulent boundary layer to follow the law $u / U = (y / \delta)^{1/5}$. [1.052 m × 1.052 m]

28. Find the ration of friction drag on the front half and rear half of the flat plate kept at zero incidence in a stream on uniform velocity, if the boundary layer is laminar over the whole plate. [2.414]
29. Boeing has a wing of chord length 6.0 m. For standard air, determine the location on the wing where boundary layer will change from laminar to turbulent when it travels at 260 km/hr. What will be the location of this transition if it travels at the same speed at 20 km altitude. For standard air: $\rho$ = 1.225 kg/m$^3$ and $u$ = 1.461 × $10^{-5}$ m$^2$/s. At 20 km altitude, $\nu = 1.599 \times 10^{-5}$ m$^2$/s. [1.101 m, 1.107 m]
30. If turbulent boundary layer on a flat plate has the velocity distribution given by $u / U = 8.74\ (Uy / \nu)^{1/7}$, obtain the relationship for variation of $\delta/x$.
31. Show that in the length of establishment for laminar flow in a circular horizontal pipe, the frictional force of he pipe wall is given by $F = \pi R^2\ (p_1 - p_2 - 1/3\,\rho U^2)$, where $U$ is the average velocity of flow in the pipe of radius $R$, and $p_1$ and $p_2$ are the pressure at sections 1 and 2 respectively.

CHAPTER 6

# Turbulent Flow

**OUTLINE**

- Introduction
- Stability of Laminar Flow
- Mean Motion and Fluctuations
- Governing Equations for Turbulent Flow
- Turbulent Boundary Layer Equations
- Velocity Scale in Turbulent Boundary Layer
- Shear Stress Model
- Universal Velocity Distribution Law in Ducts for large Reynolds Numbers
- Fully Developed Turbulent Flow in a Pipe for Moderate Reynolds Numbers
- Isotropic and Homogeneous Turbulence

## 6.1 Introduction

Air and water are easily available fluids in nature, which are flowing from one place to another. Both of these fluids flow at a very high Reynolds number and the flow is turbulent in nature. In fact, the laminar flow is an idealisation to get some analytical or numerical solution. The fact is that the flows in nature are mostly turbulent. We may see such flow in rivers, nallah, canals, channels, flow past an obstruction, the rising smoke, etc. These are a few cases just mentioned. The turbulence is responsible for the drag on the ship, aeroplanes, etc, for high rate of diffusion or heat transfer, and in many other physical phenomenons. This explains the importance of the turbulence. In some situations, turbulence may be desirable as in heat transfer or mass diffusion in a chemical reaction process whereas it may not be desirable in other situations. We may realise its importance from the fact that, had there been no turbulence then the man and other living animals probably could not have inhaled enough air, required for survival.

When the turbulence is so important then why are we afraid of it. The answer comes from our discussion in Chapter 2 and is that numerical or analytical solution of a turbulent flow is possible only with the help of certain assumptions and experimental results. This is so because the number of equations is always one less than the number of unknowns. What we call in mathematics is that the problem is not a closed one. The other problem is that the flow is not ordered like the laminar flow. The diffusion of lumps of mass across the flow introduces different type of diffusivity, whose determination is not easy because of requirement of knowledge of the history of the flow. Moreover, it is not easy to determine properties of the turbulent flow, since these are dependent on the flow characteristic that is not known *a priori*. This means if we repeat an experiment with the same flow parameters a number of times then the turbulent stress at the same point in the different runs will be different. In the laminar flow, on the contrary, the fluid property is due to molecular phenomenon and hence is same all the time.

Lastly, the turbulent flow is defined in terms of the correlation coefficients and need thorough knowledge of statistical tools. That is, solution of the turbulent flow uses the statistical based quantities. For this reason, even the super fast computer may not yield a solution to the turbulent flow or if does then it may take a large computation time. The result is that even after a lot of work done for prediction of weather, even today we are not able to predict it correctly.

Turbulence is also responsible for the wind-induced noise, jet noise, noise from flowing water and air, etc. A noise is nothing but the fluctuation of pressure at a point that travels in the media, which is sensed by us as noise. These pressure waves may be strong enough to act like a bomb and shatter the buildings and windows. The correlation of the fluctuating velocity and pressure components vary with space and time; the region in which this interaction is influential is called an eddy and is quite large in turbulent flows. The study has shown that the large eddies change to small eddies and the viscosity of the fluid ultimately disintegrates small eddies. This results in transfer of kinetic energy of the main flow into the internal energy. Thus the turbulence is responsible for internal irreversibility. This is known as eddy loss.

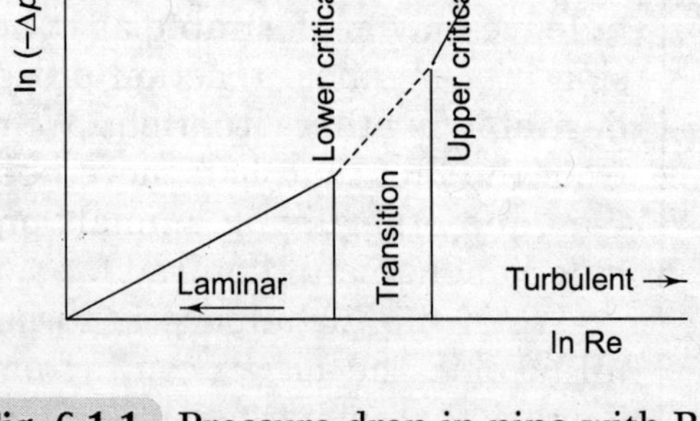


**Fig. 6.1.1** Pressure drop in pipe with Re

In Chapter 2, while describing the Reynolds experiment, it was pointed out that the flow changes to turbulent flow when the Reynolds number reaches a certain value. This experiment can also be performed with a turbulent flow and reducing the Reynolds number. Again at some Reynolds number the flow becomes laminar but this number is much less than that observed when the Reynolds number is increased to study its effect on the flow pattern. The two numbers are called the upper and the lower critical Reynold numbers. This is shown in Fig. 6.1.1 for flow through a pipe. Thus the flow region is divided into three regimes, the laminar, transition and the turbulent flow region.

We may explain onset of turbulence in a flow by considering growth of a disturbance with time or in space; the later is more important as the flow becomes turbulent while flowing and hence the history of growth with space is needed to predict commencement of turbulence. The disturbance can be expressed as components of Fourier series with sinusoidal oscillations having different frequencies and amplitudes. The analysis is based on infinitesimal amplitudes of the disturbances. If the amplitude grows with space coordinate then the flow will change to turbulent flow else it will remain laminar.

These disturbances are always present in the flow system and may be caused by the vibrations of the pump, by wind, obstruction in flow path etc, by the friction at the surface, or at the surface of two different fluids with different velocities in contact. The turbulence due to wall friction is called the wall turbulence and that due to free surfaces as free turbulence. Effect of turbulence is to make the flow more homogeneous and less dependent on the direction. Turbulence is called *homogeneous* if the correlation coefficients, which measure the turbulence structure in the flow, are same everywhere in the flow region and *isotropic* if these are independent of the direction at a point. In isotropic turbulence, which by definition is homogeneous too, the gradient of the mean velocity does not exist. If at all the gradient exists, that is, the mean velocity is changing, and then the turbulence is anisotropic.

Correlation is used in statistics to relate two variables at a point or some distance apart and is defined by the following relation:

$$Q = \overline{u'(\vec{x})\, u'(\vec{x} + \vec{r})} \quad \text{and} \quad R(r) = \frac{\overline{u'(\vec{x})\, u'(\vec{x} + \vec{r})}}{\left[\overline{u'^2(\vec{x})}\;\overline{u'^2(\vec{x} + \vec{r})}\right]^{1/2}} \tag{6.1.1}$$

Overbar shows time average, which is explained in section 6.3. Eq. 6.1.1 defines the correlation $Q$ between the function $u'$ (turbulent velocity component) at position $\vec{x}$ and $\vec{x} + \vec{r}$ where $\vec{x}$ and $\vec{x} + \vec{r}$ are the position vectors of the two points. Since there are three velocity components, there will be nine correlation coefficients. We may even have correlation between two different functions like $u'(\vec{x})$ and $v'(\vec{x} + \vec{r})$ at the two positions under consideration. Similarly we may define correlation between such functions, including even pressure, at $n$ different position in the flow region; we shall determine these correlations for isotropic and homogeneous flow later.

It is common practice to put the correlation in non-dimensional form, as shown in the equation and is known as *correlation coefficient*. Variation of the correlation coefficient is shown in Fig. 6.1.2. We note that the correlation coefficient is one when the distance of separation $r$ is zero, as is evident from Eq. 6.1.1. It decreases as $r$ increases and at some value it is zero. This decides the size of the eddy. If the two points under consideration are not falling in the eddy zone then there is no correlation between the functions at these two points. The large eddy may have size comparable to the physical dimensions of the flow whereas the small eddies may be less than 1 mm; however, the size of the small eddy is much more than the mean free path.

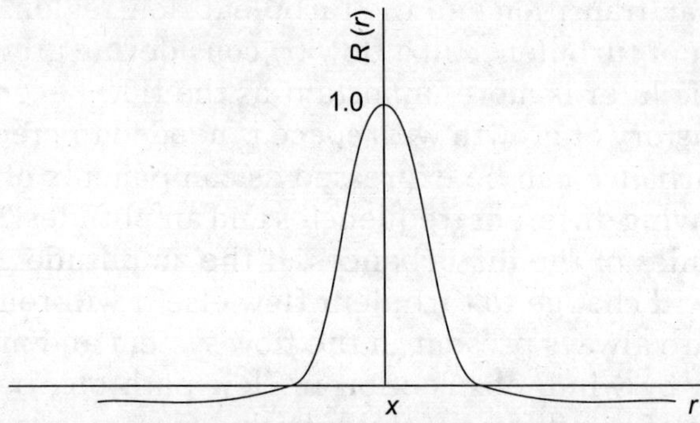


**Fig. 6.1.2** Variation of $R$ with $r$

It is also possible to relate the two variables at time $t$ and $t + \tau$ at the same position. In that case the correlation is called *autocorrelation*, which can also be expressed in non-dimensional form as before. Therefore, autocorrelation between function $u'$ at time $t$ and $t + \tau$ is

$$\overline{u'(t)\, u'(t+\tau)} \tag{6.1.2}$$

## 6.2 Stability of Laminar Flow

We have learnt under some circumstances a laminar flow changes to turbulent flow. So naturally we will be interested to know these circumstances. Same of the solutions given by Lorentz, Orr and Sommerfeld are under consideration. Lorentz calculated variation of energy of the disturbance with time and from it he concluded whether the flow would change to turbulent or remain laminar. But this energy method did not find much acceptance. On the other hand, Orr and Sommerfeld gave separately solution based on the equation of motion for small disturbances and is known as method of small disturbances. They wrote the equation of motion for the mean flow as well the actual flow considering the small disturbances as sinusoidal, which are superimposed on the main flow. The parallel flow was taken as main flow for the analysis. For such flow the component of velocity $U$ is a function of $y$-coordinate whereas pressure varies both with $x$ and $y$-coordinates. The procedure is given in the following paragraph. Thus the main flow is

$$U(y), \qquad V = W = 0, \qquad \text{and} \qquad P(x, y) \tag{6.2.1}$$

The disturbance components are assumed to be two-dimensional and may be expressed as $u'(x, y, t)$, $v'(x, y, t)$ and $p'(x, y, t)$ so that at any time $t$

$$u = U + u', \quad v = v', \quad w' = 0, \qquad \text{and} \qquad p = P + p' \tag{6.2.2}$$

The primed quantities are so small in magnitude that the product of these or their derivatives can be neglected. We may write for the velocity and pressure fields from

Eqs. 6.2.1 and 6.2.2 in the continuity and the momentum equations (*N–S* equations) to give equation of motion for the main and the actual flow. These equations can be linearised by rejecting the product terms of the small fluctuating components. Finally, we may subtract the equations for the main flow from that for the actual flow. This process results in the following set of equations.

$$\frac{\partial u'}{\partial t} + U\frac{\partial u'}{\partial x} + v'\frac{\partial u'}{\partial y} + \frac{1}{\rho}\frac{\partial p'}{\partial x} = \nu \nabla^2 u', \tag{6.2.3}$$

$$\frac{\partial v'}{\partial t} + U\frac{\partial v'}{\partial x} + \frac{1}{\rho}\frac{\partial p'}{\partial y} = \nu \nabla^2 v', \tag{6.2.4}$$

$$\frac{\partial u'}{\partial x} + \frac{\partial v'}{\partial y} = 0 \tag{6.2.5}$$

There are three equations in three unknowns for which the boundary conditions are obtained from the consideration of no slip at the wall and the physics of the flow. This means $u' = v' = 0$ at the wall (no-slip condition). Since the equation is linear partial differential equation, there can be number of solutions and their linear combination is also a solution. The velocity components $u'$ and $v'$ are function of $x$ and $y$ and can be written in terms of the stream function $\psi(x, y, t)$ of the disturbance, which for a single oscillation may be assumed to be of the form

$$\psi(x, y, t) = \phi(y)e^{i(\alpha x - \beta t)}, \tag{6.2.6}$$

where $\phi(y)$ is the amplitude function, which is assumed to depend on $y$ only because the mean flow is function of the $y$-coordinate, $\alpha$ is a real quantity such that $\alpha = 2\pi/\lambda$, $\lambda$ is the wavelength of the disturbance, and $\beta$ is a complex number. This implies that the disturbance is assumed to be expanded in a Fourier series. The quantity $\beta$ can be further assumed as

$$\beta = \beta_r + i\beta_i \qquad \rightarrow c = \beta/\alpha = c_r + ic_i \tag{6.2.7}$$

Here $\beta_r$ is the circular frequency, $\beta_i$ is the amplification factor, $c_r$ is the velocity of propagation, and $c_i$ determines the amplification/damping of the disturbance. We observe from Eq. 6.2.6 that a positive value of $c_i$ means increase in disturbance with time. We now get $u'$ and $v'$ from Eq. 6.2.6 using definition of the stream function and substitute these in Eqs. 6.2.3 through 6.2.5. The pressure term is then eliminated from Eqs. 6.2.3 and 6.2.4 by taking cross derivative and subtracting one from another. That is, differentiate Eq. 6.2.3 with respect to $y$ and Eq. 6.2.4 with respect to $x$ and then subtract the later from the former. We may further eliminate the continuity equation. (6.2.5) by invoking the stream function. This results in the equation

$$(U - c)\left(\phi'' - \alpha^2 \phi\right) - U''\phi = -\frac{i}{\alpha \mathrm{Re}}\left(\phi'''' - 2\alpha^2 \phi'' + \alpha^4 \phi\right) \tag{6.2.8}$$

Equation. 6.2.8 is the Orr–Sommerfeld stability equation that determines growth rate of the disturbance with time. Such study is called *temporal stability* analysis, since $\beta$ is taken as the complex number.

It is also possible to consider the wave number as complex number and $\beta$ a real number. In that case the solution will yield growth of the disturbance along the flow direction; the analysis is called *spatial stability*. Orr and Sommerfeld carried out temporal analysis and we will not go into more details in this primary course. Appearance of

Reynolds number Re in Eq. 6.2.8 is indicator of the fact that the equation has been cast into the non-dimensional form for which we have used the scaling factors: $U_m$ for velocity, $b$ the width of channel ($\delta$ the boundary layer thickness in case of boundary layer), and the prime over $\phi$ for the differentiation with respect to $y/b$; Re = $U_m\ b/\nu$.

A notable objection to this analysis is the assumption of two-dimensional disturbance because the disturbance in general would be three-dimensional. However, square showed that a two-dimensional disturbance leads to transition at lower critical Reynolds number as compared to a three-dimensional disturbance and hence it justifies use of two-dimensional disturbance in the analysis.

The stability equation is a fourth order total differential equation and needs four boundary conditions; two come from the no-slip condition at the wall and the remaining two from other considerations in the flow region. For example, if the flow is between the two plates then the two plate walls give as many boundary conditions, in boundary layer flow over a surface, at the boundary layer edge, in flow through a pipe, the symmetry[1] at the centre of pipe, and so on so forth.

We note that the boundary conditions are defined at different locations. This therefore requires marching technique for numerical integration of the equation. This has posed serious problems as we choose two mutually orthonormal solutions satisfying the boundary conditions at one of the wall and then march along the other boundary where we try to match the solution arrived with the given boundary conditions. In this process the two solutions try to merge into one. Therefore solution of the equation was not an easy job and required some filters to keep the two solutions always orthonormal to each other throughout the integration range. Well, we shall not delve on this matter here and come directly to the results obtained.

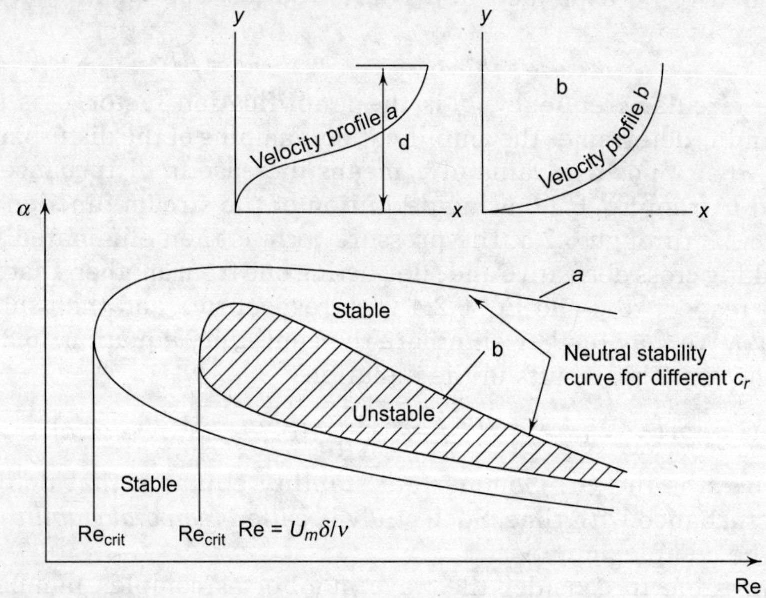


**Fig. 6.2.1** Neutral stability curve for boundary layer flow over a flat plate

[1] Anti-symmetric disturbances have also been studied.

We note that Eq. 6.2.8 has parameters $\alpha$, Re, $c_r$, and $c_i$. We may get neutral stability curve from the solution of the Eq. 6.2.8 with appropriate boundary conditions and $c_i = 0$. The neutral stability curves are plotted on $\alpha$ v/s Re in Fig. 6.2.1, as an illustration, for boundary layer flow over a flat plate for two types of velocity profiles when the flow is along the length of the plate. It is observed that (i) the flow is unstable only after a certain Reynolds number that has been termed as critical Reynolds number, (ii) the value of the critical Reynolds number depends on the type of the velocity distribution in the boundary layer, and (iii) for Reynolds number greater than the critical value, the flow is unstable only when the wave number is within certain range. Thus we conclude that the wave number and the Reynolds number are the main parameter, which decide the transition of laminar flow to turbulent flow. Thus for the flow to remain laminar the Reynolds number should be less than the critical value as then all the wave lengths will lead to a stable flow. Lord Rayleigh also showed that the velocity profiles having a point of inflexion, as in case $a$, is a necessary condition for the occurrence of instability. Later W. Tollmien showed that this is also a sufficient condition. Thus presence of a point of inflexion in velocity profile guarantees instability of the flow.

## 6.3 Mean Motion and Fluctuations

A turbulent flow is characterised by a disturbance superimposed on the mean flow and may be represented as

$$u = \overline{u} + u', \quad v = \overline{v} + v', \quad w = \overline{w} + w', \quad \text{and} \quad p = \overline{p} + p', \tag{6.3.1}$$

where the overbar represents the time average quantity for the main flow and a prime refers to the turbulence component and the symbols $u$, $v$, etc have usual meaning. Figure 6.3.1 shows two types of turbulent flows. The main flow is steady in '$a$' and unsteady in

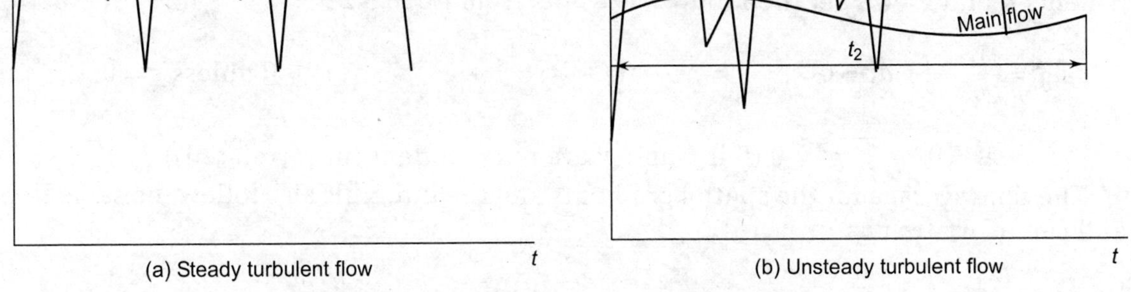


**Fig. 6.3.1** Stedy and unsteady turbulent flows

'$b$' and accordingly the turbulent flow is referred to. There are two time scales associated in turbulent flows, viz. (i) the time scale $t_1$ representing periodic time of the fluctuating components (primed variables) and (ii) the time scale $t_2$ for the periodically varying main flow; it is not necessary that the main flow has to be periodic—it may be aperiodic with main flow properties varying with time. Time scale $t_2$ is infinity for steady flow. Nevertheless the magnitude of $t_2$ is much greater than $t_1$, i.e. $t_2 >> t_1$.

The time average of a quantity $f$ at time $t_0$ is used as description of the stationary turbulence and is defined as

$$\bar{f} = \frac{1}{2T} \int_{t_0 - T}^{t_0 + T} f(x_0, t)\, dt, \tag{6.3.2}$$

where $T$ is the time of integration taken such that the average so determined is independent of these time limits. It should be large enough as compared to the fluctuating components periodic time $t_1$ but small as compared to the main flow time scale $t_2$, that is $t_1 < \tau \leq t_2$. In homogeneous turbulence, on the contrary, the space average would be desirable and can be defined exactly in the same manner. This yields

$$\bar{f}^s = \frac{1}{2x} \int_{-x}^{x} f(x, t_0)\, dx, \tag{6.3.3}$$

For a homogeneous and stationary turbulence, the two averages are assumed to be equal; this is known as *ergodic principle.*

Now consider a parallel flow to get some idea of the presence of turbulent components. For such flow, $\bar{v} = \bar{w} = 0$. Substituting the velocity components in the continuity equation, we get

$$\frac{\partial(\bar{u} + u')}{\partial x} + \frac{\partial v'}{\partial y} + \frac{\partial w'}{\partial z} = \frac{\partial \bar{u}}{\partial x} + \frac{\partial u'}{\partial x} + \frac{\partial v'}{\partial y} + \frac{\partial w'}{\partial z} = 0 \tag{6.3.4}$$

This equation shows that presence of $u'$ ensures presence of other components $v'$ and $w'$, because $\partial u' / \partial x \neq 0$. That is why, we took velocity components, as used in Eq. 6.2.2, with $w' = 0$ because, as mentioned earlier, the two-dimensional disturbance produce a more unstable flow.

We will now reproduce some of the results from statistics related to time average of the fluctuating components, their products, spatial gradients, integrals, etc. For the components $f'$ and $g'$, we get the following results from Eq. 6.3.2.

$$\bar{f} = \frac{1}{2T} \int_{t_0 - T}^{t_0 + T} f'\, dt = 0 \qquad \bar{\bar{f}} = \bar{f} \qquad \overline{(f + g)} = \bar{f} + \bar{g} \qquad \overline{f'f'} > 0 \text{ unless } f' = 0$$

$$\overline{f'} = \overline{g'} = 0 \qquad \overline{f'g'} \neq 0 \text{ (0 if } f' \text{and } g' \text{are independent (uncorrelated))} \tag{6.3.5}$$

The time average for the spatial gradients and the integrals also follow the same laws and can be written as

$$\overline{\frac{\partial f'}{\partial s}} = \overline{\frac{\partial^2 f'}{\partial s^2}} = 0, \quad \frac{\partial \overline{f'g'}}{\partial s} \neq 0 \tag{6.3.6}$$

Similarly, the time average of the quantities for the *mean flow* can be written as

$$\overline{f+g} = \bar{f} + \bar{g} \qquad \overline{f \cdot g} = \bar{f} \cdot \bar{g}$$

$$\overline{\left(\frac{\partial f}{\partial s}\right)} = \frac{\partial \bar{f}}{\partial s} \qquad \overline{\int f\, ds} = \int \bar{f}\, ds \tag{6.3.7}$$

We may also define the following ratio

$$I = \sqrt[1/3]{\left(\overline{u'^2} + \overline{v'^2} + \overline{w'^2}\right)} \Big/ U \tag{6.3.8}$$

This ratio is called the *intensity of turbulence* and is used to represent level of turbulence in the flow. For isotropic flow, $\overline{u'^2} = \overline{v'^2} = \overline{w'^2}$ and therefore $I = \overline{u'^2} / U$. We can now proceed to determine the equation of motion of a turbulent flow, discussed in the next section.

**Example 6.3.1** In a turbulent flow the instantaneous velocity components can be written as $u = \bar{u} + u'$ and $v = \bar{v} + v'$, where a bar over the variable shows the time average and $a'$, over $u$ etc. shows the turbulent fluctuation component. Determine time average of $u^2$ and $uv$.

**Solution**

$$\overline{u^2} = \overline{(\bar{u} + u')^2} = \overline{\left(\bar{u}\,\bar{u} + 2\bar{u}u' + u'^2\right)} = \overline{\bar{u}\,\bar{u}} + 2\overline{\bar{u}u'} + \overline{u'^2} = \bar{u}\,\bar{u} + \overline{u'^2}$$

$$\overline{uv} = \overline{(\bar{u} + u')\,(\bar{v} + v')} = \overline{\left(\bar{u}\,\bar{v} + \bar{u}v' + u'\bar{v} + u'\bar{v}\right)}$$

$$= \overline{\bar{u}\,\bar{v}} + \overline{\bar{u}\,v'} + \overline{u'\bar{v}} + \overline{u'v'} = \bar{u}\,\bar{v} + \overline{u'v'}.$$

**Example 6.3.2** Show that $\overline{\left(\frac{\partial F}{\partial t}\right)} = \frac{\partial \bar{F}}{\partial t}$, where $F$ is function of space and time coordinates.

**Solution** We shall evaluate ab-initio the left and right sides separately to prove this identity.

$$\text{LHS} = \overline{\left(\frac{\partial F}{\partial t}\right)} = \frac{1}{2T}\int_{-T}^{T} \frac{\partial F}{\partial t}\, dt \tag{A}$$

Differentiating $\bar{F}$ with respect to $t$ and interchanging order of the differentiation and integration process, we get

$$\text{RHS} = \frac{\partial \bar{F}}{\partial t} = \frac{1}{2T}\frac{\partial}{\partial t}\int_{-T}^{T} F\, dt = \frac{1}{2T}\int_{-T}^{T} \frac{\partial F}{\partial t}\, dt \tag{B}$$

Hence proved.

## 6.4 Governing Equations for Turbulent Flow

The continuity equation for incompressible fluids gives

$$\frac{\partial(\bar{u}+u')}{\partial x}+\frac{\partial(v+v')}{\partial y}+\frac{\partial(w+w')}{\partial z}=\left(\frac{\partial \bar{u}}{\partial x}+\frac{\partial \bar{v}}{\partial y}+\frac{\partial \bar{w}}{\partial z}\right)+\frac{\partial u'}{\partial x}+\frac{\partial v'}{\partial y}+\frac{\partial w'}{\partial z}=0 \qquad (6.4.1)$$

Time averaging Eq. 6.4.1 and using the results from Eqs. 6.3.6 and 6.3.7, we get

$$\left(\frac{\partial \bar{u}}{\partial x}+\frac{\partial \bar{v}}{\partial y}+\frac{\partial \bar{w}}{\partial z}\right)=0$$

Subtracting this result from Eq. 6.4.1, we get for the primed quantities,

$$\frac{\partial u'}{\partial x}+\frac{\partial v'}{\partial y}+\frac{\partial w'}{\partial z}=0 \qquad (6.4.2)$$

Consider a two-dimensional flow in $x-y$ plane. Since $\partial u'/\partial x \neq 0$, $\partial v'/\partial y = -\partial u'/\partial x$ is not zero. This shows that if $u'$ increases then $v'$ decreases or, $-\overline{u'v'}$ is a positive quantity. Now let us get the momentum equations in the same manner. That is, substitute the values of the variables in *N–S* equation as sum of the mean and primed value. We shall demonstrate the procedure for a two dimensional flow. The inertia term along *x*-direction for steady flow is

$$\rho\left[(\bar{u}+u')\frac{\partial(\bar{u}+u')}{\partial x}+(\bar{v}+v')\frac{\partial(\bar{u}+u')}{\partial y}\right]$$

$$=\underbrace{\rho\left(\bar{u}\frac{\partial \bar{u}}{\partial x}+\bar{v}\frac{\partial \bar{u}}{\partial y}\right)}_{(1)}+\underbrace{\rho\left(u'\frac{\partial \bar{u}}{\partial x}+v'\frac{\partial \bar{u}}{\partial y}\right)}_{(2)}+\underbrace{\rho\left(u'\frac{\partial u'}{\partial x}+v'\frac{\partial u'}{\partial y}\right)}_{(3)}+\underbrace{\rho\left(\bar{u}\frac{\partial u'}{\partial x}+\bar{v}\frac{\partial u'}{\partial y}\right)}_{(4)}+\underbrace{\rho\left(u'\frac{\partial u'}{\partial x}+u'\frac{\partial v'}{\partial y}\right)}_{(5)}$$

$$=\underbrace{\rho\left(\bar{u}\frac{\partial \bar{u}}{\partial x}+\bar{v}\frac{\partial \bar{u}}{\partial y}\right)}_{(6)}+\underbrace{\rho\left[\frac{\partial \overline{u'^2}}{\partial x}+\frac{\partial \overline{u'v'}}{\partial y}\right]}_{(7)} \qquad (6.4.3)$$

Note. We added the term 5, since by continuity equation it is zero, which when combined with term 3 gives term 7 of the equation. Terms 2 and 4 would vanish on taking the time average by Eq. 6.3.5. This leaves terms 6 and 7 in Eq. 6.4.3 that represent the inertia force along *x*-direction in case of turbulent flow.

The pressure force term is

$$\frac{\partial(\bar{p}+p')}{\partial x}=\frac{\partial \bar{p}}{\partial x}+\frac{\partial p'}{\partial x}\rightarrow\frac{\overline{\partial p}}{\partial x}=\frac{\partial \bar{p}}{\partial x}$$

We may similarly proceed to determine the viscous term of the *N–S* equation. Assembling and rearranging all the terms, we get the time averaged *N–S* equation for turbulent flow as

$$\rho\left(\bar{u}\frac{\partial \bar{u}}{\partial x}+\bar{v}\frac{\partial \bar{u}}{\partial y}\right)=-\frac{\partial \bar{p}}{\partial x}+\frac{\partial}{\partial x}\left(\mu\frac{\partial \bar{u}}{\partial x}-\rho\overline{u'^2}\right)+\frac{\partial}{\partial y}\left(\mu\frac{\partial \bar{u}}{\partial y}-\rho\overline{u'v'}\right) \qquad (6.4.4)$$

We may similarly proceed for the $y$-momentum equation. Equation 6.4.4 is same as the one for laminar flow except that the shear force terms have been modified due to the presence of the turbulence in the flow. These additional stresses are called *apparent stresses.* These momentum equations for turbulent flow are known as Reynolds equations. Since stress is a tensor of order 2, there are nine components of the apparent stress, as given below

$$-\begin{pmatrix} \rho\overline{u'^2} & \rho\overline{u'v'} & \rho\overline{u'w'} \\ \rho\overline{u'v'} & \rho\overline{v'^2} & \rho\overline{v'w'} \\ \rho\overline{u'w'} & \rho\overline{v'w'} & \rho\overline{w'^2} \end{pmatrix} \tag{6.4.5}$$

These stresses are also known as *Reynolds stress.* Their determination posses a big problem, since this not only depends on the properties of the fluid but also on the nature of the flow. Like a laminar flow we cannot be definitive to predict these stresses. However different models were developed for determining these stresses. These are centred on the existing procedures for laminar flow. Boussinesq based his model on the Newton's law of viscosity and expressed Reynolds stresses as follows

$$\tau_t = \rho\varepsilon\frac{\partial\overline{u}}{\partial y} \rightarrow \tau = \tau_l + \tau_t = \rho(\nu + \varepsilon)\frac{\partial\overline{u}}{\partial y} \tag{6.4.6}$$

$\varepsilon$ is the *eddy viscosity.* Whereas the kinematic viscosity is the phenomenon at the molecular level, as discussed in kinetic theory of gases, the eddy viscosity takes into account the momentum flux at much larger level solely due to random flow fluctuations that are not amenable to predictions.

We consider a two-dimensional, steady, and incompressible flow through a channel of height $h$ with its centre line as $x$-axis. For the mean flow $u = \overline{u}(y) \neq 0$, $\overline{v} = \overline{w} = 0$. Further, the mean turbulent terms are independent of $x$. The Reynolds equations finally give

$$\frac{1}{\rho}\frac{\partial\overline{p}}{\partial x} - \nu\frac{\partial^2\overline{u}}{\partial^2} + \frac{\partial\overline{u'v'}}{\partial y} = 0, \qquad \frac{1}{\rho}\frac{\partial\overline{p}}{\partial y} + \frac{\partial\overline{v'^2}}{\partial y} = 0, \text{ and } \quad \frac{\partial\overline{w'v'}}{\partial y} = 0$$

with the boundary conditions: $y = \pm\, h$, $\overline{u} = \overline{v} = \overline{w} = 0$, $\overline{u'v'} = \overline{u'w'} = \overline{w'v'} = 0$. The unknowns are: $\overline{p}, \overline{u}, \overline{u'v'}$, and $\overline{v'^2}$, numbering 4. Thus there are three equations in 4 unknowns. This type of problem is faced while analysing the turbulent flow.

## 6.5 Turbulent Boundary Layer Equations

We get these equations exactly in the way these were obtained for the laminar boundary layer flows using the order of magnitude of different terms and assuming that the thickness of the turbulent boundary layer is of the order of $\delta$. To simplify our analysis, we further assume a two-dimensional flow with $w = 0$ with no axial pressure gradient, such as in flow over a flat plate. In such case

$$\frac{\partial\overline{p}}{\partial y} = 0, \frac{\partial\overline{p}}{\partial x} = \frac{dp}{dx}, \frac{\partial^2\overline{u}}{\partial x^2} << \frac{\partial^2\overline{u}}{\partial y^2}, \text{ and } \frac{\partial}{\partial x}\left(-\rho\overline{u'^2}\right) << \frac{\partial}{\partial x}\left(-\rho\overline{u'v'}\right) \tag{6.5.1}$$

This leads to the following turbulent boundary layer equations after dropping the lower order terms from the *N–S* equations

$$\frac{\partial \bar{u}}{\partial x} + \frac{\partial \bar{v}}{\partial y} = 0 \tag{6.5.2}$$

$$\rho\left(\bar{u}\frac{\partial \bar{u}}{\partial x} + \bar{v}\frac{\partial \bar{v}}{\partial y}\right) = -\frac{d\bar{p}}{dx} + \frac{\partial}{\partial y}\left(\mu\frac{\partial \bar{u}}{\partial y} - \rho\overline{u'v'}\right) \tag{6.5.3}$$

That is, the velocity components in the laminar boundary layer equation have been replaced by the time average components for the mean flow and the stress term is increased by the turbulent or apparent stress.

## 6.6 Velocity Scale in Turbulent Boundary Layer

When the flow takes place over a surface, boundary layer is formed that may be turbulent at some section. In such layer there will be shear as well as turbulent stress depending on the distance of the point under consideration from the surface. On the surface, the apparent stress and the velocity fluctuating components will vanish. This implies there is no apparent stress near the surface and the shear stress is all due to laminar flow. In fact, there is a small thin layer near the surface, called laminar sublayer, where the laminar stress dominates and the viscous force dictates over the inertia force. These observations give the following equations, (using Eq. 6.5.3, for flow over the flat plate)

$$\frac{\partial}{\partial y}\left(\mu\frac{\partial \bar{u}}{\partial y} - \rho\overline{u'v'}\right) = 0,$$

$$\xrightarrow{\text{integration}} \nu\frac{\partial \bar{u}}{\partial y} - \overline{u'v'} = C$$

In the sublayer, near the wall $u'$ and $v'$ do not exist. It follows that

$$\left.\frac{\partial \bar{u}}{\partial y}\right|_{y=0} = \frac{\tau_w}{\nu} \tag{6.6.1}$$

To get some idea of the magnitude of the velocity gradient in $y$-direction, let $U_s$ be the velocity at the end of sublayer; Fig. 6.6.1 shows variation of the laminar and turbulent shear stress and velocity in the boundary layer. We observe that viscous stress dominates the near wall region where the turbulent stress is almost zero. As we move away normal to the surface, the turbulent stress first increases up to the end of buffer zone whence after it decreases. The viscous stress of course decreases with the distance. Thus, the whole boundary layer is divided into three regimes: laminar sublayer, buffer layer, and the turbulent regimes based on the dominance of the type of the stress. In laminar sublayer viscous stress dominates and turbulent stress is almost zero. In the turbulent layer, on the other hand, the viscous stress is almost zero and the turbulent stress dominates. In the buffer layer naturally both will exist.

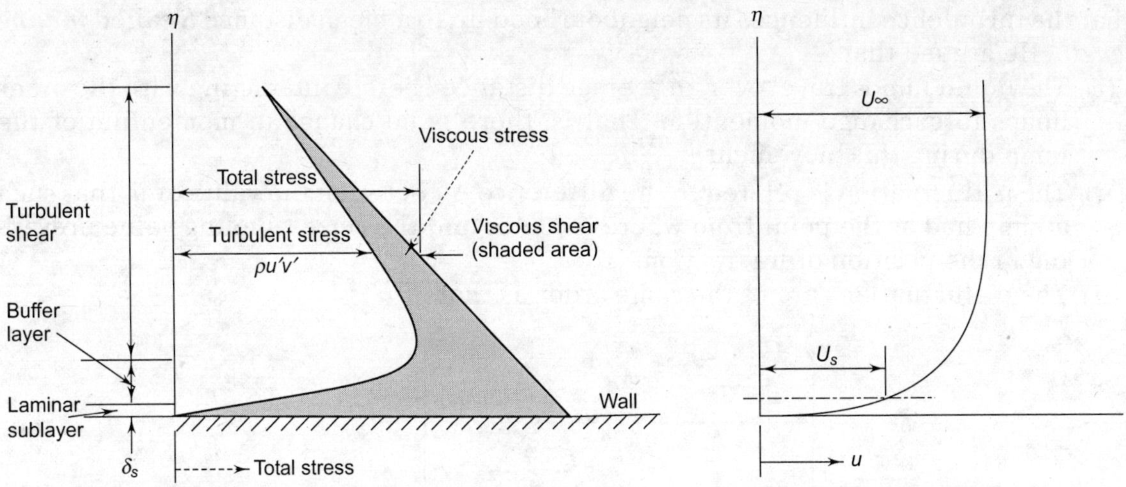


**Fig. 6.6.1** Variation of stress and velocity in turbulent region

We may find out a characteristic velocity $u_\tau$ in the sublayer by expressing the wall stress $\tau_w$ in terms of the velocity $U_s$, at the edge of the sublayer, and the sublayer thickness $\delta_s$.

$$\tau_w = \mu\left[\frac{U_s - 0}{\delta_s}\right] = \mu\frac{U_s}{\delta_s} \text{ and } u_\tau = \left(\frac{\tau_w}{\rho}\right)^{1/2} \tag{6.6.2}$$

$$\rightarrow \qquad u_\tau^2 = \nu\frac{U_s}{\delta_s} \tag{6.6.3}$$

The characteristic velocity is called the friction velocity and defines the structure of the sublayer, which has been confirmed by the experiments. $U_s$ is expected to be proportional to the free stream velocity $U_\infty$,that is, $U_s = CU_\infty$. Substituting it in Eq. 7.6.3, we get

$$\frac{\delta_s u_\tau}{\nu} = C\frac{U_\infty}{u_\tau} \tag{6.6.4}$$

Equation 6.6.4 suggests that a non-dimensional distance $\eta = \frac{yu_\tau}{\nu}$ may be defined that would help in quantifying the different zones; the viscous boundary layer is up to $\eta = 5$, turbulent and viscous for $5 \le \eta \le 70$, and turbulent for $\eta > 70$. The intermediate layer is called the buffer zone. Though we used the analysis for flow over a plate but the concept of friction velocity is true for any other flow. We shall see in the next section how does this help us in determining the velocity distribution in a turbulent flow through a pipe or duct. We know that in such geometries the entire flow region is affected by the boundary layer, unlike the flow over the surfaces where the effect is limited to a certain depth.

## 6.7 Shear Stress Model

In order to determine the apparent stress Prandtl proposed a physical model for the fluctuations in the turbulent flow. He suggested, akin to the concept of mean free path, that the turbulence influences its neighbourhood up to a mean distance $l$, called *mixing length*. He argued that

(i) The liquid lumps travel over an average distance $l$ before interacting with the other lumps to exchange momentum. That is, there is no change in momentum of the lump during this movement.

(ii) The $u'$ fluctuation is related to the difference $\Delta u$ between the value of $\bar{u}$ that such a lump had at the point from where it comes and the corresponding value prevailing in the position of interaction.

(iii) The $v'$ fluctuations are of the same order as that of $u'$.

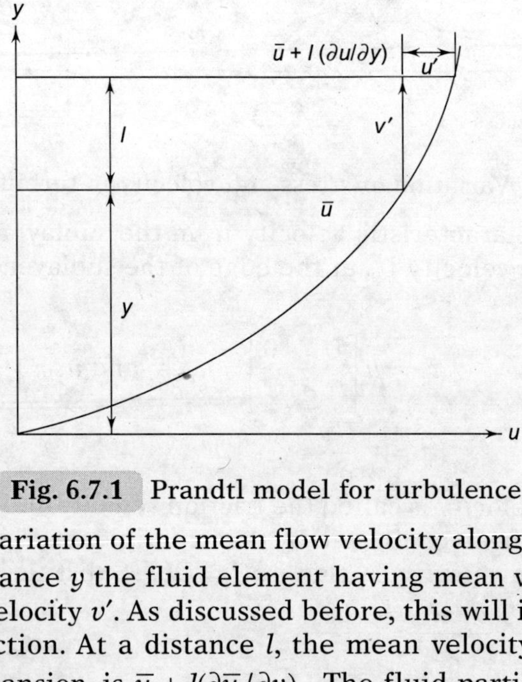


**Fig. 6.7.1** Prandtl model for turbulence

Figure 6.7.1 shows variation of the mean flow velocity along the direction normal to the surface. Let at distance $y$ the fluid element having mean velocity $\bar{u}$ in $x$-direction travels upwards with velocity $v'$. As discussed before, this will induce a negative $u'$ disturbance in the $x$-direction. At a distance $l$, the mean velocity of the fluid elements, using Taylor series expansion, is $\bar{u} + l(\partial\bar{u}/\partial y)$. The fluid particles reaching this plane from $y$ plane reduce the momentum which is proportional to $\rho l(\partial\bar{u}/\partial y)$ that results in fluctuation $u'$. Thus we may write

$$\rho l(\partial\bar{u}/\partial y) = C\rho u' \qquad \text{or,} \qquad u' = l(\partial\bar{u}/\partial y) \qquad (6.7.1)$$

$C$ is a constant that has been combined with $l$. Since $l$ depends on the magnitude of $v'$, so does the fluctuation $u'$, we may write,

$$v' \propto -u' = -C_1 u'$$

$$\rightarrow \qquad \tau_t = -\rho\overline{u'v'} = -C_1\rho l^2\left(\frac{\partial\bar{u}}{\partial y}\right) = \rho l^2\left(\frac{\partial\bar{u}}{\partial y}\right)^2 = \rho l^2\left|\frac{\partial\bar{u}}{\partial y}\right|\frac{\partial\bar{u}}{\partial y} \qquad (6.7.2)$$

The constant $C_1$ has been amalgamated with $l$ and the Reynolds stress has been expressed in the conventional derivative form by splitting the squared term. The mod (two vertical lines) is used to show that whatever the sign of the velocity gradient, its value is always positive. This is done to retain sign of $\partial\bar{u}/\partial y$ term in the expression for turbulent shear stress. Comparison between Eqs. 6.4.6 and 6.7.2 gives

$$\varepsilon = l^2\left|\frac{\partial \overline{u}}{\partial y}\right| \tag{6.7.3}$$

The mixing length $l$ is determined from the experiments. The most popularly used value of $l = 0.4y$ is obtained from the experiments of Nokuradse, and others where $y$ is the perpendicular distance measured from the surface. This theory has also been justified for the internal flows through simple geometries such as circular pipe, circular annulus, and reasonably on boundary layer flow in the absence of strong pressure gradients. This hypothesis is used most often. However, it is an empirical concept and does not help much in understanding the turbulence. Further, according to this theory, the eddy viscosity would be zero at points where $d\overline{u}/dy = 0$, as at the centre of pipe, for example, (contrary to the experimental observations). The other objections to this theory are (i) fluctuations in pressure would bring some change in the momentum of the lump and (ii) the results do not hold good for laminar sublayer and viscous buffer layer near the wall. In spite of this, this theory gives useful results.

***Taylor's Vorticity Transport Theorem***. Taylor used the vorticity equation for a two-dimensional parallel flow and in the same way he got another relation, which was exactly identical with the Prandtl relation except the constant in the mixing length. So both these theories give the same result.

***Von Karman Model***. The problem with mixing length theory is the identification of some suitable scale from the flow. Von Karman suggested the following scale that contains the time-averaged velocity in a shear flow

$$l_v = \frac{d\overline{u}/dy}{d^2\overline{u}/dy^2} \tag{6.7.4}$$

He obtained this scale by carrying out non-dimensional analysis of a two-dimensional flow along $x$-direction such that $\overline{u} = \overline{u}(y)$ and $\overline{v} = 0$. The flow can be made steady by fixing the coordinate systems with the flow itself. In such situation the mixing length $l$, Von Karman assumed, would depend on the distribution of the mean motion in the turbulent flow. We may express the velocity components at a distance $y$ around the point $P$, fixed at the origin $O$ so that $\overline{u} = 0$, using Taylor series

$$u = \left(\frac{\partial \overline{u}}{\partial y}\right)_p y + \left(\frac{\partial^2 \overline{u}}{\partial y^2}\right)_p \frac{y^2}{2} + \ldots + u' \text{ and } v = v'$$

Introducing the quantities $\xi$, $\eta$, $u'_1$, $v'_1$, $l_v$ and $A$, such that

$$x = l_v\xi,\ y = l_v\eta,\ u' = Au'_1, \text{ and } v' = Av'_1$$

We note $l$ and $A$ are the scale factors to non-dimensionalise the distance and the velocity. Substituting these in the above equation, we get

$$u = \left(\frac{d\overline{u}}{dy}\right)_p l_v\eta + \left(\frac{d^2\overline{u}}{dy^2}\right)_p \frac{l_v^2\eta^2}{2} + \ldots + Au'_1 \qquad \text{and } v = Av'_1$$

Since $u'_1$ and $v'_1$ are function of $\xi$ and $\eta$ only, this requires

$$l_v\left(\frac{d\overline{u}}{dy}\right)_p \sim l_v^2\left(\frac{d^2\overline{u}}{dy^2}\right)_p \sim A$$

or, $$l_v \sim \left(\frac{d\overline{u}/dy}{d^2\overline{u}/dy^2}\right)_p \text{ and } A \sim l_v \left(\frac{d\overline{u}}{dy}\right)_p$$

You may see from dimensional analysis that the right hand side of $l$ has the dimensions of length. Karman then related this length with the mixing length $l$ as

$$l = \kappa l_v \tag{6.7.5}$$

where $\kappa$ is universal constant, also known as Von Karman constant, whose value is approximately equal to 0.4, as mentioned before, obtained from the experimental results of Nikuradse and Reichardt. He generalised this for a three-dimensional flow too, which we would not discuss here. This theory is called the Von Karman's similarity theory. He obtained, using this scale, the velocity distribution near the wall of a flat plate boundary layer flow at large Reynolds number when the laminar sublayer would be practically non-existent. Substituting for $l_v$ from Eq. 6.7.4 in Eq. 6.7.2, we get the Reynolds stress

$$\tau_t = \frac{\rho\kappa^2 (d\overline{u}/dy)^2 (d\overline{u}/dy)^2}{(d^2\overline{u}/dy^2)^2} = \frac{\rho\kappa^2 (d\overline{u}/dy)^4}{(d^2\overline{u}/dy^2)^2} \tag{6.7.6}$$

or, $$\left(\frac{d\overline{u}}{dy}\right)^4 = \frac{\tau_t}{\rho}\frac{1}{\kappa^2}\left(\frac{d^2\overline{u}}{dy^2}\right)^2$$

$$\rightarrow \quad \left(\frac{d\overline{u}}{dy}\right)^2 = \pm\frac{u_\tau}{\kappa}\left(\frac{d^2\overline{u}}{dy^2}\right) \tag{6.7.7}$$

The last equation is obtained on the assumption that $\tau_t = \tau_w$ and using $u_\tau = \sqrt{\tau_w/\rho}$, which would be true for flows at high Reynolds number because then the *N–S* equations for turbulent flow give

$$\frac{\partial}{\partial y}\left[\mu\frac{\partial\overline{u}}{\partial y} + \tau_t\right] = \frac{\partial\overline{p}}{\partial x} = 0 \rightarrow \tau_t = \tau_{t,0},$$

where $\tau_{t,0}$ is the turbulent shear stress at the junction of turbulent and viscous zones. It is based on the suggestion from Von Karman, Nikuradse, and others that at high Reynolds number and zero pressure gradient in the $x$ direction, the viscous layer is very thin; buffer zone is neglected for convenience. Thus the turbulent stress is constant throughout the boundary layer and can be taken $\tau_w$. After integration with respect to $y$ and using the condition that $d\overline{u}/dy \rightarrow \infty$ as $y \rightarrow 0$, Eq. 6.7.7 gives

$$\frac{d\overline{u}}{dy} = \frac{u_\tau}{\kappa y}$$

One more integration with the condition $y = y_0$, $\overline{u} = 0$ and using $\eta = yu_\tau/\nu$, $\eta_0 = y_0 u_\tau/\nu$ gives

$$\frac{\overline{u}}{u_\tau} = \frac{1}{\kappa}(\ln y - \ln y_0)$$

$$= \frac{1}{\kappa}\left(\ln \eta + \ln \frac{\nu}{u_\tau} - \ln y_0\right)$$

$$= \frac{1}{\kappa}(\ln \eta - \ln \eta_0) = A \ln \eta + B \tag{6.7.8}$$

This is known as the *universal velocity profile.* Based on Nikuradse and Reichardt's data for velocity distribution on the flat plate, value of 0.4 for $\kappa$ and 0.111 for $\eta_0$ gave excellent results between the experimental measurements and the corresponding values obtained from the Universal velocity profile. Thus the velocity profile is

$$\frac{\overline{u}}{u_\tau} = 2.5 \ln \eta + 5.5 \tag{6.7.9}$$

This equation is not applicable in the region near the wall where the viscous stress dominates over the turbulent stresses. In fact in the region $5 < \eta \leq 60$, the buffer zone, it is observed that Eq. 6.7.9 does not predict the results as obtained from experiments; these results are shown in Fig. 6.8.1, in the next section.

***Reichardt's Inductive theory***. The various theories discussed above make some hypothesis and deduce the velocity law by solving the differential equations to which the equations of motion reduce to under the hypothesis. Reichardt on the other hand took the experimental data and succeeded in finding a simple formula for free turbulent flow for the case of a two-dimensional incompressible flow. The laminar stresses are negligible in the free turbulence region. Such a theory, based on formulating empirical relation from the experimental data, is termed as inductive theory. He started with the following momentum equation

$$\rho\left(\frac{\partial \overline{u^2}}{\partial x} + \frac{\partial \overline{uv}}{\partial y}\right) = -\frac{\partial p}{\partial x} \xrightarrow{\text{temporal average}} \frac{\partial \overline{u'^2}}{\partial x} + \frac{\partial \overline{u'v'}}{\partial y} = 0,$$

for the fluctuating components, since the pressure term vanishes for free turbulence. From the experimental data, Reichardt establish the following empirical relation

$$\overline{u'v'} = -\Delta(x)\frac{\partial \overline{u'^2}}{\partial y}$$

and obtained the equation, known as Reichardt's fundamental equation

$$\frac{\partial \overline{u'^2}}{\partial x} = \Delta(x)\frac{\partial^2 \overline{u'^2}}{\partial y^2}$$

The quantity $\Delta(x)$ is called the momentum transfer length and is determined experimentally.

## 6.8 Universal Velocity Distribution Law in Ducts for Large Reynolds Numbers

We will now determine the velocity distribution in turbulent flow in flows through channel taking the distance between the two plates as $2h$ and the origin of the coordinates at the surface. For large Reynolds numbers, Prandtl assumed constant shear stress throughout, i.e. $\tau_t = \tau_w$. Equation 6.7.2 gives using Eq. 6.6.2

$$\rho l^2 \left(\frac{\partial \bar{u}}{\partial y}\right)^2 = \rho u_\tau^2 \quad \text{or,} \quad \kappa^2 y^2 \left(\frac{\partial \bar{u}}{\partial y}\right)^2 = u_\tau^2$$

$$\rightarrow \qquad \frac{\partial \bar{u}}{\partial y} = \frac{u_\tau}{\kappa y} \tag{6.8.1}$$

Integration with respect to $y$ gives

$$\bar{u} = \frac{u_\tau}{\kappa} \ln y + C$$

The constant of integration is given by the condition that at $y = h$, the distance of the axis of the pipe (duct), where the velocity is maximum $\bar{u}_m$, $\bar{u} = \bar{u}_m$. This condition is based on the assumption that the condition of constant stress is applicable for the entire region. The velocity distribution is

$$\frac{\bar{u}_m - \bar{u}}{u_\tau} = \frac{1}{\kappa} k \ln\left(\frac{h}{y}\right) \tag{6.8.2}$$

Equation 6.8.2 is known as universal velocity defect law of Prandtl. We could have as well taken the boundary condition at $y = 0$ where $\bar{u} = 0$, instead of at $y = h$. But this condition cannot be satisfied by a finite constant value in Eq. 6.8.2. However, for flows at very large Reynolds number, the laminar sublayer can practically be ignored. In such case we may choose a distance $y_0$, of the order of the thickness of the sublayer, at which $\bar{u}$ can be taken 0. By writing $y$ in terms of $\eta$, Eq. 6.8.1 gives after simplifying

$$\frac{\bar{u}}{u_\tau} = \frac{1}{\kappa}(\ln y - \ln y_0)$$

$$= \frac{1}{\kappa}\left(\ln \eta + \ln \frac{\nu}{u_\tau} - \ln y_0\right)$$

$$= \frac{1}{\kappa}(\ln \eta - \ln \eta_0)$$

$$= A \ln \eta + B \tag{6.8.3}$$

Equation 6.8.3 is the same as that obtained for flow over a flat plate, which we have called universal profile. It is applicable from moderate to a very high Reynolds number. Here $A$ and $B$ are unknown and are determined from experiments. Nikuradse and Reichardt found these constants to be $A = 2.5$ and $B = 5.5$. The velocity distribution has been shown in Fig. 6.8.1.

$$\frac{\bar{u}}{u_\tau} = 2.5 \ln \eta + 5.5 \tag{6.8.4}$$

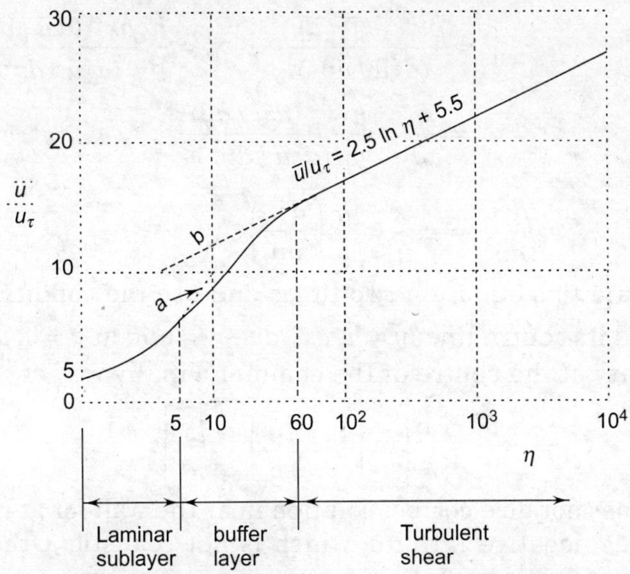


**Fig. 6.8.1** Universal velocity profile

The universal velocity profile differs from the actual data close to the wall. To be exact in the laminar sublayer and the buffer zone, as can be seen in the figure. Von Karman suggested the modifications, which are

$$\frac{\overline{u}}{u_\tau} = \eta = \frac{yu_\tau}{\nu} \quad \text{for } \eta < 5.0 \text{ (laminar sublayer)}$$

$$\frac{\overline{u}}{u_\tau} = 11.5 \log \frac{yu_\tau}{\nu} - 3.0 \quad \text{for } 5.0 < \eta < 60 \text{ (buffer zone)} \tag{6.8.5}$$

These equations are shown in Fig. 6.8.1 as curves *a* and *b*.

We may similarly find out the velocity profile in a fully developed flow in a channel using Von Karman relation with $y$ measured from its centre line and $x$ along it. For a fully developed flow, the acceleration terms vanish and the *N–S* equation gives

$$\frac{\partial \tau_t}{\partial y} = \frac{\partial \overline{p}}{\partial x}$$

or,

$$\tau_t = \frac{\partial \overline{p}}{\partial x} y + C$$

The constant of integration is obtained from the condition at the pipe axis, $y = 0$, $\tau_t = 0$. This gives $C = 0$ and $\tau_t = \frac{\partial \overline{p}}{\partial x} y$. At the wall $y = h$, the half width of the channel, $\tau_w = \frac{\partial \overline{p}}{\partial x} h$, or, $\frac{\tau_w}{\tau_t} = \frac{h}{y}$. Using the Von Karman relation in the expression for the turbulent stress, we get

$$\tau_t = \frac{\rho\kappa^2 (d\bar{u}/dy)^4}{(d^2\bar{u}/dy^2)^2}, \quad \tau_w = \frac{h}{y}\frac{\rho\kappa^2 (d\bar{u}/dy)^4}{(d^2\bar{u}/dy^2)^2}$$

$$u_\tau^2 = \frac{\tau_w}{\rho} = \frac{h}{y}\frac{\kappa^2 (d\bar{u}/dy)^4}{(d^2\bar{u}/dy^2)^2}$$

or,

$$\frac{d^2\bar{u}}{dy^2} = \pm\frac{\kappa}{u_\tau}\sqrt{\frac{h}{y}}\left(\frac{d\bar{u}}{dy}\right)^2$$

We may now integrate this equation two times and use the conditions, respectively, at the channel wall and its centre line: $y = h$, $d\bar{u}/dy \to \infty$ and at $y = 0$, $\bar{u} = u_m$, where $u_m$ is the maximum velocity at the centre of the channel. Finally we get

$$\frac{u_m - \bar{u}}{u_\tau} = \frac{1}{\kappa}\left[\ln\left(1 - \sqrt{\frac{y}{h}}\right) - \sqrt{\frac{y}{h}}\right]$$

It is seen that this does not give correct solution near the wall for at the wall where $y = h$, the velocity becomes negative infinite, which is not realistic. Otherwise, this gives a distribution very close to the actual.

## 6.9 Fully Developed Turbulent Flow in a Pipe for Moderate Reynolds Numbers

In Chapter 4, we derived a relation between the pressure gradient along the direction of flow in terms of the frictional force offered by the pipe surface. The force balance on the fluid element, as taken in Chapter 4 (see section 4.2), gives after simplification

$$\tau_w = f\rho U_{av}^2 / 8 \tag{6.9.1}$$

Different empirical relations were given at different times based on the experimental results; some of these are given below

$$\text{Blassius: } f = 0.3164\ \text{Re}^{-0.25}, \quad \text{Re} = \rho U_{av} D/\mu \text{ and Re} \le 10^5 \tag{6.9.2}$$

The other relation that is commonly used for higher range of Re is $\frac{1}{\sqrt{f}} = 2.0 \log\left(\frac{\text{Re}_D\sqrt{f}}{2.51}\right)$, which is applicable for Re up to $3.2 \times 10^6$. This relation is implicit and hence needs iteration to get the value of $f$. The dependence of $f$ on viscosity is explained from the fact that the boundary layer near the wall is laminar. It is distorted many a times by the manufacturing procedure that generates additional turbulence. Therefore friction factor is high in rough pipes as compared to smooth pipes. In experiments over a pipe with sandpaper like roughness of physical height $\varepsilon$, the measured friction factor $f$ was found by Colebrook, who expressed it as the following empirical relation

$$\frac{1}{\sqrt{f}} = -2.0 \log\left(\frac{\varepsilon/D}{3.7} + \frac{2.51}{\text{Re}\sqrt{f}}\right) \tag{6.9.3}$$

The equivalent wall roughness $\varepsilon$ for commercial pipes has been determined from few tests and is given in the following table.

**TABLE 6.9.1** Wall roughness $\varepsilon$ for commercial pipes

| Material | $\varepsilon$, m | Material | $\varepsilon$, m | Material | $\varepsilon$, m |
|---|---|---|---|---|---|
| Concrete | $0.3 - 3.0 \times 10^{-3}$ | Cast iron | $3.0 \times 10^{-4}$ | Galvanised iron | $1.50 \times 10^{-4}$ |
| Commercial steel | $5.0 \times 10^{-5}$ | Drawn tubing | $1.5 \times 10^{-6}$ | Glass, plastic | 0 |

Experimental results of Nikuradse show that the velocity profile in the pipe flow at high Re can be expressed as

$$\frac{u}{\overline{u}} = \left(\frac{y}{R}\right)^{1/n} \tag{6.9.4}$$

where $R$ is pipe radius, $y$ is the distance of the point from the wall, and $\overline{u}$ is the time mean average velocity at the pipe centre. $U_{av}$, needed in Eq. 6.9.1, can be determined in the usual way, as explained earlier in chapter 4. This gives

$$Q = \pi R^2 U_{av} = \int_0^R 2\pi R u dr$$

$$\rightarrow \quad \frac{U_{av}}{R} = \frac{2n^2}{(n+1)(2n+1)} \tag{6.9.5}$$

Substituting for $U_{av}$ and $f$ (Eq. 6.9.2) in Eq. 6.9.1, we get

$$\tau_w = \frac{0.3164}{8} \text{Re}^{-0.25} \rho U_{av}^2$$

$$= 0.03325\rho \left(\frac{U_{av}}{\overline{u}}\right)^{7/4} (\overline{u})^{7/4} (\nu / R)^{1/4}$$

Value of $n$ for such flow is found to be 7 for which $U_{av} / \overline{u} = 0.8$. Substituting these values in the above equation, we get

$$\tau_w = 0.03325\rho\,(0.8)^{7/4} (\overline{u})^{7/4} (\nu / R)^{1/4} = 0.025\ \rho(\overline{u})^{7/4} (\nu / R)^{1/4}$$

$$\rightarrow \quad u_\tau^2 = \frac{\tau_w}{\rho} = 0.025\,(\overline{u})^{7/4} (\nu / R)^{1/4}$$

$$\frac{\overline{u}}{u_\tau} = 0.74 \left(\frac{u_\tau R}{\nu}\right)^{1/7} \tag{6.9.6}$$

## 6.10 Isotropic and Homogeneous Turbulence

It will not be possible to discuss the turbulence in general in this book. But we may get some idea for the analytical approach for a simple case of isotropic and homogeneous turbulence. We mentioned in the early sections that the turbulent flow could be explained with the help of correlation tensors. Correlations involve turbulent velocity components and other fluctuating components like pressure, temperature, etc. There are three velocity components, represented by $u_1$, $u_2$, and $u_3$ in our discussions hence after. In general we use the symbol $u_i$ to represent velocity components. Let us consider $n$ points in the turbulent flow region. The $n$-point correlation of the order $k$ among the velocity components at these points is expressed as

$$C^k_{ij\ldots p}\left(\vec{X}_1, \vec{X}_2, \ldots, \vec{X}_n\right) = < u_i\left(\vec{X}_1\right) u_j\left(\vec{X}_2\right) \ldots u_p\left(\vec{X}_n\right) > \qquad (6.10.1)$$

$k$ is called the order of the tensor and equals $p$. We would restrict ourselves to a two-point 2nd order correlation. We know isotropy means that value of the correlation will not change if we take the image of the given two points about any line passing through the origin or about the origin itself that changes $\vec{r}$ to $-\vec{r}$. Reflection about the axes give

$$< u_1\, u_2 > = - < u_1\, u_2 > = 0, \text{ Similarly } < u_2\, u_3 > = 0 \text{ and } < u_1\, u_3 > = 0 \qquad (6.10.2)$$

$<u_1^2>$ is a correlation for $u_1$ component at the same point. Due to isotropy

$$<u_1^2> = <u_2^2> = <u_3^2> \rightarrow <u^2> = \frac{1}{3}\left[< u_1^2 > + < u_2^2 > + < u_3^2 >\right] \qquad (6.10.3)$$

Homogeneity demands that the correlation should depend on the relative position of the points and not on the position of the origin. This means that in the analysis one of the points can be taken as the origin and the position of other points be given with its reference. In a two-point correlation, it further simplifies the analysis because one of the axes can be coincided with the line joining the two points.

Appendix A deals with the vectors and tensors. You should read it before proceeding further. You will learn that a vector or a tensor satisfies some transformation law. Without going into details, we just mention that transformation characteristic of a homogeneous isotropic random field allows to write the following expressions for a two-point correlation tensor.

1st Order: $C_i(\vec{r}) = A_0\,(r)\, r_i$ (6.10.4)

2nd Order: $C_{ij}\,(\vec{r}) = A_1\,(r)\, r_i\, r_j + B_1\,(r)\, \delta_{ij}$ (6.10.5)

3rd Order: $C_{ijk}\,(\vec{r}) = A_2\,(r)\, r_i\, r_j\, r_k + B_2\,(r)\, r_k\, \delta_{ij} + C_2\,(r)\, r_j\, \delta_{ik} + D_2\,(r)\, r_i\, \delta_{jk}$ (6.10.6)

$= A_2\,(r)\, r_i\, r_j\, r_k + B_2\,(r)\, r_k\, \delta_{ij} + C_2\,(r)\,[r_i\, \delta_{jk} + r_j\, \delta_{ik}]$

since $c_{ijk}$ is symmetric with respect to $i, j$ so that $C_2(r) = D_2(r)$.

This means that the three components of $C_i\,(\vec{r})$ can be written in terms of one unknown $A_0(r)$, the 9 components of $C_{ij}\,(\vec{r})$ in terms of 2 unknowns and so on. The problem is still not over as we have two unknowns. To determine these, consider two points $A$ and $B$ respectively at $\vec{X}$ and $\vec{X} + \vec{r}$, as shown in Fig. 6.10.1, with the line joining these points as one of the axes $X_1$, i.e. $\vec{r} = ir$, where $r$ is the length of the line $AB$. This means $r_1 = r$ and $r_2 = r_3 = 0$. Then

$$C_{11}\,(\vec{r}) = A_1\,(r)\, r^2 + B_1\,(r),$$

$$C_{12} = C_{21} = 0,\ C_{13} = C_{31} = 0,\ C_{23} = C_{32} = 0$$

$$C_{22} = C_{33} = < u_n (\vec{x}) u_n (\vec{x} + \vec{r}) > = B_1 (r).$$

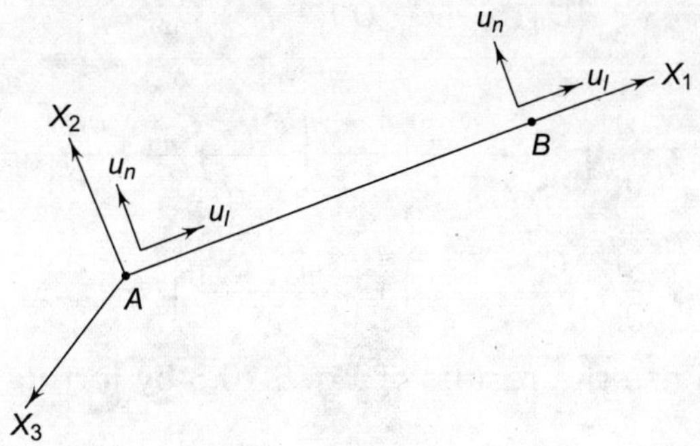


**Fig. 6.10.1** Velocity components in homogeneous and isotropic random field

That is, there will be only diagonal non-zero terms in the correlation tensor. Advantage in this form is that

$C_{11}(r)$, longitudinal correlation = $C_{ll}(r) = < u_l (\vec{x})\, u_l (\vec{x} + \vec{r})>$

and $C_{22}(r)$ = lateral correlation = $C_{nn}(r) = < u_n (\vec{x})\, u_n (\vec{x} + \vec{r})>$

can be measured experimentally. Writing these in the above equation, we get

$$B_1 (r) = C_{nn} (r), A_1 (r) = \frac{1}{r^2}\left[C_{ll}(r) - C_{nn}(r)\right]$$

and
$$C_{ij} (\vec{r}) = \frac{1}{r^2}\left[C_{ll}(r) - C_{nn}(r)\right] r_i\, r_j + C_{nn}\, \delta_{ij} \tag{6.10.7}$$

Equation 6.10.7 is further simplified in the case of incompressible fluids because the continuity equation is $\partial u_i / \partial r_i = 0$. For homogeneous field, it is immaterial as to where is the origin and we may choose the point $A$ as the origin. In such case

$$C_{ij} (\vec{r}) = < u_i (0)\, u_j (\vec{r}) > = < u_j (0)\, u_i (-\vec{r}) > = C_{ij} (-\vec{r})$$

$$\frac{\partial C_{ij} (\vec{r})}{\partial r_i} = \frac{\partial}{\partial r_i} <u_j (0)\, u_i (-\vec{r}) > = < u_j (0) \frac{\partial u_i (-\vec{r})}{\partial r_i} > = 0 \tag{6.10.8}$$

We can also write $r = |\vec{r}| = \sqrt{r_1^2 + r_2^2 + r_3^2}$, i.e. $r^2 = r_i\, r_i$. On differentiating it with respect to $r_i$, we get

$$2r \frac{\partial r}{\partial r_i} = 2r_i$$

or,
$$\frac{\partial}{\partial r_i} = \frac{r_i}{r} \frac{\partial}{\partial r}$$

$\rightarrow$
$$\frac{\partial r_j}{\partial r} = \frac{r}{r_k} \frac{\partial r_j}{\partial r_k} = \frac{r}{r_k} \delta_{jk},$$

$$\frac{\partial r_i}{\partial r_j} = \delta_{ij} \text{ and } r_i\, \delta_{ij} = r_j$$

We now differentiate Eq. 6.10.7 with respect to $r_k$ to give

$$\frac{\partial C_{ij}(\vec{r})}{\partial r_k} = \frac{r_k}{r}\frac{\partial}{\partial r}\left\{\frac{1}{r^2}\left[C_{ll}(r) - C_{nn}(r)\right] r_i r_j + C_{nn}\,\delta_{ij}\right\}$$

$$= \frac{r_k}{r} r_i r_j \frac{\partial}{\partial r}\left(\frac{C_{ll} - C_{nn}}{r^2}\right) + \frac{r_k}{r}\left(\frac{C_{ll} - C_{nn}}{r^2}\right)\frac{\partial r_i r_j}{\partial r} + \frac{r_k}{r}\frac{\partial C_{nn}}{\partial r}\delta_{ij}$$

$$= \frac{r_k}{r} r_i r_j \frac{\partial}{\partial r}\left(\frac{C_{ll} - C_{nn}}{r^2}\right) + \left(\frac{C_{ll} - C_{nn}}{r^2}\right)\left(r_i\,\delta_{jk} + r_j\,\delta_{ik}\right) + \frac{r_k}{r}\frac{\partial C_{nn}}{\partial r}\delta_{ij}$$

At this stage, we can use the results of Eq. 6.10.8 by letting $k$ = i in this equation to give

$$\frac{\partial C_{ij}(\vec{r})}{\partial r_i} = \frac{r_i r_i r_j}{r}\frac{\partial}{\partial r}\left(\frac{C_{ll} - C_{nn}}{r^2}\right) + \left(\frac{C_{ll} - C_{nn}}{r^2}\right)\left(r_i\,\delta_{ji} + r_i\,\delta_{ii}\right) + \frac{r_i}{r}\frac{\partial C_{nn}}{\partial r}\delta_{ij}$$

$$= rr_j \frac{1}{r^2}\frac{\partial}{\partial r}\left(C_{ll} - C_{nn}\right) - \frac{2rr_j}{r^3}\left(C_{ll} - C_{nn}\right) + 4r_j\left(\frac{C_{ll} - C_{nn}}{r^2}\right) + \frac{r_j}{r}\frac{\partial C_{nn}}{\partial r}$$

$$= \frac{r_j}{r}\left[\frac{\partial C_{ll}}{\partial r} + 2\frac{C_{ll}}{r} - 2\frac{C_{nn}}{r}\right] = 0$$

This gives, since $r_j / r \neq 0$,

$$\frac{\partial C_{ll}}{\partial r} + 2\frac{C_{ll}}{r} - 2\frac{C_{nn}}{r} = 0$$

$$\rightarrow \qquad C_{nn} = \frac{r}{2}\frac{\partial C_{ll}}{\partial r} + C_{ll} = \frac{1}{2r}\frac{\partial\left(r^2 C_{ll}\right)}{\partial r}.$$

$$\therefore \qquad C_{ij}(\vec{r}) = -\frac{1}{2r}\frac{\partial C_{ll}}{\partial r} r_i r_j + \left(C_{ll} + \frac{r}{2}\frac{\partial C_{ll}}{\partial r}\right)\delta_{ij} \tag{6.10.9}$$

We notice that now only one correlation $C_{ll}$ is needed to get the stress tensor and that this correlation can easily be obtained from the experimental data.

Often a question comes to our mind "what will be the size of an eddy?" Basically this question is related to some scale of the eddy size because eddies of different sizes are present in the flow. In a laminar flow, we have a feel of the size, as it can be the boundary layer thickness or some such measure we discussed in Chapter 5. We do not ask such questions as asked in turbulent flow because our senses can feel it. In general, any problem is associated with different kind of scales; these may be length, velocity, time scale, and so on so forth depending on the type of problem and the analysis. Moreover, there may be more than one scale. So is the case of turbulent flow. Different scales were proposed by the scientists to answer this question, which are derived from the analysis itself and not from the physical dimensions of the flow region. The only statement one can make is that the size of the eddy cannot be greater than the physical dimension. We will now get some turbulence scale.

It is clear that the correlation between turbulent quantities at different points determine the size of eddy. That is the correlation coefficient would be zero at points not lying within the eddy region. Using the definition of correlation coefficient, we write

$$R_{ll}(r) = \frac{C_{ll}(r)}{C_{ll}(0)} \text{ and } R_{nn}(r) = \frac{C_{nn}(r)}{C_{nn}(0)},$$

such that $R_{ll}(r) \le 1, R_{nn}(r) \le 1$, (Schwarz inequality)

$$R_{ll}(0) = 1, \text{ and } R_{nn}(0) = 1.$$

Expanding $R_{ll}(r)$ and $R_{nn}(r)$ around the origin and keeping in mind that isotropy demands the correlation coefficient to be even function, we get

$$R_{ll}(r) = R_{ll}(0) + \frac{1}{2}\frac{d^2 R_{ll}(0)}{dr^2} + O(r^4) = 1 + \frac{1}{2}\frac{d^2 R_{ll}(0)}{dr^2} + O(r^4) \approx 1 - \frac{r^2}{2\lambda_1^2}$$

$$R_{nn}(r) = 1 + \frac{1}{2}\frac{d^2 R_{ll}(0)}{dr^2} + O(r^4) \approx 1 - \frac{r^2}{2\lambda_n^2},$$

where $$\lambda_l^2 = -\frac{1}{\dfrac{d^2 R_{ll}(0)}{dr^2}} \quad \text{and} \quad \lambda_n^2 = -\frac{1}{\dfrac{d^2 R_{nn}(0)}{dr^2}} \tag{6.10.10}$$

$\lambda_l$ and $\lambda_n$ are called the longitudinal and lateral microscales of the turbulence, which are decided by the second derivative of the respective correlation coefficients at the origin.[2] For a solenoidal flow, since the longitudinal and lateral correlations are related to each other, there exists a relation between the two microscales. We proceed as follows to establish this relation.

$$C_{nn} = \frac{r}{2}\frac{\partial C_{ll}}{\partial r} + C_{ll}$$

$$\rightarrow \quad \frac{\partial C_{nn}}{\partial r} = \frac{1}{2}\frac{\partial C_{ll}}{\partial r} + \frac{r}{2}\frac{\partial^2 C_{ll}}{\partial r^2} + \frac{\partial C_{ll}}{\partial r} = \frac{3}{2}\frac{\partial C_{ll}}{\partial r} + \frac{r}{2}\frac{\partial^2 C_{ll}}{\partial r^2}$$

and $$\frac{\partial^2 C_{nn}}{\partial r^2} = 2\frac{\partial^2 C_{ll}}{\partial r^2} + \frac{r}{2}\frac{\partial^3 C_{ll}}{\partial r^3}$$

$$\Rightarrow \quad \left.\frac{\partial^2 C_{nn}}{\partial r^2}\right|_{r=0} = 2\left.\frac{\partial^2 C_{ll}}{\partial r^2}\right|_{r=0} \quad \text{or, } \lambda_l^2 = \lambda_n^2 \tag{6.10.11}$$

It is also possible to follow integral formulation to give a macro or the integral turbulence scale. It is defined as follows

$$L_1 = \int_0^\infty R_{ll}(r)\,dr \quad \text{and} \quad L_n = \int_0^\infty R_{nn}(r)\,dr \tag{6.10.12}$$

These scales are nothing but the area under the curve for the corresponding correlation coefficient when plotted against $r$. Thus these scales, unlike the microscales, cover the entire eddy region. These scales help us in studying variation of the coefficients throughout the eddy region. We know

[2] The turbulence scale is decided by the derivative of a function whence the name microscale.

$$C_{nn} = \frac{r}{2}\frac{\partial C_{ll}}{\partial r} + C_{ll}$$

Multiply both sides by $r^m$ and integrate with respect to $r$ from 0 to ∞. This gives

$$\int_0^\infty r^m\, C_{nn}\, dr = \int_0^\infty r^m\, \frac{r}{2}\frac{\partial C_{ll}}{\partial r}\, dr + \int_0^\infty r^m\, C_{ll}\, dr$$

Integrating by parts, $\int_0^\infty r^{m+1}\frac{\partial C_{ll}}{\partial r} dr = r^{m+1} C_{ll}\Big|_0^\infty - (m+1)\int_0^\infty r^m C_{ll}\, dr = -(m+1)\int_0^\infty r^m C_{ll}\, dr$

$$\therefore \quad \int_0^\infty r^m\, C_{nn}\, dr = \frac{1-m}{2}\int_0^\infty r^m\, C_{ll}\, dr$$

If $m = 0$, then $L_n = 1/2\, L_1$

Also for $m = 1$, $\int_0^\infty r\, C_{nn}\, dr = 0$

This means that $C_{nn}$ must be negative for some values of $r$, since $r$ cannot be negative in its given range, so that the integral becomes zero. Figure 6.10.2 gives an idea of variation of the correlation coefficients within the eddy.

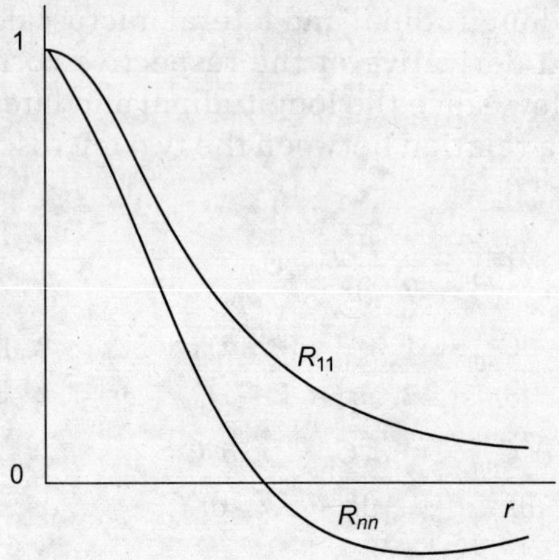


**Fig. 6.10.2** Correlation coefficient variation in homogeneous isotropic turbulent flow

Thus it is possible to define a characteristic eddy size based on the microscale $\lambda$ or integral scale $L$. further, $\sqrt{\overline{u^2}}$ is the characteristic velocity of eddy. Therefore, a characteristic Reynolds number of turbulence can be defined using these. Thus

$$\mathrm{Re}_L = \frac{L\sqrt{\overline{u^2}}}{\nu}$$

or,

$$\mathrm{Re}_\lambda = \frac{\lambda\sqrt{\overline{u^2}}}{\nu} \quad (6.10.13)$$

A few researchers were looking at this problem from the energy angle. It was an experimentally observed fact that the main flow was dissipating energy due to formation of turbulence and that the lumps of fluid move across the fluid in random manner. These lumps could be as big as the size of the tube in the Reynolds experiment on the flow through pipes. Taylor, Reynolds, Kolomogorov, etc. analysed the turbulence problem on the lines of energy spectrum as used in study of Sun radiations. We learn in physics how to express the energy flux from the Sun in terms of wavelength or frequency of radiations arriving on Earth. A wave number $k$, which is reciprocal of the eddy size $\lambda$, was defined as

$$k = \omega / \overline{u} = 2\pi / \lambda \tag{6.10.14}$$

where $\omega$ is the frequency of the turbulent velocity components and $\overline{u}$ is the time averaged mean flow velocity. The total kinetic energy of turbulent flow is

$\overline{(\overline{u} + u').(\overline{u} + u')} = \overline{u}\,.\,\overline{u} + \overline{u'\,.\,u'} = |\overline{u}|^2 + |u'|^2$ (factor ½ is not considered).

The first term represents the kinetic energy of the mean flow and the second that of turbulent kinetic energy. That is the turbulent velocity component causes transfer of energy to the eddies. Through mathematical analysis, it is possible to show dependence of the energy spectrum function $\Phi$ on the wave number, where the turbulence kinetic energy can be written in terms of energy of eddies having different wave number using Fourier integral theorem.

$$|u'|^2 = \int \Phi\,(k)\,dk$$

Such analysis leads to results as shown in Fig. 6.10.3. It shows that the energy spectrum is maximum at some eddy size corresponding to wave number $k_e$, which is towards

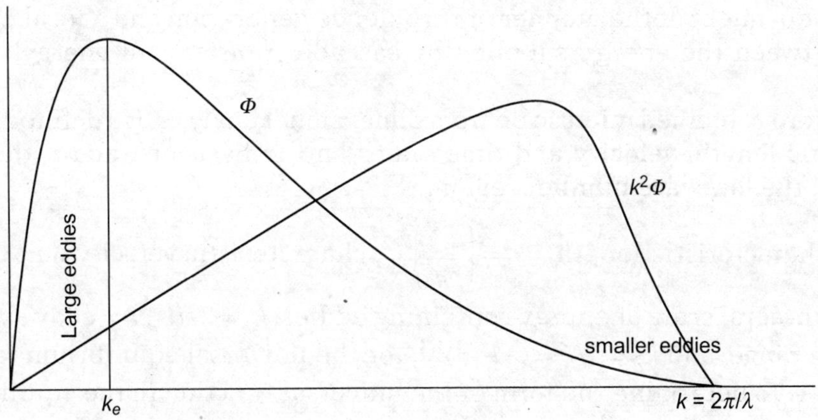


**Fig. 6.10.3** Energy spectrum as a function of wave number

the large size eddies. That is, the large sized eddy carry the large chunk of energy. The figure also shows graph for $k^2\Phi$. This factor is measure of energy dissipation that occurs at large wave number, as is seen from the figure. This is so because of the factor $k^2$. Thus the small size eddies are responsible for dissipation of kinetic energy to the main fluid.

Kolomogorov gave three hypotheses, based on dimensional analysis, about the transfer of energy from large size to the small size and ultimately conversion to internal energy of the main flow.

*Kolomogorov's first hypothesis.* For large turbulence Reynolds number, the large size eddies no longer remain isotropic in presence of the boundaries. However, for eddies of size much smaller than the characteristic eddy size $L_0$, i.e. $\lambda << L_0$, then the eddy may be considered practically isotropic in the region sufficiently away from the boundaries. In the above plot, this corresponds to the region $k >> k_e$, that is towards the right hand side of the graph. Thus in general the flow will be non-isotropic but there exists some area where it can be considered locally isotropic.

*Kolomogorov's second hypothesis.* Kolomogorov divided the space in Fig. 6.10.3 into three regions (a) $k << k_e$, (b) $k \approx k_e$, (c) $k >> k_e$, and suggested the following model of energy transfer betweeen different sized eddies in the figure.

(i) The turbulent energy produced out of the mean flow energy is fed largely to the low wave number region.
(ii) There is transfer of turbulent energy from one eddy to another in a hierarchical pattern through a cascade mechanism of energy transfer from low wave number to high wave number.
(iii) Turbulent energy is being dissipated due to friction for all eddies. This occurs in the region where $k >> k_e$. That is, the inertial and viscous forces control the motion; the manner of turbulence generation has almost no role in this region.

With these ideas in mind, Kolomogrov's second hypothesis, also called Universal Equilibrium hypothesis, states that in a locally homogeneous and isotropic region ($k >> k_e$) the motion is determined by the forces of only friction and inertia. It means spectrum shape is independant of the manner of turbulence generation and a local equilibrium is attained between the energy supplied by cascade transfer and energy dissipated by friction.

Kolomogorov obtained microscale from dimensional analysis. He defined the following characteristic length, velocity and time scales on the basis of $\nu$ and $\varepsilon$, the parameters that control the low wave number region:

$$\text{Characteristic length} = \eta = \left(\frac{\nu^3}{\varepsilon}\right)^{1/4}, \text{ characteristic velocity} = u = (\nu\varepsilon)^{1/4}.^*$$

Let $L_f$ = integral scale of energy containing eddies, $k_e = 2\pi/L_f$, $\eta$ = eddy size of univesal equilibrium range, and $k_d = 1/\eta = (\varepsilon / \nu^3)^{1/4}$; for the universal equilibrium range $k_e << k_d$, and $\text{Re}_L >> 1$. To determine the form of the universal function in the equilibrium range, let us express $\Phi$ as

$$\Phi(k, t) = \Phi_e (k / k_d)\, \eta^a u^b$$

---

* $[\nu] = \dfrac{L^2}{T}, [\varepsilon] = \left[\dfrac{d\overline{u^2}}{dt}\right] = \dfrac{L^2}{T^3}$

$$\rightarrow \quad [\eta] = \left[\frac{\nu^3}{\epsilon}\right]^{1/4} = \left(\frac{L^6}{T^3}\frac{T^3}{L^2}\right)^{1/4} = L, [u] = \left(\frac{L^2}{T}\cdot\frac{L^2}{T^3}\right)^{1/4} = \frac{L}{T}$$

$\Phi_e(k/k_d)$ is non-dimensional spectrum. $[\eta] = L$, $[u] = L/T$, and $[\Phi] = L^3T^{-2}$. Writing the dimensions on both sides of the equation and equating powers of $L$ and $T$, we get: $a = 1$ and $b = 2$. This gives

$$\Phi(k, t) = \Phi_e(k/k_d)\,\eta u^2 = (\varepsilon v^5)^{1/4}\,\Phi_e(k/k_d) \tag{6.10.15}$$

$$k_d = (\varepsilon v^5)^{1/4} \approx \mathrm{Re}_D^{3/4}\,D$$

*Kolomogorov's third hypothesis.* It states that there exists a range of wave numbers in the energy spectrum lying between $k_d$ and $k_e$ where the structure of the spectrum function depends on the inertia or spectral transfer of energy, i.e., only on the total dissipation of energy $\varepsilon$. This range of wave numbers is called universal equilibrium range. We may obtain variation of $\phi$ with $k$ demensionally for this range of wave numbers by rating that $\phi$ depends on $k$ and $\varepsilon$ and carrying out dimensional analysis, as follows:

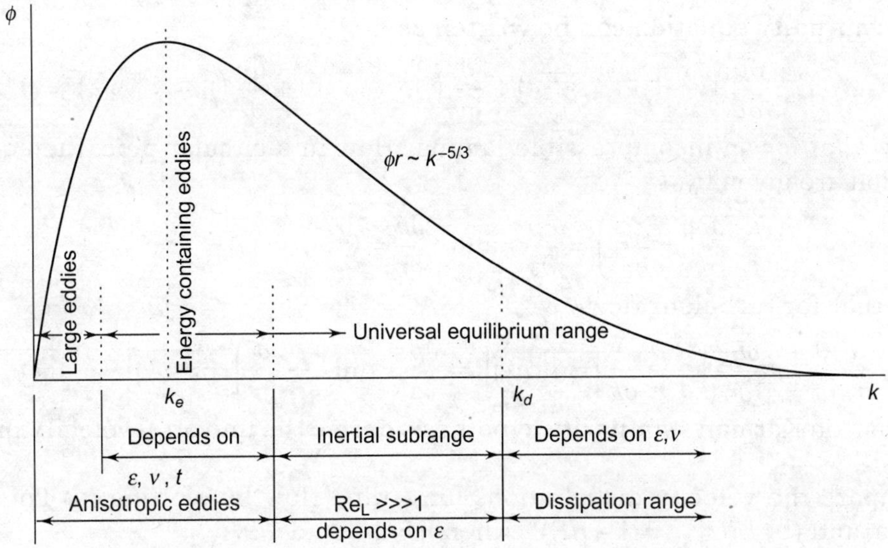


**Fig. 6.10.4** Energy spectrum based on Kolmogorov hypothesis

$$\Phi(k, t) = Ck^a \varepsilon^b \rightarrow L^3 T^{-2} = L^{-a} L^{2b} T^{-3b}.$$

$$\therefore \quad a = -5/3 \text{ and } b = 2/3.$$

$$\Phi(k, t) = C\varepsilon^{2/3} k^{-5/3} \tag{6.10.16}$$

As a result of the above discussions, the qualitative picture of $\Phi$ that over the most of the spectral range $\Phi$ varies as $k^{-5/3}$, as shown in Fig. 6.10.4, emerges.

# PROBLEMS

## Problems for Practice

1. Derive the continuity equation for compressible turbulent flow.
2. For an incompressible laminar flow in a circular pipe the equations of motion are given by

$\mu\frac{\partial}{\partial r}\left(r\frac{\partial u_z}{\partial r}\right) = r\frac{\partial p}{\partial z}$ and $\frac{\partial p}{\partial r} = 0$. Show that for turbulent flow in such pipe, the equations of motion can be written as.

$$\mu\frac{\partial}{\partial r}\left(r\frac{\partial \bar{u}_z}{\partial r}\right) - \mu\frac{\partial}{\partial r}\left[r\,\overline{(u_r'\,u_r')}\right] = r\frac{\partial \bar{p}}{\partial z} \text{ and } \rho\frac{\partial}{\partial r}\left[r\,\overline{(u_z'\,u_r')}\right] - \overline{(u_\theta'\,u_\theta')} = -\,r\frac{\partial \bar{p}}{\partial r}$$

3. Compare the velocity distribution for a pipe flow by plotting (a) Poiseuille formula, and (b) $u/u_{max} = (1 - r/R)^{1/n}$ for $n = 7$ and 9; $R$ is the pipe radius.

4. Verify that $\overline{\int_{t-T/2}^{t+T/2} u\,d\eta} = \int_{t-T/2}^{t+T/2} \bar{u}\,d\eta$

5. For compressible flows the fluctuating density is defined by $\rho' = \rho - \bar{\rho}$. Show that the continuity equation can be written as

$$\frac{\partial \bar{\rho}}{\partial t} + \frac{\partial}{\partial x}(\bar{\rho}\bar{u} + \overline{\rho' u'}) + \frac{\partial}{\partial y}(\bar{\rho}\bar{v} + \overline{\rho' v'}) + \frac{\partial}{\partial z}(\bar{\rho}\bar{w} + \overline{\rho' w'}) = 0$$

6. Show that for an incompressible laminar flow in a circular pipe, the equations of motion are given by

$$\frac{\mu}{r}\frac{\partial}{\partial r}\left(r\frac{\partial u_z}{\partial r}\right) = \frac{\partial p}{\partial z} \quad \text{and} \quad \frac{\partial p}{\partial r} = 0$$

and that for turbulent flow

$$\frac{\mu}{r}\frac{\partial}{\partial r}\left(r\frac{\partial \bar{u}_z}{\partial r}\right) - \frac{\rho}{r}\frac{\partial}{\partial r}\left[r\overline{(u_z'\,u_r')}\right] = \frac{\partial \bar{p}}{\partial z} \quad \text{and} \quad \frac{\rho}{r}\frac{\partial}{\partial r}\left[r\overline{(u_r'\,u_r')}\right] - \frac{\rho}{r}\overline{u_\theta'\,u_\theta'} = -\frac{\partial \bar{p}}{\partial r}$$

7. Using Von Karman's similarity hypothesis derive the universal velocity profile near a wall.
8. Compare the velocity distributions for a pipe flow by plotting (a) Poiseuille formula, and (b) $U/U_{max} = (1 - r/R)^n$, when $n = 1/7$ and $1/9$.
9. A 50 mm diameter (inside) horizontal smooth pipe is used to transport oil having $v = 103$ mm²/s. What power per kilometre would be required to maintain the flow when:
   (a) The flow rate of the oil is 0.005 m³/s?
   (b) The flow rate of the oil is 0.05 m³/s?
10. Oil having $v = 11.2$ mm²/s is transported through a horizontal SIT pipe line (1 km long) at a rate of 0.1 m³/s. Determine the power required to maintain the flow at a Reynolds number of 35,000.
11. Oil flows through a horizontal smooth pipe line with maximum velocity of 2 m/s. If the diameter of the pipe is 0.3 m and $v = 12$ mm²/s, find the power required per kilometre to maintain the flow at the velocity given above.
12. If the distance $2h$ between two parallel walls of a rectangular channel is taken to be much smaller than the width, show that the universal velocity distribution in a turbulent channel flow can be obtained by the Prandtl mixing length theory in the form

$$\frac{u}{u_{\max}} = 1 - \frac{1}{k}\left(\frac{u_\tau}{u_{\max}}\right)\left[\ln\left(\frac{1+\sqrt{y/h}}{1-\sqrt{y/h}}\right) - 2\sqrt{\frac{y}{h}}\right]$$

where $y$ is the distance measured from the centre line of the channel and $k$ is an empirical constant.

13. Find the average velocity of the turbulent flow in a channel in terms of the maximum velocity and the frictional velocity.
14. Using Von Karman's similarity law show that the universal velocity distribution in a turbulent channel flow is given by

$$\frac{u_m - \bar{u}}{u_\tau} = -\frac{1}{\kappa}\left[\ln\left(1 - \sqrt{\frac{y}{h}}\right) + \sqrt{\frac{y}{h}}\right]$$

15. Water at 20°C is transported through a horizontal smooth pipe line 1 km long, with maximum velocity of 3 m/s. If the Reynolds number is $10^6$, find the diameter of the pipe and the horsepower required to maintain the flow. Prandtl's universal velocity distribution may be used.
16. Using Von Karman's universal velocity distribution find the diameter of the pipe and horsepower required for Prob.
17. Show that the velocity distribution for turbulent flow in rough pipes can be expressed as $\frac{u_{\max} - u}{u_{av}} = -\sqrt{f}\ln(1 - r/R)$ and plot $(u_{\max} - u)/u_{av}$ vs. $r/R$ for the roughness Reynolds number $u_r\,\varepsilon/\nu = 100$ and Re = $10^6$.

CHAPTER

# 7 Compressible Flow

## OUTLINE

- Introduction
- Isentropic Flow Through Variable Area
- Adiabatic Frictional Flow Through Constant Area Duct— Fanno Flow
- Isothermal Flow with Friction
- Flow Through Smooth Constant Area Duct with Heat Transfer—Rayleigh Flow
- Normal Shock Wave

## 7.1 Introduction

We learnt some of the properties of the fluids in Chapter 1. However, the gases are compressible and as a result not only the density of the fluid changes but also some other properties, especially the compressibility coefficient changes. Therefore, Mach number, defined as ratio of the inertia and the elastic force, becomes an important non-dimensional number in the study of the behaviour of the compressible fluids. Moreover, as was mentioned in Chapter 2 that the complete set of mass, energy and momentum equations along with the equation of state is to be solved. These equations may also be coupled. Therefore the analytical solution is much more involved as compared to that of the incompressible fluids for which also we have seen that analytical solution in a few simple flows is available. For this reason, the subject is introduced by taking one-dimensional flow. We shall restrict ourselves to one-dimensional flows only during which we shall study isentropic flow through a variable area duct or through ducts with either presence of friction or heat transfer with the surroundings.

Compressibility of gas introduces two kinds of limitations, namely choking of flow and formation of shock wave. The former results in limiting maximum mass flow rate and latter causes loss in stagnation and static pressure as compared to isentropic flow condition. Choking is the condition arrived at while gas or vapour flows through a variable area duct and represents the condition of maximum mass flow rate, which in no case can be increased. This happens because Mach number, becomes unity at the cross section of the duct where the area is minimum; this cross section is called throat of the passage. Choking of flow through the constant area ducts may also occur due to the presence of friction or heat transfer. During the flow, entropy increases and tends to become maximum.

On the other hand, shock is the process of formation of discontinuity in the flow because of which the flow changes from supersonic to subsonic flow. In this process, there is no change in stagnation temperature but static pressure and temperature increase and stagnation pressure is decreased. The region of discontinuity is very thin.

### 7.1.1 Propagation of Infinitesimal Waves

In physics, we read about the sound waves and their propagation in a compressible or incompressible media. We shall restrict our discussions to compressible media only. Recall, when a sound is produced then the source generates alternate compression and rarefaction waves in the surrounding medium. These waves travel in all directions with a spherical wave front. Also the change in pressure results in bodily movement of the medium. This sets media particles next to it in motion and the process goes on. However, it does not mean that media particles start flowing downward. In fact, these particles vibrate about their mean position. The speed at which this disturbance propagates is called the acoustic speed.

To simplify the study, we consider that the wave front is plane so that it moves in one direction only. This type of wave is generated if some compressible fluid of density $\rho$ at pressure $p$ is placed in a tube of uniform cross sectional area '$A$' fitted with a piston at one end, as shown in Fig. 7.1.1 (a).

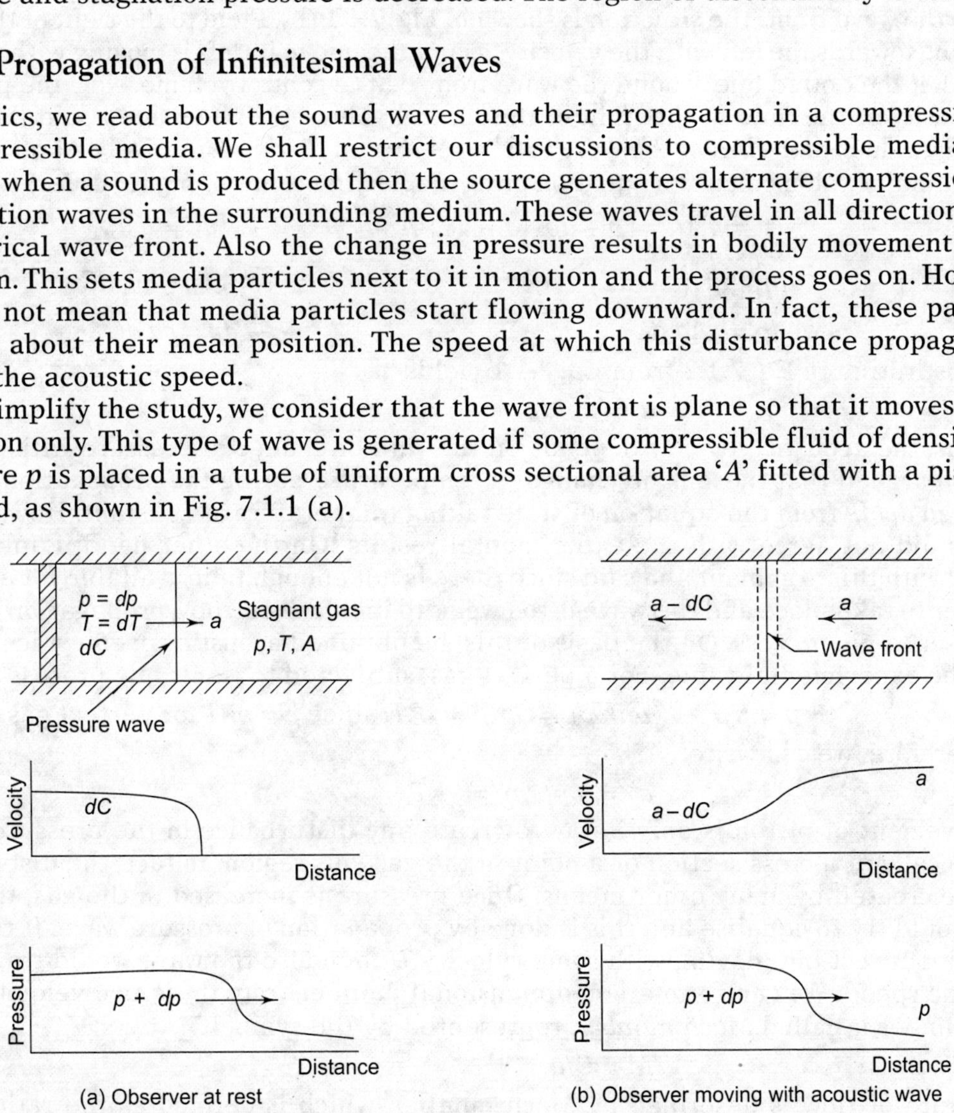


**Fig. 7.1.1** Propagation of an infinitesimal wave in constant area duct

Let a pulse be generated by pushing the piston a little to its right. This increases pressure in the film close to the piston causing movement of the fluid to its right. We assume that the intensity of the wave front is not large so that the pressure rise is not sharp. (We shall see its effect in later sections.) This way the fluid downstream is disturbed and the fluid moves with velocity $dC$, howsoever small it may be. This disturbance propagates in the medium with the speed of sound, called the *acoustic* or sound speed '$a$'. The increase in pressure and the velocity up to the wave front is shown in the figure. At some time let the fluid upstream of the wave move with velocity $dC$ (to right) whereas downstream is at rest. Thus the flow of fluid in the duct changes with movement of the wave and hence is unsteady. To analyse this phenomenon, we can change it to a steady flow by attaching the reference axes to the wave front. The situation is shown in Fig. 7.1.1 (b). Fluid to the right of the wave is moving towards the left with the velocity '$a$' and the one to its left is moving with velocity '$a$–$dC$.' Let the dotted line around the wave front be the control volume. Now the pressure and density of the fluid to the right of the control volume will be $p$ and $\rho$ and the corresponding values to the left of the wave will be $p + dp$ and $\rho + d\rho$, respectively. Application of the continuity and momentum equation gives (neglecting terms of higher order).

$$\dot{m} = \rho A a = (\rho + d\rho)(a - dC)\,A \quad \rightarrow \rho\, dC = a\, d\rho \qquad (7.1.1)$$

and $A\left[p - (p + dp)\right] = \dot{m}\left[(a - dC) - a\right] \quad \rightarrow A\, dp = \dot{m}\; dC$

or, $\quad A\, dp = \rho A a\, dC \quad \rightarrow dp = \rho a\, dC \qquad (7.1.2)$

Substituting in Eq. 7.1.2 from Eq. 7.1.1 yields

$$a^2 = dp/d\rho \qquad (7.1.3)$$

Thus, the acoustic velocity depends on the rate of change of pressure with density. Newton argued that there is no change in temperature during the propagation and he obtained $dp/d\rho$ from the equation of state taking temperature as constant. However, his results did not agree with the experimental results. On the other hand, Laplace disagreed with this argument and said since there is not enough time available for any heat transfer to take place and the wave is too weak to increase entropy, propagation of wave is an isentropic process. On the basis of this, he obtained acoustic speed, which agreed with the experimental value. For a perfect gas, subjected to isentropic process

$$p = K\rho^{\gamma} \rightarrow (\partial p/\partial \rho)_s = \gamma p/\rho = \gamma RT, \text{ since } p = \rho RT \text{ for perfect gas.}$$

Therefore, we can write

$$a^2 = (\partial p/\partial \rho)_s = \gamma RT \qquad (7.1.3)$$

Movement of piston is only a way to create any disturbance in the pressure of the fluid locally at a cross section or a point in the gaseous region. In fact, the disturbance may be created by many other means. Once pressure is increased in the gas, the pressure would try to equalise and this is done by propagation of pressure wave. If the fluid were not at rest but moving with some velocity $U$ then also the wave would travel with acoustic speed. We can form a non-dimensional number from these two velocities and this number is called *Mach number*, represented by the symbol $M$.

$$M = C/a \qquad (7.1.4)$$

Nature of flow is described by Mach number, which is defined as the ratio of the velocity of fluid to the sound velocity, $a = \sqrt{(\gamma R\,T)}$, where $\gamma$ and $R$ have usual meaning and $T$ is static temperature. Here the numerator may be relative or absolute velocity depending on location of the point where Mach number is to be determined. If determined at any point on moving part of a machine then it has to be relative velocity with respect to that part.

Equation 7.1.3 shows that the acoustic velocity in liquids will be very large, since there is a large change in pressure with almost no change in density when liquid is compressed. However, it is not so in case of gases. Now let us study the effects of a sound source (present in fluid) on the region of influence of the disturbed fluid when the fluid is also moving with some finite velocity, say from right to left. (By fixing the reference with moving fluid, we can move the source with the same velocity from left to right.) It is shown in Fig. 7.1.2. The disturbance is created at some point in the gas that spreads in all the directions with velocity $a$. Thus a disturbance travels along spherical path. When the fluid is at rest, all the successive disturbances will emanate from the same point. Let the disturbance, for convenience, be generated every second. Figure 7.1.2 (a) shows the region of influence by drawing the circles from the position of the source after 3$s$ from the first disturbance was generated. In the mean time, the first disturbance has travelled a distance 3$a$, the second 2$a$, the third $a$ and the fourth pulse is just to be generated. All these circles will be concentric, since the fluid is at rest. Thus, the disturbance will spread out uniformly in all directions.

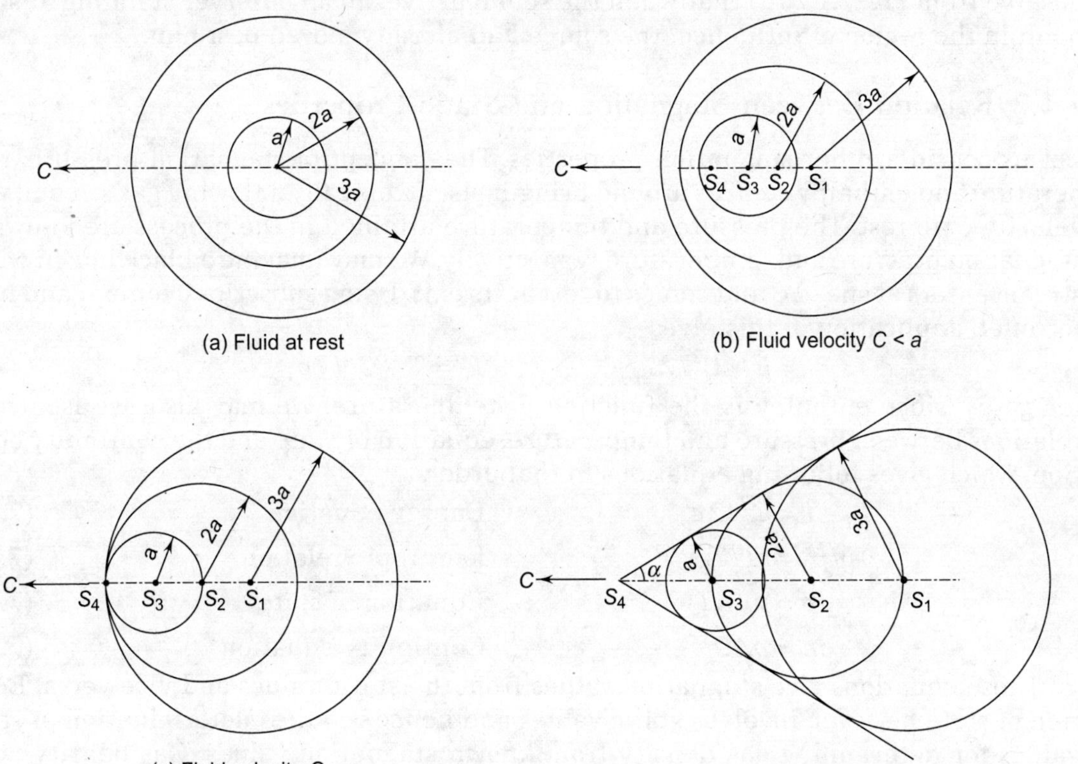


**Fig. 7.1.2** Effect of fluid flow on the region of propagation of sound in a gaseous medium

Now consider that the source of the disturbance is moving with the fluid with speed $C$. Three possibilities may occur, viz, (i) $C < a$, (ii) $C = a$, and (iii) $C > a$. The centres of the three circles of Fig. 7.1.2 (a) will move and have been shown displaced by a distance $3C$, $2C$ and $C$, respectively from the location of the first centre; these are shown after one second each. There will be sound waves during this period too. But these are shown to explain the wave propagation after lapse of 1s for clarity only. Let the sound be produced when source is at $S_1$. In one second it moves to $S_2$ by a distance $C$, in two second to $S_3$ by a distance $2C$, and so on. The source is at $S_4$ for which condition the region of media under disturbance influence is shown in the figure. We observe: when $C < a$, then disturbance affects more region downstream and less on upstream. This effect grows with increasing velocity of the source till $C = a$ when the disturbance is limited only in the region to right of the source, Fig (b) and (c). However, when $C > a$, then the region downstream takes shape of a cone. This cone is called *Mach cone* and half the angle of the cone is known as *Mach angle*. We can see from the diagram (d) that Mach angle depends on the Mach number and is related to it by the relation

$$\tan \alpha = a/C = 1/M \tag{7.1.5}$$

Increasing $M$ decreases Mach angle and hence the region of influence. Moreover, we observe from Fig. 7.1.2 (d) that when the sound arrives to an observer standing at some point in the region of influence, the source had already moved past him.

### 7.1.2 Relation Between Stagnation and Static Properties

Let us consider other stagnation properties. The concept of stagnation pressure, temperature and enthalpy centres around bringing isentropically a flowing gas stream, with velocity $C$, to rest. The pressure and temperature attained in the process are known as stagnation pressure and temperature respectively. We may imagine a black box in which stream enters at speed $C$ and comes to rest at its exit. Using subscript 0 for exit and none for inlet, application of this gives

$$h_0 = h + C^2/2 \qquad \rightarrow \qquad T_0 = T + C^2/2c_p$$

for gases, since enthalpy is the function of temperature. We may also use isentropic relations between pressure and temperature, equation of state and the continuity equation, which gives following equations in that order.

$$T_0 - T = C^2/2c_p \qquad \text{Energy equation} \tag{7.1.6}$$

$$p/p_0 = (T/T_0)^{\gamma/(\gamma-1)} \qquad \text{Isentropic relation} \tag{7.1.7}$$

$$\rho = p/(RT) \qquad \text{Equation of state} \tag{7.1.8}$$

$$m = \rho AC \qquad \text{Continuity equation} \tag{7.1.9}$$

These equations give stagnation values from the static values and vice versa. Equation of state however, involves static values and hence requires determination of static values for determining gas density from known stagnation values. Gas density is obtained using Eqs. 7.1.6 to 7.1.9. Equation. 7.1.6 involves gas velocity. If $C$ is known then these four equations give area $A$. In many situations, however, area $A$ is known but $C$ is unknown. In that case, it becomes difficult but not impossible to solve the four equations. Difficulty arises because the procedure becomes iterative. We guess a value for $C$ and proceed to get a new value of $C$ after solving Eq. 7.1.6. The procedure is repeated till a convergence criterion to reduce the difference between the chosen and calculated value of velocity to some prescribed value, is satisfied. In some situations it is easy to guess value of density and in that case iteration can be carried on density instead of velocity. In a few iterations, this procedure gives static properties for given stagnation values.

### 7.1.3 Various Regimes of Flow

Three temperatures are of importance—one is stagnation, second is absolute zero and third is temperature at the critical section, which is the section where the acloustic velocity is equal to the flow velocity. We represent these sections respectively by subscripts 0, max and superscript *. At the absolute zero temperature, the enthalpy is zero and hence the velocity is maximum that is represented by the symbol $a_{max}$. $A^*$ over a property indicate the property at critical section, which is a section where $M^* = C^*/a^*$ is always 1; we shall discuss more about it later. Equation 7.1.3 gives following relations for acoustic velocity at different reference temperature, replacing $R$ by the Victor–Mayer relation $c_p - c_v = R$. These are

$$T = \frac{a^2}{\gamma R} = \frac{a^2}{c_p\,(\gamma - 1)} \quad \text{and} \quad T_0 = \frac{a_0^2}{c_p\,(\gamma - 1)}.$$

Substituting for $T$ and $T_0$ in Eq. 7.1.6 and using relation $h = c_p\,T$, after simplification we get

$$\frac{a^2}{(\gamma - 1)} + \frac{C^2}{2} = \frac{a_0^2}{(\gamma - 1)} = \frac{C_{max}^2}{2} = \frac{\gamma + 1}{\gamma - 1}\frac{C^{*2}}{2} \tag{7.1.10}$$

4th term in Eq. 7.1.10 is obtained from the first by writing for $a$ equal to $a^* = C^*$, $C^*$ for $C$, and simplifying.

Equation 7.1.10 can be plotted on $a$ v/s $C$ graph, as shown in Fig. 7.1.3. Depending on the flow Mach number, the flow is divided into the following regimes: (i) Incompressible for $0 \le M \le 0.3$, (ii) subsonic for $0.3 \le M \le 0.9$, (iii) transonic for $0.9 \le M \le 1.1$, (iv) supersonic for $1.1 \le M \le 6$ and (v) hypersonic for $M > 6$.

Mach number at a section is obtained by dividing the flow velocity by the local acoustic velocity calculated from static temperature at that section. Instead, if we use acoustic velocity $a^*$ at the critical section to divide the flow velocity at the given section then the Mach number at the section is called the critical Mach number, being represented by $M^*$, which bears relation with $M$. That is, $M = C/a$ and $M^* = C/a^* = C/C^*$

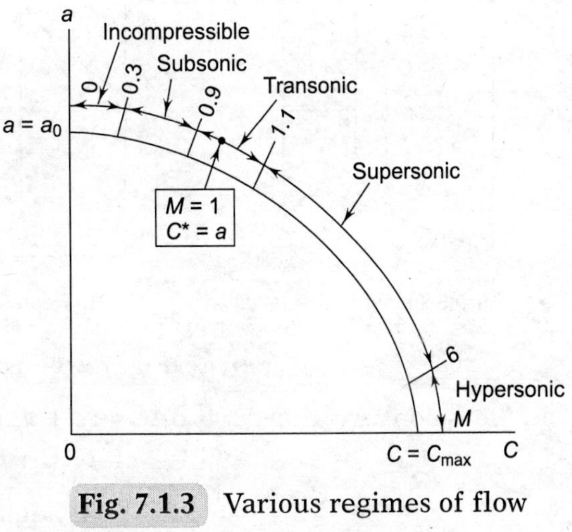


**Fig. 7.1.3** Various regimes of flow

or, $$M^{*2} = \frac{C^2}{a^{\star 2}} = \frac{C^2}{a^2}\frac{a^2}{a^{\star 2}} = \frac{a^2}{a^{\star 2}} M^2 \tag{7.1.11}$$

Considering first and last terms of Eq. 7.1.10 and dividing both sides by $C^{\star 2}/2$, we get

$$\frac{2a^2/C^{*2}}{(\gamma-1)} + \frac{C^2}{C^{*2}} = \frac{\gamma+1}{\gamma-1},$$

or, $$\frac{2}{(\gamma-1)}\frac{M^{*2}}{M^2} + M^{*2} = \frac{\gamma+1}{\gamma-1}$$

$\rightarrow$ $$M^{*2} = \frac{\frac{1}{2}(\gamma+1)M^2}{1+\frac{1}{2}(\gamma-1)M^2} \tag{7.1.12}$$

Equation 7.1.12 gives relation between Mach number and critical Mach number at a section in the isentropic flow. We can verify that when $M = 0$, $M^* = 0$ and $M = 1$, $M^* = 1$. However, when $M \rightarrow \infty$ $M^* \xrightarrow{\text{does not}} \infty$; rather its value becomes $\sqrt{\{(\gamma+1)/(\gamma-1)\}}$, which is finite.

## 7.2 Isentropic Flow Through Variable Area

A nozzle or diffuser is a passage in which area of flow is changing. In general, when flow occurs through variable area passage then pressure also changes. Whether the pressure would increase or decrease with change in area may be obtained from mass and momentum conservation equation for a differential volume $dV$, as shown in Fig. 7.2.1 (a).

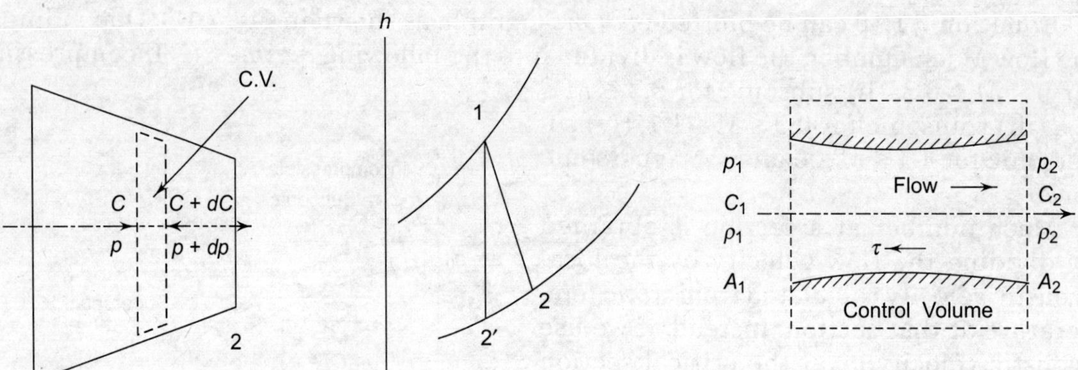


(a) Schematic representation (b) Representation in h-s chart (c) Impulse function

**Fig. 7.2.1** Expansion in nozzle

Mass conservation : $\rho AC = (\rho + d\rho)(A + dA)(C + dC)$

$= \rho AC + \rho A\, dC + \rho C\, dA + AC\, d\rho$ + higher order terms

$\rightarrow$ $\frac{d\rho}{\rho} + \frac{dA}{A} + \frac{dC}{C} = 0$ neglecting all higher order terms and

Momentum conservation : $pA - (p + dp)(A + dA) = \rho AC\{(C + dC) - C\}$
dropping higher order terms in the momentum equation gives

$$dp = -\rho C\, dC.$$

Substituting it in the mass conservation equation, we get

$$\frac{dA}{A} = \frac{d\rho}{\rho C^2}\left(1 - \frac{d\rho}{dp}C^2\right) = \frac{dp}{\rho C^2}(1 - M^2),$$

since
$$a^2 = \left(\frac{\partial p}{\partial \rho}\right)_s = \frac{dp}{d\rho} \tag{7.2.1}$$

From this equation we infer that $dA$ and $dp$ have (i) same sign if $M < 1$ and (ii) opposite sign when $M > 1$. This means area of nozzle should decrease for subsonic flow and increase for supersonic flow in order that pressure may decrease. In case a flow starts from subsonic then the nozzle would be converging in the subsonic region and diverging in supersonic flow regions. Thus there will be a section where area of nozzle is minimum. This section is called throat and Mach number at this section has to be 1. Such nozzle is called convergent-divergent or *CD* nozzle and is used in supersonic applications. In subsonic flows nozzle is convergent only.

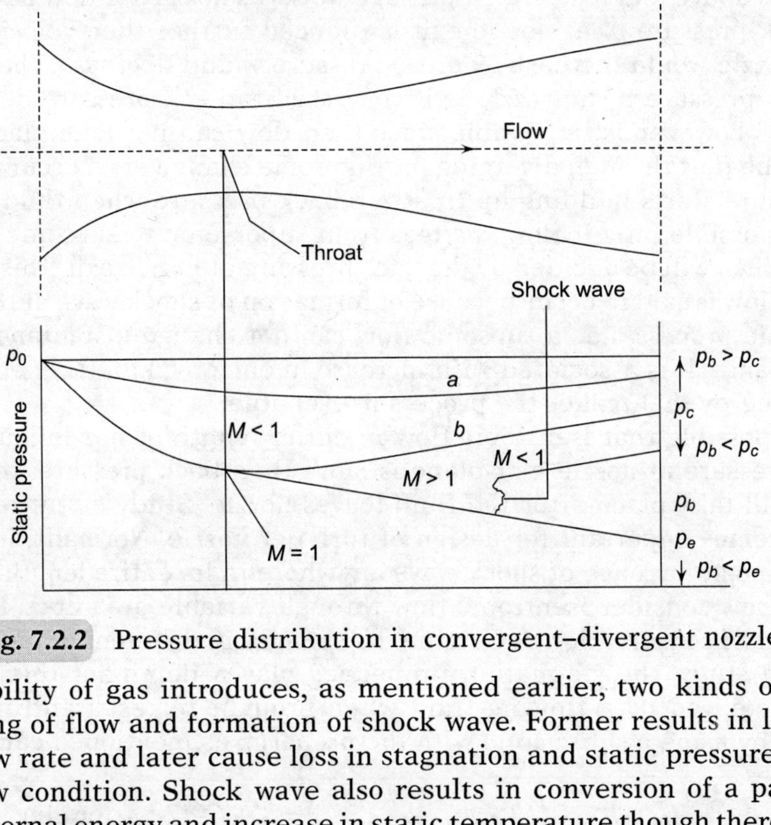


**Fig. 7.2.2** Pressure distribution in convergent–divergent nozzle

Compressibility of gas introduces, as mentioned earlier, two kinds of limitations, namely, choking of flow and formation of shock wave. Former results in limiting maximum mass flow rate and later cause loss in stagnation and static pressure compared to isentropic flow condition. Shock wave also results in conversion of a part of kinetic energy into internal energy and increase in static temperature though there is no change in stagnation temperature since no energy transfer occurs from surroundings.

This can be understood by studying flow through a converging–diverging nozzle, as shown in Fig. 7.2.2. The nozzle is supplying gas, say, to some machine through a control valve, which is initially closed. Gas enters nozzle from a reservoir at some stagnation pressure and temperature $p_0$ and $T_0$, respectively. These are stagnation values since gas is at rest initially. Now let us open the discharge valve on the down side so that flow starts. Pressure first decreases up to the nozzle throat and then increases to match the discharge pressure at exit, curve $a$ in Fig. 7.2.2. Flow through the nozzle is subsonic. Thus velocity of flow is maximum at throat. By opening control valve further, back pressure can be reduced. At some point, velocity at throat will reach velocity of sound, that is, Mach number at throat is 1. Pressure distribution in this situation is shown in Fig. 7.2.2 by curve $b$. If the control valve is opened further, the disturbance so created shall travel upstream up to throat only and no further upstream. Thus upstream flow will not notice any change downstream and hence no more increase in flow rate. This is the condition of *choking* of nozzle. The property at the throat are known as critical values and are indicated by using superscript $\star$ over the property because at this section $M = 1$ and $a = C$. Thus $M = M^\star = 1$ at this section. We shall study shortly the effect of the choking of nozzle on its performance.

We showed above that a diverging passage works as nozzle for flow having $M > 1$. That is, when back pressure $p_b$ at nozzle exit is reduced further then velocity in divergent portion of nozzle would increase whereas pressure would decrease. Therefore this will result in less pressure $p_e$ at nozzle exit compared to back pressure developed by the machine. This however, is impossible since then flow can not take place. This implies that during the flow through diverging portion some change must occur at some section so that pressure starts building up to attain back pressure when fluid reaches nozzle exit. This is possible only if flow changes from supersonic to subsonic at this section, location of which will be decided by the back pressure at nozzle exit. This sudden change in nature of flow is said to occur because of formation of shock wave at that section. It is an irreversible process, i.e., a subsonic flow can not change instantaneously to supersonic flow. Hence, it is associated with increase in entropy; kinetic energy is converted to internal energy that makes the process irreversible.

It is also possible that isentropic flow in entire length of nozzle may continue and even then pressure at nozzle exit plane is more then back pressure. In that case free expansion will take place after the fluid leaves nozzle. Study of entire range of back pressure becomes important for design of turbojet nozzle. Normally, it is designed so that there is no occurrence of shock wave anywhere in its entire length.

We shall now consider isentropic flow through variable area duct. Remember that the three laws of conservation determine the changes in different properties during the flow. In such study, the stagnation properties play a dominant role. These are the properties acquired by a flowing fluid when brought to rest isentropically and are represented by using a subscript 0 with the property, as mentioned earlier. The energy equation gives:

$$h_0 = h + C^2/2 \qquad \rightarrow \qquad T_0 = T + C^2/2c_p \text{ (for air)} \tag{7.2.2}$$

But $$a = \sqrt{\left(\frac{\partial p}{\partial \rho}\right)_s} = \sqrt{\gamma RT},\; c_p = \frac{\gamma R}{\gamma - 1}^{1}$$

$$\therefore \qquad \frac{T_0}{T} = 1 + \frac{\gamma - 1}{2}\frac{C^2}{\gamma RT} = 1 + \frac{\gamma - 1}{2}M^2 \qquad (7.2.3)$$

Since the process is isentropic, the process equation $p \propto T^{\gamma/(\gamma-1)}$ gives relation between $p$ and $p_0$. Then using equation of state, relation between corresponding densities is obtained; the relations are

$$\frac{p_0}{p} = \left[1 + \frac{\gamma - 1}{2}M^2\right]^{\frac{\gamma}{\gamma - 1}}$$

and $$\rho_0/\rho = (p_0/p)^{1/\gamma} = [1 + \{(\gamma - 1)/2\}\, M^2]^{1/(\gamma - 1)} \qquad (7.2.4)$$

Let us now consider the conservation of momentum. From the integral momentum equation, we know that a thrust acts on the passage when a fluid flowing through it accelerates or decelerates. Fig. 7.2.1 (c) shows a variable area passage with inlet and outlet sections. Control volume is also shown. Pressure force on its surface due to atmosphere will cancel out. In this case the external force on control volume is therefore due to pressure at inlet and outlet sections and the thrust $\tau$ acting on fluid due to fixed nozzle. Momentum equation gives the force on the nozzle

$$\tau = (p_2 A_2 + \rho_2 A_2 C_2^2) - (p_1 A_1 + \rho_1 A_1 C_1^2) \qquad (7.2.5)$$

But $\rho C^2 = (p/RT)\, C^2 = \gamma p\, \{C^2/(\gamma RT)\} = \gamma p M^2$.

We define the quantity in parenthesis of Eq. 7.2.5 as impulse function, $F$.

$$F = p + A\,\rho\, C^2 = p + \gamma p A\, M^2$$

$$\rightarrow \qquad \tau = F_2 - F_1 = p_2 A_2\,(1 + \gamma M_2^2) - p_1 A_1\,(1 + \gamma M_1^2) \qquad (7.2.6)$$

We observe that thrust on nozzle is due to (i) change in velocity from inlet to outlet and (ii) difference in pressure. $p_2$ however, will depend on back pressure at nozzle exit plane. Let us consider a converging nozzle. If back pressure is more than the critical pressure for the isentropic flow then pressure at nozzle exit will be back pressure. But if it is less then the nozzle exit pressure, $p_2$ will be critical pressure, which has to be used in impulse function.

We shall now determine mass flow rate at a section. From Eq. 7.2.2

$$C^2 = 2c_p\, T_0\left(1 - \frac{T}{T_0}\right) = \frac{2\gamma RT_0}{\gamma - 1}\left(1 - \frac{T}{T_0}\right)$$

Expressing $T/T_0$ in terms of $p/p_0$ by isentropic relation and using Eq. 7.2.4, we get after simplification.

[1] Remember this relation, since we shall use it often in transforming equations in desired form.

$$C^2 = 2\frac{\gamma}{\gamma-1}RT_0\left\{1-\left(\frac{p}{p_0}\right)^{(\gamma-1)/\gamma}\right\} = RT_0\left[\frac{\gamma M^2}{1+\frac{\gamma-1}{2}M^2}\right] \tag{7.2.7}$$

We may now use continuity equation to give mass flow rate across any section using Eq. 7.2.4 for $\rho$ and Eq. 7.2.7. After simplification we get

$$\frac{\dot{m}\sqrt{T_0}}{Ap_0}\sqrt{\frac{R}{\gamma}} = \sqrt{2\frac{\gamma}{\gamma-1}\left\{\left(\frac{p}{p_0}\right)^{2/\gamma}-\left(\frac{p}{p_0}\right)^{(\gamma+1)/\gamma}\right\}} \tag{7.2.8}$$

We note that Eq. 7.2.8 has been written in non-dimensional form. In this flow, stagnation values are decided by the initial conditions and are known a priori. The only variable then left is $p$, the pressure at any section. We may differentiate the equation with respect to $p$ to give condition for maximum discharge and the values of maximum mass flow density, $\dot{m}/A$. This requires two differentiations: first differentiation gives the condition for pressure and the second successive differentiation decides whether this condition is for maxima or minima. You may verify yourself that the condition obtained below by this procedure is for maximum discharge. We get

$$\left(\frac{p}{p_0}\right)_{max} = \left(\frac{2}{\gamma+1}\right)^{\gamma/(\gamma-1)} \tag{7.2.9}$$

This is the same condition as obtained for $p/p_0$ from Eq. 7.2.4 when $M = 1$ is substituted in it. In isentropic flow through a convergent divergent nozzle, Mach number becomes 1 only at the throat. Under such condition, throat will be critical section and we may use superscript $\star$, as mentioned earlier to properties at this section. It should be emphasized here that a critical section is a concept and does not dictate that it has to be present in the flow region. This means if the flow everywhere is subsonic then there is no section anywhere in the flow where $M = 1$. But critical section exists for this too. Substituting for $p/p_0$ from Eq. 7.2.9 in 7.2.8 and using $A^*$ for $A$, we get

$$\frac{\dot{m}_{max}\sqrt{T_0}}{A^\star p_0}\sqrt{\frac{R}{\gamma}} = \left(\frac{2}{\gamma+1}\right)^{(\gamma+1)/2(\gamma-1)} \tag{7.2.10}$$

We note that the critical pressure ratio and $\dfrac{\dot{m}_{max}\sqrt{T_0}}{A^* p_0}$ respectively are equal to 0.528 and 0.040418 for $\gamma = 1.4$, a value which is mostly used in calculations for air. Properties at the critical section can be obtained by substituting $M = 1$ in the relevant relations in Eqs. 7.2.3 and 7.2.4. These relations are combined with those given in Eq. 7.2.4 to give property at any section in terms of the critical property. It gives the following relations

$$\frac{T^*}{T_0} = \frac{2}{\gamma+1} = 0.833 \quad \text{and} \quad \frac{T^*}{T} = \frac{2}{\gamma+1} + \frac{\gamma-1}{\gamma+1}M^2 \tag{7.2.11}$$

$$\frac{p^*}{p_0} = \left(\frac{2}{\gamma+1}\right)^{\gamma/(\gamma-1)} = 0.528 \quad \text{and} \quad \frac{p^*}{p} = \left(\frac{2}{\gamma+1} + \frac{\gamma-1}{\gamma+1}M^2\right)^{\gamma/(\gamma-1)} \tag{7.2.12}$$

$$\frac{\rho^*}{\rho_0} = \left(\frac{2}{\gamma+1}\right)^{1/(\gamma-1)} = 0.634 \quad \text{and} \quad \frac{\rho^*}{\rho} = \left(\frac{2}{\gamma+1} + \frac{\gamma-1}{\gamma+1}M^2\right)^{1/(\gamma-1)} \tag{7.2.13}$$

$$C^* = \sqrt{2\gamma RT_0/(\gamma+1)} \tag{7.2.14}$$

We infer from Eq. 7.2.12 that in case of convergent nozzle, pressure at its exit can never be less than critical pressure corresponding to stagnation pressure[2] at its inlet. Moreover, an internal loss in convergent portion is so small that these are neglected and isentropic flow assumption gives satisfactory results.

It is seen from Eqs. 7.2.3 and 7.2.4 that the ratio of pressure, temperature and density made with the corresponding stagnation values is function of Mach number at the section in the flow and specific heat ratio. The critical Mach number also is function of $\gamma$ and $M$. Therefore, it would be desirable to put these equations in tabular form for specified value of $\gamma$. Such tables are known as isentropic flow tables and are available in market. The values at the critical section can be easily obtained, since at this section $M = M^* = 1$. Students are advised to study structure of the table. These tables have been prepared for commonly used values of $\gamma$, such as 1.33, 1.4, 1.67.

We therefore would like to develop relations so that thrust and mass flow rate can easily be obtained without solving the relevant equations. Writing continuity equation between a given section and the critical section and using Eqs. 7.2.7, 7.2.12 and 7.2.14, we get

$$\frac{A}{A^*} = \frac{1}{M}\left(\frac{2}{\gamma+1} + \frac{\gamma-1}{\gamma+1}M^2\right)^{(\gamma+1)/2(\gamma-1)} \tag{7.2.15}$$

Similarly, impulse function at the throat is obtained by substituting $M = 1$ in Eq. 7.2.6. We thus obtain

$$\frac{F}{F^*} = \frac{p}{p^*}\frac{A}{A^*}\frac{1+\gamma M^2}{1+\gamma} = \left(\frac{2}{\gamma+1} + \frac{\gamma-1}{\gamma+1}M^2\right)^{-\gamma/(\gamma-1)} \tag{7.2.16}$$

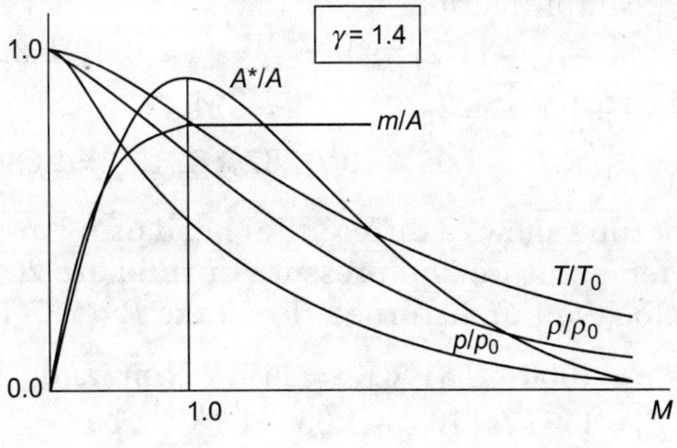


**Fig. 7.2.3** Variation of flow variables in isentropic flow

substitute for $p/p^*$ and $A/A^*$ from Eq. 7.2.12 and 7.2.15, respectively, and simplify. We note that $A/A^*$ and $F/F^*$ are functions of Mach number. These equations have been plotted on a 2-D plot in Fig. 7.2.3 for $\gamma = 1.4$.

[2] Normally the velocity of fluid at the entrance is neglected as it is small compared to that at its outlet. However, pressure at nozzle inlet is not the stagnation pressure and hence is to be calculated from gas tables.

The above relations hold true whether the area is decreasing or increasing and these are equally applicable for a nozzle or a diffuser. However, in flow through a diffuser, we keep in mind that pressure increases in the direction of the flow, which has a tendency to separate the flow from the boundary. For this reason, the degree of divergence given to a diffuser is limited to 10° – 12°.

**Example 7.2.1** The static and stagnation temperatures in a stream of argon gas are 15°C and 75°C, respectively. Assuming the flow to be reversible adiabatic, determine the velocity, Mach number and stagnation pressure of the gas stream. Also compute the percentage rise at the stagnation point. Assume $R = 209$ J/kg K and $\gamma = 1.658$ for argon.

**Solution** From given values of $R$ and $\gamma$, $c_p$ can be determined and is = 0.5266 kJ/kg K. Flow velocity $C$ can be determined from energy equation $T_0 = T + C^2/(2c_p)$.

This gives $C^2 = 2 \times (T_0 - T) \times c_p \rightarrow C = 315.9$ m/s.

$$\therefore \quad M = C/\sqrt{(\gamma RT)} = 315.9/\sqrt{(1.658 \times 209 \times 288)} = 0.796,$$

$$h_0 = c_p \times T_0 = 151.7 \text{ kJ/kg}.$$

$$p/p_0 = (T/T_0)^{\gamma/(\gamma-1)} \rightarrow p_0/p = 1.61 \text{ bar}.$$

$\therefore$ Percent rise in pressure = $(p_0/p - 1) \times 100 = 61\%$.

**Example 7.2.2** A jet aircraft flies at 1000 kmph in air at 10°C. Calculate the Mach number of flight. What are the temperature, pressure and density of air at the nose of the jet? Assume the pressure of ambient air = 1 bar. Take $R = 287$ J/kg K and $\gamma = 1.4$ for air.

**Solution** $C = 1000 \times 1000/3600 = 277.8$ m/s, $a = \sqrt{(1.4 \times 287 \times 283)} = 337.21$ and $M = C/a = 0.824$. Also $\rho = p/R\,T = 1.231$ kg/m$^3$.

The nose of the jet will become stagnation point. Thus using the isentropic relations (Eq. 7.2.3), we get, after simplification,

$$p_0/p = [1 + \{(\gamma - 1)/2\}\, M^2]^{\gamma/(\gamma-1)} = [1 + 0.2 \times 0.824^2]^{3.5} \rightarrow p_0 = 1.55 \text{ bar}.$$

$$T_0/T = 1 + \{(\gamma - 1)/2\}\, M^2 \rightarrow T_0 = 321.4 \text{ K}$$

$$\rho_0 = p_0/R\,T_0 = 1.55 \times 10^5/(287 \times 321.4) = 1.686 \text{ kg/m}^3.$$

**Example 7.2.3** A pitot tube shows a differential head of 1.2 m of water when placed in air stream. The static temperature and pressure of fluid are 20°C and 1.05 bar, respectively. Compute the velocity of undisturbed flow. Take $R = 287$ J/kg K and $\gamma = 1.4$.

**Solution** Given: $p_0 - p = 1000 \times 9.81 \times 1.2 = 11772$ N/m$^2$ and $p = 1.05$ bar.

$$\therefore \quad p_0 = 11772 \times 10^{-5} + 1.05 = 1.16772 \text{ bar}.$$

$$p_0/p = [1 + \{(\gamma - 1)/2\}\, M^2]^{\gamma/(\gamma-1)} \rightarrow M = 0.4.$$

$$a = \sqrt{(1.4 \times 287 \times 293)} = 343.11 \text{ and } C = a \times M = 137.246 \text{ m/s}.$$

**Example 7.2.4** For an isentropic flow in a channel, we are given: $M_1 = 0.3$, $T_1 = 62$°C, $p_1 = 650$ kPa, $A_1 = 0.001$ m$^2$ and $M_2 = 0.8$, where the values are given at two sections 1 and

2 some distance apart. Determine (a) the channel shape and (b) properties at section 2. Solve the problem using the relations and isentropic table.

**Solution** Eq. 7.2.3 gives $T_{01} = T_{02} = (273 + 62)\ [1 + (1.4 - 1)/2 \times 0.3^2] = 341$ K

And $p_{01} = p_{02} = 650 \times [1 + (1.4 - 1)/2 \times 0.3^2] = 692$ kPa.

Now the stagnation values and the Mach number at the sections are known. Equation 7.2.3 can be used to find out the static values. We shall now use the tables instead to solve rest part of the question. From isentropic table for $\gamma = 1.4$, we read: $T_2 / T_{02} = 0.8865$, $p_2 / p_{02} = 0.6560$, $\rho / \rho_0 = 0.74$ and $A/A^* = 1.038$.

$\therefore\ T_2 = 302$ K, $p_2 = 454$ kPa, $\rho_2 = p_2/(R\,T_2) = 5.24$ kg/m³ and $C_2 = \sqrt{(\gamma R\,T_2)} = 348$ m/s.

We may also find out $T_2 / T_1$, etc by writing $T_2/T_1 = (T_2/T_{02})/(T_{01}/T_1)$; the ratios in parantheses are obtained from isotropic flow tables at $M_1$ and $M_2$. $M_2$ is also known since $c_2$ and $T_2$ are known. It is left for the student to complete. Finally

$$\frac{A_2}{A_1} = \frac{A_2}{A^*}\frac{A^*}{A_1} = \frac{1.038}{2.035} = 0.5101 \qquad \rightarrow A_2 = 0.00051 \text{ m}^2.$$

The area of cross section has decreased and therefore the duct is converging type.

**Example 7.2.5** Air flows isentropically through a converging nozzle with $p_0 = 1$ Mpa, $T_0 = 333$ K, $p_b$ = back pressure on the nozzle = 591 kPa. Find (a) Mach number at the nozzle exit and (b) mass flow rate if the nozzle exit area is 0.001 m².

**Solution** First we have to determine whether the flow is choked or not by using Eq. 7.2.9. We have mentioned that the ratio for $\gamma = 1.4$ is 0.528. For the given problem, $p_b/p_0 = 591/1000 = 0.591 > 0.528$. Therefore, the flow is not choked. Mach number is obtained using Eq. 7.2.3.

$$\frac{1000}{591} = \left[1 + \frac{1.4 - 1}{2} M_e^2\right] \qquad \rightarrow M_e = 0.90.$$

Here subscript *e* is used for nozzle exit to differentiate it from the back pressure. We determine the mass flow rate using continuity equation rather than using Eq. 7.2.8, as it is difficult to remember it. For this we have to calculate static pressure, temperature and velocity of flow at the nozzle exit. Let us define

$$B = 1 + \frac{\gamma - 1}{2} M^2 = 1 + \frac{1.4 - 1}{2} \times 0.9^2 = 1.162.$$

$$\therefore \qquad T_e = T_0 / B = 287 \text{ K}.$$

Also $a_e = \sqrt{1.4 \times 287 \times 287} = 340$ m/s and $\rho_e = 591/(0.287 \times 287) = 7.18$ kg/m³.

$$\therefore \qquad \dot{m} = \rho_e \times (M_e \times a_e) \times A_e = 7.18 \times 0.9 \times 340 \times 0.001 = 2.20 \text{ kg/s}.$$

**Example 7.2.6** A conical diffuser has entry and exit diameter of 15 cm and 30 cm respectively. The pressure, temperature and velocity of air at entry is 0.69 bar, 340 K and 180 m/s. Determine: the (a) exit pressure, (b) exit velocity and (c) force exerted on the diffuser wall. Assume isentropic flow with $\gamma = 1.4$. Take $c_p = 1.00$ kJ/kg.

**Solution** Given $A_1 = 0.785 \times 15^2 = 176.8$ cm² and $A_2 = 706$ cm². Also $T_{01} = 340 + 180^2/(2 \times 1) = 356.2$ K $\rightarrow T_1 / T_{01} = 0.95$.

$$\therefore \qquad \rho_1 = p_1 / R\,T_1 = 0.69 \times 10^5/(287 \times 340) = 0.707 \text{ kg/m}^3.$$

From isentropic table for $\gamma = 1.4$, the corresponding Mach number is 0.5. For $T/T_0$ = 0.95, the other values are:

$$p_1/p_{01} = 0.843_1, \rho_1/\rho_{01} = 0.66044, (A/A^*)_1 = 1.34 \text{ and } (F/F^*)_1 = 1.203.$$

$$\rightarrow \quad p_{01} = p_1/0.843 = 0.818 \text{ bar} = p_{02}.$$

From Eq. 7.2.9, $p_1^* = 0.528\, p_{01} = 0.432$ bar.

From Eq. 7.2.6, $F_1^* = F_2^* = p_1^*\, A_1^*\, (1+\gamma) = 0.432 \times 10^5 \times (A_1 \times 10^{-4}/1.34)(1+1.4) = 1370$ N.

$$A_1^* = A_2^* = A_1/1.34 = 132 \text{ and } A_2/A_2^* = 706/132 = 5.37.$$

From isentropic table once again for this area ratio

$M_2 = 0.107$, $p_2/p_{02} = 0.992$, $\rho_2/\rho_{02} = 0.994$ and $F_2 = F_2^* = 4.30$ by using linear interpolation of values between $M = 0.10$ and 0.11.

$$\therefore \quad p_2 = 0.992 \times 0.818 = 0.81 \text{ bar},$$

$$\rho_2/\rho_1 = \rho_2/\rho_{02} \times \rho_{02}/\rho_1 \rightarrow \rho_2 = 0.994/0.66044 \times 0.707 = 1.065 \text{ kg/m}^3.$$

Using continuity equation, $\rho_1 A_1 C_1 = \rho_2 A_2 C_2$, we get $C_2 = 29.16$ m/s.

The force exerted on the wall is given by $F_2 - F_1 = 4.3 \times F_2^* - 1.203 \times F_1^* = 3.097 \times 1370 =$ 4243 N.

**Example 7.2.7** A nozzle is designed to expand air isentropically to atmospheric pressure from a tank in which properties are held constant at 5°C and 304 kPa. The desired flow rate is 1 kg/s. Determine the exit area of the nozzle.

**Solution** $p/p_0$ at exit = 1.0135/3.04 = 0.333. From isentropic table for $\gamma = 1.4$, we get $M = 1.36$. Thus, the flow is supersonic at the nozzle exit and the nozzle is converging–diverging. We read from the table: $T/T_0 = 0.72997. \rightarrow T_e = 203$ K. Using equation of state, we can determine density of air at this section. This comes out to 1.73 kg/m³. Also

$$C = 1.36 \times \sqrt{(1.4 \times 287 \times 203)} = 388 \text{ m/s}.$$

Using continuity equation, we get

$$1 = 1.73 \times A \times 388 \rightarrow A = 1.49 \times 10^{-3} \text{ m}^2.$$

**Example 7.2.8** In an isentropic flow of air through a duct, the static pressure is 1.3 bar, Mach number 0.7, and the area of cross-section 6.5 cm². If the mass flow rate is 0.15 kg/s, calculate the reservoir temperature and pressure. Estimate to what extent the duct area can be reduced without choking. What would be the value of the static pressure and temperature at the maximum reduced duct area condition? Use gas tables.

**Solution** Given: $M = 0.7$, $A = 6.5$ cm², $p = 1.3$ bar and $m = 0.15$ kg/s.

From gas table at $M = 0.7$, we read: $p/p_0 = 0.7209$, $T/T_0 = 0.9107$, and $A/A^* = 1.094$.

$$\therefore \quad p_0 = 1.3/0.7209 = 1.8033 \text{ bar}.$$

$$A^* = 6.5/1.094 = 5.94 \times 10^{-4} \text{ m}^2.$$

$$m/A^* = 0.15/5.94 \times 10^{-4} = 252.5 \text{ kg/s/m}^2.$$

Substituting $R = 287$ J/kg K and $\gamma = 1.4$ in Eq. 7.2.10, we get after simplification

$(m/A^*)_{max} = 0.4044\, p_0/\sqrt{T_0} \rightarrow \sqrt{T_0} = 0.40444 \times 1.8033 \times 10^5/252.5$, or, $T_0 = 834$ K.

The area of cross section can be reduced till the velocity at such section is sonic. The percentage reduction is

$$\frac{A - A^*}{A} \times 100 = \left(1 - \frac{A^*}{A}\right) \times 100 = (1 - 1/1.904) \times 100 = \mathbf{8.59\%}.$$

At such section, the pressure and temperature are $p^*$ and $T^*$, given by Eq. 7.2.11. This gives

$$P^* = 0.5282 \times p_0 = \mathbf{0.9525\ bar} \text{ and } T^* = 0.8333\ T_0 = \mathbf{695\ K}.$$

**Example 7.2.9** Compressed air at reservoir condition $p_0$ = 10 bar and 100°C is to be discharged at the rate of 15 kg/s through a converging-diverging nozzle to a back pressure of 1 bar. Calculate the nozzle cross-sectional area at the throat and at the exit plane when: (a) the nozzle flow is isentropic all along its length, (b) flow is isentropic upto the throat and with friction in the diverging portion with an efficiency of 0.95.

**Solution** Given: $p_0$ = 10 bar, $T_0$ = 373 K, $m$ = 15 kg/s, $p_b$ = 1 bar $\rightarrow p_b/p_0 = 0.1$.

From gas table, we observe that at this value of $p_b/p_0$, $M = 2.15$. Therefore, the flow is supersonic in the diverging portion and hence the throat is the critical section. Equation 7.2.11 gives for $\gamma = 1.4$

$$p^* = 0.528\ p_0 = 5.28 \text{ bar and } T^* = 0.8333 \times T_0 = 310.82 \text{ K}$$

Equation of state gives $\rho^* = p^*/RT^* = 5.82 \times 10^5/(287 \times 310.82) = 5.927$ kg/m$^3$.

$$c^* = a^* = \sqrt{(1.4 \times 287 \times 310.82)} = 353.13 \text{ m/s}.$$

$$m/A^* = \rho^*\ C^* = 5.927 \times 353.13 = 2093.0. \text{ But } m = 15 \text{ kg/s}$$

$$\rightarrow \quad A^* = 15/1093 = 0.007166 \text{ m}^2 = \mathbf{71.66\ cm^2}.$$

(a) For isentropic flow throughout the nozzle length:

At the nozzle exit, from the gas table we read corresponding to $p/p_0 = 0.1$

$$T_2'/T_0 = 0.519, \rho_2'/\rho_0 = 0.1946.$$

(Prime is used to represent the condition if the flow were isentropic.)

$$\rightarrow \quad T_2' = 373 \times 0.519 = 193.58 \text{ K},$$

$$a_2 = \sqrt{(1.4 \times 287 \times 193.58)} = 278.73 \text{ m/s},$$

$$C_2' = M_2 \times a_2 = 2.15 \times 278.73 = 599.27 \text{ m/s}.$$

From equation of state,

$$\rho_2 = 1 \times 10^5/(287 \times 193.58) = 1.80 \text{ kg/m}^3.$$

$$\therefore \quad A_2' = m/(\rho_2 \times C_2') = 0.0139 \text{ m}^2 = \mathbf{139\ cm^2}.$$

(b) For isentropic flow upto the throat only:

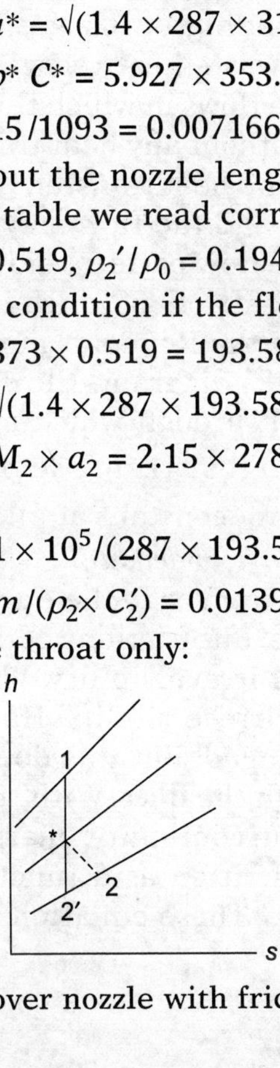


**Fig. E 7.2.9** Expansion over nozzle with friction in diverging portion

Expansion in the nozzle is shown in the figure on *h-s* diagram; expansion 1* is isentropic, 12′ is isentropic throughout the nozzle length and *2 in diverging section of the nozzle with friction. Using the definition of nozzle efficiency, we can determine temperature $T_2$.

$$0.95 = \frac{T^* - T_2}{T^* - T_2} = \frac{310.82 - T_2}{310.82 - 193.58} \quad \rightarrow T_2 = 199.4 \text{ K}.$$

Also, $h^* - h_2 = c_p\,(T^* - T_2) = 1.005 \times (310.82 - 199.4) = 111.72$ kJ/kg.
Steady flow energy equation between two points * and 2 gives

$$h_2 + \frac{C_2^2}{2000} = h^* + \frac{C^{*2}}{2000} \quad \rightarrow C_2 = 590 \text{ m/s}.$$

Equation of state gives: $\rho_2 = \dfrac{p_2}{RT_2} = \dfrac{1 \times 10^5}{287 \times 199.44} = 1.7488 \text{ kg/m}^3.$

$$\therefore \qquad A_2 = \frac{m}{\rho_2 C_2} = \frac{15}{1.7488 \times 590} \times 10^4 = \mathbf{147.37\ cm^2}.$$

## 7.3 Adiabatic Frictional Flow through Constant Area Duct—Fanno Flow

In central air conditioning systems, air is transferred from the building to the air-handling unit and then back to the building, or in a gas turbine system products of combustion are passed from turbines to waste heat recovery boiler. Like this, many other examples may be cited. These are the flows in which the length of the duct is not very large and the flow may be considered without any heat transfer. If the fluid is hot or cold then the pipe is insulated to minimise the heat transfer. In many cases, the heat transfer may not be zero but can be neglected and the flow may be considered adiabatic. However, the presence of friction affects the flow pattern and in this section we shall study how is the flow affected by the presence of friction, which is characterised by the coefficient of friction *f*, whose value depends on the Reynolds number and the roughness of the pipe and can be obtained from Moody diagram. Flow through a pipe may be assumed (idealised) to be one-dimensional through which mass flow rate across any section is $\dot{m} = \rho AC$. Since density changes from section to section, the quantity $G = \rho \mathbf{C} = \dot{m}/A$, however, will remain constant at all the sections for such constant area ducts. This quantity is called *mass flow density.*

Since there is no heat transfer from or to the duct, the stagnation enthalpy and stagnation temperature do not change. But friction reduces stagnation pressure and in turn increases entropy due to internal irreversibility. We know that entropy of any system always increases due to internal irreversibility. Heat transfer is another way that the entropy of the system can be changed. But this does not come in picture, since we are considering adiabatic flow through the duct. With this background we can now proceed to get some useful results using the continuity, energy, momentum, and state equations. Note that the state equation is written as a function of entropy for the reason that entropy is changing in the process. These equations are:

$$G = \rho C,\ h_0 = h + C^2/2 = h + G^2/(2\rho^2), \text{ and } \rho = f(s, h).$$

These equations are combined to give

$$h = h_0 - \frac{1}{2}\frac{G^2}{[f(s,h)]^2} \tag{7.3.1}$$

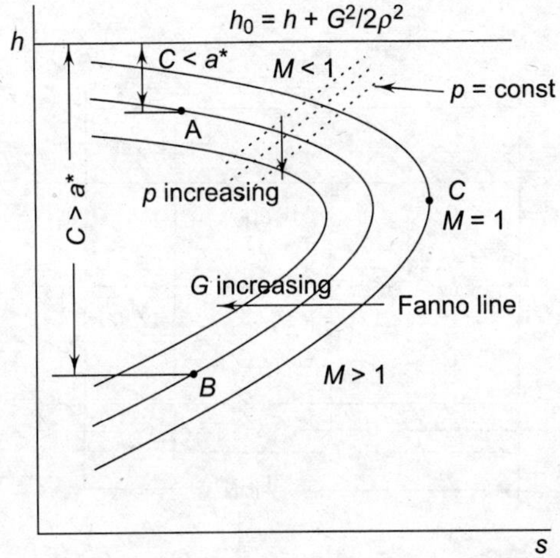


**Fig. 7.3.1** Fanno line for different mass flow densities

Equation 7.3.1 can be plotted on *h-s* diagram for given value of $h_0$. This is shown in Fig. 7.3.1. You will recognise that controlling the pressure at the nozzle exit that supplies fluid to the duct can change the velocity of the flow at the entrance to the duct. That is, the mass flow rate will depend on the conditions prevailing downward the nozzle exit in the flow. Thus, we shall get a family of curves for different values of $G$. As $G$ increases, we note from Eq. 7.3.1 that $h$ decreases. This explains the direction of increasing $G$ in the figure. The plot shows that the curve is divided into upper and the lower limb. For the lower limb, Mach number is greater than 1 whereas for the upper limb it is less than 1. The two curves meet at point $C$. At this point $ds = 0$ as the curvature is changing. Using *Tds* relation[3] and writing continuity and the energy equations in their differential form, we get

*Tds* equation : $T\,ds = c_p\,dT - T(\partial v/\partial T)_p\,dp = dh - dp/\rho = 0 \rightarrow dh = dp/\rho,$

Continuity equation : $\rho.\,dC + C.\,d\rho = 0 \rightarrow dC = -(C/\rho)\,d\rho$

Energy equation : $0 = dh + C\,dC \rightarrow dh = -C \times dC = -C \times (-C/\rho)\,d\rho = C^2\,d\rho$

or, $dp/\rho = C^2\,d\rho/\rho \rightarrow C^2 = dp/d\rho = a^2$

This means Mach number at this point is 1. So we conclude that as the flow takes place along the duct with friction, the Mach number changes and tends to 1 at some section where the entropy of the fluid becomes maximum during the flow due to friction. If the flow were initially supersonic then Mach numbers would decrease while for

[3] See Eq. 4.2.2 (c) of *Thermodynamics* by Gupta, S. C., Pearson Education.

subsonic flow it would increase; in both cases, however, the flow ultimately becomes sonic. This also means that the subsonic flow cannot become supersonic and vice-versa, since this would require entropy to decrease that is not possible. Moreover, once the flow has become sonic then there would be no more change in the flow thereafter. This means that at the end of the duct the Mach number can be $\leq 1$; if it is one then the critical section exists in the flow region otherwise it exists at some section outside the flow region.

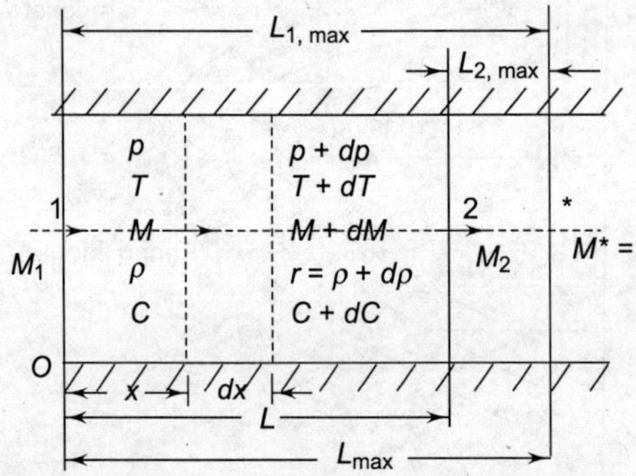


**Fig. 7.3.2** Fanno flow analysis

Now we shall develop relations so that the properties at any section down the flow may be determined if these are known at some known section. A control volume is shown in Fig. 7.3.2 with properties marked on it. We fix the origin at $O$ and measure $x$ in the direction of flow; the control volume of length $dx$ is at a distance $x$. Because of friction, a shear force $\tau_w$ would act on peripheral area $P\,dx$, where $P$ is the perimeter of the duct of cross sectional area $A$. We use a non-dimensional coefficient $f$ to express the wall shear in the following way

$$f = \frac{\text{Wall shear stress}}{\text{dynamic head}} = \frac{\tau_w}{\frac{1}{2}\rho C^2} \quad \rightarrow \tau_w = \tfrac{1}{2} f \rho C^2 \tag{7.3.2}$$

$$\text{Also } D = \text{hydraulic mean diameter} = \frac{4A}{P} \text{ and } dA_w = P\,dx = \frac{4A}{D}dx \tag{7.3.3}$$

We shall now use the different conservation relation as applicable to the control volume in their differential form and express in terms of $M^2$; the procedure is as follows:

$$\text{Continuity equation} \rightarrow \frac{d\rho}{\rho} + \frac{dC}{C} = 0 \quad \text{or,} \quad \frac{d\rho}{\rho} = -\frac{1}{2}\frac{dC^2}{C^2} \tag{7.3.4}$$

$$\text{Equation of safe} \rightarrow \frac{dp}{p} = \frac{d\rho}{\rho} + \frac{dT}{T} \tag{7.3.5}$$

$$M = \frac{C}{a} = \frac{C}{\sqrt{\gamma RT}} \text{ or, } M^2 = \frac{C^2}{\gamma RT} \xrightarrow{\text{logarithmic differentiation}} \frac{dM^2}{M^2} = \frac{dC^2}{C^2} - \frac{dT}{T} \tag{7.3.6}$$

Energy equation $\rightarrow c_p\, dT + \frac{1}{2}\, dC^2 = 0$ But $c_p = \frac{\gamma R}{\gamma - 1}$

$$\therefore \quad \frac{\gamma RT}{\gamma - 1}\frac{dT}{T} + \frac{1}{2}\, dC^2 = 0 \qquad \text{or, } a^2\,\frac{dT}{T} + \frac{\gamma + 1}{2} dC^2 = 0$$

$$\text{or,} \quad \frac{dT}{T} + \frac{\gamma - 1}{2}\, M^2\, \frac{dC^2}{C^2} = 0 \tag{7.3.7}$$

$$\text{Momentum equation} \rightarrow \dot{m}\, dC = -A\, dp - \frac{1}{2}\, \tau_w\, dA_w = -A\, dp - \frac{1}{2}\, \rho AC^2 \left(4f\, \frac{dx}{D}\right) \tag{7.3.8}$$

But $\rho C^2 = \frac{\gamma p}{\gamma RT} C^2 = \gamma p M^2$ and $\frac{\dot{m}}{A} dC = \rho C\, dC = \rho C^2\, \frac{dC}{C} = \frac{\gamma p M^2}{2}\, \frac{dC^2}{C^2}$

$$\therefore \text{Eq. 7.3.8} \rightarrow \frac{dp}{p} + \frac{\gamma M^2}{2}\, \frac{dC^2}{C^2} + \frac{\gamma M^2}{2}\left(4f\, \frac{dx}{D}\right) = 0 \tag{7.3.9}$$

Logarithmic differentiation of both sides of the Eq. 7.2.4 for $p_0/p$ gives after simplification we get

$$\frac{dp_0}{p_0} = \frac{dp}{p} + \frac{\gamma M^2}{2\left(1 + \frac{\gamma - 1}{2} M^2\right)}\, \frac{dM^2}{M^2} \tag{7.3.10}$$

Also $\qquad F = pA + \rho AC^2 = (1 + \gamma M^2)\, pA.$

Logarithmic differentiation gives

$$\frac{dF}{F} = \frac{\gamma M^2}{1 + \gamma M^2}\, \frac{dM^2}{M^2} + \frac{dp}{p} \tag{7.3.11}$$

Equations 7.3.4 to 7.3.11 give all the necessary equations to determine the properties at any section. We shall combine these equations appropriately to give following equations.

$$\text{Eqs. 7.3.7 and 7.3.6} \rightarrow \frac{dM^2}{M^2} = \left(1 + \frac{\gamma - 1}{2}\, M^2\right) \frac{dC^2}{C^2}$$

$$\therefore \quad 2\,\frac{dC}{C} = \frac{dC^2}{C^2} = \frac{1}{1 + \frac{\gamma - 1}{2}\, M^2}\, \frac{dM^2}{M^2} \tag{7.3.12}$$

$$\text{Eq.s 7.3.12 and 7.3.4} \rightarrow \frac{d\rho}{\rho} = -\frac{dC}{C} = -\frac{1}{2\left(1 + \frac{\gamma - 1}{2}\, M^2\right)}\, \frac{dM^2}{M^2} \tag{7.3.13}$$

$$\text{Eqs. 7.3.7 and 7.3.12} \rightarrow \frac{dT}{T} = -\frac{(\gamma - 1)\, M^2}{2\left(1 + \frac{\gamma - 1}{2}\, M^2\right)}\, \frac{dM^2}{M^2} \tag{7.3.14}$$

Eqs. 7.3.14, 7.3.13 and 7.3.5 $\rightarrow \dfrac{dp}{p} = -\dfrac{1+(\gamma-1)M^2}{2\left(1+\dfrac{\gamma-1}{2}M^2\right)}\dfrac{dM^2}{M^2}$ (7.3.15)

Eqs. 7.3.15 and 7.3.4 $\rightarrow \dfrac{dp_0}{p_0} = -\dfrac{1-M^2}{2\left(1+\dfrac{\gamma-1}{2}M^2\right)}\dfrac{dM^2}{M^2}$ (7.3.16)

Eqs. 7.3.15 and 7.3.11 $\rightarrow \dfrac{dF}{F} = -\dfrac{1-M^2}{2\left(1+\gamma M^2\right)\left(1+\dfrac{\gamma-1}{2}M^2\right)}\dfrac{dM^2}{M^2}$ (7.3.17)

Substituting from Eqs. 7.3.15 and 7.3.12 in Eq. 7.3.9, after simplification, we get

$$\frac{dM^2}{M^2} = \frac{\gamma M^2\left(1+\dfrac{\gamma-1}{2}M^2\right)}{1-M^2}4f\frac{dx}{D}$$

or, $$4f\frac{dx}{D} = \frac{1-M^2}{\gamma M^2\left(1+\dfrac{\gamma-1}{2}M^2\right)}\frac{dM^2}{M^2} \qquad (7.3.18)$$

Equations from 7.3.12 to 7.3.17 show dependence of the variables on $dM^2/M^2$ whereas Eq. 7.3.18 gives dependence of the later on $4f\,(dx/D)$. We can substitute it in other relations to get dependence of the variables on $4f\,(dx/D)$ as well.

These equations give a differential change. Integration of the equations from some reference value $M = 1$ to the value $M$ at the section yields the desired results. A few integrations are shown in Appendix A; students may use these to verify the following results:

$$\left(\frac{C}{C^\star}\right)^2 = M^2\frac{\gamma+1}{2\left(1+\dfrac{\gamma-1}{2}M^2\right)} \quad\text{and}\quad \left(\frac{C_2}{C_1}\right)^2 = \left(\frac{M_2}{M_1}\right)^2\frac{1+\dfrac{\gamma-1}{2}M_1^2}{1+\dfrac{\gamma-1}{2}M_2^2} \qquad (7.3.19)$$

$$\frac{\rho}{\rho^\star} = \frac{C^\star}{C} = \left[\frac{2\left(1+\dfrac{\gamma-1}{2}M^2\right)}{\gamma+1}\right]^{1/2} \quad\text{and}\quad \frac{\rho_1}{\rho_2} = \frac{C_2}{C_1} = \frac{M_2}{M_1}\left[\frac{1+\dfrac{\gamma-1}{2}M_1^2}{1+\dfrac{\gamma-1}{2}M_2^2}\right]^{1/2} \qquad (7.3.20)$$

$$\frac{T}{T^\star} = \frac{\gamma+1}{2\left(1+\dfrac{\gamma-1}{2}M^2\right)} \quad\text{and}\quad \frac{T_2}{T_1} = \frac{1+\dfrac{\gamma-1}{2}M_1^2}{1+\dfrac{\gamma-1}{2}M_2^2} \qquad (7.3.21)$$

$$\frac{p}{p^\star} = \frac{1}{M}\left[\frac{\gamma+1}{2\left(1+\dfrac{\gamma-1}{2}M^2\right)}\right]^{1/2} \quad\text{and}\quad \frac{p_2}{p_1} = \frac{M_1}{M_2}\left[\frac{1+\dfrac{\gamma-1}{2}M_1^2}{1+\dfrac{\gamma-1}{2}M_2^2}\right]^{1/2} \qquad (7.3.22)$$

$$\frac{p_0}{p_0^\star} = \frac{1}{M}\left[\frac{2\left(1+\frac{\gamma-1}{2}M^2\right)}{\gamma+1}\right]^{(\gamma+1)/2(\gamma-1)} \text{ and } \frac{p_{02}}{p_{01}} = \frac{M_1}{M_2}\left[\frac{1+\frac{\gamma-1}{2}M_1^2}{1+\frac{\gamma-1}{2}M_2^2}\right]^{(\gamma+1)/2(\gamma-1)} \tag{7.3.23}$$

$$\frac{F}{F\star} = \frac{1+\gamma M^2}{M\left[2(\gamma+1)\left(1+\frac{\gamma-1}{2}M^2\right)\right]^{1/2}} \text{ and } \frac{F_2}{F_1} = \frac{1+\gamma M_2^2}{1+\gamma M_1^2}\frac{M_1}{M_2}\left[\frac{1+\frac{\gamma-1}{2}M_1^2}{1+\frac{\gamma-1}{2}M_2^2}\right]^{1/2} \tag{7.3.24}$$

Change in entropy can also be obtained, since it is given by the relation

$$\Delta S = c_p \ln (T_2/T_1) - R \ln (p_2/p_1)$$

We may change this by using stagnation values in place of static values. As before, we again consider the sections where Mach numbers are $M$ and 1. Since $T_{02} = T_{01}$, the equation gives writing for $p_{02}/p_{01}$

$$\frac{s_2 - s_1}{R} = \ln\frac{p_{01}}{p_{02}} = \ln\frac{M_2}{M_1}\left[\frac{1+\frac{\gamma-1}{2}M_1^2}{1+\frac{\gamma-1}{2}M_2^2}\right]^{(\gamma+1)/2(\gamma-1)} \tag{7.3.25}$$

We mentioned before that at some section the flow becomes sonic and there after the flow does not change. Distance of this section from the given section depends on the Mach number at the section. If this distance is $L_{max}$, it can be obtained by integrating Eq. 7.3.18 taking $\hat{f}$ to be constant such that

$$\hat{f} = \frac{1}{L_{max}}\int_0^{L_{max}} f\,dx \tag{7.3.26}$$

Integration of Eq. 7.3.18 after simplification, gives

$$\int_0^{L_{max}} 4f\frac{dx}{D} = \int_M^1 \frac{1-M^2}{\gamma M^2\left(1+\frac{\gamma-1}{2}M^2\right)}\frac{dM^2}{M^2}$$

This equation with Eq. 7.3.26 gives

$$4\hat{f}\,L_{max}/D = \frac{1-M^2}{\gamma M^2} + \frac{\gamma+1}{2\gamma}\ln\frac{(\gamma+1)M^2}{2\left(1+\frac{\gamma-1}{2}M^2\right)} \tag{7.3.27}$$

Knowing the Mach number at the two sections, the corresponding maximum length of the duct from the section for it to become sonic can be determined and from their difference the distance between the two sections is obtained.

The relations so obtained are lengthy and time consuming to evaluate their values. However, we note that the properties depend only on $\gamma$ and $M$. Therefore these relations can be expressed in form of charts and tables; both are available in the market for different values of $\gamma$ by the name Fanno chart and Fanno tables. A Fanno table has columns for $T/T^*$, $p/p^*$, $p_0/p^*_0$, $C/C^*$, $F/F^*$ and $4fL_{max}/D$ with Mach number at the section as the independent parameter. Here * corresponds to the properties at the section where $M = 1$.

**Example 7.3.1** Air flows through a constant area pipe of diameter 7.16 mm from a tank where air temperature and pressure is 296 K and 760 mm *Hg* respectively. A converging nozzle decreases pressure to a vacuum of 18.9 mm *Hg* at the inlet to the duct. The vacuum at the other end of the duct is 412 mm *Hg*. Determine Mach number at the exit and the length of the duct. Assume adiabatic frictional flow through the pipe.

**Solution.** Let subscripts 1 and 2 be used for inlet and exit of the pipe. We are given: $T_{01} = 296$ K, $p_{01} = 760$, $p_1 = 760 - 18.9 = 741.1$ mm and $p_2 = 760 - 412 = 348$ mm of *Hg*.

$p_1/p_{01} = 741.1/760 = 0.97513$. From isentropic table at $\gamma = 1.4$, corresponding to this ratio we read $M_1 = 0.19$. We also read from the table $T_1/T_{01} = 0.99283 \rightarrow T_1 = 0.99283 \times 296 = 294$ K, $a_1 = \sqrt{(1.4 \times 287 \times 294)} = 343.68$ and $C_1 = M_1 \times a_1 = 65.3$ m/s.

To determine density at section 1, we need pressure in N/m² but it is given in mm of *Hg*. Either we may use a conversion factor or $p_1 = \rho g h = 13.6 \times 9.81 \times (741.1/1000) = 98.1$ kPa.

$$\therefore \qquad \rho_1 = p_1/(R\,T_1) = (98.1 \times 1000)/(287 \times 294) = 1.16 \text{ kg/m}^3.$$

From the table for air, $\mu$ at 294 K $= 1.82 \times 10^{-5}$ kg/ ms.

$$\therefore \qquad \text{Re}_1 = (\rho C\,D/\mu)_1 = 1.1.6 \times 65.3 \times 0.00716/(1.82 \times 10^{-5}) = 2.98 \times 10^4.$$

From the charts for friction factor, assuming smooth pipe, $f = 0.0235$ at this Re.

From Fanno table, we read for the section 1: $p_1/p^* = 5.745$ and $f\,L_{max}/D = 16.38$.[4]

$\rightarrow L_1 = L_{max} = 16.38 \times 0.00716/0.0235 = 4.99$ m.

This is the distance of the pipe section, where $M = 1$, from section 1, however, the free end of the pipe may not have $M = 1$. Therefore we proceed in the same way to determine the distance of the critical section from the free end so that length of the pipe may be determined. Since $p^*$ is constant for entire length of the pipe, we get

$$\frac{p_2}{p^*} = \frac{p_2}{p_1} \times \frac{p_1}{p^*} = \frac{348}{741.1} \times 5.745 = 2.698\,.$$

From Fanno tables corresponding to $p/p^* = 2.698$, $M_2 = 0.4$ and $f\,L_{max}/D = 2.309$.

$$\rightarrow \qquad L_2 = L_{max} = 0.704 \text{ m}.$$

$$\therefore \qquad \text{length of the pipe} = L_1 - L_2 = 4.29 \text{ m}.$$

**Example 7.3.2** In an experiment of air flowing adiabatically through a constant area duct, the experimenter measured at section 1: $p_1 = 1.5$ bar, and $T_1 = 30°$C and at section 2: $p_2 = 1.3$ bar and $T_2 = 17°$C. Verify whether the measurements can be considered accurate.

[4] While using Darcy equation, the problem of $f$ or $4f$ always is a problem as both the practices are being followed. The value of $f$ suggests that we do not have to use the factor 4 in this case. A rough guide is that if $f > 0.01$ then 4 is included in $f$ itself.

**Solution** The measurements will be correct if there is increase in entropy from the sections 1 to 2, since the flow is adiabatic and not isentropic. The air is perfect gas and the change in entropy, from thermodynamics course, is given by the equation

$$s_2 - s_1 = c_p \ln(T_2/T_1) - R \ln(p_2/p_1)$$
$$= 1.005 \times \ln(290/303) - 287 \times \ln(1.3/1.5) = -2.956 \text{ kJ/kg K}.$$

Decrease in entropy means that the measurements are not correct.

## 7.4 Isothermal Flow with Friction

In industries, the compressible fluid is transported miles away through pipes of constant diameter. Gas flow through pipes from one country to another is one such example. Such flows are idealised as isothermal rather than adiabatic. In this case, the stagnation temperature and stagnation enthalpy would change in the direction of the flow. The analysis procedure is the same as discussed above. We shall assume a perfect gas, a constant area duct, absence of body forces and one-dimensional flow, the usual assumptions.

Continuity equation $\rightarrow \dfrac{d\rho}{\rho} + \dfrac{dC}{C} = 0 \quad$ or, $\dfrac{d\rho}{\rho} = -\dfrac{1}{2}\dfrac{dC^2}{C^2}$ (7.4.1)

Equation of state $\rightarrow \dfrac{dp}{p} = \dfrac{d\rho}{\rho} = -\dfrac{dC}{C} = -\dfrac{1}{2}\dfrac{dC^2}{C^2}$ (7.4.2)

Momentum equation $\rightarrow \dfrac{dp}{p} = -\dfrac{\gamma M^2}{2}\left(\dfrac{dC^2}{C^2} + 4f\dfrac{dx}{D}\right)$

$$= -\frac{\gamma M^2}{2(1-\gamma M^2)} 4f \frac{dx}{D} \quad \text{(Using Eq. 7.4.2)} \qquad (7.4.3)$$

$$M = \frac{C}{a} = \frac{C}{\sqrt{\gamma RT}} \quad \text{or, } M^2 = \frac{C^2}{\gamma RT} \xrightarrow[\text{differentiation}]{\text{logrithmic}} \frac{dM^2}{M^2} = \frac{dC^2}{C^2} \qquad (7.3.33)$$

$$\text{Eq. 7.2.3} \xrightarrow[\text{differentiation}]{\text{logarithmic}} \frac{dT_0}{T_0} = \frac{(\gamma-1)M^2}{2\left(1+\dfrac{\gamma-1}{2}M^2\right)} \frac{dM^2}{M^2} \qquad (7.4.4)$$

These equations can be combined to give

$$\frac{dp}{p} = \frac{d\rho}{\rho} = -\frac{dC}{C} = -\frac{dC^2}{2C^2} = -\frac{dM^2}{2M^2} = -\frac{\gamma M^2}{2(1-\gamma M^2)} 4f \frac{dx}{D} \qquad (7.4.5)$$

We notice that $1 - \gamma M^2$ decides the change in the flow variables. When $M < 1/\sqrt{\gamma}$ then velocity increases and pressure decreases with increase in duct length. Thus the flow is accelerating. It is shown on a *T-s* diagram by the horizontal path *AC* through *A* in Fig. 7.4.1. Also $dM^2$ is positive. On the other hand, if $M > 1/\sqrt{\gamma}$ then the state moves along path *BC* during which the flow is decelerating and pressure is increasing. Thus $M = 1/\sqrt{\gamma}$

is the limiting value. Therefore, we will integrate Eq. 7.4.5 from $M = 1/\sqrt{\gamma}$ to $M$, representing properties at $M = 1/\sqrt{\gamma}$ section by $^\star$. We get

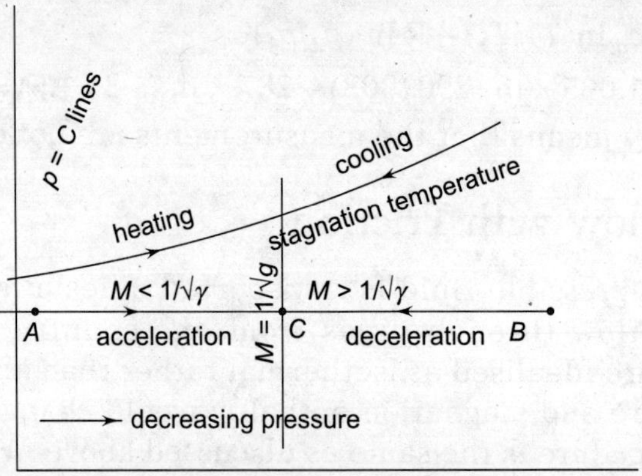


**Fig. 7.4.1** Isothermal flow with friction

$$\frac{p}{p^*} = \frac{\rho}{\rho^*} = \frac{C^*}{C} = \frac{1}{\sqrt{\gamma}\, M} \tag{7.4.6}$$

$$\frac{p_2}{p_1} = \frac{\rho_2}{\rho_1} = \frac{C_1}{C_2} = \frac{M_1}{M_2} \tag{7.4.7}$$

$$\frac{T_2}{T_1} = 1 \tag{7.4.8}$$

$$s_2 - s_1 = c_p \ln \frac{T_2/T_1}{(p_2/p_1)^{(\gamma-1)/\gamma}} = R \ln\left(\frac{M_2}{M_1}\right) \tag{7.4.9}$$

$$s - s^\star = R \ln\left(\sqrt{\gamma}\, M\right) \tag{7.4.10}$$

Logarithmic differentiation of Eq. 7.2.4 gives $dp_0/p_0$ in which we substitute for $dp/p$ from Eq. 7.4.5. This yields after simplification

$$\frac{dp_0}{p_0} = \left(\frac{\gamma+1}{2}M^2 - 1\right)\frac{dM^2}{M^2} \tag{7.4.11}$$

At $M = \sqrt{2/(\gamma+1)}$, $dp_0/p_0$ is zero. Therefore for $M < 1/\sqrt{\gamma}$ the stagnation pressure would always decrease. However, when $M > 1/\sqrt{\gamma}$ then the factor within the parenthesis in Eq. 7.4.11 is positive and the second negative (as concluded above). Therefore stagnation pressure continues to decrease till $M = \sqrt{2/(\gamma+1)}$. After this value, both the factors in Eq. 7.4.11 are negative and hence stagnation pressure would increase. This trend is shown in Fig. 7.4.2.

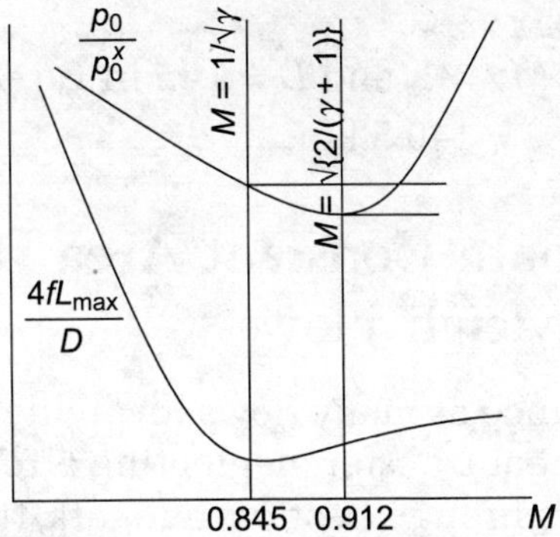


**Fig. 7.4.2** Variation of stagnation pressure and duct length

Equation 7.4.5 also relates length of the duct $dx$ with the Mach number and can be integrated in the same way as other integrals were carried out from $M = 1/\sqrt{\gamma}$ to $M$. This will give the distance $AC$ or $BC$, where $C$ represents the section at which $M = 1/\sqrt{\gamma}$. We get

$$\left(4f\frac{L_{max}}{D}\right)_M = \frac{1-\gamma M^2}{\gamma M^2} + \ln \gamma M^2 \qquad (7.4.12a)$$

The distance between the two sections 1 and 2 is given by

$$4f\frac{L}{D} = \left(4f\frac{L_{max}}{D}\right)_{M1} - \left(4f\frac{L_{max}}{D}\right)_{M2} = \frac{1}{\gamma}\left(\frac{1}{M_1^2} - \frac{1}{M_2^2}\right) + \ln\frac{M_1^2}{M_2^2} \qquad (7.4.12b)$$

where $L_{max}$ is the distance $AC$ and $L$ is the distance between two sections in the flow where Mach numbers are $M_1$ and $M_2$, respectively. We may also find out the impulse function in the same way. Figure 7.4.2 shows variation of stagnation pressure and maximum length with Mach number. This confirms our discussions about the change in stagnation pressure with Mach number.

**Example 7.4.1** A natural gas pipeline of diameter 1 m and surface roughness $\varepsilon = 5 \times 10^{-5}$ m is supplied with gas from a compressor at a pressure $p_1$ = 20 bar and temperature $T_1$ = 20°C. At this temperature, the natural gas sound speed $a_1$ = 446.1 m/s, viscosity $\mu = 9 \times 10^{-6}$ *Pa s* and specific heat ratio = 1.31 while the gas constant $R$ = 518.3 J/kg K. The pipe flow speed $C_1$ = 10 m/s. Calculate the pipe length at which the flow would be choked.

**Solution** To determine the friction coefficient from Moody diagram we need Reynolds number.

$$\rho_1 = \frac{p_1}{RT_1} = \frac{2 \times 10^6}{518.3 \times 293.15} = 13.16 \text{ kg/m}^3$$

and

$$\text{Re} = \frac{\rho_1 C_1 D}{\mu_1} = \frac{13.16 \times 10 \times 1}{9 \times 10^{-6}} = 1.463 \times 10^7$$

Darcy friction factor is given by the empirical relation

$$\frac{1}{\sqrt{f}} = -2.0 \log\left(\frac{\varepsilon / D}{3.7} + \frac{2.51}{\text{Re}\sqrt{f}}\right) \rightarrow f = 0.01076 \text{ by trial and hit method.}$$

Further, $M = 10/446.1 = 0.02242$.

Substituting $M_1 = 0.02242$, $M^2 = 1$, and $D = 1$ m in Eq. 7.4.12, we get

$$L_{max} = 140.5 \text{ km}.$$

## 7.5 Flow Through Smooth Constant Area Duct with Heat Transfer—Rayleigh Flow

This is another flow of importance as many flows involving heat transfer can be idealised as one-dimensional flow with heat transfer; neglecting effect of friction without introducing appreciable error e.g; flow through heat exchangers, furnaces, etc. due to small pipe or gas path length. We make usual assumptions of perfect gas, constant area of flow, etc. Following the procedure adopted for derivation of the equation for Fanno lines and recognising the difference that in this case that instead of stagnation enthalpy, it is stagnation pressure that would remain constant, we may write the following relations:

$$G = \rho C + p + \rho C^2 = \text{const} \rightarrow p + G^2 v = \text{const.}$$

Equation of state: $$p = p(h, s) \text{ and } \rho = \rho(h, s) \rightarrow p\,(h, s) + G^2/\rho(h, s) \qquad (7.5.1)$$

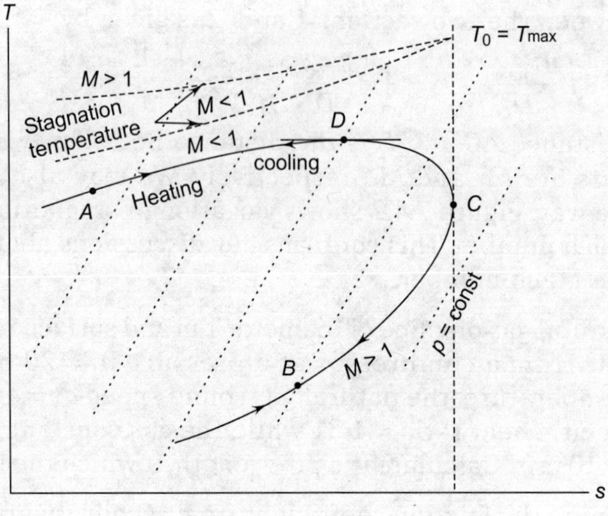


**Fig. 7.5.1** Rayleigh line

Equation 7.5.1 is shown on *T-s* diagram in Fig. 7.5.1 for one value of *G*. In fact, this equation generates a family for each value of *G* at *a* given value of $\gamma$. The constant pressure lines are also shown. We observe that

(i) There are two limbs of the curve, for the upper one $M < 1$ and for the lower one $M > 1$.

(ii) The stagnation temperature is increasing along two different paths which meet at a point where $T_0 = T_{max}$. This point lies on the vertical line tangential to the curve at the point *C*. The lower stagnation temperature curve is for subsonic flow.

(iii) Entropy is maximum at point *C*. However, the static temperature is maximum at point *D*.

(iv) Heating of the gas increases entropy and therefore the point on the curve moves to right. However, this change can take place up to the point *C* only; thereafter no more change can take place. On the contrary, cooling moves the point to the left and there is no limit on it.

Now we shall determine conditions at points *C* and *D* by determining slope of the Rayleigh line corresponding to constant entropy or constant enthalpy on *p-v* diagram. If $\theta$ is the slope of Rayleigh line at any point then from the momentum equation

$$p + G^2 v = C \rightarrow \tan\theta = (dp/dv)_R = -G^2 = -\rho^2 a^2 M^2 \tag{7.5.2}$$

For an isentropic curve on *p-v* diagram, the slope $\tan\alpha$ is

$$\tan\alpha = \left(\frac{\partial p}{\partial v}\right)_s = \left(\frac{\partial p}{\partial(1/\rho)}\right)_s = -\rho^2\left(\frac{\partial p}{\partial\rho}\right)_s = -\rho^2 a^2 \tag{7.5.3}$$

Comparison of Eqs. 7.5.2 and 7.5.3 gives same slope for the isentropic and the Rayleigh line at $M = 1$. We see from Fig. 7.5.1 that at the point *C* the Rayleigh curve has the same slope as that of constant entropy line. Therefore, at point *C*, $M = 1$. This result was expected, since the flow from upper or lower limb tends to sonic flow due to heat transfer to it. We may now determine similarly the slope of isothermal line for which $pv = C$, constant. This give slope $\tan\beta$ of the line as

$$\tan\beta = \left(\frac{\partial p}{\partial v}\right)_T = -p\rho = -\rho^2 a^2/\gamma, \tag{7.5.4}$$

Using Eq. 7.1.3, comparison with Eq. 7.5.2 gives $M = 1/\sqrt{\gamma}$ for isothermal or isenthalpic line. This means the isenthalpic line is tangential to the Rayleigh line at this point that corresponds to point *D*, since only at this point the horizontal isenthalpic lines can be tangential. This shows that the static temperature would increase up to the point *D* only after which it will decrease till the point *C*. In other words, there is limit to the heating process. Since *M* is < 1 the point will lie on the subsonic flow curve.

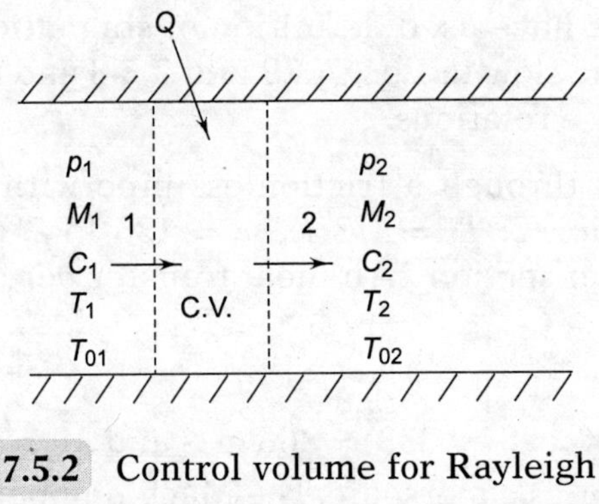


**Fig. 7.5.2** Control volume for Rayleigh flow

We may now proceed to determine the governing equations by considering a control volume, shown in Fig. 7.5.2, and writing the different equations, as done in Fanno line analysis. We get the following set of relations for constant area duct:

Continuity : $\dfrac{C_2}{C_1} = \dfrac{\rho_1}{\rho_2} = \dfrac{M_2^2}{M_1^2}\dfrac{1+\gamma M_1^2}{1+\gamma M_2^2}$ (7.5.5)

Momentum : $(1+\gamma M_1^2)p_1 = (1+\gamma M_2^2)p_2 \rightarrow \dfrac{p_2}{p_1} = \dfrac{1+\gamma M_1^2}{1+\gamma M_2^2}$ and $F_2 = F_1$ (7.5.6)

State : $\dfrac{T_2}{T_1} = \dfrac{p_2}{p_1} \times \dfrac{\rho_1}{\rho_2} = \dfrac{M_2^2}{M_1^2}\left(\dfrac{1+\gamma M_1^2}{1+\gamma M_2^2}\right)^2$ (7.5.7)

Stagnation pressure : $\dfrac{p_{02}}{p_{01}} = \dfrac{p_2}{p_1} \times \left(\dfrac{1+\dfrac{\gamma-1}{2}M_2^2}{1+\dfrac{\gamma-1}{2}M_1^2}\right)^{\gamma/(\gamma-1)}$ (7.5.8)

Stagnation temperature: $\dfrac{T_{02}}{T_{01}} = \dfrac{T_2}{T_1}\dfrac{1+\dfrac{\gamma-1}{2}M_2^2}{1+\dfrac{\gamma-1}{2}M_1^2}$ (7.5.9)

Change in entropy : $s_2 - s_1 = c_p \ln\left[(T_2/T_1)/(p_2/p_1)^{(\gamma-1)/\gamma}\right]$

$$= R\ln\left[\left(\frac{M_2}{M_1}\right)^{2\gamma/(\gamma-1)}\left(\frac{1+\gamma M_1^2}{1+\gamma M_2^2}\right)^{(\gamma+1)/(\gamma-1)}\right] \quad (7.5.10)$$

Heat transfer :

$$Q = c_p(T_{02}-T_{01}) \rightarrow \frac{Q}{c_p T_1} = \frac{(M_2^2-M_1^2)\left[2-2\gamma M_1^2M_2^2+(\gamma-1)(M_2^2+M_1^2)\right]}{2M_1^2(1+\gamma M_2^2)^2} \quad (7.5.11)$$

These relations are called Rayleigh relations and have been tabulated in the gas table by taking $M_2 = 1$ and $M_1 = M$. A little practice will help you in deriving these relations. For example, we have used definition of stagnation temperature and stagnation pressure for isentropic flow in Eqs. 7.5.9 and 7.5.8 and then substituted the values of $T_2/T_1$ or $p_2/p_1$ to give the relations.

**Example 7.5.1** Air flows through a frictionless pipe with the data given at its inlet section 1 and outlet section 2: $T_1 = 333$ K, $p_1 = 135$ kPa, $C_1 = 732$ m/s and $M_2 = 1.2$. Determine (a) properties at section 2, (b) heat transfer per unit mass, and (c) change in entropy.

**Solution.** From the data,

$$a_1 = \sqrt{(1.4 \times 287 \times 333)} = 366 \text{ m/s and } M_1 = C_1/A_1 = 732/366 = 2.0.$$

The flow at inlet section is supersonic. The Mach number is decreasing from section 1 to 2 without friction. This means there must be heat transfer between the pipe and surroundings. Using Rayleigh table for the section 1 where $M_1 = 2$, we read $T/T^* = 0.5289$, $p/p^* = 0.3636$, $v/v^* = 1.455$. The corresponding values at $M = 1.2$ are: 0.9119,

0.7958, and 1.146, respectively. Using these data and recognising that * values are same for all the sections, we can write the following relations to get properties at section 2.

$$\frac{T_2}{T_1} = \frac{T_2/T^*}{T_1/T^*} = \frac{0.9119}{0.5289} = 1.72 \rightarrow T_2 = 573 \text{ K}$$

and $$\frac{p_2}{p_1} = \frac{p_2}{p^*}\frac{p^*}{p_1} = 2.19 \qquad \rightarrow p_2 = 296 \text{ kPa.}$$

$C_2 = M_2 a_2 = 5.77$ m/s. Also $\rho_2 = p_2/(RT_2) = 1.8$ kg/m$^3$.

To determine heat transfer, we need $T_{02} - T_{01}$. These can be determined using the isentropic table at $\gamma = 1.4$ . We get $T_{01} = 599$ K and $T_{02} = 738$ K.

$$\therefore \qquad q = c_p \times (T_{02} - T_{01}) = 1.0 \times (738 - 599) = 139 \text{ kJ/kg.}$$

$s_2 - s_1 = c_p \ln(T_2/T_1) - R \ln(p_2/p_1) = 1 \times \ln(573/333) - 287 \ln(2.96/1.35) = 0.317$ kJ/kg.

From the increase in stagnation temperature, it is evident that heat is being added in the process. We may also see the effect on the stagnation pressure that can be determined at sections 1 and 2 using the isentropic table. The result is $p_{01} = 1.06$ Mpa and $p_{02} = 718$ kPa. This verifies that heat transfer results in loss of stagnation pressure, a result we got while studying Rayleigh flow.

**Example 7.5.2** A combustion chamber in a gas turbine plant receives air at 350 K, 0.55 bar and 75 m/s. The air fuel ratio is 29 and the calorific value of the fuel is 41.87 MJ/kg. Taking $\gamma = 1.4$ and $R = 287$ J/kg K for the gas, determine: (a) the initial and final Mach numbers, (b) final pressure, temperature and velocity of the gas, (c) percent stagnation pressure loss in the combustion chamber, and (d) the maximum stagnation temperature attainable.

**Solution** Let the inlet and exit sections of the combustion chamber be represented by subscripts 1 and 2 respectively. then from the given data

$$a_1 = \sqrt{1.4 \times 287 \times 350} = 375 \text{ m/s and } M_1 = C_1/a_1 = 75/350 = 0.2.$$

From isentropic tables at $\gamma = 1.4$ and $M_1 = 0.2$

$$\frac{p_1}{p_{01}} = 0.973 \rightarrow p_{01} = 0.55/0.973 = 0.565 \text{ bar}$$

and $$\frac{T_1}{T_{01}} = 0.992 \rightarrow T_{01} = 350/0.992 = 352.82 \text{ K.}$$

From Rayleigh table at $\gamma = 1.4$ and $M = 0.2$

$$\frac{p}{p^*} = 2.273, \ \frac{T}{T^*} = 0.207, \ \frac{C}{C^*} = 0.091, \ \frac{p_0}{p_0^*} = 1.235, \text{ and } \frac{T_0}{T_0^*} = 0.207.$$

Making energy balance of the combustion chamber for 1 kg of air fuel mixture, we get

$$C_p(T_{02} - T_{01}) = 41870/(29+1)$$

$$\rightarrow \qquad T_{02} = 352.82 + 1395.67/1.005 = 1741.55 \text{ K.}$$

or, $$\frac{T_{02}}{T_0^*} = \frac{T_{02}}{T_{01}}\frac{T_{01}}{T_0^*} = \frac{1741.55}{352.82} \times 0.174 = 0.859.$$

From Rayleigh table corresponding to this temperature ratio, $M_2$ = **0.64.**

Also $p/p^* = 1.525$, $T/T^* = 0.953$, $C/C^* = 0.625$, and $p_0/p_0^* = 1.061$.

$$\rightarrow \qquad p_2 = \frac{(p/p^*)_2}{(p/p^*)_1} p_1 = \frac{1.525}{2.273} \times 0.55 = 0.369 \text{ bar.}$$

Similarly $\qquad T_2 = 1611.35$ K, $C_2 = 515.11$ m/s, and $p_{02} = 0.485$ bar.

Stagnation pressure loss = $p_{01} - p_{02}$ = 0.565 – 0.485 = 0.08 bar = (0.08/0.565) × 100 = 14.16%.

Maximum stagnation temperature is $T_{01}^* = 352.82/0.174 = 2027.7$ K.

## 7.6 Normal Shock Wave

We considered generation of a wave in section 6.1 that was a very week pressure wave, which propagates in all the directions. For this reason, we imagined a tube fitted with a piston at its one end and filled with gas so that the disturbance wave travels in one direction. Now let us consider situation when the piston is producing continuous waves, each of which travel downward. This sets a temperature and pressure gradient in the fluid resulting in higher wave velocity near the piston. Situation arises when a number of waves overcome each other to reach a section simultaneously and coerce to make strong pressure waves. These strong pressure waves are called the *shock waves* and there exists a sharp pressure and temperature difference across such wave. Travel of a compression wave is shown in Fig. 7.6.1 with its thickening downstream indicating merging of more and more compression waves reaching from upstream. These wave fronts may be in any direction with respect to direction of the flow. If these are at right angles, these are called *normal shock*. Thus a normal shock wave is a very thin region of discontinuity in the flow region; the thickness is of the order $10^{-3}$ mm, which is comparable to free path. Flow across a wave shock wave is an irreversible process and is associated with increase in entropy and losses. Irreversibility of the flow imposed certain restructions on the occurrence of shock wave, which we shall investigate.

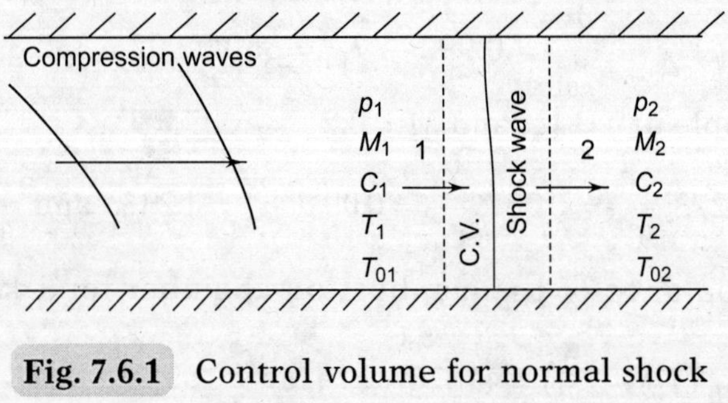


**Fig. 7.6.1** Control volume for normal shock

Consider a control volume around the normal shock with sections 1 and 2 upstream and downstream, respectively. We can write the conservation laws for one-dimensional flow as well as other equations in the same way as we did for Fanno or Rayleigh flows to give the following relations between the properties at section 2 in terms of those at section 1.

$$\text{Continuity}: \dot{m} = \rho_1 A C_1 = \rho_2 A C_2 \tag{7.6.1}$$

$$\text{Momentum}: (p_1 - p_2)\, A = \dot{m}\; (C_2 - C_1) \tag{7.6.2}$$

$$\text{Energy}: h_{01} = h_{02} = h_0 \text{ and } T_{01} = T_{02} = T_0 \tag{7.6.3}$$

$$\text{State}: h = h(s, \rho) \text{ and } s = s\,(p, \rho) \tag{7.6.4}$$

We learnt that these equations are also satisfied by the Fanno and Rayleigh lines except Eq. 7.6.2 in the former case and Eq. 7.6.3 in the latter case. This suggests that point 2 can easily be located by superimposing the Rayleigh line over the Fanno line and determining the point of intersection of the two lines for the same value of $G$, as shown in Fig. 7.6.2. Points 1 and 2 lie on the supersonic and subsonic limbs and the flow properties move from point 1 to 2. We see that entropy is increasing during this process. This means shock would occur only if the flow is supersonic and not subsonic.

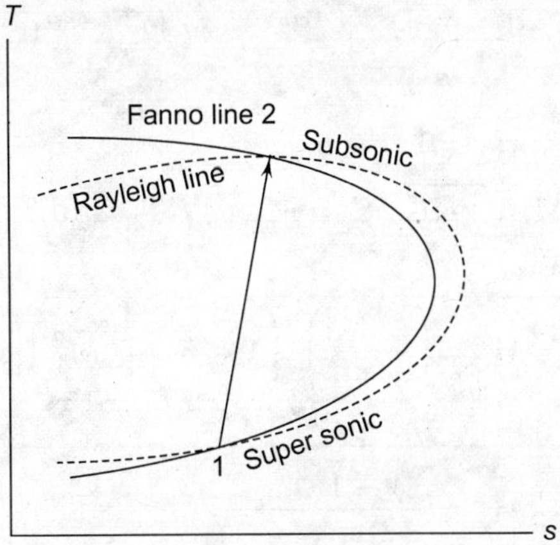


**Fig. 7.6.2** Normal shock

Because of the losses, shock wave formation is not desirable, as it results in decrease in efficiency of the machines. Examples of a strong shock wave are sonic boom created by supersonic aircrafts and the blast wave generated by the bomb explosion. We know how devastating these effects are affecting the human life and the structures. However, there are some applications wherein the shock wave is created as in pulse engine, shock tube experimental set up for study of shock wave behaviour, compressors to obtain high pressure ratio per stage, etc.

We shall now develop relations that give the properties and flow variables at the section 2 after the shock wave in terms of the values at section 1 noting that there is no change in stagnation temperature across the shock wave. We also combine Eqs. 7.6.1 and 7.6.2 to get Eq. 7.6.7.

$$\frac{a_1^2}{\gamma - 1} + \frac{C_1^2}{2} = \frac{a_2^2}{\gamma - 1} + \frac{C_2^2}{2} = \frac{1}{2}\frac{\gamma + 1}{\gamma - 1} a^{*2} \tag{7.6.5}$$

$$\rightarrow \quad \frac{a_1^2}{C_1} = \frac{\gamma+1}{2}\frac{a^{\star 2}}{C_1} - \frac{\gamma-1}{2}C_1 \text{ and } \frac{a_2^2}{C_2} = \frac{\gamma+1}{2}\frac{a^{\star 2}}{C_2} - \frac{\gamma-1}{2}C_2 \tag{7.6.6}$$

$$\frac{p_1}{\rho_1 C_1} - \frac{p_2}{\rho_2 C_2} = C_2 - C_1 \rightarrow \frac{1}{C_1}\frac{\gamma p_1}{\rho_1} - \frac{1}{C_2}\frac{\gamma p_2}{\rho_2} = \gamma(C_2 - C_1) \tag{7.6.7}$$

Writing $\frac{p_1}{\rho_1} = \frac{a_1^2}{\gamma}, \frac{p_2}{\rho_2} = \frac{a_2^2}{\gamma}$ in Eq. 7.6.7 and using Eq. 7.6.6, we get

$$\frac{\gamma+1}{2}\frac{a^{*2}}{C_1} - \frac{\gamma-1}{2}C_1 - \frac{\gamma+1}{2}\frac{a^{*2}}{C_2} + \frac{\gamma-1}{2}C_2 = \gamma(C_2 - C_1)$$

$$\rightarrow \quad C_1 \,.\, C_2 = a^{*2} \tag{7.6.8}$$

Eq. 7.6.8 is called Prandtl–Meyer relation and can be written in terms of the Mach number at sections 1 and 2, noting that

$$T_1{}^* = \frac{2}{\gamma+1}T_{01} = T_2{}^* = \frac{2}{\gamma+1}T_{02}$$

$$a_1{}^* = \sqrt{\gamma T_1{}^*}\ a_2{}^* = \sqrt{\gamma T_2{}^*} \rightarrow a_1{}^* = a_2{}^* = a^*$$

Eq. 7.6.8 gives

$$M_1^* \,.\, M_2^* = 1 \tag{7.6.9}$$

We know, $C_1 = M_1\, a_1 = M_1\sqrt{\gamma R T_1}$, $C_2 = M_2\sqrt{\gamma R T_2}$ and $T/T_0 = 2\gamma/(\gamma+1)$. Using these relations in Eq. 7.6.8, we get

$$M_1^2 \,.\, M_2^2 = \left(\frac{2}{\gamma+1}\right)^2\left(1 + \frac{\gamma-1}{2}M_1^2\right)\left(1 + \frac{\gamma-1}{2}M_2^2\right)$$

$$\rightarrow \quad M_2^2 = \frac{\frac{2}{\gamma-1} + M_1^2}{\frac{2\gamma}{\gamma-1}M_1^2 - 1} \tag{7.6.10}$$

Dividing numerator and denominator of Eq. 7.6.10 by $2(\gamma - 1)$, we get

$$M_2^2 = \frac{2/(\gamma-1) + M_1^2}{2\gamma/(\gamma-1)\, M_1^2 - 1} \tag{7.6.11}$$

Thus knowing Mach number at section 1, Mach number at section 2 is obtained from Eq. 7.6.11. We may similarly proceed to get ratio of static pressure and temperature using the above equations by substituting for $M_2$.

$$\frac{p_2}{p_1} = \frac{1+\gamma M_1^2}{1+\gamma M_2^2} = \frac{1+\gamma M_1^2}{1 + \frac{2\gamma/(\gamma-1) + M_1^2}{2\gamma/(\gamma-1)M_1^2 - 1}}$$

$$= \frac{\gamma-1}{\gamma+1}\left(\frac{2\gamma}{\gamma-1}M_1^2 - 1\right) = \frac{2\gamma}{\gamma-1}M_1^2 - \frac{\gamma-1}{\gamma+1} \tag{7.6.12}$$

$$\frac{T_2}{T_1} = \frac{1 + \frac{\gamma - 1}{2} M_1^2}{1 + \frac{\gamma - 1}{2} M_2^2} = \frac{\left(1 + \frac{\gamma - 1}{2} M_1^2\right)\left(\frac{2\gamma}{\gamma - 1} M_1^2 - 1\right)}{\frac{1}{2}\frac{(\gamma + 1)^2}{\gamma - 1} M_1^2} \tag{7.6.13}$$

$$\frac{\rho_2}{\rho_1} = \frac{p_2}{p_1}\frac{T_1}{T_2} = \frac{\left(\frac{\gamma + 1}{2}\right) M_1^2}{1 + \frac{\gamma - 1}{2} M_1^2} \tag{7.6.14}$$

Rankine-Hugoniot however expressed $\rho_2/\rho_1$ in terms of $p_2/p_1$ instead of $M_1^2$ so that we may compare the effect of occurrence of a shock wave in the flow on the development of static pressure and density change. Rewriting Eq. 7.6.12 in the following form

$$M_1^2 = \frac{\gamma + 1}{2\gamma}\frac{p_2}{p_1} + \frac{\gamma - 1}{2\gamma}$$

and substituting it in Eq. 7.6.13, we get after simplification

$$\frac{T_2}{T_1} = \frac{\frac{p_2}{p_1}\left(1 + \frac{\gamma - 1}{\gamma + 1}\frac{p_2}{p_1}\right)}{\frac{p_2}{p_1} + \frac{\gamma - 1}{\gamma + 1}}$$

and

$$\frac{p_2}{p_1}\frac{T_1}{T_2} = \frac{\frac{p_2}{p_1} + \frac{\gamma - 1}{\gamma + 1}}{1 + \frac{\gamma - 1}{\gamma + 1}\cdot\frac{p_2}{p_1}} = \frac{\rho_2}{\rho_1}$$

This gives after rearrangement of terms

$$\frac{p_2}{p_1} = \frac{\frac{\gamma + 1}{\gamma - 1}\frac{\rho_2}{\rho_1} - 1}{\frac{\gamma + 1}{\gamma - 1} - \frac{\rho_2}{\rho_1}} \tag{7.6.15}$$

We observe that $p_2/p_1 \to 1$ when $\rho_2/\rho_1 \to 1$ and $\to \infty$ when $\rho_2/\rho_1 = (\gamma + 1)/(\gamma - 1)$. Also

$$\frac{p_{02}}{p_{01}} = \frac{p_{02}}{p_2}\frac{p_2}{p_1}\frac{p_1}{p_{01}} = \left\{\frac{\frac{\gamma + 1}{2} M_1^2}{1 + \frac{\gamma - 1}{2} M_1^2}\right\}^{\gamma/\left(1 + \frac{\gamma - 1}{2} M_1^2\right)} \left(\frac{2\gamma}{\gamma + 1} M_1^2 - \frac{\gamma - 1}{\gamma + 1}\right)^{-1/(\gamma - 1)} \tag{7.6.16}$$

$$s_2 - s_1 = c_p \ln\frac{T_2}{T_1} - R\ln\frac{p_2}{p_1} = c_p \ln\left(\frac{p_{02}}{p_{01}}\right)^{-(\gamma - 1/\gamma)}$$

$$= \frac{\gamma R}{\gamma - 1}\ln\left(\frac{2}{(\gamma + 1) M_1^2} + \frac{\gamma - 1}{\gamma + 1}\right) + \frac{R}{\gamma - 1}\ln\left(\frac{2\gamma}{\gamma + 1} M_1^2 - \frac{\gamma - 1}{\gamma + 1}\right) \tag{7.6.17}$$

Equation 7.6.15 is known as *Rankine–hugoniot* relation. We have now determined all the relations[5] we need to calculate properties and Mach number at section 2 after the shok wave. These values depend on $\gamma$ and $M_1$. Therefore, the above equations have been tabulated for a few values of $\gamma$ with $M_1$ as independent variable. Such tables are available in the market and are called normal shock tables. The table gives $M_2$, $p_2/p_1$, $\rho_2/\rho_1$, $T_2/T_1$, $p_{02}/p_{01}$, $A_1^*/A_2^*$ and $p_{02}/p_1$ against $M_1$.

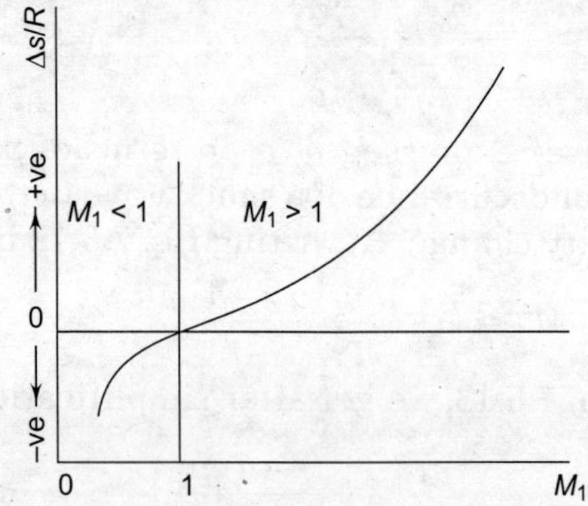


**Fig. 7.6.3** Variation of increase in entropy across normal shock

Let us come back to the validity of the statement we made in the beginning of this section that shock can occur only when the flow is supersonic. For this, Eq. 7.6.17 is plotted against $M_1$ in Fig. 7.6.3. We observe that change in entropy across the normal shock is positive only when initial Mach number is more than 1, that is, the flow is supersonic that justifies our statement. Moreover, we may compare the ratio of static pressure across the shock, given by Eq. 7.6.15, with the ratio for isentropic flow for the sections 2 and 1 to visualise the impect of a shock wave. For isentropic flow, $p \propto \rho^\gamma$, which gives

$$\frac{p_2}{p_1} = \left(\frac{\rho_2}{\rho_1}\right)^\gamma$$

Thus when $\rho_2/\rho_1 \to 1, p_2/p_1 \to 1$ .

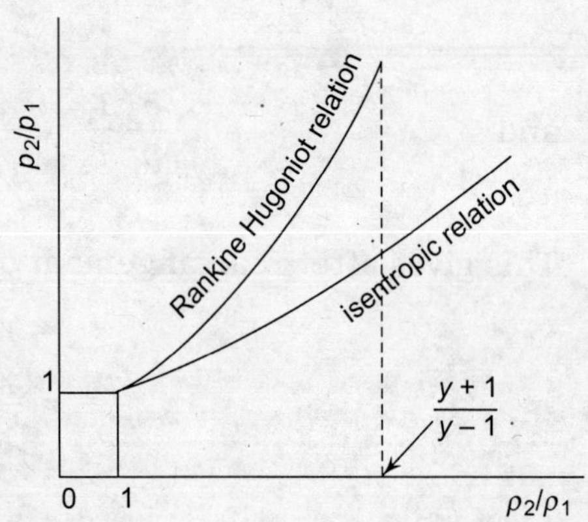


**Fig. 7.6.4** Comparison of normal shock with isentropic flow

[5] It is practice to put all the relations in terms of Mach number irrespective of the type of flow. You must have realised while deriving the relations for $p_{02}/p_{01}$ etc that we use isentropic relations and some other relations determined before. Unless we have to use these relations because the table does not exist for the value of $\gamma$ given in the problem or we have been asked to do so, it is better to determine these quantities from the appropriate table rather than from the relation. This would save a lot of time.

Figure 7.6 shows variations of $p_2/p_1$ v/s $\rho_2/\rho_1$ for two cases. We observe that as $\rho_2/\rho_1$ increases, the vertical gap between the two curves increases. However, $p_2/p_1 \rightarrow 1$ as $\rho_2/\rho_1 \rightarrow 1$ in both the cases. We define parameter shock strength as follows:

$$\xi = \frac{p_2}{p_1} - 1 = \frac{2\gamma}{\gamma+1}(M_1^2 - 1) \tag{7.6.18}$$

and express it in terms of density ratio rather pressure ratio using Rankine–Hugoniot relation.

$$\xi = \frac{p_2}{p_1} - 1 = \frac{\frac{2\gamma}{\gamma+1}\left(\frac{\rho_2}{\rho_1} - 1\right)}{\frac{2}{\gamma+1} - \left(\frac{\rho_2}{\rho_1} - 1\right)} \tag{7.6.19}$$

The factor $M_1^2 - 1$ is a measure of the strength of the shock wave. We infer from Eq. 7.6.18 that when $M_1 \rightarrow 1, p_2/p_1 \rightarrow 1$. Therefore, in transonic range, shock wave strength is very less and the flow can even be treated as isentropic flow.

**Example 7.6.1** Air at rest ($A_1$ = 347.2 m/s, $p_{01}$ = 1 bar) is accelerated by a shock wave of speed $V_s$ that is generated by a piston moving at the speed of $V_p$ = 100 m/s. Calculate the Mach number, the shock speed and the pressure on the piston.

**Solution** Let us use subscripts 1 and 2 to reperesent sections before and after the shock wave position and the reference frame to be fixed with the shock wave. Mass conservation demands

$$\rho_1 V_s = \rho_2 (V_s - V_p)$$

$$\rightarrow \quad \frac{\rho_2}{\rho_1} = \frac{M_1}{M_1 - V_p/a_1} = \frac{(\gamma+1)M_1^2}{(\gamma-1)M_1^2 + 2}$$

from Eq. 7.6.14. This gives a quadratic equation in $M_1$ and we get $M_1$ = 1.188; the other root corresponds to subsonic flow and hence not considered.

$$V_s = M_1 A_1 = 1.188 \times 347.2 = 412.5 \text{ m/s}.$$

Equation 7.6.12 gives pressure $p_2$ = 1.48 bar.

**Example 7.6.2** Air at the rate of 35 kgs flows through a nozzle in which a normal shock occurs in the divergent portion downstream of the throat. The nozzle has an area of cross section equal to 40 cm$^2$ at the location of shock. The pressure and velocity of fluid just before the shock are 2.5 bar and 480 m/s respectively. Find the Mach number, pressure and temperature after the shock.

**Solution**

Continuity equation: $\rho_1 = m/(A_1 C_1) = 1.858 \text{ kg/m}^3$,

Equation of state: $T_1 = p_1/(\rho_1 R) = 282.26$ K,

Sound speed: $a_1 = \sqrt{(1.4 \times 287 \times 282.26)} = 336.1$ m/s

and $M_1 = C_1/a_1 = 1.428$.

From normal shock table,

$$M_2 = 0.728, p_2/p_1 = 0.45317 \text{ and } T_2/T_1 = 1.2728.$$

This gives $p_2 = 3.31$ bar and $T_2 = 358$ K.

**Example 7.6.3** The ratio of the exit to entry in a subsonic diffuser is 4.0. The Mach number of a jet of air approaching the diffuser at $p_0 = 1.103$ bar and $T = 290$ K is 2.2. There is a standing shock wave at the entrance to the diffuser just outside it. The flow in the diffuser is isentropic. Determine Mach number, temperature and pressure at the exit of the diffuser. What is the stagnation pressure loss between the initial and final states of the flow?

**Solution** Let us use subscripts 1, 2 and 3 for sections before and after the shock wave and at the diffuser exit. We are given: $p_{01} = 1.013$ bar, $T_1 = 290$ K, $M_1 = 2.2$, $A_3/A_2 = 4$. From isentropic table at $\gamma = 1.4$, $p/p_0 = 0.0935$ and $T/T_0 = 0.508$ for $M = 2.2$.

This gives: $p_1 = 0.0947$ bar and $T_{01} = 570.87$ K.

From normal shock table, $M_2 = 0.547$, $p_2/p_1 = 5.48$, $T_2/T_1 = 1.857$ and $p_0/p_{01} = 0.628$. This gives: $p_2 = 0.519$ bar, $T_2 = 538.53$ K and $p_{02} = 0.636$ bar.

Again from isentropic table at $M_2 = 0.547$, $A/A^\star = 1.26$

$\rightarrow A_3/A^\star = (A_3/A_2) \times (A_2/A^\star) = 5.04.$

From isentropic table for this area ratio, $M = \mathbf{0.116}$,

$$T/T_0 = 0.997 \rightarrow T_3 = 0.997 \times 570.87 = 569.15 \text{ K},$$

and $$p/p_0 = 0.990 \rightarrow p_3 = 0.99 \times 0.636 = 0.629 \text{ bar}.$$

$$\text{Stagnation pressure loss} = p_{01} - p_{02} = 1.013 - 0.636 = 0.377 \text{ bar}.$$

**Example 7.6.4** A *CD* nozzle is provided with a pipe of constant cross section at its exit; the exit diameter of the nozzle and that of the pipe is 40 cm. The mean coefficient of friction for the pipe is 0.0025. Stagnation pressure and temperature of air at the nozzle entry is 12 bar and 600 K. The flow is isentropic in the nozzle and adiabatic in the pipe. The Mach numbers at the entry and exit of the pipe are 1.8 and 1.0 respectively. Determine: (a) the length of the pipe, (b) pressure and temperature at the pipe exit, and (c) diameter of the nozzle throat.

How the length, pressure and temperature at the pipe exit would be affected if (i) a normal shock occurs at nozzle exit or (ii) normal shock occurs in the pipe where $M = 1.22$?

**Solution** Let sections 1 and 2 be the inlet and exit of the pipe. 1 also represents the nozzle exit. From the isentropic table for $\gamma = 1.4$, we get at $M_1 = 1.8$, $p_1/p_{01} = 0.174$, $T_1/T_{01} = 0.607$ and $A_1/A_1^* = 1.439$. This gives $p_1 = 0.174 \times 12 = 2.088$ bar, $T_1 = 0.607 \times 600 = 364.2$ K and $A_1^* = 0.75 \times 0.4^2/1.439 = 0.8735$ m$^2$. This is the area of flow at the nozzle throat. If $D^\star$ is the throat diameter then

$$(\pi/4) \times D^{*2} = 0.8735 \rightarrow D^* = \mathbf{33.35\ cm.}$$

From Fanno line table ($\gamma = 1.4$) at $M_1 = 1.8$

$p_1/p^* = 0.474$, $T_1/T^* = 0.78$ and $4f\,L_{max}/D = 0.242$

$$\rightarrow \qquad L_{max} = 0.242 \times 0.4/(4 \times 0.0025) = 9.68 \text{ m}.$$

Since the flow is sonic at the pipe exit, the length of pipe = 9.68 m from the nozzle exit. Therefore the pressure and temperature at this section are: $p^* = p_1/0.474 = 4.405$ bar and $T^* = T_1/0.728 = 500$ K.

(i) *Normal shock at nozzle exit*. Let subscripts 1 and 2 be used for sections before and after the normal shock. From normal shock table at $M_1 = 1.8$, we read

$M_2 = 0.616$, $p_2/p_1 = 3.613$ and $T_2/T_1 = 1.532 \rightarrow p_2 = 7.544$ bar and $T_2 = 558$ K.

From Fanno table at $M = 0.616$, by linear interpolation, $p_2/p^* = 1.725$, $T_2/T^* = 1.116$ and $4f\,L_{max}/D = 0.45$.

$$\rightarrow \qquad L_{max} = 0.45 \times 0.4/(4 \times 0.0025) = 18 \text{ m.}$$

Also pressure and temperature at the pipe exit are 7.544/1.725 = 4.375 bar and 500 K.

We note that the length of the pipe has increased because of decrease in Mach number at its inlet. Moreover, the flow is subsonic in the pipe except at its inlet.

(ii) *Normal shock at a location where M = 1.22*. Let us use subscripts $x$ and $y$ for the sections before and after the shock location, 1 for pipe entrance and 2 for its exit. The flow is Fanno flow from pipe inlet to section 1 and after the section 1 to the pipe exit. Using Fanno table, we get the values at the section $x$ corresponding to $M_x = 1.22$.

$$\frac{p_x}{p^*} = 0.788, \quad \frac{T_x}{T^*} = 0.925 \quad \text{and } 4f\,\frac{L_{max}}{D} = 0.039$$

$$\rightarrow \qquad \frac{p_x}{p_1} = \frac{p_x}{p^*} \Big/ \frac{p_1}{p^*} = 0.788/0.474 = 1.6625$$

$$\frac{T_x}{T_1} = \frac{T_x}{T^\star} \Big/ \frac{T_1}{T^\star} = 0.925/0.728 = 1.2706.$$

$\therefore\ p_x = 1.6625 \times 2.088 = 3.471$ bar and $T_x = 1.2706 \times 364.2 = 462.75$ K.

From normal shock table, at $M = 1.22$:

$$\frac{P_y}{P_x} = 1.57 \text{ and } \frac{T_y}{T_x} = 1.141 \rightarrow p_y = 1.57 \times 3.471 = 5.45 \text{ bar and } T_y = 1.141 \times 462.75 = 528 \text{ K.}$$

We now determine length $L_1$ of the pipe between nozzle exit and shock location by assuming the maximum length $L_{max}$ up to the critical section.

$$4f\,\frac{L_1}{D} = \left(4f\,\frac{L_{max}}{D}\right)_{M_1} - \left(4f\,\frac{L_{max}}{D}\right)_{M_x} = 0.242 - 0.039 = 0.203 \rightarrow L_1 = 8.12 \text{ m.}$$

After the shock $M_y = 0.83$. From the Fanno table

$$\frac{p_y}{p^*} = 1.2375, \quad \frac{T_y}{T^*} = 1.055 \text{ and } 4f\,\frac{L_2}{D} = 0.049$$

$$\rightarrow \qquad p^* = 5.499/1.2375 = 4.403 \text{ bar and } T^* = 500 \text{ K.}$$

$$\text{and} \qquad L_2 = 0.049 \times 0.4/(4 \times 0.0025) = 1.96 \text{ m.}$$

Where $L_2$ is the length of the pipe from shock location to the pipe exit.

$$\therefore \qquad L = L_1 + L_2 = 8.12 + 1.96 = 10.08 \text{ m.}$$

*Comment.* Let us see how the length is changing in three cases. The length of the pipe comes out as 9.68 m, 18 m and 10.08 m. The nature of flow is shown in the following table

| Pipe length | Position of shock from pipe entrance | Nature of flow in pipe |
|---|---|---|
| 9.68 m | No shock | Supersonic |
| 10.08 m | Shock at $L = 8.12$ m | Supersonic/subsonic |
| 18 m | Shock at $L = 0$ m | subsonic |

We observe that pipe length is minimum when the flow is supersonic throughout the length and maximum when it is subsonic. For any length in between, the flow is supersonic initially and changes to subsonic at some distance.

# PROBLEMS

## Review Questions

1. Under what circumstances can a gas flow be treated as an incompressible flow?
2. Both isothermal and adiabatic processes are governed by the equation $pV = RT$. Comment.
3. In the steady flow of compressible fluid through a pipe of uniform diameter, it is possible to have different volumetric rates of flow at different sections?
4. In an adiabatic flow $\gamma = 1.4$, $p = 0.8$ bar (abs) and density = 0.7 kg/m$^3$. What is the sonic velocity?
5. If the compressibility of the fluid is ignored, the computed stagnation pressure would be more than the correct value. Is this true?
6. The stagnation pressure would increase with the Mach number of undisturbed flow. Is this true?
7. The ratio of stagnation to static absolute pressure in a compressible flow is 1.5. If $\gamma = 1.4$, what is the ratio of the corresponding densities? What is the ratio for corresponding temperatures?
8. A shell is fired from a field gun with a velocity of 400 m/s in air. What is the maximum rise in its skin temperature?
9. Use of separate pitot and static tubes is advisable rather than a pitot-static tube in the velocity measurement in the gas flows. Why?
10. If the Mach number is 0.9, what is the per cent error in computing the dynamic pressure when the compressibility affect of the fluid is neglected?
11. In the gas flow through a convergent nozzle, no purpose is served by reducing the back pressure below a certain limit. Is this true?
12. In a gas flow through a convergent nozzle, the gas at exit is always cooler than the gas at entrance. Explain the reason.
13. Define Rayleigh and Fanno flows.

14. At the point of maximum temperature on a Rayleigh line, what is the value of $M$? What is the value of $M$ at the point of maximum entropy on a Fanno line?
15. For what purpose is the Prandtl–Mayer supersonic expansion function used?
16. Explain the five regimes of flow of a compressible fluid with examples.
17. Derive the equation for the mass rate of flow of a compressible fluid using a venturimeter.
18. Derive the relation connecting area velocity changes and explain the nature of the velocity variation with area change in subsonic, transonic and supersonic flows of compressible fluid.
19. Explain briefly the theory of oblique shock.
20. Distinguish between (i) attached and detached shocks and (ii) compression and expansion shocks.
21. Explain the phenomenon of supersonic wave drag.

## Problems for Practice

1. A horizontal pipe conveys a gas a temperature of 5°C. At section 1 of the pipe, the diameter is 6 cm and the gas pressure is 2 bar (gauge). At section 2 of the pipe the diameter is 3 cm and the gas pressure is 1.6 bar (gauge). Assuming an isothermal change, find the velocities of the gas at these sections.
2. A body moves through standard air at 200 m/s. What is the pressure at a point on the body where the velocity of the air relative to the body is 0? Assume (a) compressible fluid, (b) incompressible fluid. For standard air, $\gamma = 1.4$, $R = 287$ J/kg K, $T = 288$ K, $p = 101$ kPa, and $\rho = 1.25$ kg/m$^3$. [127.63, 125.5 kPa]
3. Find the velocity of a bullet fired in standard air if the Mach angle is 30°. [680.35 m/s]
4. The static pressure to stagnation pressure ratio at a point in a flow stream measured with a pilot static tube is being equal to 0.82. The stagnation temperature of the fluid is 20°C. Determine the flow velocity if the fluid is air. Take $\gamma$=1.4. [180.1 m/s].
5. An object moves at 221 m/s through standard air at sea level. What are (a) the pressure and (b) the temperature at the forward stagnation point? Sea level conditions are: $p = 101$ kPa, $t = 15$°C. [134.528 kPa, 312.364°C].
6. A supersonic wind tunnel operates at test section Mach number, $M = 1.8$. Operating conditions in the tunnel test section are: 80°C and 258 kPa (gauge). A normal shock that stands in front of a total head tube reduces the stagnation pressure by 18.7%. Calculate (a) local isotropic stagnation conditions for the test section conditions and (b) the pressure sensed by the total head tube. Take barometric pressure as 101 kPa. [$p_0 = 20.6274$ bar, 16.77 bar]
7. The hot gas stream at the turbine inlet of a gas engine (JT–9D) is at 1300 K and 140 kPa. The Mach number at that location is 0.32. Calculate the critical conditions (temperature, pressure, and flow speed) that correspond to these conditions. Assume the flow properties of those of pure air. [79.395 kPa, 1105.2 K, 666.385 m/s]
8. Air at stagnation condition has a temperature of 800 K. Determine the velocity of sound and the maximum fluid velocity. What is the velocity of sound when the flow is at half the maximum velocity? [566.956 m/s, 1267.75 m/s, 491 m/s]
9. Air at rest at 90°C is accelerated isentropically. (a) What is the air speed in m/s when the Mach number becomes 0.8? (b) What is the air speed when the flow be-

comes sonic? (c) What is the Mach number when the air speed becomes 600 m/s? [287.7 m /s, 348.6 m /s, 2.2].

10. Air flows steadily through a duct from 350 kPa, 60°C, and 183 m /s at the initial state to a Mach number of 1.3 at the outlet, where the local isentropic stagnation conditions are known to be 385 kPa and 77°C. Compute the isentropic stagnation pressure and temperature at the inlet and the static pressure and temperature at the duct outlet. [415 kPa, 350 k, 139 kPa, 262 K]
11. Air in tank at 700 kPa discharges into a standard atmosphere through a converging nozzle. For isentropic flow, (a) what is the exit pressure? (b) what is the air temperature at the nozzle exit?, and (c) what nozzle exit diameter will discharge 0.8 kg/s? [370 kPa, 248 K, 5 mm].
12. Air flows through a convergent nozzle at a stagnation pressure and temperature of 300 kPa and 400 K, respectively. The exit area of the nozzle is 0.1 m$^2$. Find the mass flow rate, exit temperature, pressure, and Mach number for the discharge into an ambient of (a) 200 kPa (b) 100 kPa.
13. Air flows isentropically in a channel. At section one, the Mach number is 0.3, the area is 0.001 m$^2$, and the absolute pressure and temperature are 650 kPa and 62°C, respectively. At section two, the Mach number is 0.8. Evaluate the properties at this section. [302.3 K, 453.88 kPa, 5.23 kg/m$^3$]
14. Air in a reservoir at 825 kPa and 50°C discharges through a converging nozzle to the atmosphere. Design a nozzle for flow rate of 1.5 kg/s. Specify the exit diameter. Also design a converging diverging nozzle for the same flow rate. Specify the exit and throat diameters. [32 mm, 4 mm]
15. Air flows steadily and isentropically through a converging diverging nozzle. At the throat, the air is at 140 kPa, and 60°C. The throat cross sectional area is 0.05 m$^2$. At certain section in the divergent part of the nozzle, the pressure is 70 kPa abs. Calculate the speed and area at this section. [503.6 m /s, 0.05958 m$^2$].
16. A small rocket motor fuelled with hydrogen and oxygen is tested on a thrust stand at a stimulated altitude of 10 km. The motor is operated at chamber stagnation condition of 1500 K and 8 Mpa gauge. The combustion product is water vapour, which may be treated as an ideal gas. Expansion occurs through a converging diverging nozzle with design Mach number of 3.5 and exit area of 700 mm$^2$. Evaluate the pressure at the nozzle exit plane. Calculate the mass flow rate of exhaust gas. Determine the force exerted by the rocket motor on the thrust stand. Take $\gamma$ = 1.3 and $R$ = 461 J/kg K and ambient pressure at 10 km to be 26.5 kPa. [88.29 kPa, 0.49 kg/s, –1026.67 N].
17. A converging diverging nozzle has an exit area to throat area ratio of 2. Air enters this nozzle with a stagnation pressure of 1000 kPa and a stagnation temperature of 360 K. The throat area is 500 mm$^2$. Determine the mass rate of flow, exit temperature, exit pressure, exit Mach number, and exit velocity for the following conditions: (a) sonic velocity at the throat, diverging section acting as a nozzle and (b) sonic velocity at the throat, diverging section as a diffuser.
18. A conical diffuser has an intake area of 0.11 m$^2$ and an exit area of 0.44 m$^2$. Air enters the diffuser with a static pressure of 0.18 Mpa, static temperature of 37°C, and velocity of 267 m /s. Calculate (a) the mass flow rate of air through the diffuser, (b) the Mach number, static temperature and static pressure of air leaving the dif-

fuser, and (c) the net thrust acting upon the diffuser, assuming that its outer surfaces are acted upon by atmospheric pressure at 0.1 Mpa. [59.42 kg/s, 0.76, 344.4 K, 0.26 Mpa, 51.68 kN].

19. The stagnation pressure and temperature for a flow in a 50 mm dia pipe are 10 bar and 400 K, respectively. The average friction factor for the pipe surface is 0.008. If the Mach number changes from 3 at the entry to 1 at the exit, find (a) the length of the pipe and (b) the mass flow rate. [3.265 m, 0.93644 kg/s]
20. The pressure and temperature at the inlet of a long straight duct are 101 kPa and 298 K, respectively. The flow is considered to be adiabatic and the pipe long enough so that the flow is choked. Find the speed and temperature at the pipe exit. If the duct diameter and pressure at exit are 12 mm and 94.1 kPa respectively, find the duct length. Take $f = 0.0185$. [315.88 m /s, 248.34 K, 2.88454 m].
21. Air is drawn from the atmosphere at 20°C and 101 kPa through a converging nozzle into a long insulated 20 mm diameter tube of constant area. Flow in the nozzle is isentropic. The Mach number at the inlet to the tube is 0.15. Evaluate the mass flow rate through the tube. Calculate critical pressure and temperature for the isentropic process and for the flow through the tube. Show that these results would be changed if the Mach number at the tube inlet were raised to 0.3. [0.01896 kg/s, 53.358 kPa, 244.16 K, 13.455 kPa].
22. A circular duct passes 8.25 kg/s of air at an exit Mach number of 0.5. The entry pressure and temperature are 345 kPa and 311 K respectively. The average friction factor is 0.02. If the Mach number at entry is 0.15, determine (a) the diameter of the duct, (b) the length of the duct, (c) pressure and temperature at the exit of duct, and (d) stagnation pressure loss. [226 mm, 308.32 m, 100.78 kPa, 297.47 K, 231 kPa].
23. Air flows with negligible friction in a constant area duct. At section one, the flow properties are: $t_1 = 60.4$°C, $p_1 = 135$ kPa abs and velocity 732 m/s. Heat is added to the flow between section one and section 2, where the Mach number is 1.2. Determine the flow properties at section two, the heat transfer per unit mass and the entropy change. [295.65 kPa, 573.45 K, 576.54 m/s, 1.796 kg/m$^3$, $M = 1.2$, 139.3 kJ/kg, 0.32 kJ/kg K].
24. Air flows steadily and without friction at the rate of 1.83 kg/s through a duct with cross sectional area of 0.02 m$^2$. At the duct inlet, the temperature and absolute pressure are 260°C and 126 kPa, respectively. The exit flow discharges subsonically to atmospheric pressure. Determine the Mach number, temperature, and stagnation temperature at the duct outlet, and the heat transfer rate. [0.5, 1482 K, 1556.47 K, 1870 kJ/s].
25. In the frictionless flow of air through a 100 mm diameter duct, 1.42 kg/s enters at a temperature of 52°C and an absolute pressure of 60 kPa. Determine the amount of heat that must be added to the flow, and the fluid properties at the choked state. [17.25 kJ/kg, $T_{02} = 381.52$ K, $p_{02} = 87.67$ kPa].
26. Air is drawn from the standard atmospheric conditions through a smooth converging nozzle into the inlet manifold of a racing engine. Flow is isentropic through the nozzle. Air leaves the nozzle and enters the manifold at Mach number $M = 0.3$. In the manifold, gasoline is evaporated in a short section of constant area duct where friction is negligible. The effect is to remove 60.3 kJ/kg of thermal energy from the air stream. The fluid properties are those of pure air. Find (a) the Mach number,

and (b) the pressure at the outlet of the evaporation section. Take the standard atmospheric values as 101 kPa and 288 K. [0.26, 97.65 kPa].

27. A fuel-air mixture with the thermodynamic properties of pure air enters a combustor of constant area. The stagnation temperature of the inlet stream is constant at 400 K. Friction is negligible. When heat is added to the flow at the rate of 1.4 MJ/kg, flow at the duct exit is choked. At this condition the static pressure at the inlet is 144.6 kPa and the pressure at the duct exit is 64.7 kPa. Calculate (a) the combustor outlet temperature, (b) the combustor inlet Mach number, and (c) the loss in stagnation pressure through the combustor. [1495.4 K, 0.23, 27.56 kPa].
28. Air from an aircraft inlet system enters the engine combustion chamber where heat is added during a frictionless process in a tube with constant area of 0.01 $m^2$. The local isentropic stagnation temperature and the Mach number entering the combustor are 427 K and 0.3. The mass flow rate is 0.5 kg/s. When the rate of heat transfer is set at 404 kW, the flow leaves the combustor at 1026 K and 22.9 kPa. Determine for this process (a) the Mach number at combustor outlet, (b) the static pressure at the combustor inlet, (c) the change in local isentropic stagnation pressure during the heat addition process. [0.528, 48.8 kPa, –8.598 kPa].
29. A converging diverging nozzle has an exit to throat area ratio of 3. Air enters the nozzle with a stagnation pressure of 6.5 bar and a stagnation temperature of 360 K. Assuming a normal shock at the exit plane and considering the flow through the diverging portion of the nozzle supersonic, determine the Mach number, pressure and temperature after the shock wave. [0.5008, 2.415 bar, 342.78 K].
30. Consider the subsonic-supersonic flow through a converging-diverging nozzle. The reservoir pressure and temperature are 10 bar and 300 K. There are two locations in the nozzle where $A/A^* = 6$, one in the diverging and the other in the converging section. At each location, calculate the Mach number, $p$, $T$, and $C$. [0.097, 9.94 bar, 99.4 K, 33.6 m/s; 3.368, 0.1584 bar, 91.77 K, 646.7 m/s]
31. Air from a reservoir at a pressure $p_0$, flows through a converging diverging nozzle of throat area $A^*$ and an exit area of 2.54 $A^*$. A normal shock occurs beyond the throat where $A$ = 1.45 $A^*$. (a) What is the exit Mach number? (b) What is the receiver pressure in terms of $p_0$? and (c) What receiver pressure would produce isentropic supersonic flow without shock? [0.9382, 0.758, 0.64]
32. Air flows through a converging diverging nozzle with area ratio $A_e/A^* = 3.5$. The upstream stagnation conditions are atmospheric; the back pressure is maintained by a vacuum pump. Determine the back pressure required to cause a normal shock to stand at the nozzle exit plane and the flow speed leaving the shock. [33.4 kPa, 162.17 m/s].
33. An insulated pipe of length $L$ and diameter $D$ discharges into region where the static pressure is 4 bar. Air flows from a convergent divergent nozzle with a throat area half of the pipe. The stagnation pressure at the nozzle inlet is 10 bar. The friction factor of pipe is 0.02 and $L/D = 20$. If shock occurs in between the pipe, find the Mach number at exit from the pipe. [0.7].

CHAPTER 8

# Non-dimensional Analysis

## OUTLINE

- Introduction
- Units and Dimensions
- Non-dimensional Groups
- Non-dimensional Analysis
- Similitude

## 8.1 Introduction

We have learnt by now that any flow can be modelled in a mathematical form and that the different physical quantities governing the flow can be expressed in a functional form of other flow quantities. For example, pressure loss in laminar flow through a circular, smooth pipe depends on the flow velocity $U$, diameter $D$ and length $L$ of the pipe, and density $\rho$ and viscosity $\mu$ of the fluid. That is, $\Delta p = f(U, D, L, \rho, \mu)$. In this functional relation there are 6 variables; in general, there may be $m$ variables in such a relation. It has been our endeavour to find out such relation. We have two methods, namely, the analytical and the experimental method.

Let us examine how much effort will it require to determine such relation experimentally. In any experiment, one variable is changed at a time while others are kept constant. The experiment is conducted for a large number of this independent variable so that we may be fairly sure to study the dependence of the dependent variable(s) on this variable. Then a graph is plotted between the two. Let us assume we make 10 observations for each independent variable. In the above problem, there are 5 independent variables and therefore we have to make 50 observations, plot 5 graphs, and correlate these graphs to get a meaningful result. Morever, we shall need 5 liquids of different viscosity and other 5 liquids of different densities to perform these experiments. This is not impossible but a very time-consuming and tiring process. So we would like to reduce this work.

The solution lies in dimensional or non-dimensional analysis; the scientists have adopted the latter. We know that 5 horses cannot be added to 5 donkeys to give 10 horses—principle of *dimensional homogeneity*. In this chapter, we shall learn about the dimensions, non-dimensional groups and how to determine these so that we may be able to get the functional relationship. Finally, we shall learn how to use this information to our advantage in carrying out the experiments and to predict from these results what we need for the given situation?

## 8.2 Units and Dimensions

Each quantity describing the flow has a dimension. The number of dimensions can be limited by our choice and the dimensions so chosen are called the fundamental dimensions. These are so chosen that their number is minimum. Other dimensions can be expressed in terms of these chosen fundamental dimensions using mathematical relation. For example, dimension velocity is the rate of change of displacement. Here, displacement is length dimension and time another dimension. Similarly, dimension pressure is the ratio of force and area; the latter is the product of two lengths and the former may be taken as another dimension, force. Thus the velocity and pressure are expressed in terms of force, length and time. So there are two types of dimensions—the *fundamental dimensions* in terms of which other dimensions, the *secondary dimensions*, are expressed. However, it is still our choice to choose the fundamental dimension. Experience has taught us that mass, length and time can be taken as fundamental dimensions. These are writen as $M$, $L$ and $T$ respectively. These are measured in kg, metre and s, respectively, in the S.I. (Systemè International) units (pound, ft and s in FPS system, cm, g and s in CGS system). In these units, kg, g*, m, cm and s are the abbreviation for kilogram, gram, metre, centimetre and second, respectively, adopted by International Society of Weights and Measures having their office in Paris, France.

A given dimension is expressed in terms of these chosen fundamental dimensions using the mathematical relations. These relations may be simple, complex algebraic or differential relations. Today we have three fundamental dimensions in mechanics. Depending upon the area of application, some other variables have also been fixed as fundamental dimension. For example, current strength $I$ in a circuit has been taken as fundamental dimension in Electric Engineering or temperature $\theta$ in Thermodynamics. However, it is not necessary that there should be only three fundamental dimensions. In fact, force also had been used as fundamental dimension in *MKS* system of dimensions and units in Mechanics. Nomenclature *MLT* or $MLT\theta$, or *FMLT*, etc. is used in which the symbol indicates the fundamental dimension used in dimensional analysis.

* At many places gm has been used for g to eliminate confusion between acceleration due to gravity and unit of mass both of which have the symbol g. Note that the standard symbol of unit of mass in CGS is g and not gm.

In the dimensional system, *FMLT*, force is also taken as fundamental dimension in addition to mass, length and time. This system has been used by engineers for a long time and is known as *MKS* system as mass length, and time were measured in meter, kg and seconds, respectively. In this system, 1 kg force was defined as the force that would act on a mass of 1 kg placed in the earth's field where $g = 9.80665$ m/s$^2$; symbol kg$_f$ has also been used to differentiate it from 1 kg mass. Force '*F*' acting on a mass *m* placed in earth's gravity '*g*' is defined by Newton's second law of mechanics, which is written in mathematical form as follows:

$$F \propto mg \qquad \text{or} \qquad F = (1/g_c)\, mg \tag{8.2.1}$$

Where $(1/g_c)$ is a constant of proportionality. It can be seen that it has dimensions $F^{-1}\,MLT^{-2}$, since *g* has dimensions $LT^{-2}$, the dimensions of acceleration. Symbol $[x]$, *x* within square brackets, is used for the phrase 'dimension of *x*'. It may be viewed as conversion factor between the units of force and mass. Equation 8.2.1 then demands that numerical value of $g_c$ should be that of *g* to satisfy it and that this value will depend on the system of unit used. This system was a bit confusing as the same unit kg was used as unit for both mass and force. However, in the S.I. system of units, which is now adopted universally as the system of units to be used everywhere in the world, force is not taken as fundamental dimension. Unit of force is Newton (*N*), defined as that force when acting on 1 kg mass would accelerate it by 1 m/s$^2$. Eq. 8.2.1 is automatically satisfied if we choose $g_c$ as 1 having no dimension. Thus *MLT* system is more coherent and is known as *coherent* or consistent system whereas the *FMLT* system is *non-coherent* type. It is also possible to have less number of fundamental dimensions.

The fact that an equation is dimensionally homogeneous has been used in many useful ways. For example, we may check whether the relation is correct, at least dimensionally. If not, we may check for the error. Also if we divide one term of the equation by another then the resultant will be dimensionless quantity. This fact has been vastly used in decreasing the number of unknowns in similitude study and model development of the machine so that we are sure about its performance after its manufacture. The simulation technique proved to be a boon for us. These days the computer simulation is replacing the conventional laboratory testing and modelling techniques. But it is quite a complex process and laboratory testing has its own advantages.

A word related to dimension, but not the same, is unit. We may understand it from unit of length, which is given to length between two fixed points on a bar (kept in museum of France) at a fixed temperature. So kg, metre and second have been fixed as SI units for mass, length and time, respectively. They have some standard reference preserved in France. That is all the fundamental dimensions have been defined at some standard conditions and given some name to it. These are called the SI units. However, in practice other units are also used to meet the requirement and these differ from the SI unit by a factor called the *conversion factor*. For measurement of small distance we use cm or mm or even some other units. Idea is to keep the number associated with the size of the dimension reasonable. We would prefer 1 km instead of 1000 m for length and so on.

Each dimension has a unit, which depends on the system of the unit used. Different system of units such as FPS, CGS, MKS and SI have been in use for a long time and are still in use. However, SI system of units is most commonly used. Table 1.4.1 gives units of fundamental dimensions in these system of units.

**TABLE 8.2.1** Units for fundamental dimensions in different system of units

| Dimension | FPS | CGS | MKS | SI |
|---|---|---|---|---|
| Length | ft | cm | m | m |
| Mass | pound | g | kg | kg |
| Time | s | s | s | s |
| Force | $lb_f$ | dyne | $kg_f$ | N |
| Current | amp | amp | amp | amp |

Units for other quantities are obtained from the dimensional relation of these quantities with fundamental dimension. We have learnt some of such units in other courses. Besides these units, we also use secondary units used for practical reason. For example, length can be measured in mm, cm, m, km, and so on so forth though the unit for it in SI system is m only and all the relations relating different dimensions need these units only. Therefore, whenever, we make substitution in any relation, we have to ensure that we are using the appropriate unit and not the unit in which value is given. For example, pressure has to be in N/m$^2$ in any equation though it is given in different units like atm (read as atmosphere), bar, Pa (read as Pascal), kPa, Mpa, or m of some fluid, normally the one in use in the given problem. Table 8.2.1 gives dimensions of some physical quantities used in the study of flow.

**TABLE 8.2.2** Dimensions of physical quantities

| Physical quantity | Dimension | Physical quantity | Dimension | Physical quantity | Dimension |
|---|---|---|---|---|---|
| Mass | $M$ | Length | $L$ | Time | $T$ |
| Temperature | $\theta$ | Area | $L^2$ | Volume | $L^3$ |
| Speed, velocity | $LT^{-1}$ | Acceleration | $LT^{-2}$ | Angular speed | $T^{-1}$ |
| Density | $ML^{-3}$ | Momentum | $MLT^{-1}$ | Force, thrust | $MLT^{-2}$ |
| Diameter, radius | $L$ | Angular Momentum | $ML^2T^{-1}$ | Moment, torque | $ML^2T^{-2}$ |
| Velocity potential | $L^2T^{-1}$ | Stream function | $L^2T^{-1}$ | Circulation | $L^2T^{-1}$ |
| Weight | $MLT^{-2}$ | Work, energy | $ML^2T^{-2}$ | Power | $ML^2T^{-3}$ |
| Frequency | $T^{-1}$ | Pressure, stress | $ML^{-1}T^{-2}$ | Surface tension | $MT^{-2}$ |
| Dynamic viscosity | $ML^{-1}T^{-1}$ | Kinematic viscosity | $L^2T^{-1}$ | | |

**Example 8.2.1** The following equation is applicable in the MKS system of units: $U = 50\sqrt{(m\,i)}$, where $U$ is the velocity vector, $m$ is the hydraulic mean depth of a channel and $i$ is its longitudinal slope. Find the dimension of the constant and determine its value in FPS system.

**Solution** 50 in the given expression is a constant, say $C$, whose value depends on the units used for measuring the different quantities in the equation. That is this constant has some dimensions, which can be determined from the principle of dimensional homogeneity. Note that slope $i$ does not have any dimension and dimensions of other quantities are given in Table 8.2.1. We therefore get dimensions of the constant $C$ as

$$[C] = \left[\frac{U}{\sqrt{mi}}\right] = \frac{LT^{-1}}{L^{1/2}} = L^{1/2}T^{-1}$$

Units of $C$ in MKS system therefore is $\sqrt{m}$/s. Unit for time is same in both the units but 1 m = 3.281 ft, the conversion factor from MKS to FPS system is $(1\text{ m} / 1\text{ ft})^{1/2} = 1.81$

$\therefore$ value of $C$ in FPS system = 50 × 1.81 = 90.5

**Example 8.2.2** Newton's second law is $F \propto m\,a$, where $m$ is mass and $a$ is acceleration. This can be written as, $F = m\,a/g_c$, where $1/g_c$ is a constant of proportionality whose value and unit will depend on the system of units used. Find out its value and unit in some of the system of units used all over the world and tabulate the results.

**Solution** Following table gives the results. Note that in these systems, four fundamental dimensions, namely mass, length, time and force have been used. Difference lies in how different units are defined for each fundamental dimension for dimensional consistency of Newton's law of motion.

| System of unit | Mass | Force | Length | Time | $g_c$ |
|---|---|---|---|---|---|
| British | $lb_m$ | $lb_f$ | ft | s | 32.174 $lb_m$- ft/($lb_f$- $s^2$) |
| CGS | gm | dyne | cm | s | 1 gm-cm/dyne-$s^2$) |
| SI system | kg | N | m | s | 1 kg-m/)N-$s^2$) |
| Engineering: | | | | | |
| Metric or MKS | kg | $kg_f$ | m | s | 9.80665 kg–m/( $kg_f$-$s^2$) |
| British | $lb_m$ | poundal | ft | s | 1 $lb_m$-ft/(poundal-$s^2$) |
| | slug | $lb_f$ | ft | s | 1 slug–ft/($lb_f$-$s^2$) |

**Note:** $lb_m$ is used to represent 1 lb mass; $lb_m$ is used to differentiate it from $lb_f$ used for force in British engineering system—defined as the force acting on a mass of 1 lb when placed on equator.

We shall demonstrate the procedure taking one example of Engineering MKS system of units. In this system, 1 $kg_f$ force is defined as the force acting on 1 kg mass when placed in earth's field at equator, where $g$ = 9.80665 m/$s^2$. Using the Newton's law, we write

$$1\ kg_f = (1/g_c)\ 1\ kg \times g\ m/s^2 \rightarrow g_c = 9.80665\ kg\text{–}m/(kg_f\text{–}s^2)$$

Thus, this value of $g_c$ makes Newton's law dimensionally consistent. From the above table, we may also get the following results, using the Newton's law.

(a) Newton is a force that acts on 1 kg mass to accelerate it by 1 m/$s^2$.

(b) Slug is mass of a chunk such that when acted upon by 1 $lb_f$ force would be accelerated by 1 $ft/s^2$.

(c) 1 poundal is a force that would accelerate 1 lb mass by g $ft/s^2$ at the equator.

**Example 8.2.3** 1 kg mass is accelerated with a force of 10 $lb_f$. Determine the acceleration of mass in (i) $cm/s^2$ and (ii) $ft/s^2$.

**Solution** By definition of $lb_f$, the unit of force,
Conversion factor is 1 lb = 453.6 g and g = 32.174 $ft/s^2$ = 980.665 ≈ 981 $cm/s^2$

(i) 1 $lb_f$ = 1 lb × g $ft/s^2$ ≡ (453.6 × $10^{-3}$ kg) × 981 $cm/s^2$ = 445 kg-$cm/s^2$

∴ Acceleration of 1 kg mass = 10 $lb_f$/1 kg = 10 × 445 (kg-$cm/s^2$)/1 (kg) = 4450 $cm/s^2$

(ii) In $ft/s^2$: 1 $lb_f$ = 1 lb × g $ft/s^2$ and 1 lb = 0.4536 kg

∴ Acceleration of 1 kg mass = 10 $lb_f$ /1 kg = 10 × 32.2 (lb $ft/s^2$)/(1/0.4536) (lb) = 138 $ft/s^2$

**Example 8.2.4** If a mass is weighed on a beam at a location where $g$ = 900 $cm/s^2$, what is the reading on the balance? The spring scale originally is calibrated in a region where $g$ = 981 $cm/s^2$.

**Solution** A beam balance reads mass whereas the spring balance reads the weight. Since mass is the property of the chunk, it is the same irrespective of where it is measured. On the other hand, the force acting on this mass would change from place to place, and hence the spring balance reading, since acceleration due to gravity depends on the place.

Spring balance reading where $g$ = 900 $cm/s^2$ = 1 (kg) × 900 ($cm/s^2$) = 900 kg-$cm/s^2$

But the reading on spring balance scale will be marked 1 for the place where $g$ = 981 $cm/s^2$.

∴ Spring balance reading = 900/981 = 0.917 kg

A spring balance gives reading in kg, which is nothing but $kg_f$. That is, if a mass of 1 kg is attached to its hook then it records 1 kg though it measures force.

**Example 8.2.5** Viscosity of air in poise is given by the relation, $\mu = 1.65 \times 10^{-4} + 2.5 \times 10^{-7}$ T, where T is temperature in °*F*. Poise is used as unit of viscosity in CGS system of units where force, distance and time are measured in dyne, cm and s. Convert the above expression into its appropriate form when the units of force, distance, and time are (a) $lb_f$, ft, s and (b) $kg_f$, m/s.

**Solution** Conversion factors are: g $ft/s^2$ = 981 $cm/s^2$ and 453.6 g = 1 lb, 1 ft = 2.54 × 12 cm = 30.48 cm, or 1 cm = 1/30.48 ft. Further 32 $ft/s^2$ = 981 $cm/s^2$ → $\dfrac{cm/s^2}{ft/s^2} = \dfrac{32}{981}$

By definition, 1 dyne = 1 gm × 1 $cm/s^2$ and 1 $lb_f$ = 1 lb × g ft $/s^2$

$$\therefore \quad \frac{1\,\text{dyne}}{1\,\text{lb}_f} = \frac{1}{g}\frac{\text{gm}}{\text{lb}} \times \frac{\text{cm}/\text{s}^2}{\text{ft}/\text{s}^2} \rightarrow 1\,\text{dyne} = \frac{1}{453.6 \times 981}\text{lb}_f$$

$$1\text{ poise} = 1\text{ dyne} \times \frac{s}{\text{cm}^2} = \frac{1/(453.6 \times 981)}{1/30.48^2} = 2.09 \times 10^{-3}\ \text{lb}_f - \text{s}/\text{ft}^2$$

∴ $\mu$ in new units = $\mu$ (given) × 2.09 × $10^{-3}$ = 3.45 × $10^{-7}$ + 5.22 × $10^{-10}$ T lbf $s/ft^2$

(b) In this system, 1 kgf = 1 kg × g $m/s^2$ = 1000 (gm) × 981 ($cm/s^2$) = 981000 dyne → 1 dyne = 1/981000 $kg_f$.

$$1\text{ poise} = 1\text{ dyne} \times \frac{\text{s}}{\text{cm}^2} = \frac{1/981000}{1/100^2} = 1.02 \times 10^{-2}\ \text{kg}_f\text{-s}/\text{m}^2$$

$\therefore\ \mu = (1.65 \times 10^{-4} + 2.5 \times 10^{-7}\,T) \times 1.02 \times 10^{-2} = 1.68 \times 10^{-6} + 2.55 \times 10^{-9}\,T$ kg$_f$-s/m$^2$.

*Note* In these expressions temperature is measured in °F. If temperature is given in °C then we may use the conversion relation in the two temperature scales

C/5 = (F − 32)/9 → T°F = 1.8 C + 32°C and can be used in these relations.

## 8.3 Non-dimensional Groups

We now know what is the secondary dimension and how to determine it for a given quantity or group of quantities. By dividing a group by another group having the same dimensions, we get another group, which does not have any dimensions. For example, pressure $p$ and $\rho U^2$ have the same dimensions $ML^{-1}T^{-2}$. Therefore $p/(\rho U^2)$ will not have any dimension. Such groups are called *non-dimensional* numbers or groups. Some of these are occurring often and have been given names in honour of some scientists. In Fluid Mechanics, ratio of the terms having dimensions of force are very common, in which one of these force is inertia force and the other forces in turn are viscous force, gravity force, pressure force, elastic force, etc. We may express these forces in terms of the physical quantity of the flow or the fluid as follows:

(i) *Inertia force* $\propto m \times$ acceleration $= \rho \times$ vol. $\times (U/T) = \rho \times$ (vol./$T$) $\times U = \rho AU \times U = \rho AU \times U = \rho L^2U^2$. Here $L$ and $U$ are some characteristic length and velocity of the flow, and $\rho$ is the density of the fluid. We may write area $A$ as $L^2$ and discharge vol./ $T$ as $AU$. Similarly, we proceed to get other forces.

(ii) *Viscous force* $\propto \mu U/L \times A = \mu LU$

(iii) *Gravity force* $\propto mg = \rho L^3 g$

(iv) *Pressure force* $\propto p \times A = p\,L^2$

(v) *Elastic force* $\propto E \times A = \rho\, a^2\, L^2$, where $a$ is the sound velocity.

(vi) Surface tension force = $\sigma l$, where $\sigma$ is the surface tension.

We now take ratio of the inertia force one by one with other forces to give the following non-dimensional numbers, which have been given some name owing to their importance in flow problems.

Reynolds' number = inertia force/viscous force = $\rho LU/\mu = LU/\nu = Re$ (8.3.1)

Froude number = √(inertia force/gravity force) = $\sqrt{(U^2/Lg)} = U/\sqrt{Lg} = Fr$ (8.3.2)

Euler number = pressure force/inertia force = $p/(\rho U^2) = E$ (8.3.3)

Mach number = $\sqrt{(\text{Inertia force / elastic force}} = U/a = M$ (8.3.4)

Weber number = inertia force/surface tension force = $\rho LU^2/\sigma$ (8.3.5)

In fact, the number of such non-dimensional numbers is large. Some of them can even be combined to give some other number. We combine and use different numbers as per appearance of these numbers in the equations. Again in Eq. 8.3.1, we have given the name Reynolds number to the group $LU/\nu$ whereas $\nu/(LU)$ is also a non-dimensional number; for that matter $(LU/\nu)^p$, where $p$ is any number, is a non-dimensional number. We could have called any of these combinations as the Reynolds number. However, we have fixed our choice to $LU/\nu$ out of so many groups for Reynolds number. Similarly, we named $U/\sqrt{Lg}$ as the Froude number instead of the group $U^2/Lg$ formed by the ratio of the forces. The choice is made so that the number is greater than one and at the same time it is not too large. Since these groups carry some meaning, these are to be used in this way and not in any other way we wish. But there will be many groups that are not tagged by a particular form; we have the liberty to define them in any way, we wish. This gives flexibility and at the same time complexity, especially when the actual solution is not known and we are dependent on the non-dimensional analysis, explained in the next section.

The different non-dimensional numbers have, as mentioned earlier, importance for a specific type of flow. For example, Reynolds number has influence over the internal flows (flow through pipe, venturimeter, rotadynamic machines, etc) and external flows over the surfaces. Froude number has significance in the flows that are influenced by gravity such as flow through open channels, surface wave motion, flow of compressible flow over surfaces, jets, etc. The Mach number will exist only in compressible fluids whereas the Weber number will have its presence in situations where the surface tension force is comparable to the inertia force, such as capillary flow, manometer, etc.

Table 8.3.1 summarises a few non-dimensional numbers used in fluid flow and heat transfer analysis along with their field of application. This information helps in envisaging the types of groups likely to dominate the given problem and therefore in determining the correct number during the non-dimensional analysis.

**TABLE 8.3.1** Non-dimensional numbers in use in fluid flow and heat transfer

| Name | Symbol | Definition | Significance | Fields of use |
|---|---|---|---|---|
| Reynolds number | $Re$ | $UL\rho/\mu$ | Inertia/viscous | Viscous flows |
| Froude number | $Fr$ | $U/\sqrt{(L\ g)}$ | Inertia/gravity | Free surface flows |
| Mach number | $M$ | $U/a$ | Inertia/elastic | Compressible flow |
| Euler number | $E$ | $p/(\rho U^2)$ | Pressure/inertia | Pressure variation in flow |
| Strouhal number | $S$ | $ND/U$ | Frequencies | Vortex shedding |
| Speed number | $S$ | $NL/U$ | Peripheral/free stream speed | Fluid machines |
| Pressure coef ficient | $p_c$ | $\Delta p/(\frac{1}{2}\rho U^2)$ | Pressure drop/ dynamic pressure | Confined flows |
| Roughness coef ficient | $\varepsilon'$ | $\varepsilon/D$ | Roughness/characteristic length | Surface roughness |
| Flow coefficient ficient | $Q_c$ | $Q/ND^3$ | Discharge/characteristic discharge | Fluid mechanics |
| Drag/lift coef ficient | $C_D$ | $F/(\rho U^2 L^2)$ | Drag/inertia force | Aerodynamics |
| Efficiency | $\eta$ | $P_o/P_i$ | Output/input power | System analysis |
| Knudsen number | $Kn$ | $\lambda/L$ | Mean free path/ characteristic length | Rarefied gas flow |
| Prandtl number | $Pr$ | $C_p\ \mu/k$ | Momentum/thermal diffusivity | Forced convection Heat transfer |
| Nusselt number | $Nu$ | $hD/k$ | Total/conductive heat transfer | do |
| Grashof number | $Gr$ | $L^3\rho g\beta\Delta T/\mu^2$ | Buoyancy/viscous force | Free convection |
| Gamma | $\gamma$ | $c_p/c_v$ | Sp. heat at const $p$/const $v$ | Compressible fluid |

## 8.4 Non-dimensional Analysis

One of the important contributions of dimensions is the non-dimensional analysis. The importance can be understood from the example of laminar flow of incompressible fluid through a pipe of diameter $D$ and length $L$. The pressure loss depends on the physical dimensions $L$ and $D$ of the pipe, the fluid properties $\rho$ and $\mu$, and characteristic flow velocity $U$, as discussed in chapter 5. It may be expressed in the functional form $\Delta p = f(U, D, \rho, \mu)$. After conducting the experiment and plotting the results on graph sheet, as explained in section 8.1, it was found that the pressure loss was same whenever the group $\rho LU/\mu$ had the same value i.e; if $U$ is doubled and $L$ is halved then the *Re* will be same in two cases. We may observe the same for other experiments in which $D$ or liquid is changed. This observation gave power to the non-dimensional analysis, since what is required is to change the Reynolds number by changing any one variable or their combination to give same nondimensional number; in this case the simplest is to change the flow velocity.

The next question is how many non-dimensional numbers will influence the given flow problem? Buckingham answered this question and is known as *Buckingham $\pi$ theorem,* $\pi$ because it does not have any dimensions or is a non-dimensional entity by its definition. According to this theorem, there will be $(m–n)$ such numbers, where $m$ is the number of variables in the functional form and $n$ is the number of fundamental dimensions required to express the various quantities in the function. An analyst has the choice to have any number of fundamental dimensions which depends on the type of problem on hand. For example, we have *MFLT, MLT, FLT,* or *LT* system of fundamental dimensions.

In the above-cited example, we ignored effects of the pipe surface roughness, gravity, presence of any eddy in the flow, and so on so forth. This is to be done to simplify the study. That is, while deciding the variables that influence the flow we often do not consider such variables that do affect but to a less extent as compared to others. Or in other words, we consider only those variables that have major influence on the flow. Sometimes we even neglect such variables whose effect is not negligible but their determination or satisfying the dynamic similarity considerations may be difficult; flow around the ship is one such case. So finally, we decide upon the variables and hence the value of $m$. Next is to decide the number of fundamental dimensions. So, once the value of $m$ and $n$ are fixed then there are number of methods to determine non-dimensional numbers formed by the flow variables and geometry and the fluid properties. This analysis is known as *non-dimensional analysis.*

### 8.4.1 Rayleigh's Method

To illustrate the process, we shall consider the problem of flow of incompressible liquid through a pipe. In this method, we express the functional form for the pressure loss as

$$\Delta p = K U^a D^b L^c \rho^d \mu^e, \text{ where } K \text{ is a non-dimensional constant} \quad (8.4.1)$$

and express it in dimension form (see Table 8.2.1)

$$ML^{-1}T^{-2} = \left(LT^{-1}\right)^a (L)^b (L)^c \left(ML^{-3}\right)^d \left(ML^{-1}T^{-1}\right)^e = M^{d+e} L^{a+b+c-3d-e} T^{-a-e}$$

By the principle of dimensional homogeneity, powers of $M$, $L$ and $T$ on both sides of the equation must be same. This gives following set of three equations in 5 unknowns, namely $a$, $b$, $c$, $d$, and $e$.

$$d + e = 1, \quad a+b+c-3d-e=-1, \quad \text{and} -a-e=-2$$

We have less number of equations than the unknowns and there will be infinite number of solutions. We may determine any three of the unknowns in terms of remaining two depending upon in what form we want to express our result. However, *a priori*, these groups may not be known and it requires experience to decide these; Table 8.3.1 may be a help in such situaiton. Let us determine $b$, $d$ and $e$ in terms of $a$ and $c$. We get:

$$e = 2 - a,\ d = a - 1, \text{ and } b = -2 + a - c$$

$$\Delta p = KU^a D^{-2+a-c} L^c \rho^{a-1} \mu^{2-a} = K\left(\rho D^2/\mu^2\right)^{-1} (UD\rho/\mu)^a (L/D)^c \tag{8.4.2}$$

$$\pi_1 = UD\rho/\mu = Re, \quad \pi_2 = L/D, \quad \text{and } \pi_3 = \Delta p \times \left(\rho D^2/\mu^2\right)$$

But, $\rho D^2/\mu^2 = (UD\rho/\mu)^2 \left(\rho U^2\right)^{-1}$

$\therefore \quad \pi_3 = \Delta p / \left(\rho U^2\right)$

Note how $\pi_3$ is changed later to Euler number when we recognised that the number obtained is not a standard form. We make following observations from the above solution:

(i) In this problem three non-dimensional groups are formed by the non-dimensional analysis. This supports the Buckingham $\pi$ theorem, since for this problem, $m = 6$ and $n = 3$ ($MLT$ system being used to express dimensions of different variables).

(ii) We changed $\pi_3$ to bring it in a definite form as a combination of other non-dimensional groups in the problem; this combination is decided as to what we want.

(iii) Flow properties density and viscosity have mass in their dimensions. Only $\Delta p$ has force dimension, which however can be changed in mass dimensions. Therefore, in Mechanics, we use $MLT$ system in the majority of the cases for non-dimensional analysis.

**Example 8.4.1** Obtain an expression for the capillary rise $h$ in a small diameter tube of diameter $d$ immersed in a liquid having density $\rho$ and surface tension $\sigma$.

**Solution** The rise of liquid in a capillary tube is determined by the surface tension of the liquid and gravity, which is combined with density in specific weight $w$. We may write the functional form as

$$h = h(w, \sigma, d)$$

We may use $FLT$ or $MLT$ system to analyse this problem dimensionally. We note that this problem involves force. Therefore $FLT$ system may be preferred over the $MLT$ system. For this system we get $[\rho]$ as follows:

$$[\rho] = ML^{-3}, [F] = [m \times \alpha] \text{ or, } F = MLT^{-2}$$

$$\rightarrow \quad M = FL^{-1}T^2 \text{ and } [\rho] = FL^{-4}T^2$$

Moreover, $[h] = L$, $[w] = [\rho g] = FL^{-4}T^2 \times LT^{-2} = FL^{-3}$, $[\sigma] = FL^{-1}$, and $[d] = L$

We shall use Rayleigh's method to find out the $\pi$ groups. Proceeding in the way explained above, we get

$$h = Kw^a \sigma^b d^c$$

or, $$F^0 L^1 T^0 = \left(FL^{-3}\right)^a \left(FL^{-1}\right)^b L^c = F^{a+b} L^{-3a-b+c} T^0$$

$\rightarrow$ $$a + b = 0 \text{ and} -3a - b + c = 1$$

This gives, $b = -a$ and $c = 2a + 1$ in items of $a$

$\therefore$ $$h = Kw^a \sigma^{-a} d^{2a+1}, \text{ or } h/d = K\left(\frac{\sigma}{wd^2}\right)^{-a}$$

$\rightarrow$ $$\frac{h}{d} = f\left(\frac{\sigma}{wd^2}\right)$$

If we examine this problem from the Buckingham $\pi$ theorem then we find that $m$ = 4 and $n$ = 3, which means there must be only one non-dimensional group. Is this problem violating the Buckingham theorem? No, in fact only the fundamental dimensions $F$ and $L$ are enough for this problem, as seen from the dimensions of the variables and fundamental dimension $T$ is redundant variable. That is, we have used $FL$ system instead of $FLT$ system for the analysis. So there will be two non-dimensional groups, as obtained above.

## 8.4.2 Elimination of Fundamental Dimension One by One

This is a stepwise process involving the following steps for solution. Follow the steps as under:

(i) Write down the functional form and decide the fundamental dimensions for the analysis, say, we use $MLT$ system.
(ii) Next while eliminating say $M$, we seek a variable that has [$M$] (read as dimension mass) and divide all other variables having dimension mass with this variable to give new variables. The functional form is now changed with the newly obtained variables in place of old variables while dropping this term from the function. Now write down the dimensions below each term.
(iii) Repeat step (ii) for other dimensions till we end up with non-dimensional groups.

**Example 8.4.2** Consider a functional relation between drag force $F$, diameter of a sphere $D$, velocity $U$, and fluid properties $\rho$ and $\mu$, namely $F = F(D, U, \rho, \mu)$. Eliminate from this equation each of the dimensions $M$, $L$ and $T$ one by one and obtain the non-dimensional numbers.

**Solution**

*Step 1:* $F = F(D, U, \rho, \mu)$. Let us use $MLT$ system.

*Step 2:* First let us eliminate dimension $M$. We observe from the list of the variables in the function that $F$, $\rho$ and $\mu$ involve the dimensions of $M$ and therefore we choose the variables $F/\rho$ and $\mu/\rho$ in place of the original variable by dividing rest of the variables by $\rho$.

This gives

$$\frac{F}{\rho} = F\left(U, \quad D, \quad \frac{\mu}{\rho}\right)$$

$$\left[\frac{MLT^{-2}}{ML^{-3}}\right] \quad \left[\frac{L}{T}\right] [L\} \left[\frac{ML^{-1}T^{-1}}{ML^{-3}}\right]$$

The dimension of each terms is written below the term. Next, we eliminate $T$ by choosing $U$ and redefine the variables by dividing those variables which involve dimension $T$. For example, $F/\rho$ has dimension $L^4/T^2$ whereas that of $U$ is $L/T$, therefore we have to divide it by $U^2$ so that $T$ is gone. $D$ does not have $[T]$ and therefore we do not divide it by $U$ while $\mu$ has dimension $L^2/T$ so we divide it by $U$ and not by $U^2$ i.e. we divide suitably only those variables from which $[T]$ is to be eliminated. We thus get

$$\frac{F}{\rho U^2} = F\left(D, \quad \frac{\mu}{\rho U}\right)$$

$$L^2 \qquad L \qquad L$$

Finally, we eliminate $L$ by choosing $D$. The procedure yields

$$\frac{F}{\rho U^2 D^2} = F\left(\frac{\mu}{\rho UD}\right)$$

or, $$C_D = f(Re)$$

This way we obtain the desired non-dimensional numbers. $C_D$ is the coefficient of drag on the sphere and $Re$ is the Reynolds number.

### 8.4.3 Group Method

This is a method in which the repeating variables are identified and then we make their group with the remaining variables taking one by one. We may use Rayleigh method for the purpose, as the unknowns are determined in terms of only one unknown thus producing a non-dimensional number. This procedure, however, requires some experience in selecting the repeated variables. A rough guide may be as follows:

(i) As far a possible, the dependent variable should not be selected as repeating variable.

(ii) These variables taken together should represent all the three fundamental dimensions chosen for the analysis.

(iii) The repeating variables in three fundamental dimension system should be such that one contains the fluid property (density, viscosity, etc), the second, flow property (velocity, pressure, acceleration, etc) and the third geometric property (length, diameter, head, etc).

(iv) These three selected repeating variables should not form a non-dimensional number with themselves. For example, if we choose $L$, $U$ and $\nu$ as repeating variables, then in fact we have chosen only two variables, as the selected variables form Reynolds number by themselves.

(v) This also implies that none of these variables can have same dimensions.

We shall now demonstrate the procedure for determining the non-dimensional numbers for the propeller of an aircraft. The thrust $F$ developed by the propeller is found, from the study of the flow around it, to depend on aircraft velocity $U$, density $\rho$ and viscosity $\mu$ of the air around the propeller, its diameter $D$ and rpm. That is, the functional relation between the different variables is of the type

$$F = (U, \rho, \mu, D, N) \tag{8.4.3}$$

There are 6 physical quantities and three fundamental dimensions and therefore these form three non-dimensional numbers. These will now be determined using the different methods as discussed above.

***Rayleigh method*** Let us write Eq. 8.4.3 in the following way and then write down the dimensions of each quantity so that equation is dimensionally homogeneous. This gives

$$F = KU^a \rho^b \mu^c D^d N^e$$

$$\rightarrow \qquad MLT^{-2} = [LT^{-1}]^a\,[ML^{-3}]^b\,[ML^{-1}\,T^{-1}]^c\,[L]^d\,[T^{-1}]^e$$

or,

$$MLT^{-2} = M^{b+c}\,L^{a-3b-c+d}\,T^{-a-c-e}$$

$$\Rightarrow \qquad b + c = 1,\ a - 3b - c + d = 1 \text{ and } -a - c - e = -2$$

This way we have got three equations in 5 unknowns. Let us determine three in terms of $c$ and $e$ by solving these equations. This gives

$$b = 1 - c,\ a = 2 - c - e \text{ and } d = 2 - c + e$$

Substituting the values of the constants and arranging the numbers having same power as a group, we get

$$\frac{F}{\rho U^2 D^2} = K\left(\frac{\mu}{UD\rho}\right)^c \left(\frac{ND}{U}\right)^e$$

We notice that this problem involves 3 non-dimensional numbers as predicted by $\pi$ theorem. However, the group $\mu/(UD\rho)$ is reciprocal of Reynolds number. It does not matter and we may express the non-dimensional equation in the following way as well because any combination of non-dimensional numbers is also a non-dimensional number.

$$\therefore \qquad \frac{F}{\rho U^2 D^2} = f\left(\frac{UD\rho}{\mu}, \frac{ND}{U}\right)$$

**Elimination of fundamental dimension.** The procedure is explained above. Keeping the points in mind we proceed as follows:

$$F = (\ U,\quad D,\quad \mu,\quad \rho,\quad N)$$

*Step 1:*

$$\left[\frac{ML}{T^2}\right]\left[\frac{L}{T}\right][L]\left[\frac{M}{LT}\right]\left[\frac{M}{L^3}\right]\left[\frac{1}{T}\right]$$

*Step 2:* Eliminate $[M]$. Variables $F, \mu, \rho$ involve dimension $M$, we choose $\rho$ for elimination of $[M]$. Divide all the variables that have $[M]$ by $\rho$. This gives

$$\frac{F}{\rho} = \left(U, D, \frac{\mu}{\rho}, \quad N\right)$$

$$\frac{ML}{T^2}\frac{L^3}{M} \quad \frac{L}{T} \; L \; \frac{M}{LT}\frac{L^3}{M} \; T^{-1}$$

*Step 3:* Eliminate [$T$]. $F$, $U$, $\mu$, and $N$ have dimension $T$. We choose $U$ and divide other variables by it to give new variables

$$\frac{F}{\rho U} = \left(D, \frac{\mu}{\rho U}, \frac{N}{U}\right)$$

$$\frac{L^4}{T^2}\frac{T^2}{L^2} \quad L \; \frac{L^2}{T}\frac{T}{L} \; \frac{1}{T}\frac{T}{L}$$

*Step 4:* Eliminate [$L$]. Variables $D$, $\mu$ and $N$ have dimension $L$. We divide these by $D$ or some of its power to give the non-dimensional numbers

$$\frac{F}{\rho U^2 D^2} = \left(\frac{\mu}{\rho UD}, \frac{ND}{U}\right)$$

$$\frac{L^2}{T^2}\frac{T^2}{L^2} \quad \frac{L}{T}\frac{T}{L} \; \frac{1}{L}L$$

***Group method*** We choose three variables $U$, $D$ and $\rho$ as the repeating variables as these satisfy all the conditions mentioned in that section. $U$ represents the flow property, $D$ the geometric property and $\rho$ the fluid property while these three do not have common dimension to form a non-dimensional combination and at the same time the totality of the variables have the three fundamental dimensions, namely $M$, $L$ and $T$. Now we can take the remaining variables along with these variables one by one to form the non-dimensional groups

$$\pi_1 = FU^aD^b\rho^c, \quad \pi_2 = \mu U^aD^b\rho^c \quad \text{and} \quad \pi_3 = NU^aD^b\rho^c$$

Each of the groups. $\pi_1$, $\pi_2$, etc., can be obtained using Rayleigh method discussed before. Each of these three groups has three unknowns and there will be three equations that would get exact solution to get the constants $a$, $b$ and $c$, which will be different for different groups. Therefore we have

$$\pi_1 = \frac{F}{\rho U^2 D^2}, \quad \pi_2 = \frac{UD\rho}{\mu} \quad \text{and} \; \pi_3 = \frac{ND}{U}$$

or,
$$\frac{F}{\rho U^2 D^2} = f\left(\frac{UD\rho}{\mu}, \frac{ND}{U}\right)$$

Basically the method is same as the Rayleigh method with the difference that the single equation is split into number of equations equal to the number of groups. This makes the solution of the simultaneous equations a little easy at the cost of repeating the same steps a number of times for each non-dimensional group.

***Alternate group method*** We may also find a combination of variables that gives a non-dimensional number in a little different way. We observe that (i) $U$ and $D$ are present in all the three non-dimensional numbers, (ii) $\rho$ is present in two groups, (iii) rest variables are present in only one of the group. The Steps for solution in this method are:

(i) Choose three variables in such a way that these when taken together have all the fundamental dimensions.

(ii) These variables are chosen in the order of number of their appearances in the non-dimensional groups. For example, in this case, the variables will be chosen in the order of $U, D$ and $\rho$, which satisfies the condition (i) also.

(iii) We now choose a combination of these variables to form groups such that one has the dimension of mass, the second length and the third that of time (see the procedure given below).

(iv) If the groups so formed have the square roots, etc., then we may choose some other repeating variables.

(v) Now take the remaining variables one by one and try to find out the combination with the help of relations obtained in step (iii) so that this group has the dimension of the variable under consideration. A ratio of the variable with this group is a non-dimensional number.

In this problem, we choose $U, D$ and $\rho$. We get following relations following the steps for solution in the following equations, we use notation $[x]$ means dimension of quantity $x$.

$$[D] = L\text{: This means dimension of } D \text{ is } L$$

$$[U] = L/T = [D]/T$$

$$\rightarrow \quad [D/U] = T\text{ : read as group } D/U \text{ has dimesion } T$$

$$[\rho] = ML^{-3} = M\left[1/D^3\right] \quad \rightarrow \quad \left[\rho D^3\right] = M \text{ i.e. group } \rho D^3 \text{ has dimension } M$$

These equations are read as: $D$ has dimension $L$, $D/U$ that of $T$ and $\rho D^3$ that of $M$. Now take the remaining variables one by one and proceed as follows

$$[F] = MLT^{-2} = [\rho D^3][D][D/U]^{-2} = [\rho D^2 U^2]$$

$$\rightarrow \quad \pi_1 = \frac{F}{\rho D^2 U^2}$$

$$[\mu] = ML^{-1}T^{-1} = [\rho D^3][D]^{-1}[D/U]^{-1} = [\rho DU]$$

$$\rightarrow \quad \pi_2 = \frac{\rho DU}{\mu}$$

$$[\mathrm{N}] = T^{-1} = [D/U]^{-1} = [U/D]$$

$$\rightarrow \quad \pi_3 = \frac{ND}{U}$$

We get the same result. However, this procedure has the advantage that no simultaneous equations are to be solved. The only problem is the right selection of the repeating variables. In case we are able to conceive the non-dimensional groups likely to influence the flow, which normally would be possible with experience, then the method is straight forward and time saver.

**Example 8.4.3** The pressure difference $\Delta p$ in a pipe of diameter $D$ and length $L$ due to turbulent flow depends on the velocity $U$, viscosity $\mu$, density $\rho$, and roughness $\varepsilon$. Using Buckingham $\pi$ theorem, determine an expression for $\Delta p$.

**Solution** We may write the functional relation in different variables as follows, in which there are 7 variables and therefore there are $4\pi$ terms.

$$f(\Delta p, D, L, U, \mu, \rho, \varepsilon) = 0$$

We choose $D$, $U$ and $\rho$ as the repeating variables and form four non-dimensional groups. These are

$$\pi_1 = f(\Delta p, D, U, \rho), \quad \pi_2 = f(L, D, U, \rho), \quad \pi_3 = f(\mu, D, U, \rho), \quad \text{and } \pi_4 = f(\varepsilon, D, U, \rho)$$

We may determine the non-dimensional groups taking one by one these relations and using Rayleigh method or any other method. We shall demonstrate the procedure for the first and the last group.

$$\pi_1 = f(\Delta p, D, U, \rho) \rightarrow M^0L^0T^0 = \left(ML^{-1}T^{-2}\right)(L)^b\left(LT^{-1}\right)^c\left(ML^{-3}\right)^d$$

or $\quad 1 + d = 0, \; -1 + b + c - 3d = 0 \text{ and } -2 - c = 0$

$\rightarrow \quad d = -1, \; c = -2 \text{ and } b = 0$

$$\therefore \quad \pi_1 = \frac{\Delta p}{\rho U^2}$$

Similaraly, consider $\pi_4 = f(\varepsilon, D, U, \rho) \rightarrow M^0L^0T^0 = (L)(L)^a(LT^{-1})^b(ML^{-3})^c$
or, $1 + a + b - 3c = 0$, $c = 0$, and $b = 0 \rightarrow a = -1$

$$\therefore \quad \pi_4 = \frac{\varepsilon}{D}$$

The other two relations give non-dimensional numbers

$$\pi_2 = \frac{L}{D} \text{ and } \pi_3 = \frac{\mu}{DU\rho}$$

$$\therefore \quad \frac{\Delta p}{\rho U^2} = f\left(\frac{L}{D}, \frac{\mu}{DU\rho}, \frac{\varepsilon}{D}\right)$$

This shows that the drop in pressure depends on the Reynolds number, the distance between two points between which the pressure drops and the roughness of the pipe. The ratio $\varepsilon/D$ is called the Knudsen number.

To summarise, a physical phenomenon can be described with the help of equations like

$$\pi_1 = K\pi_2^a \, \pi_3^b \ldots$$

or, $\quad \ln \pi_1 = A + a \ln \pi_2 + b \ln \pi_3 + \ldots$

Normally, two of the non-dimensional groups dominate the flow and are considered first keeping other fixed, as far as possible. Let us take the first two namely $\pi_1$ and $\pi_2$. We note that the plot on log-log graph will be a straight line. If the slope changes at any point then it reflects some major change in the flow condition. This can be seen from the Moody diagrams for a pipe where the slope of the curve $C_d$ v/s *Re* changes on the log-

log graph at the instant the flow is changed to turbulent flow. Thus log-log graph have special role in dimensional analysis.

**Example 8.4.4** A jet of liquid is discharged with initial velocity $V$ at an angle $\theta$ with the horizontal. Determine the range $R$ of the jet.

**Solution** Range is the distance travelled along the horizontal direction, which will be affected by the earth's gravity $g$. We may write the functional relation in the form: $R = f(V, \theta, g)$. We choose $V$ and $g$ as repetitive variables and then take other variables in turn. Taking $R$, writing the functional equation $f(R, V, g) = 0$ and using Rayleigh method, we get $R = A\, V^2/g$, where $A$ is a constant, $\theta$ is itself a non-dimensional number and there is no need to proceed with this.

We observe that $\theta$ does not appear in this solution, simply because it does not have any dimension. Therefore, it is not clear what is the role of $\theta$ in this problem. Is it that $A$ includes effect of $\theta$? Nothing can be said from the analysis we carried out. Such types of problems are faced when we have vector quantity forming non-dimensional group.

Instead let us use the components $u$ and $v$ of the vector $V$, where $u = V\cos\theta$ and $v = V\sin\theta$ and express $R$ as $R = f(u, v, g)$ and get the non-dimensional form using any of the methods. We shall get,

$$R = A\frac{uv}{g} = A\frac{V^2 \sin\theta\cos\theta}{g} = C\sin 2\theta\frac{V^2}{2g}$$

It is obvious that the coefficient of $V^2/g$ is a function of $\theta$, which could not be inferred from the analysis done above. This happens many a times and working with the components of a vector rather than the vector itself may give more information many times.

**Example 8.4.5** The frictional torque $T$ on a rotor rotating at a speed of $N$ rpm in a fluid of density $\rho$ and viscosity $\mu$ in a turbulent flow is given by $T = \rho D^5 N^2 f\left[\frac{\mu}{D^2 N\rho}\right]$. Prove this by non-dimensional technique. (AMIE 1975)

**Solution** There are five variables and therefore 2 $\pi$'s. (We have to be careful about $T$ as it is used for torque as well as the dimension $T$). Let us choose $D$, $N$ and $\rho$ as the repeating variables and proceed as follows:

$$[D] = L, [N] = T^{-1} \rightarrow [1/N] = T$$

and
$$[\rho] = ML^{-3} \rightarrow [\rho D^3] = M$$

$$[T] = ML^2T^{-2} = [\rho D^3 \times D^2 \times N^2] = [\rho D^5 N^2] \rightarrow \pi_1 = \frac{T}{\rho D^5 N^2}$$

$$[\mu] = ML^{-1}T^{-1} = [\rho D^3 \times D^{-1} \times N] = [\rho D^2 N] \rightarrow \pi_2 = \frac{\mu}{\rho D^2 N}$$

$$\therefore \quad \frac{T}{\rho D^5 N^2} = f\left(\frac{\mu}{\rho D^2 N}\right)$$

**Example 8.4.6** A partially submerged body (ship) is towed in water. The resistance $R$ to its motion depends on the density $\rho$, the viscosity $\mu$ of water, length $L$ and velocity $U$ of the body, and the acceleration due to gravity $g$ of the earth field. Determine the non-dimensional numbers for this kind of motion. (AMIE, Summer, 1978)

**Solution** The functional form of the problem is

$$f(R, \rho, \mu, L, U, g) = 0$$

It has 6 dimensions and therefore there are 3 non-dimensional numbers. Let us choose $\rho$, $L$ and $U$ as the repeating variables. Then

$$[L] = L, [U] = LT^{-1} \rightarrow [L/U] = T$$

and $$[\rho] = ML^{-3} \rightarrow [\rho L^3] = M$$

$$[R] = MLT^{-2} = [\rho L^3 \times D \times (L/U)^{-2}] = [\rho L^2 U^2] \rightarrow \pi_1 = \frac{R}{\rho L^2 U^2}$$

$$[\mu] = ML^{-1}T^{-1} = [\rho L^3 \times L^{-1} \times (L/U)^{-1}] = [\rho UL] \rightarrow \pi_2 = \frac{\mu}{\rho UL}$$

$$[g] = LT^{-2} = [L(L/U)^{-2}] = [U^2/L] \rightarrow \pi_3 = \frac{U^2}{Lg} \text{ or } \frac{U}{\sqrt{Lg}}$$

$$\therefore \quad \frac{R}{\rho L^2 U^2} = f\left(\frac{\mu}{\rho UL}, \frac{U}{\sqrt{Lg}}\right)$$

Resistance to towing of the submerged body depends on the Reynolds and Froude numbers. This result was expected, since the water in contact with the body offers viscous force whereas the water surface waves produce surface drag that depends on the gravity.

**Example 8.4.7** The resistance $R$ to a supersonic plane during the flight can be considered as function of its length, velocity of flight $U$, and density $\rho$, viscosity $\mu$ and bulk modulus $K$ of air. Establish this relation. (AMIE, Summer, 1997)

**Solution** The functional form of the problem is

$$f(R, \rho, \mu, L, U, K) = 0$$

It has 6 dimensions and therefore there are 3 non-dimensional numbers. Let us choose $\rho$, $L$ and $U$ as the repeating variables. Then

$$[L] = L, [U] = LT^{-1} \rightarrow [L/U] = T$$

and $$[\rho] = ML^{-3} \rightarrow [\rho L^3] = M$$

$$[R] = MLT^{-2} = [\rho L^3 \times D \times (L/U)^{-2}] = [\rho L^2 U^2] \rightarrow \pi_1 = \frac{R}{\rho L^2 U^2}$$

$$[\mu] = ML^{-1}T^{-1} = [\rho L^3 \times L^{-1} \times (L/U)^{-1}] = [\rho UL] \rightarrow \pi_2 = \frac{\mu}{\rho UL}$$

$$[K] = ML^{-1}T^{-2} = \left[\rho L^3 \times L^{-1} \times (L/U)^{-2}\right] = \left[\rho U^2\right] \rightarrow \pi_3 = \frac{K}{\rho U^2}$$

$$\therefore \qquad \frac{R}{\rho L^2 U^2} = f\left(\frac{\mu}{\rho U L}, \frac{K}{\rho U^2}\right)$$

**Example 8.4.8** The efficiency of a fan depends on density $\rho$, viscosity $\mu$ of the fluid, angular velocity $\omega$ and diameter $D$ of the fan, and discharge $Q$. Determine the functional relationship for the efficiency. (AMIE, Winter, 1976)

**Solution** The functional form of the problem is

$$f(\eta, \rho, \mu, D, \omega, Q) = 0$$

It has 6 dimensions and therefore there are 3 non-dimensional numbers. Let us choose $\rho$, $D$ and $\omega$ as the repeating variables. Then

$$[D] = L, \ [\omega] = T^{-1} \rightarrow [1/\omega] = T$$

and

$$[\rho] = M L^{-3} \rightarrow [\rho D^3] = M$$

$$[\mu] = M L^{-1}T^{-1} = \left[\rho D^3 \times D^{-1} \times (1/\omega)^{-1}\right] = \left[\rho\omega D^2\right] \rightarrow \pi_2 = \frac{\mu}{\rho\omega D^2}$$

$$[Q] = L^3T^{-1} = \left[D^3 (1/\omega)^{-1}\right] = \left[\omega D^3\right] \rightarrow \pi_3 = \frac{Q}{\omega D^3}$$

$$\therefore \qquad \eta = f\left(\frac{\mu}{\rho\omega D^2}, \frac{Q}{\omega D^3}\right)$$

**Example 8.4.9** If viscosity $\mu$ of a gas depends on its mass density $\rho$, mean velocity of molecules $\overline{C}$, diameter of the molecule $d$ and mean free path $\lambda$, determine expression for $\mu$.

**Solution** We shall use *MLT* system of fundamental dimensions. Let $\mu = f(\rho, \overline{C}, d, \lambda)$. Let us use group method and choose the variables $d$, $\rho$, and $\overline{C}$ as the repeating variables. Then the non-dimensional group taking other variable $\mu$ can be expressed as

$$\pi_1 = \mu\, d^a \rho^b\, \overline{C}^{\,e}$$

$$\rightarrow \qquad M^0L^0T^0 = \left[\frac{M}{LT}\right][L]^a\left[\frac{M}{L^3}\right]^b\left[\frac{L}{T}\right]^e$$

We have also written the dimensions of the each term. Equating powers of $M$, $L$ and $T$, we get: $b + 1 = 0$, $a - 3b + e - 1 = 0$, and $-e - 1 = 0$. This gives, $a = -1$, $b = -1$ and $e = -1$. Therefore,

$$\pi_1 = \frac{\mu}{\overline{C}\, d\rho}, \text{and similarly, } \pi_2 = \frac{\lambda}{d}$$

$$\Rightarrow \qquad \frac{\mu}{\overline{C}\, d\rho} = f\left(\frac{\lambda}{d}\right) \quad \text{or,} \quad = \frac{\mu}{\overline{C}\, d\rho} = A\left(\frac{\lambda}{d}\right)^q \text{ (say)}$$

The nature of the function is determined from the experimental data or physics of the problem. We have assumed a functional relation of a polynomial type of degree $q$ with a single term only and $A$ is a constant. Maxwell observed that $\mu$ does not depend on $\lambda$. This is possible if $\rho \sim 1/\lambda$. Take $p$ =1 in the assumed function, then $\mu = A\,\overline{C}$.

## 8.5 Similitude

Similitude is the science of study of the flow by performing experiment over a model, which has been designed so that the experimentation may be carried out on it in a laboratory, and the performance of the prototype be predicted from the results obtained on the model. This is also known as modelling of the prototype and has proved a very useful experimental method to ensure performance of the prototype before we start manufacturing it. The actual prototype may be very big or small in size, or may be using a fluid that may be replaced by a more convenient fluid during the experimentation. This would not only facilitate an easy handling of the experimental set up but will also reduce the cost of the experimentation, which is a critical factor. These days we are shifting towards computer simulation as the variables can be changed easily and the results obtained promptly. This has reduced the experimentation time considerably. The other advantage of such packages is that we may change the variables to any extent and see the effect on the flow pattern. This way it helps in giving more information to us. However, simulation of the actual flow and the geometry of the flow pose many challenges.

This kind of study requires that the model and the flow pattern over it should be exactly similar to that for the prototype; the degree of variation in the flow pattern in the prototype and the model would determine the error in the predicted results. The two flow patterns will be similar when these have geometric, kinematic and dynamic similarity.

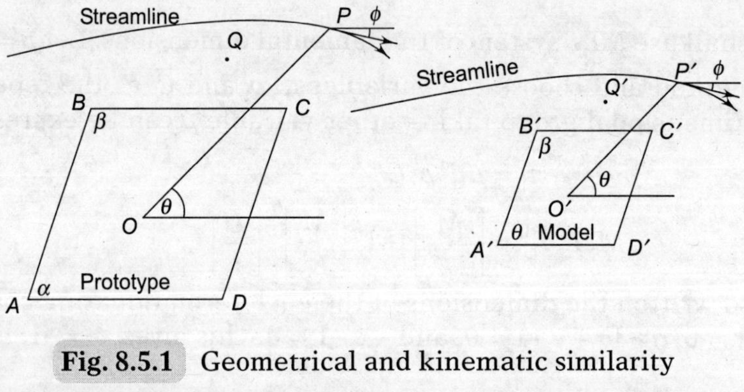


**Fig. 8.5.1** Geometrical and kinematic similarity

*Geometric similarity*. Figure 8.5.1 shows a streamline, representing the flow pattern, around a prototype and its model reduced in size to 60% from all the sides. We observe that

(i) The model retains the angles of the prototype during the reduction (or enlargement) in size of the prototype; reduction is done when the prototype is big and enlargement when it is small.

(ii) The size is reduced by the same ratio in all the directions. This implies that

$$\frac{AB}{A'B'} = \frac{AD}{A'D'} = \frac{BC}{B'C'} = \frac{CD}{C'D'} = L_r, \text{ the scale ratio} \tag{8.5.1}$$

(iii) The point $P$ and $P'$ are geometrically similar points in the two flow regions if $OP/O'P' = S$ and the two lines $OP$ and $O'P'$ are at the same angle $\theta$ from the reference axis. The geometrically similar points are also referred to as the corresponding points.

A model is said to be geometrically similar to its prototype when the first two conditions are satisfied. However, models have also been made which do not satisfy the above conditions and are called distorted models. These are required in the study of flows when equal reduction in size leads to errors in the experiential results. For example, while studying movement of ships in the shallow but broad rivers like Ganges, an equal reduction of the width and depth of flow will develop a model in which the boundary layer effects may dominate during the experiments and the predicted results will be far from correct. Otherwise, as far as possible, we use geometrically similar models for the study.

Since the prototype has a scale factor of $L_r$ in length dimension, we can find out the scale ratio of area and volume too; these are:

$$\frac{\text{Area of prototype}}{\text{Area of model}} = L_r^2 \quad \text{and} \quad \frac{\text{Volume of prototype}}{\text{Volume of model}} = L_r^3 \tag{8.5.2}$$

*Kinematic similarity.* We read in Chapter 2 that in kinematics we study displacement, velocity and acceleration of the fluid at a point in the flow region. By kinematic similarity, we mean that displacement, velocity or acceleration at the geometrically similar points, in the flow around the prototype and its model, bear the same ratio. Let $P$ and $Q$ are two points in the flow around the prototype and $P'$ and $Q'$ are their corresponding geometrically similar points then kinematic similarity means

$$\frac{U_P}{U_{P'}} = \frac{U_Q}{U_{Q'}}, \text{ etc,} \quad \text{and} \quad \frac{\alpha_P}{\alpha_{P'}} = \frac{\alpha_Q}{\alpha_{Q'}}, \text{ etc,} \tag{8.5.3}$$

Expressing velocity as displacement per unit time in Eq. 8.5.3 gives

$$\frac{U_m}{U_p} = \frac{(L/T)_m}{(L/T)_p} = \frac{L_m}{L_m} \times \frac{T_p}{T_m} = L_r/T_r, \text{ where } T_r = \frac{T_m}{T_p} \tag{8.5.4}$$

$T_r$ is the time scale and can be obtained from Eq. 8.5.4.

*Dynamic similarity* It demands that the ratio of the force at the geometrical points in the flow around the prototype and the model must bear the same ratio at all the geometrical similar points. Let us consider the geometrical points $P$ and $P'$. The forces acting on the fluid element are viscous, inertia, pressure force, etc, which we discussed in section 8.5.4. Dynamic similarity requires, using subscript $p$ for prototype and $m$ for the model, that at the geometrical similar points

$$\frac{(\text{Inertia force})_p}{(\text{Inertia force})_m} = \frac{(\text{Viscous force})_p}{(\text{Viscous force})_m} = \frac{(\text{Gravity force})_p}{(\text{Gravity force})_m} = \ldots$$

or, $$\left(\frac{\text{Inertia force}}{\text{Viscous force}}\right)_p = \left(\frac{\text{Inertia force}}{\text{Viscous force}}\right)_m, \left(\frac{\text{Inertia force}}{\text{Gravity force}}\right)_p = \left(\frac{\text{Inertia force}}{\text{Gravity force}}\right)_m = \ldots$$

or, $$Re_p = Re_m, \qquad Fr_p = Fr_m, \ldots \tag{8.5.5}$$

This means, all non-dimensional numbers for the two flows must be same at the corresponding points. In practice, it is very difficult to satisfy all these conditions. For example, consider the movement of a ship in the sea. Its velocity and power requirements were shown to be the function of the Reynolds and the Froude numbers. Their equality requires to choose the model and the fluid for experimentation so that

$$\left(\frac{UL}{\nu}\right)_p = \left(\frac{UL}{\nu}\right)_m \quad \text{and} \quad \left(\frac{U^2}{Lg}\right)_p = \left(\frac{U^2}{Lg}\right)_m$$

These two conditions are satisfied for a model when

$$\frac{U_m}{U_p} = \left(\frac{\nu_m}{\nu_p}\right) \times \frac{L_p}{L_m} \tag{i}$$

$$\frac{U_m}{U_p} = \sqrt{\frac{L_m}{L_p}} \tag{ii}$$

Therfore, both the conditions will be satisfied when

$$\frac{\nu_m}{\nu_p} \times \frac{L_p}{L_m} = \sqrt{\frac{L_m}{L_p}}$$

$$\rightarrow \qquad L_r = \frac{L_p}{L_m} = \left(\frac{\nu_p}{\nu_m}\right)^{2/3} \tag{iii}$$

Satisfying these two conditions is not an easy job, rather almost impossible because it will be difficult to select fluid that may have kinematic viscosities to satisfy this condition. This means that it is not possible to satisfy dynamic similarity in toto in such cases, as both the Reynolds and Froude numbers for a model and the prototype cannot be the same.

*Modelling criterion* While making a model, we choose a scale so that the size of the model is reasonable. Yet another factor to be considered is that creation of exactly identical conditions for the model is not an easy job, since we often neglect the effects of many variables such as surface roughness, capillary action, surface tension, eddies, etc. Their effect depends on the size of the body. Thus, while modelling such situations, we have to be extra careful about the model size so that it may not significantly change the effect of such unaccounted variables in the similitude study. From this discussion, it is evident that complete similitude has been ruled out and we make the following concessions to ease out the modelling problems.

(i) Assume the surface roughness profile similarity.
(ii) Neglect the real flow effects.
(iii) Consider only the most predominant forces, at the most two, so that effect of one is determined through dimensional analysis and that of another in some other way

that would be dictated by the given problem. We shall explain the procedure with the help of some solved problem later.

(iv) As far as possible, a distorted model be avoided. But we should consider it seriously if the model study is affected severely by doing so. This is important when one of the physical dimensions is too large as compared to other dimensions.

We shall now consider flow through a rotamachine to develop a non-dimensional model. The fluid is assumed to be incompressible to avoid consideration of compressibility effects. In such flows, the performance or dependent variables are power $P$, specific energy $gH$, and efficiency $\eta$, which are found to depend on density $\rho$ and viscosity $\mu$ of the fluid, rotor diameter $d$, its rpm $N$, and the volume flow rate of the fluid over the rotor. The specific energy is defined as $gH$, where $H$ is the head acting on turbine or developed by the pump. This statement can be written in the functional form as given below.

$$P = f_1(\rho, d, \mu, N, Q),$$

$$gH = f_2(\rho, d, \mu, N, Q),$$

and $$\eta = f_3(\rho, d, \mu, N, Q)$$

Each of these equations has 6 variables, 5 of which are common. Thus for each of these variables there are three non-dimensional groups; two groups will be same and only one group would change with choice of these variable in the analysis. These can be determined by any of the methods discussed above. Let us choose $N$, $d$ and $\rho$ as the repeating variables. Then

$$[d] = L, [N] = T^{-1} \rightarrow [1/N] = T$$

and $$[\rho] = ML^{-3} = M[d^{-3}] \rightarrow [\rho d^3] = M$$

The remaining independent variables $Q$ and $H$ appear in all the three equations. These give the following non-dimensional numbers.

$$[Q] = L^3T^{-1} = [d^3N] \rightarrow \pi_2 = \frac{Q}{Nd^3} \text{ and similarly}$$

$$[\mu] = ML^{-1}T^{-1} \rightarrow \pi_3 = \frac{\rho N d^2}{\mu}$$

Now let us take other variables to get the non-dimensional numbers:

$$[P] = ML^2T^{-3} = [\rho d^3 \times d^2 \times N^3] = [\rho N^3 d^5] \rightarrow \pi_1 = \frac{P}{\rho N^3 d^5}$$

and $$[gH] = LT^{-2} \times L = L^2T^{-2} = [d^2][N^2] = [N^2d^2] \rightarrow \pi_4 = \frac{gH}{N^2d^2}$$

$\eta$ is a non-dimensional number. We can therefore write

$$\frac{P}{\rho N^3 d^5} = f_1\left(\frac{Q}{Nd^3}, \frac{\rho N d^2}{\mu}\right),$$

$$\frac{gH}{N^2 d^2} = f_2\left(\frac{Q}{Nd^3}, \frac{\rho Nd^2}{\mu}\right),$$

$$\eta = f_3\left(\frac{Q}{Nd^3}, \frac{\rho Nd^2}{\mu}\right)$$

You may identify the number $\rho Nd^2/\mu = \rho Ud/\mu$ as the Reynolds number. Normally the flow in rotamachines is turbulent when the viscous effects are negligible and may be ignored during the dimensional analysis. The condition to be satisfied for dynamic similarity therefore is to satisfy the flow parameter. That is, the flow parameter $Q/(Nd^3)$ must be same for the model and the prototype. When the tests are carried out under these conditions on a geometrically similar model, then the efficiency, the power parameter and the head parameter will also be the same for the two. So from the measurement of these on a model, it will be possible to predict power requirements for the prototype.

Though not related to the non-dimensional analysis, it would be of interest to note that the square of the non-dimensional number $\frac{Q}{Nd^3}$ can be divided by the cube of $\frac{gH}{N^2 d^2}$ so as to give a new number without the $d$ term. This number is used more commonly in pumps and is called the specific speed of the pump.

$$\pi = \sqrt[1/4]{\left(\frac{Q}{Nd^3}\right)^2 \bigg/ \left(\frac{gH}{N^2 d^2}\right)^3} = \frac{N\sqrt{Q}}{(gH)^{3/4}}$$

We may similarly make some other combination. Try and get specific speed of turbine, which is $\frac{N\sqrt{p/\rho}}{(gH)^{5/4}}$

**Example 8.5.1** The characteristics of a spillway are to be studied by means of a geometrically similar model constructed to the scale ratio of 1:10.

(a) If the maximum rate of flow in the prototype is 28.3 cumec, what will be the corresponding flow in model?
(b) If the measured velocity in the model at a point on the spillway is 2.4 m/s, what will be the corresponding velocity in the prototype?
(c) If the hydraulic jump at the floor of the model is 50 mm high, what will be the height of jump in the prototype?
(d) If the energy dissipated per second in the model is 3.5 J, what energy is dissipated in the prototype? (AMIE Summer, 1988)

**Solution** Flow in a spillway has free water surface and hence is dominated by the Froude number. In this problem, we are given the length scale ratio. Using the Froude similarity, we can determine the scale ratio for other quantities in the following way:.

This similarity requires

$$\left(\frac{U}{\sqrt{L}}\right)_m = \left(\frac{U}{\sqrt{L}}\right)_p \Rightarrow \frac{U_p}{U_m} = \sqrt{\frac{L_p}{L_m}} = \sqrt{L_r},$$

$$T_r = \frac{T_p}{T_m} = \frac{L_p / U_p}{L_m / U_m} = \sqrt{L_r}$$

and $$Q_r = \frac{Q_p}{Q_m} = \left(\frac{L_p}{L_m}\right)^2 \frac{U_p}{U_m} = \left(\frac{L_p}{L_m}\right)^{2.5}$$

(a) Here $L_r$, $T_r$ and $Q_r$ are the scale ratio for length, time and discharge. [1] From the given data ($L_r$ = 10 and $Q_p$ = 28.3 m³/s), $Q_m$ = 28.3 / (10)$^{2.5}$ = **0.0895 m³/s**

(b) The velocity scale is shown to be

$$\frac{U_p}{U_m} = \sqrt{\frac{L_p}{L_m}} = \sqrt{10} \rightarrow U_p = 2.4 \times \sqrt{10} = 7.589 \text{ m/s}$$

(c) Hydraulic jump has length dimension. This gives

$$\frac{H_p}{H_m} = \frac{L_p}{L_m} = 10 \rightarrow H_p = 50 \times 10 = 500 \text{ mm}$$

(d) We may determine scale ratio for power using the relation $P$ = force × distance/time and assuming that the same fluid is used for prototype and the model.

$$\text{Force} = \text{mass} \times \text{acceleration} = \rho L^3 \times (U/T) = \rho L^2 (L/T)U = \rho L^2 U^2$$

$$F_r = \frac{F_p}{F_m} = \frac{(\rho L^2 U^2)_p}{(\rho L^2 U^2)_m} = L_r^3$$

and $$P_r = \frac{P_p}{P_m} = \frac{(FL/T)_p}{(FL/T)_m} = \frac{(\rho L^2 U^3)_p}{(\rho L^2 U^3)_m} = L_r^{3.5}.$$

This gives: Power dissipated in the prototype = 3.5 × (10)$^{3.5}$ = 11068 W.

**Example 8.5.2** A 7.2 m high and 15 m long spillway discharges 94 m³/s discharge under a head of 2.0 m. If a 1:9 scale model is to be made, determine the model dimensions, head over spillway model and the model discharge. If the model experiences a force of 7.5 kN, determine force on the prototype. (AMIE, Summer, 1991)

**Solution** Given $L_r$ = 9

∴ Dimensions of the model = 7.2/9 and 15/9 = 0.8 and 1.67 m, respectively.

The head over the spillway model is 2/9 = 0.222 m

Also, we have learnt in Example 8.5.1 that the flow in spillway is controlled by Froude number for which the discharge scale ratio is $L_r^{2.5}$.

This gives discharge for the model = 94/(9)$^{2.5}$ = 0.387 m³/s. Finally

$$F_r = \frac{F_p}{F_m} = L_r^3 \rightarrow F_p = 7.5 \times 9^3 = 5467.5 \text{ kN}$$

[1] This points out that the scale ratio for other quantities can be expressed in terms of the scale ratio for the length using the similarity law.

**Example 8.5.3** The ratio of a submarine and its model is 30:1. The speed of submarine is 10 m/s, the model is to be tested in a wind tunnel. Find the speed of air in wind tunnel. Also determine the ratio of drag between the model and the prototype. Take the value of kinematic viscosity for sea water and air as 0.012 and 0.016 stokes respectively. The density for sea water and air is given as 1030 and 1.24 kg/m$^3$ respectively.

**Solution** The Reynolds number determines resistance to the motion of a submarine when submerged fully under water. This requires

$$\left(\frac{UL}{\nu}\right)_m = \left(\frac{UL}{\nu}\right)_p$$

$$\rightarrow \quad U_m = \left(\frac{L_p}{L_m}\right) \times \left(\frac{\nu_p}{\nu_m}\right) \times U_p = 30 \times \frac{0.012}{0.016} \times 10 = 400 \text{ m/s}$$

$$\frac{R_p}{R_m} = \frac{(L^2 \times \rho U^2)_p}{(L^2 \times \rho U^2)_m} = (30)^2 \times \frac{1030}{1.24} \times \left(\frac{10}{400}\right)^2 = 467.22$$

Symbol $R$ is used for drag on the submarine.

**Example 8.5.4** The pressure drop in an aeroplane model of size 1/10 of its prototype is 80 N/cm$^2$. The model is tested in water. Find the corresponding pressure drop in the prototype. Take density of air = 1.4 kg/m$^3$, viscosity of water = 0.01 poise and that of air 0.00018 poise.

**Solution** This problem involves pressure and viscous forces. Therefore, for model testing, the Reynolds and Euler numbers must be the same for the flow around the model and the prototype. This gives:

$$\left(\frac{\rho UL}{\mu}\right)_m = \left(\frac{\rho UL}{\mu}\right)_p$$

$$\rightarrow \quad \frac{U_m}{U_p} = \left(\frac{L_p}{L_m}\right) \times \left(\frac{\mu_m}{\mu_p}\right)\left(\frac{\rho_p}{\rho_m}\right) = 40 \times \frac{0.01}{0.00018} \times \frac{1.24}{1000} = 2.755$$

$$U_m^2 / (p_m/\rho_m) = U_p^2 / (p_p/\rho_p)$$

$$\rightarrow \quad \frac{p_m}{p_p} = \left(\frac{U_m}{U_p}\right)^2 \times \frac{\rho_p}{\rho_m} = (2.755)^2 \times \frac{1000}{1.24}$$

or, $$p_p = 0.01306 \text{ N/cm}^2$$

**Example 8.5.5** A 1/20 model of a naval ship having a submerged surface of 5 m$^2$ and length 8 m has a total drag of 20 N when towed through water at a velocity of 1.5 m/s. Calculate the total drag on the prototype when moving at the corresponding speed. Use the relation $F_f = 1/2\, C_f \rho A U^2$ for calculating the skin resistance. The value of $C_f = 0.0735/Re^{1/5}$. Take kinematic viscosity of water as 0.01 stoke and density as 1000 kg/m$^3$.

**Solution** We discussed this problem before and found that the Froude and Reynolds numbers determine the total resistance. Unfortunately, it is difficult to satisfy equality of both these numbers for the two flows, as was pointed out earlier. In such situation, we satisfy one of the conditions and handle the other non-dimensional separately. Normally the dominant number is satisfied. In this case it is the Froude number. For this reason the skin coefficient is given in the problem.

The procedure is to determine the viscous force using the skin coefficient and the wave or surface drag using the similarity criterion. Before proceeding, since the kinematic viscosity is given in stoke and the conversion factor to change it to the SI units is $10^{-4}$.

Reynolds number for the model = $\dfrac{UL}{\nu} = \dfrac{1.5 \times 8}{0.01 \times 10^{-4}} = 1.2 \times 10^{7}$

Therefore, the skin coefficient for the model = $\dfrac{0.0735}{\left(1.2 \times 10^{7}\right)^{1/5}} = 2.82 \times 10^{-3}$

And the viscous force on the model = $\dfrac{1}{2} \times 2.82 \times 10^{-3} \times 1000 \times 5 \times (1.5)^{2} = 15.862$ N

→ Wave resistance on the model = 20 – 15.862 = 4.138 N

Using similarity consideration, the wave resistance on the prototype is calculated. Accordingly

$$\left(\frac{U}{\sqrt{Lg}}\right)_p = \left(\frac{U}{\sqrt{Lg}}\right)_m$$

$$\rightarrow \qquad U_p = \sqrt{\frac{L_p}{L_m}}\, U_m = \sqrt{20} \times 1.5 = 6.708 \text{ m/s}$$

$$\left(\frac{R_w}{\rho U^2 L^2}\right)_p = \left(\frac{R_w}{\rho U^2 L^2}\right)_m$$

$$\rightarrow \qquad R_{wp} = \frac{\rho_p}{\rho_m} \times \left(\frac{L_p}{L_m}\right)^2 \times \left(\frac{U_p}{U_m}\right)^2 \times R_{wm} = 33104 \text{ N}$$

From the given empirical relation, we can find out the viscous force on the prototype, since the Reynolds number can now be determined. Note that the length scale is 20 and therefore the area scale will be 20 × 20 and accordingly the length and area in the empirical relation would change.

Reynolds number for the model = $\dfrac{UL}{\nu} = \dfrac{6.708 \times 160}{0.01 \times 10^{-4}} = 1.073 \times 10^{9}$

Therefore the skin coefficient for the prototype = $\dfrac{0.0735}{\left(1.073 \times 10^{9}\right)^{1/5}} = 1.1486 \times 10^{-3}$

And the viscous force on the prototype $R_{fp} =$

$$\frac{1}{2} \times 1.1486 \times 10^{-3} \times 1000 \times (5 \times 400) \times (6.708)^2 = 51683.8 \text{ N}$$

$\therefore$ Total drag on the prototype $= R_{wp} + R_{fp} = 84787.8$ N

**Example 8.5.6** An aircraft is to fly at a height of 9 km (where the temperature and pressure are –45°C and 30.2 kPa respectively) at 400 m/s. A 1/20 scale model is tested on a pressurised wind tunnel in which the air is at 15°C. For complete dynamic similarity, what pressure and velocity should be used in the wind tunnel? Given for air: $\mu = T^{3/2}/(T + 117)$, where $T$ is in Kelvin.

**Solution** Similarity of the flow around the model and the prototype demands equality of the Mach and Reynolds number, since the compressibility and viscous forces affect the flow pattern. Moreover, we are given in the problem variation of viscosity with temperature and therefore it is known for the two situations. Using the similarity laws, we get

$$\left(\frac{U}{a}\right)_m = \left(\frac{U}{a}\right)_p$$

$$\rightarrow \qquad U_m = \frac{a_m}{a_p} \times U_p = \sqrt{\frac{\gamma R T_m}{\gamma R T_p}} \times U_p = 450 \text{ m/s}$$

$$\left(\frac{\rho UL}{\mu}\right)_m = \left(\frac{\rho UL}{\mu}\right)_p \text{ or, } \left(\frac{p}{RT}\frac{UL}{\mu}\right)_m = \left(\frac{p}{RT}\frac{UL}{\mu}\right)_p$$

$$\rightarrow \qquad \frac{p_m}{p_p} = \frac{U_p}{U_m} \times \frac{L_p}{L_m} \times \frac{T_m}{T_p} \times \frac{\mu_m}{\mu_p} = \frac{U_p}{U_m} \times \frac{L_p}{L_m} \times \left(\frac{T_m}{T_p}\right)^{5/2} \times \left[\frac{T_p + 117}{T_m + 117}\right]$$

and
$$p_m = \frac{p_m}{p_p} \times p_p = \frac{400}{450} \times 20 \times \left(\frac{288}{228}\right)^{5/2} \frac{288 + 117}{228 + 117} \times 30.2 = 821 \text{ kPa}$$

## PROBLEMS

### Review Questions

1. Drive the SI units of (a) Chezy's coefficient, (b) dynamic viscosity, and (c) specific speed of a centrifugal pump as used in practice.
2. Derive the dimensions of (a) vorticity, (b) velocity potential function, and (c) circulation in *MLT* system.
3. Verify the dimensional homogeneity of the following shear stress formula.

(a) $\tau_0 = \dfrac{0.664}{\sqrt{R_{ex}}}\dfrac{\rho U^2}{2}$ (b) $\dfrac{dy}{dx} = \dfrac{S_0 - S_i}{1 - \dfrac{Q^2 T}{gA^3}}$ (used in gradually varying flow)

4. What dimensionless parameters can be formed from ($Q$, $L$, $H$, $\mu$, $B$, $\rho$, $g$)? The discharge $Q$ is dependent variable.
5. Identify the $\pi$ terms in the following problem of shear stress in sheet flow: $\tau = f(y, V, \rho, g, \sigma)$.
6. In group method, the dependent variable should not be selected as a repeating variable. Is this true?
7. List out the objectives of model studies.
8. The pattern of stream lines in two flow systems is similar. What is the nature of similarity between the two systems?
9. What is the nature of similarity that is to be obtained in dynamic similarity between two systems?
10. It is virtually impossible to achieve two dynamic similarities in the design of the models. Comment.
11. What is the height of a hydraulic jump?
12. The wave resistance encountered by a ship is 81 kN. What is the force on a 1:30 reduced scale model? Assume that the fluid in both the cases is same.
13. A 3 m diameter penstock supplies 6 m$^3$/s of water to a hydraulic turbine. For dynamic similarity, what should be the rate of flow through a 10 cm diameter pipe?
14. Derive the scale ratio for (a) time, (b) discharge, and (c) energy using the Froude law of similarity.
15. Write a brief note on the selection of scale in hydraulic modelling.
16. Why is it necessary to use distorted models? List out the merits and limitations of distorted models.
17. Explain the scale effects in the hydraulic model analysis.
18. List out practical problems in which each of the following similitude criterions is applicable: Froude, Reynolds, Weber, and Mach.

## Problems for Practice

1. A model of rectangular pier 1.5 m wide and 4.5 m long in the river is built to a scale of 1/25. The average depth of water in the river is 3 m. The model was tested in a laboratory where the velocity of the flow was maintained constant at 0.65 m/s. It was observed that the force acting on the model was 3.2 N and the height of the standing wave was 3.5 cm. Determine for the prototype: (i) the corresponding speed, (ii) the force acting, (iii) the height of the standing wave at the nose, and (iv) the coefficient of drag resistance. [3.25 m/s, 61.25 kN, 87.4 cm, 2.57]
2. Assuming that the viscous force $F$ exerted by a fluid on a sphere of diameter $D$ depends on the viscosity $\mu$, mass density of the fluid $\rho$, and the velocity of the sphere $v$, obtain expression for the viscous force.

3. The resisting torque against the motion of a shaft in a lubricated bearing depends on the viscosity $\mu$, the rotational speed $N$, the diameter $D$ and the bearing intensity $P$. Show that $T = \mu N D^3 f\left(\frac{p}{\mu N}\right)$
4. The pressure $\Delta p$ generated by a pump is a function of the impeller dia $D$, the rotational speed $N$, the fluid density $\rho$, viscosity $\mu$, and the rate of discharge $Q$. Show that $\Delta p = \rho N^2 D^2 f\left(\frac{Q}{ND^3}, \frac{\rho ND^2}{\mu}\right)$
5. The resisting force $R$ of a supersonic plane during flight can be considered to be dependent upon the length of the aircraft $l$, velocity $V$, air viscosity $\mu$, air density $\rho$, and the bulk modulus of air $K$. Express the functional relationship between the variables and the resisting force.
6. A partially submerges body is towed in water. The resistance $R$ to its motion depends on the density $\rho$, the viscosity $\mu$ of water, length $L$ of the body, velocity $V$ of the body and the acceleration due to gravity $g$. Show that the resistance to the motion can be expressed in the form

   $R = \rho L^2 V^2 \phi\left(\frac{\mu}{\rho LV}, \frac{Lg}{V^2}\right)$ AMIE, Summer 78
7. The efficiency of a fan depends on density $\rho$, dynamic viscosity $\mu$ of the fluid, angular velocity $\omega$, diameter $D$ of the rotor, and the discharge $Q$. Express $\eta$ in terms of dimensionless parameters. AMIE, Winter 76
8. Using Buckingham $\pi$ theorem, show that the velocity through a circular orifice is given by $V = \sqrt{2gH}\ \phi\left[\frac{D}{H}, \frac{\mu}{\rho VH}\right]$, where $H$ is head causing flow, $D$ is the diameter of the orifice, $\mu$ is the coefficient of viscosity, $\rho$ is the density and $g$ is the acceleration due to gravity. AMIE, Winter 77
9. The pressure difference $\Delta p$ in a pipe of diameter $D$ and length $L$ due to turbulent flow depends on the velocity $V$, viscosity $\mu$, density $\rho$ and roughness $k$. Using $\pi$ theorem, obtain the expression for $\Delta p$.
10. Derive on the basis of dimensional analysis suitable parameters to present the thrust developed by a propeller. Assume that the thrust $P$ depends upon the angular velocity $\omega$, speed of advance $V$, diameter $D$, dynamic viscosity $\mu$, mass density $\rho$, elasticity of the fluid medium which can be denoted by the speed of sound in the medium $C$. AMIE, Winter 81
11. The frictional torque $T$ of a disc of diameter $D$ rotating at a speed $N$ in a fluid of viscosity $\mu$ and density $\rho$ in a turbulent flow is given by $T = D^5 N^2 \rho \phi\left[\frac{\mu}{D^2 N \rho}\right]$. Prove this by the method of dimensions. AMIE, Summer 75
12. Using $\pi$ theorem, show that the discharge $Q$ consumed by an oil ring is given by

$$Q = ND^3\phi\left[\frac{\mu}{\rho Nd^2}, \frac{\sigma}{\rho N^2 d^3}, \frac{w}{\rho N^2 d}\right]$$

where $d$ is the internal diameter of the ring, $N$ is the rotational speed, $\rho$ is the density, $\mu$ is the viscosity, $\sigma$ is the surface tension, and $w$ is he specific weight of oil.

13. Find non-dimensional factor for a weir.
14. In a Couette flow, the velocity $u$ at a point depends on its location $y$ (measured perpendicularly from one of the plates), the distance of separation $h$ between the plates, the relative velocity $U$ between the plates, the pressure gradient $dp/dx$ imposed on the flow and the viscosity of the fluid. Find a relation in dimensionless form to express $u$ in terms of the independent variables as described above.
15. In the study of vortex shedding phenomenon due to the presence of a bluff body in a flow through a duct the following parameters are found to be important: velocity of flow $V$, density of liquid $\rho$, coefficient of dynamic viscosity $\mu$, hydraulic diameter of the duct $D$, the width of the body $B$, and the frequency of vortex shedding $n$. Obtain the dimensionless parameters governing the phenomenon.
16. The capillary rise $h$ of a liquid of density $\rho$ and surface tension $\sigma$ in a tube of diameter $D$ depends on the contact angle $\theta$ and acceleration due to gravity $g$. Find the expression for $h$ in terms of dimensionless variables using Rayleigh's indicial and group methods.
17. A pipe of diameter 1.5 m is required to transport an oil of specific gravity. 0.9 and viscosity 0.03 poise at the rate of 3000 lps. Tests were conducted on a 15 cm diameter pipe using water at 20°C. Find the velocity and rate of flow in the model. Viscosity of water at 20°C = 0.01 poise. [5.091 m/s, 89.9 lps]
18. The ratio of lengths of a submarine and its model is 30:1. The speed of submarine (prototype) is 10 m/s. The model is to be tested in a wind tunnel. Find the speed of air in wind tunnel. Also determine the ratio of the drag between the model and its prototype. Take the value of kinematic viscosities for sea water and air as 0.012 stokes and 0.016 stokes respectively. The density for sea water and air is 1030 kg/m$^3$ and 1.24 kg/m$^3$ respectively. [400 m/s, 467.22]
19. A ship model of scale 1/50 is towed through sea water at a speed of 1 m/s. A force of 2 $N$ is required to tow the model. Determine the speed of ship and the propulsive force on the ship, if prototype is subjected to wave resistance only. [7.07 m/s, 250 kN]
20. In the model test of a spillway the discharge and velocity of flow over the model were 2 m$^3$/s and 1.5 m/s respectively. Calculate the velocity and discharge over the prototype, which is 36 times the model size. [9 m/s, 15552 m$^3$/s] AMIE Summer 80
21. In a geometrically similar model of a spillway, the discharge per meter length is 1/6 m$^3$/s. If the scale of the model is 1/36, find the discharge per meter run of the prototype. [36 m$^3$/s] AMIE Summer 79
22. A spillway model is to be built to a geometrically similar scale of 1/50 across a flume of 600 mm width. The prototype is 15 m high and maximum head on it is expected to be 1.5 m. (i) what height of model and what head on the model should be used? (ii) If the flow over the model at particular head is 12 lps, what flow per meter length of the prototype is expected? (iii) If the negative pressure in the model is 200 mm, what is the negative pressure in prototype? Is it practicable? [0.3 m, 0.03 m, 7071 m$^3$/s, –10.0 m, not practicable] AMIE Winter 77

23. In a 1/20 model of stilling basin, the height of the hydraulic jump in the model is observed to be 0.2 m. What is the height of the hydraulic jump in the prototype? If the energy dissipated in the model is 0.1 kW, what is the corresponding value for the prototype? [4 m, 3577.7 kW] AMIE Summer 77
24. A 1:64 model of an open channel in concrete which has Manning's $n = 0.014$ is constructed. Find the value of $n$ for the model. [0.007] AMIE Winter 87
25. The pressure drop in an aeroplane model of size 1/10 of its prototype is 80 N/cm$^2$. The model is tested in water. Find the corresponding pressure drop in the prototype. Take density of air = 1.24 kg/m$^3$. The viscosity of water is 0.01 poise and that of air 0.00018 poise. [0.01306 N/cm$^2$]
26. An agitator of diameter $D$ requires power $P$ to rotate at a constant speed $N$ in a liquid of density $\rho$ and viscosity $\mu$.
    (i) Show with the help of $\pi$ theorem that $P = \rho N^3 D^5 \phi\left[\rho N D^2 / \mu\right]$
    (ii) An agitator of 225 mm diameter rotating at 23 rev/s in water requires a driving torque of 1.1 Nm. Calculate the corresponding speed and the torque required to drive a similar agitator of 675 mm diameter rotating in air (Viscosities: air $1.86 \times 10^{-5}$ poise, water $1.01 \times 10^{-3}$ Pas. Densities: air 1.2 kg/m$^3$, water 1000 kg/m$^3$). [39.22 rps, 0.933 Nm]
27. A 1/25 mode of ship, when towed in water at a velocity of 2 m/s, experiences a resistance of 2 kg. The wetted area of the model is 1 m$^2$. Calculate for the prototype: (a) the corresponding speed and (b) the total drag, if the skin-friction coefficient for model is 0.004 and that for the prototype is 0.016. [10 m/s, 69483 kg]

CHAPTER

# 9 Rotamachines

OUTLINE

- Introduction
- Pump
- Axial Flow Pumps
- Hydraulic or Water Turbine
- Governing of Turbines

## 9.1 Introduction

Water that flows down to the sea from the mountains possess a huge amount of energy with it. Man had always been thinking of using this energy to produce electricity before consuming it for irrigation, drinking or in industries, etc. The flowing water produces a force on a body when placed in its passage. This force may be due to many reasons that we have studied in this course on fluid mechanics. We shall study in this chapter the force which is useful for production of power.

To harness this energy, a solution was thought of for storing water in high altitude areas and use it at the place of need by transporting it through open channels or pipes and impinging on a surface moving in a certain direction to get the work. Recall from mechanics that a force does a work when acting upon a body and moves it with some velocity. Thus we need a source of force and the displacement of the point of application of the force. In this manner both these requirements were met with. Remember that the water sources at different geometrical sites will vary in many ways. For example, the quantity of water stored, the height of the free surface of water, the area of catchment, etc. will not be same as these depend on the local conditions of rain and topography of the area. This has given a wide choice for us and requires different types of machines so that the water source may be harnessed with maximum efficiency throughout the year. In this chapter, we shall study the methods of conveying water through long distances and the precautions to be taken to avoid water hammer, which we discussed in Chapter 4 and the basic principles of transfer of energy from water to the electricity generator.

We also need to raise pressure of the liquid so that it can be used for different applications in the industry. For example, hydraulic press, cranes, etc. The pressure may be very high in such applications. Whether the pressure is to be raised or energy of water is to be used for power generation, there is energy transfer between the water and the system. The system may be reciprocating or rotary type. We shall cover both of these, including the applications of high pressure liquids, in the following chapters.

### 9.1.1 Impact of Jets

An easy way to convert kinetic energy of water to shaft work is to make it flow in a form of a jet of diameter $d$ and strike the moving surface that results in change in its momentum. We discussed a jet striking a flat surface placed in its path at some angle in Example 2.4.6. We observed that

1. The pressure is atmospheric everywhere and therefore the speed at the point of impinging and of leaving the surface by the jet is same if frictional loss is neglected; however, there is change in its direction.
2. The flow is divided in two streams such that $\dfrac{Q_2}{Q_3} = \dfrac{1 + \cos\theta}{1 - \cos\theta}$ where $Q_2$ and $Q_3$ are the volume flow rate and $\theta$ is the angle made by the plate with the direction of the jet, as shown in Fig. 9.1.1. This is obtained on the consideration that the frictional force between the surface and water is zero. When $\theta = 90°$ then the two volume flow rates are equal.
3. The normal force on the surface is given as $F_n = \rho QC \sin\theta$, where $Q$ is the volume flow rate through the jet such that $Q = Q_2 + Q_3 = \rho aC$, where $C$ is the velocity and $a$ is the area of cross section of the jet, $\rho$ is the density of the water striking the surface.
4. When the plate is moving with velocity $u$ then some of the mass will fill in the space created by the movement of the plate and will not strike the surface. Thus the force produced per unit of time will decrease. However, such systems can be brought to rest by fixing the reference axes with the moving plate. In that case the jet velocity is the relative velocity $(C - u)$ with respect to the plate. That is write $(C - u)$ for $C$ when the axes are also moving with the plate with its speed.

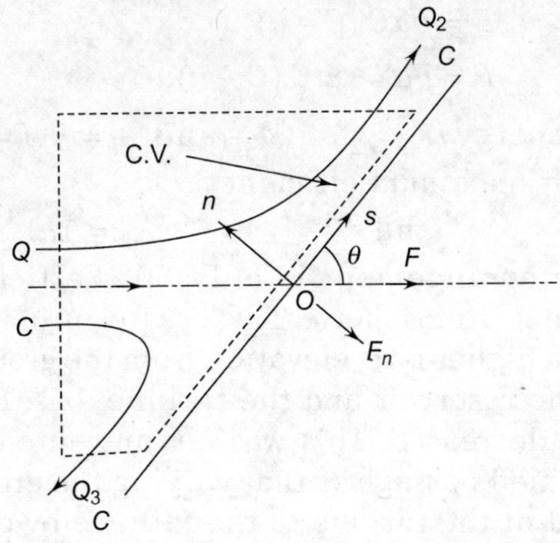


**Fig. 9.1.1** Jet impinging on an inclined plane

The force $F_n$ can be resolved in the direction of the plate motion and perpendicular to it. The component in the direction of $u$ will produce power $P$ given by, writing $(C - u)$ for $C$,

$$F = [\rho Q\,(C - u) \sin \theta] \sin \theta = \rho\, a\,(C - u)^2 \sin^2 \theta$$

$$P = Fu = \rho a\,(C - u)^2\, u \sin^2 \theta \tag{9.1.1}$$

$$I = \text{Input energy}/s = \tfrac{1}{2}\,\rho a C^3 \quad \text{and} \quad \eta = P/I = \frac{2u}{C}\left(1 - \frac{u}{C}\right)^2 \tag{9.1.2}$$

Efficiency of this energy conversion is function of $u$ for given $C$. Differentiating it with respect to $u$ and equating to zero gives the condition for maximum efficiency. We get

$$(C - u)\,(C - 3u) = 0 \rightarrow u = C/3 \quad \text{and} \quad \eta_{\max} = 8/27 \tag{9.1.3}$$

The other root of the equation $C = u$ is not acceptable because then water will not strike the plate and hence this corresponds to the condition of no work transfer or zero efficiency.

Such arrangement can at the most convert 1/3rd of the energy input into the shaft work—a very highly inefficient process. Therefore we have to think of an alternate method to increase the efficiency. The first point that comes to our mind is that whole of the mass of water is not striking the plate. Another problem with it is the implementation of such a scheme, as the plate cannot be allowed to travel in a direction continuously. This can be done for a short distance but not continuously. The solution lies in fixing number of plates over a drum that rotates about its axis so that one plate moves ahead and its position is taken by the new plate. In this way, the space created due to movement of the plate from its original position is minimised. The water trapped between the two plates travels and hits the leading plate before some part of it escapes striking it, which is small enough to the extent that it can be neglected in the calculation. So the whole water $\rho a C$ strikes the surface.

We may now apply the integral momentum theorem or what we call as the impulse momentum theorem. $(C - u)$ in Eq. 9.1.1 is now $C$ and $\theta = 90°$. Thus we shall get

$$F = \rho a C (C - u)$$

$$P = Fu = \rho a C (C - u)u \quad (9.1.4)$$

$$I = \text{Input energy}/s = \tfrac{1}{2} \rho a C^3 \quad \text{and} \quad \eta = P/I = 2u (C - u)/C^2 \quad (9.1.5)$$

Equation 9.1.3 gives for maximum efficiency

$$u = C/2 \quad \text{and} \quad \eta_{max} = 1/2 \quad (9.1.6)$$

We conclude that the arrangement is only 50% efficient. Moreover, pressure throughout the flow region is atmospheric. This creates problem in placing the turbine on the ground because then higher is its elevation from the ground lower is the difference of free water surface in the reservoir and the turbine axis. Thus the velocity of water issued from the nozzle is decreased. This was circumvented by flowing water over a curved surface in the confined passage so that varying the area of the flow suitably and then connecting the outlet of the turbine to the tailrace by diverging pipe can control the pressure at the turbine outlet.

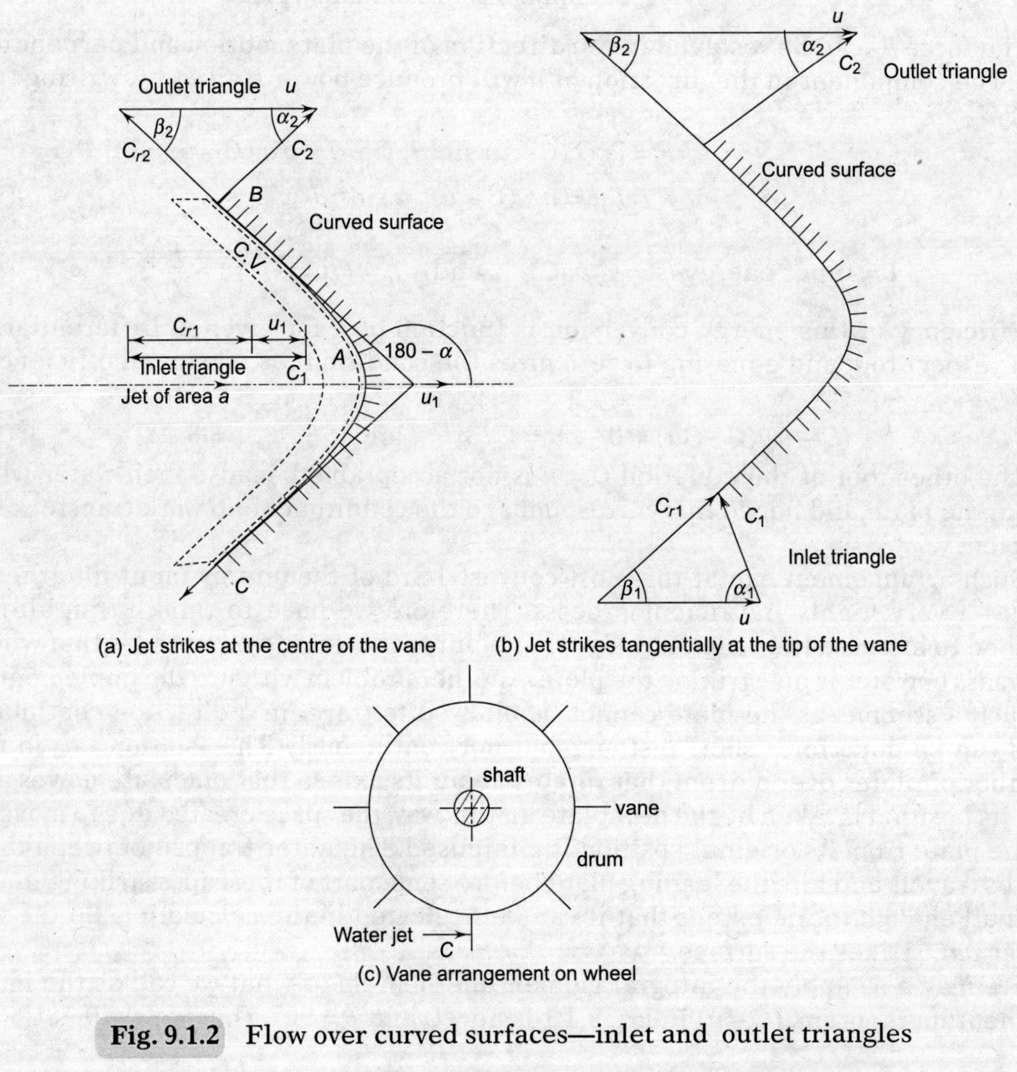


**Fig. 9.1.2** Flow over curved surfaces—inlet and outlet triangles

Figure 9.1.2 shows two ways of executing this solution. In (a) the jet, shown striking the vane at its centre $A$, glides over the surface and leaves the vanes at its edges $B$ or $C$ along the tangent to the vane whereas in (b) it enters from one end and comes out from the other end. The pressure is all atmospheric throughout. Hence the relative velocity of the water with respect to vane will be same at the inlet and outlet for smooth flow. At point $A$ the velocity of flow and the vane velocity is in the same direction but at the edge, the direction of water and vane velocity is different. The vanes are mounted on a wheel periphery, such as shown in Fig. 9.1.2(c). Water jet of area $a$ arrives at the vane with velocity $C_1$ along the direction of motion of the vane.

For analysis of such machines we may use a reference attached either to ground or to the moving blade. Three mutually perpendicular lines passing through the point under consideration on the moving blade are taken as reference axes. One of the reference axis is along the direction of rotation in tangential plane, second along the radial in meridonial plane and third along shaft axis in axial plane. We use following notations in our discussions:

| | | |
|---|---|---|
| $C$ = Absolute velocity | $C_r$ = Relative velocity of fluid desked may we write w.r.t. a point on blade | |
| $C_f$ = Flow or mass velocity | $C_w$ = Whirl or tangential velocity | $u$ = Blade velocity |
| $\alpha$ = Nozzle angle | $\beta$ = Blade angle | $m$ = Mass flow rate |
| Subscripts 1 and 2 | for inlet and outlet of flow | |

In this notation, $u$ is the blade velocity of the point $P$ where moving reference axes are fixed. Therefore, it is also the absolute velocity of moving reference. Then absolute velocity of fluid moving over the surface at the point $P$ can be written as vector sum of its relative velocity with respect to moving reference and absolute velocity of moving reference. That is, $C = C_r + u$. This equation represents a triangle, called *velocity triangle.*

When the plate and the jet move in the direction of the plate then the inlet triangle will be a straight line, as shown in Fig. 9.1.2(a). The jet is deffected by angle $(180 - \alpha)$ from its path when it glides over the curved surface and leaves it in the direction of $C_{r2}$ along the tangent to the surface at exit end, Fig. 9.1.2(a). We note that $\beta_2 = \alpha$, $u_1 = u_2 = u$ (say), and the pressure is everywhere atmospheric pressure, since the flow over the surface is open to atmosphere. The impinging water jet splits in two directions at the striking point. Since it strikes at right angle, half of the water goes on upper side and another half on the lower surface and leave the plate surface at two different sections with same velocity magnitude but in different directions. We may consider the control volume shown in Fig. 9.1.2(a). The momentum equation will be same as if whole of the water leaving the plate is assumed to leave at one section only. We thus get the following relations from the conservation of mass and momentum.

$$m = \rho a C$$

$$F = m\,[(C - u) - \{-(C - u)\cos\alpha\}] = m\,(C - u)\,(1 + \cos\alpha)$$

$$P = Fu = m\,(C - u)\,(1 + \cos\alpha)\,u \tag{9.1.7}$$

Here $C_1$ is $C$ and $F$ is the force exerted by the liquid on the surface. This force is along the vane motion, since the momentum is considered in this direction in the above equation. This gives the amount of work done. We mentioned earlier that mass $m$ of water striking the plate will decrease as the plate velocity increases and thus it is a function of $u$. However, if we bring in a new plate as the old one moves some distance than $m$ can be treated independent of $u$. Moreover, the input energy to the blade is kinetic energy of striking fluid, which is 1/2 $mc^2$. Now the efficiency can be determined and from it the condition for maximum efficiency is obtained. The result is

$$u = C/2 \quad \text{and} \quad \eta_{max} = (1 + \cos\alpha)/2 \tag{9.1.8}$$

This shows that efficiency can be 100% if $\alpha = 0^0$. That is the jet is deflected back. This requires a semicircular vane. For practical reasons, the angle of deflection is kept from 160 to 170°.

In the same way, we may analyse the flow in case (b). In this case, we have to ensure that the relative velocity vector is along the tangent to the blade for smooth entry of the liquid. This is essential as otherwise some energy will be lost in the impact made by the fluid jet with blade and blade may also be set to vibrate. This forms an important guiding principle for rotamachines. The results will be the same if the pressure is atmospheric everywhere and we take $u_1 = u_2 = u$. This assumption is satisfied because the height of the vane is much less as compared to the radius of the drum over which these are mounted. Moreover, it is also possible that water moves in horizontal plane after striking the surface so that the two points of the blade surface—one where the water strikes and another where it leaves—are at the same distance from the axis of rotation. In that case $u_1 = u_2$, since $u = \pi DN/60$.

However, the situation is different if the pressure is also different; this is discussed in next section. The machines so developed are called *rotamachines*.

### 9.1.2 Rotamachines

Reciprocating systems were used as thermal systems for power production or pumping fluids under pressure to different destinations. However, a reciprocating system has inherent disadvantage that the piston is heavy. It accelerates from rest and then decelerates to come again to rest. This results in large variation of inertia force. Such system therefore cannot be operated at high rpm. Further, small cylinder is used as far as possible to keep weight of cylinder and piston low for dynamic and commercial reasons. This results in low swept volume. Thus, both low rpm and low cylinder volume reduces capacity of the reciprocating system. Moreover, flow rate also varies during discharge stroke.

A rotating system overcomes these limitations and is therefore preferred over reciprocating system. These machines are dynamically balanced and produce less noise. *We use word rotamachines for rotating machines in which fluid flows over the rotating element formed from curved surfaces called blades or vanes.* Rotamachine requires power to raise pressure of liquid. Work transfer from rotor to fluid takes place that increases fluid pressure. Such a device is called a pump. A part of input work is also transformed into kinetic energy and therefore fluid comes out from rotor at high velocity. Diffuser and/or volute casing are used to transform[1] part of kinetic energy of outgoing fluid to

[1] Two terms, transfer and transformation, are used while studying rotamachine. The former refers to the energy exchange between rotor and the fluid whereas the latter means change of form of energy such as pressure to kinetic energy.

pressure energy since the purpose of a pump is to raise pressure and not velocity. It is also possible to transfer energy of the fluid to rotor connected to a machine so that it gives driving power. A rotamachine developing power is called a *turbine.*

In rotamachines, fluid may flow along radial, axial or in between the two directions. Accordingly, the machine is called radial, axial or mixed flow rotamachine. In radial machines fluid flows in radial direction at right angles to shaft axis; flow may be from outer periphery to the centre or vice versa. In both the cases, centrifugal force plays a vital role the in transfer of energy between rotor and fluid, as there is finite change of radius at entry and exit of fluid. For this reason, pump and compressor are outward flow type machine because centrifugal force helps in building pressure. For the same reason radially inward turbines are preferred. Rotor of radial machine is made integral part of the hub and shaft, which makes it mechanically strong and allows it to run at high peripheral velocity. The two factors, namely, higher peripheral velocity and added effect of centrifugal force in building pressure, make such machine very useful for applications, which require high delivery pressure. However, such machine is not suitable for handling large volume flow rate, as area of flow is restricted. This can be seen from Fig. 9.1.3 where area of flow at eye is $\pi\left(d_t^2 - d_r^2\right)/4$, where $d_t$ and $d_r$ are the diameter of the tip and root of the eye. Therefore energy transfer is also restricted, as it is proportional to mass flow rate. Hence, radial turbine is not suitable for applications requiring large volume flow rate.

In axial flow type machines, flow is along the shaft axis and many of the shortcomings of a radial machine are overcome. Rotating blades are fixed at equal distance at given section on a drum (known as rotor) along its periphery forming space in between the blades through which the fluid glides smoothly. Since the fluid must enter the passage along the tangent to the rotating blades at the inlet, the fluid has to be directed accordingly. This is done by placing row of fixed guide vanes before the row of rotating vanes along the rotor periphery. The fixed vane is called stator. The set of stator and rotor is known as stage. In multi-stage machine, number of such set is repeated along the shaft axis; the distance between rows is kept as small as possible. Figure 9.1.3(c) shows schematic view of an axial flow turbine for a compressible fluid.

Figure 9.1.3(b) shows axial flow machine for incompressible fluids. In such machines only one row of vanes is placed on the rotor whereas guide vanes are placed in a different way. Normally (not shown in the figure), the water enters the guide vane at right angle to the shaft axis and is turned along the axis by the guide vane so that water enters the moving blade in the axial direction. In these machines, distance of the point on the moving vane changes. Therefore, we take the mid points *A*, *B*, *C* etc. on each blade, which are at same distance from the shaft axis. The flow over the blades is along the axis. Turbine and pump are normally single stage. However, multistage axial flow pumps are used also.

The hydraulic systems are simple because of the fact that liquid is incompressible and its density remains constant. Since the density is already known, it is not an unknown variable. But in compressible fluids, it is a function of the state of gas and hence changes from section to section during the flow and need to be determined. Further, density of water is about 900 times more than that of air and therefore amount of volume of fluid to be handled is much less requiring less area of flow and comparatively much lower velocities. This reduces weight, size and maintenance of hydraulic equipment.

Further, pressure and density are related to each other for incompressible fluids by the relation $p = w\,H$, where $H$ is called pressure head and $w$ = specific weight = $\rho g$. It is common practice to work in terms of pressure head rather than pressure in hydraulic pumps and turbines.

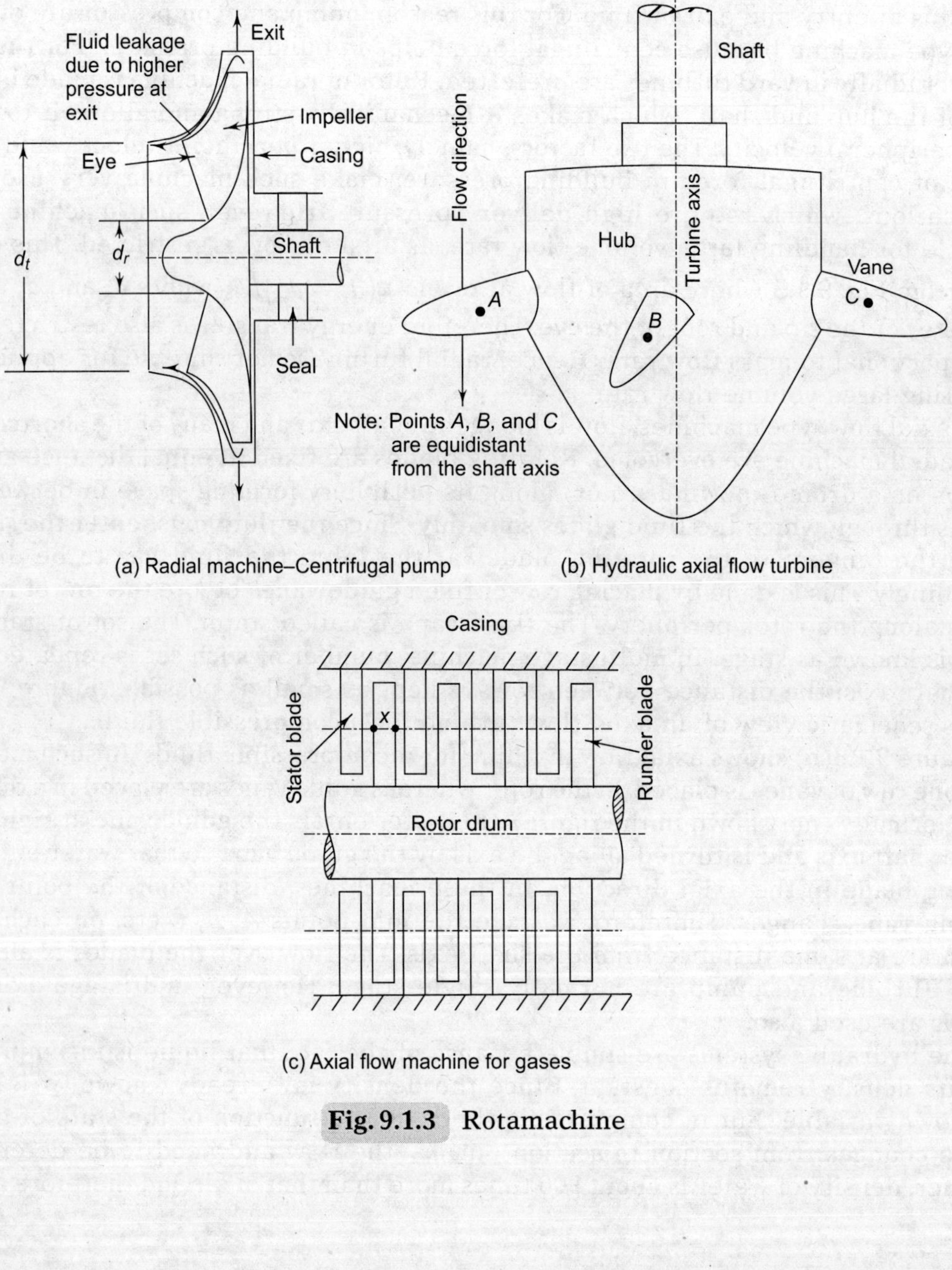


(a) Radial machine–Centrifugal pump (b) Hydraulic axial flow turbine

(c) Axial flow machine for gases

**Fig. 9.1.3** Rotamachine

### 9.1.2.1 Work Exchange in Rotamachines

Radial machines have a number of equi-spaced blades fixed along the periphery of a hub such that water flows through the space formed by these vanes; it is called rotor or runner. The vanes are surrounded from sides by shrouds, which prevents leakage. Figure 9.1.4(a) shows one blade of a rotor. Fluid enters at section 1 and leaves at 2, as shown; circles passing through these points are pitch circles with centre *O*. Water glides over the vanes and produces a force that rotates the blades. The flow is same, at least assumed so, through the spaces formed between any two successive vanes on the rotor.

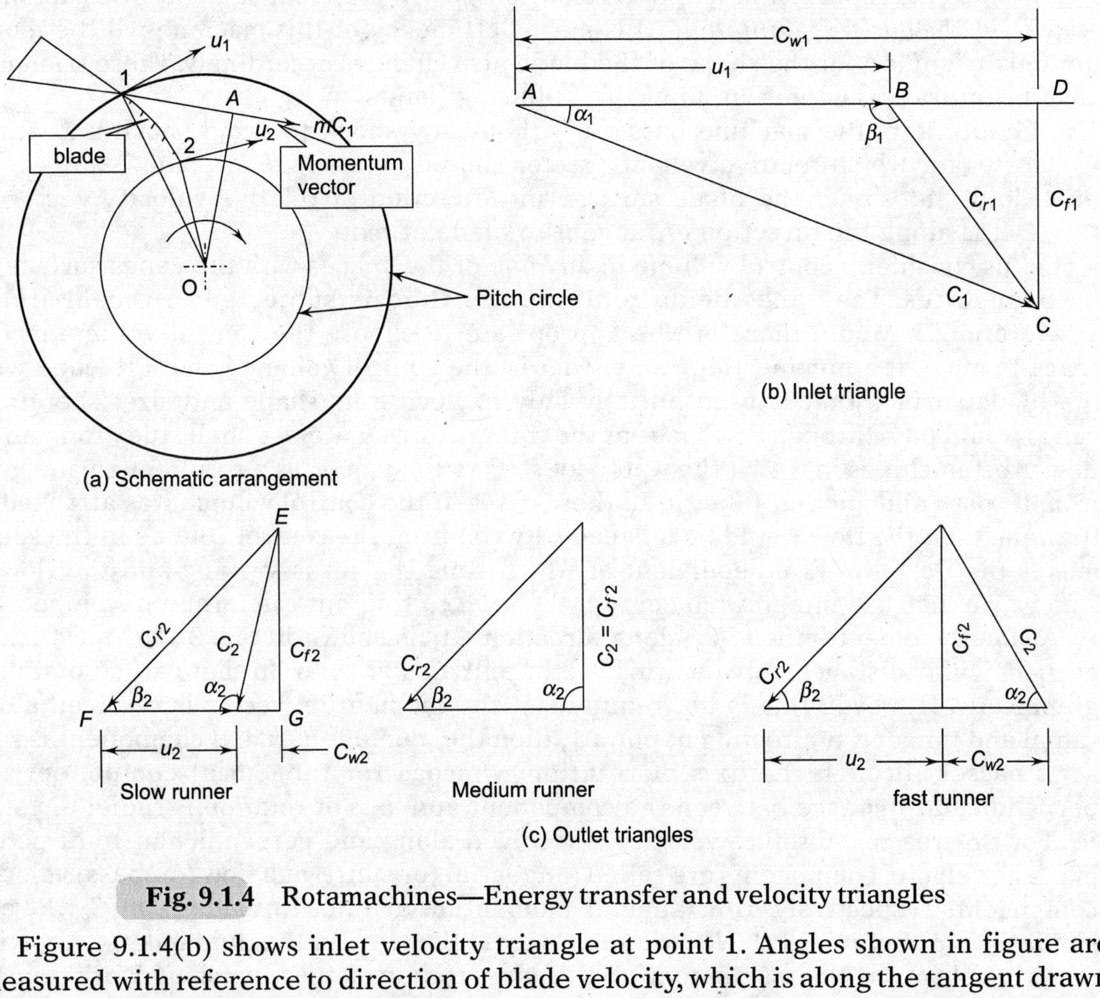


**Fig. 9.1.4** Rotamachines—Energy transfer and velocity triangles

Figure 9.1.4(b) shows inlet velocity triangle at point 1. Angles shown in figure are measured with reference to direction of blade velocity, which is along the tangent drawn at the pitch circle passing though the point of consideration, for example point 1 for inlet triangle. Fluid jet comes from the nozzle and hence angle made by this velocity vector with $u$ is called *nozzle angle*, $\alpha$. As per convention used to represent angle between two vectors, this angle is included between positive directions of two vectors. However, the angle made by relative velocity vector is measured with –ve direction of $u$ instead of its + ve direction because then it becomes interior angle of the triangle. Some books have taken exterior angles of triangle for it but both are in use. It is represented by symbol $\beta$ and called *blade angle* since for smooth flow fluid must enter along the tangent to the blade.

Figure 9.1.4(c) shows the outlet velocity triangle. Three outlet velocity triangles have been shown as possible triangles depending on the type of rotamachine. We observe that blade velocity at the exit is increasing from left to right in these triangles and accordingly the rotor is named as slow, medium and fast runner. Thus, shape of the triangle depends on the rotational speed of the rotor. *However, medium runner is mostly used for the reason that for other given conditions, when blade outlet angle is 90° absolute velocity at the outlet is minimum*. This means efficiency of this machine will be more. Not only triangles but the shape of the blade also changes accordingly. These triangles have been drawn keeping in mind the following points:

i. Fluid enters the machine passage without any shock to avoid shock and entry losses for which relative velocity vector should be along $\beta_1$ at point 1, and
ii. Fluid glides over the blade surface and direction of relative velocity vector at exit is along the direction of tangent to blade at exit.

Let us consider a control volume in any one of the spaces with the vanes surface as its boundaries. The vanes being rotating, the fluid is subjected to a centrifugal acceleration; it would therefore be appropriate to choose the control volume in the space to make it a non-inertial frame. That is, the control volume is not attached with the blade surface that is used initially only to decide its shape and size. Therefore, vanes would pass through this space as the rotor rotates. Later we shall study tangential flow type machines in which the water jet strikes the vanes, as a result, one vane goes from its path and another takes it. In these cases, if the control volume was attached to the vane then the flow would be unsteady. By choosing the control volume in the space makes the flow always independent of which vane the jet is striking and the flow is steady. We may use integral momentum theorem to find out the torque developed.

At inlet, momentum is $m\ C_1$ along direction $C1$, as shown in Fig. 9.1.4(a). $OA$ is the perpendicular distance between this vector and axis of rotation that can be obtained graphically. However, this is not required. If the momentum vector is resolved along radial and tangential direction at point 1, then the moment of radial component is zero as it passes through the axis of rotation whereas for tangential component the perpendicular distance between the component and axis of rotation is radial distance $R_1$. For this reason, absolute velocity is resolved along and perpendicular to direction of $u$. The velocity components are called tangential (or whirl) and flow (or mass) velocity components respectively. Thus angular momentum at inlet to blade is $m\ C_{w1}\ R_1$ and similarly at outlet it is $m\ C_{w2}\ R_2$, where $m$ is mass flow rate of the fluid in kg/s. Applying the principle of angular momentum[2] to C.V. gives torque on the rotor shaft. Work transfer to or from fluid is

$$T = m(C_{w1}R_1 - C_{w2}R_2) \quad \text{and}$$

$$\text{Work transfer/s} = T \times \omega = m\ (C_{w1}u_1 - C_{w2}u_2) \tag{9.1.9}$$

[2] Angular momentum is the moment of momentum about the axis of rotation.

Here $\omega$ is angular speed of shaft in radian/s. It is seen that the tangential component of absolute velocity is responsible for transfer of energy. In case of turbine, it causes rotation of the rotor whence the name whirl velocity is assigned to it. However, the radial component of velocity will produce thrust, which is equal to the rate of change of linear momentum in radial direction. Momentum at the inlet and exit are $m\ C_{f1}$ and $m\ C_{f2}$, respectively. Therefore thrust $F$ on the support for the shaft is

$$F = m\ (C_{f1} - C_{f2}) \tag{9.1.10}$$

In axial machines, radial distance of a point on the blade changes along its height. Therefore inlet and outlet velocity triangles are drawn at mid point of the blade height (marked $x$, in Fig. 9.1.3(c)) for the calculation of work transfer between rotor and fluid assuming uniform flow conditions over the blade height. These two points are at the same distance $R$ from the axis of rotation. Therefore $u_1 = u_2 = u$ (say). Axial velocity of flow is usually taken same for all the stages of the machine so that there is no axial force on the rotor otherwise a thrust bearing has to be provided to balance this force.

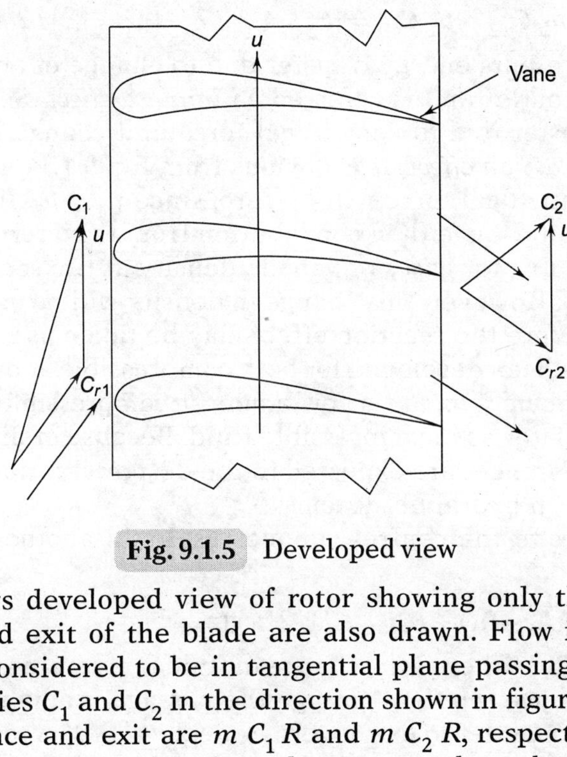


**Fig. 9.1.5** Developed view

Figure 9.1.5 shows developed view of rotor showing only three blades[3]; velocity triangles at inlet and exit of the blade are also drawn. Flow is assumed to be two-dimensional and is considered to be in tangential plane passing through mid point '$x$' with absolute velocities $C_1$ and $C_2$ in the direction shown in figure. In this case, angular momentum at entrance and exit are $m\ C_1\ R$ and $m\ C_2\ R$, respectively. It is to be noted that component of $C$ in tangential plane do not pass through shaft axis about which angular momentum is obtained. Therefore both the tangential and axial (or flow) component will contribute towards angular momentum. If there is no change in axial velocity of flow, as usually the case is, then there will be no change in angular momentum

[3] Area of the flow in rotor blade row will increase, decrease, or remain the same depending on the type of machine and is discussed at appropriate place. While drawing Fig. 9.1.3, no consideration is given for it.

of this component from inlet to exit and therefore its contribution to total change in angular momentum of fluid is zero. Even if the axial velocities change from the blade inlet to its outlet section, the angular momentum of the axial component of velocity will be in plane normal to the rotational plane and hence it would not contribute to power transfer; however it would effect thrust on bearings. Therefore,

$$T = m\,(C_{w1} - C_{w2})R$$

$$\text{and work transfer/s} = m\,(C_{w1} - C_{w2})u \qquad (9.1.11)$$

Equation 9.1.11 is same as Eq. 9.1.1, since $u_1 = u_2$. From the inlet and outlet velocity triangles for rotamachine, Fig. 9.1.4(b), we can write using law of triangles

$$C_{r1}^2 = C_1^2 + u_1^2 - 2u_1C_1 \cos\alpha_1 = C_1^2 + u_1^2 - 2u_1C_{w1}, \text{ since } C_1 \cos\alpha_1 = C_{w1}$$

or, $$u_1C_{w1} = \left(C_1^2 + u_1^2 - C_{r1}^2\right)/2$$

Similarly, $$u_2C_{w2} = \left(C_2^2 + u_2^2 - C_{r2}^2\right)/2$$

$$\therefore \quad w = u_2\,C_{w2} - u_1\,C_{w1} = \left(C_2^2 - C_1^2\right)/2 + \left(u_2^2 - u_1^2\right)/2 + \left(C_{r1}^2 - C_{r2}^2\right)/2 \qquad (9.1.12)$$

The first term represents energy transfer due to change of absolute kinetic energy of fluid from inlet to outlet and is referred as *impulse effect*. Second and third terms represent energy transfer due to centrifugal force and change in relative velocity of fluid; later is called *reaction effect*. It is obvious from Eq. 9.1.11 that for axial machine, contribution from centrifugal force will be zero, since $u_1 = u_2$. If area of flow through blades passage does not change then contribution from third term is zero for liquids as $Cr_2$ is exactly equal to $Cr_1$; for gases or vapour, density at two sections will be different and therefore $C_{r2} \approx C_{r1}$. However, this change in density will be negligibly different in a single stage and therefore the reaction effect may be taken as zero in such case too.

Rotamachines have been developed for both compressible and incompressible fluids. Air, carbon dioxide, steam, etc. are a few amongst compressible fluids used in gas or vapour turbines and water as incompressible fluid. Because of difference in the nature of the fluid, many differences are expected in the design; it is not covered in this book, since our emphasis is on hydraulic systems.

Application of *sfee* to the control volume results in another expression for work transfer.

$$w = (h_2 - h_1) + \left(C_2^2 - C_1^2\right)/2 + g(z_2 - z_1) \qquad (9.1.13)$$

*For incompressible fluid*, there is no change in fluid temperature and hence in internal energy of the liquid, since the flow through the passage is adiabatic. Normally change in potential energy from inlet to exit is zero or neglected. Moreover, $v_2 = v_1 = 1/\rho$. Thus work transfer results in change in pressure and velocity in incompressible fluids. But in case of compressible fluids, enthalpy is function of temperature, which changes along the flow. So change in enthalpy is equally responsible for work exchange.

Though both the rotary compressor and the turbine operate on the same principle, there is a difference in their operation. In a turbine, fluid velocity is increased by decreasing its pressure in a nozzle. Fluid then enters the passage formed by rotor vanes through which relative velocity may remain constant or change depending on the type of turbine in use. This device transfers various energies of the moving fluid to rotor as shaft work. But in a centrifugal pump (or compressor) both pressure and velocity increase at exit of rotor and work is transferred to the fluid from the pump rotor. High velocity at the exit is converted into pressure by passing fluid through diverging passage.

In reaction stage, the stator, unlike impulse stage, also contributes towards change in pressure and velocity. This fact is expressed by the term *degree of reaction* defined as ratio of contribution of rotating elements to that of stage in change of prime quantity. This prime quantity may be change in velocity or pressure depending on whether the machine is hydraulic turbine or pump. In gas and steam turbine or air compressors, change in enthalpy is taken as prime quantity since it is easy to deal with enthalpy with such fluids. An insight into the first $T\,ds$ equation $T\,ds = dh - v\,dp$ for a reversible process ($ds = 0$) reveals that change in enthalpy is same as change in pressure provided density is constant. Whereas this is exactly true for water, it is approximately true for steam or air for change in density of compressible fluids is not large in a stage. So it is immaterial which of them is taken as prime quantity in the definition of degree of reaction. We write the following equation for degree of reaction

$$A = \frac{\text{Static enthalpy change in rotor}}{\text{Static enthalpy change in stage}} \tag{9.1.14}$$

### 9.1.3 Similarity Relations

Having designed a machine, we need to study its performance under design and off-design conditions before a green signal is given to manufacture it. A model is prepared so that experiments on it may be performed in laboratory and performance of actual machine be predicted from the laboratory tests. Correct prediction of results requires careful study of the variables, which affect the performance. We would simultaneously like to reduce number of these variables so that minimum number of experiments are performed and the results are presented on a graph to show meaningful trend. We know that if there is only one independent variable then a graph can easily be plotted on a two dimensional graph sheet. If there are two independent variables then a parametric family keeping one independent variable constant can be drawn. However, if the number of independent variables is more, then it increases our work to plot number of graphs for drawing any meaningful conclusion. We discussed the similitude in Chapter 5 and you should read it first.

Coming to hydraulic rotamachines, power and efficiency are dependent variables, which are found to depend on density $\rho$ of fluid, volume flow rate $Q$, rpm $N$ and diameter $D$ of rotor, acceleration due to gravity $g$, and head $H$[4]. In fact, $gH$ instead of $g$ and $H$ affect the performance and $gH$, known as *specific energy*, is taken as one variable. In pump $gH$ also becomes dependent quantity since purpose of a compressor is to raise pressure. However, it does not make much difference since the non-dimensional numbers may be grouped in any way to give another non-dimensional number, which

[4] Head has different meaning in turbine and compressor and shall be discussed at the appropriate place later.

we shall consider later. We may write a functional relation among these variables as follows:

$$\eta, P = f(\rho, Q, N, D, gH) \tag{9.1.15}$$

There are six variables in this equation. By Buckingham $\pi$ theorem, there exist three non-dimensional numbers among six dimensional variables, namely,

$$\pi_1 = \frac{N\sqrt{P}}{\sqrt{\rho}(gH)^{5/4}}, \quad \pi_2 = \frac{ND}{\sqrt{gH}}, \quad \text{and } \pi_3 = \frac{Q}{ND^3} \tag{9.1.16}$$

which can be determined using non-dimensional techniques. $\pi_1$ may be combined with $\pi_2$ as follows to give

$$\pi_1^2 = \frac{N^2 P}{\rho} \frac{\pi_2^5}{(ND)^5} = \frac{P}{\rho N^3 D^5}$$

$\pi_1{}^2$ is also used as non-dimensional number because it does not involve square root though the most commonly used form is $\pi_1$. The functional relationship among different non-dimensional numbers is expressed as follows :

$$\eta, \frac{N\sqrt{P}}{\sqrt{\rho}(gH)^{5/4}} = f\left[\frac{ND}{\sqrt{gH}}, \frac{Q}{ND^3}\right], \quad \text{That is, } \eta, \pi_1 = f(\pi_2, \pi_3) \tag{9.1.17}$$

If viscosity of fluid is also considered then Reynolds number will also appear on right hand side. However, it has a little effect on performance of such machines and hence not included in this analysis. It is worth mentioning at this stage that by combining these non-dimensional groups several other non-dimensional groups were obtained for specific purpose with the result that there are number of such non-dimensional groups in use. This happened because of different demands on turbine, pump and compressor.

For example, performance curves of any machine were developed so that these help user in making selection to meet his needs. This gave number of graphs referred as main characteristic, that help in their selection. For a machine in operation, we need operating curves so that the machine may run optimally at different load conditions or machine may be used at some other place. In both the cases appropriate graphs will be needed to predict its performance. So, we have a number of coefficients, specific quantity and unit quantity in use. We shall now discuss these. $\pi_1$ is known as *specific speed*. In practice however specific speed is defined as follows:

$$(N_s)_{\text{turbine}} = N\sqrt{P}/H^{5/4} \quad \text{and} \quad (N_s)_{\text{pump}} = N\sqrt{Q}/(gH)^{3/4} \tag{9.1.18}$$

Since $P = \rho g Q H$ (for incompressible fluid),

then $N\sqrt{(P/\rho)}/(gh)^{5/4} = N\sqrt{Q}/(gH)^{3/4}$

Clearly, these are not non-dimensional numbers. Here $P$ is in kW, $Q$ in m$^3$/s, $H$ in m and $N$ in rpm. Also both the expressions of Eq. 9.1.18 represent same non-dimensional number $\pi_1$ but is expressed in terms of $P$ or $Q$ because of their special importance in turbine and pump, respectively. Thus, specific speed of a turbine is that speed at which turbine will run when it produces unit power under unit head. Similarly, we may define specific speed of a pump.

$\pi_2 = ND/\sqrt{(gH)}$ is identified as *head coefficient* for incompressible fluids. For fluids, $gH = \Delta p/\rho$, where $\Delta p$ is the difference in pressure between two points. $ND$ represents blade velocity and is expressed as ratio of jet velocity $C$. Thus $ND \propto C$. Therefore, head coefficient may be written as

$$\pi_6 = 1/\pi_2^{\,2} = gH/(ND)^2 \propto gH/C^2 = \Delta p/(\rho\, C^2)$$

Ratio of blade speed $u$ to impinging jet velocity $C$ from the nozzle is called speed ratio, $K_u$. Thus,

$$K_u = u/C \tag{9.1.19}$$

$\pi_3 = Q/(ND^3)$ is referred as *capacity coefficient*. But $Q = C_f A \propto D^2 C_f$, where $C_f$ is velocity of flow across area $A$. And $ND^3 = ND \times D^2 \propto C\,D^2$. We may rewrite $\pi_3$ as

$$\pi_3 \propto D^2\, C_f / C\, D^2 = C_f/C = \phi \tag{9.1.20}$$

$\phi$ is called *flow coefficient*. Symbol $K_f$ is also used for it.

*Unit quantities* are used for plotting *main characteristic curves of a given hydraulic turbine*. The reason for use of unit quantity is that a consumer will be interested to know the performance of the available machine to operate under available head and discharge conditions before installing it. Hence, for a particular machine the dimension and the specific speed are fixed but discharge, head and rpm at the location would be different from the one where it was tested and performance curves were developed. The answer to the problem comes in the form of unit quantities used for performance curves. The operating variables such as discharge, rpm and power developed by turbine are reduced to unit head. From the main characteristic curves the performance of the turbine can be predicted knowing available $H$.

Unit quantities are obtained from corresponding specific quantity and coefficients as follows, taking $D$, $\rho$ and $g$ constant:

$$\pi_2 = ND/\sqrt{(gH)} \quad \Rightarrow \quad N_u = N/\sqrt{H} \tag{9.1.21}$$

$$\pi_3 = \frac{Q}{ND^3} = \frac{Q}{\sqrt{gH}}\,\frac{\sqrt{gH}}{ND}\,\frac{1}{D^2} \Rightarrow Q_u = \frac{Q}{\sqrt{H}} \tag{9.1.22}$$

and
$$\pi_1 = N\sqrt{\frac{P}{\rho}}\,(gH)^{5/4}$$

$$\rightarrow \quad \pi_1^2 = N^2\frac{P}{\rho}(gH)^{5/2} = \frac{P}{H^{3/2}}\,\frac{N^2D^2}{gH}\,\frac{1}{D^2 g^{3/2}}$$

$$\Rightarrow \quad P_u = P/H^{3/2} \tag{9.1.23}$$

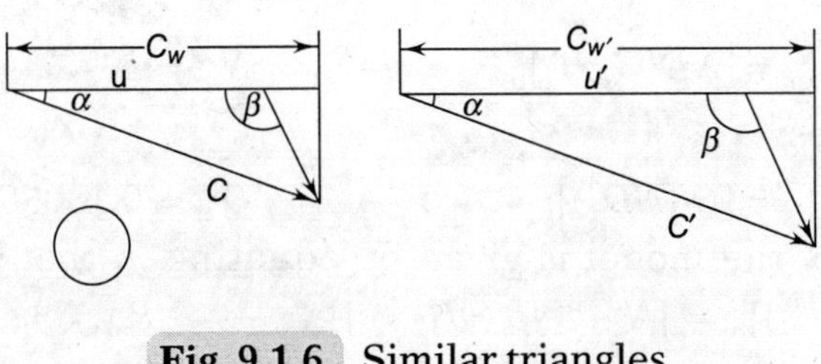


**Fig. 9.1.6** Similar triangles

Yet another important feature of unit quantity is discovered if velocity triangles are drawn for two different operating heads say $H$ and $H'$, as shown in Fig. 9.1.6. These triangles are similar, since all the three angles of the triangles are equal. Thus, we can write

$$u/u' = C/C' = C_w/C'_w = C_f/C'_f = \sqrt{H/H'}$$

The kinematic similarity is automatically satisfied. Thus a unit quantity satisfies all the conditions for model testing provided model is geometrically similar to its prototype.

It is common practice to represent blade velocity and flow velocity as percentage of head, which respectively are speed ratio $K_u$ and flow ratio $K_f$. In pumps, this head is manometric head, $H_m$

$$K_u = u/\sqrt{2gH_m} \qquad \text{and} \qquad K_f = C_f/\sqrt{2gH_m} \tag{9.1.24}$$

Where values of $K_u$ vary from 0.95 to 1.25 and that of $K_f$ from 0.1 to 0.25; values increase with increase in specific speed. Also ratio $D_1/D_2$ varies from 1/3 to 2/3. With this information, velocity triangle at exit can be drawn to find out the angles.

Scaling of the prototype brings some changes in the flow pattern because of change in size and hence there is some shift in the operating point of the model from that of prototype and requires a careful study while designing experiments for the model. For example, in case of centrifugal pump, the discharge at the best operating point is lower than that obtained from the similarity relations. We shall not be able to deal these aspects in details.

**Example 9.1.1** For a given hydraulic site, a Francis turbine is designed to have runner diameter of 3.0 m while operating at 300 rpm under head of 45 m with an overall efficiency of 82% to generate 6.75 MW. Before starting its manufacturing, it is desirable to perform test on its geometrically similar model having scale ratio 1:8 under head of 9 m. Suggest the rpm and discharge and predict power of the model. What is the specific speed of the model.

**Solution** Given: $P = 6.75$ MW, $H = 45$ m, $N = 300$ rpm, $D = 3.0$ m, $\eta_o = 0.82$ (for prototype)

Model: scale ratio = 1/8 i.e. $D_m/D_p = 1/8$ and so on, $H_m = 9.0$ m, $\eta_o = 0.82$ (assumed)

For model analysis, the three non-dimensional numbers given in Eq. 9.1.16 are to be satisfied. A look at these numbers tell us that $N$ and $Q$ can be determined for model by ensuring $\pi_2$ and $\pi_3$ to be same. However, discharge for prototype is not given which can be determined as follows:

$$P = (\eta_o\, wQH)_p \quad \rightarrow \quad 6750 = 0.82 \times (9.81 \times 1000) \times Q_p \times 45 \text{ or } Q_p = 18.685 \text{ m}^3/\text{s}$$

We shall use subscript $p$ and $m$ for prototype and model respectively. Equality of $\pi_2$ gives:

$$[ND/\sqrt{(gH)}]_m = [ND/\sqrt{(gH)}]_p \quad \rightarrow \quad N_m = 1073.3 \text{ rpm.}$$

Equality of $\pi_3$ gives:

$$[Q/(ND^3)]_m = [Q/(ND^3)]_p \quad \rightarrow \quad Q_m = 0.13056 \text{ m}^3/\text{s.}$$

Power produced by the model is given by equating $\pi_1$ that gives

$$[N\sqrt{P}/\{\sqrt{\rho}\,(gH)^{5/4}\}]_m = [N\sqrt{P}/\{\sqrt{\rho}\,(gH)^{5/4}\}]_p \quad \rightarrow \quad P_m = 9.4336 \text{ kW}$$

We may also obtain power developed by the model as follow:

$$P_m = (\eta_0 wQH)_m = 0.82 \times 9.81 \times 0.13056 \times 9 = 9.452 \text{ kW}$$

This value is little more than that obtained from equality of $\pi_1$, which means overall efficiency of the model is more than the assumed value of 0.92. Also the specific speed comes out to be

$$N_{s,m} = 211.5 \qquad \text{and} \qquad N_{s,p} = 211.5$$

**Example 9.1.2** The Francis turbine of example 9.1.1 is to be used at a place where available head is 60 m. What should be its operating rpm and the flow rate at this new place? Determine the power produced by the machine. Will you recommend use of the machine at this place?

**Solution** Since the same machine is to be used at different locations and hence the unit quantities are to be same and not the non-dimensional numbers of Eq. 9.1.16, as was done in Example 9.1.1. These unit quantities are defined by Eqs. 9.1.13 to 9.1.15. Satisfying these conditions and using subscripts 1 and 2 for the two sites, we get

$$N_u \quad \rightarrow \quad (N/\sqrt{H})1 = (N/\sqrt{H})_2 \qquad \text{or} \qquad N_2 = 346.4 \text{ rpm} > N_1$$

Since rpm has increased, the centrifugal stress developed in rotating elements will be more and needs to be checked. The increase in rpm is (346.4 – 300) = 15.47%, which is allowed as the rotary systems are designed to take 200% rotational speed in case of failure of power supply.

$$Q_u \quad \rightarrow \quad Q_2 = Q_1\sqrt{(H_2/H_1)} = 20.82 \text{ m}^3/\text{s}$$

$$P_u \quad \rightarrow \quad P_2 = P_1\,(H_2/H_1)^{3/2} = 10.4 \text{ MW}$$

We can recommend the use of this machine if the quantity of water available is large enough to meet the requirement as obtained above.

**Example 9.1.3** In the Hirakund power station, there are four Kaplan and two Francis turbines, as in 1985, operating under an average head of 26.5 m. The overall efficiency of the Kaplan turbine is 90% and of the Francis turbine is 88%. Both types have a mechanical efficiency of 94.4% between the turbine and the alternator. (a) The design flow rate for the Kaplan turbine is 170 m$^3$/s and the operating speed 150 rpm. Calculate the shaft power, the electric power output, and the specific speed. (b) For Francis turbine, the specific speed is 2.25, and an electrical power rating of $P_e$ = 24 MW. Determine the flow rate and the rotational speed. Take $g$ = 9.8 m/s$^2$.

**Solution**

(a) For Kaplan turbine:

$$\text{Water power} = P_w = wQH,$$

$$\text{Shaft power} = P_M = \eta_o \times P_w = 0.9 \times 9.81 \times 170 \times 26.5 \times 10^{-3} \text{ MW} = 39.734 \text{ MW}$$

$$\text{Input to alternator} = P_e = \eta_m \times P_M = 0.944 \times 39.734 = 37.5 \text{ MW}$$

$$\omega = 2\pi N/60 = \pi \times 150/30 = 15.7 \text{ rad/s}$$

From Eq. 9.1.15

$$\pi_1 = \omega_s = \frac{\omega\sqrt{P_M/\rho}}{(gH)^{5/4}} = \frac{15.7 \times \sqrt{39.7 \times 10^6 / 10^3}}{(9.8 \times 26.5)^{5/4}} = 3$$

Note Definition of specific speed is different from what is used in practice. The difference lies in that it is (i) non-dimensional (ii) Watt is used instead of kW, and (iii) angular speed is used instead of revolutions per second.

(b) For Francis turbine:

$$P_M = \frac{P_e}{\eta_m} = \frac{24}{0.944} = 25.42 = \eta_o wQH$$

$$\rightarrow \quad Q = \frac{P_M}{\eta_o wH} = \frac{25.42 \times 10^6}{0.88 \times 9810 \times 26.5} = 111.2 \text{ m}^3/\text{s}$$

$$\pi_1 = \omega_s = \sqrt{\eta_o} \times \frac{\omega\sqrt{Q}}{(gH)^{3/4}}$$

$$\rightarrow \quad \omega = \frac{\omega_s (gH)^{3/4}}{\sqrt{\eta_o Q}} = \frac{2.25 \times (9.81 \times 26.45)^{3/4}}{\sqrt{0.88 \times 111.2}} = 14.71 \text{ rad/s} = \frac{30 \times 14.71}{\pi} = 140.5 \text{ rpm}$$

## 9.2 Pump

Pumps (and compressors) are mechanical systems used in our day to day life. For example, a pump is used to raise pressure of liquids as in municipal water works and drainage system, agricultural and irrigation system, tube well, oil pumps in lubrication system, crude oil in petroleum industry, etc. In fact, a pump is used to increase pressure of liquid and can handle any kind of liquid, even contaminated with solid particulates to some degree. In power plants pump is used to handle ash slurry made with water, which is highly erosive in nature. In chemical industries, it is used to handle corrosive fluids, and so on. But its performance is affected by the increase in the viscosity of the liquid and the presence of contamination in it.

The function of the pump is to raise pressure of fluid and impart desirable velocity to it so that fluid may be transported from one place to another; more is the velocity the more is the flow rate and hence less time required for transportation. However, we have learnt that the head loss occurring in transportation increases as square of the flow velocity. It therefore requires pressure of the fluid at the outlet of the pump to be more to overcome these losses. Thus, a compromise is required between the pressure and the velocity at the pump exit. It must, however, be emphasised that the purpose of a pump is mainly to raise the pressure and not the velocity of the fluid. The Following chart gives a broad spectrum of the pumps available in the market.

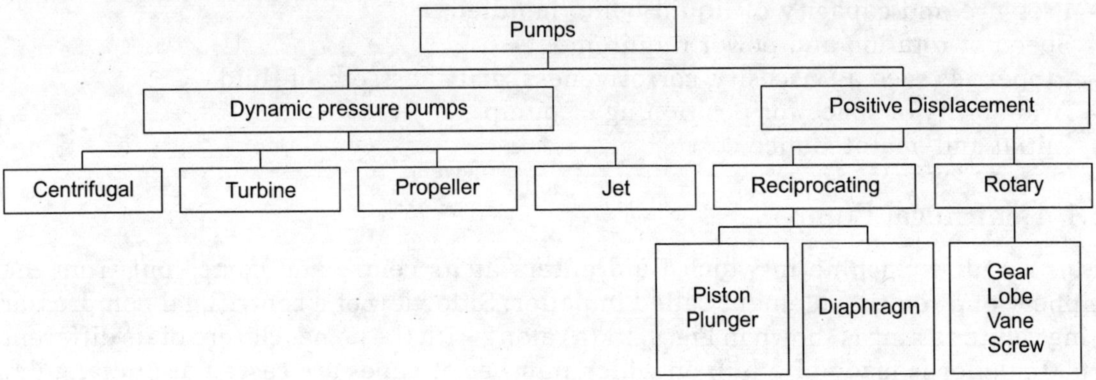


In dynamic pressure pump, work is transferred to fluid flowing over rotating blades of the impeller by changing direction of flow. This increases both pressure and velocity of fluid; mechanism of increase in pressure will be discussed later. Therefore fluid flows in the direction of increasing pressure and chances of separation and back flow are quite high, which results in decrease in efficiency and stable operation of the pump. For this reason their use is limited to those applications where moderate pressure rise is required. These are preferred as compared to reciprocating pump when required volume flow rate is high and/or pressure rise is not large. Dynamic pressure type systems include both turbine and rotary pump, which can handle compressible as well as incompressible fluids. Further, fluid may flow from outer to inner side or vice versa, as in radial flow machines or along the shaft axis of such machines. Methods like multi-staging have been developed to use these for high-pressure applications also. It is a technique in which discharge of the first stage is fed to the next stage and so on. This way a high-pressure ratio can be obtained. Here pressure ratio is defined as ratio of discharge pressure to suction pressure.

Amongst positive displacement type, reciprocating pump is commonly used. This system can deliver very high pressure since the cylinder piston surface does not allow back flow, whence the name. However, capacity, which is defined as mass flow rate of the device per minute, is low because of limited size of the cylinder. Plunger—barrel arrangement is also a reciprocating system like piston cylinder. It is used for diesel fuel pumps as quantity of fuel required is very low but discharge pressure is quite high. Plunger has length comparatively large with respect to its diameter; cylinder is called barrel. Other techniques include gear, lobe, vane, and screw pumps, which are rotary devices, in which back flow does not occur. The rotary type gear pump has wide application in industry mainly for raising pressure of lubricating oil. This pump can be used for extremely large pressure ratio. For this reason, these are used extensively in hydraulic controls.

If the pressure ratio is high then the pumps are placed in series with each other. Or if large quantity is required then many pumps operate with same pressure ratio and the discharge is fed to a common pipe. This arrangement is called parallel arrangement. Both the arrangements are often used in industrial application.

Some of the parameters that decide the selection of a pump are:

- Pressure and capacity of liquid being handled
- Speed of rotation and power requirement
- Properties such as viscosity, corrosiveness, grittiness, etc. of fluid
- Availability of space for positioning of pump
- Initial and maintenance cost

## 9.2.1 Centrifugal Pump

This is a radial machine in which fluid enters at its centre and comes out from the periphery of a rotating element called impeller. Side view of a centrifugal compressor having volute casting is shown in Fig. 9.2.1 (a) along with the nomenclature of its different parts. Impeller is made of a hub on which number of vanes are casted at equiangular positions either on its one or both faces making passage for the fluid to glide through it for which the curvature of vanes is so designed that fluid flows without separation. Pump is known as single or double suction pump accordingly. A double suction pump is used where large mass flow rate is required. A shaft connects hub to motor and passes through the pump casing, which provides bearing for the shaft, see Fig. 9.2.1(b). Seal is provided between casing and pump shaft to stop leakage of fluid.

Pressure across the passage between two consecutive vanes is assumed to be same but due to rotation of the vane pressure on forward face becomes more and less on trailing face. That is, on one side of a vane there will be positive pressure and on its other side (in another passage) it will be negative, as shown in figure. Therefore, fluid will have tendency to move around the edge in the space between casing and impeller causing losses. Fixing a ring around the vanes can prevent this. This ring is called shroud. Further, pressure at impeller exit is more than that at its entrance and that the exit and entrance are connected through space between casing and blade edge. Therefore, there are chances of back flow from exit to suction side of impeller through this casing space, see Fig. 9.1.3(a). Reducing clearance space can decrease it.

Fluid coming out of impeller has sufficiently high velocity and is converted to pressure using any or a combinations of (i) diffuser (ii) spiral (also known as volute) casing. A diffuser is a diverging passage having a suitable angle of divergence so that separation of flow does not occur. Pressure loss at its entry is reduced by designing the entrance angle of the diffuser so that liquid coming from impeller enters into this passage smoothly. For this reason,

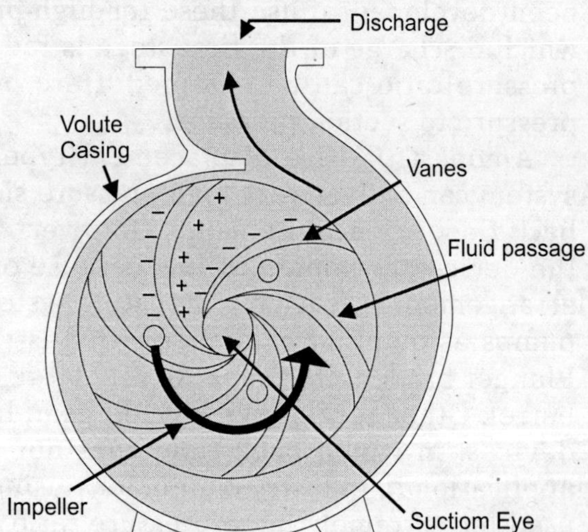


**Fig. 9.2.1(a)** centrifugal pump

velocity of the entering fluid should be in the direction of tangent to diffuser blade at its entrance. It should be noted that the number of diffuser vanes is kept less than that of moving vanes to make the flow uniform along peripheral direction. Therefore, a diffuser passage is fed by more than one moving vane and the streamlines at the entrance are curved. We know from fluid mechanics that such flows are associated with secondary flow, which cause loss of head. Therefore, initial portion of the passage is designed to straighten the streamlines. This process is completed within a short distance during which the diffuser passage is curved and fluid angles go on changing. The section where streamlines are almost straight is called throat and after this section the diffuser passage is a straight diverging section so that the area of flow goes on increasing. Normally a rectangular section is used with constant depth and varying width. Angle of divergence is around 8° to 10° to avoid separation of fluid from the wall of the passage.

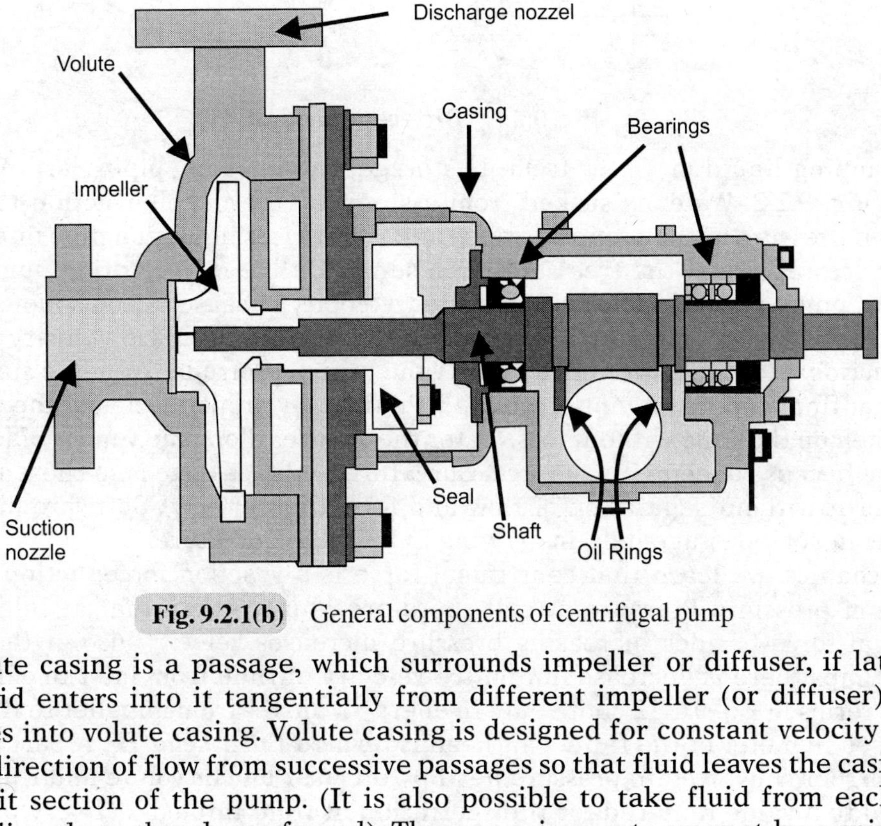


**Fig. 9.2.1(b)** General components of centrifugal pump

A volute casing is a passage, which surrounds impeller or diffuser, if later is also used. Fluid enters into it tangentially from different impeller (or diffuser) passages and moves into volute casing. Volute casing is designed for constant velocity of flow. It changes direction of flow from successive passages so that fluid leaves the casing finally at the exit section of the pump. (It is also possible to take fluid from each diffuser channel directly to the place of need). These requirements are met by a spiral casing, whence the name spiral casing. Normally a volute casing is enough, since magnitude of velocity in pump is not high. The volute casing is also called the scroll casing.

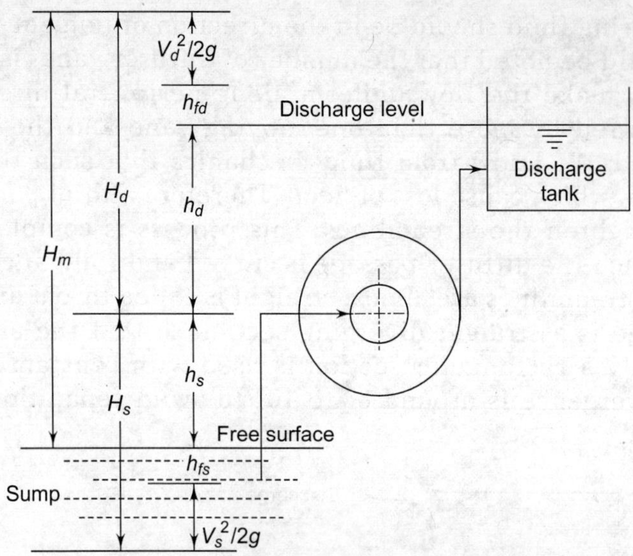


**Fig. 9.2.2** Different heads

For pumping liquid or water from its storage, called sump, piping arrangement is shown in Fig. 9.2.2. Water is sucked from a sump by the impeller action that causes decrease in pressure at the pump entrance and water rises in suction pipe due to higher atmospheric pressure at the free surface in sump. Unlike reciprocating pump, flow in centrifugal pump is continuous with constant velocity because of continuous pumping of fluid. Suction pipe diameter is normally kept large to decrease velocity in pipe to reduce frictional losses. Lower edge of the vane, which protrudes from the surface, is so shaped that fluid enters tangentially at all its points. By proper design of the edge, fluid can be taken on the vane without loss. Water then moves along the vane surface towards outer edge because of centrifugal force. Curvature of the passage is so chosen that fluid glides over it without separation of flow and formation of eddy. Software packages to study flow in such passage help in arriving at suitable curvature.

In Mechanics, we learn that centrifugal force is a reaction force; action for which comes from pressure. Pressure at exit therefore is more than that at inlet to vane. Centrifugal force besides increasing pressure increases velocity also at the impeller exit. In pumps there is decrease in relative velocity of fluid from inlet to exit. Thus all the three terms in Eq. 9.1.12 contribute in energy transfer from impeller to fluid, which is expressed in meter of the fluid. This head is termed ***Euler head, $H_e$***. A part of velocity head is further converted to pressure head by diffuser and/or volute and fluid velocity is reduced to a value required for transportation of fluid through delivery pipe.

## 9.2.1 Terms Used in Centrifugal Pump

Figure 9.2.2 shows a centrifugal pump placed at a height $h_s$ from the free surface of liquid. Here the suction pipe is shown as vertical pipe but in practice it need not be so. However, the length of the suction pipe will be minimum if it is vertical and in that case the frictional losses will also be minimum. Liquid is stored in a tank placed at a height $h_d$ from the centre line of the pump. We use the following terms in centrifugal pump analysis, see Fig. 9.2.2:

- ***Suction lift head, $h_s$***. It is the vertical distance between free surface of liquid level in the sump and pump centre line. It should not be confused with the length of suction pipe, which will be more if not vertical.
- ***Discharge lift head, $h_d$***. It is the vertical distance between the top surface of liquid level in discharge tank and pump centre line. The sum of suction lift and discharge lift is called ***total static*** or ***vertical lift***, *H*.
- ***Suction head, $H_s$***. In addition to suction lift head, pump also has to generate head for overcoming friction loss $h_{fs}$ in suction pipe, entry loss $h_i$ at entrance to suction pipe and running fluid in suction pipe with velocity $V_s$. Thus

$$H_s = h_s + h_i + \frac{V_s^2}{2g} + h_{fs} = \frac{p_s}{w} + \frac{V_s^2}{2g} \tag{9.2.1}$$

where $p_s = w\,(h_s + h_i + h_{fs})$ is the vacuum below atmospheric pressure that would be recorded by a pressure gauge placed at the entrance of the pump. This can be verified by applying modified Bernoulli's equation between two points one at the entrance section of the suction pipe and another at its discharge end where the pressure gauge is fitted and taking free surface in sump as datum for measuring height. If $h$ is depth of suction end of the pipe from free surface then pressure at this end = $p_a + wh$. This gives:

$$\frac{p_a}{w} + h + \frac{V_s^2}{2g} = \frac{p_s}{w} + \frac{V_s^2}{2g} + (h_s + h) + (h_i + h_{fs})$$

or,

$$p_s = w\,(h_s + h_i + h_{fs})$$

- ***Discharge head, $H_d$***. In addition to discharge lift head, pump has to generate head for overcoming friction loss $h_{fd}$ in discharge pipe and imparting discharge velocity $V_d$. Thus

$$H_d = h_d + \frac{V_d^2}{2g} + h_{fd} = \frac{p_d}{w} + \frac{V_d^2}{2g} \tag{9.2.2}$$

where $p_d$ is the pressure measured by a pressure gauge placed just at exit of the pump. We may show it in the way it was shown for suction side.

- ***Manometric head, $H_m$*** and ***manometric efficiency***. Manometric head is defined as change in total energy head produced by the pump when fluid moves through it. If we neglect the change in potential head from inlet to exit of pump then

$$H_m = H_d + (-H_s) = \left(\frac{p_d}{w} - \frac{p_s}{w}\right) + \frac{V_d^2}{2g} - \frac{V_s^2}{2g} \tag{9.2.3}$$

Since the impeller transfers energy to the fluid, manometric head is nothing but Euler head except when there are energy losses associated in energy transfer process. These losses are discussed later in this chapter. The loss is expressed in terms of manometric efficiency defined as manometric head divided by Euler head.

Manometric efficiency, $\eta_{man} = H_m / H_e$ (9.2.4)

Manometric head may be seen as the head required to provide suction and lift head plus the losses in pipe from inlet of suction pipe to the discharge point of the delivery pipe plus velocity head at the delivery pipe end. That is,

$$H_m = h_s + h_d + h_{fs} + h_{fd} + \frac{V_d^2}{2g} = H_s + H_d - \frac{V_s^2}{2g}, \qquad (9.2.5)$$

using Eqs. 9.2.1 and 9.2.2.

- ***Net positive suction head, NPSH***. Concept of net positive suction head is that vapours should not be formed in the flow as these form bubbles, which are trapped inside liquid. These bubbles collapse when reach with flow in regions of high pressure where actual temperature is lower than the condensation temperature. This results in formation of cavity and flow of liquid from all the sides resulting in tremendous rise in pressure. This phenomenon results in mechanical loss as well as loss in efficiency of the machine. It is called *cavitation* and has to be avoided for better performance and long life of the machine. Formation of vapour depends on the surrounding temperature[5]. Therefore the minimum pressure in suction line should not fall below vapour pressure of the liquid, $p_v$. It is seen from Fig. 9.2.2 that minimum pressure in suction line will be at the impeller eye. Pressure at this point can be obtained by applying modified Bernoulli equation between the points on free surface in sump and near eye, neglecting velocity of liquid on the free surface and taking this surface as datum. We get

$$\frac{p_a}{w} = \frac{p_1}{w} + h_s + h_{fs} + \frac{V_s^2}{2g}$$

$$\Rightarrow \qquad \frac{p_1}{w} = \frac{p_a}{w} - \left( h_s + h_{fs} + \frac{V_s^2}{2g} \right) > \frac{p_v}{w}$$

$$\text{or,} \qquad \frac{p_a}{w} - \frac{p_v}{w} - \left(h_s + h_{fs}\right) > \frac{V_s^2}{2g} \qquad (9.2.6)$$

In literature, term $(p_a/w - p_v/w - h_s - h_{fs})$ *is named as net positive suction head, NPSH.* Thus from Eq. 9.2.6, NPSH > $V_s^2/2g$, which is assumed 3m of water. Eq. 9.2.6 helps in deciding positioning of the centre line of pump above sump level. Normally a large diameter pipe with minimum length between sump and pump is used to reduce the losses so that vertical height of the pump from sump may be more. However, in practice, suction lift head is not kept more than 7m of water as vibration and cavitation has been reported for higher values.

## 9.2.2 Effect of Vane Shape and Operating Variables

Figure 9.2.3 shows three types of vanes used in practice. They differ in the way the vanes are curved with reference to direction of rotation, whence the name. For example,

[5] All liquids start changing their phase to vapour when they reach saturated liquid state. This depends on the pressure and temperature of the liquid. If pressure is decreased at a given temperature then at a particular pressure the liquid will start boiling. This pressure is called vapour pressure.

a backward curved vane is curved opposite to direction of rotation whereas forward curved in the direction of rotation. We can draw velocity triangles at exit in three cases for the same peripheral velocity $u_2$, which are shown in Fig. 9.2.4. We observe from Fig. 9.2.4 that (i) $\beta_2$, $Cw_2$, and $C_2$ increase as we go through backward, radial to forward curved vanes and (ii) $\beta_2$ is < 90° for backward, $\beta_2 = 90°$ for radial and $\beta_2 > 90°$ for forward curved vane. Therefore energy transfer to fluid and absolute velocity at pump exit is least in backward curved and maximum in forward curved case. However, chances of back flow in backward curved pump are less. Moreover, when the impeller rotates at high rpm, as in rotating compressor, centrifugal force acting on vane tries to straighten it out. For these reasons, radial type is mostly preferred; it is easy to manufacture too. But

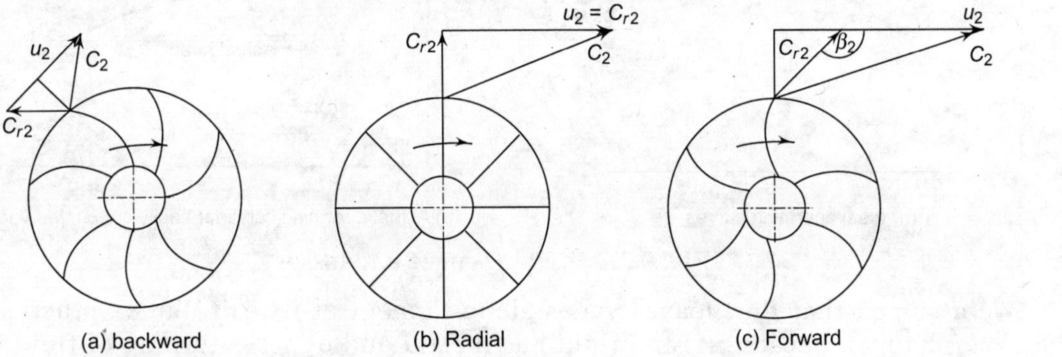


**Fig. 9.2.3** Different vane shapes

backward curved pump is finding more and more application with improvement in manufacturing techniques because of their higher efficiency. In all the three type, entry is along axial direction having zero angular momentum in the plane of rotation. Euler head developed using Eq. 9.1.1 and Fig. 9.2.4 is

$$H_e = C_{w2}\, u_2/g = (u_2 - C_{f2} \cot \beta_2)\, u_2/g = (u_2 - Q \cot \beta_2/A_2)\, u_2/g \qquad (9.2.7)$$

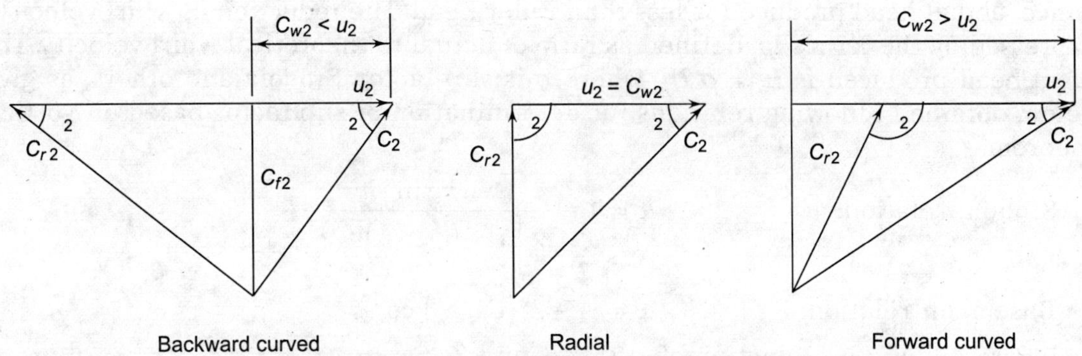


**Fig. 9.2.4** Velocity diagram for different vanes

Where $Q$ is the volume flow rate and $A_2$ is the area of flow at impeller exit, which gives, $C_{f2} = Q/A_2$. The rest of the symbols have the usual meaning. This equation shows the effect of vane angle at its exit on pressure head development. For a given rpm of impeller, variables in Eq. 9.2.7 are $H_e$, $Q$ and $\beta_2$. Therefore, we get a parametric family for different $\beta_2$ between $H_e$ and $Q$; this is shown in Fig. 9.2.5(a). Curve for radial blade is horizontal which means there is no effect of discharge on the head developed.

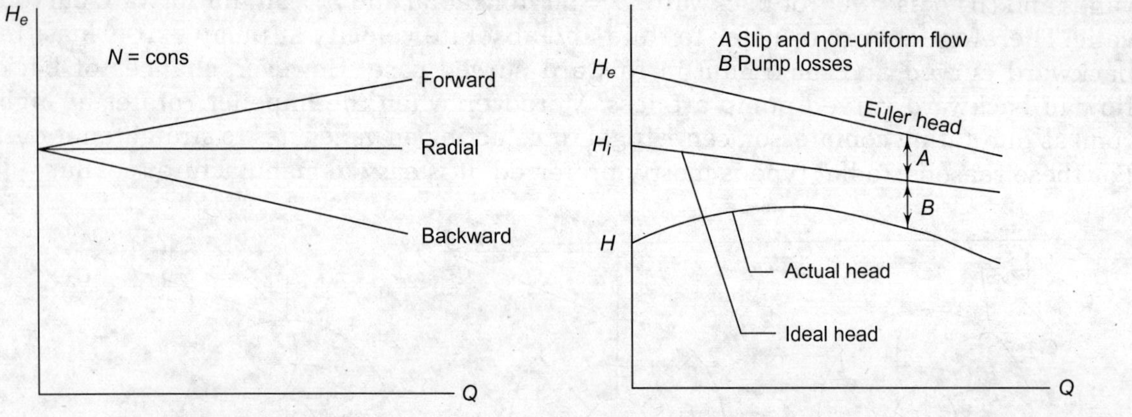


(a) Ideal operating curves (b) Actual operating curve for backward curved vane pump

**Fig. 9.2.5** *H* v/s *Q* curve and losses

We assumed that flow leaves vanes along tangent at its exit. But in practice, this does not happen because (i) the fluid has inertia and opposes turning of fluid in the direction of rotation and (ii) that the fluid rotates in direction opposite to the direction of rotation in the impeller passage to have zero vorticity as it was zero at entrance to the impeller, by vorticity theorem. Though no change in vorticity is true only for frictionless flow but for the small distance travelled by fluid, an assumption of no change in vorticity is justified. Also velocity distribution along impeller periphery is non-uniform even at the design operating conditions[6]. That is, $C_2$ varies along the impeller periphery, both in magnitude and direction resulting in decrease in whirl velocity at exit. This is shown in Fig. 9.2.6. It results in less energy transfer to fluid and hence actual head produced is less than Euler head. The reduction in whirl velocity is expressed by the term *slip*, defined as ratio of actual to theoretical whirl velocity. Thus ideal head produced is $H_i = \sigma H_e$, where $\sigma$ is slip factor. Stodola and others, as given below, obtained following relations for determination of slip factor based on vorticity theorem.

Stodola relation: $$\sigma = 1 - \frac{\pi \sin \beta_2}{Z\left[1 - \left(C_f / u_2\right) \cot \beta_2\right]}$$

Busemann relation: $$\sigma = A - B\left(C_f / u_2\right) \cot \beta_2,$$

where $A$ and $B$ are function of $r_2 / r_1$, $\beta_2$ and $Z$

[6] A pump is designed for a given pressure head , mass flow rate and rpm. Under these conditions, it has maximum efficiency. These conditions are known as design operating conditions.

Stanitz relation: $$\sigma = 1 - \frac{0.63 \sin \beta_2}{Z\left[1 - \left(C_f / u_2\right) \cot \beta_2\right]}$$

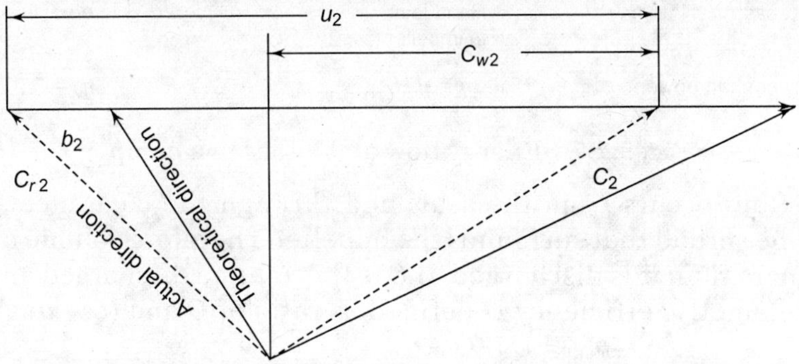


**Fig. 9.2.6** Effect of inertia in backward curved pump

Slip factor improves if the number of vanes is increased. However, this increases the blockage ratio, that is, area of flow reduces and hence the mass flow rate too. Therefore a compromise is to be made.

The result of slip and non-uniform flow is to reduce the pressure head produced at impeller exit and the effect is shown by vertical distance *A* in Fig. 9.2.5(b). This loss decreases with increase in *Q*. Losses occurring in diffuser further decreases actual head *H*; vertical distance *B* is measure of this loss and we observe from Fig. 9.2.5(b) that this loss first decrease upto some *Q* and then starts increasing with increasing *Q*. The loss consists of friction loss in diffuser passage and the aerodynamic loss at the entrance to it because of deviation between actual flow angle and diffuser blade angle. The later loss changes with *Q* and hence explains the trend of the diffuser loss. So the losses in diffuser will be minimum only at the design point and therefore the actual head *H* varies with *Q* as shown in Fig. 9.2.5(b). Ratio of *H* to $H_i$ is known as hydraulic efficiency,

$$\eta_h = H / H_i \qquad \text{or} \qquad H = \eta_h H_i = \eta_h \sigma H_e \tag{9.2.8}$$

Besides the losses mentioned earlier, other losses also occur. Fig. 9.2.7 gives a picture of flow of energy and the loss occurring in different sections. Power from electric motor is known as shaft power. But due to mechanical losses in bearing, power that reaches pump rotor is less. Thus

$$\text{rotor power} = w\,(Q + q)\,H_i,$$

where $(Q + q)$ is the dischange from the sump. This loss is expressed by mechanical efficiency, defined as

$$\eta_{\text{mech}} = \text{rotor or impeller power/ shaft power} \tag{9.2.9}$$

Mechanical efficiency usually is 0.95 to 0.98.

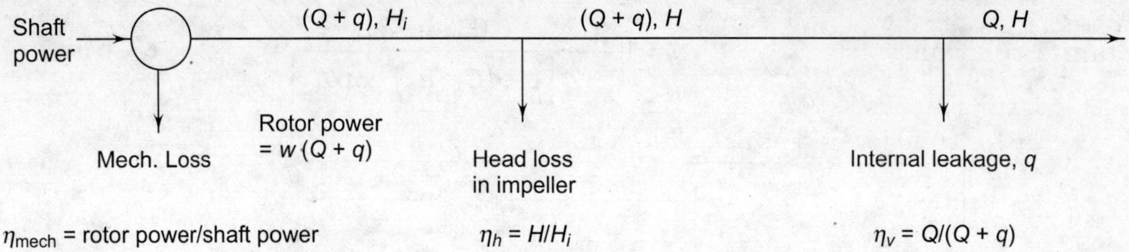


**Fig. 9.2.7** Energy flow and losses in a pump

Leakage of fluid occurs from the seals and through the clearance space between the casing and the shroud that surrounds the impeller. Therefore volume of fluid handled by impeller is more than it is discharged. If $Q$ is the volume discharged and $q$ the volume leakage then volumetric efficiency is defined as ratio of $Q$ and $(Q + q)$.

$$\eta_{vol} = Q\,/\,(Q + q) \tag{9.2.10}$$

A pump may be considered as a black box, which is fed with shaft power and water power $wQH$ leaves it. We may express performance by introducing the term overall efficiency,

$$\eta_o = \text{water power/shaft power} = wQH\,/\,\text{shaft power}$$

We multiply and divide this equation by rotor power while writing expression for the rotor power in the denominator. This gives *overall efficiency*

$$\eta_o = \frac{wQH}{\text{Shaft power}} \times \frac{\text{Rotor power}}{w(Q+q)H_i} = \frac{H}{H_i} \times \frac{Q}{Q+q} \times \frac{\text{Rotor power}}{\text{Shaft power}} = \eta_h\,\eta_{vol}\,\eta_{mech} \tag{9.2.11}$$

We may also express manometric head in terms of $Q$ and $N$ to see their effect on it, since these are operating variables which are changed in actual set up. Let suffix 1 and 2 represent inlet and exit of impeller and say a fraction $k$ of the kinetic energy head at exit of impeller $V_2^2/2g$ is recovered as pressure head in diffuser and/or volute casing. Then manometric head is,

$$H_m = (p_2 - p_1)/w + k\,V_2^2/2g$$

Assuming no loss in impeller, Bernoulli's equation gives:

$$p_1/w + V_1^2/2g + Cw_2u_2/g = p_2/w + V_2^2/2g$$

or, $$(p_2 - p_1)/w = \left(V_1^2 - V_2^2\right)/2g + Cw_2u_2/g$$

But from outlet velocity triangle,

$$C_{w2} = u_2 - C_{f2}\cot\beta_2 \qquad \text{and}$$

$$C_2^2 = C_{w2}^2 + C_{f2}^2 = C_{f2}^2\,\text{cosec}^2\,\beta_2 + u_2^2 - 2u_2\,C_{f2}\cot\beta_2$$

$$\therefore \qquad (p_2 - p_1)/w = \left(C_{f1}^2 + u_2^2 - C_{f2}^2\,\text{cosec}^2\,\beta_2\right)/\,2g$$

Here it is assumed that the entry to pump is radial, which means $C_1 = C_{f1}$, and has been used in above relation. We further assume that there is no change in velocity of flow, that is $C_{f1} = C_{f2}$. Substituting for $(p_2 - p_1)/w$, $C_2^2$ in the equation for manometric head and simplifying, we get

$$H_m = (AN^2 + BNQ + CQ^2)/2g, \tag{9.2.12}$$

where $A = (1 + k)\,(\pi D_2/60)^2, \quad B = -2k \cot \beta_2/(\pi D_2 \beta),$

and $C = (1 + k \cot^2\beta_2 - \text{cosec}^2\beta_2)\,/(\pi D_2 b)^2$

Here $b$ is depth of impeller passage at its exit.

## 9.2.3 Cavitation, Minimum Speed of Operation and Priming in Pump

Cavitation and priming are two important phenomena associated with pump. We have explained earlier as to why cavitation occurs. It is specific characteristic of hydraulic machines caused by vaporisation of liquid. We mentioned earlier that the pressure is minimum at impeller eye. Let the pressure at the section is $p_v$, which is saturation pressure of the liquid at the temperature at this section. Any further decrease in pressure would result in formation of vapour. Under this critical condition, the flow velocity $V_s$ at eye can be obtained by applying modified Bernoulli equation. This gives

$$\frac{V_s^2}{2g} = \frac{p_a}{w} - \frac{p_v}{w} - h_s - h_{fp} = NPSH$$

However, $V_s^2/2g$ should be less then this values so that cavitation does not occur. Thoma gave cavitation number $\sigma$, defined as

$$\sigma = V_s^2/2gH \le NPSH/H, \tag{9.2.13}$$

where $H$ is head produced by pump; $\sigma$ should therefore be less than $NPSH/H$ for cavitation free flow. Critical value of Thoma's cavitation factor $\sigma_c$, depends on specific speed, efficiency of the pump and number of vanes and is given by empirical relation

$$\sigma_c = 1.042 \times 10^{-3}\,(N_s)^{4/3}, \text{ where } N_s \text{ is pump specific speed} \tag{9.2.14}$$

Thus cavitation factor of given installation should always be greater then $\sigma_c$. That is,

$$\sigma_c < \sigma < NPSH/H \tag{9.2.15}$$

Manufacturer of pump gives the required *NPSH*, which depends on the design of the pump. When in use, available *NPSH* is obtained for those conditions for which the pump operates at maximum efficiency without any objectionable noise. Therefore *available NPSH must be more than the NPSH prescribed by manufacturer for cavitation free operation.* Thus Thoma's factor decides the positioning of the pump.

On the other hand, pump fails to lift liquid at the time of its starting. This happens because suction pipe is filled with air when liquid flows back to sump due to failure of one-way valve fitted at the entrance of suction pipe. Since air is lighter than liquid, rotation of impeller does not create enough suction head to raise liquid from sump. It is overcome by filling the suction pipe and the space around the impeller with liquid by pouring it through a funnel fixed in delivery pipe or by sucking air from the pump using an extraction pump that also pulls liquid from sump and fills the whole space around impeller. This process is called *priming*.

A pump can operate only when its rpm is sufficient to move liquid over the vane. This will happen when head developed by pump at the time of start is more than the manometric head. But at this condition, since flow is to start, the only force to develop head is centrifugal head = $(u_2^2 - u_1^2)/2g$. Therefore,

$$(u_2^2 - u_1^2)/2g \geq H_m \quad \text{or} \quad (u_2^2 - u_1^2)/2g \geq \eta_m \, H_e = C_{w2} \, u_2/g$$

$$\Rightarrow \qquad N^2 \geq 2g\left(\frac{60}{\pi}\right)^2 \frac{H_e \eta_m}{D_2^2 - D_1^2} \tag{9.2.16}$$

Where $D_1$ and $D_2$ are diameter of mid point of eye and impeller outlet respectively. Equation 9.2.16 gives the minimum rpm for a given pump. Normally an electric motor is used to run the pump and therefore value of $N$ is determined by motor rpm. In that case $D_2$ is checked from Eq. 9.2.16; actual pump diameter should be more than this diameter so that pump may be used at that motor rpm.

### 9.2.4 Performance Characteristics of Centrifugal Pump

Three types of performance characteristic curves, namely, main, operating and iso-efficiency characteristic curves are used for a centrifugal pump. Experiments are performed on the given pump running at a given (fixed) rpm for different values of the discharge through it while measuring power consumption and manometric head

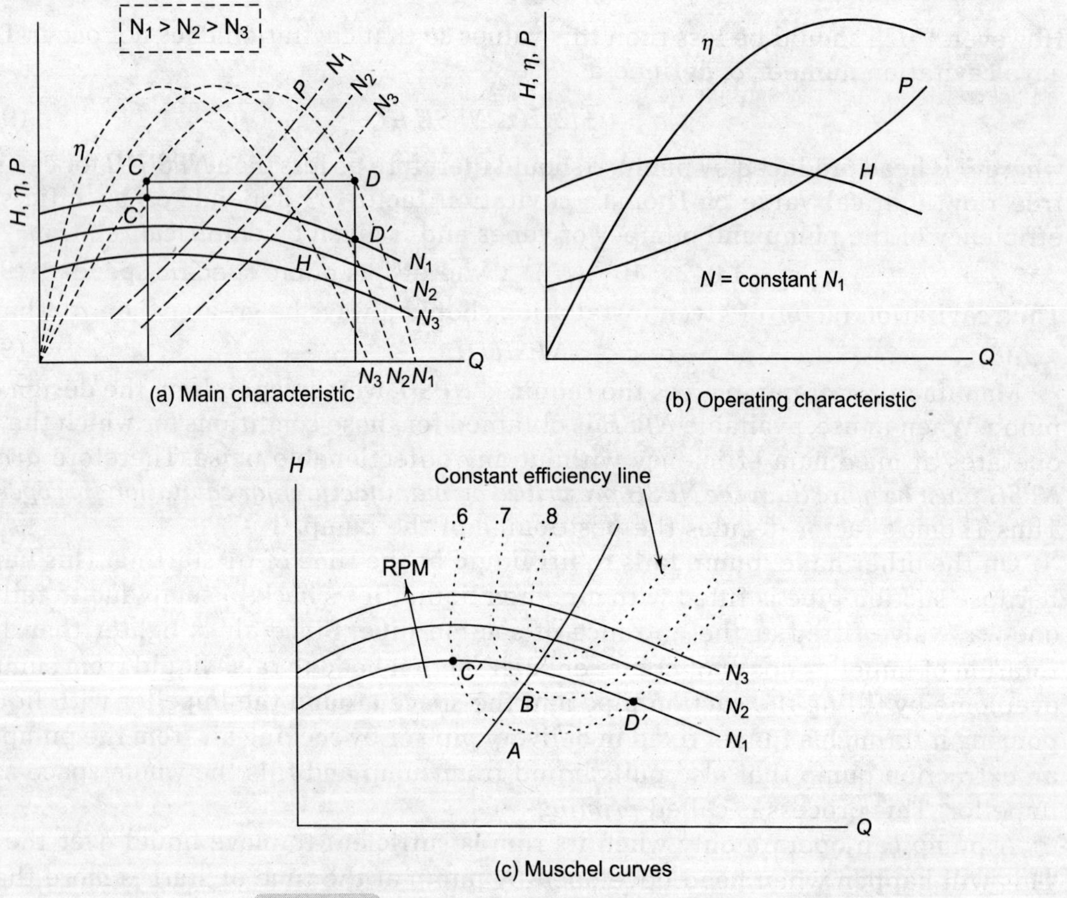


**Fig. 9.2.8** Pump's performance curves

produced for each setting. From these observations, efficiency is obtained for a given value of $Q$ and the results are plotted on a graph with $Q$ along abscissa and $H$, $P$ and efficiency along ordinate. This experiment is repeated at a few other rpm and the results plotted on the same graph. This gives the parametric family in the parameter rpm. Curves so obtained are main characteristic, shown in Fig. 9.2.8(a). This chart helps in selection of the pump for given operating conditions.

Like any other system, pump will also operate at its maximum efficiency when run at its design rpm. Therefore, manufacturers give operation graph, which is nothing but the graph from main characteristic at the design rpm. The normal operation condition of pump corresponds to the point at which the efficiency is maximum. Since the efficiency curve is quite sharp near the operating point, it is desirable to run the pump at the design flow rate as far as possible.

Yet another graph of importance is the iso-efficiency graph, also known as Muschel graph because discharge requirement during operation may vary and we would like to run the pump almost at the same efficiency. For this, Muschel chart is very useful since it gives the desired operating speed. It is a graph between Head and flow rate with the efficiency as the parameter of the family of the curves. These curves are obtained from main characteristic at a given efficiency by drawing a horizontal line $CD$, see Fig. 9.2.8(a). This line meets the efficiency curve for a particular rpm $N_1$ (say) at $C$ and $D$. Now draw vertical lines through points, $C$ and $D$ to meet the $H$ = constant line corresponding to $N = N_1$ at $C'$ and $D'$ and read values of $H$ and $Q$ corresponding to points $C'$ and $D'$. This gives two points $C'$ and $D'$ on Muschel graph. The horizontal line will also meet the constant efficiency curves at different rpm from which we get other points in the same manner and plot on the Muschel graph. This gives an iso-efficiency curve. The procedure is then repeated for different values of efficiency. We get these curves from the main characteristics only. We can also join the points for which rpm is the same. The results are shown in Fig. 9.2.8(c).

It is observed that as efficiency increases, width of the curve decreases and ultimately at some efficiency it will shrink to a point that corresponds to the point of maximum efficiency for that pump. Also for each RPM there is a point of maximum efficiency and hence there will be only one point for it, which is shown as points $A$, $B$, etc. in the graph. Thus, the line joining these points is the line at which the pump is to be operated for that efficiency.

All the three types of characteristics is shown in Fig. 9.2.8.

**Example 9.2.1** The impeller of a centrifugal pump has an outer diameter of 250 mm and an affective area of 0.017 m$^2$. The blades are bent backwards so that the direction of outlet relative velocity make an angle of 148° with the tangent drawn in the direction of impeller rotation, the diameters of suction and delivery pipes are 150 mm and 100 mm respectively. The pump delivers 0.013 m$^3$/s at 1450 rpm when the gauge pressure on the suction and delivery pipes close to the pump show heads of 4.6 m below and 18.0 m above atmosphere respectively. The head losses in the suction and delivery pipes are 2.0 m and 2.9 m respectively. The motor, driving the pump delivers 8.67 kW. Assuming that water enters the pump without shock and whirl, determine (i) the solution efficiency and (ii) the overall efficiency of the pump.

**Solution**

**Given:** $D_2 = 0.25$ m $\quad A_f = 0.017\ m^2 \quad \beta_2 = 180 - 148 = 32° \quad D_s = 0.15$ m

$D_d = 0.1$ m $\quad N = 1450$ rpm $\quad Q = 0.031\ m^3/s \quad h_{fs} = 2.0$ m

$h_{fd} = 2.9$ m $\quad P = 8.67$ kW $\quad p_s/w = -4.6$ m $\quad p_d/w = 18.0$ m

$C_{w1} = 0$

For determining manometric efficiency, we calculate manometric and Euler heads. Manometric head is given by Eq. 9.2.3 for which velocity of flow in suction and discharge pipe is obtained using continuity equation. This gives:

$$V_s = \frac{4Q}{\pi D_s^2} = 1.754 \text{ m/s}, \qquad V_d = \frac{4Q}{\pi D_d^2} = 3.947 \text{m/s, and}$$

$$H_m = \left(\frac{p_d}{w} - \frac{p_s}{w}\right) + \left(\frac{V_d^2}{2g} - \frac{V_s^2}{2g}\right) = \frac{(18 + 4.6) + (3.947^2 - 1.754^2)}{2 \times 9.81} = 23.3 \text{ m} \qquad \text{(A)}$$

To determine Euler head, see Fig. 9.2.4 for backward curved vane.
From given data, we get

$$u_2 = \pi D_2 N/60 = 18.98 \text{ m/s} \quad \text{and} \quad C_{f2} = Q/A_f = 1.82 \text{ m/s}$$

From outlet velocity diagram:

$$C_{w2} = u_2 - C_{f2} \cot \beta_2 = 16.07 \text{ m/s and} \quad H_e = C_{w2}\, u_2/g = 31.1 \text{ m}$$

$$\therefore \quad \eta_{man} = H_m/H_e = \mathbf{74.71\ \%}$$

$$\eta_o = w\, Q\, H_m/P = 9.81 \times 0.031 \times 23.23/8.67 = \mathbf{81.48\%}$$

**Example 9.2.2** A centrifugal pump, in which water enters radially, delivers water to a height of 165 m. The impeller has a diameter of 360 mm and width 180 mm at inlet and the corresponding dimensions at the outlet are 720 mm and 90 mm respectively; its rotational speed is 1200 rpm. The blades are curved backwards at 30° to the tangent at exit and the discharge is 0.389 $m^3/s$. Determine (a) theoretical head developed, (b) manometric efficiency, (c) pressure rise across the impeller assuming losses to be 12% of velocity head at exit, (d) pressure rise and the loss of head in the volute casing, (e) the vane angle at inlet, and (f) power required to drive the pump assuming an overall efficiency of 70%. What would be the corresponding mechanical efficiency?

**Solution** Refer Fig. 9.2.2 for nomenclature and Fig. 9.2.4(a) for velocity triangle at outlet.

**Given:** $H_m = 165$ m $\quad D_1 = 0.36$ m $\quad D_2 = 0.72$ m $\quad \beta_2 = 30°$

$B_1 = 0.18$ m $\quad B_2 = 0.09$ m $\quad Q = 0.389 \quad N = 1200$ rpm

$\therefore \quad u_2 = \pi D_2 N/60 = 45.24$ m/s $\quad$ and $u_1 = u_2/2 = 22.62$ m/s

Further $\quad Q = \pi D_1 B_1 C_{f1} = \pi D_2 B_2 C_{f2}$

But $\quad D_1 B_1 = D_2 B_2$, (check from given data)

$\therefore \quad C_{f1} = C_{f2} = 0.389/(\pi \times 0.72 \times 0.09) = 1.91$ m/s

**Assumption**. (i)Radial entry at inlet and (ii) 100% volumetric efficiency, since no information is given.

(a) Theoretical head developed is Euler head = $H_e = C_{w2}u_2/g$. $C_{w2}$ is obtained from velocity triangle at outlet. From Fig. 9.2.4(a),

$$C_{w2} = u_2 - C_{f2}/\tan\beta_2 = 45.24 - 1.91/\tan 30 = 41.93 \text{ m/s}$$

$$\therefore \quad H_e = 41.93 \times 45.24/9.81 = 193.4 \text{ m}$$

(b) Manometric efficiency = Manometric Head/ Euler Head = 165/193.4 = **85.3%**

(c) Applying modified Bernoulli equation at points inlet (1) and outlet (2) of the pump and neglecting the change in potential energy, we get

$$\text{Pressure rise in impeller} = H_e + \frac{C_1^2 - C_2^2}{2g} - h_{\text{loss}} \quad \text{(A)}$$

But, $C_1 = C_{f1} = 1.91 \rightarrow C_1^2/(2g) = 0.186$ m and $\dfrac{C_2^2}{2g} = \dfrac{C_{f2}^2 + C_{w2}^2}{2g} = 89.79$ m

$$\rightarrow \quad h_{\text{loss}} = 0.12\frac{C_2^2}{2g} = 0.12 \times 89.79 = 10.77 \text{ m}$$

Substituting the different values in the Eq. (A), we get

Pressure rise in impeller = 193.4 + (0.1862 – 89.79) – 10.77 = 93.02 m

(d) $H_m$ = pressure rise in impeller + pressure rise in volute casing. This gives

Pressure rise in volute casing = 165 – 93.02 = 71.98 m

Loss of head in pump = loss in impeller + loss in volute casing

This gives, loss of head in volute casing

$$= H_e - H_m - h_{\text{loss}} = 193.4 - 165 - 10.77 = 17.63 \text{ m}$$

(e) Refer Fig. 9.1.4 for inlet velocity triangle. From this triangle,

$$\tan\beta_1 = C_{f1}/u_1 = 1.91/22.62 \quad \rightarrow \quad \beta_1 = 4.8°$$

(f) Power required = $w\,Q\,H_m/\eta_o$

$$= (9.81 \times 1000) \times 0.389 \times 165 / 0.7 = 899.5 \times 1000 \text{ W} = \mathbf{899.5\ kW}$$

$$\eta_o = \eta_{\text{mano}} \times \eta_v \times \eta_{\text{mech}} \quad \rightarrow \quad \eta_{\text{mech}} = 0.7/0.853 = \mathbf{82\%}$$

**Example 9.2.3** A centrifugal pump lifts water under a static head of 36 m of water of which 4 m is suction lift. Suction and delivery pipes are both 150 mm in diameter. The head loss in suction pipe is 1.8 m and in delivery pipe 7 m. The impeller is 380 mm in diameter and 25 mm wide at mouth and revolves at 1200 rpm. Its exit blade angle is 35°. If the manometric efficiency of the pump is 82%, determine (a) the discharge through the pump, and (b) the pressure at the suction and delivery branches of the pump.

**Solution** Refer section 9.2.1 for various heads used in the following solution.

**Assumption.** Radial entry at the pump inlet.

**Given:** $H = 36$ m $\quad h_{fs} = 1.8$ m $\quad h_{fd} = 7 \text{ m} \rightarrow H_m = 36 + 1.8 + 7 = 44.8$ m

$D_s = D_d = 0.15$ m $\quad D_2 = 0.38$ m $\quad N = 1200 \rightarrow u_2 = 23.87$ m/s

$B_2 = 0.025$ m $\quad \beta_2 = 35°\quad h_s = 4$ m $\rightarrow h_d = H - h_s = 32$ m

$\eta_{mano} = 0.82 \rightarrow H_e = C_{w2}\, u_2/g = H_m/\,\eta_{mano}$ or $C_{w2} = 22.45$ m/s

(a) $Q = (\pi D_2 B_2)C_{f2}$, where $C_{f2} = (u_2 - C_{w2})\tan\beta_2 = (23.87 - 22.45)\tan 35 = 0.99$ m/s
This gives

$$Q = \pi \times 0.38 \times 0.025 \times 0.99 = \mathbf{0.0295\ m^3/s}$$

Also $Q = \pi D_s^2\, V_s/4$, where $V_s$ is velocity of water in suction pipe. This gives, $V_s = V_d$ = 1.67 m/s, since pipe diameter is same for suction and discharge pipes.

$$\therefore \quad V_s^2/2g = V_d^2/2g = 0.142 \text{ m}$$

Pressure on suction side of the pump

$$= -\left(h_s + h_{fs}\, V_s^2/2g\right) = -5.942 \text{ m} = 10 - 5.942 = 4.058 \text{ m}$$

$$= (4.058 \times 9.81 \times 1000)/1000 = \mathbf{39.8\ kN/m^2}$$

Similarly, pressure on delivery side of the pump

$$= \left(h_d + h_{fd}\, V_d^2/2g\right) = 39.14 \text{ m} = \mathbf{3.84\ bar\ (gauge)}$$

## 9.3 Axial Flow Pump

In these pumps, as the name indicates, liquid flows along the shaft axis. These pumps have high specific speed and are used where large discharge at low delivery head is required. Also the static suction lift has to be small. These restrictions are put because of high specific speed, which requires the pump to operate at high rpm. These are suitable for irrigation, drainage, sewage, etc. These are designed for large capacities and low heads, generally under 12 m and for vertical shaft. Eqs. 9.1.1 or 9.1.12 can be used to find out power requirement. This pump is often called the propeller pump because its impeller somewhat resembles a marine propeller. Figure 9.3.1 shows a schematic view of this pump.

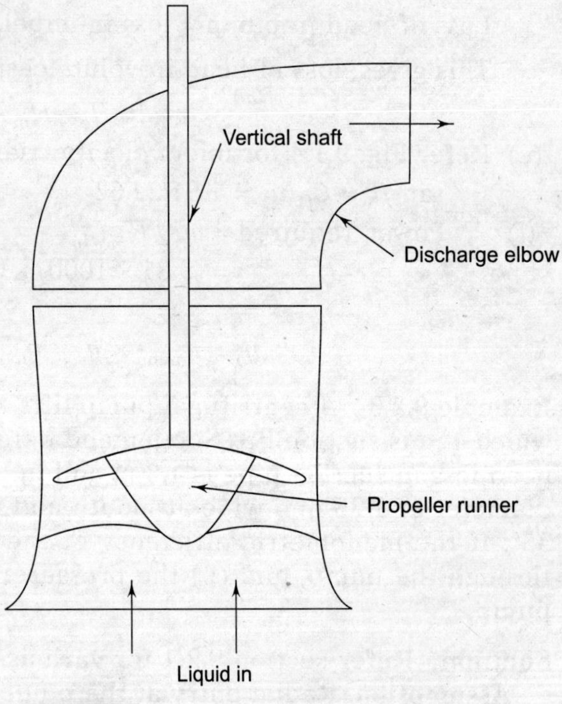


**Fig. 9.3.1** Axial flow pump

The impeller is the place close to the sump, about 1 m above, and therefore chances of cavitation are minimum.

These pumps, unlike centrifugal pumps, cannot be operated throughout the range of operation. Thus the operation is confined close to the point of maximum efficiency or the design point. Otherwise the operation becomes noisy and chances of cavitation are increased. Anderson gave the following empirical relation for the cavitation factor.

$$\sigma = \frac{8.8 \times 10^{-4}}{\eta_Q^2} N_s^{4/3} \tag{9.3.1}$$

Wislicenus introduced another cavitation criterion, known as suction specific speed or cavitational specific speed

$$s = \frac{N\sqrt{Q}}{H_{sv}^{3/4}}, \text{ where } H_{sv} = \sigma_{\text{crit}} \cdot H \tag{9.3.2}$$

Stepanoof gave the value of $s$ as 167.5 for single suction pump and 237 for double suction pump. The maximum value of $s$ attained is 300 for single and 400 for double suction pumps; the higher value is obtained for a more rational construction and higher standard of workmanship for the pumps.

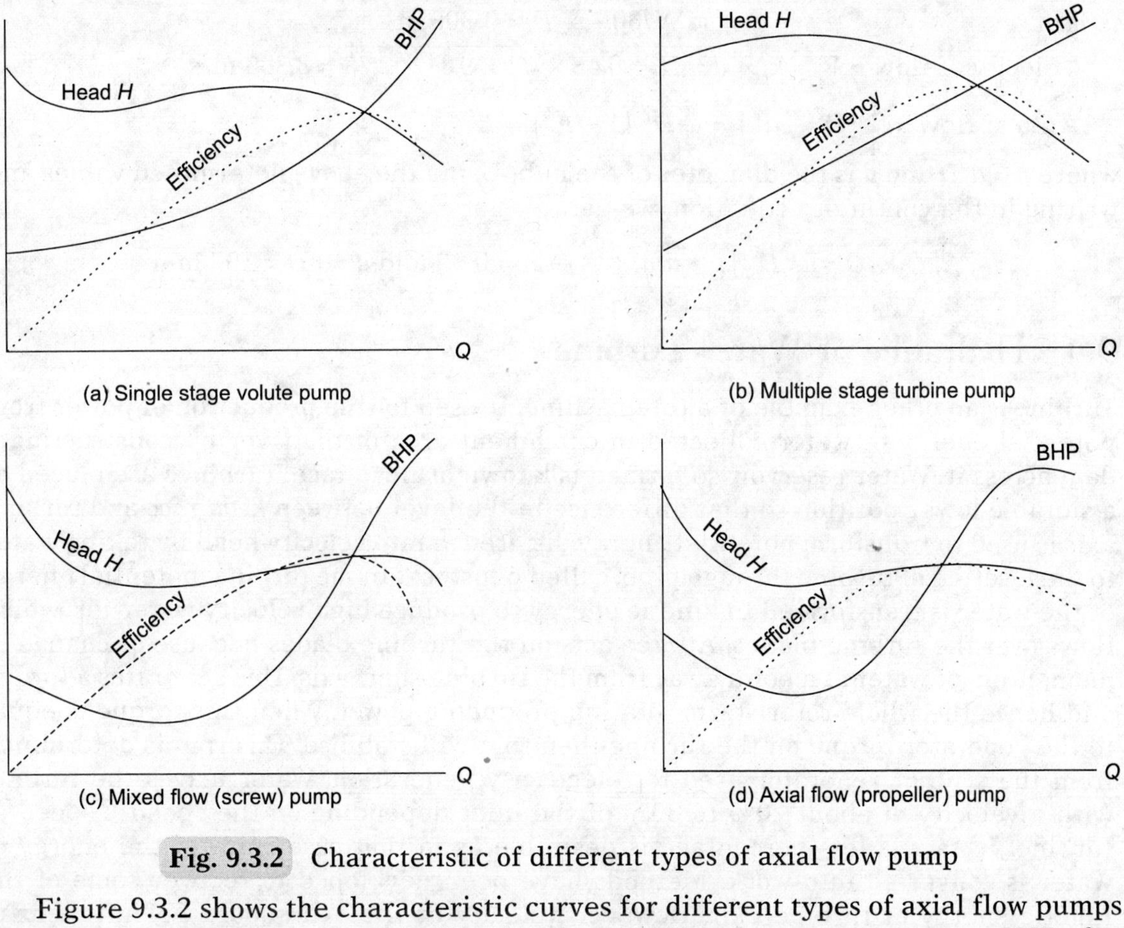


**Fig. 9.3.2** Characteristic of different types of axial flow pump

Figure 9.3.2 shows the characteristic curves for different types of axial flow pumps. The head v/s discharge curves, unlike that for centrifugal pump, are steeper. BHP of an axial flow pump increases with head and therefore overloading of motor is a possibility in such case. So we have to be careful while operating an axial flow pump. The precaution is also taken by the manufacturer by way of providing a device to stop the motor before overloading.

**Example 9.3.1** Find the main dimensions of a propeller pump which is required to deliver 210 l/s of water against a head of 2.75 m. Assume specific speed of the pump as 200 units, speed ratio 2.1 and flow ratio 0.45.

**Solution**

**Given:** $Q = 210$ l/s $= 0.210$ m$^3$/s, $H = 2.75$ m, $K_u = 2.1$, and $K_f = 0.45$

Using the expression for the specific speed of a pump, we get

$$N = N_s \frac{H^{3/4}}{\sqrt{Q}} = 200 \times \frac{2.75^{3/4}}{\sqrt{0.21}} = 935 \text{ rpm}$$

Induction motor operating at 960 rpm is available in the market. Therefore, we choose $N = 960$.

Peripheral speed $u = K_u \times \sqrt{2gH} = 2.1 \times \sqrt{2 \times 9.81 \times 2.75} = 15.4$ m/s

But $u = \pi DN/60 \rightarrow D = 0.305$ m

Velocity of flow $= V_f = k_f \times \sqrt{2gH} = 0.45 \times \sqrt{2 \times 9.81 \times 2.75} = 3.305$ m/s

Area of flow $= \dfrac{\pi}{4}\left(D^2 - d^2\right) = \dfrac{\pi}{4} D^2 \left(1 - n^2\right)$

where $n = d/D$ and $d$ is the diameter of the hub. Using the above determined values and writing in the continuity equation, we get

$$Q = \frac{\pi}{4} D^2 \left(1 - n^2\right) \times V_f \rightarrow n = 0.35 \text{ and } d = nD = 107 \text{ mm}$$

## 9.4 Hydraulic or Water Turbine

Turbine is an other example of a rotamachine. It used for the production of power from potential energy of water collected in catchment area of the river by constructing a dam across it. Water reservoir so formed is known as head race. Turbines are placed at a suitable lower position so that difference in the level between *head race* and turbine axis is used to transform potential energy difference into velocity head by taking water to the location of turbine through pipe, called penstock. In the turbine, potential energy of the water is transformed to kinetic energy to produce high velocity water jet, which flows over the turbine blades. A force acts on the turbine blades because of change in momentum of water at a point away from the turbine shaft axis. This generates a torque and hence the whole rotor starts rotating producing power. When this torque is equal to the generator torque on the turbine then rpm will stabilise. This rpm is determined from the type of generator used for electricity generation. Water leaves the turbine with a velocity of about 20% to 30% of the inlet depending on the specific speed of turbine. In fact, a less percentage is desirable as in that case more of the energy of water is converted into work. Methods have been developed to recover some of the kinetic energy at turbine exit. However, velocity at the exit should be sufficient to move water further downstream. Velocity triangles at inlet and exit may be drawn and power produced can be calculated using Euler turbine equation.

Water is collected at different parts of the land whose topography and quantity of water flowing through the rivers differ greatly and therefore the conditions at different places will be different. In some rivers, water may flow throughout the year whereas in others it flows only during the rainy season. Not only this, the height of the dam and the location of the turbine house will vary with location and hence the available head. Therefore, a proper study of the data available for water runoff in the area where a water turbine plant is to be erected is made for the last 20 to 25 years or even more and the average value of the discharge available is calculated. Also from the topological study of the terrain, safe height of the dam and submersible area (area around river drowned under water as a result of storage) is calculated.

In a nutshell, there are so many variables that a single turbine cannot be thought of to match all the requirements. Different inventors like Pelton, Francis, etc. developed different types of turbines, named after their names. It should be noted that in India, hydraulic power plants are in use as base load as well as peak load plants. *Base load* plants are those which produce power throughout the year and for full 24 hours whereas a *peak load* plant runs only at the time when electrical load is more than that produced by a base load plant. Since hydraulic plant can be started in no time, it suits as peak load plant very well.

This also gives an opportunity to use a storage plant at those places where water is not available continuously for power generation and there exists another base load plant in that area. Idea is to use a turbine pump — a rotamachine that works as turbine when rotated in one direction else as pump in another direction of rotation. When there is demand on the base load plant more than its average load then water flows through turbine and electricity is generated. Exit water is stored in a tank. At a time of the day when there is not much load on the base plant then the machine is run as a pump and water is lifted back to the head race. No doubt, power required for lifting water is more than that is generated by turbine. But since the base load plant is able to run at constant load, it runs at its maximum efficiency that offsets the power loss in rotamachine and in overall situation there is a gain. Such plant is known as *storage plant.*

It will be desirable at this stage to study the types of turbines being in use so that we may get a better picture of these machines. Following chart gives its classification:

**Classification of Turbine**

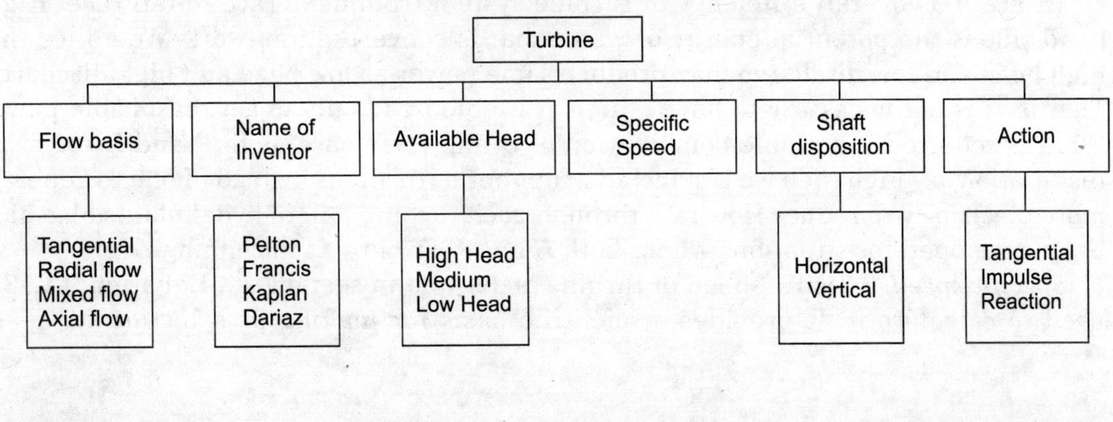

***Reaction and Impulse action***. While discussing pump, we mentioned about impulse action. The classification is based on whether fluid expands or not while passing through the runner space. Accordingly it is classified as impulse or reaction turbine. Turbine in which pressure is atmospheric everywhere like Pelton wheel, expansion does not occur and hence is impulse turbine. Pressure decrease is possible only in turbines completely closed from atmosphere. Therefore it is possible to keep water pressure above atmospheric pressure at the turbine inlet or below it at the turbine outlet. Thus in a reaction turbine pressure decreases and relative velocity of fluid increases while fluid flows over the runner blades. During flow, kinetic energy is transferred to work and hence velocity at exit is less. We may write

$$Cr_2 > Cr_1 \quad \text{for reaction turbine and} \quad Cr_2 \leq Cr_1 \quad \text{for impulse turbine}$$

Whereas $C_2 < C_1$ for both turbines (9.4.1)

For impulse turbine, $Cr_2$ may be less than $Cr_1$ because of frictional losses over the blade. Ratio $Cr_2/Cr_1$ is called *blade velocity coefficient*, which depends on the kind of surface, manufacturing process, and Reynolds number of the flow and may be around 0.9.

***Low, Medium or High Head***. Water leaving the turbine is left in tailrace. Difference between mean sea levels of free surface in head race and tail race in ideal case is known as *gross available head* $H_g$. Ideal case means no losses during flow from head race to tail race. However in Pelton turbine available gross head is difference of level of water level in head race and turbine shaft axis, since pressure everywhere is atmospheric. (Thus positioning of such turbine matters a lot as height of turbine above tailrace is lost as useful head). Furthermore, some head is lost due to friction when flow takes place through pipe (or channel) and turbine. Therefore the head actually available for work is less and is called net available or simply *net or available head*, *H*. Thus

$$H = H_g - h_f - h$$

where $h$ is height of turbine shaft axis from tailrace level, in case of Pelton turbine. Turbine is classified on the basis of gross head as high, medium or low head turbine. *Conventionally high means above 250 m, medium 60–250 m and low below 60 m of average head.* However, these numbers give an idea of the range. If discharge $Q$ is available from a source at a head of $H$ m then power developed by the turbine will be:

$$P = wQH \times \eta_o / 1000 \text{ kW} \tag{9.4.2}$$

Where $\eta_o$ is overall efficiency of turbine system from head race to tail race. Right hand side is the potential energy of water that is converted into work. We notice that high head and low discharge may produce same power as low head and high discharge. That is, it is not necessary to have a high available head only to get reasonable power. What is required is suitable combination of $Q$ and $H$. In case, both $Q$ and $H$ for given installation are high then we can install a number of turbines instead of one to generate more electricity to reduce flow rate through each turbine. Micro-hydel plants also have been developed for situations where both $H$ and $Q$ are low as in running river.

***Specific speed***. Specific speed of turbine is defined in section 9.1. Equation 9.1.18 is used to calculate it. It provides a scientific basis for making its selection for given situation and comparing performance of different kind of turbines. Table 9.4.1 gives idea of specific speeds and available head for which commonly used turbine is suitable. Inspection of this table reveals that:

(i) Specific speed increases from radial to axial turbine.
(ii) Speed ratio $K_u$ and flow ratio $K_f$ increase with specific speed.
(iii) Rotor angles at its inlet and outlet, measured from + ve $u$ direction, increase with specific speed.
(iv) Fluid leaves rotor blade radially and then flows to tail race.
(v) Number of blades decreases drastically with increase in specific speed.
(vi) Runner diameter also decreases with specific speed (results are not tabulated here as its actual size would depend on other factors too). However, ratio of width to runner diameter increases to offset reduction in area of flow due to decrease in diameter.
(vii) Flow direction changes from radial to axial with specific speed.

Consider a place where 67 m³/s of water is available at head 12.8 m. We have to decide the type of turbine to be used and its number required. Let us assume overall efficiency is 90% and that the generator coupled with turbine runs at 1610.47 rpm. Equation 9.4.2 and 9.1.18 give total power and specific speed

$$P = 1000 \times 9.81 \times 67 \times 12.8 \times 0.9/1000 \text{ kW} = 7571.75 \text{ kW}$$

and $$N_s = 1610.47 \times \sqrt{7571.75}/(12.8)^{5/4} = 599$$

Thus any turbine with this specific speed can be used. As per Table 9.4.1, Kaplan turbine is suitable. We may even decide to choose 2, 3 or 4 turbines. In that case each turbine will produce power $P' = P/n$, where $n$ stands for the number of turbines. Specific speed of turbine is calculated and tabulated in the following table:

| No. of turbine | $N_s$ | No. of turbine | $N_s$ | No. of turbine | $N_s$ | No. of turbine | $N_s$ |
|---|---|---|---|---|---|---|---|
| 1 | 599 | 2 | 423.57 | 3 | 345.85 | 4 | 299.5 |

This table provides many choices and one can take appropriate decision regarding the type of turbine and its number. This should be kept in mind that efficiency increases with specific speed and therefore a turbine with higher specific speed will be preferred. It will also be possible to see if any turbine available with us or easily available in the market may be used and does not require ordering a tailor-made turbine. Idea of available head also guides in making a suitable choice.

This plant generates around 7.5 MW of electricity. If we have 6–7 m head available at a place then power produced is of the order of 3 to 4 MW. In our country where number of canals are constructed to take water from river to far off places, this provides a good source of energy. Fortunately, discharge through these canals is much more. The added advantage of such a plant is that electricity produced may be distributed in surrounding areas to reduce distribution cost and power loss.

**TABLE 9.4.1.** Important features of most commonly used hydraulic turbines

| Variable | Pelton | Francis | | | Kaplan/Propeller |
|---|---|---|---|---|---|
| | | slow | medium | fast | turbine |
| Efficiency | | Increasing | ⟶ | ⟶ | |
| Diameter | | Decreasing | ⟶ | ⟶ | |
| Specific Speed | 10–50 | 50–100 | 100–150 | 150–250 | 250–850 |
| | | | ⟶ | ⟶ | |
| Head, m | >250 | 250 | | 50 | < 50 |
| Flow direction | Tangential | Radial | Mixed | Mixed | Mixed to axial |
| Action | Impulse | Reaction | Reaction | Reaction | Reaction |
| $K_u$ | 0.45–0.47 | 0.6–0.69 | 0.69–0.75 | 0.75–0.88 | 0.88–2.2 |
| $K_f$ | 0.97–0.99 | 0.12 | 0.12–0.28 | 0.28 –0.35 | 0.35–0.70 |
| $\alpha_1$ | 0 | 12–25 | 25–32 | 32–40 | 40–75 |
| $\beta_1$ | 0 | 60–90 | 90 | 90–120 | 120–165 |
| $\alpha_2$ | | 90 | 90 | 90 | 90 |
| $\beta_2$ | | 15 | 15 | 15 | 15 |
| $n = B/D$ | | 0.04 | 0.12–0.25 | 0.25–0.5 | 0.35–0.6 |
| Z | | 20–26 | 14–20 | 10–14 | 3–6 |

***Speed ratio and flow ratio***. Speed ratio $K_u$ and flow ratio $K_f$ are non-dimensional blade speeds and flow velocity, defined as

$$K_u = u / \sqrt{2gH} \qquad \text{and} \qquad K_f = C_f / \sqrt{2gH} \tag{9.4.3}$$

Their value for different turbines is given in Table 9.4.1. These are useful variables for designing a given turbine. In hydraulic systems, $\sqrt{2gH}$ is commonly used as reference velocity for making a given velocity non-dimensional.

***Flow Direction***. As the name suggests, this is the classification based on the direction of the flow with respect to the axis of the turbine. This means that flow in a radial machine is along normal to shaft axis, in axial machine along the axis and in mixed flow turbine, it is some where in between. The fluid enters radially and leaves axially in the mixed flow turbines. From Table 9.4.1, it is observed that as the nature of flow shifts from radial to axial, the specific speed of the turbine increases and the number of runners decreases. Other methods of classification do not need any discussion, since their title gives the idea.

## 9.4.1 Pelton Turbine

Pelton turbine consists of a Pelton wheel around whose periphery a number of double hemispherical cups, called buckets, are fitted at equiangular positions. Water strikes cup at its centre. Figure 9.4.1(a) shows the arrangement of buckets around the periphery the and (b) shows its details. The cup is divided into two halves by a sharp edge that divides entering water equally in two streams, one moving left and the other to right; each side of the cup is almost a hemisphere. Water then turns sideways leaving out from the outer side of the cup rim so that the bucket turns it by 160–165°. Various dimensions of the cup with reference to the jet diameter $d$ are:

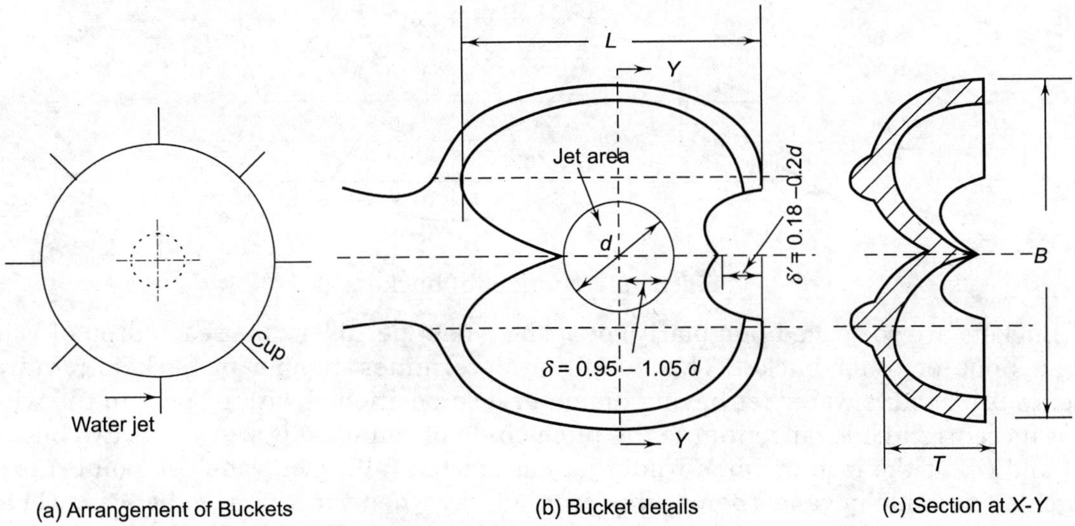


**Fig. 9.4.1** Pelton wheel and cup details

radial length $L = 2.3$ to $2.8\ d$, depth $T = 0.6$ to $0.9\ d$ axial width $B = 2.8$ to $3.2\ d$,

$\beta_1 \approx 5°$ to $8°$, $\beta_2 \approx 10°$ to $20°$.

Ideal inlet velocity diagram is a straight line but in practice, cup angle at inlet is around 5° to 8° because of the dividing surface. Water flows from head race through pipe, called penstock, to the nozzle fitted at the end of pipe from where it travels through atmosphere and strikes cup at its centre tangentially to the pitch circle. Consequently, this turbine is called tangential flow type. Water flows smoothly over the cup and develops power, which rotates turbine shaft in its bearings. At a time water strikes only a few buckets, 2 or at the most 3, and the rest of them rotate empty with wheel. The empty bucket interferes with jet, the moment it enters jet path and splinters water reducing the quantity of water striking the cup and hence loss of power produced. To avoid it, material from bottom of the bucket is notched out so that water may pass through it without any interference; its depth $\delta'$ is such that the bucket in perpendicular position to the jet direction gets whole water and the incoming bucket is just ready to take water with further rotation. $\delta' = kd$, where $k$ is normally taken as 0.18 to 0.2 $d$. Notch point is at a distance $\delta \approx 0.95$ to $1.05\ d$ from the centre of cup.

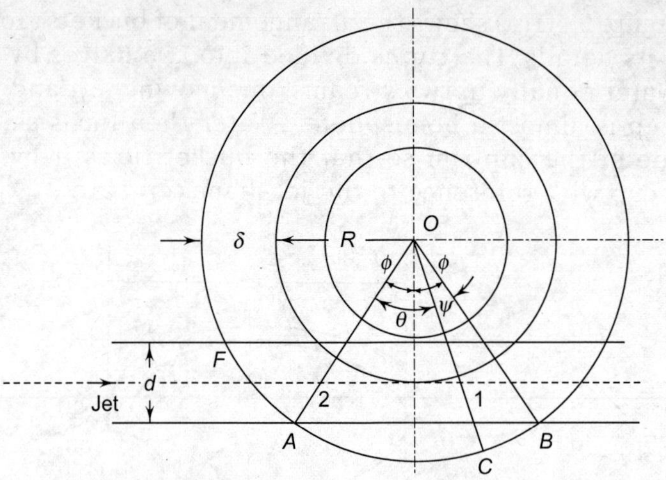


**Fig. 9.4.2** Number of buckets

Buckets are so spaced (angularly) over the wheel periphery that each drop of water strikes one or other bucket. This condition determines number of buckets required. Figure 9.4.2 shows water jet having diameter $d$ so positioned with respect to the wheel that its centre line is tangential to the pitch circle of radius $R$. It also shows two buckets $OA$ and $OB$ at the instant when whole jet has started falling on vane $OA$; point $A$ is the notch point made in vane. (Before this position, say when vane 2 was in between $OA$ and $OF$ positions, part of the jet falls on it and part of it passes to right.). However, water to its right, which passed earlier, travels with velocity $C$ the distance up to bucket 1 so that no drop escapes striking surface to produce work. For this, time taken for travel of water should be equal to or less than that taken by bucket 1 to reach point $B$, else some of water will not hit the surface. For minimum number of blades, this time should be equal. Equating two times give the angle of rotation of the bucket in the mean time. Time taken by the wheel to rotate angle $\psi$ at the angular speed $\omega = u/R$ is $\psi/\omega$. We calculate the time taken by the jet to travel distance $AB$ as follows. From Fig. 9.4.2,

Distance to travel by jet = $AB = 2\,(R + \delta)\,\phi \approx 2(R + \delta)\sin\phi$,

where $\cos\phi = (R + d/2)/\,(R + \delta) = (m + 1)/(m + 2k);\ m = D/d,\ k = \delta/d$

$\therefore \quad 2\{(R + \delta)\sin\phi\}/\,C = \psi/\omega$, where $\omega = u/R$

$\rightarrow$
$$\psi = (2u/C)(1 + 2k/m)\sin\phi \tag{9.4.4}$$

$m$ is called *jet ratio* defined as ratio $D/d$. This gives, angle between buckets

$$\theta = 2\phi - \psi \quad \text{and number of bucket, } Z = 360/\theta \tag{9.4.5}$$

In practice it is found that Eq. 9.4.5 gives more number of buckets and therefore empirical relations are used to determine the number. Tygun relation is one among many relations. It is

$$Z = 0.5\text{ m} + 15 \tag{9.4.6}$$

For production of more power, more than one jet can issue water jet. This also ensures that more buckets remain full with water at a time. These nozzles are spaced equi-angularly around the wheel. Up to 6 nozzles are in use but the most common installations have only 2 placed diametrically opposite. Naturally, specific speed will increase as $\sqrt{n}$, where $n$ is number of nozzles used.

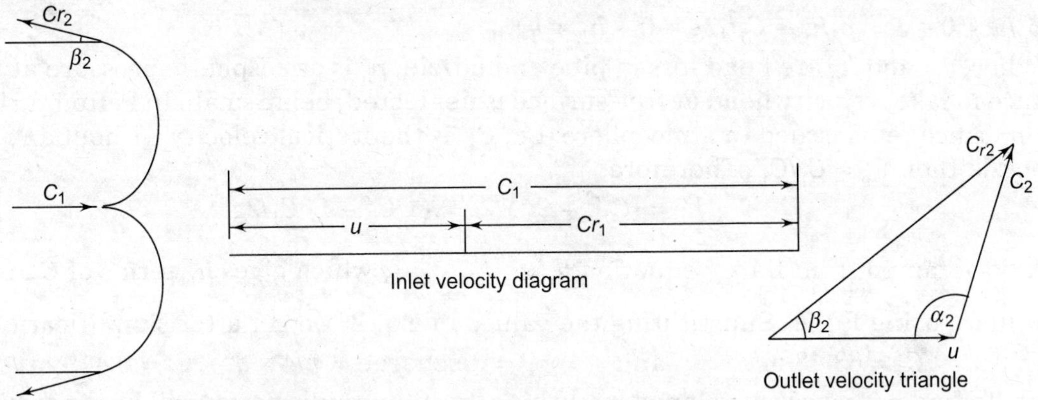


**Fig. 9.4.3** Velocity diagram for Pelton wheel

The inlet velocity triangle will be a line, as shown in Fig. 9.4.3; outlet velocity triangle is also shown. Theoretical speed ratio for maximum efficiency is ½ but in practice its value is taken around 0.47. Normally angle $\beta_2$ is around 10° to 15°. With speed ratio ½ and such a small angle $\beta_2$, velocity triangle will take shape as shown in figure. We may determine work done by turbine, volumetric efficiency, mechanical and overall efficiency using Eqs. 9.1.1, 9.2.10, 9.2.9, and 9.2.11 respectively. However, for turbine hydraulic efficiency expression will change to

$$\eta_h = H_e / H \tag{9.4.7}$$

where $H_e$ is Euler head and $H$ is available head.

**Example 9.4.1** A Pelton wheel nozzle, for which velocity coefficient $C_v = 0.97$, is 400 m below water level of a lake. The jet diameter is 80 mm, the pipe diameter $D$ is 60 cm, its length is 4 km and $f = 0.0032$. The bucket deflects the jet trough 165° and they run at 0.48 times the jet speed, bucket friction reducing the relative velocity at outlet by 15% of the relative velocity at inlet. Mechanical efficiency = 90%. Determine (a) flow rate and (b) shaft power developed by the turbine.

**Solution**

**Given:** $d = 80$ mm, Pipe dia, $D = 60$ cm Pipe length, $L = 4$ km $f = 0.0032$

$K_f = 0.97$ $K_u = 0.48$ $\eta_{\text{mech}} = 0.9$ $K = 1 - 0.15 = 0.85$

$z$ = height of water surface in lake above nozzle axis = 400 m

**Steps:** This problem explains how to take into account the head loss in pipe, nozzle and to calculate shaft horse power. Following steps will do the needful.

1. We have to first determine jet velocity so that discharge through nozzle may be determined while taking into account the friction loss in pipe. It can be obtained by using (i) mass conservation between two sections one in pipe and another at nozzle

outlet (ii) modified Bernoulli's equation between points on free surface in lake and exit of nozzle. This gives, if $A$ and $C$ are pipe area and velocity of flow in pipe respectively,

$$AC = a\,C_1 \text{ (Mass conservation )} \quad \Rightarrow \quad C = (d/D)^2\,C_1 = 0.0177\,C_1 \qquad \text{(A)}$$

$$p_a/w + 0 + z = p_1/w + C_1^{\,2}/2g + 0 + h_L + h_n, \qquad \text{(B)}$$

Where $h_L$ and $h_n$ are head loss in pipe and nozzle, $p_a$ is atmospheric pressure at free surface in lake, velocity head at free surface is neglected, being small. In Pelton turbine $p_1 = p_a$, since jet is issued in atmosphere. Let $C_1'$ is theoretical velocity without any loss in nozzle then $C_v = C_1/C_1'$. Therefore,

$$h_n = (C_1'^2 - C_1^2)/2g = (1/C_v^2 - 1)\,C_1^2/2g \qquad \text{(C)}$$

$h_c$ is obtained from Darey equation $h_L = \dfrac{4fL}{D} \times \dfrac{C^2}{2g}$, which gives $h_L$ terms of $C$ and in turn in $C_1$, using Eq. A. Substituting the values in Eq. B, we get after simplification

$$C_1 = 83.48 \text{ m/s} \qquad \text{and} \qquad Q = \text{discharge} = \pi/4 \times d^2 \times C_1 = 0.419 \text{ m}^3\text{/s}$$

2. To find out power developed by turbine, first hydraulic power will be determined from the velocity diagram, which is shown in Fig. 9.4.3. Let us first find out all that is required to draw it. $C_1$ has been obtained and therefore $u = K_u \times C_1 = 40.07$ m/s, since $K_u$ is given. Inlet triangle gives $Cr_1 = C_1 - u = 43.4$ m/s and then $Cr_2 = 0.85 \times Cr_1 = 310.89$ m/s. It may be noted that bucket velocity coefficient is not given rather loss in relative velocity is given. From this bucket velocity coefficient comes out to 0.85. Also angle of deflection of jet is 165°, which means $\beta_2 = 180 - 165 = 15°$. With this data, outlet velocity triangle can be completed. This gives, assuming runner is fast ( see Fig. 9.4.3):

$$Cw_2 = u - Cr_2 \cos \beta_2 = 4.44 \text{ m/s},$$

$$\text{Hydraulic power} = (wQ/g)\,(Cw_1 - Cw_2)\,u = 1327 \text{ kW}$$

$$\text{Shaft power} = 0.9 \times 1327 = 1194.3 \text{ kW}$$

Note a positive value of $Cw_2$ means that our assumption is correct. It would not affect our calculations had it been zero or negative as the value was to be used in turbine equation taking the sign as well.

**Example 9.4.2** A Pelton wheel is to be designed for shaft power 9560 kW, head 350 m, speed 750 rpm, overall efficiency 0.85, and jet diameter not to exceed 1/6 of the wheel diameter. Determine: (i) wheel diameter, (ii) jet diameter, and (iii) Number of jets required. Take nozzle velocity coefficient $C_v = 0.985$ and speed ratio = 0.45

**Solution**

**Given:** $P = 9560$ kW $\quad N = 750$ rpm $\quad H = 350$ m

$\eta_o = 0.85$ $D/D \le 1/6$ $\quad C_v = 0.985$ $\quad K_u = 0.45$

**Steps:**

1. Wheel diameter can be obtained if $u$ is known since $N$ is given. But $C_1 = C_v \times \sqrt{2gh}$ = 81.62

$\therefore \quad u = K_u \times C_1 = 37.3$ m/s $\quad$ and $\quad D = 60\,u/(\pi N) = 0.95$ m

2. Taking $d/D = 1/6$, we get $d = 0.158$ m

3. To find out number of jets, what we require is the total discharge and discharge of each nozzle. Discharge of each nozzle is = $\pi d^2/4 \times C_1$ = 1.6 m³/s. Further,

$$wQH/1000 = \eta_o \times P \quad \Rightarrow \quad Q = .85 \times 9560/(9.81^7 \times 350) = 3.27 \text{ m}^3/\text{s}$$

and number of jets required = 3.27/1.6 = 2

Now we have to recalculate diameter of nozzle so that each of the nozzle has discharge = 3.27/2 = 1.635 m³/s.

This gives: $d = \sqrt{\dfrac{4 \times 1.635}{\pi \times 81.62}} = 0.1597$ m

**Example 9.4.3** A single jet Pelton turbine is required to drive a generator to develop 10 MW. The available head at the nozzle is 760 m. Assuming electric generation efficiency 95%, Pelton wheel efficiency 87%, coefficient of velocity for nozzle 0.97, mean bucket velocity 0.46 of jet velocity, outlet angle of bucket $15^0$ and the relative velocity of the water leaving the buckets 0.85 of that at inlet, find: (i) the flow in m³/s, (ii) the diameter of jet, (iii) force exerted by the jet on the bucket, (iv) the best synchronous speed for generation at 50 Hz and (v) the corresponding mean diameter if the ratio of the mean bucket circle diameter to the jet diameter is not to be less than 10.

**Solution**

**Given:** $P = 10000$ kW $\quad H = 760$ m $\quad \eta_{gen} = .95 \quad \eta_h = .87$

$C_v = 0.97 \quad K_u = .46 \quad \beta_2 = 15 \quad K = 0.85$

$m \geq 10 \quad f = 50$ Hz

1. Here Pelton wheel efficiency is nothing but the hydraulic efficiency. To find $Q$, Eq. 9.4.2 may be used as all variables except $Q$ are known. Otherwise also power generated will depend on discharge and hence this information should be used for determination of $Q$. Moreover, hydraulic power of turbine will be more than what is available at generator outlet because of mechanical losses, accounted for by mechanical efficiency, and generator losses, taken care of by generator efficiency. This gives,

$$P = 10000/(\eta_{gen} \times \eta_{mech}) = 9.81 \times Q \times 760 \Rightarrow$$
$$\mathbf{Q = 1.62\ m^3/s}$$

2. Now jet diameter can be obtained from $Q = \pi d^2/4\ C_1$, where $C_1 = C_v \times \sqrt{2gH}$ This gives,

$$C_1 = 0.97 \times \sqrt{(2 \times 9.81 \times 760)} = 118.45 \quad \text{and} \quad d = \mathbf{0.132\ m}$$

3. Force exerted on bucket is given by $\rho\, Q\, (C_{w1} - C_{w2})$, for which velocity triangle, as shown in Fig. 9.4.3 are drawn. From given data:

At inlet, $\quad C_{w1} = C_1 = 118.45$, $u = K_u \times C_1 = 54.5$, $C_{r1} = C_1 - u =$ 64 m/s

and at outlet, $\quad C_{a2} = K \times Cr_1 = 54.4$, $C_{w2} = u - C_{r2} \cos \beta_2 = 1.95$

[7] $w/1000 = 1000 \times 9.81/1000 = 9.81$. That is, when power is in kW then $w$ can be taken as 9.81.

a + ve sign means that $\alpha_2 < 90°$

$\therefore$ Force on bucket = $1000 \times 1.62 \times (118.45 - 1.95) = 188.8$ kN

4. We mentioned earlier that turbine speed is chosen according to that of generator for which it depends on the number of poles. Assuming minimum jet ratio 10, wheel diameter = $10 \times d = 1.32$ m and for this diameter, rpm of turbine is $N = 60u/\pi D = 788$. For generator, $N_g$ = generator rpm = $120\ f/p = 120 \times 50\ /p = 6000/p$, where $p$ is number of poles, which will be always even number as poles are present in pair in a generator. For its rpm nearest to that of turbine, $p = 8$ and $N = 750$. Therefore diameter of the wheel is to be changed to give desired bucket velocity. Thus $D = 60 \times 54.5/(\pi \times 750) = 1.387$ m

### 9.4.2 Francis Turbine

Pelton wheel is used for high heads. But there are many places where available head may not be that high as required for Pelton turbine. In that case, as mentioned earlier, we can still generate reasonable waterpower by running a machine with high discharge. But this is not that simple as it seems to be. Lower $H$ means lower $u$ for same optimum value 0.47 of $K_u$ (for Pelton wheel) and therefore lower runner diameter $D$, since rpm of a turbine is controlled by the generator. This in turn will reduce area of flow as it is equal to $\pi DB$, where $B$ is width of runner and is chosen as fraction of $D$. Thus for a machine designed for $K_u = 0.47$, both diameter and width of the wheel will be less and hence it will produce lower power, which we do not intend to. So a solution lies in increasing $K_u$ and $K_f$.

Since the available head is reduced by the height of the wheel from tail race, positioning the pelton wheel above tail race is a problem. A solution to put the turbine under water is not feasible because of maintenance reason. Therefore, solution was thought in the form of a closed turbine in which fluid expands on the runner blades inside a closed casing so that pressure at exit is lower than atmospheric pressure. Water is then taken to tail race by a diverging tube, called draft tube. This gives flexibility to place turbine anywhere above tailrace. Fig. 9.4.4 shows plan and front view of Francis turbine in first angle method of orthographic projection. Water, supplied to scroll casing by Penstock from head race, enters fixed guide vanes placed round the runner. Water comes out from guide vane in a direction along tangent to runner blade at entry so that fluid enters smoothly without any shock at designed operating conditions. Francis turbine shaft is normally vertical and therefore plane of runner rotation is horizontal; water comes out from runner in radial direction and falls to draft tube by gravity.

Guide vanes are hinged on a ring such that angular movement of ring by external mechanism connected to governing system rotates guide vanes about their hinge in a manner that the area of flow changes and the passage through vanes may even be closed completely. Thus changing mass flow rate through these guide vanes control load on turbine.

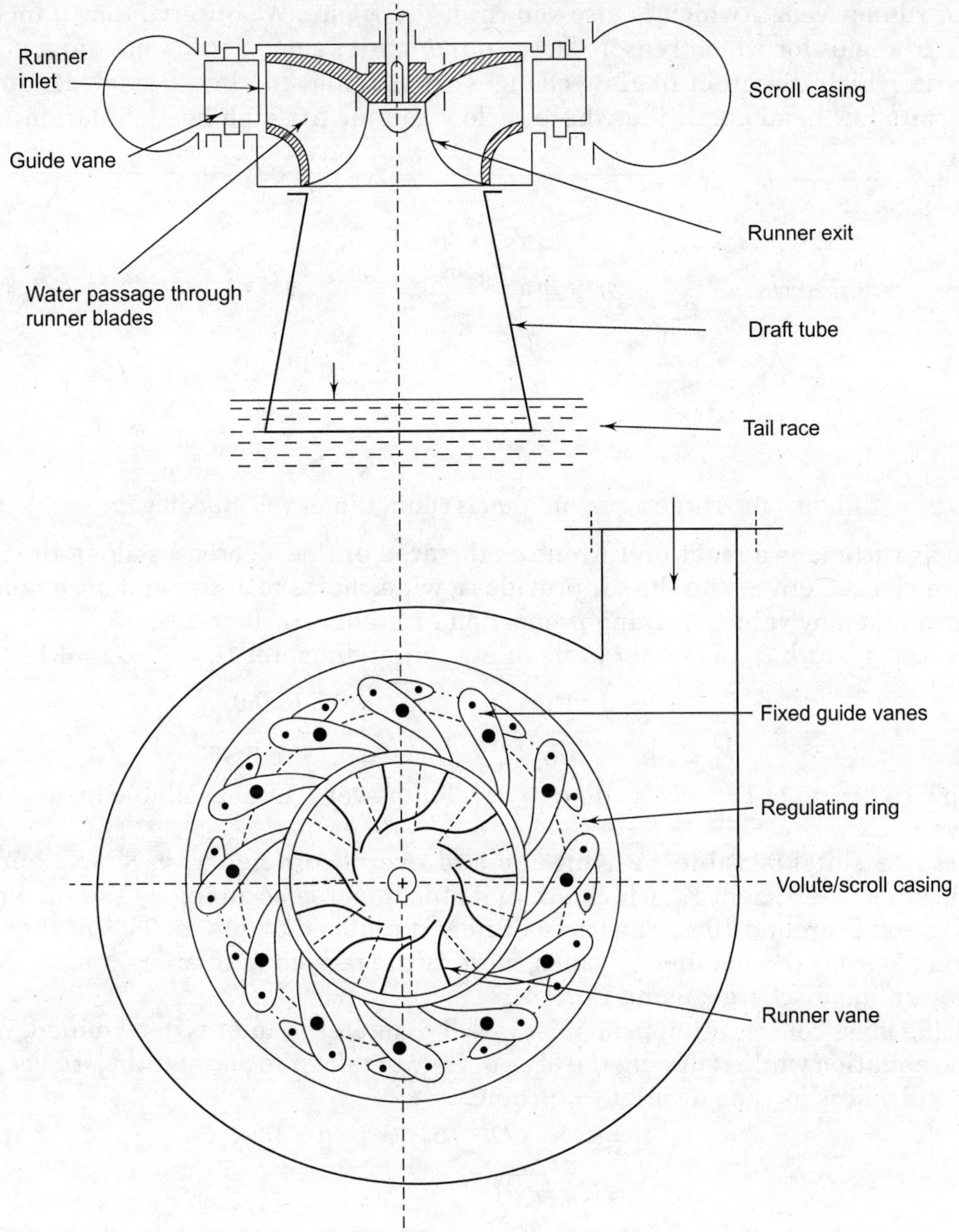


**Fig. 9.4.4** Francis turbine

Francis developed three types of runners with increasing $K_u$, $K_f$, specific speed and decreasing number of vanes (see Table 9.4.1). For this reason, a Francis turbine can be used for a wide range of available heads. The three inlet triangles are shown in Fig. 9.4.5; proportion of runner speed at inlet and absolute velocity is such that blade angles in three cases are > 90°, = 90° and < 90°. Consequently, there is a change in the shape of runner vanes, which is also shown in the figure. We observe that $u$ increases in these triangles for which reason these runners are named as slow, medium and fast runners in which direction of flow changes from radial to almost axial. High speed runner suits low head condition whereas slow runner fits high head situation.

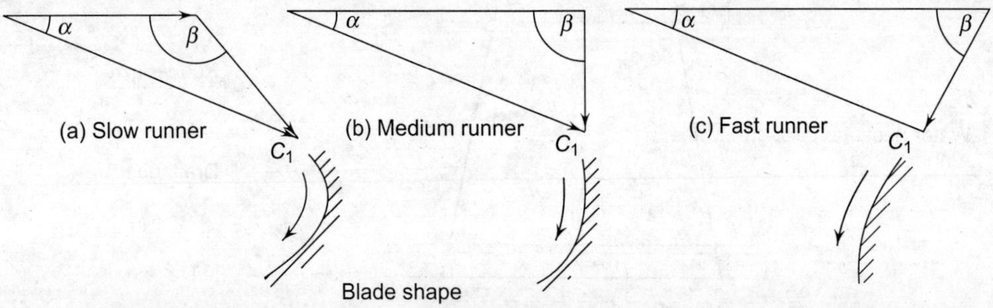


**Fig. 9.4.5** Three types of Francis runner-inlet velocity diagam

Francis turbine is a reaction turbine as the area of flow decreases along the runner vanes which accelerates the flow. It provides a wide choice to designer since blade inlet angle can take any value. Working proportion of Francis turbine are:

$n$ = ratio of width $B_1$ of runner at its inlet to inlet diameter $D_1 = B_1/D_1 = 0.10$ to 0.45,

$$K_u = \text{speed ratio} = u_1/\sqrt{2gH} = 0.6 \text{ to } 0.9,$$

and $$K_f = \text{Flow ratio} = C_{f1}/\sqrt{2gH} = 0.15 \text{ to } 0.30.$$

Stepwise design of Francis turbine to develop power $P$ from available head $H$ at rpm $N$ is given below:

1. Assume suitable value of hydraulic and overall efficiency, $n$, $K_u$, $K_f$, and vane thickness coefficient $K_t$, which accounts for the area occupied by the thickness of blade; it is around 10%. Thus area of flow is reduced to 90% to 95% of theoretical area. More is the number of vanes, more is the reduction in area.
2. Determine discharge, using Eq. 9.4.2.
3. Using mass conservation principle, wheel diameter at inlet is determined. Writing the equation while reducing the area of flow to take into account the area occupied by thickness of vane at inlet to turbine,

$$Q = \pi (K_t \times D_1 \times B_1)\, C_{f1} = n\, \pi\, D_1^2 K_t\, C_{f1}$$

$$C_{f1} = K_{f1}\sqrt{2gH}$$

$$\Rightarrow \quad D_1 = [Q/\{K_{f1}\, K_t\, \pi\, n\sqrt{2gH}\,\}]^{1/2} \tag{9.4.8}$$

Once $D_1$ is known $B_1$ can be obtained.

4. Now $u_1$ can be obtained from equation: $u_1 = \pi D_1 N/60$.
5. Assuming radial flow at exit, i.e., $C_{w2} = 0$, $C_{w1}$ can be obtained from the Euler turbine Eq. 9.1.1. Inlet triangle can be drawn now and angles $\alpha_1$ and $\beta_1$ can be obtained using trigonometric relations.
6. Normally $D_2$ is $D_1/2$ and flow velocity is assumed to remain constant. Continuity equation gives $B_2 = B_1$.

7. Draw outlet velocity diagram and determine angle $\beta_2$.

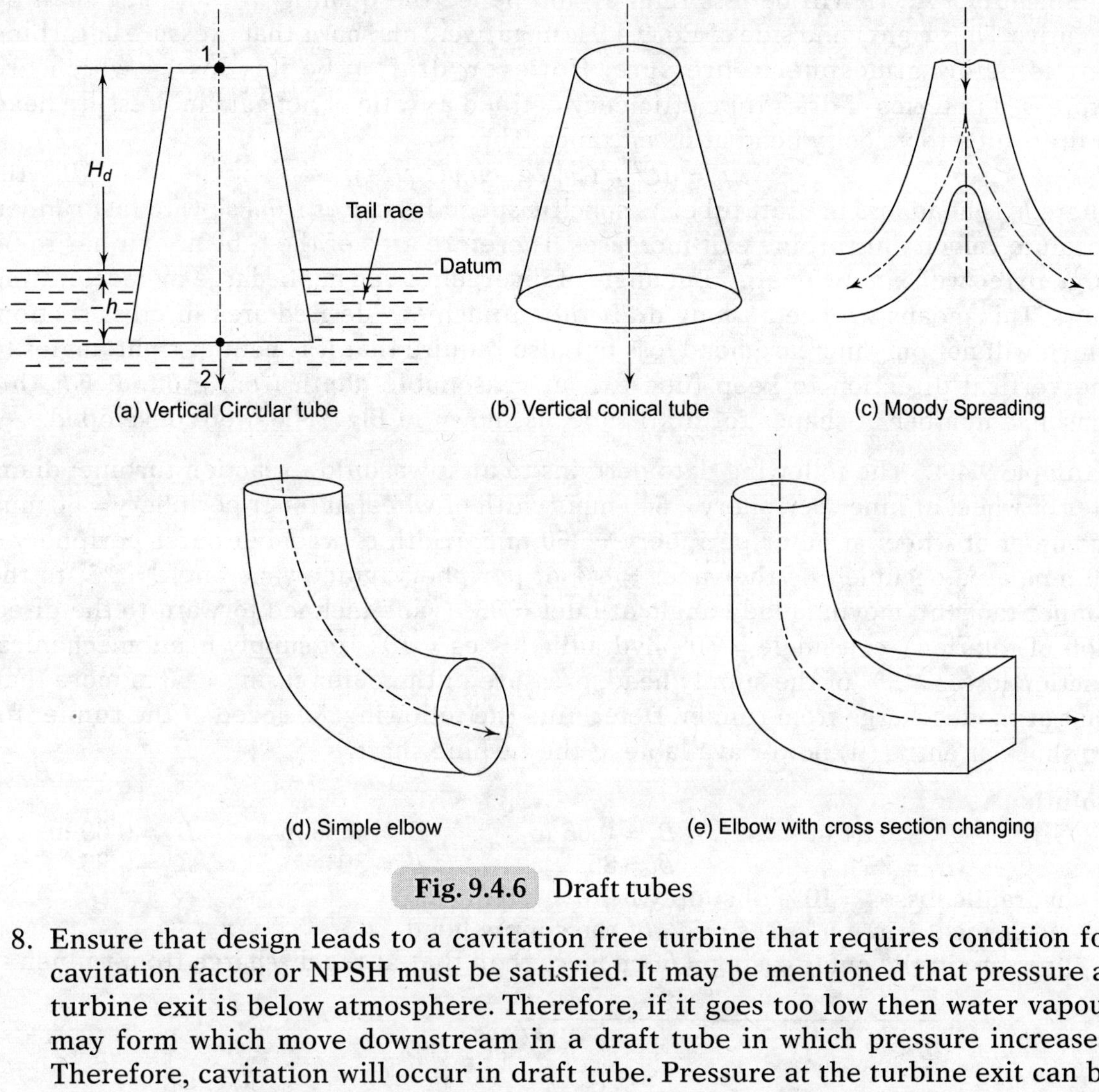


**Fig. 9.4.6** Draft tubes

8. Ensure that design leads to a cavitation free turbine that requires condition for cavitation factor or NPSH must be satisfied. It may be mentioned that pressure at turbine exit is below atmosphere. Therefore, if it goes too low then water vapour may form which move downstream in a draft tube in which pressure increases. Therefore, cavitation will occur in draft tube. Pressure at the turbine exit can be obtained by considering draft tube as diverging pipe, as shown in Fig. 9.4.6(a) with one end 1 at turbine exit and other end 2 dipped under water at depth h in tailrace. We assume no hydraulic losses from inlet of draft tube to its outlet. Applying Bernoulli's and continuity equations at these two points taking free surface of water in tail race as datum, we get

$$p_1/w + C_1^2/2g + H_d = (p_a/w + h) + C_2^2/2g + (-h) \quad \text{and} \quad A_1C_1 = A_2C_2$$

Here $H_d$ is height of turbine exit from tailrace level *and not the length of draft tube.* Tube may be of any length, it affects only if head losses are considered. Solving these two equations, we get

$$(p_1 - p_a)/w = \left(C_2^2 - C_1^2\right) / 2g - H_d \qquad (9.4.9)$$

Since $A_2 > A_1$, $C_2$ will be less than $C_1$ and hence the quantity $\left(C_2^2 - C_1^2\right)/2g$ shall be negative. Thus right hand side of Eq. 9.4.9 is negative. This shows that pressure at turbine exit is below atmospheric pressure. However, draft tube has losses which are expressed in terms of draft tube efficiency defined as ratio of net gain in pressure head in draft tube to velocity head at its entrance. That is,

$$\eta_d = \{\left(C_1^2 - C_2^2\right)/2g - h_f\}/(C_1^2/2g), \qquad (9.4.10)$$

where $h_f$ is head loss in draft tube. As specific speed increases from slow to fast runner, absolute velocity at turbine exit increases. Therefore area of the tube at exit has to be large to recover kinetic energy. But angle of divergence can not be large in decelerating flows. This means we need a long draft tube to achieve desired area of cross section, which will not only increase head loss but also require that it is bent at right angles to the vertical direction to keep tube exit at reasonable depth from ground. For this reason, a number of shapes for draft tube, as shown in Fig. 9.4.6, were developed.

**Example 9.4.4** The following data pertains to an inward flow reaction turbine; diameter of wheel at inner periphery = 540 mm, width of wheel at inner periphery = 60 mm, diameter of wheel at outer periphery = 360 mm, width of wheel at outer periphery = 90 mm, area occupied by the vanes = 8% of periphery, guide vane angle = 25° to the runner tangent, moving blade angle at inlet = 95° (vane inclined forward to the direction of rotation), exit angle = 30°, hydraulic losses = 10% of supply head, mechanical friction losses = 5% of the supply head, pressure in the outer casing = 66 m more than that at the discharge from runner. Determine the following: (i) speed of the runner for no shock at entry, (ii) power available at the turbine shaft.

**Solution**

**Given:** $D_1 = 0.54$ m $\quad B_1 = 0.06$ m $\quad D_2 = 0.36$ m $\quad B_2 = 0.09$ m

$\alpha_1 = 25°$ $\quad \beta_1 = 85°$ $\quad \beta_2 = 30°$ $\quad K_t = 0.95$

hydraulic losses = 10% of supply head

mechanical friction losses = 5% of the supply head

pressure in the outer casing = 66 m more than that at the discharge from runner.

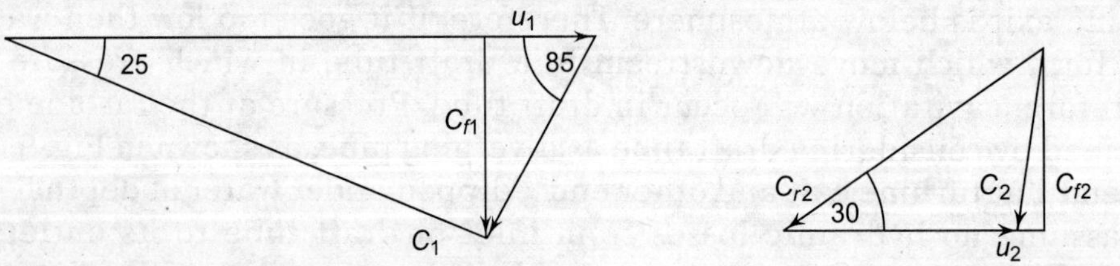


(a) Inlet velocity triangle (b) Outlet velocity triangle

**Fig. E9.4.4** Inward reaction turbine

Note that vane inlet angle is given with reference to direction of rotation of vane whereas in the convention we have taken its supplementary angle. Therefore $\beta_1 = 180° - 95° = 85°$. Since $N$ is not given, we can not draw velocity triangles. But the shape of these triangles may look like that given in Fig. E9.4.4. Difference of pressure head between casing, that is inlet of guide vane to the outlet of runner vane is given. That is, available head is 66 m. Considering the hydraulic losses of 10%, the actual or Euler head is 0.9 × 66 m. This head is used in generating Euler work and kinetic energy of mass leaving runner vane. Therefore it requires to determine kinetic energy at exit first. Taking $K_t$ same at inlet and outlet of runner, continuity equation gives:

$$K_t\,(\pi D_1 B_1)\,C_{f1} = K_t\,(\pi D_2 B_2)\,C_{f2} \quad \rightarrow \quad C_{f1} = C_{f2}$$

From inlet velocity triangle:

$$u_1 = C_{f1}\,(\cot 25 + \cot 85) = 2.23\,C_{f1} \quad \text{and} \quad C_{w1} = C_{f1}\cot 25 = 2.144\,C_{f1}.$$

Also $u_2 = D_2/D_1 \times u_1 = 1.49\,C_{f1}$

From outlet triangle:

$$C_{f2} = C_{f1} = (u_2 + C_{w2})\tan 30$$

$$\rightarrow \quad C_{w2} = 0.243\,C_{f1} \qquad \text{and} \quad C_2 = \sqrt{(C_{w2}^2 + C_{f2}^2)} = 1.03\,C_{f1}$$

$$\therefore \quad 0.9 \times 66 = (C_{w2}\,u_2 + C_{w1}\,u_1)/g + (1.03\,C_{f1})^2/2g$$

$$\rightarrow \quad C_{f1} = 2.14 \text{ m/s and } C_2^2/2g = 5.56 \text{ m}$$

Note that here $C_{w2}$ is negative for the outlet velocity triangle shown in figure. Substitution of $C_{f1}$ yields.

$$u_1 = 2.144\,C_{f1} = 22.61 \text{ m/s, which gives} \quad N = 60\,u_1/(\pi D_1) = \mathbf{800\ rpm}$$

$$Q = \text{discharge} = (K_t\,\pi D_1 B_1) \times C_{f1} = 0.949 \text{ m}^3\text{/s , and}$$

$$P = \text{power} = (w/1000) \times Q \times (C_{w2}\,u_2 + C_{w1}\,u_1)/g \times \eta_{\text{mech}} = \mathbf{4710.3\ kW}$$

**Example 9.4.5** In an inward flow reaction turbine (vertical shaft) the sum of the pressure and kinetic heads at entrance to the spiral casing is 132 m and vertical distance between this section and the tail race level is 3.3 m. The peripheral velocity of the runner at entry is 33 m/s, discharge from the runner is without whirl. The hydraulic losses are: (a) losses between turbine entrance and discharge from guide vanes = 4.95 m, (b) losses in the runner = 8.8 m, (c) losses in draft tube = 0.88 m, (d) kinetic energy head rejected to the tail race = 0.55 m. Determine (i) the guide vane angle and the runner blade angle at inlet, (ii) the pressure heads at entry to and discharge from the runner.

**Solution**

**Given:** Sum of the pressure and kinetic heads at entrance to the spiral casing is 132 m. vertical distance between this section and the tail race level is 3.3 m.

$u_1 = 33$ $\qquad C_{f1} = C_{f2} = 11$ $\qquad C_{w2} = 0$ (i.e. discharge is radial)

Losses between turbine entrance and discharge from guide vanes = 4.95 m,

Losses in the runner = 8.8 m $\qquad$ losses in draft tube = 0.88 m

Kinetic energy head rejected to the tail race = 0.55 m

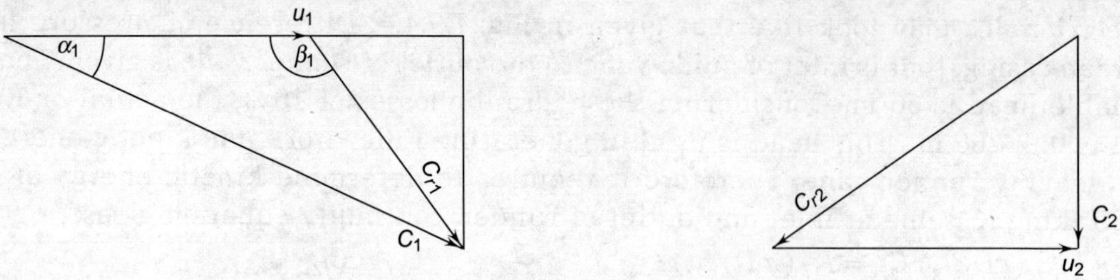


**Fig. E9.4.5** Inward reaction turbine

1. Inlet and outlet velocity diagrams are drawn in Fig. E9.4.5. Since $H$ and $u_1$ are known, $C_{w1}$ can be obtained by using Euler turbine equation, noting that discharge is radial.

   Head utilised by runner, $H = C_{w1}\, u_1/g = 132 + 3.3 - (4.95 + 8.8 + 0.88 + 0.55) = 120.12$

   $\rightarrow \qquad Cw_1 = 35.71$

   From velocity triangle: $\tan \alpha_1 = C_{f1}/C_{w1} \rightarrow \qquad \alpha_1 = 17.12°$

   and $\tan (180 - \beta_1) = C_{f1}/\,(C_{w1} - u_1)$

   $\rightarrow \qquad \beta_1 = 103.84°$

2. Take tail race as datum and apply modified Bernoulli's equation between entrance to turbine and exit of guide vane, we get from given data:

   $$132 + 3.3 = C_1^2/2g + p_1/w + z_1 + 4.95 = C_1^2/2g + p_1/w + 3.3 + 4.95,$$

   since turbine is horizontal

   $\therefore$ $p_1/w$, pressure head at runner inlet = 55.87 m

3. Similarly applying modified Bernoulli's equation between turbine entrance and runner outlet, we get

   $$132 + 3.33 = C_2^2/2g + p_2/w + z_2 + H + 4.95 + 8.8$$

   But $H$ = 120.12 m (determined above), $C_2 = C_{f2} = 11$ and $z_2 = 3.3$

   Substitution and simplification gives:

   pressure head at runner outlet, $p_2/w = -8.04$ m

   Negative sign shows that pressure is negative and is greater than NPSH, which means there are chances of cavitation.

### 9.4.3 Kaplan Turbine

Francis turbine was suitable for medium heads. Therefore turbine with high specific speed were required for low heads to generate power efficiently. That means, flow has to be almost axial and the number of blades to be reduced. Further, Francis turbine perform very well at the design load. However, their performance was degraded at partial loads. This was because the governing mechanism used in this machine was changing the gate opening, as mentioned earlier, which changed direction of absolute velocity of water from guide vane and hence of relative vector too results in entry losses. Kaplan considered both these points and designed a turbine, which is called Kaplan turbine. He developed link mechanism to change direction of runner at different turbine loads, which improved its performance significantly over part loads. In next section, we shall discuss performance characteristics of different turbines. This turbine is very commonly used in planes where there is network of canals since this can operate for heads even less than 10 m; this head can easily be available in actual flow conditions and discharge in such canal is quite high.

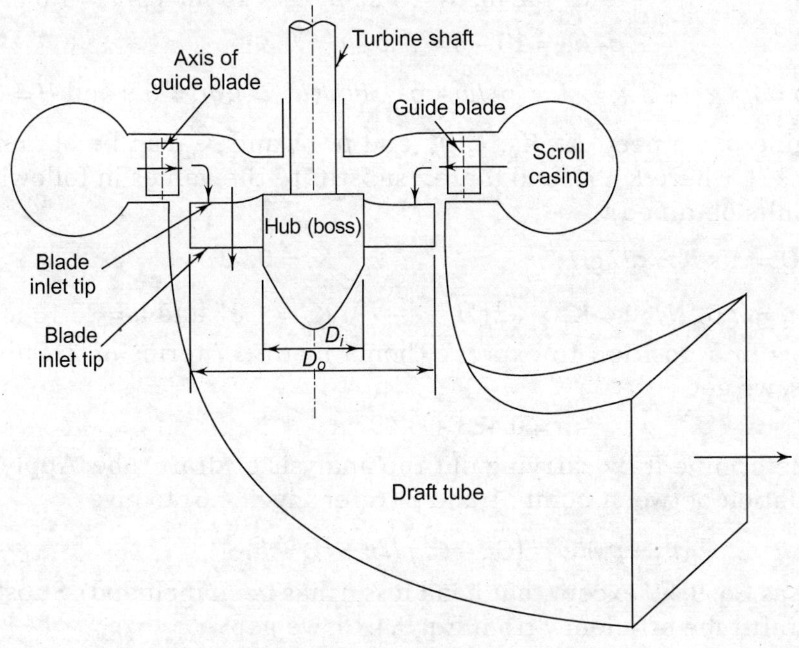


**Fig. 9.4.7** Kaplan turbine

Arrangement for the turbine is shown in Fig. 9.4.7. Water enters the scroll casing from the penstock and is directed over the guide blades, which can be rotated by the governing mechanism about its axis. Guide vane is so designed that water comes out in the axial direction at its exit and falls over the moving blade fitted on the hub. It then goes to the tailrace through the draft tube. The draft tube is very long and runs even in kilometer. The shaft is normally vertical.

**Example 9.4.6** A Kaplan turbine working under a head of 10 m and at a design speed of 250 rpm has a flow rate of 24 m$^3$/s. The diameter of the runner and hub are 2 m and 1 m, respectively. The inlet and outlet diameter of the draft tube are 2 m and 3 m, respectively. The pressure recorded at inlet of the draft tube is 3 m vacuum. The vapour and barometric pressure are 1.6 m and 10 m, respectively. Efficiency of draft tube is 80%. The Thoma's cavitation factor for the turbine is given by the relation $\sigma = \eta_d\ K_f^2 + \lambda\ K_u^2$, in which $\eta_d$ is the draft tube efficiency, $K_f$ is the flow ratio, $K_u$ is the speed ratio and $\lambda$ is a dimensionless number defined by $\lambda = (p_2/w - p_{min}/w)/(u^2/2g)$ in which $u$ is the tangential velocity, $(p_2/w)$ is the pressure head at inlet to draft tube, and $(p_{min}/w)$ is the minimum pressure head at a point on the blade. If overall efficiency is 90%, determine the minimum pressure on the blade.

**Solution**

**Given:** $Q = 24\ m^3/s, N = 250\ rpm, D_o = 2\ m, D_i = 1\ m$

$\rightarrow$ $A = \pi/4(D_o^2 - D_i^2) = 3/4\pi$

Draft tube: $d_1 = 2\ m, d_2 = 3\ m, p_a/w = 10\ m,\ p_v/w = 1.6\ m,$

$$p_2/w = 10 - 3 = 7\ m,\ \eta_d = 0.8$$

$$\sigma = \eta_d K_f^2 + \lambda K_u^2,\ \lambda = (p_2/w - p_{min}/w)/(u^2/2g),\ \eta_o = 0.9 \text{ and } H = 10\ m$$

To determine $\sigma$, we need $K_f$, $K_u$, $\lambda$, $\eta_d$ and $u$. $K_f$ and $K_u$ can be obtained from the relations (9.1.27), where $K_u$, $K_f$ and $u$ are: (substitute the values in following steps and verify the results obtained).

$$Q = A \times K_f \times \sqrt{2gH} \quad \rightarrow \quad K_f = 0.727$$

$$u = \pi D_o N/60 = K_u \times \sqrt{2gH} \quad \rightarrow \quad K_u = 1.87 \text{ and } u = 26.18\ m/s$$

We are now in a position to express Thoma factor in terms of $\lambda$. Substituting the known values, we get

$$\sigma = 0.423 + 3.497\,\lambda \qquad (A)$$

We shall determine it by carrying out the analysis of draft tube. Applying modified Bernoulli equation between points 1 and 2 (refer Fig. 9.4.6) to give

$$(p_1 - p_a)/w = \left(C_2^2 - C_1^2\right)/2g - H_d + h_f$$

(it is same as Eq. 9.4.9 except that head loss $h_f$ has been included). Substituting for $h_f$ in terms of draft tube efficiency from Eq. 9.4.10, we get

$$p_1/w = p_a/w - H_d - \eta_d\,(C_1^2/2g)$$

$C_1$ can be determined as diameter of draft tube at it inlet and $Q$ are given.
This gives: $C_1 = Q/(\pi/4\ d_1^2) = 7.64\ m/s$
Substituting for $\eta_d, C_1, p_1/w, p_a/w$, we get:
$H_d$ = height of the draft tube inlet from the tail race = 0.62 m. Cavitation factor is defined as

$$\sigma = (p_a/w - p_v/w - H_d)/H_d = (10 - 1.6 - 0.62)/10 = 0.778 \qquad (B)$$

Equating (A) and (B), we get, $\lambda = (p_2/w - p_{min}/w)/(u^2/2g) = 0.1015$

$\rightarrow$ $p_{min}/w = -\ 6.545$ m or 6.545 (vacuum)

$\rightarrow$ or $p_{min} = 9.81 \times 6.545 = 64.2\ kN/m^2$ (vacuum)

## 9.4.4 Performance Characteristic of Hydraulic Turbines

***Main Characteristic***. Earlier, while discussing non-dimensional analysis, the importance of unit quantities was mentioned. In turbines, the independent variable is speed which is changed during experiment at a given gate opening (that is, for given load on turbine).

Therefore, gate opening is taken as parameter in the performance curves in which abscissa is $N_u$. Quantities of interest for prediction of performance of machine are discharge, power and overall efficiency. Hence $Q_u$, $P_u$ and $\eta_o$ are taken along ordinate or normally called $y$ axis. Such curves are known as *main characteristic* of turbines and are shown in Fig. 9.4.8 for Pelton, Francis and Kaplan turbine. We observe that

1. At a given gate opening, while unit discharge remains almost same for Pelton wheel, it decreases for Francis turbine and increases for Kaplan turbine with unit speed. A decease in Francis turbine is the result of centrifugal force that opposes inward flow of water in Francis turbine. However, as gate opening decreases, decrease in unit discharge occurs in all turbines.
2. Trend of unit power and overall efficiency is same for all turbines and there is increase in their value with increase in gate opening at any unit speed. Also both of these increase from zero at zero speed (naturally) with unit speed, reach maximum at some value and then falls off to zero finally.

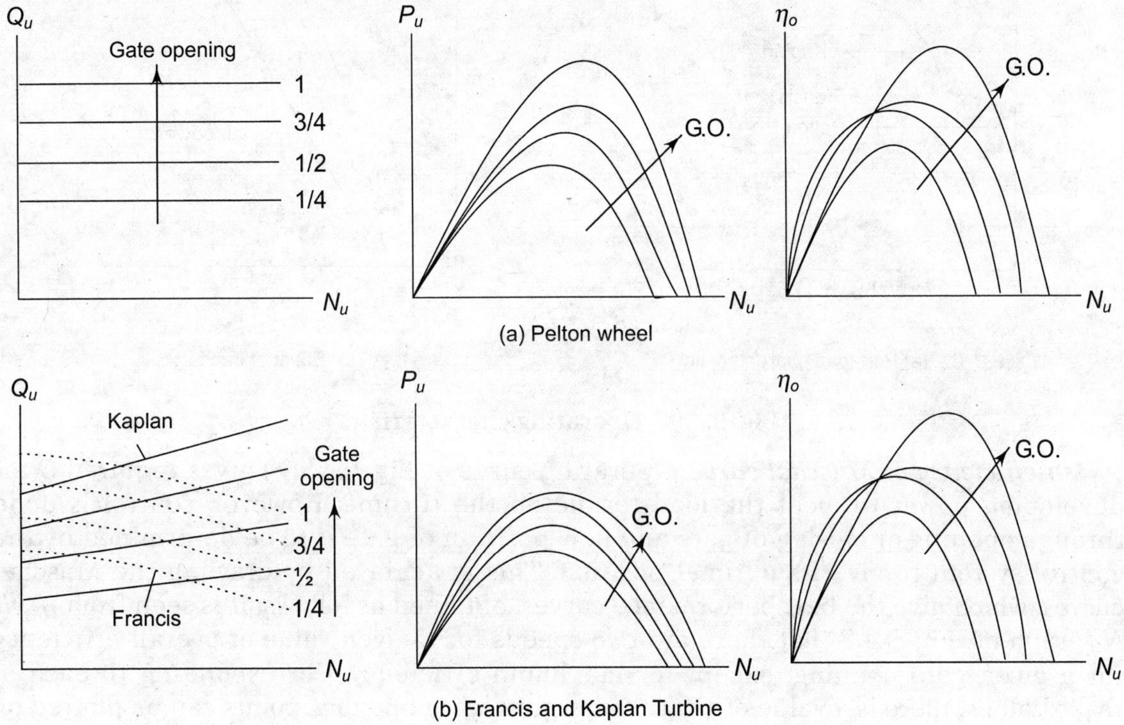


**Fig. 9.4.8** Main characteristic for different turbines

***Operating characteristic.*** The main characteristic helps in selecting a turbine for a given situation or predicting its behaviour at different places where the available head and discharge may be different. However, once a turbine is installed then for day-to-day operation, we need the operating characteristic, which is a plot of overall efficiency v/s percentage of full load. These are shown for Pelton, Francis and Kaplan turbines in Fig. 9.4.9(a). We observe that performance of Francis turbine is worst at part loads and that of Kaplan is the best. In fact, efficiency of Kaplan turbine remains almost equal to that at the design condition for a load upto 30% or so of the full load. That has made this turbine very popular.

Graph between $P$ v/s $Q$ at the design speed may be used for control purpose, manual or automatic. Because of friction, turbine will start rotating only after some minimum quantity of water has been supplied to it. That explains the shift of starting point of the curve from the origin of the graph shown in Fig. 9.4.9(b). At a given load, flow rate is to be changed to keep rpm constant and is obtained from this graph. This generates correcting signal to operate the governing mechanism to adjust for the new flow rate requirement. We still need some more information to decide the gate opening, which is obtained from Muschel curve.

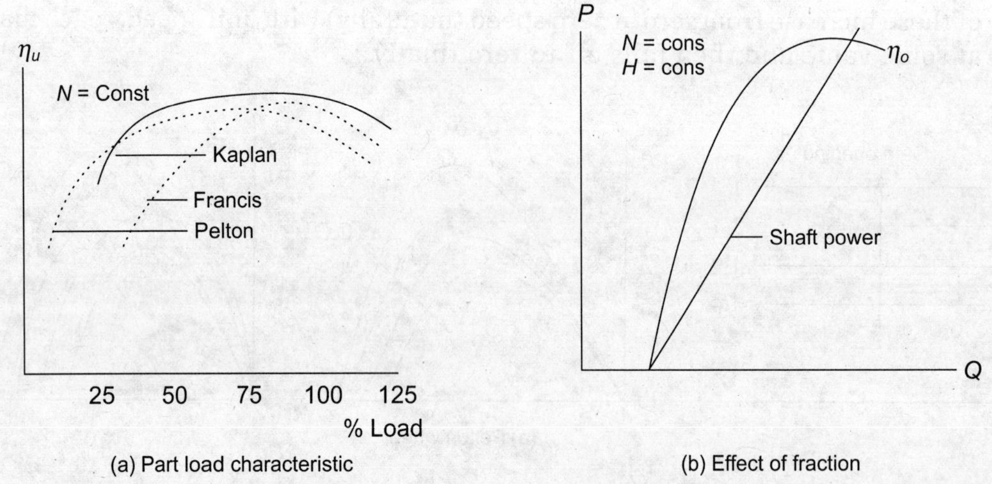


**Fig. 9.4.9** Operating characteristic

***Iso-efficiency or Muschel curve***. Operating curve of Fig. 9.4.9(b) gives required $Q$ for developing power to meet the load torque on the turbine. However, control is done through opening or closing of gate and new position of gate is to be determined by the control system to give its optimal position. This is achieved by drawing the Muschel curves, which give the best performance curves, obtained as follows: It is seen from $\eta_o$ v/s $N_u$ graph of Fig. 9.4.8 that there are two speeds for a given value of overall efficiency for a given gate opening, except for maximum efficiency. Corresponding to each of these values, there is a value of $Q_u$. Thus for given gate opening, points can be plotted on $Q$ v/s $N$ graph for different efficiency values. The process is repeated for different gate openings. All the points corresponding to same efficiency are now joined to give Muschel curves, as shown in Fig. 9.4.10, which is a graph between $Q$ and $N$ with gate opening as parameter. Now for given $N$ and value of $Q$ obtained from Fig. 9.4.9, gate opening is determined. Therefore, both graphs as shown in Figs. 9.4.9 and 9.4.10 are required by any control system to determine gate opening position. Also corresponding to maximum efficiency, the single points when joined give the best performance curve. This means that any point on this curve gives the gate opening for which maximum efficiency is

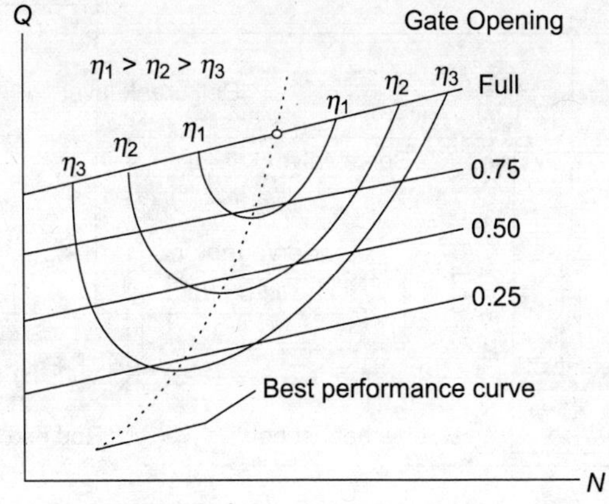


**Fig. 9.4.10** Muschel curves

the value of the iso-efficiency curve at $Q$ and $N$ corresponding to the point. However, turbine can run only at constant rpm or with some regulation fixed by the electricity regulations. Hence, it may not be possible to operate the machine with that efficiency.

## 9.5 Governing of Turbines

Turbine is used as prime mover for running electric generator on which the amperage load changes throughout the day and the turbine speed fluctuates. When load increases speed goes down and vice versa. This affects the frequency of A.C. thus generated. Fluctuation in the frequency is permitted within certain limits, called regulation, which is fixed in India at ± 4% of nominal frequency of *AC* produced. Therefore, the frequency has to be kept within 48 and 52 hertz. This requires control of the output of the turbine as per the load on it and is done by changing flow rate of the working fluid over its blades. This requires the gate valve to be regulated, which is done by using a governor.

We use all-mechanical or hydro-mechanical governing systems that used the power of the liquid to move the valve. A governor is mounted on the turbine shaft. The change in speed of the turbine shaft is sensed by governor sleeve that moves up or down with the speed. This shift in the position of the sleeve is then transferred to the control circuit using link mechanism in all-mechanical and hydraulic pressure in hydro-mechanical servo system, see Fig. 9.5.1. In both the cases, the sleeve motion is transferred to a double beat spool fitted in a cylindrical casing. The casing carries two service ports *A* and *B* (see Fig. 9.5.1 a) on which rests the spool. Service ports *A* and *B* are connected to the blank and rod end of the servo motor and the ports are covered in normal position of the spool. Piston rod in the servomotor actuates the gate valve opening when the spool is moved from its normal position on either side. We shall discuss the two systems in the next para. These days electronic controls are being used and you have to study it in a course on controls.

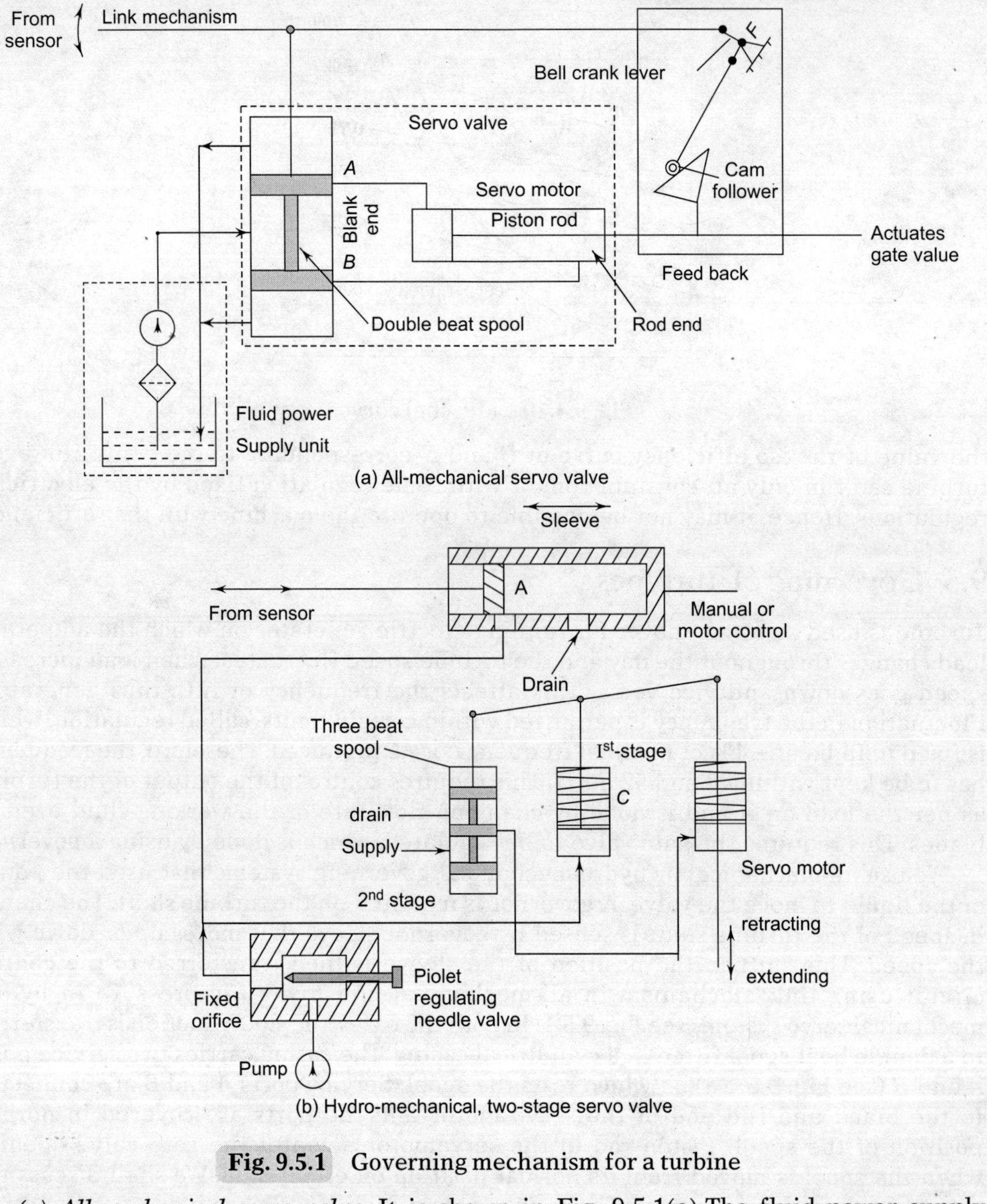


**Fig. 9.5.1** Governing mechanism for a turbine

*(a) All-mechanical servo valve*. It is shown in Fig. 9.5.1(a).The fluid power supply, servo valve and the feed back mechanism are marked in the figure. Spool valve has two spools, also called double beat valve. The space between the spool is connected to a pump that supplies liquid under high pressure; the space outside the spool is

connected to the drain so that liquid can move back to the sump. Let the speed go up that moves up the governor sleeve and hence the spool from its mean position, opening both the service ports. Consequently, high pressure fluid flows to the blank end of the cylinder through port *A*, while the liquid from the rod end flows back to the sump via port *B*, extending the piston rod and closing the gate valve to reduce mass flow rate. Simultaneously, the cam follower of the feed back mechanism lowers down. This rotates the bell-crank lever end down resulting in lowering of the spool. Here, the self-correcting movement is given to the spool and not to its sleeve. Soon the spool returns back to its non-operating position to stop flow of the liquid to or from the servo motor. In actual practice, the spool oscillates about this position for a few number of cycles.

*(b) Two-stage Hydro-mechanical Servo Valve.* This is shown in Fig. 9.5.1(b). In this case the speed sensor moves the piston *A*, which is fitted in a sleeve, covering partially the port in the sleeve, as shown in Eq. 9.5.1(b). This port gets supply from the pump through a fixed orifice (shown next to the pump) whose area can be changed by a needle valve, if desired. Liquid entering the cylinder goes back to the pump through the drain ports. The moving piston changes area of flow and hence the flow rate. This results in change in pressure at the discharge side of the fixed orifice. Liquid is also supplied to the spring controlled hydraulic cylinder *C*, the first-stage of the valve, and to the three beat spool, the 2nd stage of the valve. Variation in pressure moves piston of 1st stage value up or down. This motion is transmitted to three-beat spool by the lever attached thus disturbing spool position of, 51 stage valve. If the spool moves up then high pressure liquid is passed on to the servomotor and if it moves down, the liquid from servomotor goes to drain. Lever connecting spool is attached to servomotor that acts as a feedback to bring spool to the normal position. Arrangement is also made to change relative position of the sleeve by moving it axially either manually or using an electric motor.

Once the motion is obtained from the actuator, it is used in different ways according to the type of turbine used.

*Pelton Turbine.* A nozzle is fitted at the end of the penstock in which a spear can move axially. The spear is connected to the servo motor and reduces water supply during extending motion. Precaution is also taken to ensure no water hammering takes place. This is a process that is very dangerous and results because of a very large amount of moving water in the penstock is brought to rest in a very short time if the spear quickly closes the nozzle completely. Due to this closure, the whole kinetic energy is converted into pressure energy that will be in thousands of bar. Thus a pressure wave is produced that travels back and forth in the penstock. It leads to complete failure and a very loud sound. This process is called hammering. However, by controlling deceleration, the hammering process can be avoided. In Pelton turbine, water is deflected from its path so that it does not strike the turbine vanes when the spear moves in; this way the gate valve may be closed slowly without any risk of the speed to cross the danger limit. Figure 9.5.2(a) shows the arrangement.

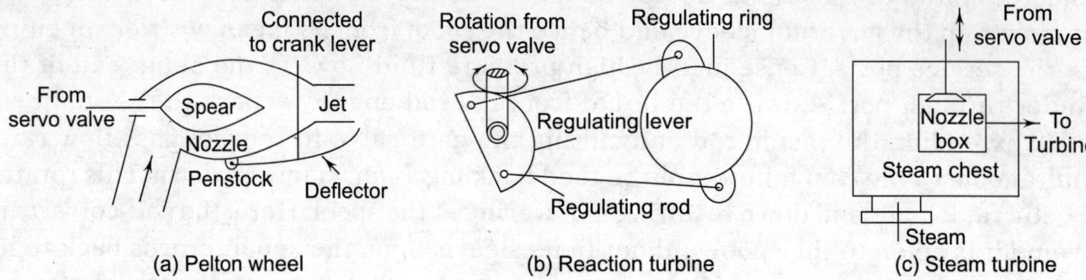


**Fig. 9.5.2** Mechanism of transferring servo motor motion to different turbines

*Francis and Kaplan turbine.* In Francis and Kaplan turbine, water enters the runner blades through guide vanes hinged on regulating ring around the runner vane ring. Regulating ring is connected to the regulating lever by two regulating rods at opposite corners on the ring, as shown in Fig. 9.5.2(b). The lever can be turned in both directions about its axis by the motion of the servo motor using suitable linkage (not shown). This rotates the ring in either direction around its axis. Ring movement is transferred to fixed vanes in both turbines by using appropriate linkage (again not shown) so that the free end of the fixed vane moves around the hinge and thus changing the area of flow and in turn the water flow rate. In Kaplan turbine, this motion is also passed on to the runner vanes so that water enters it smoothly. A pressure relief valve is attached to the turbine casing so that in case of sudden closure of the water flow in the event of grid failure water is by passed to the tail race.

# PROBLEMS

## Review Questions

1. When water flows over a series of curved vanes, the energy transferred by the water to the vane is maximum when the velocity of the whirl at exit of vanes is zero or negative. Is it true?
2. The work done by the fluid on a curved vane per second is 1000 J. If the angular velocity is 20 rad/s, what is the torque?
3. State the advantage of making flow over a curved vane over that on plane plate.
4. How are pumps classified? State the advantages and disadvantages of rotary pumps over reciprocating pumps. Name the important considerations when selecting a pump for a particular application.
5. Explain how atmospheric pressure pushes hydraulic oil up into the inlet port of a pump?
6. What is the importance of various non-dimensional numbers used in rotamachine? Name them and explain their role.

7. Define specific speed of turbine and compressor, as used in practice. Are these non-dimensional numbers? Determine the SI units whether it is non-dimensional or not for both the turbine and compressor.
8. Explain why the rotamachine runner changes from radial to axial type as its specific speed increases.
9. How does the number of vanes, vane shape, efficiency, angle of deflection of jet, and hydraulic efficiency of any rotamachine vary with increasing specific speed?
10. Which points should be considered for selecting right type of hydraulic turbine for a hydraulic plant?
11. Define (a) Euler head, (b) static head, (c) manometric head, and (d) total head.
12. Enumerate the losses that occur when a centrifugal pump operates and how are the losses accounted in calculations. Draw energy flow diagram to explain your answer.
13. Discuss the influence of the blade exit angle on the performance of a centrifugal pump. Which one will you prefer from the three types of centrifugal pumps and why?
14. How does discharge influence efficiency of centrifugal pump? What do you understand by iso-efficiency curves? Explain their importance.
15. What is the significance of characteristic curves and why different types are used?
16. Why are centrifugal pumps used sometimes in series and sometimes in parallel? Draw the characteristic curves for a centrifugal pump when tested at constant speed.
17. What is a positive displacement pump? How does it differ from a turbo-pump in its basic principle of operation?
18. What is the primary function of the diffusers in a turbine pump?
19. Why are centrifugal pump impeller vanes are backward curved?
20. What is the pump turbine? Is it the same as a turbine pump?
21. What is the difference between radial flow type, mixed flow type, and axial flow type turbo-pumps?
22. The difference between the water levels in the sump and overhead tank is $H$. What additional head a pump should generate?
23. The speed ratio of a centrifugal impeller increases with specific speed. Is this true?
24. What is the minimum starting speed of a centrifugal pump? A centrifugal pump impeller has outlet and eye diameters equal to 30 cm and 18 cm, respectively. If $H_m = 36$ m, what is the minimum starting speed?
25. What is cavitation, and what is its cause? At which section of the pump, it is likely to occur?
26. Explain significance of NPSH and Thoma factor.
27. On what factors does the cavitation in water turbines depend? Enumerate some methods to avoid cavitation in water turbines.
28. Why priming is required in centrifugal pumps but not in centrifugal compressors? What methods are used for it?
29. What is NPSH of a centrifugal pump? How is it related to the cavitation index of the pump?

30. If NPSH = 2 m and $H_m$ = 48 m, what is the Thoma number of the pump?
31. The rate of flow of a fluid delivered by a centrifugal pump must be reduced in order to obtain more head. Is this true?
32. If the centrifugal pump is delivering less quantity than the designed capacity, what are the possible reasons?
33. In axial flow machines, we assume negligible centrifugal force though the fluid flows over a rotating element. Justify it.
34. What is the principle of a hydraulic turbine?
35. Explain logically why power output of any turbine first increases from zero to some maximum value and then becomes zero again as its speed increases?
36. How are hydraulic turbines classified? Draw a comparison between impulse and reaction turbines.
37. Draw a general layout of a hydroelectric plant using an impulse turbine and define the following: (a) gross head, (b) net head, (iii) hydraulic efficiency, and (d) overall efficiency of impulse turbine.
38. What are the disadvantages of an outward flow radial turbine as compared to a radial inward flow turbine?
39. Why cannot reaction turbines be operated by impact of a jet on the vanes like impulse turbine?
40. Draw a schematic diagram of a Francis turbine and explain briefly its construction and working.
41. What is the primary objective of using more than one jet in a Pelton wheel installation?
42. What is the chief function of a deflector in a Pelton wheel installation?
43. Why do Pelton wheel bucket have a notch at the centre of their outer periphery?
44. The cross-sectional area of a spiral casing of a reaction turbine reduces in the direction of flow. What is the reason?
45. The B/D ratio of a Francis turbine increases with specific speed. Substantiate the reason.
46. Why the three Francis runners are named so? What are these names? Show the effect of this variation on the inlet and outlet velocity triangles.
47. The working head on a water turbine is 400 m and its unit speed is 33. What is its operating speed?
48. A reaction turbine consumes 120 $m^3/s$ while working under a head of 64 m. What is the unit discharge?
49. A model Francis runner while running at 1500 rpm consumes 1/200 of the prototype discharge. If the speed of the prototype machine is 300 rpm, what is the model scale ratio?
50. A model Kaplan turbine while working under 1/5 of prototype head consumes 1/1000 of the prototype rate of flow of water. If the output of the model is 4 kW, what is the output of the prototype?
51. Why are draft tubes not used in impulse wheel installations?
52. State the advantage of draft tube and justify not using it in Pelton wheel.
53. What factors would decide length of draft tube? Justify its bending after some depth instead of a vertical draft tube.

54. A fast Francis runner has a specific speed of 200 and unit power of 16. What is its unit speed?
55. What is the purpose of constructing the main characteristic curves of hydraulic turbine against variable speed?
56. In operating characteristic of a turbine, graph $p$ v/s $q$ is shown as a straight line. Give your comments.

## Problems for Practice

1. How much hydraulic power would a pump produce when operating at 140 bar and delivering 0.001 $m^3/s$ of oil? What power-rated electric motor would be selected to drive the pump if its overall efficiency is 85%?
2. A centrifugal pump rotating at 1500 rpm delivers 0.2 $m^3/s$ at a head of 15 m. Calculate the specific speed of the pump and the power input. Assume overall efficiency of the pump as 0.68. If this pump were to operate at 900 rpm, what would be the head, discharge and power required for homologous conditions? Assume overall efficiency remains unchanged at new rpm. {Hint: use unit quantities} [88.05, 43.3 kW, 5.4 m, 0.12 $m^3/s$, 9.35 kW] (IES 98).
3. A centrifugal pump delivers water against a net head of 14.5 m and a design speed of 1000 rpm. The vanes are curved back to an angle of 30° with the periphery. The impeller diameter is 300 mm and outlet width 50 mm. Determine the discharge of the pump if manometric efficiency is 95%. 90.1675 $m^3/s$]. AMIE Winter 1983.
4. The impeller of a centrifugal pump having external and internal diameters 500 mm and 250 mm respectively, width at outlet 50 mm and running at 1000 rpm, works against a head of 40 m. The velocity of flow through the impeller is constant and equal to 2.5 m/s and the vanes are set back at an angle of 40° at outlet. Determine (a) inlet vane angle, (b) work done by the impeller on water per second, and (c) manometric efficiency. [10.81°, 119.2 k, 64.6%].
5. A centrifugal pump impeller whose external and internal diameters are 400 m and 200 mm respectively, is running at 950 rpm. The rate of flow through the pump is 0.035 $m^3/s$ and the suction and delivery pipes ate 120 mm and 80 mm respectively. If the outlet vane angle is 450, the flow velocity is constant and equal to 1.8 m/s, determine (a) inlet vane angle, (b) the overall efficiency, and (c) manometric efficiency. [10.260, 61.76%, 73.65%].
6. Find the number of pumps required to take water from a deep well under a total head of 120 m. All the pumps are identical and are running at 800 rpm. The specific speed of each pump is given as 25 while the rated capacity of each pump is 0.16 $m^3/s$. [4].
7. It is required to deliver 0.048 $m^3/s$ of water to a height of 24 m through a 150 mm diameter and 120 m long pipe by a centrifugal pump. If the overall efficiency of the pump is 75% and coefficient of friction is 0.01 for the pipeline, find the power required to drive the pump. [22.76 kW].
8. The following data refer to a centrifugal pump: the diameter of the impeller at inlet and outlet = 180 mm and 360 mm respectively, the width of the impeller at

inlet and outlet = 144 mm and 72 mm respectively, the rate of flow through the pump = 17.28 litre/s, speed of the impeller = 1500 rpm, vane angle at the outlet = 45°, and the water enters the impeller radially at inlet. Neglecting losses through the impeller, find the pressure rise in the impeller. [40.5 m].

9. A centrifugal pump lifts water under a static head of 36 m of water of which 4 m is suction lift. Both suction and delivery pipes are 150 mm in diameter. The head loss in suction pipe is 1.8 m and in delivery pipe 7 m. The impeller is 380 mm in diameter and 25 mm width at mouth and revolves at 1200 rpm. Its exit angle is 35°. If the manometric efficiency of the pump is 82%, determine (a) the discharge through the pump, and (b) the pressure at the suction and delivery branches of the pump. [0.0295 $m^3$/s, – 5.942 m, 39.142 m].

10. Losses in a Pelton wheel may be modelled as follows:
Loss due to bucket friction and shock = $K1\ (V - u)^2/2g$, and loss due to bearing friction and windage loss = $K2\ u^2/2g$, where $V$ and $u$ are jet and bucket velocity; $K1$ and $K$ are constants.
Prove that the maximum efficiency of the Pelton wheel occurs when the ratio $u/V$ is

$$\frac{u}{V} = \frac{1 - \cos\theta + K1}{2\,(1 - \cos\theta) + K1 + K2}$$

$\theta$ = bucket angle at outlet. Volumetric losses are neglected.

11. For the data given for a Pelton turbine, calculate the friction factor for the buckets, hydraulic efficiency and mechanical efficiency. Overall efficiency 0.82, speed ratio 0.45, coefficient of velocity 0.98, jet deflection angle 160, windage and bearing losses $\propto N^2$, and volumetric efficiency 1. (IES 97).

12. Pelton wheel is to be designed for the following specifications: brake power = 9.56 MW, head = 350 m, speed = 750 rpm, overall efficiency = 85%, jet diameter not to exceed 1/6 of the wheel diameter. Determine: (i) the wheel diameter, (ii) diameter of jet, and (iii) the number of jets required. Take $K_f$ = 0.985 and speed ratio = 0.45. [0.95 m, 0.158 m, 2].

13. The water available for Pelton wheel is 4 $m^3$/s and the total head from the reservoir to the nozzle is 250 m, the turbine has two runners with two jets per runner. All the four jets have the same diameters. The pipe is 3 km long. The efficiency of transmission through the pipeline and the nozzle is 91%. The velocity co-efficient of each nozzle is 0.975 and $4f$ of the pipe = 0.0045. Determine: (a) the power developed by the turbine, (b) the diameter of the jet, and (iii) the diameter of pipeline. {Hint: efficiency of power transmission is = $(H - h_f)/H$, where $H$ is total head and $h_f$ is frictional head loss in pipe.} [7640 kW, 0.14 m, 0.955 m].

14. The following data relates to a double overhung Pelton wheel: output of generator = 25 MW, generator efficiency = 93%, effective head at the base of nozzle = 300 m, Pelton wheel efficiency = 85%, coefficient of velocity = 0.97, speed ratio = 0.46, jet ratio = 12. Determine the following: (i) size of the jet, (ii) mean diameter of runner, and (iii) synchronous speed. [0.3 m, 3.6 m].

15. A Pelton wheel of 1.1 m mean bucket diameter works under a head of 500 m. The deflection of jet is 165° and its relative velocity is reduced over the bucket by

15% due to friction. If the diameter of jet is 100 mm and the water is to leave the bucket without any whirl, determine: (i) rotational speed of wheel, (ii) ratio of bucket speed to jet velocity, (iii) impulsive force and power developed by the wheel, (iv) available power, (v) power input to the buckets, and (vi) efficiency of the wheel with power input to bucket as reference input. Take $C_v = 0.97$. [75 rpm, 0.45, 7250 N, 3.7 MW, 3481.8 kW, 90%].

16. The following data are the design particulars of a large Pelton wheel: head at distributor = 630 m, discharge = 12.5 m$^3$/s, power = 65 MW, speed of rotation = 500 rpm, runner diameter = 1.96 m, number of jets = 4, jet diameter = 192 mm, angle through which the jet is deflected by the bucket = 165°, and the mechanical efficiency = 96%. Determine the hydraulic power losses in the distributor nozzle assembly and the buckets.
17. A reaction turbine works at 450 rpm under a head of 120 m. Its diameter at inlet is 1.2 m and the flow area is 0.4 m$^2$. The angles made by absolute and relative velocities at inlet are 20° and 60°, respectively with the tangential velocity. Determine: (i) the volume flow rate, (ii) the power developed, and (iii) the hydraulic efficiency. [5.211 m$^3$/s, 5272 kW, 86%].
18. A Francis turbine supplied through a 6 m diameter penstock has the following particulars: output of installation = 63.5 MW, flow = 117 m$^3$/s, speed = 150 rpm, hydraulic efficiency = 92%, mean diameter of turbine at entry = 4 m, mean blade height at entry = 1 m, entry diameter of draft tube = 4.2 m, velocity in tailrace = 2.4 m/s. The static pressure head in the penstock measured before entry to the runner is 57.4 m. The point of measurement is 3 m above the level of the tailrace. The loss in the draft tube is equivalent to 30% of the velocity head at entry to it. The exit plane of the runner is 2 m above the tailrace and the flow leaves the runner without swirl. Determine: (i) overall efficiency, (ii) direction of flows relative to the runner at inlet, and (iii) pressure head at entry to the draft tube. [90.7%, 146°, – 4.25 m].
19. An in ward flow reaction turbine has an external diameter of 1 m and its breadth at inlet is 250 mm. If the velocity of flow at inlet is 2 m/s, find weight of water passing through the turbine per second. Assume 10% of the area of flow is blocked by blade thickness. If the speed of the runner is 210 rpm and guide blades make an angle of 10° to the wheel tangent, draw the inlet velocity triangle and find: (i) the runner vane angle at inlet, (ii) the velocity of wheel at inlet, (iii) the absolute velocity of water leaving the guide vanes, and (iv) the relative velocity of water entering the runner blade. [13.68 kN/s, 80.07°, 11.5 m/s, 2.03 m/s].
20. An inward flow reaction turbine is supplied 0.233 m$^3$/s of water under a head of 11 m. The wheel vanes are radial at inlet and the inlet diameter is twice the outlet diameter. The velocity of flow is constant and equal to 1.83 m/s. The wheel makes 370 rpm. Determine: (i) guide vane angle, (ii) inlet and outlet diameters, and (iii) the width of the wheel at inlet and outlet. Assume that the discharge is radial and there are no losses in wheel. Take speed ratio as 0.7. Neglect thickness of the vanes. [10.09°; 0.53 m, 0.265 m; 76.5 mm, 153 mm].
21. In an inward flow reaction turbine the diameter of the outer periphery is two times the diameter of inner one, and the turbine operates under a head of 20 m. The

turbine has radial tips at the inlet while at the exit the blades make an angle of 30° with the forward tangent. Assuming a constant radial velocity of flow and that the blade friction accounts for a dissipation energy equivalent to 10% of kinetic energy at the inlet, determine: (i) runner velocity at the rim, and (ii) hydraulic efficiency of the turbine. Assume the turbine discharges radially at the outlet. [13.7 m/s, 95.66%].

22. A Francis turbine with an overall efficiency of 76% is required to produce 150 kW. It is working under a head of 8 m. The peripheral velocity = 0.25 $\sqrt{(2gH)}$ and the radial velocity of flow is 0.95 $\sqrt{(2gH)}$. The wheel runs at 150 rpm and the hydraulic losses in the turbine are 20% of the available energy. Assuming radial discharge, determine: (i) the guide blade angle, (ii) the wheel vane angle at inlet, (iii) diameter of wheel at inlet and (iv) width of the wheel at inlet. [30.75°, 35.18°, 0.398 m, 0.179 m].
23. A Francis turbine works under a head of 25 m and produces 11760 kW while running at 120 rpm. The turbine has been installed at a station where atmospheric pressure is 10 m of water and vapour pressure is 0.20 m of water. Calculate the maximum height of the straight draft tube for the turbine.
24. The propeller reaction turbine of runner diameter 4.5 m is running at 48 rpm. The guide blade angle at inlet is 145° and the runner blade angle at outlet is 25° to the direction of vane. The axial flow area of water through the runner is 30 m$^2$. If the runner blade angle at inlet is radial, determine: (i) hydraulic efficiency of the turbine, (ii) discharge through the turbine, and (iii) power developed by the runner. [57.5%, 237.6 m$^3$/s, 15.156 MW].
25. A turbine is to operate under a head of 25 m at 200 rpm. The discharge is 9 m$^3$/s. If the efficiency is 90%, determine the performance of turbine under a head of 20 m. [$N$ = 178.88 rpm, $Q$ = 8.05 m$^3$/s, $P$ = 1421.4 kW].
26. A hydraulic turbine is to develop 1015 kW when running at 120 rpm under a net head of 12 m. Work out the maximum flow rate and specific speed for the turbine if the overall efficiency at the best operating point is 92%. In order to predict its performance, a 1:10 scale model is tested under a head of 7.2 m. What would be the speed, power output and water consumption of the model if it runs under the conditions similar to the prototype? {Hint: the model is designed so that head coefficient, power coefficient and flow coefficient are same as for prototype.} [$Q$ = 9.372 m$^3$/s, $Ns$ = 171.2, 929.5 rpm, 0.0726 m$^3$/s, 4.72 kW].
27. In a project of hydroelectric scheme, 300 m$^3$/s of water is available under a head of 15 m. Alternative schemes are proposed using Francis turbines with specific speed 395 and Kaplan turbine with specific speed 680. If normal running speed is 150 rpm, determine the number of machines needed in the two schemes. The units in any one installation are to be of equal power and the turbine efficiency may be assumed 90%. Also work out the power output of the machine in each case.
28. A Pelton wheel develops 5880 kW shaft power at 210 rpm when operating under a head of 225 m. If overall efficiency of the wheel at the best operating point is 85%, determine the unit speed, unit discharge and unit power. It is now desired to test this wheel at a site where the maximum supply head is 15 m and the available precision tachometer has a maximum range of 200 rpm. Work out the speed,

discharge and power output of the wheel under changed head.

29. In a hydroelectric generating plant, there are 4 similar turbines of total output 220 MW. Each turbine is 90% efficient and runs at 100 rpm under a head of 65 m. It is proposed to test the model of the above turbines in a flume where discharge is 400 litres/s under a head of 4 m. Work out the size (scale ratio) of the model. Also calculate the model speed and power results expected from the model.
30. A hydroelectric power station required to generate 60 MW at 50 cycles per second frequency at the available head of 20 m suggests the use of Kaplan turbine for which specific speed is 600 units. Make calculations for (a) the speed and discharge through the turbine assuming that it has an overall efficiency of 90%, (b) leading dimensions of the turbine assuming flow ratio 0.6 and that hub diameter of the wheel is 0.35 times the outside diameter, (c) maximum safe elevation of runner to avoid cavitation taking vapour pressure 0.26 m of water and critical value of Thoma's cavitation factor.

CHAPTER

10

# Other Hydraulic Systems

OUTLINE

- Reciprocating Pump
- Pressure Intensifier
- Accumulator
- Hydraulic Crane
- Hydraulic Lift
- Hydraulic Press
- Hydraulic Coupling
- Hydraulic Ram
- Hydraulic Controls

## 10.1 Introduction

There are a number of hydraulic systems, other than the hydraulic turbines and pumps that are of great service to the human beings. These devices may be reciprocating or rotary type or even the movement may be intermittent. In this chapter, we shall look at such devices.

Reciprocating systems have been used for a long time, since the industrial revolution. It is made of a piston reciprocating in a cylinder between the two extreme piston positions, known as dead centres. These are used in steam engine, internal combustion engine, reciprocating pumps and many fluidic actuators. The main reason for using reciprocating pump and compressor is their capability of producing high pressure without any back flow and for that reason these are known as positive displacement pump. However, the volume and hence mass handled per cycle or minute, inertia forces, dynamic balancing, vibration problem, requirement of a good and heavy foundation, etc. are forcing us to go for rotary kind of systems. But as mentioned in the beginning, these have an importance of their own and form a very important place in fluidic controls.

You shall learn about the working of reciprocating pump as well as the effect of inertia and friction on its performance. Use of air vessel and how it helps in reducing of power consumption and discharge fluctuations is dealt with in details.

Air compressor is also used extensively. These may be reciprocating or rotary type. In these devices, the compressibility of fluid plays important role and mass of air-discharged decreases with increase in discharge pressure. Moreover, increase in density with the pressure introduces analytical complexity as it is a variable and has to be determined at the different sections. In addition, air is lighter as compared to water. Therefore, the mass of the fluid handled becomes significantly less. In order to have sufficient mass flow rate, the flow velocity is very large in rotary compressors. However, the inertia effect will be pronouncedly less in such systems. We shall not cover compressor in this book.

## 10.2 Reciprocating Pump

A reciprocating pump is made from a cylinder and piston arrangement, as shown in Fig. 10.2.1. Two spring loaded plates[1] (also called disc) valves are fitted in the cylinder head. Suction valve *S* allows fluid to enter into the cylinder and not the other way. It opens due to vacuum produced inside cylinder by movement of piston when pulled outside by the crank mechanism, consisting of connecting rod, crank arm and crank shaft. Liquid comes from the sump through the suction pipeline whose free end is dipped in liquid in sump and other end to the suction side of the cylinder. A

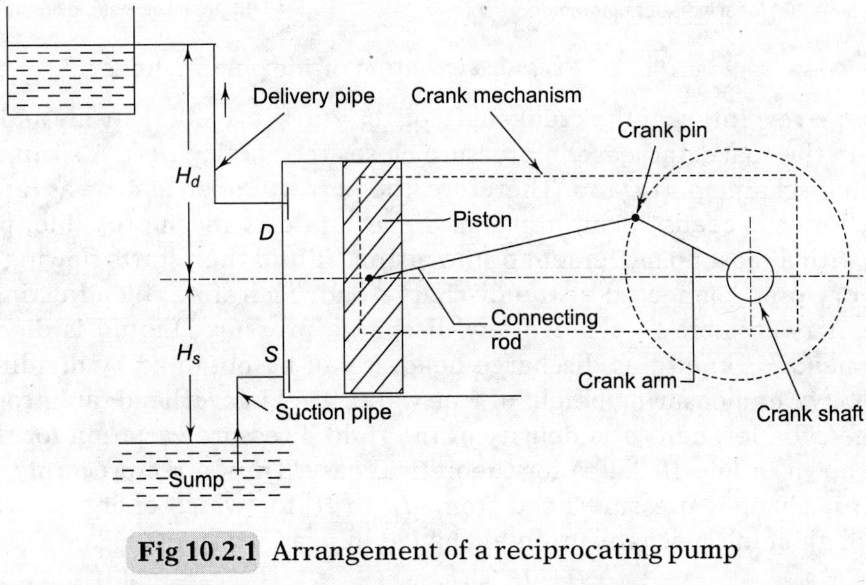


**Fig 10.2.1** Arrangement of a reciprocating pump

[1]A plate port is a disc, which is fitted along with spring in the space made in the cylinder head. This arrangement allows fluid to flow only in one direction, the direction of the pressure gradient. Hence, they are one-way valves.

filter is fitted at the dipped end of the pipe to remove dust and foreign material from the liquid. Cylinder is placed at a suitable position such that its centre line is at a height $H_s$ from the free surface of liquid in the sump, Fig. 10.2.1. $H_s$ is called suction head and is negative since pressure on free surface of water in the sump is atmospheric. This results in maximum vacuum at the suction port. If pressure loss in the suction valve is neglected then theoretical vacuum inside the cylinder during outward piston movement may be taken as $H_s$ gauge. Figure 10.2.2(a) shows variation of pressure inside cylinder for the suction stroke.

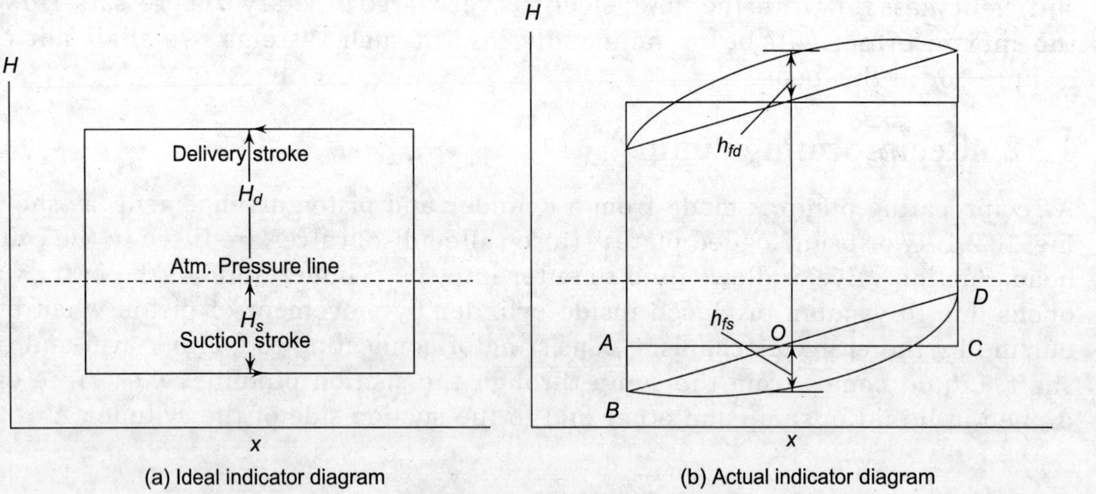


**Fig. 10.2.2** Pressure variation inside pump cylinder

After half revolution of the crankshaft, piston starts travelling inside and compresses the fluid in the space. Increase in pressure closes the suction port. A spring-loaded disc is fitted in discharge port *D* too. Therefore, the port will open as soon as liquid pressure inside cylinder exceeds spring tension. Since liquid is incompressible, pressure will increase with almost no movement of the piston.[2] Fluid then starts flowing out through the delivery pipe connected at the discharge end. Therefore, by adjusting the spring pressure, it is possible to achieve any discharge pressure. Liquid is discharged at a pressure head $H_d$, known as discharge head. It can be obtained by dividing discharge pressure by $\rho g$ or measuring height of free water level in overhead tank from the centre line of the cylinder; here $\rho$ is density of the fluid. Pressure variation for the complete cycle is shown in Fig. 10.2.2(a) for theoretical case. Thus, a reciprocating pump works such that it changes pressure head from $-H_s$ to $H_d$ for which energy is transferred to fluid through crank mechanism. Total change in head

$$H = H_s + H_d \tag{10.2.1}$$

is called *total static head.* These heads are measured in metres of the liquid being pumped.

[2]In case of compressible fluids, discharge port will open late and not at the *ODC* as in case of liquid.

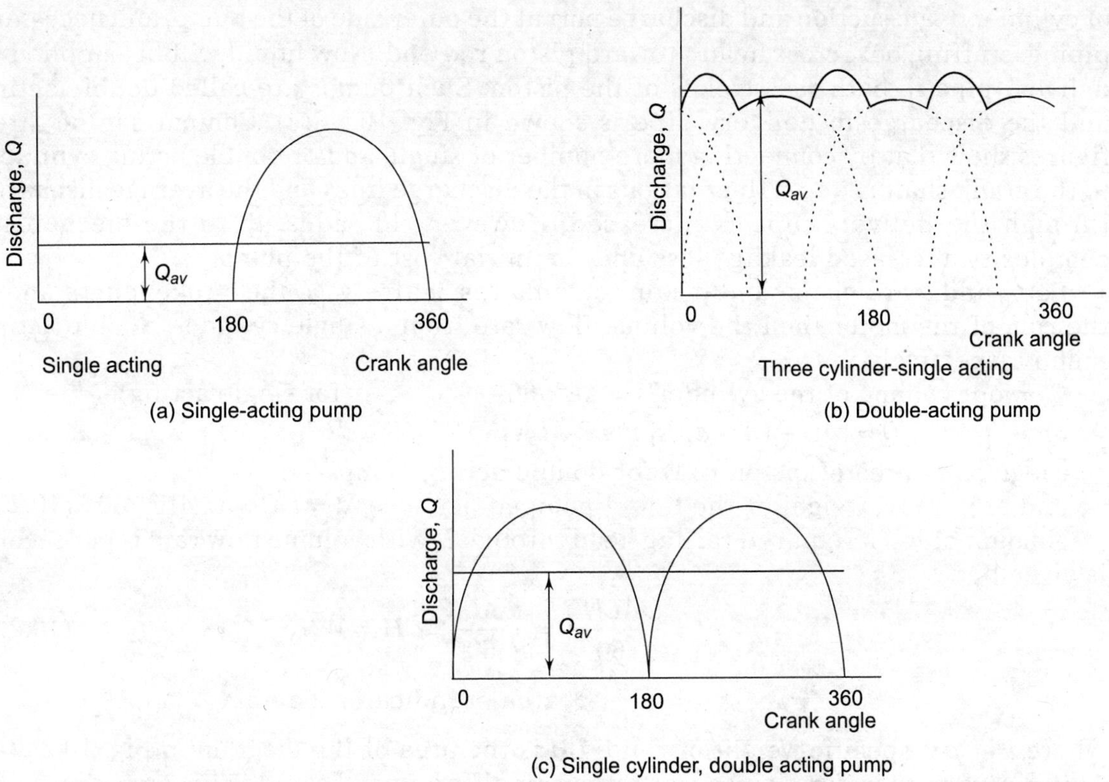


**Fig. 10.2.3** Variation of discharge for different pump arrangements

We observe that fluid does not flow from cylinder continuously. Fluid flow starts when piston is at out dead centre (*ODC*), during its inward motion, reaches the maximum at some position of piston and finally becomes zero at the end of delivery stroke. Pump discharge $Q$ v/s crank angle is shown in Fig. 10.2.3(a). Flow fluctuations are the result of changing speed of piston. Hence, flow velocity also fluctuates in both suction and delivery pipes. It, therefore, brings in inertia effect due to acceleration or deceleration of a large liquid mass present in the pipe. We would prefer to decrease the discharge fluctuation level in order to have uniform velocity and less inertia effects. One of the methods adopted is to supply liquid to the discharge pipe from two or more cylinders. The cylinders may have their individual drive or run by the same crankshaft of the crank mechanism. In the latter case, the crankshaft shall have as many crank arms as the number of cylinders. These crank arms will be spaced angularly with equal angle between them. That is, if there are three cranks then the angle between crankarms will be 120°. Figure 10.2.3(b) shows variation of discharge with crank angle over one cycle for 3 single acting pumps, called three-throw pump. (Flow discharge curve for the other 2 cylinders is shown shifted right by 120° and 240°, respectively). As seen from the figure, the average discharge is increased and discharge fluctuations decreased by this arrangement. We may also make the pump double acting by providing another set of cylinder head, suction and discharge port at the outer side of the pump, and necessary piping, stuffing box, cross head[3] toward piston rod end. Now liquid will be supplied to delivery pipe in both the strokes of the piston. Such pumps are called double acting and the discharge in delivery pipe is shown in Fig. 10.2.3(c). Comparison of three figures show that by connecting more number of single and/or double acting cylinders to the crank shaft reduces fluctuations in the discharge lines and the average discharge through the delivery pipe is increased. However, this adds up to the mechanical complexity, increased leakage loss and the initial cost of the pump.

Let $A$ and $a$ are piston and piston rod area respectively, $L$ the stroke length and $N$ the rpm of the motor then the volume flow rate from a single cylinder, double acting pump respectively is

$Q$, swept volume of the cylinder/s = $ALN/60$ (for single acting)

or $Q = \{AL + (A - a) L\}\, n \cong 2\, ALN/60$

(neglecting area of piston rod) for double acting pump,

and $W$, weight of the liquid pumped per second = $wQ = w\, ALN/60$ (10.2.2)

Amount of work required raising head through $H$ with volume flow rate $Q$ per second is given by

$$P = \frac{wALNH}{60} = \frac{wALN}{60} \times H = WH, \qquad (10.2.3)$$

$$= \frac{wAN}{60} \times \text{area of indicator diagram.}$$

where $w = \rho g$, specific weight of fluid. $LH$ is the area of the diagram in Fig. 10.2.2(a). This diagram, therefore, gives work input to different scale and is known as *indicator diagram.*

## 10.2.1 Inertia Effect on Pressure Head

During outward or inward movement, piston starts from rest, accelerates and attains maximum velocity at some position after which decelerates to stop at other end of travel. Linear acceleration of the piston can be determined by expressing displacement of piston from IDC, see Fig. 10.2.4. Let $l$ and $r$ be the length of connecting rod and crank arm, $\omega$ the angular speed of the crank arm, $\theta$ the angle turned by the crank arm at some part of the stroke as shown in the figure and $x$ the displacement of piston from dead centre, then

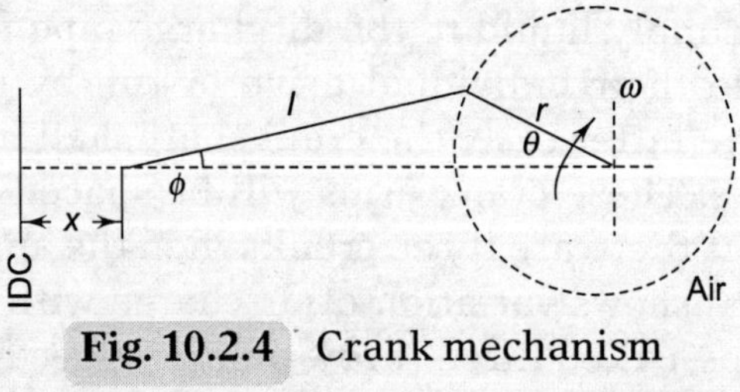


**Fig. 10.2.4** Crank mechanism

[3]Since piston rod has to move along cylinder axis because of stuffing box, a device used to stop leakage, connecting rod cannot be fixed to piston. In this case, additional machine element called cross head is used; this contains a slider fitted at one end with the piston rod and connecting rod is hinged with it. This slide reciprocates in guides.

$$x = (l + r) - (l \cos \phi - r \cos \theta) = l\,(1 - \cos \phi) + r\,(1 - \cos \theta) \cong r\,(1 - \cos \theta), \quad (10.2.4)$$

$$l \sin \phi = r \sin \theta \qquad \text{or} \qquad \sin \phi = r / l \sin \theta$$

In most applications, the ratio of the length of connecting rod to the crank rod, $l/r$ is large and therefore $\phi$ is so small that $\cos \phi$ approximates to 1 and therefore the first term on R.H.S. of Eq. 10.1.4 has been neglected. Writing $\omega t$ for $\theta$ and differentiating with respect to $t$, we get velocity and acceleration of piston,

$$V_P = \omega r \sin \omega t \qquad \text{and} \qquad \alpha_p = \omega^2 r \cos \omega t \quad (10.2.5)$$

However, our interest is to find out velocity and acceleration of the liquid in suction and discharge pipe since the inertia force of the fluid in these pipes shall determine pressure head at inlet or exit ports of the pump. This is obtained from conservation principle of mass, which requires mass of liquid entering the cylinder equal to that supplied by the suction pipe. Similarly, mass leaving the cylinder has to be equal to that entering delivery pipe. If $A$ represents area of flow of any of the pipes, $V$ and $\alpha$ the velocity and acceleration of the fluid in pipe and $L$ the length of pipe (subscript $s$ or $d$ may be used to represent suction or discharge pipe whenever required) then conservation of mass or volume flow rate demands,

$$V = V_P \times A_p / A = (A_p/A)\ \omega r \sin \omega t, \qquad \alpha = (A_p/A)\ \omega^2 r \cos \omega t \quad (10.2.6)$$

Inertia force = mass of liquid in pipe × acceleration = $(\rho A\, L) \times (A_p / A)\ \omega^2 r \cos \omega t$

Intensity of this force = Force/Area of flow)

$$= \frac{(\rho AL) \times \left(A_p / A\right) \omega^2 r \cos \omega t}{A} = (\rho L) \times \left(A_p / A\right) \omega^2 r \cos \omega t$$

From Fig. 10.2.4, we note that $x = (l + r) - (l \cos \phi + r \cos \theta) \approx r - r \cos \theta$, since $\phi$ is very small and hence $\cos \phi \approx 1$. This gives: $r \cos \theta = r \cos \omega t = r - x$. We may now determine the acceleration head corresponding to the intensity of the inertia force.

$H_a$ = Acceleration head = Intensity of force $/(\rho g)$

$$= (L / g) \times \left(A_p / A\right) \omega^2 r \cos \omega t = (L / g) \times \left(A_p / A\right) \omega^2 r\,(r - x) \quad (10.2.7)$$

Pressure head at suction port

$$= -H_s - H_{as} = -H_s - (L_s / g) \times (A_p / A_s)\ \omega^2 r \cos \theta \qquad 0 \le \theta \le \pi \quad (10.2.8)$$

Pressure head at discharge port

$$= H_d - H_{ad} = H_d - (L_d / g) \times (A_p / A_d)\ \omega^2 r \cos \theta \qquad \pi \le \theta \le 2\pi \quad (10.2.9)$$

Negative sign with $H_s$ in Eq. 10.2.8 has been taken because pressure at suction port is below atmospheric pressure. Equations 10.2.8 and 10.2.9 are plotted against $x$ in Fig. 10.2.2(b) where it is shown as inclined straight line. Term $(r - x)$ in Eq. 10.2.7 for acceleration head shows that variation of acceleration head with $x$ is linear. In actual installations, length of suction pipe and delivery pipe is large enough so that the swept volume of the pump is negligible. In that case, inertia pressure head of the mass contained within the cylinder may be neglected. Therefore pressure head inside the cylinder is the same as that at its port. Equation 10.2.8 gives acceleration head at $\theta = 0$ and $\theta = \pi$ as

$$-L_s \times (A_p / A_s)\ \omega^2 r \qquad \text{and} \qquad L_s \times (A_p / A_s)\ \omega^2 r$$

respectively. Now consider triangles *OAB* and *OCD* in Fig. 10.2.2(b). *AB* = *CD* from Eq. 10.2.8 and *O* is mid point of *AC*. Therefore area of two triangles is same. Similarly, this may be verified for the discharge end. We, therefore, conclude that area of the indicator diagram is not changed due to liquid inertia. Or in other words there is no effect of inertia of the liquid in pipe on power requirement.

A negative acceleration head at $\theta = 0$ may create problem as vacuum at the suction port will increase and may reach close to vapour pressure corresponding to the temperature of the liquid at which it starts boiling. Moreover, the acceleration head varies as square of the angular velocity of crank arm. This limits the rpm of the pump in order to avoid separation of flow. Therefore, positioning of the pump above sump level and its rpm are governed by onset of the caviatation. The separation problem is a likely to occur at the beginning of the suction stroke or end of the discharge stroke, as can be seen from Fig. 10.2.2, since there is a possibility only at these two points for vacuum to reach the danger point. Example 10.2.1 illustrates the procedure to calculate the maximum speed of operation of the pump.

An other problem may arise if acceleration head near the other end of suction stroke is high enough to exceed delivery head even before completion of suction stroke. In that situation the discharge port will open before piston completes its suction stroke and the liquid will flow from the discharge pipe as well into cylinder thus increasing the total volume flow rate beyond the capacity of cylinder. However, in most situations, leakage of liquid from joints result in Low volume flow rate. We use term *slip* to express the difference in theoretical and actual discharge of the pump and is expressed as percentage deviation of discharge with respect to theoretical discharge .

$$\text{Slip} = (Q_{\text{th}} - Q) / Q_{\text{th}} \times 100 \tag{10.2.10}$$

It is, therefore, possible to have negative slip in reciprocating pump if inertia effect is too large. Such condition is likely to occur when rpm of the pump is high.

From Fig. 10.2.2(b), it is seen that pressure at the exit port is lower than discharge head by an amount equal to the acceleration head at the end of discharge stroke. It is a possibility that the acceleration head is so large that the pressure at exit port may reach saturation pressure corresponding to temperature of fluid. In that case vapour formation may take place there too and cavitation occurs. Therefore absolute pressure at the end of delivery stroke should be more or in limit equal to vapour pressure $p_v$. This gives

$$p_a/w - (H_d - H_{ad}) \geq p_v \quad \text{or} \quad H_d - H_{ad} \leq (p_a - p_v)/w$$

where $p_v$ is saturation pressure at atmospheric temperature and the right hand side is normally taken equal to 7.8 m of water.

## 10.2.2 Effects of Friction and Inertia on Pressure Head

The other factor that influences performance of a pump is the property of fluid, mainly the viscosity and compressibility. Whereas viscosity results on head loss during flow of liquid through pipe, the compressibility leads to change in density of the fluid along the rising pressure. Head loss in a pipe is given by Darcy equation. Using Eq. 10.2.6 for velocity of flow through pipe of length $l$ and diameter $d$ ($l = L_s$ and $d = d_s$ for section pipe or $L_d$ and $d_d$ for delivery pipe)

$$h_f, \text{ friction head} = \frac{fl}{d}\frac{V^2}{2g} = \frac{fl}{d}\frac{(\omega r \sin\theta)^2}{2g}\left(\frac{A_p}{A}\right)^2 \tag{10.2.11}$$

We note from this equation that (i) friction or viscous head loss is proportional to $\sin^2\theta$ and therefore varies parabolically and (ii) is always positive. Combined effect of acceleration and viscous effect is shown in Fig. 10.2.2(b). Table 10.2.1 summarises all these effects at three positions, namely at *IDC*, mid point and *ODC* of piston.

**TABLE 10.2.1.** Pressure head at three piston positions

| | Suction stroke | | | Discharge or delivery stroke | | |
|---|---|---|---|---|---|---|
| $\theta$ | 0 | $\pi/2$ | $\pi$ | $\pi$ | $3\pi/2$ | $2\pi$ |
| Static head | $-H_s$ | $-H_s$ | $-H_s$ | $H_d$ | $H_d$ | $H_d$ |
| Acceleration head | $B\,L_s/A_s$ | 0 | $-B\,L_s/A_s$ | $B\,L_d/A_d$ | 0 | $-B\,L_d/A_d$ |
| Friction head | 0 | $KL_s/d_s$ | 0 | 0 | $KL_d/d_d$ | 0 |
| Where $K = (f/2g)(A_p/A_d)^2(\omega r)^2$ and $B = A_p\,\omega^2 r/g$ | | | | | | |

We also observe from Fig. 10.2.2(b) that area of the indicator diagram is increased. That is, more power is required. Recalling from geometry course that area of a parabola is 2/3rd of that of the rectangle with height equal to maximum ordinate of parabola, which is $h_f$ at $\theta = \pi/2$ for suction stroke and $3\pi/2$ for discharge stroke,

Total area of indicator diagram = $(H_s + H_d + 2/3\ h_{fs} + 2/3\ h_{fd})\ L$ and

$$\text{Average work done} = w\,A\,N/60 \times \text{area of indicator diagram}$$
$$= w\,A\,N/60 \times (H_s + H_d + 2/3\ h_{fs} + 2/3\ h_{fd})\ L$$
$$= W(H_s + H_d + 2/3\ h_{fs} + 2/3\ h_{fd}), \tag{10.2.12}$$

$$P_1 = \text{power required for overcoming friction only}$$
$$= \text{Area of parabolas} \times w\,A\,N/60$$
$$= 2/3\ h_f \times w\,A\,N/60 = (W fl/3gd)\ (A_p/A)^2\ (\omega r)^2 \tag{10.2.13}$$

after substituting for $h_f$ from Eq. 10.2.11 and simplifying the expression.

**Example 10.2.1** The diameter and stroke length of a single-acting reciprocating pump are 75 mm and 150 mm respectively. It takes supply of water from a sump 3 m below the pump through a pipe 5 m long and 40 mm dia. It delivers water to a tank 12 m above the pump through a pipe 15 m long, 30 mm dia. If separation occurs at 75 kN/m$^2$ below the atmospheric pressure, find the maximum speed at which pump may be operated without separation. Assume that the piston has a simple harmonic motion. Also find out the power of motor to be used to run this pump.

**Solution** We learnt that separation is due to inertia head at the beginning of suction stroke or end of discharge stroke, since at these two points vacuum may be enough within the pump to cause separation of the flow. Inertia head at these points is obtained from Equation 10.2.8 and 10.2.9, respectively, by substituting $\theta = 0$ or $2\pi$ in these equations. We are given all the data to evaluate these terms.

**Given:** Pump $D = 0.075$ m $\rightarrow A_p = \pi/4\ D^2 = 0.0044$ m$^2$

$$L = 0.15 \text{ m} \rightarrow r = L/2 = 0.075 \text{ m}$$

Suction line $H_s = 3$ m, $L_s = 5$ m, $d_s = 0.04$ m $\rightarrow A_s = 0.001256$ m$^2$

Discharge line $H_d = 12$ m, $L_d = 15$ m, $d_d = 0.03$ m $\rightarrow A_d = 0.0007068$ m$^2$

Separation head, $H_{sep} = 75$ kN/m$^2$ = $75 \times 1000/(9.81 \times 1000)$

= 7.645 m of water below atmosphere

$\therefore H_{as}$ at beginning of suction stroke = $(L_s/g) \times (A_p / A_s)\ \omega^2 r = 0.134\ \omega^2$, and

$H_{ad}$ at end of discharge stroke = $-(L_d/g) \times (A_p / A_d)\ \omega^2 r = -0.7168\ \omega^2$

To avoid separation, gauge pressure should be less than or equal to separation pressure. This gives:

During suction stroke, $(H_s + H_{as}) \le H_{sep} = H_{sep} = 7.645$ in limit

$\rightarrow$ $3 + 0.134\ \omega^2 = 7.645 \Rightarrow \omega = 5.88$ rad/s $= 2\pi N/60 \rightarrow N = 56.15$ rpm

During delivery stroke, $(H_d + H_{ad}) \le H_{sep} = H_{sep} = 7.645$ in limit

$\rightarrow$ $12 - 0.7168\ \omega^2 = 7.645 \Rightarrow \omega = 5.235$ rad/s $= 2\pi N/60 \rightarrow N = 50.0$ rpm

The pump therefore can operate at maximum speed of 50 rpm and separation would occur at the end of discharge stroke. Power of motor will be chosen on the basis of pump rpm = 50. Eq 10.2.12 gives area of the indicator diagram, which is the work done per cycle. Therefore

$H$ = total head to be developed by pump = $(H_s + H_d + 2/3\ h_{fs} + 2/3\ h_{fd})$,

where $h_{fs} = A_p\ L_s/d_s$ and $h_{fd} = A_p\ L_d/d_d$. Substituting the given values, we get

$h_{fs} = A_p\ L_s/d_s = 0.0044 \times 5/0.04 = 0.55$ m and $h_{fd} = 0.0044 \times 15/0.03 = 2.2$ m

$\rightarrow$ $H = 3 + 12 + 2/3 \times 0.0044 + 2/3 \times 2.2 = 20.5$ m

$Q$ = discharge/s = $A_p \times L \times N/60 = 0.00055$ m$^3$/s

$\therefore$ $P = wQH/1000$ kW $= 9.81 \times 0.00055 \times 20.5 = 0.11$ kW

## 10.2.3 Reduction of Discharge Fluctuation and Acceleration Head

Inertia effect is basically because of acceleration or deceleration of large mass of liquid in pipe and the easiest way is to reduce this mass. This is achieved by providing an air vessel just at the inlet or at outlet to the pump; arrangement is shown in Fig. 10.2.5. An air vessel is a device that traps air above the liquid surface in a closed space. Air is compressed when liquid level rises and vice versa. So during suction stroke, liquid is supplied from the pipe as well as the vessel on suction side during which liquid level in vessel is lowered. Air in the space above the water surface expands in the process. Whereas during discharge stroke, liquid continues to flow in suction pipe and is stored in the air vessel, as pressure above the liquid surface in air vessel is lower than atmospheric pressure. Thus flow velocity in the suction line remains constant throughout the pump cycle. Similar events take place on discharge side whereby fluctuation in discharge is practically eliminated. This continues alternately resulting in flow with almost uniform velocity in the discharge and suction pipes except a small portion that connects the air vessels with the cylinder. However, the mass in this length is too small to take any cognisance of.

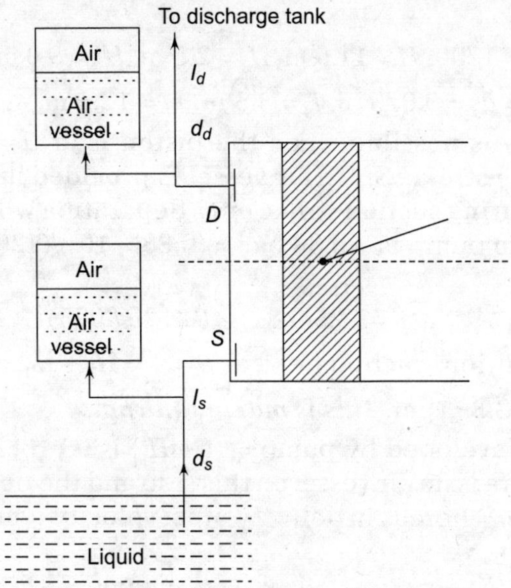


**Fig. 10.2.5** Air vessels provided at suction ans delivery side

Incidentally, there is not only reduction in inertia head and fluctuations in discharge but saving in power consumption also. This can be verified by comparing power requirement in two situations viz. with and without air vessel. Equation 10.2.13 gives power wasted to overcome friction loss. For the second case with air vessel, using Eq. 10.2.11 for piston velocity:

$V$, Velocity of flow in pipe = $A_p V_p / A = \pi \omega r \, A_p / A$

$$h_f = \text{Loss of head due to friction} = \frac{fl}{d}\frac{V^2}{2g} = \frac{fl}{2gd}\left(\frac{A_p}{A}\right)^2\left(\frac{\omega r}{\pi}\right)^2$$

$$P_2, \text{ Power wasted in overcoming friction} = W \times h_f = \frac{Wfl}{2gd}\left(\frac{A_p}{A}\right)^2\left(\frac{\omega r}{\pi}\right)^2 \quad (10.2.14)$$

$\therefore$ Percentage saving of work = $(P_1 - P_2)/P_1 \times 100 = 1 - 3/(2\pi^2) = 84.8\%$, (10.2.15)

for a single cylinder. Similarly saving for a double acting pump arrangement is 39.2%. As fluctuations in the discharge pipe decreases necessity of air vessel is not felt. Power saving too decreases.

**Example 10.2.2** A single acting reciprocating pump is to raise a liquid of density 1200 kg/m³ through a vertical height of 11.5 m, from 2.5 m below pump axis to 9 m above it. The plunger moves with simple harmonic motion, has diameter 125 mm and stroke 225 mm. The suction and delivery pipes are of 75 mm diameter and 3.5 and 13.5 m long respectively. There is a long vessel placed on the delivery pipe near the pump exit but there is no air vessel on the suction pipe. If separation takes place 0.88 bar below atmospheric pressure, find: (i) maximum speed with which the pump can run without separation taking place, (ii) power required to drive the pump, if $f = 0.02$. Neglect slip for pump.

**Solution**

**Given:** $\rho = 1200, H = 11.5$ m, $H_s = 2.5$ m, $H_d = 9.0$ m, $D = 0.125$ m, $L = 0.225$ m

$d_d = d_s = 0.075$ m, $l_s = 3.5$ m, $l_d = 13.5$ m, and $f = 0.02$

1. Separation of flow is possible when the piston is at $\theta = 0$ for suction stroke and $\theta = 2\pi$ for discharge stroke. Since air vessel is provided in delivery pipe, separation is likely to occur during suction stroke only. Separation will occur when total suction head at entrance to pump is > 0.88 bar = $(0.88 \times 10^5)/(1200 \times 9.81) = 10.97$ m of the liquid. But

$$H_{as} = (L_s/g) \times (A_p/A_s)\,\omega^2 r = (3.5/9.81) \times (0.125/.075)^2 \times \omega^2 \times (0.225/2) = 44.49\,\omega^2$$

$\therefore$ In limiting condition, suction head at $\theta = 0 = H_s + H_{as} = 44.49\,\omega^2$

$\Rightarrow$ $N = 63.39$ rpm. *This is maximum rpm.*

2. Total head to be developed by pump = $H + H_{fd} + 2/3\, H_{fs}$. Factor 2/3 on suction side comes because there is no air vessel on this side and the fictional losses are parabolic, as discussed above. Whereas in delivery pipe, velocity remains constant. For a single acting pump,

$$Q = \pi D^2 L/4 \times N = 0.00294 \text{ m}^3/\text{s}$$

$\rightarrow$ $$V_s = V_d = Q/A_d = 0.665 \text{ m/s}$$

$$H_{fd} = \frac{f l_d}{d_d}\frac{V_d^2}{2g} = 0.081 \text{ m}$$

and $$H_{fs} = \frac{f l_s}{2 g d_s}\left(\frac{A\omega r}{A_s}\right)^2 = 0.207 \text{ m}$$

Maximum friction head occurs at $\theta = \pi/2$

$\therefore$ $P = 1200 \times 9.81 \times 0.00294\,(11.5 + 0.081 + 2/3 \times 0.207) = 406$ kW

## 10.3 Pressure Intensifier

In actual applications, requirement of pressure and volume flow rate depends on the type of operation and normally in the same operation cycle these condition will in general differ for different operations. As an example, consider blanking of a hole in a sheet. The operation cycle consists of (i) fast forward motion of the punch towards the sheet, (ii) punching and (iii) fast retracting of the punch; in the first and last operations large volume flow rate at a very low pressure whereas in the second process a small volume flow rate at high pressure is required. Such situations are very common in industrial applications. Using a pump to generate pressure required for second process is waste of energy, since it is required for a very small time of the whole cycle. In such a case a pressure intensifier with an accumulator and sequential directional control valves (DCV) may be used.

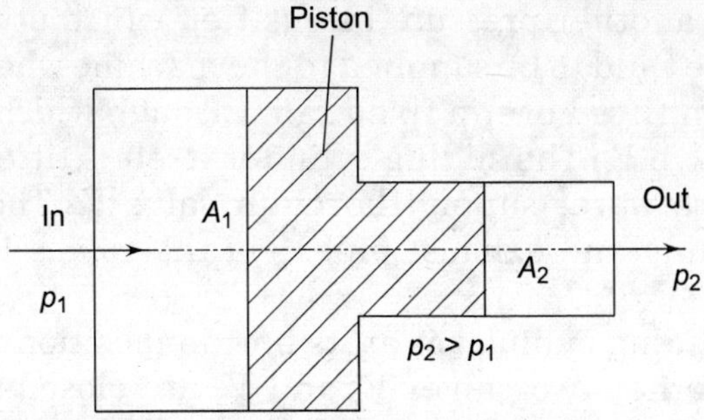


**Fig. 10.3.1** Pressure intensifier

Figure 10.3.1 shows the pressure intensifier in which a stepped cylinder and stepped piston are used. Liquid at low pressure enters at its big end side and is discharged at high pressure from the smaller end. Since the forces on both its faces balance the piston, it gives the relation

$$p_1 A_1 = p_2 A_2 \quad \text{or} \quad p_2 = p_1 A_1 / A_2 \qquad (10.3.1)$$

Here $A_1$ and $A_2$ are the area of the piston faces on the big and small ends, as shown in figure. Commercial intensifiers are available with pressure ratio up to 7. But the intensifiers are very costly.

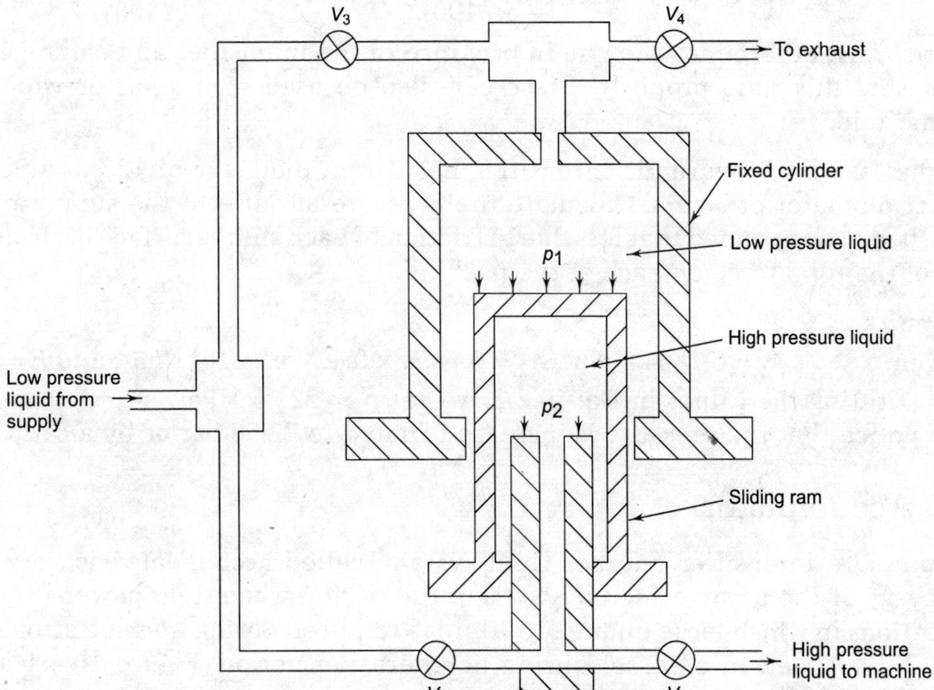


**Fig. 10.3.2** Another hydraulic intensifier

Figure 10.3.2 shows another pressure intensifier, which utilises low pressure liquid to produce high pressure fluid to be supplied to the machine when needed. The intensifier consists of a sliding ram fitted over a fixed ram through which water can enter into its inner space through valve $V_1$. The sliding cylinder itself is fitted inside a fixed cylinder which gets low-pressure water supply thorough valve $V_3$. The two cylinders are also connected to the machine and exhaust pipes via the valves $V_2$ and $V_4$ respectively, as shown in the figure.

Initially when the sliding cylinder is at its bottom position then the valves $V_1$ and $V_4$ are opened while the other two valves $V_2$ and $V_3$ are closed so that the low pressure water enters into the inside space of the sliding cylinder and the water in the fixed cylinder space flows out through valve $V_4$. Once the sliding cylinder has reached the top position, the valves $V_1$ and $V_4$ are closed while the valves $V_2$ and $V_3$ are opened. Water now enters the upper space from the supply and pushes down the sliding cylinder causing delivery of high-pressure water to the machine. The cycle is repeated automatically by using the control valves in the hydraulic circuit.

Let $A_1$ and $A_2$ be the cross sectional areas of sliding and fixed ram and $p_1$ and $p_2$ the pressure of low and high pressure liquids respectively. Further let $k\%$ be the friction loss at each sliding surface, then for equilibrium of the sliding ram

$$p_1 A_1 (1 - k / 100) = \frac{p_2 A_2}{(1 - k/100)} \rightarrow p_2 = \frac{p_1 A_1}{A_2 \times (1 - k/100)^2} \qquad (10.3.2)$$

Ratio $A_1/A_2$ determines the rise in pressure in the intensifier and can be controlled by choosing this ratio properly. These can also be used with a gas or vapour as the working fluid.

**Example 10.3.1** A hydraulic intensifier has a ram diameter of 12 cm and a sliding cylinder diameter of 60 cm. Calculate the pressure at outlet if the supply pressure is 1 Mpa. It may be assumed that loss due to friction at each sliding surface of the intensifier is 5% of the total force on each of the packing.

**Solution**

**Given:** $A_1 = 0.7854 \times 0.6^2$, $A_2 = 0.7854 \times 0.12^2$, $p_1 = 10^6$ N/m$^2$, and $k = 0.05$

Substituting the values in Eq. 10.2.2, we get $p_2$ = 22.56 MPa

We notice that the pressure is increased from 1 to 22.56 bar or by a ratio 22.56

## 10.4 Accumulator

Like pressure intensifier another fluid system, called accumulator, is very useful to use energy of the pump effectively during the period when fluid power is not used in applications in which large quantity of fluid is required during working stroke. For this, the fluid is stored in a vessel during no-operation period. Figure 10.4.1 shows two types of hydraulic accumulator. In simple accumulator, a sliding ram moves up in a fixed cylinder when the machine does not consume fluid. A load is placed on the ram so that it is equal to the product of the fluid pressure from pump and area of the ram. Thus accumulator pressure can be fixed by using appropriate weights on the ram.

When machine needs liquid then it is supplied partly by the pump and the rest by the accumulator during which the ram moves down. The volume of the liquid collected in the cylinder is equal to that supplied during no operation period. If $p$, $a$ and $h$ are liquid pressure, ram area and displacement of ram from minimum to maximum position, then

Energy supplied by fluid

$$= \text{energy stored by accumulator} = p \times a \times h = p \times \text{cylinder volume} \qquad (10.4.1)$$

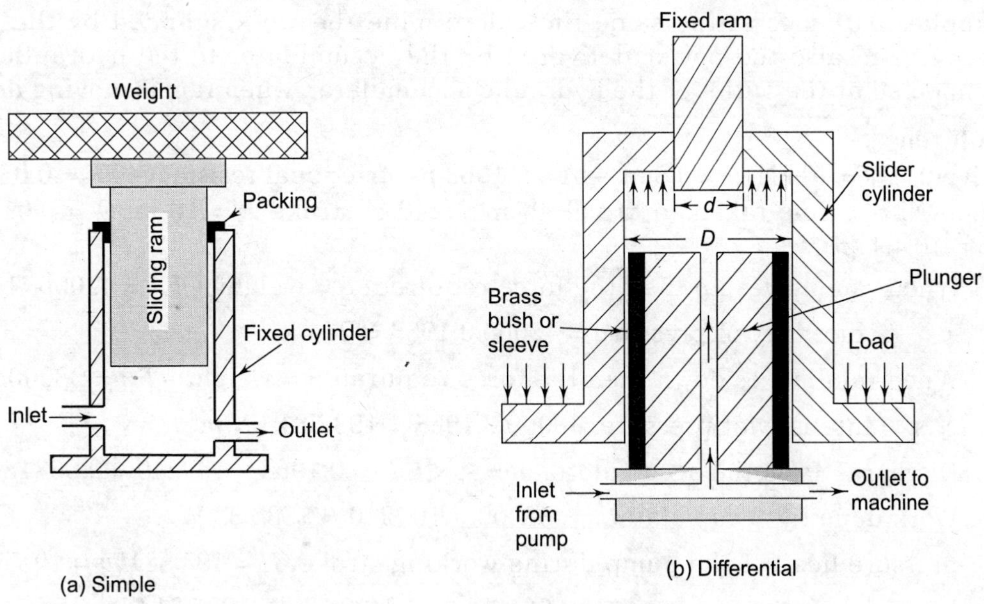


**Fig. 10.4.1** Hydraulic accumulator

Weight required in simple accumulator is large. This is reduced in differential accumulator, which is shown in Fig. 10.4.1(b). It consists of a central plunger around which inverted cylinder is placed that slides up and down along the plunger axis; a brass sleeve is used between them to prevent wear and tear of the cylinder's surfaces. Load is placed on the flange of the sliding cylinder. To reduce this load, a fixed ram is provided, as shown in the figure. During non-operation period, water enters the accumulator and exerts pressure on the cylinder surface around fixed ram and pushes it up. The cylinder moves down when liquid flows from it. We may write following equations considering equilibrium of the cylinder and relation for work transfer:

$$W = \pi/4\ (D^2 - d^2) \times p \qquad \text{and Work done/s} = W\, h/t = \pi/4\ (D^2 - d^2) \times p \times h/t \qquad (10.4.2)$$

Where $W$, $p$, $D$, $d$, $h$, and $t$ are weight on cylinder (including its own weight), liquid pressure, cylinder diameter, ram diameter, and vertical displacement of cylinder in time $t$, respectively. Load can be applied by sliding standard weights around the inverted cylinder.

**Example 10.4.1** The total weight including the self-weight of ram placed on the sliding ram of a hydraulic accumulator is 40 kN. The diameter of the ram is 500 mm. If the frictional resistance against the movement of the ram is 5% of the total weight, determine the intensity of pressure of water when (i) the ram is moving up with a uniform speed, and (ii) the ram is moving down with uniform speed. If the stroke of the ram is 10 m and the ram falls through the full stroke in 4 min steadily, find the work done by the accumulator per second. If the pump, connected to the inlet of the accumulator, supplies 0.01 m³/s at the same time, determine the work supplied by the pump per second and also the power delivered by the accumulator to the hydraulic machine, connected at the outlet of the hydraulic accumulator, when ram is moving downwards.

**Solution**

**Given:** $W = 40$ kN, $D = 0.5$ m $\rightarrow A = 0.1963$ m², frictional resistance $FR = 0.05 \times 40000 =$ 2000 N, $t$ = time taken to travel 10 m = 240 s, stroke $L = 10$ m, $Q = 0.01$ m³/s. See Fig. 10.4.1 (a).

When ram moves up then the force required, $F$ = weight + $FR$ = 42000 N

$\therefore$ $p$, intensity of pressure = 42000/0.1963 = 214 kN/m²

When ram moves down then the force required $F$ = weight – $FR$ = 38000 N

$\therefore$ $p$ = intensity of pressure, 38000/0.1963 = 193.581 kN/m²

Discharge from the accumulator, $q = A \times L / t = 0.1963 \times 10 / 240 = 0.00818$ m³/s

Work done by accumulator/s = 38000 × 10 /240 = 1583.33 W

pressure head of the pump during working stroke, $H = 193.581/w = 19.733$ m

Work done by pump= $w\,Q\,H = 9810 \times 0.01 \times 19.733 = 1935.81$ W

Total work delivered to the machine = 1583.33 + 1935.81 = 3519 W or 3.519 kW

**Example 10.4.2** It is required to transmit 36.75 kW power from an accumulator through a pipeline of 10 cm diameter and 1500 m long. The ram is loaded with a weight of 1250 kN and the friction loss in the pipeline equals 2.5% of the total power being transmitted. Work out the diameter of the ram if the friction factor $f = 0.01$ for the pipeline.

**Solution**

**Given:** $P = 36.75$ kW, $d = 0.1$ m, $L = 1500$ m, $W = 1250$ kN

$$\text{power loss} = 0.025 \times 36.75 = 0.9187 \text{ kW}$$

$$\text{Friction head loss } h_f = \frac{4fl}{d}\frac{U^2}{2g} = \frac{4 \times 0.01 \times 1500}{2 \times 9.81 \times 0.1} U^2 = 30.58U^2$$

Power loss in friction = $w\,Q\,h_f = 9810 \times 0.7854 \times 0.1^2 \times U \times 30.58U^2 = 2353\,U^3 = 918.7$

$\rightarrow$ $U$, flow velocity in pipe = 0.731 m/s and $Q = 0.7854 \times 0.12 \times U = 5.738 \times 10^{-3}$ m³/s

This discharge comes form the accumulator. If $H$ is the pressure head of water in the accumulator then

Power developed by accumulator = $w\,Q\,H = 36.75 \times 1000$ (given)

or, $9810 \times 5.738 \times 10^{-3} \times H = 36.75 \times 1000 \rightarrow H = 652.87$ m of water

Pressure intensity = $w\,H = 9810 \times 652.87 = 6.4$ MN/m$^2$

But pressure intensity = W/ram area

or, $$6.4 \times 10^6 = \frac{125 \times 1000}{\pi/4 \times D^2} \rightarrow D = 0.499 \text{ m}$$

## 10.5 Hydraulic Crane

A hydraulic crane is used for lifting load to move it from one place to another and is very useful in the industries. It works on the principle of simple machines that the mechanical advantage is high. This also means that the velocity ratio is also high. We study these two terms in chapter on simple machines in Mechanics course. Basic or simple machines were discovered in our effort to push a large mass using a small force a man or animal could exert. Lever, pulley, etc are a few examples of such machines. In these machines, a large weight (load) could be moved because of large displacement of the effort. Let $P$ and $W$ be the effort and the load and their respective displacement be $y$ and $x$. Then making energy balance, we get

$$W\,x = Py \qquad \text{or,} \qquad W/P = y/x \tag{10.5.1}$$

Ratio $W/P$ is called mechanical advantage and $y/x$ the velocity ratio. For 100% efficient machine the two are equal, otherwise their ratio is the efficiency. Thus

$$\eta = \frac{Wx}{Py} = \frac{MA}{VR} \tag{10.5.2}$$

A crane consists of a hydrualic jigger shown inside the dotted rectangular box and the craning arrangement shown in Fig.10.5.1. A hydraulic jigger is the basic unit that can be used in any application. It consists of a fixed cylinder inside which a sliding ram is fitted; the open end of the ram is fitted to the pulley block that moves with the ram. On the contrary, the fixed cylinder is fastened to a fixed pulley block. Both the pulley blocks have equal number of pulleys around which a rope passes with one of its free end fixed at $A$, as shown in the figure, and the other end attached to the craning attachment. The craning mechaninsm consists of a vertical post mounted on the pedastal; the post and the pedastal supports the jib which is fitted with two pulleys, as shown in Fig. 10.5.1. The free end of the rope passes around these pulleys and then hooked to the load. Water can be supplied or taken out from the fixed cylinder by controlling the inlet and outlet valves.

When high-pressure water is fed to the cylinder, it pushes the ram out increasing the centre-to-centre distance. Therefore the load end of the rope rises; the rise stops if the water supply is cut off. When the load is to be lowered then the outlet valve is opened resulting in the outage of the water due to the pressure on the ram due to load. This explains the working of the crane.

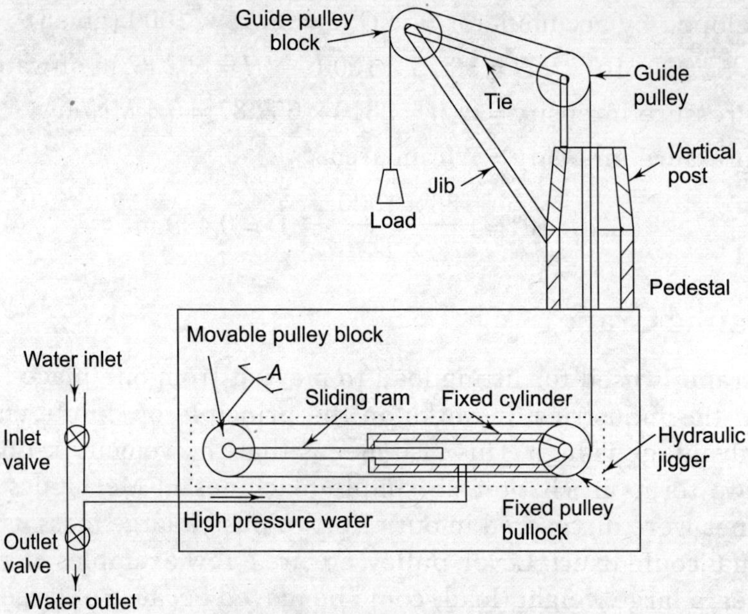


**Fig. 10.5.1** Hydraulic crane

We will not go into details of these simple machines. If there are $n$ pulleys at each end of the pulley blocks, then the vertical distance moved by the crane will be $2n$ times the travel of the ram. Let $F$ be force on the ram, ram movement be $y$, load be $W$ and the vertical distance moved by load $x$ and $\eta$ be the efficiency of the crane then from the basic machine principle

$$F.y.\eta = W.x \rightarrow \frac{W}{F} = \eta.\frac{y}{x} \text{ or, } M.A. = \eta \times V.R \tag{10.5.3}$$

Note that $x$ is equal to $2n$ times the ram movement. Thus more is the number of the pulleys less is the velocity ratio, *V.R.* In other words for given displacement of the load, the movement of the effort will be less. In hydraulic crane and lift, the load displacement is more important and not the amount of load as we can generate enough effort using hydraulic power. In such machines, mechanical advantage is always less than 1, since electric motor is being used as power source. Therefore the main aim is to have large load movement for a small displacement of the effort.

**Example 10.5.1** The efficiency of a hydraulic crane, which is supplied water under a pressure of 70 N/cm² for lifting a weight through a height of 10 m, is 60%. If the diameter of the ram is 150 mm and velocity ratio 1/6, find (i) the weight lifted, and (ii) the volume of water required in litres to lift the weight.

**Solution**

**Given:** $\eta_c = 0.6$, $p = 7 \times 10^5$ N/m², $x = 10$ m, $D = 0.15$ m $\rightarrow A = 0.0176$ m² and $F = p \times A = 12369$ N, $VR = 1/6$

$$\eta_c = \frac{W \times \text{distance moved by weight}}{F \times y} = \frac{W}{F \times VR} \rightarrow W = 1236.9 \text{ N}$$

Volume of water required for lift $= A \times y = 0.0176 \times 10/6 = 0.0945$ m³ $= 29.45\ l$

## 10.6 Hydraulic Lift

Figure 10.6.1 shows another way in which a hydraulic jigger can be used. As pointed out in the last section, jigger is the main unit that is capable of producing movement of the rope at the desired tension. Now it depends on us how do we utilise it. If we attach a cage of the lift to the rope end then it is the hydraulic lift.

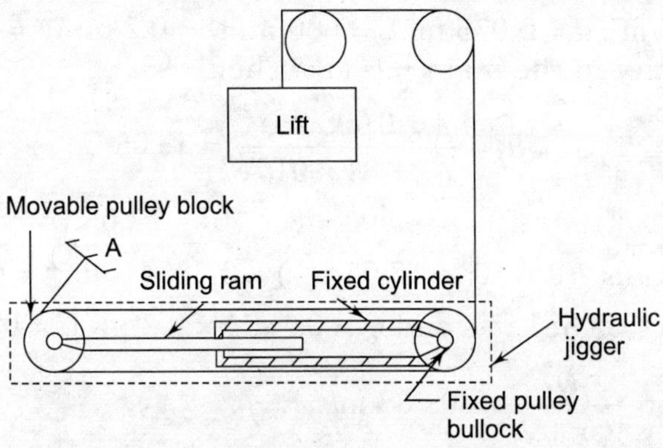


**Fig. 10.6.1** Hydraulic lift

**Example 10.6.1** A hydraulic lift is required to lift a load of 12 kN through a height of 10 m, once in every 1.75 minutes. The speed of the lift is 0.75 m/s. During the working stroke of the lift, water from the accumulator and the pump at a pressure of 400 N/cm$^2$ is supplied to the lift. If the efficiency of the pump is 80% and that of the lift 75%, find the power required to drive the pump and the minimum capacity of the accumulator. Neglect frictional losses in the pipe.

**Solution**

**Given:** $W$ = 12 kN, $H$ = 10 m, $t$ = 1.75 min = 105 s, $v$ = lift speed = 0.75 m/s, $p = 400 \times 10^4$ N/m$^2$, $\eta_p = 0.8$, $\eta_l = 0.75$

Total work done in the lift = useful work = $\dot{W} \times v$ = 9000 W or, 9 kW

Power supplied to lift per second = useful work/$\eta_l$ = 9/0.75 = 12 kW

Time taken by lift to travel through 10 m = $t' = H/v$ = 10/0.75 = 13.33 s

Idle period of lift = $t - t'$ = 105 – 13.33 = 91.67 s.

The accumulator stores the energy in this idle period that is supplied during the working stroke. Let $P_1$ be the power output of the pump then power supplied by accumulator to the lift = 12 – $P_1$

Energy stored in the accumulator during idle period = 91.67 × $P_1$ kJ

= energy supplied by the accumulator in $t'$s = (12 – $P_1$) × 13.33 → $P_1$ = 1.523 kW

Input to the pump = $P_1/\eta_p$ = 1.523/0.8 = 1.90375 kW

Energy stored in accumulator = accumulator capacity = $P_1$ × 91.67 = 13.96 kJ

**Example 10.6.2** An accumulator maintains a steady pressure of 8 Mpa in a 7.5 cm hydraulic main. A hydraulic lift is supplied with pressure water from the main and the point at which the supply to the lift is drawn off is at a distance of 600 m from the accumulator. The ram of the lift is 20 cm in diameter and the load on it, inclusive of its own weight, is 120 kN. Assuming the friction of the ram and cage etc. to be equivalent to an addition of 5% of the gross load on the ram, determine the speed with which the lift will ascend. Take $f = 0.008$ for the hydraulic main and neglect the minor losses.

**Solution**

**Given:** $p = 8 \times 10^6$ N/m², $d = 0.075$ m, $L = 600$ m, $D = 0.2$ m, $W = 120$ kN, $f = 0.008$

Let the flow velocity in the main = $U$ m/s. Then

$$h_f = \frac{4 \times 0.008 \times U^2}{2 \times 9.81 \times 0.075} = 13.05\, U^2$$

pressure intensity at the lift $= p - w\, h_f = 8 \times 10^6 - 13.05\, U^2$

and force on the ram= $(8 \times 10^6 - 13.05\, U^2) \times 0.7854 \times 0.2^2 = (120 + 0.05 \times 120) \times 10^3$

$$\rightarrow \quad U = 5.581 \text{ m/s and } Q = 0.7854 \times 0.2^2 \times U = 0.0246 \text{ m}^3/\text{s}$$

$$\therefore \quad \text{velocity of lift } \frac{Q}{\pi/4 \times 0.2^2} = 0.783 \text{ m/s}$$

## 10.7 Hydraulic Press

It is another useful simple machine that is used in industries. It has many application in the sheet metal working and forging and is shown in Fig. 10.7.1. It is a simple machine that is based on the Pascal law that pressure in the fluid at rest is same everywhere if the gravity effect is neglected. To generate a large force at the load end we use two pistons, one is the plunger of area $a$ and the other is called the ram, which has area $A$ such that $A >> a$. A simple lever is used to take further advantage. For lever, taking moment about the fulcrum $F$, we get

$$F \times l = P \times L \rightarrow F = P \times L/l$$

But $\quad p$ = pressure of liquid = $F/a$

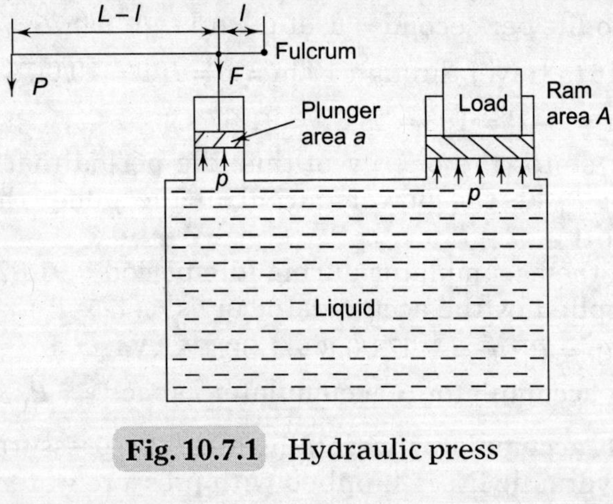


**Fig. 10.7.1** Hydraulic press

$$\therefore \quad W = \text{Load lifted} = p \times A = P \times \frac{L}{l} \times \frac{A}{a} \tag{10.7.1}$$

Thus, by proper choice of the lengths and the areas, we can achieve any mechanical advantage. However, it should be kept in mind that the velocity ratio will also be high. This means a large movement of the effort shall move the load by small distance. Therefore, a proper combination has to be taken looking into the capacity of applying the effort.

We get another equation from the fact that the volume displaced by the plunger is equal to that goes to the ram area. From this consideration the rise in the load can be determined. If $h$ is the stroke of the plunger and $N$ is the number of strokes made per second by it then the displaced volume

$$Q = a\,h\,N = A\,H \tag{10.7.2}$$

$H$ is the lift of the ram in a second.

**Example 10.7.1** A hydraulic press has a ram of 200 mm diameter and a plunger of 30 mm diameter. It is used for lifting a weight of 3 kN. Find the force required at the plunger. How much force will be required at the end of a lever, if used, having the ratio $l/L = 1/10$. If the stroke of the plunger is 100 mm find the distance travelled by the weight in 100 strokes. Determine the work done.

**Solution**

**Given:** $A = 0.7854 \times 0.2^2 = 0.0314\ \text{m}^2$, $a = 0.7854 \times 0.03^2 = 7.068 \times 10^{-4}$, $W = 3000$ N, $l/L = 0.1$

Force required at the plunger = $(a/A) \times W = 67.52$ N

From Eq. 10.6.1, force required at the end of the lever = $P = 6.752$ N

Displacement of the load is obtained by calculating the displacement of the liquid by movement of the plunger in 100 strokes. This gives

Volume of the displaced liquid = $100 \times 0.1 \times a = 0.007068\ \text{m}^3$

$\therefore$ Load displacement = $0.007068/A = 0.2251$ m

**Example 10.7.2** The ram and plunger of a hydraulic press are 25 cm and 3 cm, respectively, and the leverage of the handle is 10:1. With a plunger stroke of 25 cm, the press is able to lift a load of 200 kN through 1.25 m in 2 minutes. Determine (i) force applied at the end of the lever (ii) power required to drive the plunger, and (iii) number of stroke to be performed by the plunger in one second. Assume that the packing friction of the plunger as well as the ram is 4% of the load.

**Solution**

**Given:** $d = 0.03$ m, $D = 0.25$ m, $L/l = 10$, $h = 0.25$ m, $W = 200$ kN, $H = 1.25/120$ m, $k = 0.04$

See Fig. 10.6.1. From the given data

$$p \times a = F - kF \quad \text{and} \quad W = p \times A\,(1 - k)$$

$$\rightarrow \quad F = \frac{Wa}{A(1-k)^2} = \frac{200 \times 10^3 \times \pi/4 \times 0.03^2}{\pi/4 \times 0.25^2 \times 0.96^2} = 3120 \text{ N}$$

Effort required at the end of the lever = 3120/10 = 312 N

$$\text{Number of strokes} = \frac{\text{Total volume of liquid to be supplied}}{\text{Volume of liquid displaced / stroke in plunger}}$$

$$= \frac{AH}{ah} = \left(\frac{0.25}{0.03}\right)^2 \frac{1.25/120}{0.25} = 2.2 \approx \textbf{3 strokes}$$

Work done at the plunger $= F \times h \times n = 3120 \times 0.25 \times 2.2 = 2262$ W or 2.262 kW

## 10.8 Hydraulic Coupling

Figure 10.8.1 shows a hydraulic coupling. A coupling is used to transmit power from driving to driven shaft. Therefore a hydraulic coupling is a special hydraulic system in which there is direct transfer of fluid power from pump to motor. The pump impeller is fixed to the driving shaft, the motor impeller with the driven shaft and the two are placed face to face in a casing; contact between them is through fluid that circulates between the vanes provided around their periphery. Due to centrifugal force exerted by the rotation of the driving shaft, fluid is forced to go towards the vanes in motor where it moves around vane and returns back to the pump vane. While fluid moves over the motor cups, its momentum is changed and hence force is exerted on the cups that rotates the driven shaft. Thus, the motor works as turbine and whence the name turbine runner. Initially because of friction driven shaft takes some time to start rotation. But once the torque overcomes this friction, the shaft picks up the speed. However, due to flexibility in the coupling, speed of the driven shaft is a little less than that of the driving shaft, which is expressed by the term slip.

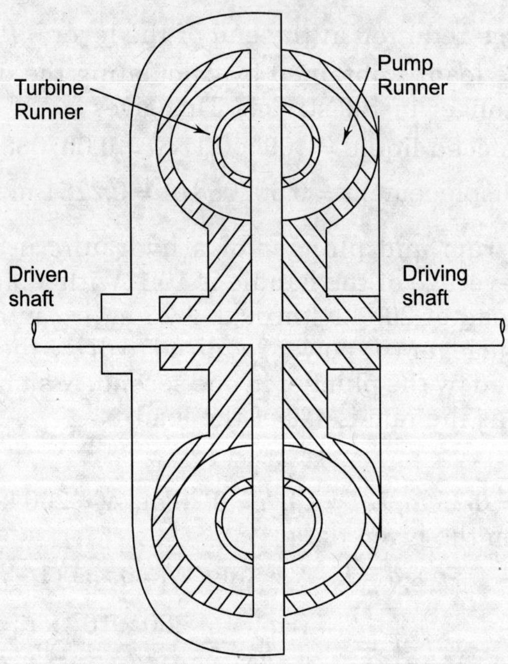


**Fig. 10.8.1** Hydraulic coupling

Let the subscripts $p$ and $m$ are used for pump and motor, $\tau$ for the torque, $P$ for power and $\omega$ for angular speed then we may write following equations:

$$P_m = (\tau\omega)_m \qquad P_p = (\tau\omega)_p$$

$\rightarrow$ Efficiency of fluid coupling $= P_m / P_p = (\tau\omega)_m / (\tau\omega)_p$ (10.8.1)

and slip $= (\omega_p - \omega_m)/\omega_p$ (10.8.2)

Slip normally varies from 2% to 4% under steady state operation while coupling efficiency increases steadily from 0% to about 90% with speed ratio of motor to that of pump. In day-to-day language, we may say this works like as if two fans are placed face to face. Let one of them is supplied electricity. This fan rotates and we observe that soon the second fan also starts rotating though no electricity is supplied to it. This is the fluid coupling action.

The fluid coupling is used in applications like automobiles, marine engines, ropeway cable drive units, agricultural machinery and such applications where driven shaft is supposed to run close to the speed of the driving shaft and where chances of shock loads are possible, since those situations can be handled easily by fluid coupling.

## 10.9 Hydraulic Ram

Hydraulic ram is shown in Fig. 10.9.1. This is a very useful static machine that is able to lift water above the level of the water in the supply tank utilising the momentum of incoming water by closing the valve for water hammer type phenomenon to occur. However, the fraction of water lifted to that supplied is much less than 1; rest of the water spills out at lower level.

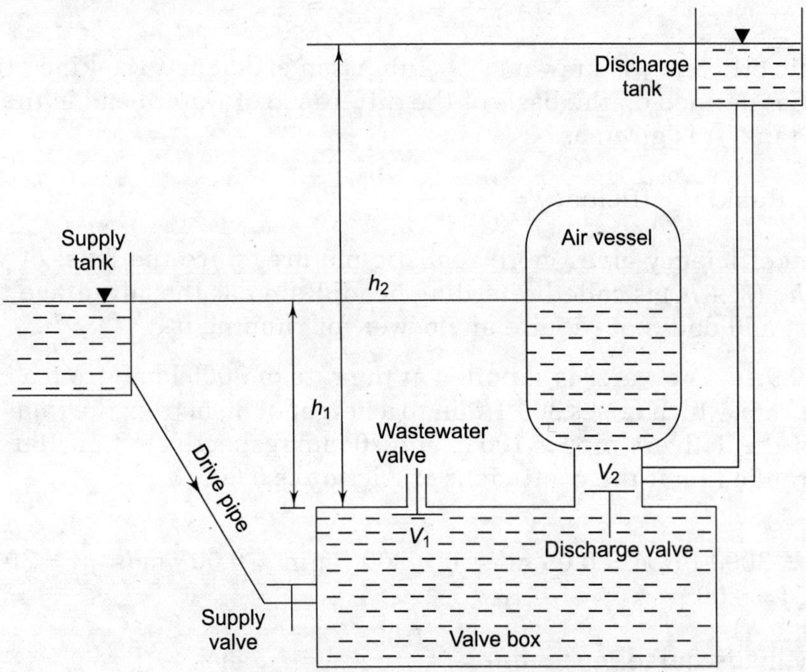


**Fig. 10.9.1** Hydraulic ram

Supply tank is connected to the valve box by pipe, called the drive pipe. The valve box contains the wastewater valve $V_1$ and the discharge valve $V_2$, as shown; both these valves are one-way valve. On the delivery side an air vessel is fitted in parallel to the water delivery pipe. Initially the valve $V_1$ is open and when the supply valve is opened then water starts moving through the drive pipe and goes as waste through $V_1$. However, as the speed picks up, the dynamic thrust too increases that closes valve $V_1$ suddenly at some time. The closure raises pressure in the valve box opening the valve $V_2$. Now the water flows to the delivery pipe as well as to the air vessel. This compresses air and therefore increases the pressure is the vessel. At some time the pressure is enough to force the water to rise to the water tank. Thus water is lifted. After some time the momentum in the valve box will be destroyed then the delivery valve will close and the waste water valve $V_1$ will open thus resetting the valve box to repeat the cycle. Now water will be supplied to the discharge tank from the air vessel as air in the vessel is at high pressure. Let

$Q$ = discharge from the supply tank $\quad$ $q$ = discharge from the valve box to the discharge tank

$h_1$ = height of free surface in supply tank above the valve box

$h_2$ = height of free surface in discharge tank above the valve box

Then energy supplied by the supply tank = $w\,Q\,h_1$

and energy received by the discharge tank = $w\,q\,h_2$

Therefore hydraulic ram efficiency is

$$\eta = \frac{wqh_2}{wQh_1} = \frac{q}{Q} \times \frac{h_2}{h_1} \tag{10.9.1}$$

Hydraulic efficiency is known as $D'$ Aubuisson efficiency. Rankine efficiency is also used, which is defined on the basis of the difference of water head in the discharge and the supply tank. It is given as

$$\text{Rankine efficiency} = \frac{q}{Q} \times \frac{h_2 - h_1}{h_1} \tag{10.9.2}$$

Ramkine efficiency gives more realistic picture, since the level of water is raised from $h_1$ to $h_2$; $(h_2 - h_1)$ is called retarding head. Ram has the advantage that there is no moving part and does not require any power for running it.

**Example 10.9.1** The water is supplied at the rate of 3000 l/min from a height of 4 m to a hydraulic ram, which raises 300 l/min to a height of 30 m from the ram. The length and diameter of the delivery pipe is 100 m and 70 mm respectively. Calculate the efficiency of the hydraulic ram if the coefficient of friction is 0.009.

**Solution**

**Given:** $Q$ = 3000 l/min = 0.05 m³/s, $q$ = 300 l/min = 0.005 m³/s, $h_2$ = 30 m, $L$ = 100 m, $d$ = 0.07 m, $f$ = 0.009

$$\text{Velocity in discharge pipe} = q/a = \frac{0.005}{\pi/4 \times 0.07^2} = 1.3 \text{ m/s}$$

$$\text{Head loss due to pipe friction} = \frac{4 \times 0.009 \times 100}{0.07} \times \frac{1.3^2}{2 \times 9.81} = 4.43 \text{ m}$$

Effective pressure head developed at the discharge valve = $h_2 + 4.43 = 34.43$ m

From Eq. 10.8.1

$$\text{D' Aubuisson efficiency} = \frac{q \times \text{effective head}}{Q \times h_1} = \frac{0.005 \times 34.43}{0.05 \times 4} = 0.8607 \text{ or, } \mathbf{86.07\%}$$

$$\text{Rankine efficiency} = \frac{q \times (\text{effective head} - h_1)}{Q \times h_1} = \frac{0.005 \times (34.43 - 4)}{0.05 \times 4} = 0.8453 \text{ or,}$$

**84.53%**

**Example 10.9.2** The supply pipe in a hydraulic ram installation is 60 mm in diameter and 4 m in length. The waste valve is 130 mm diameter having an effective lift of 6.5 mm and a weight of 13.25 N; it makes 125 beats per minute. Find the discharge through the delivery pipe per minute against a delivery head of 8 m above the water level in the supply reservoir. Neglect all losses.

**Solution**

**Given:** $d = 0.06$ m, $L = 4$ m, valve weight = 13.25 N, $h = 6.5$ mm

Waste valve diameter $D = 130$ mm

→ Base area of the waste valve = $0.7854 \times 130^2 = 13273 \text{ mm}^2$

Upward pressure required for closing the valve = weight of the valve/area

$$p = 13.25/13273 = 0.000998 \text{ N/mm}^2$$

→ pressure head $= p/w = 0.000998 \times 10^6/9810 = 0.102$ m of water

∴ Flow velocity $U$ through the waste valve = $\sqrt{2gh} = \sqrt{2 \times 9.81 \times 0.102} = 1.414$ m/s

Area of flow through which the waste water flows = $\pi \times D \times h = 2655 \text{ mm}^2$

Area of the supply pipe $= 0.7854 \times 60^2 = 2827 \text{ mm}^2$

∴ maximum velocity in the supply pipe $U = 2655 \times 1.414/2827 = 1.328$ m/s

$$\text{Retarding head} = \frac{LU}{gt} = \frac{4 \times 1.328}{9.81 \times t} = 8$$ (difference of water level in the discharge and supply tanks)

→ $t = 0.0677$ s

During this time the water is supplied to the discharge line from the supply line. However, the velocity changes and can be taken as the average velocity = $U/2 = 0.664$ m/s

∴ Quantity of water supplied/beat = $2827 \times 10^{-6} \times 0.664 = 1.27 \times 10^{-4} \text{ m}^3/\text{s}$

quantity of water supplied per minute = $1.27 \times 10^{-4} \times 125 = 0.15625 \text{ m}^3/\text{min}$ = 15.625 l/min

## 10.10 Hydraulic Controls

Controls of any system is very important to regulate its output irrespective of any change on its input or the environment. Even the demand on the system may change and then the system has to change to adjust its output accordingly. We may not be able to offset these changes completely but what can be done is to offset the effects by bringing back the system to its normal or expected behaviour in the least possible time and with least deviations from expected output by adding some mechanism to it.

Imagine we are sending a rocket in earth orbit and fail to control its trajectory during the flight then what would happen? Or if a power plant generates electric current with variable frequency or voltage, will that be acceptable to us? Answer is BIG no. We make sure that a power plant on which electrical load varies throughout the day and around the year, quality of the coal varies because of supply from different coal fields, and so on, produces electric current with frequency and voltage within limits permitted by the electricity generation authority of the country. Designing proper controls does this.

Consider an internal combustion engine. If it is used for power generation then the situation is same as for other power plant. But if it is used in vehicle then arrangement has to be made so that the vehicle speed may be varied at the will of the operator. An air conditioning system has to take care of variation of heat load on the space that would vary with season and occupancy so that temperature and humidity in the space may be maintained within the permissible limits. Like this, we can make a comprehensive list of controls required in an industry. A simple example we see in our daily life is that of automatic pumping of water to overhead tank. The control circuit takes care of running or stopping the motor to keep water level in the tank between the two defined levels. Again these limits can easily be changed as per our wish. Automation is the need of the day and is possible because of reliability of the control circuits. Like this we may quote innumerable applications needing controls of some or other parameters. The process of controlling the output parameter as per our requirement against all variations on the machine is called control. In this section we will concentrate on the basic aspects of hydraulic circuits only, as it is just impossible to this topic in its entirety. Hydraulic control is very commonly used in automobile, power plants and machine tools because of its capacity to handle large power with ease. We shall cover the hydraulic components and hydraulic circuits used in the control of thermal engineering systems rather than going to the detailed analysis of control circuits.

### 10.10.1 Hydraulic Pump and Motor

Power required for actuating the mechanism to control system input, which usually is quite large, is obtained from the fluid pressure. For this fluid power has been used in industrial applications because of its many inherent advantages. Hydraulic systems require a positive displacement type pump; in such applications volume flow requirement is low but pressure may be as high as 100–150 bar. For this reason positive displacement pumps like gear pump, vane pump, etc have been used. A gear pump is shown in Fig. 10.10.1(a), in which both the gears are external gears of equal size and carry involute teeth numbering 5 to 15. The gear pump is placed within the sump and therefore there is no problem of cavitation. One of the gears is driven by external power source and other rotate because of gearing action. The whole casing of the pump is filled with liquid. When the driver rotates then both gears take the liquid with them on the outer side through the spacing between gears and casing. This sucks liquid from the sump through the suction port. The continuous feeding of the liquid to the delivery port side raises the pressure and the fluid starts flowing when the no-return valve in the down line opens due to increased pressure. Pressure developed affects the force exerted between the meshing teeth that in turn will determine the power required for the pump. Meshing teeth of two gears also prevent flow of liquid from high-pressure side to low-pressure side. Volume of liquid pushed per revolution, called *volumetric displacement* is given by

$$V_d = \pi / 4\left(D_0^2 - D_i^2\right)L \tag{10.10.1}$$

where $D_0$ and $D_i$ are the outer and inner diameter and $L$ is width of the gear teeth. The actual discharge is less because some space is provided around the gear teeth to avoid interference with the casing through which some of the liquid returns or leaks back. We use *volumetric efficiency* to express it in the same way as used in other rotary pumps studied in Chapter 9; it is about 90%. Volumetric efficiency decreases with increase in discharge pressure. Moreover, in such pumps volume flow varies directly as rpm of the pump.

Theoretical discharge from a gear pump can be obtained using following relations:

$$Q = 2\pi b\, m^2\, N\,[z + (1 + \pi^2 \cos \alpha/12)]\ \text{m}^3/\text{min} \tag{10.10.2}$$

where $Q$, $b$, $m$, $N$, $z$, and $\alpha$ are theoretical discharge, width of gear, module of gear, rpm, number of teeth on gear, and pressure angle, respectively. The following empirical relation gives the quick discharge from the gear pump:

$$q = 0.95\,\pi C\,(D - C) \qquad \text{and } Q = qN/60\ \text{m}^3/\text{s} \tag{10.10.3}$$

where $C$ and $D$ are centre to centre distance between the gear wheels and outside diameter of the gear, respectively.

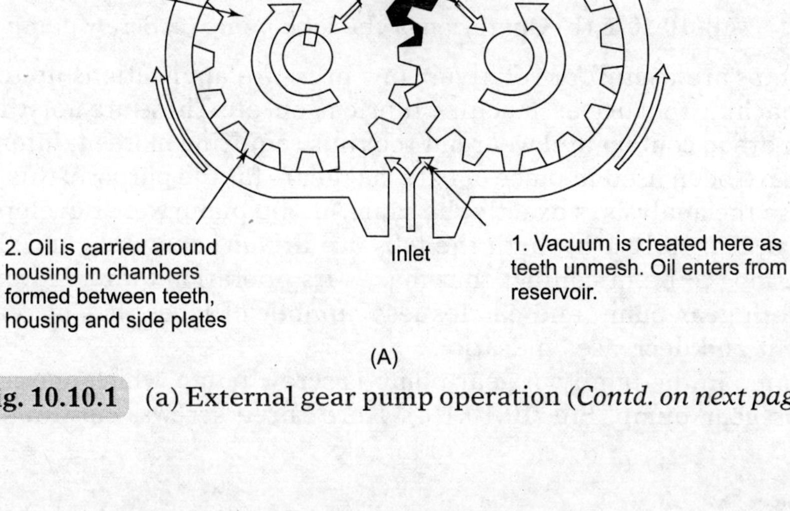


**Fig. 10.10.1** (a) External gear pump operation (*Contd. on next page*)

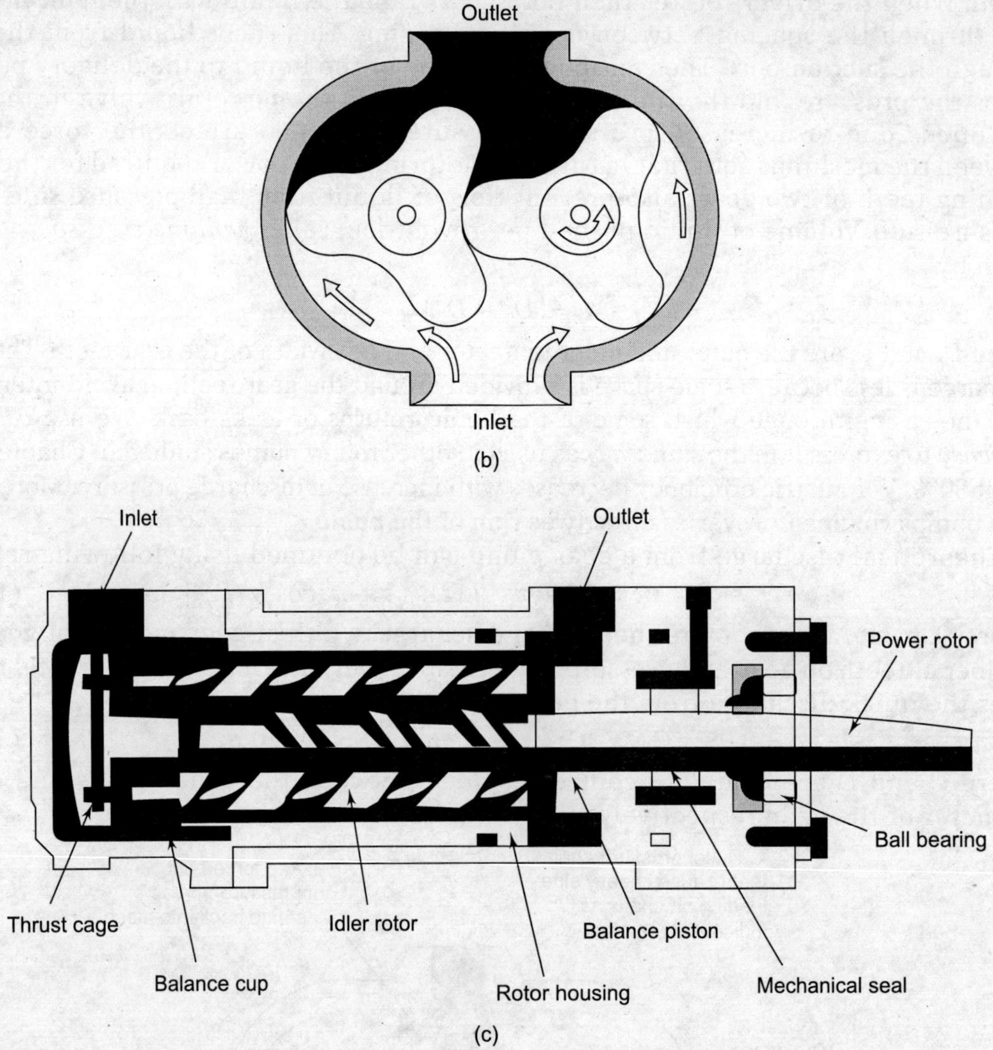


**Fig 10.10.1** (b) Operation of the lobe pump (c) Screw pump.

Gear pumps are useful for relatively low pressure applications up to 10 bar such as high duty machine tool drives, machine lubrication, etc. These are noisy in the operation as the teeth are in contact and wear out too thus requiring more maintenance. (Internal gears also have been used in place of external gears for the purpose, this is not discussed here because the analysis is exactly the same.) Lobe pump were developed to overcome this problem, Fig. 10.10.1(b); both the lobs are driven separately and therefore do not require the lobs to be in contact that makes its operation quite. It works exactly like external teeth gear pump and carries less number of lobes that increases volumetric displacement and decreases pulsation.

Other pump in the family of gear pump is screw pump, which operates on the same principle as gear pump, Fig. 10.10.1(c). Here three screws, meshing within a close-fitting housing, deliver non-pulsating flow quietly and efficiently. The central screw is power driven and the other two are the idler rotors, which also act as rotating seal. The idler rotors are in rolling contact with the central power rotor and are free to float in their respective housing on a hydrodynamic oil film.

**Vane pump**. Figure 10.10.2 shows a vane pump. It consists of a rotor having equi-spaced slots each fitted with a sliding vane. The rotor is placed inside cam ring having eccentricity *e* with the rotor, as shown in the figure. Eccentricity between the shaft and ring can be adjusted by external mechanism and in that case this pump is called variable displacement pump. When rotor rotates, centrifugal force forces the vanes to remain always in contact with the inner surface of the cam ring. Moreover the space formed by two successive vanes, the rotor and cam ring surfaces goes on changing during the rotation due to eccentricity between rotor and cam ring. During one half rotation of the shaft this space increases and the liquid is sucked in from the sump through the inlet port and in the later half it decreases and the liquid pressure is increased; liquid is then discharged through the discharge port. Vanes do not allow back flow and in that respect their performance is better than the gear pump.

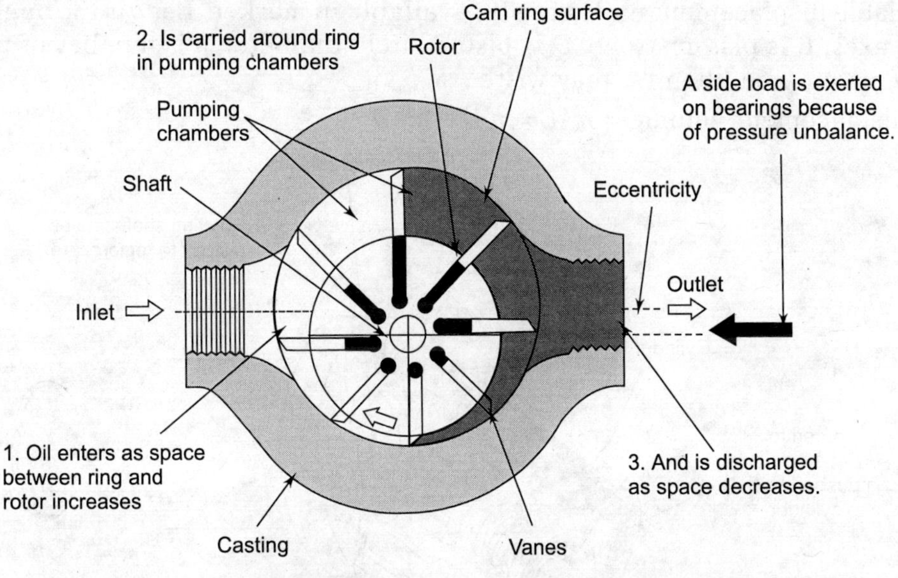


**Fig. 10.10.2** Vane pump operation

Let $N$ is rotor rpm, $D_c$ is cam ring diameter, $D_r$ is rotor diameter, $L$ is width of the rotor, $e$ is eccentricity whereas $e_{max}$ is the maximum possible eccentricity of the arrangement and $V_{d,\,max}$ is maximum possible volumetric displacement. Then

$$e_{max} = \frac{D_c - D_r}{2} \text{ and } V_{d,\,max} = \frac{\pi}{4}\left(D_c^2 - D_r^2\right) L \qquad (10.10.4)$$

At any time the volumetric displacement is obtained by substituting $e$ for $e_{max}$. In such arrangement pressure force acts on the shaft because pressure on discharge port is more than that on the suction port. This is taken care of in balanced vane pump by providing two discharge ports opposite to each other; similarly two suction ports opposite to each other are placed at right angles to the direction of the discharge ports.

A vane pump can built up pressures up to 700 bar and is a popular pumping unit used in machine tool drives and other industrial applications.

**Piston Pump.** A piston pump is a reciprocating pump with the arrangement for driving number of pistons simultaneously in a cylinder block. Basically two types of arrangements are used to move the pistons. In axial arrangement, a cylinder block contains number of cylinders bored in it along a circle each of which houses a piston. Thus the piston axes and cylinder axis are parallel. Cylinder block is fixed at an angle (and not in line) to the driving shaft, as shown in Fig. 10.10.3(a), which is connected to the swash plate that provides drive to the cylinder block. Each piston rod is connected to the flange on the driving shaft by ball-and-socket joint along the pitch circle matching the cylinders made in cylinder block, see Fig. 10.10.3(a). When shaft rotates, the pistons move in and out according to the distance between the driving shaft flange and the cylinder block. The cylinder head carries two ports separated from each other; one is connected to the suction side and other to the discharge side. Displacement volume of the pump depends on the angle $\theta$ between the axes of driving shaft and cylinder; it is zero when $\theta = 0$. Normally, angle $\theta$ is kept 23° to 30°. Arrangement can also be made to change this angle and variable displacement units are also available in market. Let $\theta$ is the angle between the two axes, $L$ is piston stroke, $D$ is piston circle diameter, $n$ is number of pistons and $A$ is the piston area, then we may write

$V_d$, displacement volume $= nAL = nAD \tan \theta$ (10.10.5)

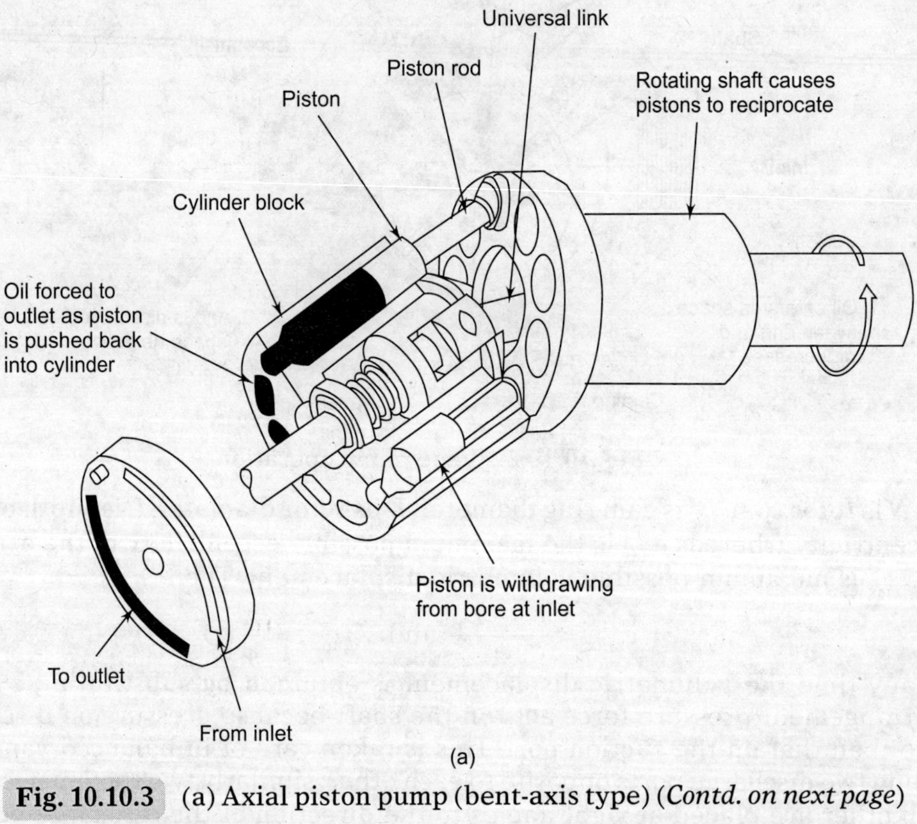


**Fig. 10.10.3** (a) Axial piston pump (bent-axis type) (*Contd. on next page*)

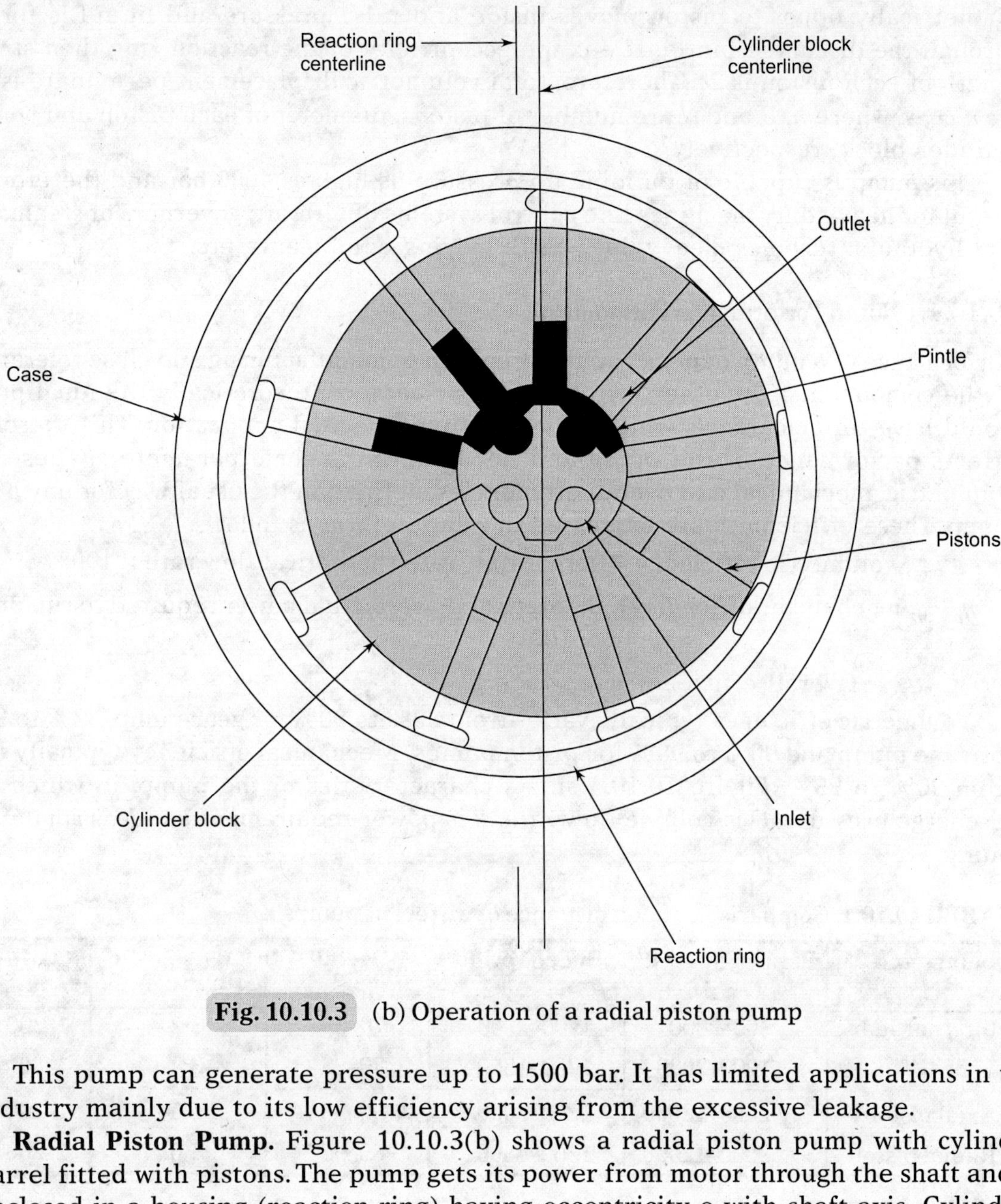


**Fig. 10.10.3** (b) Operation of a radial piston pump

This pump can generate pressure up to 1500 bar. It has limited applications in the industry mainly due to its low efficiency arising from the excessive leakage.

**Radial Piston Pump.** Figure 10.10.3(b) shows a radial piston pump with cylinder barrel fitted with pistons. The pump gets its power from motor through the shaft and is enclosed in a housing (reaction ring) having eccentricity e with shaft axis. Cylinders open in fixed pintle; one is suction and the other is delivery pintle. Pistons remain in touch with reaction ring due to back pressure of the fluid. As the shaft rotates, the piston on the suction side moves away from the pintle and sucks liquid while the diametrically opposite piston moves inside and raise pressure and liquid is forced through the discharge pintle. If e is the eccentricity of the reaction ring then stroke length of each piston is $2e$. Therefore, total volumetric displacement per minute is $0.5\ \pi\, e\, n\, D^2 N$, where $n$, $D$ and $N$ are number of pistons, diameter of each piston and rpm of cylinder block, respectively.

This pump is capable of building up pressure as high as 4000 bar and therefore is useful for heavy duty machines like control systems of aircraft; governors of steam, gas and hydraulic turbines; heavy duty earth moving equipments, etc.

#### 10.10.1.1 Pump Performance Parameters

Performance of a pump depends on the precision of manufacturing and close tolerances of the components. The other factor includes clearance; theoretically the ideal pump would have zero clearance, which is not possible in actual application. This therefore affects performance of the pump and needs defining some parameters. These are volumetric, mechanical and overall efficiencies, defined in the usual way for any other pump. These efficiencies are expressed in relation form as follows:

$$\eta_{vol}, \text{ volumetric efficiency} = \text{actual flow rate/theoretical flow rate} \times 100\%$$

$$\eta_{mech}, \text{ mechanical efficiency} = \text{theoretical power/actual power required to run pump} \times 100\%$$

$$\text{Overall efficiency} = \eta_{mech} \times \eta_{vol}$$

Volumetric efficiency typically varies from 80% to 90% for gear pump, 82% to 92% for vane pump and 90% to 98% for piston pumps. Mechanical efficiency typically runs from 90% to 95%. Figure 10.10.4 shows characteristics of the pump, in which $p$ is discharge pressure, $Q$ is volume flow rate, $P$ is power requirement and $N$ is rpm of the pump.

**TABLE 10.10.1.** Comparison of performance of different pumps

| Pump type | Pressure rating, psi | Speed rating | Overall efficiency | KW/wt. ratio | $Q$, GPM |
|---|---|---|---|---|---|
| Internal gear | 500–2000 | 1200–2500 | 70–85 | 3.2 | 1–200 |
| Vane | 1000–2000 | 1200–1800 | 80–95 | 3.2 | 1–80 |
| Axial piston | 2000–12000 | 1200–3000 | 90–98 | 6.5 | 1–200 |
| Radial piston | 3000–12000 | 1200–1800 | 85–95 | 5.0 | 1–200 |

Table 10.10.1 compares performance of different pumps. We note that gear pumps have low-pressure rating and the least performance. The maintenance cost is high due to wear and tear of the teeth, which also decreases volumetric efficiency during use. However, these are simple in design, compact in size and economic in price. These are therefore most commonly used in industry. On the other hand piston pumps provide the best performance, can operate at high rpm generating a very high pressure and even

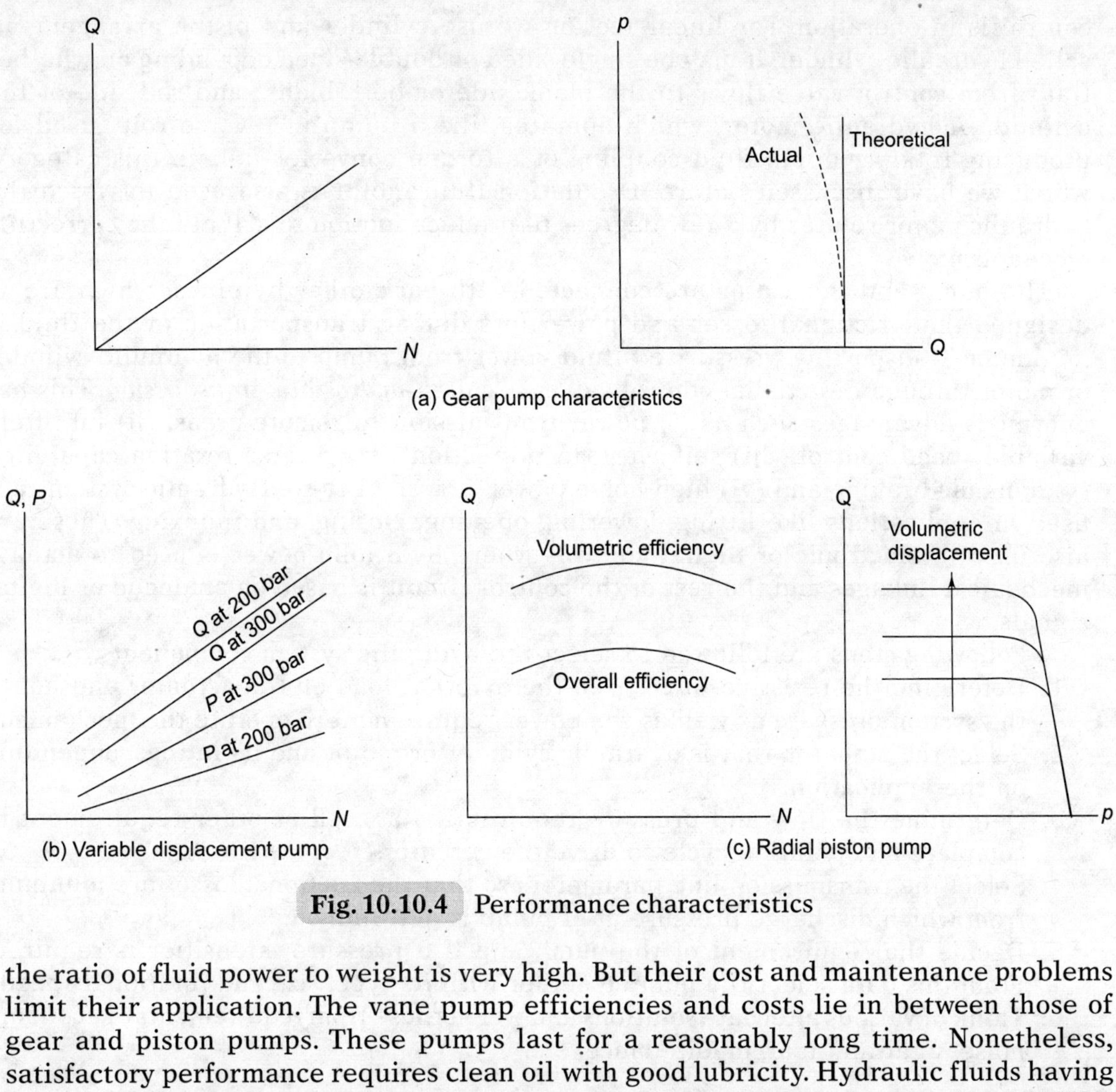


**Fig. 10.10.4** Performance characteristics

the ratio of fluid power to weight is very high. But their cost and maintenance problems limit their application. The vane pump efficiencies and costs lie in between those of gear and piston pumps. These pumps last for a reasonably long time. Nonetheless, satisfactory performance requires clean oil with good lubricity. Hydraulic fluids having Saybolt number 70–200 at temperature varying from 50°–70°C are recommended for use. Another factor affecting the performance is cavitation and care has to be taken that the pressure in the suction line does not fall below the saturation pressure of the fluid at the working temperature. These pumps on the other hand do not require priming like a centrifugal pump.

## 10.10.2 Hydraulic Cylinders and Motors

We discussed various pumps used in hydraulic controls. These pups are required so that the pressure of the liquid can be used to actuate a hydraulic cylinder or a motor. These devices are called actuators. The motion is then used to bring change in the system so that its output agrees with the user requirement or the set point, if feed back control is in operation. For linear motion we use cylinder and piston arrangement, called hydraulic cylinder. It may be single-sided or double-sided depending on whether fluid from control valve flows to the blank side or both blank[4] and rod side of the cylinder. A hydraulic motor, which operates like a pump in reverse role, used for producing rotary motion; fluid coupling or a torque converter falls in this category, which we have discussed before. Oscillation fluid motor is also used in which the hydraulic motor rotates by a few degrees to produce motion to actuate the correcting mechanism.

The pump and the motor are connected with each other by pipes, which are so designed that frictional losses and power loss during transportation of the fluid is minimum. Transporting pressurised fluid power from pump to the hydraulic cylinder or motor through directional control valves is called *hydrostatic transmission*. This has enormous advantages such as (i) power transmission to remote areas, (ii) infinitely variable speed control, (iii) self-overload protection, (iv) reverse rotation capability, (v) dynamic braking and (vi) high horse power to weight ratio. Hydraulic systems are used in applications like lifting, lowering, opening, closing, and indexing. These are also used in electronic or digital controls where hydraulic power is used to actuate mechanical linkages and the rest of the control circuit is based on analogue or digital signals.

Following steps are followed to select the hydraulic system components.

1. Determine the resistive force from the external load on the actuator and hence the system pressure as well as the power requirement to operate the mechanism.
2. Select the actuator, that is hydraulic cylinder or motor and its ratings, depending on the application.
3. Determine the flow and pressure requirement as well as other requirement to complete an operation cycle to drive the actuator.
4. Select the transmission line parameters so that the frictional losses are minimum from which discharge pressure at of pump is determined.
5. Decide the requirement of the pump and if a pressure intensifier is required. From this data select the pump together with its type, size and rotational speed.
6. Think of various alternate solutions and weigh these from requirements of economy, noise, wear and tear, maintenance, etc.

#### 10.10.2.1 Hydraulic Cylinder

This is just like the reciprocating pump or compressor, which we have studied and therefore will not go into its details. We shall however mention of spring loading and cushioning. In single cylinder system a spring is placed on the piston road side. Oil enters from the blank side, pushes piston out during *extension stroke* and compresses the spring while moving out. Piston returns back due to compressive force of the spring during the *retraction stroke* of the piston during which the oil on the blank side flows back to the sump. Flow of liquid to and from the cylinder is controlled by directional

[4] In a piston and cylinder arrangement, the cylinder head side is known as blank side while the other side on which there is piston rod, is called rod end.

control valve (DCV). Thus force can be exerted during the extension stroke in a single-sided cylinder. But in double-sided cylinders, force is exerted in both the strokes. Let $F$, $p$, $A_p$, $A_r$, and $V$ represent the force, pressure, piston area, piston rod area, and velocity the moving piston; the subscripts $p$, ext and ret are used for piston, extension and retraction strokes, respectively; and $Q_{in}$ is the volume flow rate to the cylinder. We may write following equations from force and volume consideration:

$$F_{ext} = p \times A_p \quad \text{and} \quad V_{ext} = Q_{in} / A_p \tag{10.10.6}$$

$$F_{ret} = p \times (A_p - A_r) \quad \text{and} \quad V_{ret} = Q_{in} /( A_p - A_r) \tag{10.10.7}$$

$$P\text{, power produced} = v_p \times F = Q_{in} \times p \tag{10.10.8}$$

$V_p$ is the velocity of piston.

### 10.10.2.2 Hydraulic Motor

These are the components to which hydraulic fluid is supplied under pressure that causes the rotating element to either turn by some angle as in limited rotation and oscillation fluid motor or rotate like in gear and other kind of motors. In all these devices, liquid pressure acts on one side of a surface that exerts a force and rotates that element, since it is mounted on a surface connected to a rotating shaft. Figure 10.10.5(a) shows a limited rotation and gear type motor. Liquid enters one side of the moving vane and exerts force on the whole surface it comes in contact. This rotates the shaft in one direction. Next time the liquid enters another side and brings back the vane to the original position. If $R$, $L$, $p$ and $T$ are used for radius, axial length, pressure and torque and suffix $r$ and $o$ for the rotor and casing end of the rotating vane then

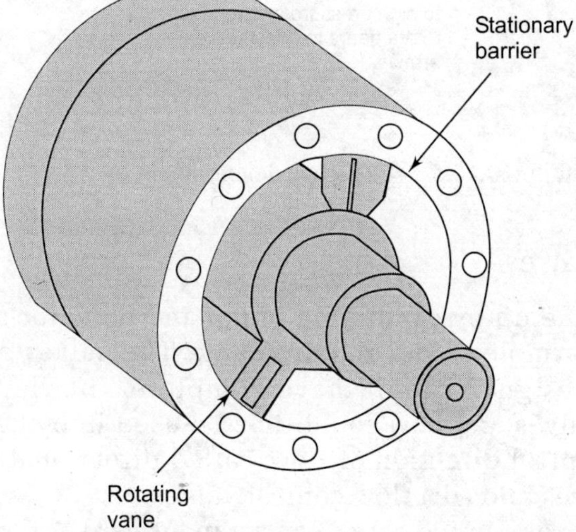


**Fig. 10.10.5** (a) Limited rotation hydraulic actuator. (*Courtesy of Rexnord Inc., Hydraulic Components Division, Racine, Wisconsin.*)

$A$, area of the vane surface on which pressure acts = $(R_o - R_r)\ L$

$F$, force on the vane surface = $p\ (R_o - R_r)$

$T$, torque on the shaft = $F \times$ average distance form axis of rotation

$$= p\ (R_o - R_r)\ L \times \{(R_o + R_r)/2\} = p\ L\left(R_o^2 - R_r^2\right)/2 \qquad (10.10.9)$$

In gear motor, which works exactly like a gear pump but in reverse mode, the force acting on the gear tooth face rotates the gear continuously. This is shown in Fig. 10.10.5 (b).

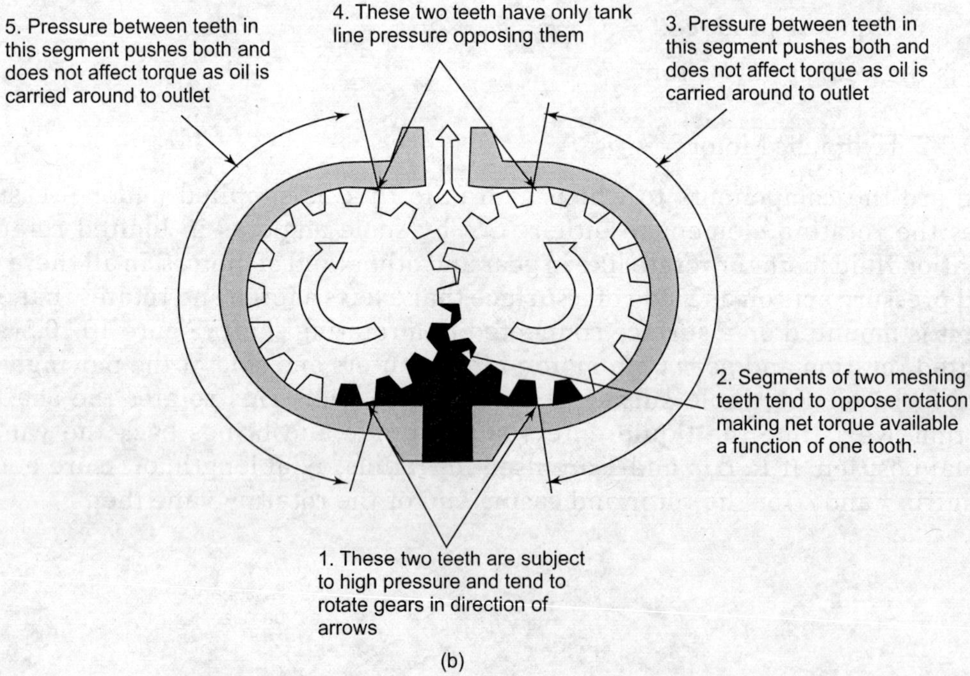


**Fig. 10.10.5** (b) Torque development by a gear motor

## 10.10.3 Control Valves

A fluid system is made up by combining pump and actuator through proper valves to control the output variable as per requirements. The valves play an important role in such systems and designer should have complete knowledge about the type and functioning of the valves and how could they be used in hydraulic circuits. Purpose of any valve is (i) to control direction of flow, called directional control valve (DCV), (ii) pressure control valves and (iii) flow control valves.

Hydraulic actuator, as studied above can act in forward or reverse direction depending on whether fluid enters it from blank or rod end and that the speed of actuation of the piston rod depends on rate of volumetric flow of liquid. Therefore it is essential to control both the direction and volume flow rate, which is implemented by designing the hydraulic circuits (that act as brain of the control system) using the directional and volume control valves. We also have to ensure safety of the control system that may be jeopardised due to excessive build-up of pressure in the circuit for any reason. This will happen when piston has reached extreme end of its travel. It may affect the safety of the hydraulic system as well as continued pressure increase may not be desirable on the load. The problem is overcome in two ways: (i) using a pressure relieving valve that limits maximum pressure or (ii) the output of the pump is diverted somewhere else. There are many more such requirements and shall be unfolded as we proceed further.

We shall first discuss in the following sections about the valves and then how are these used to make different kind of hydraulic circuits to take care of various situations and industrial demands.

### 10.10.3.1 Directional Control Valves

A directional control valve has number of ports, called ports or ways. A valve having '*n*' port is termed as *n*-way valve whether all of them are used or not. Say there are four ports in a valve and one of them has been plugged (blocked) using a plug then it is refereed as four-way valve used as a three way valve and not as three way valve. Simplest valve is a two way valve in which flow can take place only in one direction. Such a valve is called a check valve and is used frequently not to allow back flow. For example, a feed check valve used in boiler. The check valve will not allow steam or water to flow to feed pump if the pump fails to work for any reason or say there is electricity failure that would stop feed pump working.

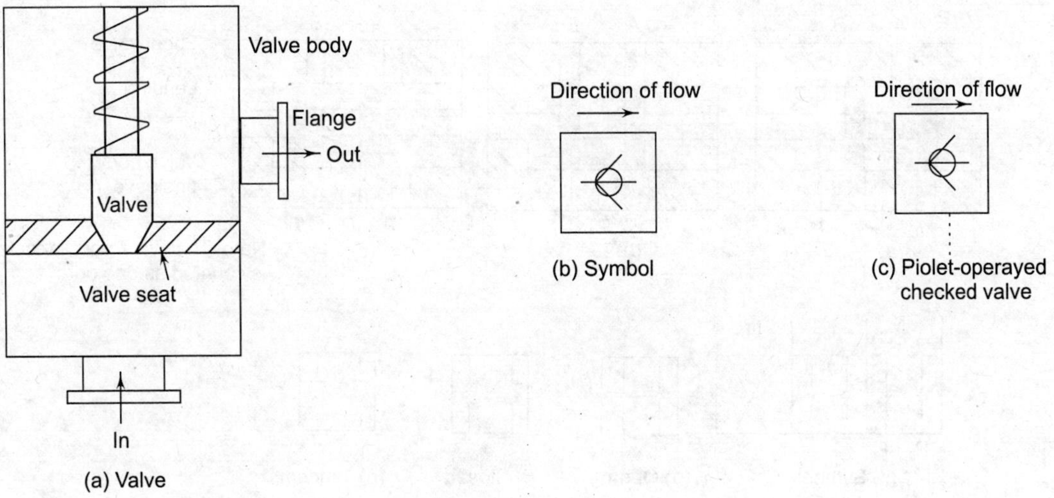


**Fig. 10.10.6** (a) Check valve (b) Symbol (c) Pilot operated check value

Figure 10.10.6 shows one such valve with the symbol used to represent it in a hydraulic circuit; the arrow shown with the symbol is not its part and is drawn to lay emphasis on the direction of flow. A check valve sits in its seat due to gravity or light spring pressure to keep it in its position. In the forward direction of flow the pressure would lift valve to allow the flow. But in opposite direction of flow, the fluid pressure would force it to sit in its seat and no flow can take place. Yet, it is possible that a pilot signal may lift the valve, as in pilot operated check valve, to allow back flow if required; its symbol is shown in Fig. 10.10.6(c). The control signal is shown as dotted line and may be generated manually, mechanically, using a solenoid or hydraulic pressure from the circuit. A mechanical type signal is generated using foot operated lever or push button. In a solenoid, electric current is used whereas pilot signal is obtained from the hydraulic system itself. Appendix D lists different symbols used in hydraulic circuits, including the pilot signals used to actuate the valve. You have to learn these symbols so that you may understand the hydraulic circuit; more details can be obtained from Thermal Engineering by the author published by Pearson Education, New Delhi. Valve is brought back to its normal position or other position using similar signals; spring is commonly used.

*Spool Valve.* Directional control valves have more than one port as shown in Fig. 10.10.7—one is connected to the pump shown as *P*, other to the reservoir shown as *T* and there may be then one, two or more ports connected to actuator, shown as *A*, *B*, etc; normally there exists two such ports. A four-way valve is very common in hydraulic circuits, one of which may be plugged to use it as three-way valve. The valve body is called sleeve and has a spool, which can be moved in two or three positions. The spool valve is shown in Fig. 10.10.7(c) with the terminology used for it. It is having two landings of same size as that of the port provided in the sleeve and can be moved from one position to another by applying a pilot signal. There are spools having three or more landings too.

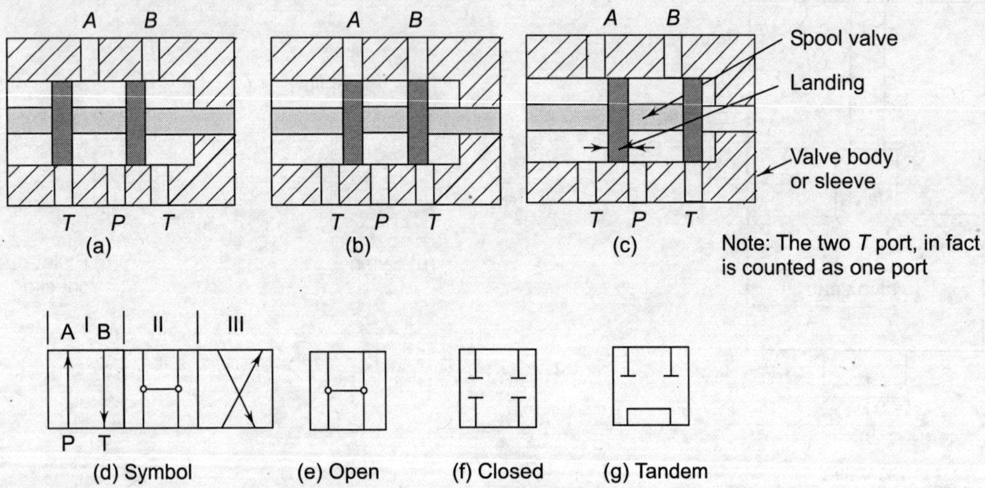


**Fig. 10.10.7** Three position four-way valve

Figure 10.10.7(a-c) shows a DCV with 5 ports but is called four-way valve because two *T* ports are counted as one. In arrangement (a), the spool is shifted to left; flow is from port *P* to *A* and from *B* to *T*. When spool is in mid position, figure (b), then no flow is taking place among the ports and when in extreme right position (c) flow is from *P* to *B* and from *A* to *T*. Thus position of spool decides direction of pressurised fluid, which may be to *A*, *B* or none. That is why these valves are called directional valve. In this valve, spool takes three positions. These three positions are symbolically represented in Fig. 10.10.7(d); the arrow indicates the direction of flow and the ports being connected. Each position of valve is shown by a square with ports *P* and *T* on bottom side and ports *A* and *B* on top side. This is shown in Fig. (a) though by convention their positions are fixed. That is why, *P*, *T*, etc. have not been indicated in other figures (e) (f) etc. Port *A* is always connected to the actuator and port *B* to the reservior, by convention. The three squares side by side show the three positions of the spool valve in the order left, centre and right position. Normally the spool valve is in centre position and is shifted to left or right depending on the type of action needed for that hydraulic circuit by applying appropriate control signal. Here the central position of spool is used to hold the actuator is no-operation mode, since there is no flow among the different ports. Three arrangements for the centre position of spool are shown in Fig. 10.10.7(e), (f) and (g) but there may be other arrangements too. In (e) all the four ports are connected to each other and therefore the output of the pump as well as through the ports *A* and *B* is directed back to the reservoir. Such arrangement is termed open. This and the tandem arrangement shown in (g) unload the valve since pump is connected to the sump and pressure developed by it is minimum, almost none. In tandem position, Fig. 10.10.7(g), ports *A* and *B* are not communicating with other ports but *P* is connected to *T*. In arrangement (f), called the closed arrangement, all the four ports are closed to each other; the arrangement shown in Fig. 10.10.7(b) is of this type.

As mentioned before different methods are in use to move the spool valve. Fig. 10.10.8 shows the symbol used for actuating devices. So long the force exerted by the actuating device acts on the spool, it remains in that position. As soon as the force is released, the spool returns to the original position due to return spring. This return force even may be produced by a solenoid current or a pilot signal from the fluid circuit.

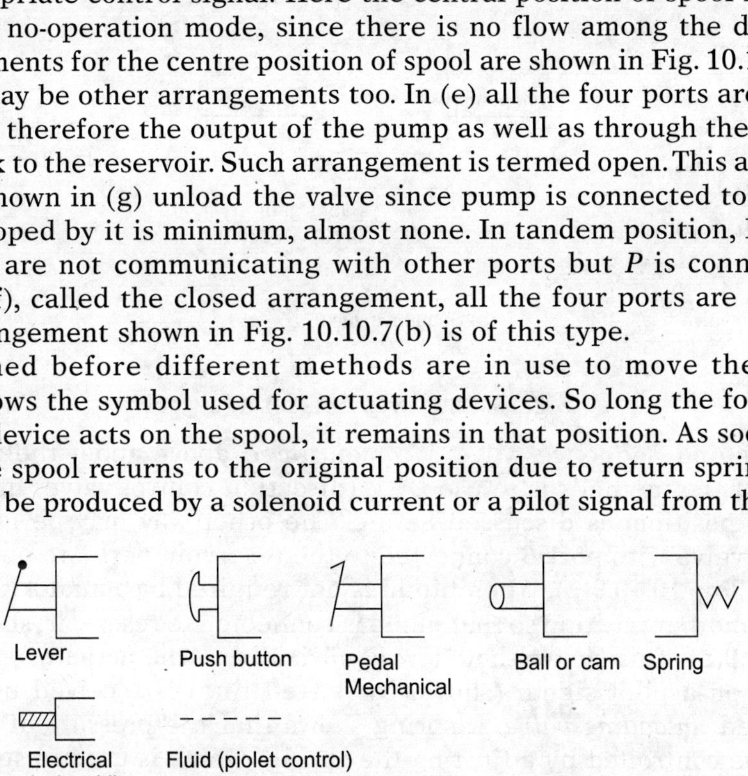


**Fig. 10.10.8** Symbol for actuating devices

Above para describes working of a linear spool. Rotary spool is also used. We shall omit the details and can be read from any book on fluidics or hydraulic systems.

### 10.10.3.2 Pressure Control Valves

In hydraulic control system, the pump rotates continuously to generate pressure required to meet the load during working stroke. If this pressure remains at its maximum even when not required then we are wasting not only energy in developing the pressure but also the energy input builds up the heat and raises temperature affecting the properties of the fluid. Besides, there may be local or other reason that results in building up of pressure in some pockets crossing dangerous value. Also continuous rise in pressure like in pressure welding is undesirable. That is, the pressure either has to be released completely to save the components or not allowed to increase after a certain value during the operation. For this reason different pressure control valves are available in market. We shall discuss some of these and show their use in hydraulic circuit for a particular function. However, their use in a circuit depends on the innovative nature of the designer, which is true for any kind of hydraulic circuits.

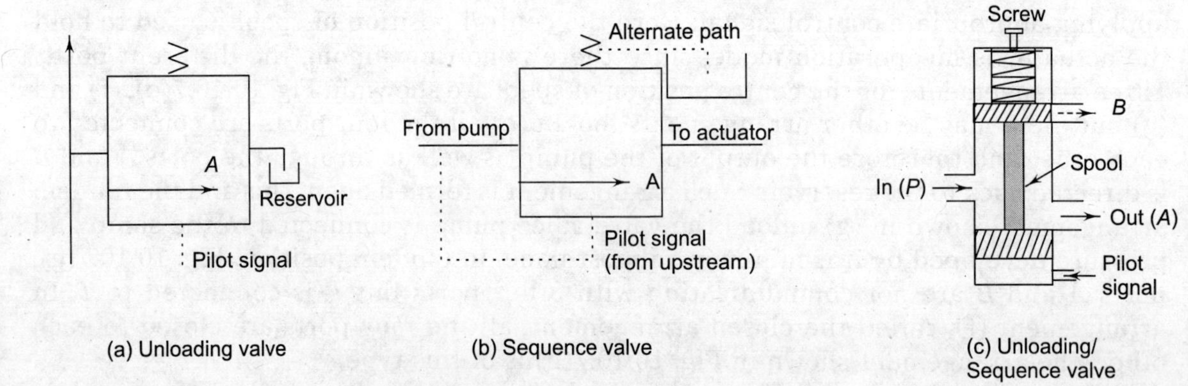


**Fig. 10.10.9** Symbols for a few valve

**Unloading and Sequence valve.** We mentioned above about the unloading of the pump. One of the methods is to use DCV (direction control valve) in open or tandem centre valve position, as discussed before. The other way may be using a three-way, two-position valve with port $B$ connected to the reservoir, port $P$ to pump and port $A$ to the actuator, Fig. 10.10.9(c). When liquid is not required by actuator the valve position is shifted to another position so that pump is connected to reservoir; shown symbolically in Fig. 10.10.9(a). This shows that flow is normally to the actuator and is diverted to reservoir when a pilot signal (shown by dotted line) is received by the valve. This valve is called *unloading valve.* A spring shows that the pressure at which unloading occurs can be controlled by adjusting the spring, which is used in such valves.

It is also possible to connect port $B$ to some another hydraulic circuit instead of reservoir to direct the hydraulic fluid for another operation. Consider the blanking operation of a sheet, for example. For this process, first fluid at moderate pressure is passed from port $A$ to the blank end of the cylinder to move the ram to the work piece in the minimum time after which high pressure fluid flows to it by diverting the liquid from pump through the pressure intensifier to start blanking operation, since large force is required during punching. So the port $B$ is connected to pressure intensifier to supply high pressure fluid to the machine. Such valve, which routes fluid to the actuator through different hydraulic circuits in sequence, is called *sequence valve.* Fig. 10.10.9(b) shows symbol for sequence valve. Here the pilot signal is shown coming from pump or upstream side; it is also possible that the signal may be from downstream side or external. Position of arrow *A* shows that the flow in alternate direction is closed normally and a spring indicates spring control of pressure at which the control is transferred to alternate path from *B*. The valve operates only when the pilot signal comes.

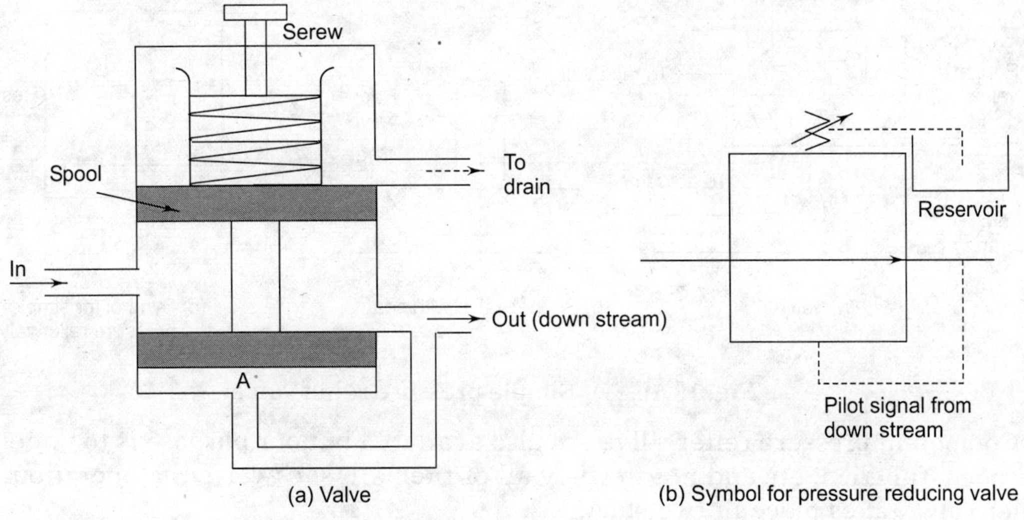


**Fig. 10.10.10** Pressure reducing valve

**Pressure Reducing Valve.** If for any reason pressure builds up downstream of the control valve then for the safety of the system, flow of pressurised fluid is stopped. For this purpose pressure-reducing valve, as shown in Fig. 10.10.10, is used. If pressure down stream increases then the pressure below the spool at A will also increases. This will move the spool up closing the fluid line to down stream and opening the line to the drain. Once the pressure becomes normal then again the flow starts to the original port. Figure also shows the symbol used for this valve. It is seen that when downstream pressure is more than the set value of the spring, the valve acts as unloading valve.

**Simple Pressure Relief Valve.** We provide relief valve in the circuit so that if pressure increases beyond the preset value it operates to release the pressure for safety. Fig. 10.10.11 shows schematic arrangement of a simple pressure relief valve (you may improve it) with its symbol. It consists of a poppet valve seated in its seat, Fig. 10.10.11(b). The poppet valve is a valve with its sides tapered like the one used in check valve, which sits in the seat made in the valve body. Flow will take place from the pump to the actuator via this valve till the pressure is build-up for some reason that would raise the valve from its seat against the spring pressure and the fluid then flows out of the other passage to the reservoir, since the flow takes the path of least resistance.

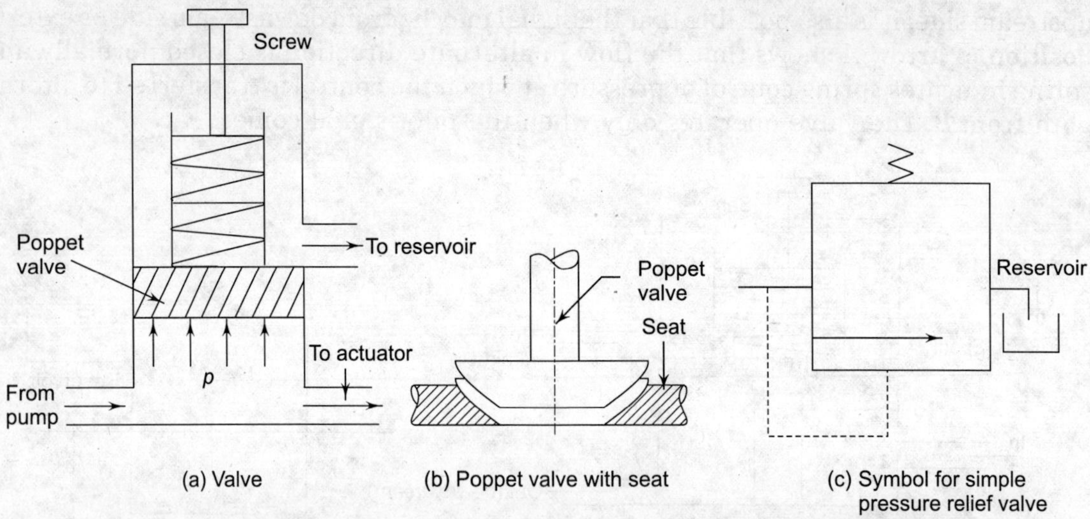


**Fig. 10.10.11** Simple pressure relief valve

Compound pressure relief valve are also available but our purpose is to understand the need, function, etc and not to study all of them. In such valve the operation of the relief valve takes place in two stages.

*In a nut-shell, the unloading valve, sequence valve, pressure-reducing valve or simple pressure-relief valve are more or less the same as their operation is same; difference is in the way these are used and the pilot signal is taken from.*

**Counterbalance Valve.** It is shown in Fig. 10.10.12. When hydraulic cylinder is vertical then piston moves down due to gravity. To keep it in the top position, this valve is used. Vertical cylinders are used in vertical press, lifts, trucks, etc. The counterbalance valve is used with a four-way, two-position control valve. Bottom of the cylinder is connected to the primary port of the counterbalance valve whose secondary port is connected to port *A* of DCV, as shown in figure. Spring pressure is made large so that the spool is not able to move due to liquid pressure in the cylinder on its blank side (above piston). When the cylinder is to be operated then liquid flows to the cylinder via DCV setting it in left wing position. This increases pressure on the piston end, moves position in the cylinder to downword increasing pressure on the rod end and therefore at A below the spool. The spool therefore is lifted so that the liquid flows through the secondary port and 4-way valve to the reservoir. While in the other valve position fluid flows through the check valve and primary port of the counterbalance valve to the bottom of the cylinder to lift the piston. This arrangement ensures that the piston in cylinder does not move to bottom due to gravity. The figure also shows the symbol used to represent this valve.

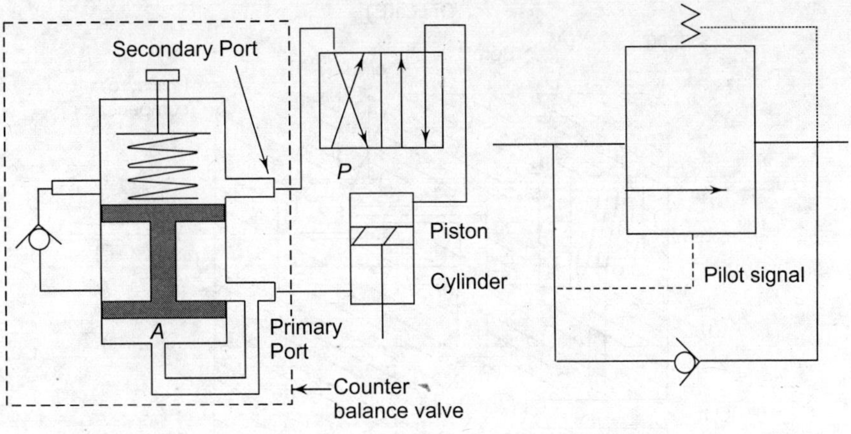


**Fig. 10.10.12** Counterbalance valve with hydraulic circuit

### 10.10.3.3 Flow Control Valves

Speed of actuation of the hydraulic cylinder or motor depends on the volumetric flow rate for which we use valves whose area of opening determines the volume flow rate. These valves may be non-pressure compensated such as needle, globe, gate valve, etc. (Fig. 10.10.13 (a)) or may be pressure-regulated using pressure signals from downstream

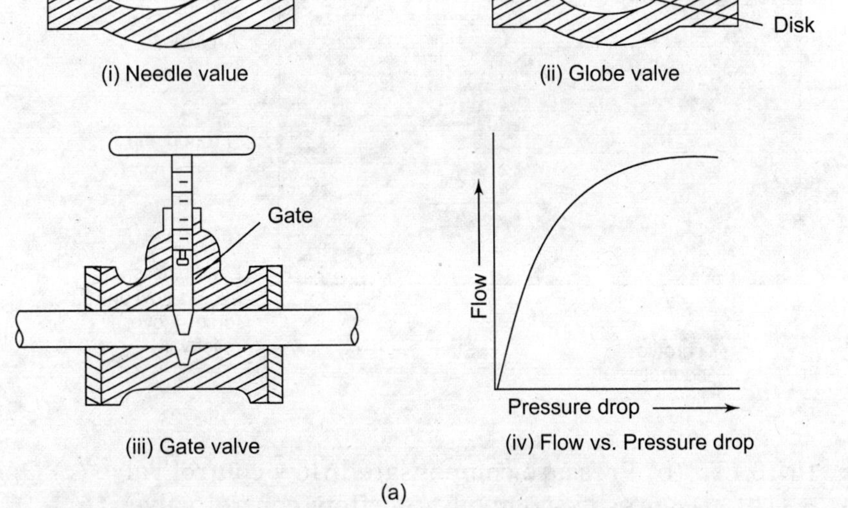


**Fig. 10.10.13** (a) Non-pressure compensated flow control valves

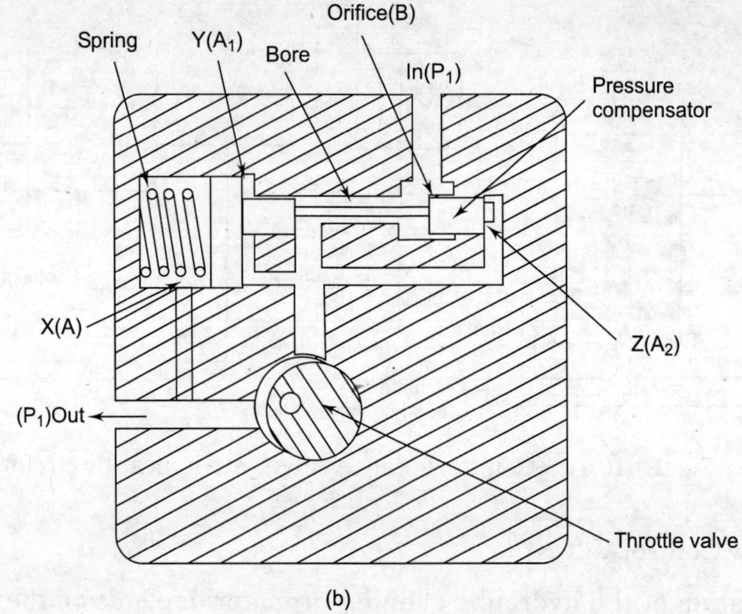


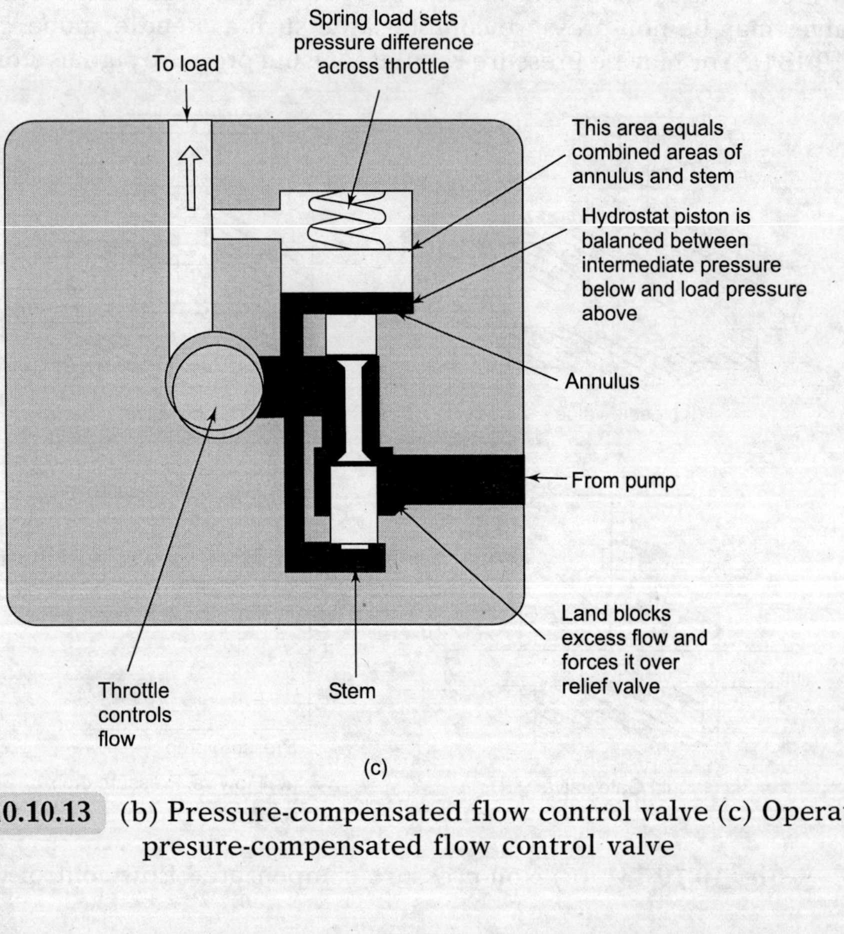


**Fig. 10.10.13** (b) Pressure-compensated flow control valve (c) Operation of presure-compensated flow control valve

or upstream (Fig. 10.10.13(b)). In the latter type of valve, the area of flow changes automatically as pressure changes either on upstream side or downstream side whereas in the non-pressure compensated type the area is controlled manually or using an actuator. Fig. 10.10.13 gives idea of some non-pressure compensated valves and their working is evident from the figure itself. Different names are given because of the shape of the valve. For example needle valve is so called because of the needle type shape of its end controlling flow.

Figure 10.10.13(c) shows the pressure compensated flow control valve. If the load on the cylinder fluctuates significantly then system pressure will also fluctuate that would change the volume flow rate, since it depends on the pressure difference across the valve orifice. In such circumstances, a pressure compensated flow control valve is recommended for use, as it will maintain constant pressure across the orifice. Liquid flows from pump to the load via the orifice and this valve. Forces on the piston between the spring and the spool act from (i) spring, (ii) load pressure $p_1$ at its upper surface having area $A_2$ and (iii) supply pressure $p_2$ on the stem of the spool of area $A_1$ at bottom and the annulus below the piston of area $(A_1–A_2)$. It is simple to make balance of liquid forces to see that the force acting on the piston is equal to $(p_2–p_1)\,A_2$, which is equal to the spring force that remains constant. This way a constant pressure can be maintained across the orifice. When there is a change, the spool moves up and down to control the variation in pressure.

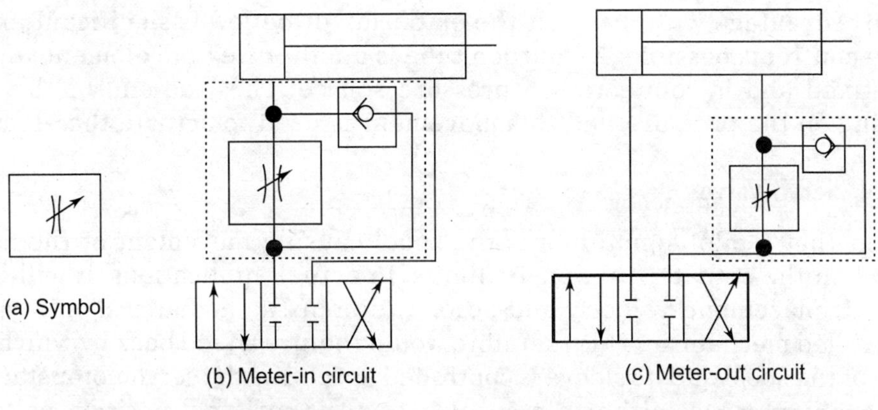

**Fig. 10.10.14** Flow control schemes

Figure 10.10.14(a) shows the symbol to represent a flow control valve; an arrow shows that the area of flow is variable. This valve may be used in (i) meter-in, (ii) meter-out and (iii) bleed-off position. In these circuits, a check valve is provided for free reverse flow. Fig. 10.10.14(b) shows the hydraulic circuit for meter-in arrangement, wherein the flow control valve is placed in the pressure line leading to the cylinder. This circuit is used where the load is constant and positive as in feed to the surface grinder table. Since the flow is restricted to the cylinder, pump is required to pump more fluid and the excess is returned to the sump. On the other hand the flow control valve is placed in the return line, Fig. 10.10.14(b), in meter-out circuit. This arrangement is used where a constant force is needed during the reverse operation like drilling. In drilling, the work piece tries to drag the drill after the hole is made. To avoid this a back pressure is to be maintained. In such situations the meter-out arrangement gives satisfactory results. The third arrangement is not shown. As the name suggests, this arrangement acts like bleeding. The valve is neither placed in the feed nor in the reverse line. Rather, the feed line is connected to the sump through this valve. Thus this valve regulates the flow to the cylinder through feed line by controlling the bleeding amount to the sump. This arrangement allows maintaining a constant flow to the cylinder and hence is suitable for constant load applications.

**Other Flow Control Valves.** In controls, pressure and temperature controlled switches are often used to control flow rate through the orifice by changing the area of the orifice. Normally a diaphragm or a bellow is used. In diaphragm type, a diaphragm is fixed around its periphery and a differential pressure exerts on its two surfaces exposed to the downstream and upstream pressures respectively. The difference in pressure moves the diaphragm up or down. This displacement may be used to control the area of the flow or to operate the electrical switch to make it on or off. In what mode the switch is to be used depends on the choice of the designer. A temperature switch is more or less like a pressure switch except that it is designed to control the temperature. Moreover, it also has provision to offset variation in ambient temperature. All these switches have arrangement to set the desired cut-off point. You shall read about such valves in the course on refrigeration.

We may use a bellow in place of diaphragm. The bellow contracts or expands with pressure. Advantage with it is that the extension of bellow is significant and therefore accurate control is possible. A Bourdon tube is another flexible element whose one end is kept fixed and is connected to pressure source. The free end of the tube moves depending on the pressure and this movement is used to actuate the electric switch.

### 10.10.3.4 Servo Valve

We learnt that a pilot signal moves the spool causing movement of the actuator that ceases when the actuator reaches its limits. In actual applications, it will be preferred that the displacement, velocity and acceleration of the actuator should match that of the controlled mechanism. This therefore would require a feed back by which the relative position of the spool and its sleeve is controlled as soon as the servomotor starts actuating. The mechanism moves either the spool back to its original position or the sleeve to cover the service port such that DCV sends no more fluid to the servomotor. Such valve has infinitely variable positioning capabilities and is called servo valve.

This may be done mechanically or using electrical means. Accordingly the servo valve is called all mechanical type or electro-hydraulic servo valve. These are shown in Fig. 10.10.15. In figure(a) the sleeve of the valve moves with the piston of servomotor. For example, when spool is moved to its left, as shown in the figure, by a pilot signal then fluid flows from port $A$ to the servo motor pushing the piston to right. Piston rod is connected to the lever having hinge at its bottom and connected to the sleeve of the DCV, as shown in figure; right movement of the piston rod moves the valve sleeve in the direction of the movement of the spool and thus in the final position the spool landings again cover the ports and no more liquid flows to servomotor. It is also possible to move the spool back to its original position by having some other linkage mechanism. The idea is that the original relative position of the spool with respect to sleeve is to be restored back. Such a valve is called *single-stage valve* and is used where the spool signal is strong enough to move the spool of the DCV.

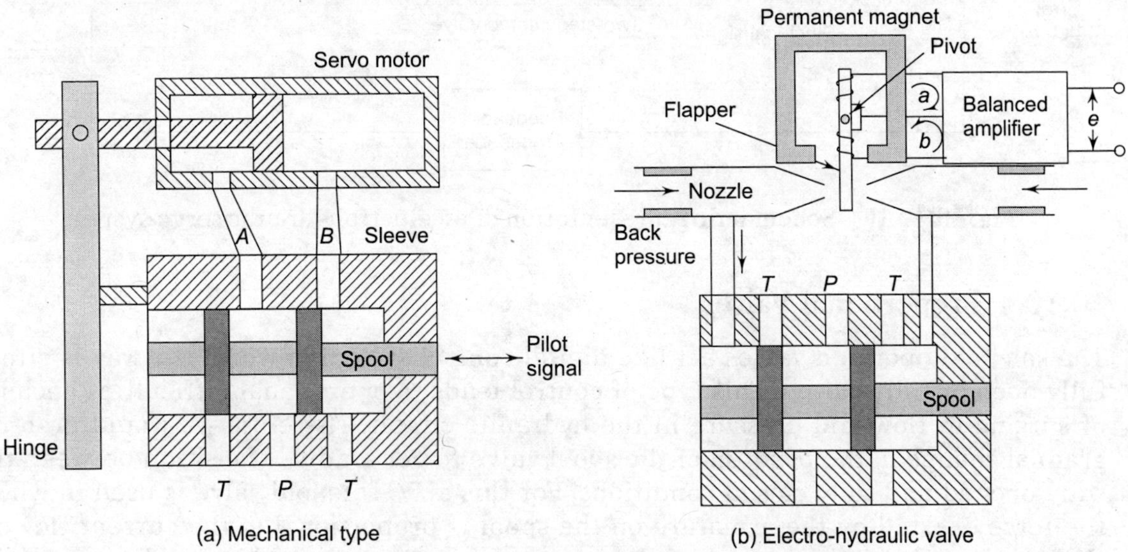


**Fig. 10.10.15** Servo valve

In large flow rating systems, the accurate positioning of the spool using a pilot signal or a solenoid valve becomes difficult due to large amount of forces required for movement of the spool. In such a case, miniature hydro-mechanical actuators, internal to the servo valve, perform this function. This can be seen in Fig. 10.10.15(b). A magnetic torque motor having permanent magnet is used with its armature extending into the air gap between the two nozzles through which fluid flows continuously. The armature acts as a flapper of the valve and has two windings, which get energised from the current supplied by the balanced amplifier to which an input '*e*' is supplied when some action is required. In normal flapper position, this electric potential is zero, current in the two arms *a* and *b* is equal and the flapper takes position such that the back pressure in both the nozzles on its two sides is same. Let some voltage *e* is applied to the armature which moves flapper to say right restricting the flow passage of the right nozzles. This increases the back pressure in the right nozzle and the signal is transferred to the spool of the main valve, since nozzle is connected to the DCV, as seen in the figure. The signal so generated is amplified and is able to shift position of the spool. Rest of the function is same as discussed before. Such valve is called *two-stage valve* since the control signal is given first to the electric motor, which generates and passes the signal to the spool. The signal is thus passed to the spool in two stages and not directly. Since electrical energy is used for the first stage valve, the servo valve is called electro-hydraulic valve. An electrohydraulic servo system is shown schematically in Fig. 10.10.16.

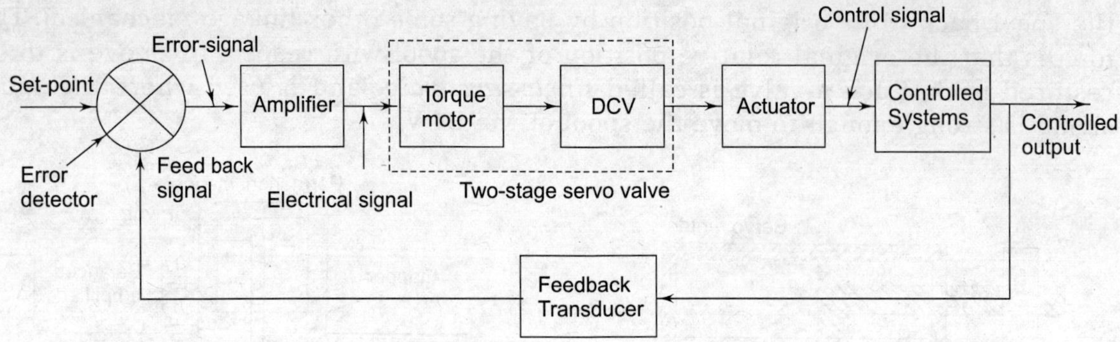


**Fig. 10.10.16** Schemactic representation of an electrohydraulic servo system

## 10.10.4 Proportional Valves

The solenoid operated valves act like digital control system in which the way is either fully open or fully closed. This type of control leads to operational difficulties because of surging in flow and pressure in the hydraulic circuit. Therefore we would prefer a gradual change in the position of the spool valve to get gradual transition between the fully opened and fully closed conditions. For this a *DC* solenoid valve is used in which the force exerted by the armature on the spool is proportional to the current flowing through it and is independent of the armature movement over the working range of the solenoid. In such a valve, we prefer to have a variable current control rather than variable voltage because current may change even if there is no change in the voltage.

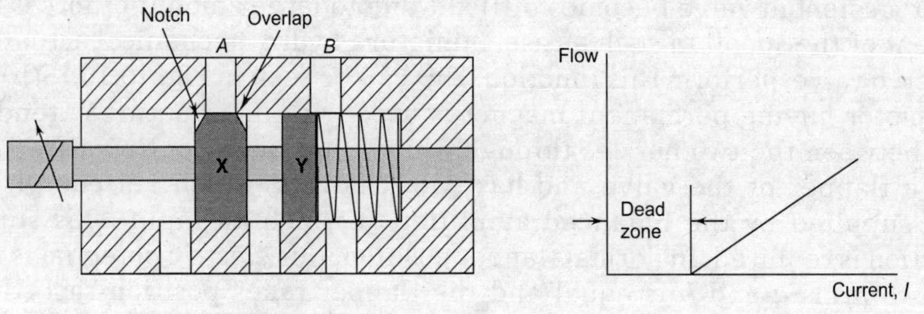


**Fig. 10.10.17** Proportional control valve

We have mentioned earlier that the valve landing is just enough to cover the port width. In the proportional valves, notched spool or overlap spool valve having landing more than the port width are used. Proportional valve allows a better control of the flow rate because the orifice (port) is progressively opened. Figure 10.10.17 shows a proportional valve and variation of flow with solenoid current. It is seen that flow starts only after certain value of current has been passed to the solenoid to move the spool by a distance equal to the overlap and that it is increasing linearly with the current due to notch provided. This gives a dead zone, as shown in the figure, as a result of the overlap.

### 10.10.5 Hydraulic Circuits Design and Analysis

We have now learned about the various control valves, actuators and servo valves. These components are selected according to the requirements and are connected to meet the requirements for control of the operation. The ultimate goal is to control the motion of servomotor to change the input so as to maintain output of the system in consonance with the applied load and conditions. Therefore, a careful design of the circuit is necessary to achieve (i) safety of operations, (ii) good performance, (iii) efficiency of operation, (iv) minimum deviation in output from the set point value and (v) the operation in minimum possible time with minimum amount of energy requirements. The circuits are shown using symbols shown in Appendix *D* and some have already been discussed before. These symbols have been adopted by American National Standards Institute (ANSI).

We will now consider different hydraulic circuits shown in Fig. 10.10.18 and learn how to read these and what functions they are used for. Our discussions are with reference to linear actuators. Before we take up study of the circuits, we may point out some basic requirements in any hydraulic circuit for the safe and efficient operation of the control system.

- ✓ A maximum pressure relief valve (MPRV) is essential in any hydraulic circuit for its safety because fluid supply to cylinder is stopped at the end of the stroke of the piston and consequently pressure build-up begins. Therefore the MPRV will relieve the pressure and the circuit will remain safe.
- ✓ It will be a better idea to use unloading valve (Fig. 10.10.9 a) in the circuit so that the pump runs at no load and its output is discharged back to the sump during no operation. While using an unloading valve, a check valve may also be used in conjunction with it to check flow from the main circuit to the sump so that the pressure in hydraulic circuit is not affected.

***Single-stroke Hydraulic Cylinder***

Figure 10.10.18(a) shows implementation of the DCV for control of single stroke hydraulic cylinder so that the piston moves outward, called extension stroke, to exert force on the load. Name of the hydraulic components is given but you have to learn these symbols and will not be given in the circuits. Liquid is sucked through a filter to remove impurities and is fed to 3-way, 2-position DCV by a pump. MPRV is placed just at the discharge end of the pump for the reasons already discussed. The valve is pedal operated and spring-offset. When the operator pushes the pedal, the valve operation is shown by its left wing. In this position fluid moves to the cylinder pushing the piston out to extend it. Piston reaches the extreme end of its stroke and then the MPRV opens so that a constant force on the load is maintained. When the operation is over, operator leaves the pedal and the spring brings back the control valve in the position shown in right side. This is the position of retracting of cylinder in which the spring force on the piston moves it towards the cylinder head and the liquid flows back to sump via the valve. The valve remains in this condition when no operation is taking place, which is the spring-offset position.

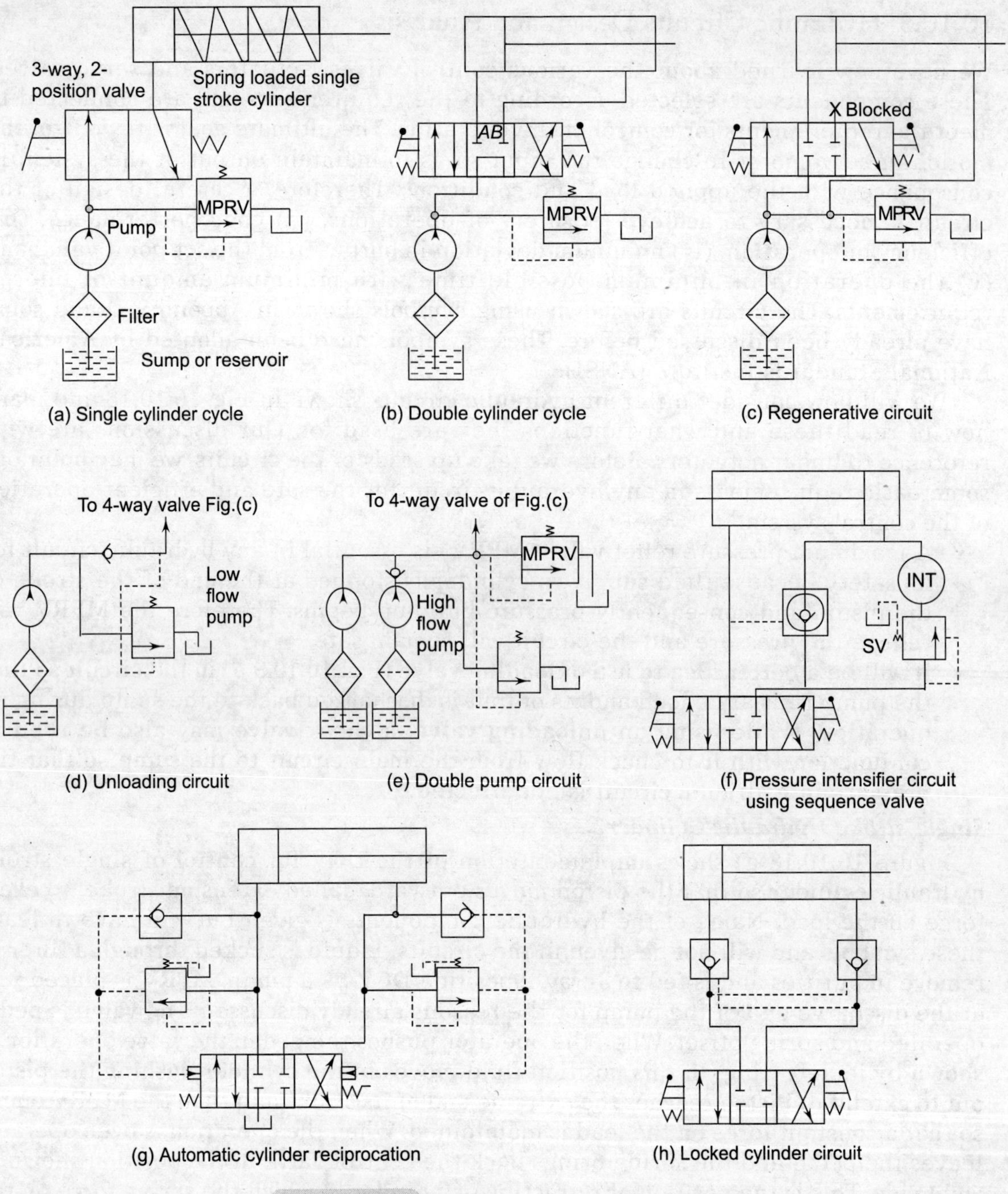


(a) Single cylinder cycle (b) Double cylinder cycle (c) Regenerative circuit

(d) Unloading circuit (e) Double pump circuit (f) Pressure intensifier circuit using sequence valve

(g) Automatic cylinder reciprocation (h) Locked cylinder circuit

**Fig. 10.10.18** Different hydraulic circuits

***Double-stroke Hydraulic Cylinder, Regenerative Cycle***

Figure 10.10.18(b) shows the flow directions for a double stroke cylinder. In this case the fluid is sent to the piston rod side also during retracting operation of the piston. Therefore, there is no need of spring and a 4-way, 3-position DCV is used. Port *B* is connected to the piston rod end. Rest of the diagram is same as discussed just before. In the spring-offset centring position (middle wing of valve positions), no operation is taking place and the liquid from pump flows back to sump thus providing unloading of the pump. When the left pedal is pressed, liquid flows to cylinder end and liquid from the rod end returns to sump via the valve. The process is reversed when the right hand paddle is pressed; the pedals are provided on the front end of the control. This way the piston can be moved out or in and pressure can be maintained for any length of time just keeping the right pedal pressed otherwise the spring will restore the valve position to the middle position.

In the arrangement of Fig. 10.10.18(b), piston moves with same speed both in extension and retracting operation. There are many processes where fast movement of the piston is desirable in extension, retracting or both. Figure 10.10.18(c) shows the hydraulic circuit to achieve this requirement for extension stroke. This circuit is called *regenerative circuit*. We know that speed can be increased by increasing flow rate to the cylinder and also that the fluid from the rod end is returned to the sump. Nevertheless it is possible to redirect liquid from rod end to the blank end by connecting the return line to the pump. It is simple to write down the following equation using the subscripts *P*, *T* and *R* to refer for pump, total and rod end and the subscripts *ext* and *ret* for extension and retracting stroke, respectively:

$$Q_T = Q_P + Q_R \quad \rightarrow Q_P = Q_T - Q_R \text{ (since liquid from rod end flows to blank end)}$$

$$\Rightarrow \quad Q_P = A\, v_{\text{ext}} - (A - A_R)\, v_{\text{ext}}$$

$$\rightarrow \quad v_{ext} = Q_P / A_R \tag{10.10.10}$$

A is the area of the piston. It is seen that extending velocity now depends on the area of the piston rod, which can be made small to give large velocity. There is law of nature that nothing comes free. So what price are we paying for large extending velocity? It is the load carrying capacity. In regenerative cycle, pressure difference on two sides of the piston will be less as compared to the simple circuit and hence the force exerted on the load is decreased. However, there will be no load on the piston rod during such movements.

In the circuit shown in Fig. 10.9.18(c), only pump discharge is supplied to rod end and hence the retracting velocity is

$$v_{\text{ret}} = Q_P / (A - A_R) \tag{10.10.11}$$

Comparing Eq. 10.9.10 and 10.9.11, we note that the velocity of retracting and extending is same when ratio of piston is double of the piston rod area. When it is so and fluid pressure in the cylinder is p then force on the load is

$$F = p(A - A_R) = p\, A_R \tag{10.10.12}$$

Regenerative circuit is very useful in situations where quick motion at low load is needed during the cycle of operations, may it be forward or return motion and have proved to be boon for saving time of operation of the cycle. This may be compared with the quick-return mechanism used in shapers.

***Unloading and pressure Intensifying Circuit***

We discussed the unloading and sequence valves and their functions in section 10.10.2. How could these be used in hydraulic circuit is shown in Fig. 10.10.18(d) and (f). In both figures, Fig. 10.10.18(c) is the base figure and the changes that are to be made are shown; rest of them will remain same. While using unloading valve the MPRV is removed. As pressure builds up in the circuit after the piston reaches the extreme position, unloading valve comes in action relieving the pump. However, the check valve does not allow pressure to fall in the main circuit and constant force exerts on the load while the pump just runs at no load.

Figure 10.10.18(f) shows another modification in which a pressure intensifier, shown as INT, is also used to increase the pressure through a sequence valve. Initially when the valve is in its left wing position then low pressure fluid is passed to the cylinder. When load pressure increases on the piston rod beyond the set pressure of the sequence valve then liquid flows through the intensifier. Cylinder end is connected to port *B* through pilot check valve so that retracting of piston may take place when a pilot signal is given to it and otherwise not. Such circuit is used in spot welding where first the electrodes are brought in contact with the weld piece that requires fast movement of the electrodes to the work piece and not much force and after completion of this process high force is required to weld the pieces. Advantage is that a low pressure pump meets all the requirements of the process cycle.

This may also be done without using a pressure intensifier. Instead, two pumps—one having high volume flow rate but capable of working at low pressure and another having low volume flow rate but can work at high discharge pressure, operate in parallel, as shown in Fig. 10.10.18(e). Here a check valve is placed in low pressure circuit so that when pressure builds up the low pressure pump is automatically unloaded because of the pilot signal and the flow is from high pressure pump only. Advantage is that the very expensive pressure intensifier is eliminated and this arrangement is more economical.

***Some Other Hydraulic Circuits***

There are number of hydraulic circuits for different purposes. We shall discuss only a few more so that we have more exposure of power of such circuits. But that is not the end and it depends on the imagination of the person designing the circuit.

*Reciprocating Motion of Piston.* In applications like reciprocating motion of the work bench in surface grinder, the actuator should reciprocate continuously. Fig. 10.10.18(g) shows a circuit for this purpose. Position of DCV is shifted from right to left wing and vice versa of the valve due to building up of the pressure in the hydraulic circuit after the piston reaches its extreme end. This requires for each side of the cylinder a check, sequence valve and hydraulic pilot signal to operate DCV. Consider the rod end where the piston has retracted completely and pressure has started building up, at that moment the corresponding sequence valve operates. The signal is sent to the DCV, which shifts it to other wing and extension stroke starts. This goes on continuously.

*Locking of the Cylinder.* We may need to lock the cylinder and want extending or retracting motion to start as and when we wish. In that case pilot check valve may be used, as shown in Fig. 10.10.18(h). For example, when the DCV is in its left wing of operation then fluid goes to the blank end but the liquid from rod end can flow only if some signal is given to it. Similarly for other side of operation. So the circuit operates only when pilot signal is passed on to the check valve.

# PROBLEMS

## Review Questions

1. A three-throw pump has cylinders. How do pistons in these cylinders work: one after the other or simultaneously?
2. A double acting reciprocating pump without air vessel delivers maximum discharge during the middle of the every stroke. What is the maximum value and what is the average value?
3. What is slip of a reciprocating pump? Why does it occur? Under what circumstances is the slip negative?
4. Why is acceleration head zero at the middle of every stroke of a reciprocating pump?
5. Acceleration head of a reciprocating pump is considered while computing the cylinder pressure but is ignored in power computations. Why?
6. In a double acting reciprocating pump the cylinder area is 0.03 $m^2$ and stroke length = 0.5 m. If the crank speed is 30 rpm, what is the theoretical discharge?
7. In a single acting reciprocating pump the cylinder area is 0.03 $m^2$ and $L = 0.5$ m. If the crank speed is 30 rpm, what is the theoretical discharge?
8. In a single acting reciprocating pump without air vessels gives an average rate of flow equal to 100 lpm. What is the maximum instantaneous rate of flow?
9. Actual $p$ v/s $q$ chart varies in many respects from a theoretical chart. Explain and how do we take into account these deviations?
10. In pumps either a minimum or maximum operating speed is required. Do you agree with the statement? Justify your answer for (a) reciprocating pump (b) centrifugal pump.
11. A single acting piston pump runs at 30 rpm and has an overall efficiency of 80%. The total head developed is 40 m of water. If the effective cylinder volume is 0.015 $m^3$, what is the power requirement of the pump?
12. What are the two primary functions of air vessels fitted to reciprocating pumps? What is the per cent power saved in overcoming pipe friction by fitting air vessels to (i) a single acting pump and (ii) double acting pump?
13. What is the nature of fluid that enters in air vessel: air or liquid? When does it enter and leave air vessel during the operation of the pump?
14. What is a hydraulic circuit?
15. What is the difference between meter-in and meter-out circuits?
16. Why are rotary pumps used in hydraulic circuits of the positive displacement type?
17. How does a vane pump differ from a centrifugal pump?

18. What is the principle merit of a (i) screw pump and (ii) radial piston pump?
19. What purpose does a (i) speed control valve and (ii) pressure compensating valve serve?
20. A pressure compensating valve always precedes a speed control valve. Is this true?
21. What type of valve changes the direction of flow to a hydraulic motor?
22. What is the function of a hydraulic accumulator? Define its capacity in terms of pressure $p$, area of ram $A$, and stroke of ram $L$.
23. An accumulator has a ram of area 0.06 $m^2$ and stroke 5 m. If the oil supply pressure is 60 bar, what is the capacity of the machine?
24. What is the primary function of a hydraulic pressure intensifier?
25. A hydraulic pressure intensifier has rams of diameter 12 cm and 30 cm. What pressure multiplication can the instrument achieve?
26. In a hydraulic press, the load lifted is 3 times the effort. The velocity ratio is 4. What is the efficiency of the press?
27. A hydraulic lift raises the lift cage of gross weight 120 kN once in every 100 seconds through a height of 30 m. what should be the output power of the pump that supplies the high pressure fluid to the lift?
28. A fluid coupling transmits 320 metric HP. The torque of the driving shaft is 50 Nm. If the speed of the driven shaft is 1450 rpm, what is the efficiency?
29. What is the distinction between fluid coupling and the fluid torque converter?
30. In a fluid torque converter, the speeds of the driving and driven shafts are 1400 and 700 rpm respectively. If the efficiency of the machine is 90%, what is the torque multiplication obtained?
31. A hydraulic ram consumes water at the rate of 500 lps under a head of 4 m and delivers at the rate of 15 lps under a total head of 30 m. Compute the Rankine and D' Abuisson's efficiencies of the ram.
32. Name the basic components required in hydraulic system. Describe importance of each of these systems.
33. What is a positive displacement pump and in what ways does it differ from a centrifugal pump? Is it necessary that a positive displacement pump will be a reciprocating type only? Why such pumps need a relief valve?
34. Explain working of any three positive displacement type pumps used commonly in hydraulic circuits. Which one of these would you recommend to use and on what basis?
35. Name the important considerations when selecting a pump for a particular application.
36. What is pressure compensated vane pump and how does it work?
37. When pump delivery side is connected to sump then it develops pressure just a little more than atmosphere. Then how does it develop pressure when supplying fluid to the actuator?
38. What parameters affect the noise level of a pump?
39. What is cavitation and where is it likely to occur in pump? Name four rules that will control or eliminate cavitation.
40. May we use a gear pump as variable displacement pump? When are the variable displacement pumps used?

41. Name two types of piston pumps and explain their working with the help of sketches. Discuss their merits and demerits.
42. What is the difference between single acting and double acting cylinder? Explain importance of cushioning.
43. What is a limited rotation hydraulic actuator? How does it differ from a hydraulic motor?
44. Can a piston pump be used as a piston motor?
45. Why does a hydraulic motor use more flow and deliver less torque then its corresponding theoretical value
46. Why does the rod of a double acting cylinder retract at a greater velocity then it extends for the same input flow rate? What determines the speed of hydraulic actuator?
47. What determines the pump size in a hydraulic system?
48. Explain how the vanes are held in contact with the cam ring in high performance vane motor?
49. What type of hydraulic motor is generally most efficient?
50. Explain working of a fluid coupling and determine its efficiency.
51. What is the function of directional control valve and how are these named?
52. When is a pressure-reducing valve used in hydraulic circuit? How is it different from pressure relief valve? Mention application of each.
53. How are the DCV actuated? Describe all the methods used for the purpose.
54. How does a pilot check valve differ from a simple check valve?
55. Explain difference between closed centre and open centre system.
56. What is a four-way, spring-centred, three-position valve?
57. What is the difference between meter-in and meter-out controls?
58. Explain working and functioning of sequence valve, counter balance valve, and unloading valve.
59. What is the difference between pressure intensifier and accumulator? Explain their working. May we have an accumulator like air vessel used in reciprocating pump?
60. What is servo valve and what is its function in hydraulic system? What are their common applications?
61. What is the difference between a mechanical-hydraulic and electrical-hydraulic servo valve?
62. What is the purpose of feed back in a servo system? What are the various feed back strategies used and how are these decided?
63. Explain what is meant by a closed loop system by drawing a block diagram.
64. What is the difference between single stage and two-stage servo valve?
65. Compare electro-hydraulic servo valve and proportional hydraulic valves.
66. Draw a hydraulic circuit for a drilling machine incorporating the following features: (a) slow drill feed and (b) rapid movement of drill to and from the work place.
67. What is meant by synchronisation of a cylinder motion? Name the various methods used to obtain it.
68. What is the purpose of regenerative circuit? Why is the load carrying capacity of regenerative circuit is small if its piston rod area is small?

## Problems for Practice

1. The bore and stroke of a single acting reciprocating pump are 100 mm and 200 mm respectively and the plunger has simple harmonic motion. The suction pipe is 80 mm in diameter and 4 m long and the centre of the pump is 3.5 m above the water level in the sump. Find the maximum speed at which the pump can run. Assume separation to occur at an absolute pressure head of 2.5 m of water. Take atmospheric pressure head = 10.3 m of water. [78.45 rpm]
2. A single acting pump has a bore of 75 mm diameter and a stroke of 150 mm. It draws water from a sump, whose water level is 3 m below the pump axis, through a pipe 30 mm in diameter and 4.25 m in length. The pump delivers the water to a tank through a 15 m long pipe 25 mm in diameter, to a height of 12 m above the pump axis. If separation occurs at 2.5 m (abs), find the maximum speed of the pump. Take atmospheric pressure head = 10.3 m of water. [41.8 rpm]
3. A single acting pump has a plunger of diameter 150 mm and a stroke 300 mm. The length of the suction and delivery pipes are 5 m and 20 m, respectively, and their diameter is 75 mm. The suction and delivery heads are 3.5 m and 8 m, respectively. Find the pressure head in the cylinder at the beginning, middle, and end of the suction and delivery strokes. Take $f = 0.0075$ and speed of the pump = 30 rpm. Find also the power required to drive the pump. Take atmospheric pressure = 10.3 m of water.
4. A double acting reciprocating pump has a bore of 150 mm and a stroke of 300 mm. The suction pipe has a diameter of 100 mm and is fitted with an air vessel. Find the rate of flow onto or from the air vessel when the crank makes angles of 30°, 90°, and 120° with the inner dead centre. Determine also the crank angles at which there is no flow into or from the air vessel. Assume that the pump runs at 120 rpm and that the piston has simple harmonic motion.
5. A single acting reciprocating pump has cylinder 100 mm in diameter and a stroke of 400 mm. The delivery pipe is 75 mm in diameter and 40 m long. Taking $f = 0.01$, find the power saved by fitting a large air vessel to the delivery pipe. The pump runs at 6 rpm. The air vessel is fitted 2 m away from the cylinder. [5.756 kW]
6. A double acting reciprocating pump has a plunger of diameter 150 mm and a stroke of 450 mm. The suction and delivery pipes are both 100 mm dia, the former being 6 m long. If $f = 0.008$, find the work saved by fitting an air vessel 1.5 m away from the cylinder on both the pipes. Assume simple harmonic motion of the plunger and that the crank performs 60 rpm. [0.424 kW]
7. A single acting reciprocating pump has a plunger diameter of 150 mm and stroke length of 300 mm. The pump runs at 35 rpm and lifts water through a height of 20 m. Calculate the theoretical discharge and the theoretical power required. If the actual discharge is 2.885 l/s, find the percentage of slip. If the delivery pipe is 15 m long and has a diameter of 100 mm, find the acceleration pressure head at the beginning of the delivery stroke. If a large air vessel is fitted close to the cylinder on the delivery side, find the pressure head necessary in the cylinder to overcome friction in the delivery pipe. Take $f = 0.0075$. [4.77%, 6.932 m, 0.034 m]

8. A double acting reciprocating pump has a plunger of 150 mm dia and a stroke of 450 mm. The suction lift is 3.5 m, the length of the suction pipe is 6 m, the pipe dia is 100 mm. Calculate the maximum speeds at which the pump could run under the following conditions: (a) no air vessel on the suction side, (b) a very large air vessel on the suction side close to the cylinder. Assume pipe friction $f = 0.006$, separation pressure head = 2.5 m (abs). Comment. [35.6, 174.3]
9. A double acting reciprocating pump of bore 200 mm and stroke 400 mm lifts water from a sump 1 m below and delivers to a height of 30 m above the cylinder axis. The suction and delivery pipes are 100 mm in dia, and are 2.5 m and 40 m long respectively. The pump runs at 35 rpm. The piston rod is 75 mm in diameter. Determine the load on the piston when it has moved through a distance 100 mm from the inner dead centre. Take $f = 0.0075$. [1595.2 N]
10. The bore and stroke of a double acting reciprocating pump are 350 mm and 450 mm, respectively. The pump runs at 60 strokes per minute. The suction pipe is 10 m long while the delivery pipe is 400 m long. The diameter of the delivery pipe is 15 mm. The pump is situated at a height of 2.75 m above the sump water level, the outlet of the delivery pipe being 72.25 m above the pump. Calculate the diameter of the suction pipe for the condition that separation is just avoided at the commencement of the suction stroke. Assume separation occurs at an absolute pressure head of 2.5 m of water. Find also the power required to drive the pump neglecting other friction losses in the pipes. Take $f = 0.0075$. [234.3 mm, 42.26 kW]
11. A reciprocating pump has three cylinders (single acting), the pistons of which are operated by cranks at 120°. The pistons are 75 mm dimmer and the stroke is 150 mm. The cylinders discharge water into a single pipe 50 mm dia and 64 m long. The pump speed is 60 rpm and there is no air vessel on the delivery side. Give diagrams on a crank angle basis, showing how the water velocity and acceleration in the delivery pipe differ from those obtained with a single acting reciprocating pump of the above dimensions and speed. If the pump discharges to atmosphere 30 m above the level of the cylinders, calculate the range of pressure in the pipe just before the pump. [Hint: draw the acceleration diagram and see how much variation takes place; this is the range] [11.67 m]
12. The supply pipe in a hydraulic ram installation is 60 mm in dia and 4 m in length. The waste valve is 130 mm in diameter having an effective lift of 6.5 mm and a weight of 13.25 N. It makes 125 beats per minute. Find the discharge through the delivery pipe per minute against a delivery head of 8 m above the water level in the supply reservoir. Neglect all losses. [15.625 l/min]
13. A hydraulic crane has to lift 22270 N through 14 m at 1.4 m/s once every two minutes. The velocity ratio of the pulley system is 6 and its efficiency is 0.75. The crane cylinder is 220 mm dia and the efficiency is 0.95 allowing for friction of the packing. The crane is fed from an accumulator, which is supplied at a steady rate from a pump. Calculate (a) the necessary pressure in the crane cylinder during lifting, (b) the duty of the pump in $m^3/s$, and (c) the minimum capacity of the accumulator. [4.93 $N/mm^2$, 0.000739 $m^3/s$, 0.08131 $m^3$]

14. A hydraulic crane is operated by a ram fitted with a jigger. The crane is required to lift 59840 N at 1.25 m/s through 80 m once every 5 minutes. The diameter of the ram is 750 mm, the pressure of water is 4 N/mm$^2$ and the mechanical efficiency is 80%. A pump supplies water continuously to the cylinder via an accumulator, which can store water at the required ram pressure. Find (i) the required minimum pump output, (ii) the minimum volumetric capacity of the accumulator, and (iii) the stroke of the jigger ram. [0.00508 m$^3$/s, 1.199 m$^3$, 3.387 m]
15. The ram of a hydraulic crane has a diameter of 150 mm and the ratio between the movement of the load and the ram is 6:1. Water is supplied through a 37.5 mm dia pipe 480 m long, the pressure at the inlet end of the pipe being 7.7 N/mm$^2$. The coefficient of friction for the pipe is 0.01. A pressure of 0.6 N/mm$^2$ on the ram is required to overcome mechanical losses. Determine (i) the maximum speed with which a load of 11350 can be lifted and (ii) the load and speed of lifting which corresponds to the maximum power obtainable from the crane. [1.32 m/s, 13350.8 N, 1.187 m/s]
16. How much hydraulic power would a power pump produce when operating at 140 bars and delivering 0.001 m$^3$/s of oil? What power rated electric motor would be selected to drive this pump if its overall efficiency is 85%?
17. A pump has a displacement volume of 98.4 cm$^3$. It delivers 0.0152 m$^3$/s of oil at 1000 rpm and 70 bars. If the prime mover input torque is 124.3 N-m, (a) what is overall efficiency of the pump? (b) what is the theoretical torque required to operate the pump?
18. Find the actual delivery of a gear pump of following specifications: number of teeth 16, gear width 0.065 m, module 6 mm, pressure angle 200, rpm 1600, and volumetric efficiency = 88%.
19. A gear pump with following specifications run at 1400 rpm. Module 3 mm, number of teeth 12, gear width 15 mm. Calculate the discharge of the pump. Also calculate the horsepower required for the electric motor to drive the pump, when working against a pressure of 150 bar.
20. Find out the delivery of gear pump of following specifications: number of teeth 16, gear width 0.065 m, module 6 mm/tooth, pressure angle 20°, gear outer diameter 0.108 m, rpm 1600, and volumetric efficiency 88%.
21. A radial piston pump has the following specifications: maximum pressure 300 bar, number of plunger 7, diameter of plunger 50 mm, maximum eccentricity 10 mm, speed of rotation of the shaft 1500 rpm. Calculate the theoretical and actual discharge and power required to drive the pump. Mechanical efficiency 80% and volumetric efficiency 90%. Specific gravity of oil is 0.9.
22. Oil at 0.001 m$^3$/s and 70 bar enters the low-pressure inlet of a 3:1 booster. Find the discharge flow rate and pressure.
23. A pump supplies oil at 0.0016 m$^3$/s to a 40 mm diameter double acting cylinder. If the load is 5000 N (extending and retracting) and the rod diameter is 20 mm, find the (a) hydraulic pressure during the extending stroke, (b) piston velocity during the extending stroke, (c) cylinder kW power during the extending stroke, (d) hydraulic pressure during the retracting stroke, (e) piston velocity during

the retracting stroke, and (f) cylinder kW power during the retracting stroke. [3.98 Mpa, 1.27 m/s, 6.37 kW, 5.31 Mpa, 1.70 m/s, 8.50 kW]

24. It is require to transmit 36.75 kW power from an accumulator through a pipeline 10 cm in diameter and 1500 m long. The ram is loaded with a weight of 1250 kN and the friction loss in the pipeline equals 2.5% of the total power being transmitted. Work out the diameter of the ram if the friction factor $f$ = 0.01 for the pipeline. [0.499 m]
25. An accumulator with 30 cm diameter ram and 6 cm stroke length is loaded with 400 kN total weight. The packing friction amounts to 5% of load on the accumulator and the ram falls steadily throughout its full range in 120 s. During this downward movement of the ram at the stated speed, the pump delivers 7.5 litres/s of water to a hydraulic machine. Work out the total power being delivered to the hydraulic machine. [59.355 kW]
26. A hydraulic intensifier gets the low pressure liquid at a pressure of 50 bar and delivers it to a machine at a pressure of 200 bar. If the intensifier has a capacity of 0.025 $m^3$ and stroke 1.25 m, calculate the diameters of the fixed ram and sliding cylinder to be used for the intensifier. [0.319 m]

CHAPTER

11

# Experimental Methods

OUTLINE

- Introduction
- Measurement Techniques
- Design of Experimental Set-up

## 11.1 Introduction

Experimental methods have been used to study the physics of a given problem to determine the variables that affect the performance of a given machine or system. Once a certain theory is developed then it is verified from the data obtained experimentally. An example of determining velocity of sound through air will tell more than the words. Newton assumed that air temperature may be taken as constant during the distance a sound wave travels and obtained the sound velocity. However, it did not agree with the experimental results. Laplace then argued that the disturbance is so week and time elapsed is so small for travel of disturbance that the changes in air during the sound wave propagation may be assumed to be isentropic. He obtained another relation for the velocity of sound. The result obtained from this analysis matched very well with experimental results. Ultimately, the scientists over the world agreed for the Laplace results. So the experimental techniques have proved to be very valuable for scientific world in developing the data bank with which the results of any theoretical analysis may be compared to check its validity and applicability. It has been adopted universally a standard technique and requires very accurate measurement.

Design of an experimental set-up by itself is a challenging job because there are a number of variables involved in any physical problem. In Rotamachines, power or efficiency depends on $D$, $m$, $p_{01}$, $p_{02}$, $RT_{01}$, $RT_{02}$; $p_0$ and $T_0$ are the stagnation pressure and temperature, $D$ is the rotor diameter, $m$ the mass flow rate and subscripts 1 and 2 correspond to the inlet and outlet sections of the rotor. In fact, there are other variables too like viscosity, surface roughness, etc, that also affect the performance of the machine. But their effect is comparatively small and therefore these were not considered. If we combine the gas constant $R$ and $T_0$ to form a single variable $RT_0$ then we need not to bother which gas is used in the rotamachine and about its density, since $p_0$ is another variable in the above list of independent variables. Now only 6 physical variables are left that form three non-dimensional numbers. This way it is possible to reduce number of variables and therefore our experimental efforts.

Any experimental set-up for such machine would require measurement of stagnation pressure and temperature, and velocity at different positions in the machine that would depend upon the type of investigation. This example stresses the need of identifying first the important variables to be measured and then to select the right instrument for its measurement to desired accuracy.

Let us take another example of load test on *IC* engine. Purpose of such a test is to predict performance of the engine at different load on it or running at different rpm, both the parameters are important depending upon the application. If engine is used in automobile then the rpm is important and load is important when used for power generation in which case it runs at a predefined speed. The parameters that may be used for measuring performance can be listed as brake specific fuel consumption, thermal efficiency, brake mean effective pressure, specific brake power, torque, etc. These are obtained from measurement of (i) fuel consumption per unit time (conventionally hour) (ii) rpm, (iii) power produced by engine using any dynamometer, and (iv) frictional power or indicated power using any of the standard methods. Having identified the variables to be measured and their range, we next select the instruments needed for such measurements.

In measurements the most important part is selection of right type of instrument and therefore needs a brief discussion about it. We know that a variable is measured using some property of the working substance used in the instrument that changes with the variable to be measured, called the measurand. For example, volume, resistance, and so on are the properties of a working substance used for measurement of temperature. Accordingly a large variety of thermometers are available for measuring temperature and each variety is again available for different range. For example, mercury thermometers are available to measure temperature up to 100°C and more with different resolution. Some are suitable for low range while others for high temperature range. Same is the situation for other measurands too. A designer therefore should know the advantages and disadvantages of each of the instrument and the range of measurand to be measured during the experiment so that he can make proper selection.

We know that length can be measured by a cm scale, a Vernier calliper, a micrometer, or some other method reading length to different number of decimal places. For example a simple scale reads accurately up to 1 mm, a Vernier scale up to 0.1 mm, a micrometer reads up to 0.01 mm. We decide their suitability upon the accuracy required for the purpose. Another important observation about the instruments is that the property of its working substance does not vary linearly with the measurand for entire range the instrument is used for. The instrument, nevertheless, is normally calibrated linearly. This means that the value of the property of the instrument is determined at two different states using the standard source and then the range in between is divided linearly. This linear division is an approximation, since the variation of the property with the measurand is not linear. Recall in a thermometer based on Celsius scale, two points at 0°C and 100°C of ice and steam are used for the purpose. Then the interval is divided into 100 equal parts. Since the change in property is unfortunately not linear, the designer has to take care of all these factors as well before deciding the instrument for measuring a particular measurand. This requires some experience in experimentation. So the actual values involved in measurement guide in their proper selection.

## 11.2 Measurement Techniques

The physical quantities that ought to be measured in science and engineering laboratories are: temperature, pressure, force, torque, mass, rpm, flow rate, current strength, velocity, stagnation and static pressure, etc. These instruments may be contact or non-contact type and may be based on different principles. We cannot discuss all this in this chapter and the student has to learn about these basic instruments from their laboratory course. However, a few of them of importance for fluid mechanics laboratory shall be described in brief.

*Pressure and Velocity Measurement.* For accurate measurement of pressure in laboratory, piezometric tubes or tapings perpendicular to flow direction are used. A piezometric tube measures piezometric head, which is the sum of pressure and potential head. It is a tube placed in the fluid at right angle to the direction of flow with other end open to atmosphere. While placing tube, care is to be taken that it does not disturb the main flow for which reason its end fixed in pipe be flushed with pipe's inner surface. The fluid rises in the tube to a height so that it exerts pressure at the bottom of the tube, which is equal to that at immediate point on its outside in the main flow. This rise of liquid in tube gives gauge pressure at the point of measurement. Tube height should be large enough that the liquid rising in it does not flow out. A taping made at right angle to the pipe will read the static pressure. But, for accuracy, four equi-spaced holes are drilled in pipe along its circumference and connected. This arrangement takes care of variation of pressure at the section of measurement and gives average pressure at the section. In industries, pressure gauges of different type are used, most of which operate a diaphragm or a bellow due to pressure difference and the movement of this sensing element is recorded. Pitot tube is used for measurement of stagnation pressure. Principally, the tube is place opposite to the flow direction so that the fluid comes to rest at its nose, shown in Fig. 11.2.1. Therefore pressure at the Pitot tube nose is stagnation pressure and the manometer attached with it would measure this.

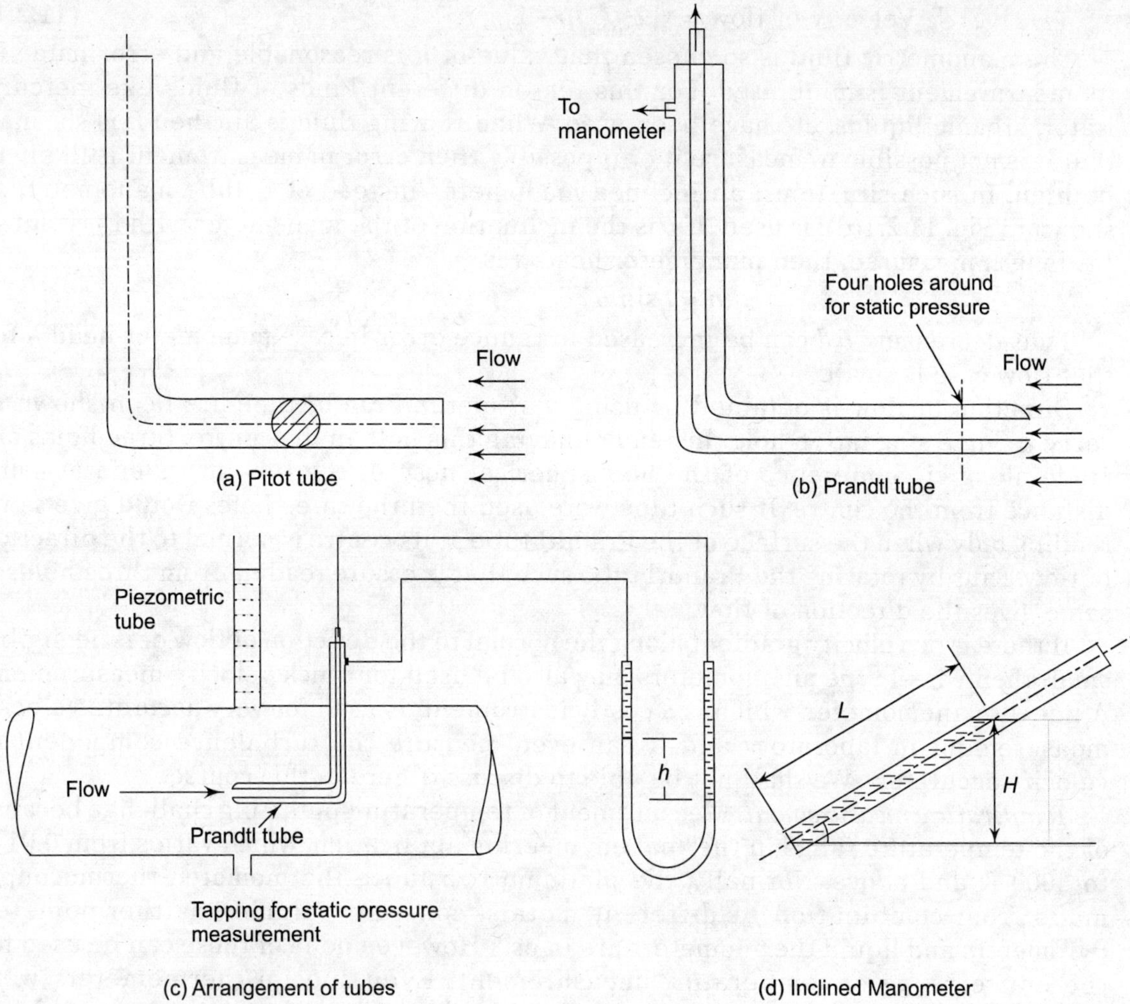


**Fig. 11.2.1** Pressure and velocity measurement

For measurement of velocity, Prandtl tube is used that measures both stagnation and static pressure simultaneously. Its two tapings are connected to the two limbs of the $U$ tube manometer and the difference in head of the manometric liquid gives velocity. However, it will be better to use a pitot and static pressure tube separately for measurement of stagnation pressure and the static pressure because the Prandtl tube measures these two parameters at different sections. Moreover, the stream lines at the section where the static pressure is measured are not parallel that also introduces some error. Thus the overall effect is that there is bound to be some error in measurement of velocity if measured by a Prandtl tube.

If $h$ and $\rho_m$ are the height and density of the manometric liquid in the $U$ tube of manometer connected to the two tapings of Prandtl tube and $\rho$ is density of the fluid flowing through the pipe then

$$\text{Velocity of flow} = \sqrt{[2(\rho_m/\rho - 1)\, g\, h]} \quad (11.2.1)$$

The manometric fluid is so chosen that value of $h$ is reasonable and error made in its measurement is minimised. For this reason different kinds of fluids like mercury, water, organic liquids, etc have been used. When flowing fluid is air then $h$ is so small that it is not possible to measure it or if possible then error in measurement is likely to be high. In such situations, an inclined manometer instead of $U$ tube manometer, as shown in Fig. 11.2.1(d), is used. If $\theta$ is the inclination of the manometer with horizontal, $l$ is length measured, then manometer head $h$ is

$$h = l \sin \theta$$

Thus decreasing $\theta$, $l$ can be increased to reduce error in case manometer head $h$ for that flow case is small.

Direction of flow is obtained by using yaw meter. Prandtl tube has been shown to carry a single stagnation hole. Instead of having this hole in the centre, three holes are made along circumference of the hemispherical nose of the tube at its end at some distance from the centre. If such tube were used then the three holes would give same reading only when the surface of the Prandtl tube at its centre is normal to the direction of flow. Thus by rotating the Prandtl tube such that pressure reading from three holes is same gives the direction of flow.

If there is no velocity gradient along the normal to the direction of flow or is negligibly small then wheel type anemometers may also be used for quick velocity measurement. A hot wire anemometer, which is a costly instrument, is used for very accurate velocity measurement in laboratory and it can even measure the turbulence component of velocity accurately. We shall not be able to discuss it here in this course.

*Temperature measurement.* Measurement of temperature puts a big challenge because of the temperature range in thermal engineering application, which varies from 0.01 K to 3000 K and above. Normally, the platinum resistance thermometer, thermocouple made from combination of different metals, constant volume gas thermometer, pyrometers and liquid thermometers are in use. However, none of these can be used for the entire range of temperature measurement. Even special thermometers were developed to measure very low temperature. This is a very exhaustive subject and will not be covered here.

*Discharge measurement.* Venturimeter, orifice plate and nozzle are used for measurement of discharge for internal flows whereas weir or notch (rectangular or $V$ notch), are used for measurement of external flows. These methods were discussed in chapter 4. In these methods, a $U$ tube manometer measures pressure drop for flow through pipes whereas for weir or notch a Vernier scale is good enough to measure head $H$. Though all these methods have one or another limitations, but the results obtained are satisfactory because of the calibration procedure adopted to determine value of coefficient of discharge. Experiments are performed whereby discharge is measured by the given apparatus as well as standardised instrument. From this, value of $h$ is obtained for the instrument at known discharge. The results are then plotted and the graph is used for finding discharge. This is known as calibration of the instrument. The calibration is done time to time to ensure that the results are accurate. This is required because of the changes that occur in the instrument due to its use. Frequency of calibration depends on the type of flowing fluid.

There are other methods too. You have to learn about these in your laboratory course work. A very simple method that is very common in laboratories is to measure time taken for prefixed quantity of flow using a stopwatch from which discharge is calculated.

*Shaft Power.* It is measured by using dynamometer. Basically a dynamometer is either absorption or transmission type. In the former case power generated by an engine is absorbed while in later it is utilised fruitfully. Naturally the later would be preferred, however, both are in use. Rope brake and hydraulic dynamometer are commonly used as absorption type whereas *DC* or *AC* generator is used as transmission type dynamometer.

For low power engines, rope brake is commonly used because of its low cost. A large pulley of diameter $D_p$ is fixed on the shaft and a rope of diameter $d_r$ is placed around its periphery. Lower end of the rod carry a fixed dead weight $W$ whereas the upper end is fixed through a spring balance. Spring tension can be changed by hand wheel. This exerts a torque on the shaft = $(W - S) R$, where $R$ is radius at which the force on rope acts from the axis of rotation, and $S$ is the spring balance reading. This load torque opposes the shaft motion due to friction between rope and pulley surface. Spring tension is adjusted so that the engine runs steadily at fixed rpm. Steady state is reached when the torque applied by engine on shaft equals the load torque due to dynamometer. Power developed by engine in Watt (kW) is

$$P = [2\,\pi N\,(W - S)\,R/60]/1000 \text{ kW}; \; R = (D_p + d_r)/2 \qquad (11.2.2)$$

Hydraulic dynamometer is used in situations where power from engine is large, since the heat developed in rope brake dynamometer at the drum periphery would be too large. (Though arrangement to cool the drum is made but it is not sufficient to absorb large power.) A hydraulic dynamometer works on the principle of hydraulic coupling in which there are fixed and moving cups opposite to each other so that water circulates in them. Moving cups are mounted on the dynamometer shaft coupled to the engine shaft that pushes water to come out from its periphery where it enters the fixed cups. The fixed cups are integral part of the dynamometer stator, which is placed on the bearing and carry loading arrangement. Increasing this load creates the torque on the stator and in turn on rotor since the circulating water tries to rotate the fixed cups. For hydraulic dynamometer,

$$P = K\,N, \qquad (11.2.3)$$

where $K$ is a constant supplied by the manufacturer and $N$ is the rpm. Student should see carefully the manual whether $K$ is given for $N$ in rps or rpm. The rpm can be measured using a contact type or non-contact type tachometer; the later gives better results.

For generator type dynamometer, we have to measure voltage, current strength, and phase. We have also to find out overall efficiency of the generator at that load so as to find out the power generated by the engine. Therefore it is essential to have load characteristic curves for that dynamometer and needs to be calibrated time to time. Moreover a resistive load is normally put across the generator in laboratory as electric load.

*Volumetric efficiency.* It is determined by finding out the actual volume flow rate per second. Air is drawn into a big tank to reduce the turbulence from where it is supplied to the engine through a pipe fitted with an orifice plate. Pressure drop is measured across the orifice using $U$ tube manometer filled with manometric fluid and then actual volume flow rate of air is obtained using Eq. 3.4.16. Theoretical or maximum volume that may be inhaled per unit time is obtained from the knowledge of swept volume, which is given as pump or engine capacity. If $D$, $L$ and $n$ are the bore, stroke length and number of cylinders, respectively, then Theoretical volume flow rate = $(\pi D^2/4) \times L \times n \times N / k$, where $k$ is one for two-stroke engine and two for four-stroke engine. Ratio of two volumes so obtained is volumetric efficiency.

## 11.3. Design of Experimental Set-up

Having selected the parameters to be measured and the instruments to be used for the purpose the next step is to design the experimental set up. Designing of an experimental set up depends on the type of the problem and therefore we shall illustrate the procedure by taking example for plotting the main characteristic curves of Pelton wheel. We learnt in chapter 2 that these are the curves for unit power, unit discharge and efficiency against unit speed. We wish to design the experimental set up for this. Normally the experiment is carried out at constant head in the laboratory. We make following observations:

1. Power, discharge, speed and head acting on the Pelton wheel is to be measured at different load on the turbine for the same head and given gate opening. Unit quantities are calculated from these measurements using Eqs. 9.1.22 to 9.1.23.
2. For applying load on the turbine and measuring power developed by it, a dynamometer is used. A rope brake dynamometer is normally used with laboratory model of Pelton wheel. If the rope brake arrangement is not provided, this has got to be arranged.
3. Water is supplied from a water source to Pelton wheel and collected after it comes out from it. A valve is to be provided in the supply line to fix the head at some desired value.
4. Since pressure in Pelton wheel everywhere is atmospheric, we shall need an open channel to transfer water from turbine casing to the tank where water is collected. (Pumping it back to the supply tank by a centrifugal pump can recirculate water collected in the tank.) For measurement of discharge[2] , we would put a $V$ notch in the channel and a vertical Vernier scale to measure depth of water level above the crest.
5. For measuring rpm, normally a tachometer is used and pressure gauge is placed on the upstream side of the turbine to know the head.

[2] For measurement of discharge in internal flows, we may use any method discussed in section 11.2. For example, an orifice plate may be fixed in the pipe carrying water from turbine to the tank with a $U$ tube manometer filled with mercury and connected to the tapings in the orifice plate. Discharge is obtained from the relation $Q = \pi/4\,(d/100)^2 \times Cd \times \sqrt{\{2g\,(\rho m/\rho - 1)\}}\sqrt{(h_m/100)} = K\sqrt{h_m}$; $h_m$ is difference of height of liquid column in two limbs of the manometer, in cm, $d$ is pipe diameter in $m$ and $Cd$ is the coefficient of discharge. It is good practice to put constant $K$ in normalised way. That is if $K = .000012$ then it should be written as $1.2 \times 10^{-5}$; or to put in this form such that the numbers in the related column are in normalised form.

With this in mind, we may develop an experimental set up, as shown in Fig. 11.3.1.

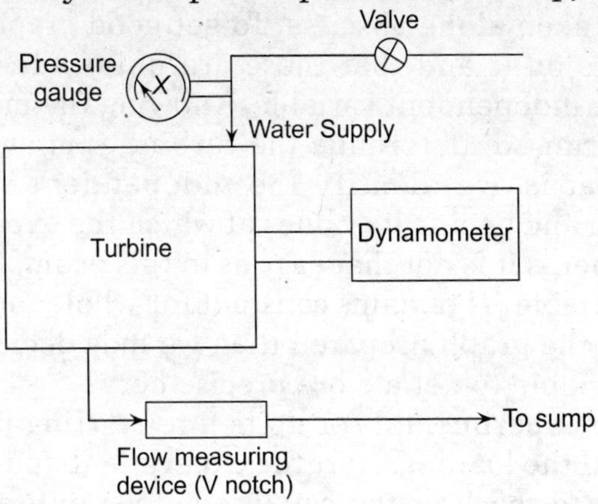


**Fig. 11.3.1** Experimental set-up for Pelton wheel

*Performing Experiment.* Having developed the experimental set up, next thing is to write down the stepwise details to perform it. In this example the steps will be:

1. Arrange all the instruments needed for experiment. In this case, we shall need tachometer (and may be a stopwatch) as other instruments are fixed with the set up.
2. Take those readings, which remain constant during the experiment. We may either measure these from the set up or see from the records provided with the set up. That is, we shall record diameter of the brake pulley and the rope, initial water level (if any) in the flow measuring device, etc.
3. Prepare an observation table in which different readings needed for calculation be recorded. For this purpose develop an observation table, since you know what measurement are to be taken for different set of experimentation. A typical observation table may look like this:
   Brake diameter, $D$ ------$m$, Rope diameter, $d_r$ -----$m$,
   Set No.---- Gate Opening -----
   $Q = 1.42 \times 10^{-2.5} \times h^{5/2}$ or $K\, h^{5/2}$
   ($h$ is height of water level above crest in the notch in cm.)
   pressure gauge reading, $p$ bar, $H = p/0.0981$ m of $H_2O$
   $P_u = [\, 2\pi N(W_1 - W_2) \times (D + d_r)\, /60]/H^{3/2} \times 10^{-3} = K_1 N(W_1 - W_2)$; $P$ is in kW

| S. No. | N, rpm | Dynamometer | | $h$ | $Q = K\,h^{5/2}$ | $Q_u = Q/\sqrt{H}$ | $N_u$ | $P_u = K_1 N$ |
|---|---|---|---|---|---|---|---|---|
| | | $W_1$ | $W_2$ | cm | m³/s | | $= N\sqrt{H}$ | $\times (W_1 - W_2)$ |
| | | | | | | | | |

This is how the observations and calculation table can be organised. Here calculation table is integrated with the observation table. However, it may be desirable to have the two separate tables in which case a primary key be fixed to connect the two.

4. Note that $N$ is the first column for the reason that unit speed is the independent variable and is taken along abscissa. To get good graph, it is desirable that there are 5 to 6 points on it and that these are uniformly distributed along abscissa representing the independent variable. Knowing the maximum rpm of the turbine at which it can run, we determine the turbine rpm at which it has to be run for experiment. That is, we identify the independent variable and its range for experimentation and decide its values at which the experiment is to be performed. In some experiments it is not that easy as in this example, since it is the rpm that is independent variable ($H$ remains constant for whole range of $N$). Once experiment is complete and the graph prepared then we may decide some more values of the independent variable to get a more precise curve.
5. Now inspect the experimental set up before starting it and see every thing is all right. Remove all the loading, start the experiment and wait for some time to allow the steady state to reach or the conditions to stabilise and then set the value of independent variable to the decided value. In most of the experiments, such as in heat transfer or thermal systems, some time is required to attain the steady state and this time should be allowed for accuracy and repeatability of the results. Thus after each setting of the independent variable, we must wait till all the readings have become constant.
6. We must also develop list of precautions to be taken while performing experiment to minimise the error. For example, while performing this experiment, you should run the turbine till water level reaches the notch crest in the channel so that the initial reading of the scale is taken and should be recorded. Scale is initially set at its zero at this point. But due to use some error creeps in and then the scale reading is corrected.

   [In our experiment, we first set the gate opening, run the turbine at no load and take the readings. Now tighten the spring by the hand wheel so that the tachometer reads the value we fixed before starting the experiment. Allow the machine to run for some time till the water level in the channel becomes constant. Now take the reading. Repeat this experiment for all the values of $N$ initially decided till we reach the last set. The whole series of experimental steps may be repeated for another gate opening].

*Calculations and Conclusion.* After performing the experiment, calculation is done using the appropriate relations. It is a good practice to develop calculation table so that whatever is needed is available in that table. Again, as seen from the observation table, the relations should be reduced in terms of unknowns that vary during the experiment. Though tabular method is good but there is no match to the visual impact. That is, if these results are presented in graphical form then it speaks much more than the numbers in the table. A look at the graph tells you the density of the point. If some more experimentation is required at some other value of independent variable, or if some of the experimental points are far away from the curve due to some error that might have been made during taking the reading or making calculation, it is quite evident from the curve. Therefore a graph should be plotted and then decision be taken if some calculations need to be checked for error or that set of experiment is to be performed again or some more data is needed to get more precise information, etc. Lastly, the results obtained from such experiment should be compared with the standard results provided by the manufacturer for that system or the standard results available from theory. If there is large deviation then experiment should be repeated to be sure that no error has been made. If results are the same then the difference in the expected and actual results should be explained for.

### 11.3.1 Performance of Centrifugal Pump

We shall take a few more example of design of experimental set up. Like this you may take any experiment and design yourself. Try the questions given at the end and compare your set up with the one available in your laboratory.

*Objective.* Purpose of this experimental set up is to draw operating characteristic curves for pump when the pump operates at constant rpm.

*Measurements required.* Operating characteristic is the plot of power required, pressure head developed across pump and pump efficiency at different rate of flow of the fluid over its impeller. The measurements are same as that for a turbine except that an electric motor will be required to transfer energy to the fluid. Therefore we do not need a dynamometer, the power input to the motor can be determined using a wattmeter or voltmeter and ammeter in the input line to the motor; if three phase line is supplying the current then we have to measure power factor of the line also. For measurement of pressure, we put two pressure gauges, one on the top of suction line and another at the discharge side just after the pump. Water is supplied to an overhead tank; position of which (height from pump) will fix the discharge pressure. A valve is provided in the discharge line to change the flow rate. Noting the total number of turns required for opening it full, it is easy to know the valve opening as fraction of full gate opening at any position. We use an orifice plate to measure the flow rate, which is placed on the discharge side. Tachometer is used to measure and maintain the speed of the pump constant by changing power supply at different gate opening. Find out if some special arrangements are required to be made to change the power supply or does it happen automatically; if yes, how will it be done and also design it?

### 11.3.2 Performance of Centrifugal Compressor

Since compressor and pump look same in many ways, do you think the set up for a pump may be used for it as well.

1. Find out the variables used for plotting the performance characteristic and whether these are dimensional or non-dimensions. If these are non-dimensional then what are their dimensions?
2. Pressure and temperature can be read at the suction and delivery side of the compressor using instruments.
3. What arrangement do you suggest for measurement of discharge? If we use the orifice plate will it make any difference and what difference does this make?
4. How would you vary the flow discharge and simulate the condition of choked flow? Are the sufficient work shop facilities available in your college to fabricate such arrangement?

5. You have studied three types of impellers, namely backward curved, radial and forward curved. Let us say the supplier is in a position to supply the three runners of the same impeller diameter. Will it be possible for you to use the same test rig for the three impeller?
6. Think of all the possible measurements needed to get the characteristics with speed as parameter and make list of instruments needed with the type and range for measuring each of these parameter.
7. Now change the relations in workable formula to reduce the multiplication and division of constants appearing in these. Then suggest the observation and calculation tables.
8. List the precautions to be taken and determine the values of independent variable at which reading is to be taken.
9. Draw the curves and show the surge and choke line.

### 11.3.3 Flow Measurement in Pipe and Open Channels

Prandtl tube, Venturimeter, orifice plate and nozzles have been used for measurement of discharge; we discussed these in section 11.2. The experimental set up needs

1. A pipe with arrangement for fitting any of the device so that one can be replaced by other in the same pipe, a centrifugal pump to make water flow (a blower in case of air) through the pipe and a tank to collect water. We may also fix all the arrangements one after another along the length of pipe. Study the advantages and disadvantages of providing all of these together. What special arrangement is needed for Prandtl tube?
2. How would you arrange changing the flow measuring device in the pipe?
3. For measurement of pressure, we shall use $U$ tube. For this we make one tapping on upstream side at a distance $D$, the diameter of pipe. (Do you suggest one or four tapings along the circumference at the same axial location?) Decide position of another taping. What considerations you shall make for it?
4. Decide the independent parameter and its values for experimentation. Obtain workable formula, develop the observation and calculation table and plot the results on 2$D$ graph. Also sketch the experimental set up. Do you have the confidence to fabricate it in your workshop? If yes, what precautions you have to take to ensure that the relations developed in section 11.2 can be used?
5. Is it possible to use the test rig as such to find out coefficient of discharge of the measuring device and their dependence on the Reynolds number or some additions are to be made? Discuss.
6. Also design the calibration procedure.

*Open Channel.* Difference in pipe and channel is that in former flow is due to pressure difference between inlet and outlet of the pipe whereas in channels it is due to gravity. Therefore methods like Venturimeter, orifice plate could not be used. However, the Prandtl tube or better a piezometer may be used to measure velocity and using graphical integration technique, discharge can be obtained. However, this is time consuming and is not normally used in field.

# PROBLEMS

## Review Problems

These problems are based on laboratory course and you may not be able to get the answer for all the questions given below from this book. You have to search it from other sources.

1. What steps will you take before performing any experiment?
2. How do we choose the instruments required for experimentation? Study different temperature measuring instruments available in market and make a comparative study. For accurate measurement of temperature, which instrument would you suggest?
3. Platinum resistance thermometer is used for accurate scientific measurement. May we use it for any temperature range. Find out the range of temperature for which its use is recommended.
4. Yet another very popular method of temperature measurement is thermocouple. On which principle is it based and what are the sources of error in measurement?
5. May we use any two dissimilar metals to make thermocouple? What is the criterion for selection of these two materials to be used for thermocouple? Are we able to take care of Thomson effect in measurement?
6. Dial gauges, recorder or indicators are placed at remote place from the place of temperature measurement. How is it done and does this affect accuracy of temperature measurement?
7. Study different types of rpm measuring instruments and write down their principle. Explain what is stroboscopic effect and how is it meaningfully used for measurement of rpm? Are the results obtained reliable?
8. What is the difference between analog and digital instruments? Learn about these in your laboratory course. While using digital instruments, you must have observed that the reading fluctuates. Then how do you decide the reading? Are you sure that the reading you have taken is error-free?
9. In Prandtl tube, holes for measurement of static and stagnation pressure are not at the same location. How do you decide location of holes for static pressure measurement? Justify under what conditions this tube would give velocity and whether it is at stagnation or static holes?
10. When fluid direction is changed, as in orifice plate, pressure loss occurs. Why? Can you list various ways in flow that will result in pressure loss and what is the way to get rid of these?
11. Can we use flow measuring instruments used for internal flow for measuring discharge in open channel? Justify your answer.
12. In any experiment how do you decide values of the independent variable? If there are more than one independent variable, what strategy is adopted?
13. How do you make sure that your results are correct? Explain the procedure.
14. You have a two-stage air compressor and its performance is to be determined. Write down the complete procedure including the figure of experimental set up, observation and calculation tables, and precautions.

15. An axial-flow air compressor is to be used with gas turbine. For this test is to be carried out so that it can be matched with the turbine for which characteristic curves are available. Suggest the experimental set up and what parameters need to be measured for the purpose.
16. Design a simple experimental rig for (i) verification of Bernoulli equation, (b) demonstration of ram principle, and (c) flow measurement in pipe using different methods studied in this book. How would you fabricate and test it in your laboratory? Will you also develop calibration test rig in these cases? Discuss the procedure.
17. Variable speed test is also performed in laboratory on centrifugal pump. Think of the purpose of such experiment and develop the experimental procedure using the set up for the fixed rpm rig.
18. There are three types of vanes used in centrifugal pumps. Would you recommend different test rig for each or use the same making an arrangement for changing the impeller only?

APPENDIX

# A Review of Mathematical Concepts

## A1 Vector Algebra, Calculus and Integrals

Vectors and tensors have provided a great relief in mathematical analysis of fluid flow through or around the bodies and therefore some space is provided here to recapitulate their important aspects. The quantities like, mass, density, temperature, energy are scalar quantities. For scalar quantity, knowledge of the magnitude is enough to describe that quantity. The velocity and forces are the vector quantities because the information about the direction in which these act is also needed in addition. Thus a vector has both magnitude and direction. These have unique components when resolved along the reference axes which change with rotation of the axes. However, the compo- nents in two different sets of axes are related to each other through mathematical relations that can easily be obtained.

On the other hand the stresses and strains are the tensors of order 2 because the surface on which these act also need to be specified. In many other cases we need some more information. Luckily, we deal with quantities upto tensors of order 2. The scalar and vector are also called tensor of order 0 and 1, respectively.

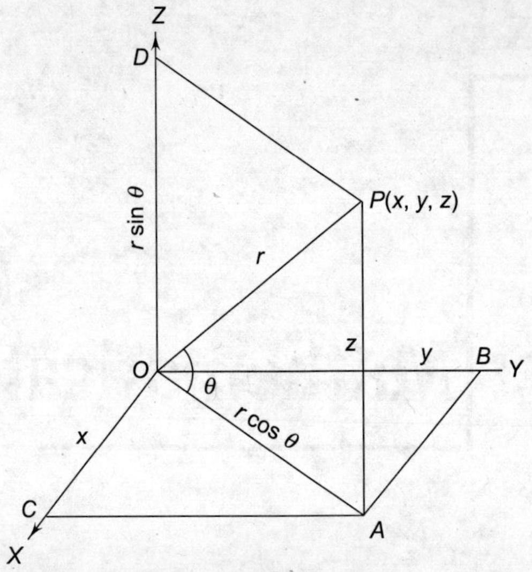


**Fig. A1.1** Vectors in Cartesian coordinates

The first thing we learn in vectors is that it can be resolved along the mutually perpendicular directions. There can be $n$ number of such directions in hyperspace. But we work out the equations only in three-dimensional space, which require a fixed reference from where the distances are to be measured. Normally these references are assumed to be fixed with the ground and are treated as inertial reference frames.[1] The frame has three axes, which meet each other at its origin, the point from which the distance of a point in space is measured. These axes may be linear or curvilinear. We shall discuss about the Cartesian coordinates that are shown in Fig. A1.1. In this figure, $O$ is the origin, $OX$, $OY$ and $OZ$ are the three axes in Cartesian system and $P$ is a point in space. The distance $OP$ is called position vector of the point, represented as $r$[2], inclined at an angle $\theta$ with the horizontal plane made by the axes $OX$ and $OY$. The coordinates $(r, \theta, z)$ are called cylindrical polar coordinates.

$OP$ has projection $OA$ on the horizontal plane and $OD$ along $z$-direction. $OA$ further has components $OC$ and $OB$ along $OX$ and $OY$, respectively. Thus the vector $OP$ has three components of lengths $x$, $y$ and $z$ and is represented by the coordinates $(x, y, z)$. Unit vectors are used to represent position vector, in which $\vec{r} = ix + jy + kz$, where $i$, $j$ and $k$ are the unit vectors along the three axes. A unit vector is a vector along its line of action having unit magnitude.

[1]An inertial frame is the one, which is either stationary or moving with the constant velocity. When the frame itself moves with velocity as function of time, it has its own acceleration and is called non-inertial frame.

[2] Different notations are in use to represent a vector. In this book we are using notations, the bold and an arrow over the vector like $\vec{r}$.

## A1.1 Vector Algebra

Two Vectors are added following the law of parallelogram. Draw the lines parallel to the vector with length equal to the size of the vector, to some scale, and then complete the parallelogram. The length of the diagonal gives the magnitude and the direction of the resultant of the two vectors added. The addition process is cumulative and associative. This means **A** + **B** = **B** + **A** and **A** + (**B** + **C**) = (**A** + **B**) + **C**

Let **A** and **B** are the two vectors $(A_x, A_y, A_z)$ and $(B_x, B_y, B_z)$, $\theta$ is the angle between the two vectors, and $\phi$ be a scalar. We may write the following relations from vector algebra:

$$\mathbf{A}\,.\,\mathbf{B} = A_x B_x + A_y B_y + A_z B_z = \mathbf{B}.\,\mathbf{A} = AB\cos\theta \tag{A.1}$$

$$\mathbf{A}^2 = \mathbf{A}.\,\mathbf{A} = A_x^2 + A_y^2 + A_z^2 \tag{A.2}$$

$$\vec{A}\times\vec{B} = \begin{vmatrix} i & j & k \\ A_x & A_y & A_z \\ B_x & B_y & B_z \end{vmatrix} = i\left(A_y B_z - A_z B_y\right) + j\left(A_z B_x - A_x B_z\right) + k\left(A_x B_y - A_y B_x\right) \tag{A.3}$$

$$= AB\sin\theta = -\,\vec{B}\times\vec{A}$$

We note that the dot product of the two vectors give a scalar quantity but the cross product yields another vector whose direction is determined by the right hand screw rule while rotating vector $A$ towards $B$. That is why, $\mathbf{A}\times\mathbf{B} \neq \mathbf{B}\times\mathbf{A}$ because the direction of the two resultant vectors will be 180° opposite to each other. Further, $\mathbf{A}\times\mathbf{B} = \mathbf{0}$ (null vector), if the two vectors are parallel to each other. In fluid mechanics, a cross product $\mathbf{r}\times\mathbf{F}$ is of common use and is called torque of the force **F** acting at a point **r**. If any of the vector is multiplied by a scalar $\phi$ then the magnitude of the vector is increased by $\phi$ times without changing the line of action and direction of the vector.

The dot product is both cumulative and associative but the cross product is not cumulative as changing the position of the vectors in the product changes the sign of the resultant vector. The dot and cross products of the unit vectors are often used. Therefore we would look into the results of such operation.

$$i.i = j.j = k.k = 1, \qquad i.j = j.k = k.i = 0 \tag{A.4}$$

$$i\times i = j\times j = k\times k = 0, \qquad i\times j = k,\ j\times k = i, \text{ and } k\times i = j \tag{A.5}$$

$$j\times i = -k,\ k\times j = -i \quad \text{and} \quad i\times k = -j$$

We have triple products too. **A** . (**B** × **C**) is of importance. **B** × **C** gives a vector perpendicular to both **B** and **C** and pointing in the direction of right hand screw rotated from **B** to **C**. This product represents the area on the plane occupied by the parallelogram formed by these two vectors. When we take the cross product with other vector **A** then this gives the volume of the parallelepiped formed by the three vectors. There exists the following relations in these vectors:

$$\mathbf{A}\,.\,(\mathbf{B}\times\mathbf{C}) = \mathbf{B}\,.\,(\mathbf{C}\times\mathbf{A}) = (\mathbf{A}\times\mathbf{B})\,.\,\mathbf{C} = \begin{vmatrix} A_x & A_y & A_z \\ B_x & B_y & B_z \\ C_x & C_y & C_z \end{vmatrix} \tag{A.6}$$

We note that the vectors **A**, **B**, and **C** are taking their position in the cyclic order in the cross product. This is important as if we change this order then a negative sign appears due to non-associative nature of the vector product.

## A1.2 Vector Calculus

The vector differential operator *del*, denoted as $\nabla$, is defined as $\nabla = i\partial/\partial x + j\partial/\partial y + k\partial/\partial z$. It operates on scalars, vectors, etc, like any other differential operator and also form a dot or cross product with other vectors. When it operates on a scalar $\phi$, it results in $\nabla\phi$ which is known as gradient. This is a vector that acts along the outward normal drawn at the $\phi$ = constant surface. Let us consider two constant $\phi$ surfaces and determine the change $d\phi$ in the direction $d\vec{r}$, which is different from the direction of the normal $\vec{n}$ on the surface $\phi = C$.

$$d\phi = \frac{\partial\phi}{\partial n}\vec{n}.d\vec{r} = \frac{\partial\phi}{\partial x}dx + \frac{\partial\phi}{\partial y}dy + \frac{\partial\phi}{\partial z}dz = (\nabla\phi).d\vec{r} \tag{A.7}$$

From this equation, we conclude that $\Delta\phi$ gives the maximum change in $\phi$ along the normal direction. The other results are summarised below:

$$\nabla.\vec{B} = \partial B_x/\partial x + \partial B_y/\partial y + \partial B_z/\partial z \tag{A.8}$$

$$(\vec{A}.\nabla) = A_x\partial/\partial x + A_y\partial/\partial y + A_z\partial/\partial z \tag{A.9}$$

$$(\vec{A}.\nabla)\phi = A_x\partial\phi/\partial x + A_y\partial\phi/\partial y + A_z\partial\phi/\partial z \tag{A.10}$$

$$(\vec{A}.\nabla)\vec{B} = A_x\partial\vec{B}/\partial x + A_y\partial\vec{B}/\partial y + A_z\partial\vec{B}/\partial z \tag{A.11}$$

and

$$\nabla.\nabla \equiv \nabla^2 = \frac{\partial^2}{\partial x^2} + \frac{\partial^2}{\partial y^2} + \frac{\partial^2}{\partial z^2} \tag{A.12}$$

We observe that if $\nabla$ is used in the position of **A** in Eq. A.1 then the dot product gives a scalar quantity. That is, it acts like any other vector. However, if it is used in position **B** then it gives an another operator, which is normally written in parenthesis as (**A**.$\nabla$) to avoid any confusion when operating on any scalar or a vector quantity. Moreover, the dot product of $\nabla$ with itself is another operator $\nabla^2$ called the *Laplacian operator*. This operator is of much use in study of flows, vibration of strings, propagation of sound, light and electromagnetic wave through a medium, and many other fields.

The vector product of $\nabla$ with a vector **A** is called the *curl* of **A**. Curl has importance in the continuously deformable media as it symbolised with the rotation of the vector field **A** about the mass axis of the particle and is given the name Vorticity of **A**. We may write the following relations:

$$\nabla\times\vec{B} = \begin{vmatrix} i & j & k \\ \partial/\partial x & \partial/\partial y & \partial/\partial z \\ B_x & B_y & B_z \end{vmatrix}$$

$$= i\left(\partial B_z/\partial y - \partial B_y/\partial_z\right) + j\left(\partial B_x/\partial_z - \partial B_z/\partial_x\right) + k\left(\partial B_y/\partial_x - \partial B_x/\partial_y\right)$$

or,

$$\vec{\omega} = i\omega_x + j\omega_y + k\omega_z \tag{A.13}$$

$\vec{\omega}$ is the curl of vector **B**, normally called the Vorticity vector. We may verify from Eq. A.13 that $\nabla \times (\nabla \phi)$ and $\nabla .(\nabla \times \vec{A})$ result in null vector. We may also need following vector operations:

$$\nabla(\phi\psi) = \phi\nabla\psi + \psi\nabla\phi \tag{A.14}$$

$$\nabla.(\phi\vec{A}) = (\nabla\phi).\vec{A} + \phi(\nabla.\vec{A}) \tag{A.15}$$

$$\nabla \times (\phi\vec{A}) = \phi(\nabla \times \vec{A}) + (\nabla\phi) \times \vec{A} \tag{A.16}$$

$$\nabla.(\vec{A} \times \vec{B}) = \vec{B}.(\nabla \times \vec{A}) - \vec{A}.(\nabla \times \vec{B}) \tag{A.17}$$

$$\nabla \times (\vec{A} \times \vec{B}) = \vec{A}(\nabla.\vec{B}) + (\vec{B}.\nabla)\vec{A} - \vec{B}(\nabla.\vec{A}) - (\vec{A}.\nabla)\vec{B} \tag{A.18}$$

$$\nabla(\vec{A}.\vec{B}) = (\vec{B}.\nabla)\vec{A} + (\vec{A}.\nabla)\vec{B} + \vec{B} \times (\nabla \times \vec{A}) + \vec{A} \times (\nabla \times \vec{B}) \tag{A.19}$$

$$(\vec{A}.\nabla)\vec{A} = \nabla(A^2/2) - \vec{A} \times (\nabla \times \vec{A}) \tag{A.20}$$

$$\nabla \times (\nabla \times \vec{A}) = \nabla(\nabla.\vec{A}) - \nabla^2\vec{A} \tag{A.21}$$

Note that $(\vec{A}.\nabla)\phi$ is not the same as $\vec{A}.\nabla\phi$; in the former $\vec{A}.\nabla$ is the operator on $\phi$ whereas in the second it is dot product of two vectors **A** and $\nabla\phi$. In vectors, one has to take such care and this shows the importance of the parentheses ( ). It is difficult to remember these vector identities. A thumb rule is that if the quantity on the left is vector then every term on the right has to be a vector.

Second order differential functions.

$$\nabla.\nabla\phi = \nabla^2\phi, (\nabla.\nabla)\vec{A} = \nabla^2\vec{A} \tag{A.22}$$

$$\nabla \times \nabla\phi = 0, \nabla.(\nabla \times \vec{A}) = 0 \tag{A.23}$$

$$\nabla \times (\nabla \times \vec{A}) = \nabla(\nabla.\vec{A}) - \nabla^2\vec{A} \tag{A.24}$$

## A1.3 Vector Integrals

Integral approach is also of equal importance in the analysis wherein we come across with the line, surface and the volume integrals. As the name indicates these are the integration processes respectively along a prescribed line in the flow region, over the control surface and over the volume under consideration. The following theorems play a vital role in changing one integral form to another.

Gauss theorem: $$\iiint_{CV} \nabla\phi dV = \iint_{CS} (\phi.\hat{n})dS \tag{A.25}$$

Divergence theorem: $$\iiint_{CV} \nabla.\vec{A}dV = \iint_{CS} \vec{A}.\hat{n}dS \tag{A.26}$$

Stokes' theorem: $$\iint_{CS} \hat{n}.(\nabla \times \vec{A})dS = \int_{C} \vec{A}.d\vec{c} \tag{A.27}$$

Here $\hat{n}$ is the unit vector along the normal to the control surface, $dS$ is the surface area of the element, $d\vec{c}$ is the vector line element of the closed curve $C$ drawn in the flow region and rest of the symbols have usual meaning. $(\phi.\hat{n})$ is the component of $\phi$ along the normal $\hat{n}$ to the surface element $dS$. In many situations the curve is not closed. In that case, the line integration of $\nabla\phi$, the gradient of $\phi$, from the point 1 to the point 2 along the curve $C$ is given as

$$\int_1^2 (\nabla\phi).d\vec{c} = \int_1^2 \left(\frac{\partial\phi}{\partial x}\right) dx + \left(\frac{\partial\phi}{\partial y}\right) dy + \left(\frac{\partial\phi}{\partial z}\right) dz = \int_1^2 d\phi = \phi_2 - \phi_1 \quad \text{(A.28)}$$

This integral relation will be used in several cases.

## A2 Orthogonal Coordinate Systems

In the above discussions, we restricted ourselves to the Cartesian coordinates, which is one of the many orthogonal coordinate systems. In fact, the vector approach has the advantage that it is independent of the coordinate system used in the analysis; the difference comes because of the different ways of expressing the gradient, divergence and curl in different coordinate systems. Method of stretching factors $h_i$ is developed to relate the different curvilinear systems to the Cartesian system by expressing length $ds$ of a line in different system of coordinates, which should be same irrespective of the coordinates.

$$ds^2 = dx^2 + dy^2 + dz^2 = (h_1\, dx_1)^2 + (h_2\, dx_2)^2 + (h_3\, dx_3)^2 \quad \text{(A.29)}$$

We observe that $h$ is 1 for all the three directions in Cartesian coordinates whereas for other coordinates it is to be obtained by determining the length of a line in that system. That is, in general $h$'s will be function of the coordinate system. The components of gradient, divergence and curl are given as

$$\text{Gradient:} \quad \frac{\partial\phi}{h_1\,\partial x_1}, \frac{\partial\phi}{h_2\,\partial x_2} \text{ and } \frac{\partial\phi}{h_3\,\partial x_3} \quad \text{(A.30)}$$

$$\text{Divergence:} \quad \text{Div } \vec{A} = \frac{1}{h_1 h_2 h_3}\left[\frac{\partial}{\partial x_1}(h_2 h_3 A_1) + \frac{\partial}{\partial x_2}(h_3 h_1 A_2) + \frac{\partial}{\partial x_3}(h_1 h_2 A_3)\right] \quad \text{(A.31)}$$

$$\text{Componenet of curl } \vec{A} \text{ along } \hat{k} = \frac{1}{h_2 h_3}\left[\frac{\partial}{\partial x_2}(h_3 A_3) - \frac{\partial}{\partial x_3}(h_2 A_2)\right], \text{ etc} \quad \text{(A.32)}$$

***Cylindrical coordinates***. In Cartesian coordinates the unit vectors remain constant. However, in the cylindrical coordinates, the radial coordinate is measured along the line joining the point with the origin and hence its direction changes with $\theta$. Therefore the unit vectors defined along the $r$ and $\theta$ directions change with $\theta$, though the magnitude remains unchanged. For this reason, their derivative with respect to $\theta$ will account for variation with the direction; this variation is given as follows:

$$\frac{\partial\hat{i}_r}{\partial\theta} = \hat{i}_\theta, \frac{\partial\hat{i}_\theta}{\partial\theta} = -\hat{i}_r \quad \text{(A.33)}$$

$$\rightarrow \quad \frac{\partial(A_\theta \hat{i}_\theta)}{\partial \theta} = \hat{i}_\theta \frac{\partial A_\theta}{\partial \theta} - A_\theta \hat{i}_r \text{ and } \frac{\partial(A_r \hat{i}_r)}{\partial \theta} = \hat{i}_r \frac{\partial A_r}{\partial r} + A_r \hat{i}_\theta \tag{A.34}$$

It may be noted that for the Cartesian coordinates, the cross derivatives of the unit vector will be zero; that is, $\partial i/\partial y = \partial i/\partial z = 0$ and so on.

Further, from Fig. 2.1.1 (b),

$$x = r\cos\theta,\ y = r\sin\theta,\ z = z \text{ and } ds^2 = dr^2 + (rd\theta)^2 + dz^2$$

Therefore, $h_1 = 1$, $h_2 = r$ and $h_3 = 1$ for the cylindrical coordinates. We may now write the following relations.

$$\vec{A} = A_r\hat{i}_r + A_\theta\hat{i}_\theta + A_z\hat{i}_z \text{ and } \vec{B} = B_r\hat{i}_r + B_\theta\hat{i}_\theta + B_z\hat{i}_z \tag{A.35}$$

$$\vec{A}.\vec{B} = A_r B_r + A_\theta B_\theta + A_z B_z \tag{A.36}$$

$$\vec{A}\times\vec{B} = \begin{vmatrix} \hat{i}_r & \hat{i}_\theta & \hat{i}_z \\ A_r & A_\theta & A_z \\ B_r & B_\theta & B_z \end{vmatrix} \tag{A.37}$$

$$\nabla = \hat{i}_r \frac{\partial}{\partial r} + \hat{i}_\theta \frac{1}{r}\frac{\partial}{\partial \theta} + \hat{i}_z \frac{\partial}{\partial z} \tag{A.38}$$

$$\nabla.\vec{A} = \frac{1}{r}\frac{\partial(rA_r)}{\partial r} + \frac{1}{r}\frac{\partial A_\theta}{\partial \theta} + \frac{\partial A_z}{\partial z} \tag{A.39}$$

$$\vec{A}.\nabla = A_r \frac{\partial}{\partial r} + \frac{A_\theta}{r}\frac{\partial}{\partial \theta} + A_z \frac{\partial}{\partial z} \tag{A.40}$$

$$\nabla\times\vec{B} = \frac{1}{r}\begin{vmatrix} \hat{i}_r & r\hat{i}_\theta & \hat{i}_z \\ \frac{\partial}{\partial r} & \frac{\partial}{\partial \theta} & \frac{\partial}{\partial z} \\ B_r & rB_\theta & B_z \end{vmatrix} = \hat{i}_r\left(\frac{1}{r}\frac{\partial B_z}{\partial \theta} - \frac{\partial B_\theta}{\partial z}\right) + \hat{i}_\theta\left(\frac{\partial B_r}{\partial z} - \frac{\partial B_z}{\partial r}\right) + \hat{i}_r\left(\frac{1}{r}\frac{\partial (rB_\theta)}{\partial r} - \frac{1}{r}\frac{\partial B_r}{\partial \theta}\right) \tag{A.41}$$

$$\nabla^2\phi = \frac{1}{r}\frac{\partial}{\partial r}(r\frac{\partial \phi}{\partial r}) + \frac{1}{r^2}\frac{\partial^2\phi}{\partial \theta^2} + \frac{\partial^2\phi}{\partial z^2} \tag{A.42}$$

***Spherical coordinates.*** The spherical coordinate system is shown in Fig. 2.1.1, in which $r$ is measured along the radius, angle $\theta$ is measured from the vertical direction that is the $z$-direction and angle $\phi$ from some reference direction, designated as $x$-direction corresponding to the Cartesian coordinates. This system of coordinates is also refereed to as $(r, \theta, \phi)$ system. The axis in the plane of the $x$-axis and perpendicular to the $x$-axis is the $y$-axis. Relation between the Cartesian and spherical coordinates is shown in the figure. We are using $\phi$ as a scalar function of coordinates, hence in the following relations, we shall use $\lambda$ in place of $\phi$. In spherical coordinate system,

$$x = r\sin\theta\cos\lambda,\ x = r\sin\theta\sin\lambda,\ z = \cos\theta \text{ and } ds^2 = dr^2 + (r\,d\theta)^2 + (r\sin\theta\,d\lambda^2)$$

Therefore, $h_1 = 1$, $h_2 = r$ and $h_3 = r\sin\theta$

The above relations take the following form in spherical coordinates in which there is one more curvilinear coordinate.

$$\nabla = \hat{i}_r \frac{\partial}{\partial r} + \hat{i}_\theta \frac{1}{r}\frac{\partial}{\partial \theta} + \hat{i}_\lambda \frac{1}{r \sin\theta}\frac{\partial}{\partial \lambda} \tag{A.43}$$

$$\nabla . \vec{A} = \frac{1}{r^2}\frac{\partial\left(r^2 A_r\right)}{\partial r} + \frac{1}{r \sin\theta}\frac{\partial(A_\theta \sin\theta)}{\partial\theta} + \frac{1}{r\sin\theta}\frac{\partial A_z}{\partial \lambda} \tag{A.44}$$

$$\vec{A}.\nabla = A_r \frac{\partial}{\partial r} + \frac{A_\theta}{r}\frac{\partial}{\partial\theta} + \frac{A_z}{r\sin\theta}\frac{\partial}{\partial\lambda} \tag{A.45}$$

$$\nabla^2\phi = \frac{1}{r^2}\frac{\partial}{\partial r}\left(r^2\frac{\partial\phi}{\partial\rho}\right) + \frac{1}{r^2\sin\theta}\frac{\partial}{\partial\theta}\left(\sin\theta\frac{\partial\phi}{\partial\theta}\right) + \frac{1}{r^2\sin\theta}\frac{\partial^2\phi}{\partial\lambda^2} \tag{A.46}$$

***Body-fitted coordinates.*** In this system, one of the coordinates is along the normal to the surface and the other two are fitted to the surface of the body. To know why we use such system, let us consider a few cases. We studied in Chapter 5 that the boundary layer thickness varies as $\sqrt{x}$, where $x$ is the distance of the section from the leading edge. Thus the $y$-coordinate of the boundary layer is a function of $x$. However, we choose the variable $\eta = y/\delta$ instead for the boundary layer thickness, since then it is same at all the sections.

Importance of this system of coordinates is appreciated very well when we want to discretise the differential equations in terms of the value of the function at different grid points. We face the problem to carry out this operation when the grid point is close to the boundary, as then the grid points around the point are not equidistant. We witnessed this problem in Chapter 3 while discussing plotting of the flow net. This problem can be handled if the grid is parallel to the surface all along it. In flow problems the constant $\psi$ and constant $\phi$ lines, which are orthogonal to each other, provide the best coordinate system. Fortunately, the streamline near the surface follows the surface. In a majority of cases, the $\psi$ and $\phi$ coordinate system provides a right kind of coordinate system. Moreover, the conformal mapping that we studied in Chapter 3 permits to change such coordinate system in a rectangular system using proper transformation function. Therefore, the body-fitted coordinate system is used frequently in analysing flow problems using computational methods. We will not discuss this in details for which you may consult Rao, JS.

## A3 Tensor Notation

A tensor of order 2 is represented as $A_{ij}$, where $j$ and $i$ are the independent indices having values 1, 2, 3. Here independence means $j$ and $i$ can take any value independent of each other. Thus $A_{ij}$ will have nine components, which can conveniently be shown in the matrix form as follows:

$$\begin{vmatrix} A_{11} & A_{12} & A_{13} \\ A_{21} & A_{22} & A_{23} \\ A_{31} & A_{32} & A_{33} \end{vmatrix} \tag{A1.5.1}$$

In the first row, $i = 1$ and then $j$ is taken in succession 1, 2 and 3 to give the elements of the matrix. In other words $j$ changes along the column and $i$ along the row. If we fix $j$ equal to $i$ then we have the element $A_{ii}$ of the matrix. However, $A_{ii}$ represents the sum $A_{11} + A_{22} + A_{33}$ in the summation convention that is used in tensor notation; $i$ is called repeating variable. Unless specified, the summation convention applies. Acording to the convention, the repeating variable takes valuse 1 through 3, in three-dimensional space, and such terms are added.

There are two special tensors $\delta_{ij}$, called Kronecker delta, and the alternating unit tensor $\varepsilon_{ijk}$, which are defined as follows:

$$\begin{aligned} \delta_{ij} &= 1 \text{ if } i = j \text{ and } 0 \text{ if } i \neq j \\ \varepsilon_{ijk} &= 1 \text{ if } ijk = 123, 231, \text{ or } 312 \text{ (123 are in cyclic order)} \\ &= -1 \text{ if } ijk = 321, 213 \text{ or } 132 \text{ (123 are in anti-cyclic order)} \\ &= 0 \text{ if } i = j, j = k, \text{ or } k = i \text{ (any two indices are same)} \end{aligned} \quad \text{(A1.5.2)}$$

We have the following important identities relating these special tensors with the unit vetors:

$$\begin{aligned} \boldsymbol{e}_i \cdot \boldsymbol{e}_j &= \delta_{ij} \\ \boldsymbol{e}_i \times \boldsymbol{e}_j &= \varepsilon_{ijk}\, \boldsymbol{e}_k \\ \varepsilon_{ijk}\, \varepsilon_{hjk} &= 2\, \delta_{ih} \\ \varepsilon_{ijk}\, \varepsilon_{mnk} &= \delta_{im}\, \delta_{nj} - \delta_{in}\, \delta_{jm} \end{aligned} \quad \text{(A1.5.3)}$$

Let us verify the validity of the second and third equations of Eq. A1.5.3. Expanding the RHS of second equation, we get

$$\varepsilon_{ijk}\, \boldsymbol{e}_k = \varepsilon_{ij1}\, \boldsymbol{e}_1 + \varepsilon_{ij2}\, \boldsymbol{e}_2 + \varepsilon_{ij3}\, \boldsymbol{e}_3 \quad \text{(A1.5.4)}$$

We now choose $i$ and $j$ from 1 to 3 in this equation. Alternating tensor in the terms for which $i = j$ will be zero, –1 where the indices are in anti-cyclic order and +1 when these are in cyclic order. This leaves, for these combinations, the terms on the RHS of A1.5.4, respectively as $\boldsymbol{e}_3 - \boldsymbol{e}_2$, $\boldsymbol{e}_1 - \boldsymbol{e}_3$, $\boldsymbol{e}_2 - \boldsymbol{e}_1$, which is $\boldsymbol{e}_i \times \boldsymbol{e}_j$. Now we consider the third equation of Eq. A1.5.3. The LHS

$$\begin{aligned} \varepsilon_{ijk}\varepsilon_{hjk} = {} & \cancelto{0}{\varepsilon_{i11}\varepsilon_{h11}} + \varepsilon_{i12}\,\varepsilon_{h12} + \varepsilon_{i13}\,\varepsilon_{h13} + \varepsilon_{i21}\,\varepsilon_{h21} + \cancelto{0}{\varepsilon_{i22}\varepsilon_{h22}} \\ & + \varepsilon_{i23}\,\varepsilon_{h23} + \varepsilon_{i31}\,\varepsilon_{h31} + \varepsilon_{i32}\,\varepsilon_{h32} + \cancelto{0}{\varepsilon_{i33}\varepsilon_{h33}} \\ = {} & \varepsilon_{i12}\,\varepsilon_{h12} + \varepsilon_{i13}\,\varepsilon_{h13} + \varepsilon_{i21}\,\varepsilon_{h21} + \varepsilon_{i23}\,\varepsilon_{h23} + \varepsilon_{i31}\,\varepsilon_{h31} + \varepsilon_{i32}\,\varepsilon_{h32} \end{aligned} \quad \text{(A1.5.4)}$$

If in Eq. A1.5.4 $i \neq h$ one of the two terms of each pair is zero, making the RHS zero; if $i = h$, two terms are equal to unity, which add to make the RHS equal to 2, thus the RHS becomes $2\delta_{ih}$. This not only proves the identity of the two equations we considered but also demonstrates how to use these special tensors and their power in the field of tensors.

We use notation $w_{i,j}$ for the derivative $\partial w_i / \partial x_j$ in tensors. We note that in the notation $w_{i,j}$ a subscript after, represents differentiation with respect to that subscript. For example, $\dfrac{\partial w_i}{\partial x_i} = \dfrac{\partial w_1}{\partial x_1} + \dfrac{\partial w_2}{\partial x_2} + \dfrac{\partial w_3}{\partial x_3} = \nabla . \vec{w} = w_{i,j}$. Thus both the summation and derivative conventions have been used while expanding $\partial w_i / \partial x_i$, since $i$ is repeating variable.

Equation A1.5.1 is the matrix notation of tensor $A$. We come across such notations in stress, strain, etc., which we discussed in chapter 1. These have nine components as a stress represents the change in a vector (velocity) along any of three directions; there being three velocity components and three directions and consequently have 9 components. One other application of importance and need to be discussed here is direction cosines of the new axes of reference with the old axes in the rotated axes of frame. We often face such situation. For example the stress system for a given set of axes $X_1$, $Y_1$, and $Z_1$ is given. We learn in Mechanics that the stresses would be different on different planes. Thus a system of axes $X_2$, $Y_2$, and $Z_2$ may be defined such that $X_2$ is perpendicular to the plane under consideration. Then $Y_2$ and $Z_2$ will be perpendicular to the face $X_2$ and these three planes will be mutually perpendicular to each other. Each of the axes will make angle with the old axes whose cosine is named direction cosine. That is each new axis has three direction cosines and thus there are 9 direction cosines. We use letters $l$, $m$, and $n$ to represent the direction cosines and the nine components can be represented in the matrix form as

$$\begin{vmatrix} l_1 & m_1 & n_1 \\ l_2 & m_2 & n_2 \\ l_3 & m_3 & n_3 \end{vmatrix}$$

In this form, we have used row instead of column to represent a vector quantity.[3] In this form, the form row represents direction cosines of $\mathbf{X}_2$ system of axes with $X_1$, $Y_1$, and $Z_1$ respectively and so on.

Consider for example, a vector having its components $\vec{X} \equiv (X, Y, Z)$ in $X_1$, $Y_1$, and $Z_1$ and $\overrightarrow{X'}$ $(X', Y', Z')$ in $X_2$, $Y_2$, and $Z_2$ systems. We normally represent the components of a vector by a column vector but it has been shown as row vector in the above representation that may be written as $\overrightarrow{X}^T$. Without going into the rigorous details of deriving the transformation relation in the rotated axes of reference, we write the final result.

$$\begin{Bmatrix} X' \\ Y' \\ Z' \end{Bmatrix} = \begin{vmatrix} l_1 & m_1 & n_1 \\ l_2 & m_2 & n_2 \\ l_3 & m_3 & n_3 \end{vmatrix} \begin{Bmatrix} X \\ Y \\ Z \end{Bmatrix} \text{ or, } \overline{X'}^T = D \times \overrightarrow{X'}^T \tag{A1.5.2}$$

This will also apply to a tensor of order 2, used in place of vector. However, the transformation is to be applied twice. So we have the transformation law for tensors as

$$\begin{vmatrix} \sigma'_{11} & \sigma'_{12} & \sigma'_{13} \\ \sigma'_{21} & \sigma'_{22} & \sigma'_{23} \\ \sigma'_{31} & \sigma'_{32} & \sigma'_{33} \end{vmatrix}^T = \begin{vmatrix} l_1 & m_1 & n_1 \\ l_2 & m_2 & n_2 \\ l_3 & m_3 & n_3 \end{vmatrix} \left[ \begin{vmatrix} l_1 & m_1 & n_1 \\ l_2 & m_2 & n_2 \\ l_3 & m_3 & n_3 \end{vmatrix} \begin{vmatrix} \sigma_{11} & \sigma_{12} & \sigma_{13} \\ \sigma_{21} & \sigma_{22} & \sigma_{23} \\ \sigma_{31} & \sigma_{32} & \sigma_{33} \end{vmatrix}^T \right]^T \tag{A1.5.3}$$

We observe the pattern in the transformation while using row notation. Pre-multiply the transpose of the vector by the direction cosine of the new axes. For a second order tensor this process is to be done two times with the transpose of the last result obtained

[3] It is common practice to represent a vector by a column matrix. This is not used here because it is convenient to write it in a row form to save a few lines. Even those following the former practice use transpose of the matrix while writing it to save space. Recall transpose of a matrix changes column into a row and vice versa.

as the new matrix to be operated upon. Eq. 1.5.3 gives component of stresses in the new set of axes.

For orthogonal coordinate system, the direction cosines are not completely independent of each other. They satisfy the relations of the following type

$$l_1^2 + l_2^2 + l_3^2 = 1 \tag{A1.5.4}$$

$$l_1^2 + m_1^2 + n_1^2 = 1 \tag{A1.5.5}$$

$$l_1\, l_2 + m_1\, m_2 + n_1\, n_2 = 0 \tag{A1.5.6}$$

$$l_1\, m_1 + l_2\, m_2 + l_3\, m_3 = 0 \tag{A1.5.7}$$

We may similarly write other relations taking different combinations. These relations could have been written in a very simple form had we used double subscript notation, say $l_{ij}$, for the direction cosines and used the summation convention for repeated variables. This means we had used $l_{11}$, $l_{12}$ and $l_{13}$ in place of $l_1$, $m_1$ and $n_1$, respectively. Eq. A1.5.4 through A1.5.7 take the form (you may verify it).

$$l_{ir}\, l_{jr} = \delta_{ij} \tag{A1.5.8}$$

the law of transformation of vector and tensor may be expressed as follows

$$X'_i = l_{ir}\, x_r,\ A'_{ij} = l_{ir}\, l_{js}\, A_{rs},\ A_{ijk} = l_{ir}\, l_{js}\, l_{kt}\, A_{rst},\ \text{etc.} \tag{A1.5.9}$$

Though we can work using matrices yet we notice that working in the subscripted notation is much easier. In this analysis, the Kronecker delta has special importance. We would therefore write a few relation related to it. These are (verify these)

$$\delta_{rs}\, a_s = \delta_{r1}\, a_1 + \delta_{r2}\, a_2 + \delta_{r3}\, a_1 = a_r,\ \delta_{ir}\, \delta_{rj} = \delta_{ij} \text{ and } x_r\, x_s\, \delta_{rs} = x_r \tag{A1.5.10}$$

We will now demonstrate the power of the tensor notations in deriving Eq. A1.5.8. Let us consider two points in a reference frame at points $\vec{X}$ and $\vec{Y}$ whose position vector in the new frame of axes, obtained by rotating the axes about the origin, represented by a bar over the old coordinates. We know in this process the length of the lines joining these points with the origin would not change. This means, using summation convention, $X_i X_i = X_j X_j$ and $Y_i Y_i = Y_j Y_j$. Now consider transformation of a tensor $A$. According to the transformation law for position vectors, $X_i Y_i = l_{ir} X_r l_{is} Y_s = l_{ir} X_r Y_s$. But $X_i Y_i = X_r Y_r = \delta_{rs} X_r Y_s$, so that we must have $(l_{ir} l_{is} - \delta_{rs}) X_r Y_s = 0$. This equation must hold for any arbitrary vectors $\vec{Y}$ and $\vec{Y}$. Therefore, $l_{ir} l_{is} = \delta_{rs}$, which is Eq. A1.5.8.

Before closing the discussions on the properties of tensor, let us mention of the process of contraction, which reduces the order of a tensor by 2 by summing over a pair of its indices. Consider a third order tensor $A_{ijk}$ and take $j = k = r$. Then the tensor is $A_{irr} = B(i)$, say. Will $B(i)$ be a vector? To answer this, apply the transformation law to the tensor that gives $A_{ijk} = l_{im}\, l_{jn}\, l_{ko}\, A_{mno}$, so that $A_{irr} = l_{im}\, l_{rn}\, l_{ro}\, A_{mno}$. But $l_{rn}\, l_{ro} = \delta_{no}$.

Thus $$A_{irr} = l_{im}\, \delta_{no}\, A_{mno} = l_{im}\, A_{mnr}$$

or, $$B(\text{i}) = l_{im}\, B(m)$$

Thus $B$ (i) is a vector. This process of reducing the order of a given tensor by summing over a pair of its indices is called contraction.

## A4 Integration of Euler Equation

In case gravity is the only body force acting on the control volume and the fluid is incompressible fluid for which $\rho = C \rightarrow \nabla\rho = 0$, the pressure force and the body force term at a point $\vec{r}$ can be combined and defined as new variable say $p^*$, where

$$p^* = p - \rho\, \vec{g}\,.\vec{r} \tag{A1.6.1}$$

This changes the Euler equation 2.4.15 to (written in vector form)

$$\frac{\partial \vec{U}}{\partial t} + (\vec{U}.\nabla)\,\vec{U} = -\frac{1}{\rho}\nabla(p^* + \rho\vec{g}.\vec{r}) + \vec{g} = -\frac{1}{\rho}\nabla p^* \tag{A1.6.2}$$

We may write Euler equation, Eq. 2.4.15, in the following form using Eq. A.20.

$$\frac{\partial \vec{U}}{\partial t} + \nabla\left(\frac{U^2}{2}\right) + \frac{1}{\rho}\nabla p - \vec{g} = \vec{U}\times\left(\nabla\times\vec{U}\right)$$

Let us integrate this equation between the points 1 and 2 along a line $C$ in space, whose element of length is dc. We get

$$\int_1^2 \frac{\partial \vec{U}}{\partial t}.d\vec{C} + \int_1^2 \nabla\left(\frac{U^2}{2}\right).d\vec{C} + \int_1^2\left(\frac{\Delta p}{\rho}\right).d\vec{C} - \int_1^2 \vec{g}.d\vec{C} = \int_1^2 \vec{U}\times(\nabla\times\vec{U}).d\vec{C} \tag{A1.6.3}$$

But $\int_1^2 \nabla\left(\frac{U^2}{2}\right).d\vec{C} = \frac{U_2^2}{2} - \frac{U_1^2}{2}$ and

$$\int_1^2 \vec{g}.d\vec{C} = \int_1^2 \nabla(\vec{g}.\vec{r}).d\vec{C} = \vec{g}.\vec{r}_2 - \vec{g}.\vec{r}_1$$

Other terms is Eq. A.1.6.3 can be integrated if curve $C$ happen to be stream line or the flow is irotational, in both the case the right hand side of the equation results to zero. The pressure integral becomes (along the stream line)

$$\int_1^2 \frac{\Delta p}{\rho}\,.\,d\vec{s} = \frac{1}{\rho}\int_1^2 \Delta p\,.\,d\vec{s} = \frac{1}{\rho}(p_2 - p_1),$$

if density remains constant along the steam line. Inserting these integrands in Eq. A.1.6.3 gives the Bernoulli equation

$$\frac{p_1}{\rho} + \frac{U_1^2}{2} + gz_1 = \frac{p_2}{\rho} + \frac{U_2^2}{2} + gz_2 \tag{A1.6.4}$$

However, if $\rho$ is not constant then it cannot be taken out of the integral sign and then

$$\int_1^2 \frac{\Delta p}{\rho}\,.\,d\vec{s} = \int_1^2 \frac{dp}{\rho}$$

whose integration depends on the relation between $p$ and $\rho$ that we have discussed in Chapter 1.

## A5 Conformal Mapping

Conformal mapping has been a very useful tool in Fluid Mechanics in decomposing a complex flow problem into a simple one so that the velocity and pressure fields may be determined conveniently. This analysis works on a complex plane $z = x + iy$, where $x$ and $y$ are the ordinary $x$ and $y$-axes in the Cartesian coordinate system and $i = \sqrt{-1}$. Thus any point on a plane can be represented by $z = r\,e^{i\theta}$, where $r$ is the polar distance of the point $P$ from the origin and $\theta$ is the angle made by the line $OP$

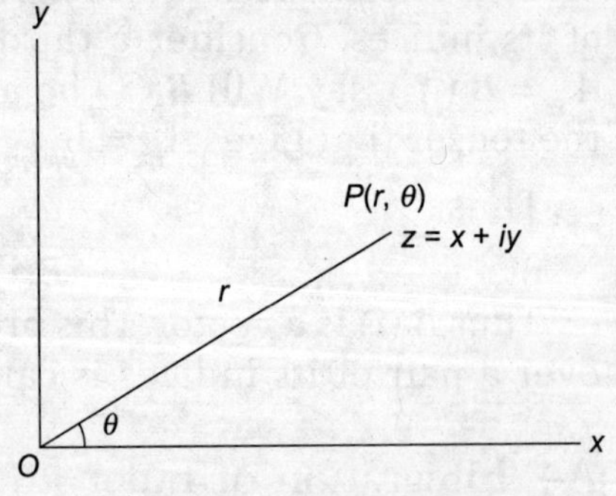


**Fig. A1.7.1** A point on a complex plane

from the + ve $x$-axis, see Fig. A1.7.1. $x$ and $y$ are called the real and the imaginary parts of the complex number $z$. Image of the point in the $x$-axis will lie at $z = re^{-i\theta} = \bar{z}$ and is called complex conjugate of $z$.

Any curve on $x$-$y$ plane is therefore function of $z$ and can be differentiated or integrated like the real numbers. However, as mentioned, in chapter 3, that for differentiation to be unique, the Cauchy–Riemann conditions should be satisfied. Only then the function is analytical. Thus the functions we are going to deal with are analytical and single-valued having unique value at a point. This is so because we have learnt that a single streamline can pass through a given point in the flow field. Thus the stream function and hence the velocity potential are single-valued continuous functions. The value of the integral along a closed curve in space too depends on the presence of singularity in the region confined by the curve. These singularities were recognised as pole and zeros. We shall now discuss conformal mapping in more details.

Let us consider an analytic complex function $w = z^2$. The complex function $w$ is taken as $\phi + i\psi$, where $\phi$ and $\psi$ are the velocity potential and stream function, respectively, as defined in chapter 2. This function transforms a flow in $x$-$y$ plane to $w$ or $\psi - \phi$ plane. To see how the flow would appear in $z$ and $w$ planes, we can separate out the function into its real and imaginary part giving

$$\phi + i\psi = (x + iy)^2 = (x^2 - wy^2) + i\,(2xy) \rightarrow \phi = (x^2 - y^2) \text{ and } \psi = 2xy \qquad \text{(A1.7.1)}$$

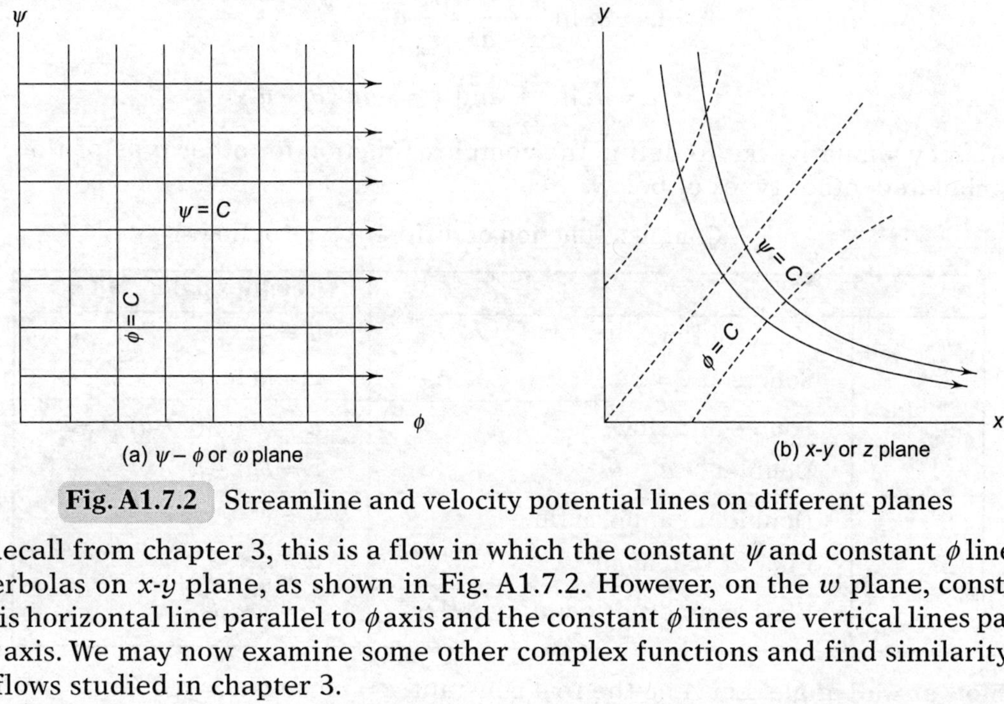


**Fig. A1.7.2** Streamline and velocity potential lines on different planes

Recall from chapter 3, this is a flow in which the constant $\psi$ and constant $\phi$ lines are hyperbolas on $x$-$y$ plane, as shown in Fig. A1.7.2. However, on the $w$ plane, constant $\psi$ line is horizontal line parallel to $\phi$ axis and the constant $\phi$ lines are vertical lines parallel to $\psi$ axis. We may now examine some other complex functions and find similarity with the flows studied in chapter 3.

**Uniform flow**. Such a flow is represented by the function $w = \phi + i\psi = Az$, where $A$ is a complex constant. Let $A = a + ib$, then substituting it and separating out the right hand side into real and imaginary parts, we get

$$\phi + i\psi = (a + ib)\,(x + iy) \rightarrow \phi = ax - by \text{ and } \psi = ay + bx$$

Also $\frac{dw}{dz} = u - iv = a + ib \rightarrow u = a$ and $v = -b$ (A1.7.2)

Therefore $w = Az$, where $A$ is a complex constant, represents flow in $x$-$y$ plane having constant $u$ and $v$ velocity components. That is, it represents a uniform flow moving from left to right.

**Source at $z = a$.** A line source is the one in which the radial velocity of flow decreases with $r$ and there is no tangential velocity. Let the source is place at a point $z = a$ and is represented by the function $w = m \ln (z - a)$, where $m$ is a real constant. Taking $z - a = r' e^{i\theta'}$ and separating into real and imaginary parts, we get

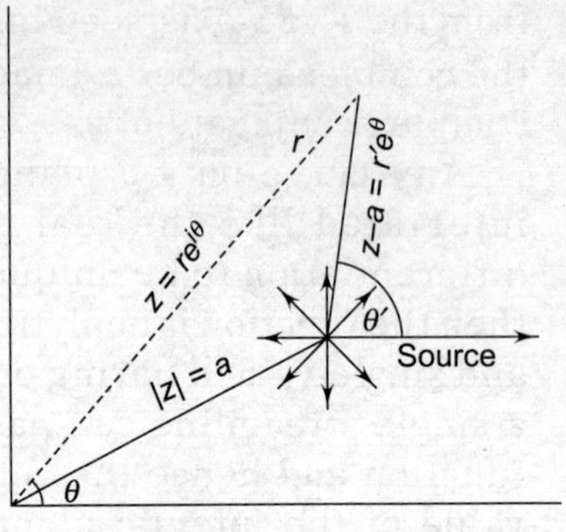


**Fig. A1.7.3** Source flow

$$\phi = m \ln r', \ \psi = m\theta', \ dw/dz = m/(z - a)$$

and $|dw/dz| = |m/(z - a)| = m/r' = q/(2\pi r')$ (A1.7.3)

The streamlines in $x$-$y$ plane are shown in Fig. A1.7.3.

**Source and sink pair.** Let the source and the sink be at $r = \pm a$ from the origin in $x$-$y$ plane. Then

$$w = m \ln \frac{z + a}{z - a} = \frac{r_1}{r_2} e^{i(\theta_1 - \theta_2)}$$

$$\rightarrow \qquad \phi = m \ln \frac{r_1}{r_2} \text{ and } \psi = -m(\theta_1 - \theta_2)$$

We may similarly like to define the complex function for other type of flow, which are tabulated other types of below.

Complex function of different type of flows

| S. No. | Type of flow | Complex potential |
|---|---|---|
| 1. | Uniform flow | $w = Az$ |
| 2. | Source at $z = a$ | $w = m \ln (z - a)$ |
| 3. | Source and sink | $w = m \ln \{(z + a)/(z - a)\}$ |
| 4. | Doublet at $a$ | $w = \mu/(z - a)$ |
| 5. | Doublet in uniform flow | $w = Uz + \mu/z$ |
| 6. | Flow at wall angle | $w = Az^n$ |
| 7. | Flow around stationary cylinder | $w = z + 1/z$ |

**Flow at wall angle.** Let $A$ be the real constant, $z = re^{i\theta}$ and $w = r_1 e^{i\theta_1}$

$$\rightarrow \qquad r = (r_1 / A)^{1/n} \text{ and } \theta = \theta_1/n$$

This shows that the distance and the angle in $w$ plane become $1/n$ in the $z$ plane. $n = 1$ gives the uniform flow, as discussed before.

So far, we chose a complex function to see what type of flow does it represent in the physical plane. However, it is required to determine the complex velocity potential too for flows around any complex shape in the physical plane $x$-$y$. Conformal mapping allows transformation of this surface to a circle in another plane $\varsigma$, where $\varsigma = f(z)$, $\varsigma = \xi(x,y) + i\eta)$x,y) is an analytical function. This requires that both $\xi$ and $\eta$ must satisfy the Laplace equation and the Cauchy–Reimann conditions $\xi_x = \eta_y$ and $\xi_y = -\eta_x$. The advantage of the conformal transformation is that the complex potential will remain same in both the planes. We may prove this statement as follows.

$$\phi_x = \phi_\xi \xi_x + \phi_\eta \eta_x$$

$$\phi_{xx} = \phi_{\xi\xi}\xi_x^2 + \phi_\xi \xi_{xx} + 2\phi_{\xi\eta}\eta_x\xi_x + \phi_\eta \eta_{xx} + \phi_{\eta\eta}\eta_x^2$$

$$\nabla^2\phi = (\eta_x^2 + \eta_y^2)(\phi_{\eta\eta} + \phi_{\xi\xi})$$

after determining the second derivative w.r.t. $y$, using Cauchy-Riemann conditions and simplifying the expression for $\nabla^2\phi$. Nevertheless, $\nabla^2\phi$ is zero. This means that $\phi_{\eta\eta} + \phi_{\xi\xi} = 0$ in the transformed plane too. Similarly, we may start with $\psi$. That is, if $w(\varsigma)$ is the complex velocity potential in $\varsigma$ plane then the complex potential in the physical plane is given by $w(f(z))$.

The complex velocity field w in the $\varsigma$ plane is $w(\varsigma)$. The relation

$$w(\varsigma) = w(z)\frac{dz}{d\varsigma} = w(z)g' \quad \text{(A1.7.4)}$$

then gives its corresponding value in the physical plane, where g is inverse function of $f$, that is, $z = g(\varsigma)$. We conclude that an analytical function $f(z)$ can help in transforming a contour on the physical plane to some surface on another plane on which the complex velocity potential is known. This gives immediately the complex velocity field on the physical plane just replacing the variable by $f(z)$. We have learnt above that the transformation function $z = \varsigma + c^2/\varsigma$ transforms an ellipse of semi-major axis $a + c^2/a$ and semi-minor axis $a - c^2/a$ into a cirlce of radius $a$ from $z$ plane to $\varsigma$ plane, where $a$ and $c$ are constants and the analyst can choose. If we choose $a = c$ then ellipse reduces to a line of length 4a. In other words, a plate of lenght 4a generates a circle of radius $a$ in the transformed plane. Similarly, a function $\varsigma = ze^{-i}$ effects anti-clockwise rotation of the flow field through an angle $\alpha$.

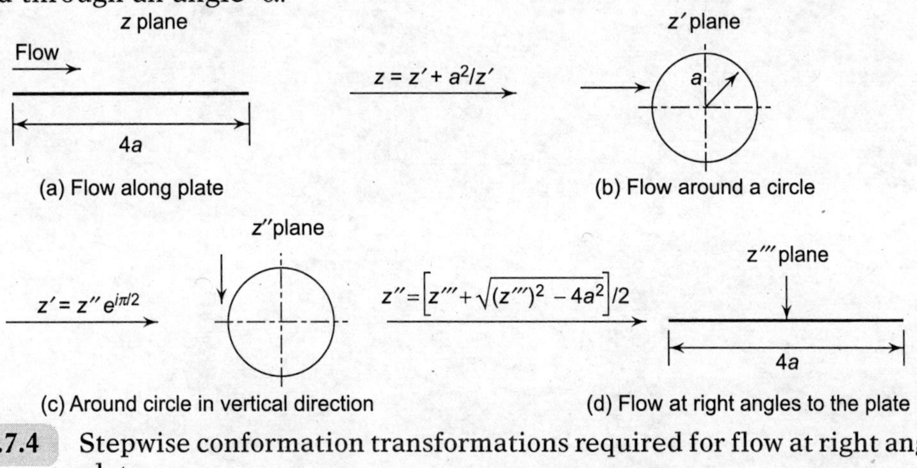


(a) Flow along plate (b) Flow around a circle

(c) Around circle in vertical direction (d) Flow at right angles to the plate

**Fig. A1.7.4** Stepwise conformation transformations required for flow at right angles to plate.

Many a time, a number of transformations need to be implemented one after another to achieve the aim. For example, consider the flow striking the horizontal plate along the vertical direction. This requires the transformations, as shown in Fig. A1.7.4. We observe that to generate flow at right angles to the plane plate, we can start from the flow along the plate with three successive transformations that change the flow pattern as shown in the figure. The figure also shows the transformation function that results in different planes.

APPENDIX

# B Properties of Standard Atmosphere

Atmosphere is divided into five regions shown in Fig. 1.3.3. These regions are: troposphere, Stratosphere, Mesosphere, ionosphere, and exosphere. The limits of the various regions shown in figure are approximate and vary with altitude, time and season.

Approximate physical properties of some common fluids at standard atmospheric pressure 760 mm Hg and temperature 0°C are given at table B.1 and B.2.

**Troposphere**. This region extends up to 11 km from the mean sea level in which the temperature decreases with altitude uniformly at the rate of 6.5°C per km. This is the region in which normal weather and clouds occur and contains about 75% by weight of the atmosphere. The composition of troposphere is given in the following table.

| Gas | Composition% | Gas | Composition% |
|---|---|---|---|
| Oxygen | 21.00 | Nitrogen | 78.00 |
| Argon | 0.94 | Hydrogen | 0.01 |
| Carbon dioxide | 0.03 | Others | 0.02 |

**Stratosphere**. The region next to troposphere upto 32 km is the stratosphere in which temperature of atmosphere remains constant at 216.5 K. It contains a little less than

25% by weight of the atmosphere. The two regions are separated by the tropopause, which is at an altitude of 11 km. Composition of air does not differ much from the troposphere.

**Mesosphere**. This region is characterised by the concentration of ozone that absorbs ultra violet rays from Sun to save life on the earth and whose depletion is sending sense of insecurity in human beings. It ranges from 32 to 76 km. The temperature first rises to 350 K, remains at that value for some height and then decreases to 240 K.

**Ionosphere**. The region beyond mesosphere up to 640 km is called ionosphere. This region contains electrically conducting layers, the composition of which changes during day and night. The temperature increases with altitude.

**Exosphere**. Region beyond 640 km is exosphere and density in this region is very low. There is no sharp boundary between ionosphere and exosphere.

**TABLE.** Properties of the international standard atmosphere at sea level

| | | | |
|---|---|---|---|
| Pressure | 1.0133 bar | Temperature | 288.2 K |
| Density | 1.225 kg/m$^3$ | Sound velocity | 340.3 m/s |
| Coefficient of viscosity | $1.789 \times 10^{-5}$ kg/ms | Kinematic viscosity | $1.461 \times 10^{-5}$ m$^2$/s |

Let subscript 0 refer to the property at the sea level. Then variation of the properties with altitude is given by the following relations:

*Troposphere.*

$$T = T_0 - mz, \quad \frac{p}{p_0} = \left(1 - \frac{mz}{T_0}\right)^{g/mR} = \left(\frac{T}{T_0}\right)^{g/mR} \quad \frac{\rho}{\rho_0} = \left(\frac{T}{T_0}\right)^{(g - mR)/mR}$$

$$\frac{\mu}{\mu_0} = \left(\frac{T}{T_0}\right)^{0.75}$$

At *tropopause*: $z = 11$ km, $p = 0.227$ bar, $T = 216.5$ K,
$\rho = 0.365$ kg/m$^3$, and $\mu = 1.42 \times 10^{-5}$ kg/ms.

*Stratosphere.* We may obtain pressure at any height $z$ from the sea level using Eq. 1.3.4. Remember that the stratosphere starts from tropopause at which height the various properties are known. Let subscript $t$ represent this region. Then $z$ in Eq. 1.3.4 is $z - z_t$. Writing these values, we get

$$\frac{p}{p_t} = e^{-g(z - z_t)/(RT_t)}, \quad \frac{\rho}{\rho_t} = \frac{p}{p_t}, \quad \mu = \mu_t$$

**TABLE B-1** Approximate physical properties of some common fluids at standard atmospheric pressure 760 mm Hg and temperature 0°C

| Substance | Density $kg/m^3$ | Dynamic Viscosity kg/ms | Thermal conductivity j/ms deg C |
|---|---|---|---|
| *Gases* | | | |
| Air | 1.293 | $1.71 \times 10^{-5}$ | $2.41 \times 10^{-2}$ |
| Argon | 1.784 | $2.12 \times 10^{-5}$ | $1.62 \times 10^{-2}$ |
| Helium | 0.1785 | $1.86 \times 10^{-5}$ | $1.415 \times 10^{-1}$ |
| Hydrogen | 0.08988 | $8.4 \times 10^{-6}$ | $1.684 \times 10^{-1}$ |
| Nitrogen | 1.251 | $1.66 \times 10^{-5}$ | $2.43 \times 10^{-2}$ |
| Oxygen | 1.429 | $1.92 \times 10^{-5}$ | $2.44 \times 10^{-2}$ |
| *Liquids* | | | |
| Benzene | 879.2** | $0.757 \times 10^{-3}$* | 0.147** |
| Carbon Tetrachloride | 1632.0 | $1.348 \times 10^{-3}$ | 0.112 |
| Mercury | 13595.1 | $1.681 \times 10^{-3}$ | 8.0 |
| Water | 999.84 | $1.787 \times 10^{-3}$ | 0.561 |

*$T = 10°C$ **$T = 20°C$

Data taken from G.W.C. Kays and T.H. Laby, Tables of Physical and Chemical Constants, London: Longmans Green, 1966.

**TABLE B-2** Properties of the standard atmospheric*

| Altitude km | Temperature °C | Pressure mm Hg | Density $kg/m^3$ | Viscosity g/ms | Sound Velocity m/s |
|---|---|---|---|---|---|
| 0 | 15.00 | 760.0 | 1.2250 | 0.01789 | 340.3 |
| 0.5 | 11.75 | 716.0 | 1.1673 | 0.01774 | 338.4 |
| 1.0 | 8.50 | 674.1 | 1.1117 | 0.01758 | 336.4 |
| 1.5 | 5.25 | 634.2 | 1.0581 | 0.01742 | 334.5 |
| 2.0 | 2.00 | 596.3 | 1.0066 | 0.01726 | 332.5 |
| 2.5 | –1.24 | 560.2 | 0.9570 | 0.01710 | 330.6 |
| 3.0 | –4.49 | 529.5 | 0.9093 | 0.01694 | 328.6 |
| 3.5 | –7.74 | 493.4 | 0.8634 | 0.01678 | 326.6 |
| 4.0 | –10.98 | 462.5 | 0.8194 | 0.01661 | 324.6 |
| 4.5 | –14.23 | 433.2 | 0.7770 | 0.01645 | 322.6 |

(*Cont....*)

| Altitude km | Temperature °C | Pressure mm Hg | Density kg/m$^3$ | Viscosity g/ms | Sound Velocity m/s |
|---|---|---|---|---|---|
| 5.0 | –17.47 | 405.4 | 0.7364 | 0.01628 | 320.5 |
| 5.5 | –20.72 | 379.1 | 0.6975 | 0.01612 | 318.5 |
| 6.0 | –23.96 | 354.2 | 0.6601 | 0.01595 | 316.5 |
| 6.5 | –27.21 | 330.6 | 0.6243 | 0.01578 | 314.4 |
| 7.0 | –30.45 | 308.3 | 0.5900 | 0.01561 | 312.3 |
| 7.5 | –33.69 | 287.3 | 0.5572 | 0.01554 | 310.2 |
| 8.0 | –36.94 | 267.4 | 0.5258 | 0.01527 | 308.1 |
| 8.5 | –40.18 | 248.7 | 0.4958 | 0.01510 | 306.0 |
| 9.0 | –43.42 | 231.0 | 0.4671 | 0.01493 | 303.8 |
| 9.5 | –46.66 | 214.4 | 0.4397 | 0.01475 | 301.7 |
| 10.0 | –49.90 | 198.8 | 0.4135 | 0.01458 | 299.5 |
| 11.0 | –56.38 | 170.3 | 0.3648 | 0.01422 | 295.2 |
| 12.0 | –56.50 | 145.5 | 0.3119 | 0.01422 | 295.1 |
| 13.0 | –56.50 | 124.4 | 0.2666 | 0.01422 | 295.1 |
| 14.0 | –56.50 | 106.3 | 0.2279 | 0.01422 | 295.1 |
| 15.0 | –56.50 | 90.85 | 0.1948 | 0.01422 | 295.1 |
| 20.0 | –56.50 | 41.47 | 0.08910 | 0.01422 | 295.1 |
| 25.0 | –51.60 | 19.12 | 0.02516 | 0.01448 | 298.4 |
| 30.0 | –46.64 | 8.978 | 0.01841 | 0.01475 | 301.7 |
| 40.0 | –22.80 | 2.154 | 0.003996 | 0.01601 | 317.2 |
| 50.0 | –2.50 | 0.5984 | 0.001027 | 0.01704 | 329.8 |

APPENDIX

# C Derivation of Von Karman Integral Equation

We shall now get the von Karman integral equations for a steady, two-dimensional and incompressible flow for which the Euler equation in Cartesian coordinates are

$$u\frac{\partial u}{\partial x} + v\frac{\partial u}{\partial y} = -\frac{1}{\rho}\frac{\partial p}{\partial x} + \nu\frac{\partial^2 u}{\partial y^2}$$

We integrate this equation from $y = 0$ to $\delta$, the edge of the boundary layer. This gives

$$\int_0^\delta \left(u\frac{\partial u}{\partial x} + \nu\frac{\partial u}{\partial y}\right)dy = \int_0^\delta \left(-\frac{1}{\rho}\frac{\partial p}{\partial x} + \nu\frac{\partial^2 u}{\partial y^2}\right)dy \qquad \text{(C.1)}$$

$$\text{or, } \int_0^\delta u\frac{\partial u}{\partial x}dy + \int_0^\delta \nu\frac{\partial u}{\partial y}dy = -\int_0^\delta \frac{1}{\rho}\frac{\partial p}{\partial x}dy + \nu\int_0^\delta \frac{\partial^2 u}{\partial y^2}dy$$

Integration by parts of the second term on the left hand side of Eq. C.1 gives

$$\int_0^\delta \nu\frac{\partial u}{\partial y}dy = [u\nu]_0^\delta - \int_0^\delta u\frac{\partial \nu}{\partial y}dy = U_\infty \nu_\delta + \int_0^\delta u\frac{\partial u}{\partial x}dy$$

(by continuity equation)

$$= -U_\infty\int_0^\delta \frac{\partial u}{\partial x}dy + \int_0^\delta u\frac{\partial u}{\partial x}dy$$

Substituting this result in Eq. C.1 and writing $dp/dx$ for $\partial p/\partial x$, we get

$$2\int_0^\delta u \frac{\partial u}{\partial x}\, dy - U_\infty \int_0^\delta \frac{\partial u}{\partial x}\, dy = -\int_0^\delta \frac{1}{\rho}\frac{dp}{dx}\, dy + \nu \left.\frac{\partial u}{\partial y}\right|_{y=0}$$

or,
$$\int_0^\delta \left( 2u \frac{\partial u}{\partial x} - U_\infty \frac{\partial u}{\partial x} - U_\infty \frac{dU_\infty}{dx} \right) dy = -\frac{\tau_w}{\rho}$$

or,
$$\int_0^\delta \frac{\partial}{\partial x}\left[ u\,(U_\infty - u)\right] dy + \frac{dU_\infty}{dx}\int_0^\delta (U_\infty - u)\, dy = \frac{\tau_w}{\rho}$$

or,
$$\int_0^\infty \frac{\partial}{\partial x}\left[ u\,(U_\infty - u)\right] dy + \frac{dU_\infty}{dx}\int_0^\infty (U_\infty - u)\, dy = \frac{\tau_w}{\rho}$$

$dp/dx$ is changed to $-du_\infty/dx$ using the Bernoulli's equation and the upper limit $\delta$ is changed in the last to $\infty$ because for values of $y$ from $\delta$ to $\infty$, the integrand is zero. Now this can be changed into the standard form as given in the text.

APPENDIX D

# Fluid Power Symbols

The symbols shown conform to the American National Standards Institute (ANSI) specifications. Basic symbols can be combined in any combination. No attempt is made to show all combinations.

| Lines and Line Functions | |
|---|---|
| Line working | 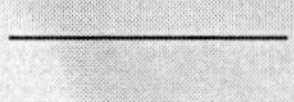 |
| Line, Pilot (L > 20W) | 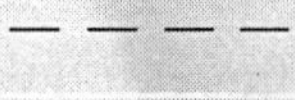 |
| Line, Drain (L < 5W) | 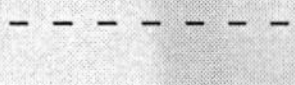 |
| Connector |  |
| Line, Flexible |  |
| Line, Joining | 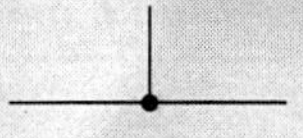 |
| Line, Passing | 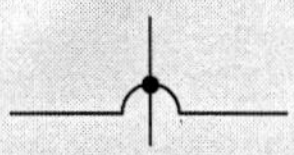 |

(*Contd*)

| Direction of flow hydraulic pneumatic | 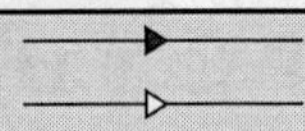 |
|---|---|
| Line to reservoir above fluid level below fluid level | 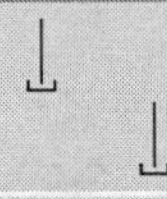 |
| Line to vented manifold | 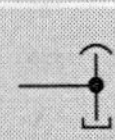 |
| Plug or plugged connection |  |
| Restriction fixed | 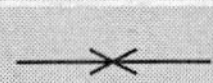 |
| **Pumps** | |
| Pump, Single fixed displacement |  |
| Pump, Single variable displacement |  |
| **Motor and Cylinders** | |
| Motor, Rotary fixed displacement |  |
| Motor, Rotary variable displacement |  |
| Motor, Oscillating |  |
| Cylinder, Single-acting | 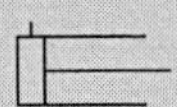 |
| Cylinder, Double-acting | 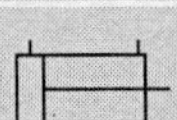 |
| Cylinder, Differential rod | 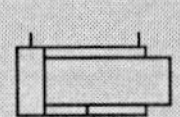 |
| Cylinder, Double-end rod | 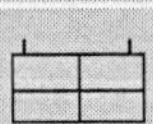 |
| Cylinder, Cushions both ends | 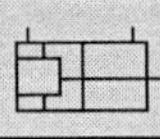 |

*(Contd)*

| Miscellaneous Units | |
|---|---|
| Direction of Rotation (Arrow in front of staff) |  |
| Component enclosure |  |
| Reservoir, Vented |  |
| Reservoir, Pressurized |  |
| Pressure gage |  |
| Temperature gage |  |
| Flow meter (flow rate) | 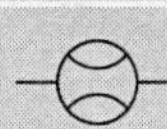 |
| Electric motor |   |
| Accumulator, Spring-loaded | 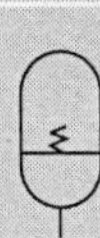 |
| Accumulator, Gas-charged |  |
| Filter or strainer | 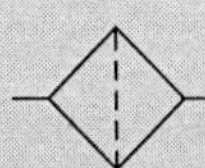 |
| Heater | 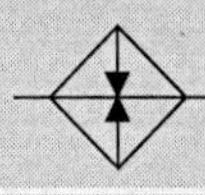 |
| Cooler | 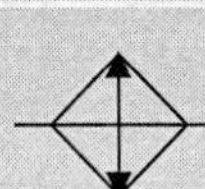 |
| Temperature controller | 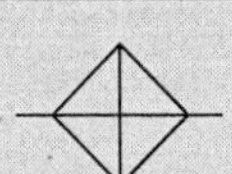 |
| Intensifier | 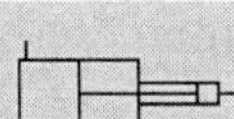 |
| Pressure Switch |  |

(*Contd*)

| Basic Value Symbols | |
|---|---|
| Check valve |  |
| Manual shut-off valve | 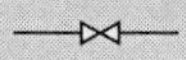 |
| Basic valve envelope |  |
| Valve, Single-flow path, Normally closed | 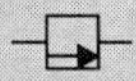 |
| Valve, Single-flow path, Normally open |  |
| Valve, Maximum pressure (relief) | 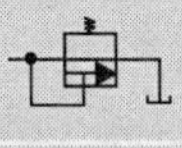 |
| Basic valve symbol, Multiple flow paths | 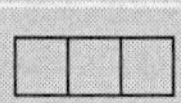 |
| Flow paths blocked in center position | 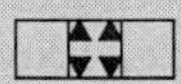 |
| Multiple flow paths (arrow shows flow direction) |  |
| **Valve Examples** | |
| Unloading valve, Internal drain, Remotely operated | 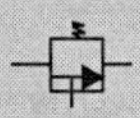 |
| Deceleration valve, Normally open | 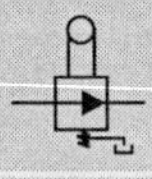 |
| Sequence valve, Directly operated, Externally drained | 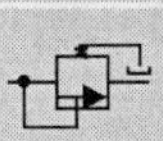 |
| Pressure-reducing valve | 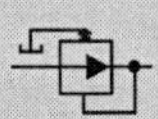 |
| Counterbalance valve with integral check | 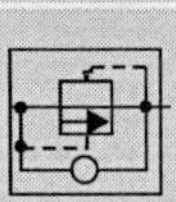 |
| Temperature and pressure-compensated flow control with integral check | 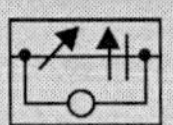 |
| Directional valve, Two-position, Three-connection |  |
| Directional, valve, Two-position, Three-connection |  |
| Valve, Infinite positioning (indicated by horizontal bars) | 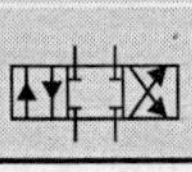 |

(*Contd*)

| Methods of Operation | |
|---|---|
| Pressure compensator |  |
| Detent | 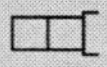 |
| Manual | 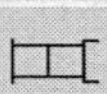 |
| Mechanical |  |
| Pedal or Treadle |  |
| Push button | 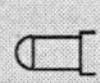 |
| Lever | 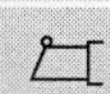 |
| Pilot pressure |  |
| Solenoid |  |
| Solenoid-controlled pilot-pressure-operated |  |
| Spring |  |
| Servo |  |

APPENDIX

# E Some Important Relations at a Glance

## Chapter 2

Streamline equation:

$dx/u = dy/v = dz/w$ (in Cartesian coordinates) (2.2.1)

Kinematic relations:

Velocity and acceleration fields:

Vector form:

$$\vec{U} = f(\vec{r}, t) \rightarrow \vec{\alpha} = d\vec{U}/dt = \partial\vec{U}/\partial t + \partial\vec{U}/\partial\vec{r}.\partial\vec{r}/\partial t = \partial\vec{U}/\partial t + (\vec{U}.\nabla)\vec{U} \quad (2.2.2)$$

Cartesian Coordinate *s*:

$$\alpha_x = Du / Dt = \partial u / \partial t + u\partial u / \partial x + v\partial u / \partial y + w\partial u / \partial z,$$

$$\alpha_y = Dv / Dt = \partial v / \partial t + u\partial v / \partial x + v\partial v / \partial y + w\partial v / \partial z,$$

$$\alpha_z = Dw / Dt = \partial w / \partial t + u\partial w / \partial x + v\partial w / \partial y + w\partial w / \partial z,$$

Cylindrical polar Coordinate *s*:

$$\alpha_r = Du / Dt = \partial u / \partial t + u\partial u / \partial r + (v / r)\partial u / \partial\theta + w\partial u / \partial z - v^2 / r, \quad (2.2.3a)$$

$$\alpha_\theta = Dv / Dt = \partial v / \partial t + u\partial v / \partial r + (v / r)\partial v / \partial\theta + w\partial v / \partial z + uv / r,$$

$$\alpha_z = Dw / Dt = \partial w / \partial t + u\partial w / \partial r + (v / r)\partial w / \partial\theta + w\partial w / \partial z.$$

Deformation of fluid element:

$$\dot{\varepsilon}_{xx}=\frac{u+(\partial u/\partial x)\,\delta x-u}{\delta x}=\frac{\delta u}{\delta x}.\ \text{Similarly}\ \dot{\varepsilon}_{yy}=\frac{\delta v}{\delta y}\ \textit{and}\ \dot{\varepsilon}_{zz}=\frac{\delta w}{\delta z} \tag{2.3.3a}$$

$$\dot{\alpha}=\frac{v+(\partial v/\partial x)\,\delta x-v}{\delta x}=\frac{\delta v}{\delta x},\ \dot{\beta}=-\frac{u+(\partial u/\partial y)\,\delta y-u}{\delta y}=-\frac{\delta u}{\delta y}.$$

$$\therefore\quad \dot{\varepsilon}_{xy}=\dot{\alpha}-\dot{\beta}=\frac{\delta v}{\delta x}+\frac{\delta u}{\delta y} \tag{2.3.3b}$$

Stream function, velocity potential and Vorticity, etc:

$$u=\partial\psi/\partial y\ \text{and}\ v=-\partial\psi/\partial x \tag{2.3.1}$$

$$\rho u=\rho_0\partial\psi/\partial y\ \text{and}\ \rho v=-\rho_0\partial\psi/\partial x \tag{2.3.2}$$

$$\omega_z=\frac{1}{2}\left(\dot{\alpha}+\dot{\beta}\right)=\frac{1}{2}\left(\frac{\delta v}{\delta x}-\frac{\delta u}{\delta y}\right) \tag{2.3.4}$$

$$\vec{\omega}=\frac{1}{2}\,\text{curl}\,\vec{U}\ \text{and}\ \vec{\Omega}=\text{curl}\,\vec{U},\ \text{where}$$

$$\text{curl}\,\vec{U}=\nabla\times\vec{U}=\begin{vmatrix} i & j & k\\ \partial/\partial x & \partial/\partial y & \partial/\partial z\\ u & v & w\end{vmatrix}\ \text{in (Cartesian) and}\ \frac{1}{r}\begin{vmatrix} e_r & re_\theta & e_z\\ \partial/\partial r & \partial/\partial\theta & \partial/\partial z\\ u_r & ru_\theta & u_z\end{vmatrix}\ \text{(cylindrical)}$$

$$\rightarrow\omega_x=\frac{1}{2}\left(\frac{\delta w}{\delta y}-\frac{\delta v}{\delta z}\right),\ \omega_y=\frac{1}{2}\left(\frac{\delta u}{\delta z}-\frac{\delta w}{\delta x}\right),\ \text{and}\ \omega_z=\frac{1}{2}\left(\frac{\delta v}{\delta x}-\frac{\delta u}{\delta y}\right)\ \text{(Cartesiann)} \tag{2.3.5}$$

$$\omega_r=\frac{1}{2}\left(\frac{\delta u_r}{r\delta\theta}-\frac{1}{r}\frac{\delta(u_\theta r)}{\delta z}\right),\ \omega_\theta=\frac{1}{2}\left(\frac{\delta u_r}{\delta z}-\frac{\delta u_z}{\delta r}\right),\ \text{and}\ \omega_z=\frac{1}{2r}\left(\frac{\delta(ru_\theta)}{\delta r}-\frac{\delta u_r}{\delta\theta}\right)\ \text{(Cylindrical)}$$

$$\Gamma=\oint_C\vec{U}.d\vec{l}=\iint_S\left(\nabla\times\vec{U}\right).d\vec{S}=\iint_S\vec{\Omega}.d\vec{S}, \tag{2.3.9}$$

Continuity, momentum and energy equations:

$$\text{Vectorform}:\frac{\partial\rho}{\partial t}+\vec{U}.\nabla\rho+\rho.\nabla\vec{U}=\frac{\partial\rho}{\partial t}+\nabla.(\rho\vec{U})=0 \tag{2.4.1}$$

$$\nabla.\vec{U}=0,\quad\text{or}\ \left(\frac{\partial u}{\partial x}+\frac{\partial v}{\partial y}+\frac{\partial w}{\partial z}\right)=0\ \text{(Cartesian coordinates)} \tag{2.4.3}$$

Cylinderical coordinates: $\frac{\partial \rho}{\partial t}+\frac{1}{r}\frac{\partial(\rho r u_r)}{\partial r}+\frac{\partial(\rho u_\theta)}{r\partial\theta}+\frac{\partial(\rho u_z)}{\partial z}=0,$ (2.4.4)

Spherical coordinates: $\frac{\partial \rho}{\partial t}+\frac{1}{r^2}\frac{\partial(\rho r^2 u_r)}{\partial r}+\frac{1}{r\sin\theta}\frac{\partial(\rho u_\theta \sin\theta)}{\partial\theta}+\frac{1}{r\sin\theta}\frac{\partial(\rho u_\phi)}{\partial\phi}=0$ (2.4.5)

Streamline coordinate: $\frac{\partial \rho}{\partial t}+\frac{1}{A}\frac{\partial(\rho u_s A)}{\partial s}=0$ and $\rho u_s A=\text{Const.}$ (2.4.6)

Average velocity

$$\bar{u}=\frac{1}{A}\int_A u\,dA=\frac{1}{H}\int_0^H u\,dy\,(\text{Rectangular})\quad \bar{u}=\frac{1}{A}\int_0^R 2\pi r u\,dr\,(\text{cylindrical}) \tag{2.4.7}$$

Integral form of continuity equation: $\frac{\partial}{\partial t}\iiint_V \rho\, dV+\iint_S \rho\vec{U}.d\vec{A}=0$ (2.4.8)

Reynolds transport theorem: $\left(\frac{dN}{dt}\right)_{CM}=\frac{\partial}{\partial t}\iiint_{CV}\eta\rho dV+\iint_{CS}\eta\rho(\vec{U}.d\vec{A})$ (2.4.9)

$$\sum \vec{F}=\left(\frac{dm\vec{U}}{dt}\right)_{CM}=\frac{\partial}{\partial t}\iiint_{CV}\vec{U}\rho dV+\iint_{CS}\vec{U}\rho(\vec{U}.d\vec{A}) \tag{2.4.11}$$

and $\tau=\frac{\partial}{\partial t}\iiint_{CV}(\vec{r}\times\vec{U})\rho dV+\iint_{CS}(\vec{r}\times\vec{U})\rho(\vec{U}.d\vec{A})$ (2.4.12)

$$\frac{d(m\vec{U}_r)}{dt}=m\frac{d(\vec{U}_r)}{dt}=\frac{d(\vec{U}-\vec{U}_c)}{dt}=m\frac{d\vec{U}}{dt}-m\,\bar{\alpha}_c=\sum\vec{F}-m\,\bar{\alpha}_c \tag{2.4.13}$$

Euler equation

Cartesian Coordinate s:

$$\begin{aligned}&\partial u/\partial t+u\partial u/\partial x+v\partial u/\partial y+w\partial u/\partial z=-(1/\rho)\partial p/\partial x+B_x\\&\partial v/\partial t+u\partial v/\partial x+v\partial v/\partial y+w\partial v/\partial z=-(1/\rho)\partial p/\partial y+B_y\\&\partial w/\partial t+u\partial w/\partial x+v\partial w/\partial y+w\partial w/\partial z=-(1/\rho)\partial p/\partial z+B_z\end{aligned} \tag{2.4.15}$$

Cylindrical polar Coordinates:

$$\begin{aligned}&\partial u_r/\partial t+u_r\,\partial u_r/\partial r+(u_\theta/r)\partial u/\partial\theta+u_z\,\partial u_r/\partial z-u_\theta^2/r=-(1/\rho)\partial p/\partial r+B_r\\&\partial u_\theta/\partial t+u_r\,\partial u_\theta/\partial r+(u_\theta/r)\partial u_\theta/\partial\theta+u_z\,\partial u_\theta/\partial z+u_r u_\theta/r=-(1/\rho)\partial p/\partial(r\theta)+B_\theta\\&\partial u_z/\partial t+u_r\,\partial u_z/\partial r+(u_\theta/r)\partial u_z/\partial\theta+u_z\,\partial u_z/\partial z=-(1/\rho)\partial p/\partial z+B_y\end{aligned} \tag{2.4.16}$$

Navier-Stoke equation

$$\partial u/\partial t + u\partial u/\partial x + v\partial u/\partial y + w\partial u/\partial z = -(1/\rho)\partial p/\partial x + B_x + \nu\nabla^2 u$$

$$\partial v/\partial t + u\partial v/\partial x + v\partial v/\partial y + w\partial v/\partial z = -(1/\rho)\partial p/\partial y + B_y + \nu\nabla^2 v \qquad (2.4.20)$$

$$\partial w / \partial t + u\partial w / \partial x + v\partial w / \partial y + w\partial w / \partial z = -(1/\rho)\partial p / \partial z + B_z + \nu\nabla^2 w$$

# Chapter 3

Uniform flow: $\psi = Uy$ and $\phi = Ux$ (3.2.1)

Source and sink flow: $\psi = k\theta$ and $\phi = k\ln r$ (3.2.3)

Anticlockwise free vortex: $\psi = -k\ln r$ and $\phi = k\theta$ (3.2.5)

Doublet flow: $\psi = -\dfrac{\mu\sin\theta}{2\pi r}$ and $\phi = \dfrac{\mu\cos\theta}{2\pi r}$ (3.3.3)

Flow around cylinder:

$$\psi = U\left(r - \frac{R^2}{r}\right)\sin\theta \text{ and } \phi = U\left(r + \frac{R^2}{r}\right)\cos\theta \qquad (3.3.9)$$

$$u_r = \frac{\partial\psi}{r\partial\theta} = U\left(1 - \frac{R^2}{r^2}\right)\cos\theta \text{ and } u_\theta = -\frac{\partial\psi}{\partial r} = -U\left(1 + \frac{R^2}{r^2}\right)\sin\theta \qquad (3.3.10)$$

$$C_p = \frac{p - p_s}{\tfrac{1}{2}\rho U^2} = \left(1 - 4\sin^2\theta\right) \qquad (3.3.14)$$

Flow around cylinder with clockwise free vortex:

$$\psi = U\left(r - \frac{R^2}{r}\right)\sin\theta + \frac{\Gamma}{2\pi}\ln\frac{r}{R} \text{ and } \phi = U\left(r + \frac{R^2}{r}\right)\cos\theta - \frac{\Gamma}{2\pi}\theta \qquad (3.3.11)$$

$$u_r = \frac{\partial\psi}{r\partial\theta} = U\left(1 - \frac{R^2}{r^2}\right)\cos\theta \text{ and } u_\theta = -\frac{\partial\psi}{\partial r} = -U\left(1 + \frac{R^2}{r^2}\right)\sin\theta - \frac{\Gamma}{2\pi r} \qquad (3.3.12)$$

$$C_p = \frac{p - p_s}{\tfrac{1}{2}\rho U^2} = 1 - \left(-2\sin\theta - \tfrac{\Gamma}{2\pi RU}\right)^2 \qquad (3.3.15)$$

$$C_D = \frac{\text{Pressure force along flow direction}}{\tfrac{1}{2}\rho U^2 \times \text{projected area of cylinder}} = \frac{1}{2R\,x\,1}\int_0^{2\pi} C_p\, R\cos\theta\, d\theta = 0$$

$$\text{and } C_L = \frac{\text{Pressure force} \perp \text{to flow direction}}{\tfrac{1}{2}\rho U^2 \times \text{projected area of cylinder}} = \frac{1}{2R\,x\,1}\int_0^{2\pi} C_p\, R\sin\theta\, d\theta = \frac{\Gamma}{RU} \qquad (3.3.16)$$

The flow around the cylinder of radius $R$ in the complex plane $z$ is given by the function

$$W = U\left(z + R^2/z\right) \qquad (3.3.20)$$

Conformal transformation that transforms cylinder into aerofoil in z plane

$$z = \tfrac{1}{2}\left(\tilde{z} + c^2/\tilde{z}\right) \tag{3.3.21}$$

(Choice of $c$ and displacement of the cylinder centre of the from the origin determines the shape of the aerofoil)

Stagnation pressure: $p_0 = p + \tfrac{1}{2}\rho U^2$ (3.4.4)

Flow measurement:

Venturimeter:

$$Q_a = C_d\, Q = C_d A_2 U_2 = \frac{C_d\, A_1 A_2}{\sqrt{A_1^2 - A_2^2}}\sqrt{2gh} = K\sqrt{2gh}\,, h = h_m\left(\frac{\rho_m}{\rho} - 1\right) \tag{3.4.11}$$

Nozzle and orifice plate: $Q_a = C_d A_2 \sqrt{2gh}$ (3.4.12)

Rectangular weir: $Q_a = C_d \int_0^H u\, b\, dh = \tfrac{2}{3} C_d b\sqrt{2g}\left[\left(H + U^2/2g\right)^{3/2} - \left(U^2/2g\right)^{3/2}\right]$ (3.4.15)

Triangular weir: $Q = Cd \int_0^H u\, b\, dh = 8/15\ Cd \tan\theta\sqrt{(2g)}\ H^{5/2}$ (3.4.16)

## Chapter 4

Flow through pipe:

Major loss, Darcy equation: $h_f = \dfrac{fL}{d}\cdot\dfrac{U^2}{2g} = \dfrac{fLQ^2}{12\, d^5}$ (4.2.1)

Minor losses $h_{\text{minor loss}} = K\, U^2/(2g)$ (4.2.2)

Sudden enlargement of pipe: $h_L = \left(1 - \dfrac{A_1}{A_2}\right)^2 \dfrac{U_1^2}{2g}$ (4.2.3)

Sudden contraction of pipe: $h_L = 0.5\dfrac{U_2^2}{2g}$ (4.2.4)

Flow through channel:

Chezy equation $U = C\sqrt{mi}$ (4.5.7)

Manning equation $C = \tfrac{1}{n} m^{1/6}$ $\rightarrow U = \tfrac{1}{n} m^{2/3}\sqrt{i}$ (4.5.8)

$$E_s = d + \frac{U^2}{2g} = d + \frac{q^2}{2gd^2} \tag{4.5.22}$$

$$d_c = \left(\frac{q^2}{g}\right)^{1/3} \quad \text{and} \quad E_{\min} = \frac{3}{2} d_c \tag{4.5.23}$$

## Chapter 5

Couette flow: $\frac{u}{U}=B\frac{y}{h}\left(1-\frac{y}{h}\right)+\frac{y}{h}$, where $B=-\frac{h^2}{2\mu U}\frac{dp}{dx}$ (5.2.2)

$$q=\int_0^h u\,dy=-\frac{h^3}{12\mu}\frac{dp}{dx}\rightarrow\overline{U}=\text{average velocity}=q/h=-\frac{h^2}{12\mu}\frac{dp}{dx} \tag{5.2.3}$$

Or. $\frac{dp}{dx}=-\frac{12\mu\overline{U}}{h^2}$. But $(p_1-p_2)/L=-dp/dx$

$$\therefore\ p_1-p_2=L\,dp/dx=12\mu\overline{U}L/h^2 \tag{5.2.4}$$

Hagen- Poiseuille flow

$$u_{\max}=-\frac{R^2}{4\mu}\frac{dp}{dx}\ \text{at}\ r=0$$

$$Q=\int_0^R 2\pi r u_x\,dr=-\frac{\pi R^4}{8\mu}\frac{dp}{dx}=\pi R^2U\rightarrow U=-\frac{R^2}{8\mu}\frac{dp}{dx}=u_{\max}/2 \tag{5.2.7}$$

$$\text{and }-dp/dx=\frac{8\mu Q}{\pi R^4}=\frac{128\mu Q}{\pi D^4} \tag{5.2.8}$$

Stokes flow past sphere: $F=6\pi\mu aU_\infty$ (5.3.4)

$$\text{Loss in mass flow rate}=\rho U\delta^\star=\int_0^\infty\rho(U-u)\,dy\ \rightarrow\delta^\star=\int_0^\infty(1-u/U)\,dy \tag{5.4.1}$$

$$\delta_m=\int_0^\infty(u/U)(1-u/U)\,dy\quad\text{and}\quad\delta_e=\int_0^\infty(u/U)^2(1-u/U)\,dy \tag{5.4.2}$$

B.L. equation: $\frac{\partial u}{\partial x}+\frac{\partial v}{\partial y}=0$, and $u\frac{\partial u}{\partial x}+v\frac{\partial u}{\partial y}=-\frac{1}{\rho}\frac{dp}{dx}+\nu\frac{\partial^2u}{\partial y^2}$ (5.4.5)

$$\tau_w=\rho\frac{d}{dx}\left(\theta U^2\right)+\rho U\frac{dU}{dx}\delta^\star \tag{5.5.1}$$

## Chapter 7

Isentropic relations:

$$a^2=(\partial p/\partial\rho)_s=\gamma RT \tag{7.1.3}$$

$$\frac{a^2}{(\gamma-1)}+\frac{C^2}{2}=\frac{a_0^2}{(\gamma-1)}=\frac{C_{\max}^2}{2}=\frac{\gamma+1}{\gamma-1}\frac{C^{\star2}}{2} \tag{7.1.10}$$

$$M^{\star 2} = \frac{C^2}{a^{\star 2}} = \frac{C^2}{a^{\star 2}}\frac{a^2}{a^{\star 2}} = \frac{a^2}{a^{\star 2}} M^2 \quad (7.1.11)$$

$$T_0/T = 1 + \{(\gamma - 1)/2\}\{C^2/((\gamma RT)\} = 1 + \{(\gamma - 1)/2\} M^2 \quad (7.2.3)$$

$$p_0/p = [1 + \{(\gamma - 1)/2\} M^2]\gamma/(\gamma - 1)$$

$$\rho_0/\rho = (p_0/p)^{1/\gamma} = [1 + \{(\gamma - 1)/2\} M^2]1/(\gamma - 1) \quad (7.2.4)$$

$$F = p + A\,\rho\, C^2 = p + \gamma p A\, M^2$$

$$\tau = F_2 - F_1 = p_2\, A_2\,(1 + \gamma M_2^{\,2}) - p_1\, A_1\,(1 + \gamma M_1^{\,2}) \quad (7.2.6)$$

$$C^2 = 2\frac{\gamma}{\gamma - 1} RT_0 \left\{1 - \left(\frac{p}{p_0}\right)^{(\gamma-1)/\gamma}\right\} = RT_0\left[\frac{\gamma M^2}{1 + \frac{\gamma - 1}{2} M^2}\right] \quad (7.2.7)$$

$$\frac{\dot{m}\sqrt{T_0}}{Ap_0}\sqrt{\frac{R}{\gamma}} = \sqrt{2\frac{\gamma}{\gamma - 1}\left\{\left(\frac{p}{p_0}\right)^{2/\gamma} - \left(\frac{p}{p_0}\right)^{(\gamma+1)/\gamma}\right\}} \quad (7.2.8)$$

$$\left(\frac{p}{p_0}\right)_{\max} = \left(\frac{2}{\gamma + 1}\right)^{\gamma/(\gamma-1)} \quad (7.2.9)$$

$$\frac{\dot{m}_{\max}\sqrt{T_0}}{A^{\star}p_0}\sqrt{\frac{R}{\gamma}} = \left(\frac{2}{\gamma + 1}\right)^{(\gamma+1)/(2(\gamma-1))} \quad (7.2.10)$$

$$\frac{T^{\star}}{T_0} = \frac{2}{\gamma + 1} = 0.833 \quad \text{and} \quad \frac{T^{\star}}{T} = \frac{2}{\gamma + 1} + \frac{\gamma - 1}{\gamma + 1} M^2 \quad (7.2.11)$$

$$\frac{p^{\star}}{p_0} = \left(\frac{2}{\gamma + 1}\right)^{\gamma/(\gamma-1)} = 0.528 \quad \text{and} \quad \frac{p^{\star}}{p} = \left(\frac{2}{\gamma + 1} + \frac{\gamma - 1}{\gamma + 1} M^2\right)^{\gamma/(\gamma-1)} \quad (7.2.12)$$

$$\frac{\rho^{\star}}{\rho_0} = \left(\frac{2}{\gamma + 1}\right)^{1/(\gamma-1)} = 0.634 \quad \text{and} \quad \frac{\rho^{\star}}{\rho} = \left(\frac{2}{\gamma + 1} + \frac{\gamma - 1}{\gamma + 1} M^2\right)^{1/(\gamma-1)} \quad (7.2.13)$$

$$C^{\star} = \sqrt{2\gamma RT_o / (\gamma + 1)} \quad (7.2.14)$$

$$\frac{A}{A\star} = \frac{1}{M}\left(\frac{2}{\gamma + 1} + \frac{\gamma - 1}{\gamma + 1} M^2\right)^{(\gamma+1)/2(\gamma-1)} \quad (7.2.15)$$

For other relations refer the chapter.

## Chapter 9

Work exchange in rotamachines:

$T = m(C_{w1}R_1 - C_{w2}R_2)$ and

Work transfer/s = $T \times w = m\,(C_{w1u1} - C_{w2u2})$. (9.1.9)

Non-dimensional groups: $\pi_1 = \dfrac{N\sqrt{P}}{\sqrt{\rho}\,(gH)^{5/4}}$, $\pi_2 = \dfrac{ND}{\sqrt{gH}}$, and $\pi_3 = \dfrac{Q}{ND^3}$ (9.1.16)

Specific speed: $(Ns)_{\text{turbine}} = N\sqrt{P}/H^{5/4}$ and $(Ns)_{\text{pump}} = N\sqrt{Q}/(gH)^{3/4}$

Speed ratio: $Ku = u/C$

Flow ratio: $= \phi$ or $Kf = Cf/C$

Head Coefficient: $ND/\sqrt{(gH)}$

Unit quantity:

Unit speed, $Nu = N/\sqrt{H}$ (9.1.21)

Unit discharge, $Q_u = \dfrac{Q}{\sqrt{H}}$ (9.1.22)

Unit power, $P_u = P/H^{3/2}$ (9.1.23)

# Bibliography


Ames, W. F. (Ed.). *Nonlinear PDE—A Symposium of Method of Solution.* Academic Press, 1967.

Anderson, Jr, J. D. *Fundamental of Aerodynamics.* New York: McGraw-Hill. 1986.

Batchelor, G. K. *The Theory of Homogeneous Turbulence.* Cambridge: Cambridge University Press. 1953.

Batchelor, G. K. *Fluid Dynamics.* Cambridge: Cambridge University Press. 1967.

Binder, R. B., W. E. Stewart, and E. N. Lightfoot. *Transport Phenomena.* Wiley. 1960.

Biswas & Eswaran (Eds). *Turbulent Flows Fundamentals.* New Delhi: Narosa. 2002.

Bose, T. K. *Numerical Fluid Dynamics.* New Delhi: Narosa. 1997.

Carslaw, H. S. and J. C. Jaeger, *Conduction of Heat in Solids.* 2nd edition. Oxford, 1959.

Cebeci, T. and A. M. O. Smith. *Analysis of Turbulent Boundary Layers.* Academic Press. 1974. Douglas, J. F. et al. *Fluid Mechanics.* New Delhi: Pearson Education. 2001.

Durand, W. F. *Fluid Mechanics Vol. 1 Aerodynamic Theory.* Julius Springer. 1934.

Esposito, A. *Fluid Power with Applications.* New Delhi: Pearson Education. 2005

Evans, H. L. *Laminar Boundary Layer Theory.* Addison Wesley. 1968.

Faber, T. E. *Fluid Dynamics for Physicists.* Cambridge: Cambridge University Press. 1995.

Fay, J. A. *Introduction to Fluid Mechanics.* New Delhi: PHI. 1994.

Fox, R. W. and A. T. McDonald. *Introduction to Fluid Mechanics.* John Wiley & Sons. 2001.

Garde, R. J. Fluid Mechanics through Problems New Age Int. New Delhi. 1997.

Goldstein, S. (Ed.) *Modern Developments in Fluid Mechanics.* New York: Oxford University Press. 1938.

Gupta, A. K. and M. M. Oberai (Eds). *QIP Course on Turbulent Fluid Flow.* Kanpur: IIT Kanpur. 1997.

Hinze, J. O. *Turbulence.* New York: McGraw Hill. 1987.

Hughes, W. F. and J. A. Brighton. *Fluid Dynamics—Schaum's Series.* New Delhi: TMH. 2004.

H. W. Liepmann and A. Roshko. *Elements of Gas Dynamics.* New York: John Wiley & Sons. 1956.

John, J. E. A. and W. L. Habermann. *Introduction to Fluid Mechanics.* 3rd edition. PHI. 1988.

Karamcheti, K. *Principles of Ideal Fluid Aerodynamics.* John Wiley & Sons. 1966.

Kaufmann, W. and E. G. Chilton. *Fluid Mechanics.* TMH.

Kumar, D. S. *Fluid Mechanics and Fluid Power Engineering.* New Delhi: S.K Kataria & Sons. 2003.

Kumar, K. L. *Engineering Fluid Mechanics.* New Delhi: Eurasia Publishing House. 2000.

Lal, J. *Fluid Mechanics.* New Delhi: Metropolitan Book Co.

———. *Hydraulic Machines.* New Delhi: Metropolitan Book Co.

Lamb, H. *Hydrodynamics.* New York: Dover Publications. 1945.

Langlois, W. F. *Slow Viscous Flow.* New York: Macmillan. 1964.

L. D. Landau and E. M. Lifshitz. *Fluid Mechanics.* New York: Pergamon Press. 1959.

Ligget, J. A. *Fluid Mechanics.* Singapore: McGraw Hill Int. 1994.

Lin, C. C. *The Theory of Hydrodynamic Stability.* Cambridge: Cambridge University Press. 1955.

Massey, B. S. *Mechanics of Fluid.* Van Nostrand. 6th ed. 1989.

Meksyn, D. *New Methods in Boundary Layer Theory.* New York: Pergamon Press. 1961.

Milne Thomson, L. M. *Theoretical Hydrodynamics.* New York: The Macmillan Co. 1955.

Modi, P. N. and S. M. Seth. *Hydraulics and Hydraulic Machines.* New Delhi: Standard Book Co. 1971.

Mohanty, A. K. *Fluid Mechanics.* New Delhi: PHI. 2004.

Murliudhar, K. and G. Biswas, *Advanced Engineering Fluid Mechanics.* New Delhi: Narosa. 1999.

Myers, G. E. Analytical Methods in Conduction Heat Transfer. McGraw Hill.

Nagaratnam, S. *Fluid Mechanics and Systems.* New Delhi: TMH. 1971.

Olmsted, M. H. *Advanced Calculus.* New Delhi: Eurasia Pub. House. 1970.

Ozisik, M. N. *Heat Conduction.* New York: John Wiley & Sons. 1993.

Pai. S. I. *Introduction to the Theory of Compressible Flows.* D. Van Nostrand. 1959.

Pai, S. I. *Viscous Flow Theory—Laminar Flow.* D. Van Nostrand. 1956.

Pankhurst, R. C. and D. W. Holder. *Wind Tunnel Techniques.* London: Pitman. 1952.

Philips, E. G. *Functions of a Complex Variable with Applications.* Oliver & Boyd. 1958.

Prakash, R. and J. S., Saini (Ed). *Topics in Advanced Fluid Mechanics*, QIP Course. Roorkee: Roorkee University. 1975.

Ramadurgaih, D. *Fluid Mechanics and Machinery.* New Delhi: New Age Int. 2002.

Ramamrutham, S. *Hydraulics, Fluid Mechanics and Fluid Machines.* New Delhi: Dhanpat Rai Publishing Co. 2004.

Rao, J. S. *Turbomachine Unsteady Aerodynamics.* New Delhi: Wiley Eastern Ltd. 1994.

Roberson J. A. and C. T. Crowe. *Engineering Fluid Mechanics.* Jaico Pub. 1998.

Rosenhead L. (Ed.). *Laminar Boundary Layer Theory.* Oxford University Press. 1963.

Roy, D. N. *Applied Fluid Mechanics.* Ellis Horwood. 1988.

Schlichting, H. *Boundary Layer Theory.* New York: McGraw Hill. 1968.

Sears, W. R. *Theoretical Aerodynamics.* New York: Ithaca. 1970.

Shames, I. H. *Mechanics of Fluid.* McGraw Hill. 1982.

Shapiro, A. H. *The Dynamics and Thermodynamics of Compressible Fluid Flow*. New York: Ronald Press. 1958.

Som, S. K. and G. Biswas. *Fluid Mechanics and Fluid Machines*. New Delhi: TMH. 2004.

Streeter, V. L. and E.B. Wylie. *Fluid Mechanics*. McGraw Hill. 1983.

Streeter, V. L. et al. *Fluid Mechanics*. McGraw Hill. 1998.

Tennekes, H. and J.L. Lumley. *A First Course in Turbulence*. MIT Press. 1956.

Townsend, A. A. *The Structure of Turbulent Shear Flow*. Cambridge: Cambridge Univ. Press. 1956.

White, F. M. *Viscous Fluid Flow*. New York: McGraw Hill. 1974.

Yaha, S. M. *Turbomachine Compressors and Fans*. New Delhi: TMH. 2002.

Yuan, S.W. *Foundations of Fluid Mechanics*. New Delhi: Prentice Hall. 1976.

# Index